The Ecological Status of European Rivers: Evaluation and Intercalibration of Assessment Methods

Developments in Hydrobiology 188

Series editor
K. Martens

The Ecological Status of European Rivers: Evaluation and Intercalibration of Assessment Methods

Edited by

Mike T. Furse[1], Daniel Hering[2], Karel Brabec[3], Andrea Buffagni[4], Leonard Sandin[5] & Piet F.M. Verdonschot[6]

[1]Centre for Ecology and Hydrology, CEH Dorset, Winfrith Technology Centre, Winfrith Newburgh, Dorchester, Dorset DT2 8ZD, United Kingdom
[2]University of Duisburg-Essen, Institute of Hydrobiology, Universitätsstr.5, 45117 Essen, Germany
[3]Masaryk University, Department of Zoology and Ecology, Kotlárská 611 37, Brno, Czech Republic
[4]CNR-Water Research Institute, Via della Mornera, 25 I-20047 Brugherio (Milano), Italy
[5]Swedish University of Agricultural Sciences, Department of Environmental Assessment, P.O. Box 7050, S-750 07 Uppsala, Sweden
[6]Alterra, Department of Ecology and Environment, Droevendaalsesteeg 3, 6700 AA Wageningen, The Netherlands

Reprinted from Hydrobiologia, Volume 566 (2006)

Springer

Library of Congress Cataloging-in-Publication Data

A C.I.P. Catalogue record for this book is available from the Library of Congress.

ISBN 1-4020-5160-3

Published by Springer,
P.O. Box 17, 3300 AA Dordrecht, The Netherlands

Cite this publication as Hydrobiologia vol. 566 (2006)

Cover illustration: Astrid Schmidt-Kloiber (Vienna)
Photos: W. Graf, A. Schmidt-Kloiber, K. Pall, G. Zauner

Printed on acid-free paper

Printed in the Netherlands

TABLE OF CONTENTS

Hydrobiologia (2006) 566:1–2
M.T. Furse, D. Hering, K. Brabec, A. Buffagni, L. Sandin & P.F.M. Verdonschot (eds), The Ecological Status of European Rivers: Evaluation and Intercalibration of Assessment Methods
DOI 10.1007/s10750-006-0113-4

The ecological status of European rivers: evaluation and intercalibration of assessment methods

Mike T. Furse[1,*], Daniel Hering[2], Karel Brabec[3], Andrea Buffagni[4], Leonard Sandin[5] & Piet F. M. Verdonschot[6]

[1]*Centre for Ecology and Hydrology, CEH Dorset, Winfrith Technology Centre, Winfrith Newburgh, Dorchester, Dorset DT2 8ZD, UK*
[2]*Institute of Hydrobiology, University of Duisburg-Essen, Universitätsstr. 5, 45117 Essen, Germany*
[3]*Department of Zoology and Ecology, Masaryk University, Kotlárská, 611 37 Brno, Czech Republic*
[4]*CNR-Water Research Institute, Via della Mornera, 25 I-20047 Brugherio (Milan), Italy*
[5]*Department of Environmental Assessment, Swedish University of Agricultural Sciences, P.O. Box 7050, S-750 07 Uppsala, Sweden*
[6]*Department of Ecology and Environment, Alterra, Droevendaalsesteeg 3, 6700 AA Wageningen, The Netherlands*
*(*Author for correspondence: E-mail: mtf@ceh.ac.uk)*

In this special issue we present the major results of the EU funded research project STAR (Standardisation of River Classifications: Framework method for calibrating different biological survey results against ecological quality classifications to be developed for the Water Framework Directive; contract number EVK1-CT-2001-00089).

The aims of STAR were to develop methodologies, tools and background information to assess rivers throughout Europe using diatoms, macrophytes, invertebrates, fish and hydromorphological features. The project's research questions and structure are described in detail by Furse et al. (2006).

STAR has generated results over a wide spectrum of topics, ranging from river typologies and new methodologies for assessing the condition of rivers using macrophytes to the uncertainty of assessment approaches. This special issue is structured to reflect the broad scope of the project and is sub-divided into seven sections. Each contains up to six papers describing specific results and each is introduced by a summary paper reviewing the main findings of the papers in the section. Individually, these sections are:

- Stream and river typologies
- Linking organism groups
- Macrophytes and diatoms
- Hydromorphology
- Tools for assessing European streams with macroinvertebrates
- Intercalibration and comparison
- Errors and uncertainty in bio-assessment methods

Acknowledgements

We would like to express our gratitude to those researchers from inside and outside the consortium who contributed to the review process: Kåre Aagaard (Trondheim, Norway), Rick Battarbee (London, UK), Jean-Nicolas Beisel (Metz, France), Sebastian Birk (Essen, Germany), Jürgen Böhmer (Kirchheim/Teck, Germany), Matthias Brunke (Flintbek, Germany), John Davy-Bowker (Dorchester, UK), Hugh Dawson (Dorchester, UK), Francois Edwards (Dorchester, UK), Stefania Erba (Brugherio, Italy), Christian K. Feld (Essen, Germany), Nikolai Friberg (Silkeborg, Denmark), Jeroen Gerritsen (Owings Mills, USA), Peter Goethals (Gent, Belgium), Peter Haase (Biebergemünd, Germany), Mattie O'Hare (Dorchester, UK), Charles Hawkins (Utah, USA), Anna-Stiina Heiskanen (Ispra, Italy), Nigel Holmes (Huntingdon, UK), Bob Hughes (Corvallis, USA), Štěpán Husák (Třeboň, Czech Republic), Jiri Jarkovsky (Brno, Czech Republic), Jochem Kail (Bonn, Germany), Ellen Kiel (Vechta, Germany), Morten Lauge-Pedersen (Silkeborg, Denmark), Sovan Lek (Tolouse, France),

Manuela Morais (Evora, Portugal), John Murray-Bligh (Exeter, UK), Rebi Nijboer (Wageningen, The Netherlands), Thomas Ofenböck (Vienna, Austria), Isabel Pardo (Vigo, Spain), A. Pasteris, Steffen Pauls (Biebergemünd, Germany), Edwin Peeters (Wageningen, The Netherlands), Paulo Pinto (Evora, Portugal), Didier Pont (Lyon, France), Karel Prach (České Budějovice, Czech Republic), Martin Pusch (Berlin, Germany), Paul Raven (Bristol, UK), Bruno Rossaro (Milan, Italy), Astrid Schmidt-Kloiber (Vienna, Austria), Mario Sommerhäuser (Essen, Germany), Gunta Springe (Salaspils, Latvia), Katerina Sumberova (Brno, Czech Republic), Philippe Usseglio-Polatera (Metz, France), Wouter van de Bund (Ispra, Italy), Hanneke Vlek (Wageningen, The Netherlands), Jean-Gabriel Wasson (Lyon, France), Geraldene Wharton (London, UK) and Thom Whittier (Corvallis, USA).

Reference

Furse, M., D. Hering, O. Moog, P. Verdonschot, R. K. Johnson, K. Brabec, K. Gritzalis, A. Buffagni, P. Pinto, N. Friberg, J. Murray-Bligh, J. Kokes, R. Alber, P. Usseglio-Polatera, P. Haase, R. Sweeting, B. Bis, K. Szoszkiewicz, H. Soszka, G. Springe, F. Sporka & I. Krno, 2006. The STAR project: context, objectives and approaches. Hydrobiologia 566: 3–29.

Hydrobiologia (2006) 566:3–29
M.T. Furse, D. Hering, K. Brabec, A. Buffagni, L. Sandin & P.F.M. Verdonschot (eds), The Ecological Status of European Rivers: Evaluation and Intercalibration of Assessment Methods
DOI 10.1007/s10750-006-0067-6

The STAR project: context, objectives and approaches

Mike Furse[1,*], Daniel Hering[2], Otto Moog[3], Piet Verdonschot[4], Richard K. Johnson[5], Karel Brabec[6], Kostas Gritzalis[7], Andrea Buffagni[8], Paulo Pinto[9], Nikolai Friberg[10], John Murray-Bligh[11], Jiri Kokes[12], Renate Alber[13], Philippe Usseglio-Polatera[14], Peter Haase[15], Roger Sweeting[16], Barbara Bis[17], Krzysztof Szoszkiewicz[18], Hanna Soszka[19], Gunta Springe[20], Ferdinand Sporka[21] & Il'ja Krno[22]

[1]*Centre for Ecology and Hydrology, CEH Dorset, Winfrith Technology Centre, Winfrith Newburgh, Dorchester, Dorset DT2 8ZD, UK*

[2]*Institute of Hydrology, University of Duisburg-Essen, Universitaetsstr. 5, 45117 Essen, Germany*

[3]*Institute for Hydrobiology and Aquatic Ecosystem Management, University of Natural Resources and Applied Life Sciences Vienna, Max Emanuel Strasse 17, A-1180 Vienna, Austria*

[4]*Department of Ecology and Environment, Alterra, Droevendaalsesteeg 3, 6700 AA Wageningen, The Netherlands*

[5]*Department of Environmental Assessment, Swedish University of Agricultural Sciences, P.O. Box 7050, S-750 07 Uppsala, Sweden*

[6]*Department of Zoology and Ecology, Masaryk University, Kotlárská, 611 37 Brno, Czech Republic*

[7]*Hellenic Centre for Marine Research, Institute of Inland Waters, 46.7 km Athens-Sounion Avenue, 190 13 Anavyssos, Greece*

[8]*CNR-Water Research Institute, Via della Mornera, 25 I-20047 Brugherio (Milano), Italy*

[9]*Centre of Applied Ecology, University of Evora, Apartado 94, Lago dos Colegiais 2, 7002–554 Evora, Portugal*

[10]*Department of Freshwater Ecology, NERI, National Environmental Research Institute, Vejlsøvej 25, P.O. Box 314, DK-8600 Silkeborg, Denmark*

[11]*South West Region, Manley House, Kestrel Way, Environment Agency, EX2 7LQ Exeter, Devon, UK*

[12]*Vyzkumny Ustav Vodohospodarsky T.G. Masayka, Drevarska 12, 657 57 Brno, Czech Republic*

[13]*LABBIO, Unterbergstrasse 2, 39055 Laives, Italy*

[14]*Centre of Ecotoxicology, Biodiversity and Environmental Health, University of Metz, Campus Bridoux, Rue de Général, 57070 Metz, France*

[15]*Senckenbergische Naturforschende Gesellschaft, Lochmuehle 2, D-63599 Biebergemünd, Germany*

[16]*Freshwater Biological Association, The Ferry House, Far Sawrey, LA22 0LP Cumbria, UK*

[17]*Institute of Ecology and Nature Protection, Department of Applied Ecology, University of łodź, Banacha 12/16, 90-237 łodź, Poland*

[18]*Department of Ecology and Environmental Protection, Agricultural University of August Cieszkowski, ul. Piątkowska 94C, 61-691 Poznan, Poland*

[19] *Lake Protection Laboratory, Instytut Ochrony Środowiska, Kolektorska 4, 01-692 Warsaw, Poland*

[20]*Institute of Biology, University of Latvia, Miera 3, 2169 Salaspils, Latvia*

[21]*Institute of Zoology, Department of Hydrobiology, Slovak Academy of Sciences, Dubravska cesta 9, 84206 Bratislava, Slovakia*

[22]*Faculty of Science, Department of Ecology, Comenius University Bratislava, Mlynská dolina B-2, 842 15 Bratislava, Slovakia*

*(*Author for correspondence: E-mail: mtf@ceh.ac.uk)*

Key words: Water Framework Directive, ecological status, biological quality elements, intercalibration, uncertainty, software

Abstract

STAR is a European Commission Framework V project (EVK1-CT-2001-00089). The project aim is to provide practical advice and solutions with regard to many of the issues associated with the Water

Framework Directive. This paper provides a context for the STAR research programme through a review of the requirements of the directive and the Common Implementation Strategy responsible for guiding its implementation. The scientific and strategic objectives of STAR are set out in the form of a series of research questions and the reader is referred to the papers in this volume that address those objectives, which include: (a) Which methods or biological quality elements are best able to indicate certain stressors? (b) Which method can be used on which scale? (c) Which method is suited for early and late warnings? (d) How are different assessment methods affected by errors and uncertainty? (e) How can data from different assessment methods be intercalibrated? (f) How can the cost-effectiveness of field and laboratory protocols be optimised? (g) How can boundaries of the five classes of Ecological Status be best set? (h) What contribution can STAR make to the development of European standards? The methodological approaches adopted to meet these objectives are described. These include the selection of the 22 stream-types and 263 sites sampled in 11 countries, the sampling protocols used to sample and survey phytobenthos, macrophytes, macroinvertebrates, fish and hydromorphology, the quality control and uncertainty analyses that were applied, including training, replicate sampling and audit of performance, the development of bespoke software and the project outputs. This paper provides the detailed background information to be referred to in conjunction with most of the other papers in this volume. These papers are divided into seven sections: (1) typology, (2) organism groups, (3) macrophytes and diatoms, (4) hydromorphology, (5) tools for assessing European streams with macroinvertebrates, (6) intercalibration and comparison and (7) errors and uncertainty. The principal findings of the papers in each section and their relevance to the Water Framework Directive are synthesised in short summary papers at the beginning of each section. Additional outputs, including all sampling and laboratory protocols and project deliverables, together with a range of freely downloadable software are available from the project website at www.eu_star.at.

Context

The Water Framework Directive

Europe has a hundred years of experience of using biological assemblages to assess the condition of streams and rivers. The first procedures were developed early in the 20th century in central Europe and were based on the concept of saprobity (Sladecek, 1973). Saprobic systems varied in their design and application but could use both micro- and macroscopic plant and animal communities in order to evaluate sites. A wide diversity of techniques blossomed throughout the 20th century (Hellawell, 1978, 1986) and, whilst a range of different biological groups continued to be used, the use of benthic macroinvertebrates became by far the commonest approach (Metcalfe, 1989; Metcalfe-Smith, 1994). Each country or, sometimes, region of a country tended to develop their own methodological procedures (Knoben et al., 1995). These incorporated a common internal approach to sampling, sample processing, indexation and quality classifications (Birk & Hering, 2002).

Whilst a range of specific monitoring traditions was evolving in individual states, the formation of the European Union resulted in a growing convergence of the legislative infrastructure of its Member States and the strategies adopted to implement this legislation. The mechanism commonly used to implement common community practices has been the issue of a directive from the European parliament. In the 1990's pressure grew for the rationalisation of these 'water quality' directives into a single overarching directive to meet this objective (Mandl, 1992). The resultant directive, commonly known as the Water Framework Directive or WFD, was published in 2000 (European Commission, 2000).

Significantly, the directive embraced the concept of the 'Reference Condition' (Hughes, 1995) as a unifying concept for aiding the harmonization of results obtained in a variety of different countries/regions using a variety of their own 'traditional' assessment protocols. This concept had already been applied successfully in the United Kingdom through the development and application of RIVPACS (Wright et al., 1989, 2000) and had subsequently been taken up outside Europe in Australia (Norris, 1994) and Canada (Reynoldson et al., 1995, 2000; Rosenberg et al., 2000).

The WFD recognised type specific biological reference conditions based on a physical and chemical typology of surface water bodies in each European eco-region sensu Illies (1978). For this purpose Member States were expected to develop a reference network for each stream type containing a sufficient number of sites of high ecological status to provide a sufficient level of confidence about the values for the reference condition.

The term 'Ecological Status' was the overarching term coined by the WFD to represent the 'quality of the structure and functioning of aquatic ecosystems associated with surface waters'. Five categories of Ecological Status are recognised by the directive; High, Good, Moderate, Poor and Bad. The WFD provides normative definitions of the biological community structure associated with the High, Good and Moderate status classes. (European Commission, 2000). Member States are required to implement programmes of measures in order that all surface water bodies achieve at least 'good Ecological Status' within a defined timetable.

Whereas only macroinvertebrate data were required for the application of most prediction and assessment systems, the WFD required the sampling and interpretation of data on a broader suite of 'biological quality elements' (BQEs). These included phytoplankton, other aquatic flora, macroinvertebrates and fish. Parameters to be considered for each element are the composition and abundance of its biotic assemblages. In addition the age structure of fish populations shall be taken into consideration.

In common with systems such as RIVPACS (Wright et al., 2000), the WFD required that observed metric values for BQEs in a water body undergoing monitoring were mathematically compared with expected values for reference condition sites based on predictive modelling, hindcasting or expert judgement. The WFD presumed that the ratios so-calculated would be in the range 0–1 and the numerical value derived by such a comparison was termed the Ecological Quality Ratio (EQR). The division of the value range of an EQR into classes provides a mechanism for categorising the ecological status of sites.

The precise BQEs to be monitored will be dependant on the type of monitoring to be undertaken. The WFD recognises three forms of monitoring: surveillance (to provide an assessment of the overall surface water status within each catchment), operational (to establish the status of water bodies identified as being at risk of failing to meet environmental objectives) and investigative (the source and magnitude of a specific pollutant). In surveillance monitoring, parameters indicative of all biological elements shall be monitored except where it is not possible to establish reference conditions for a particular element due to that element's high degree of natural variability in the water body being monitored. In contrast, operational and investigative monitoring may be restricted to one or two BQEs.

In addition to the direct monitoring of the biological assemblages, the other quality elements to be monitored for the classification of Ecological Status comprise hydromorphological, chemical and physiochemical elements supporting the biological elements.

Common Implementation Strategy

The WFD sets the framework for future monitoring of surface waters and sets out the mechanisms for reporting on the results of monitoring programmes and the formulation of river basin management plans, based upon the information gathered by monitoring and other sources. However, it is not prescriptive of the methodologies to be used to collect and process biological samples nor the specific metrics or multi-metrics to be used to calculate the Ecological Quality Ratios or the class value limits of these EQRs for each of the five classes of Ecological Status. It also provides no specific guidance on how the results of monitoring of the many and diverse quality elements shall be integrated in order to provide a single classification of the water body's status nor on how estimates of the required level of confidence and precision should be made.

For these reasons (European Commission, 2001), a Common Implementation Strategy (CIS) was established in order to develop common understanding of the technical and scientific implications of the directive and, in so doing, to achieve its harmonised implementation. Amongst the many guidance documents emanating from CIS working groups are reports on the establishment of the Intercalibration Network and on the intercalibration exercise (European Commission, 2002), on establishing reference conditions and ecological status class boundaries (European Commission, 2003a), on monitoring for the WFD (European Commission, 2003b) and on the overall approach to the classification of Ecological Status (European Commission, 2003c).

A series of Geographical Intercalibration Groups (GIGS) have been set up to agree on the intercalibration strategy to be adopted in discrete geographical areas of the European Union. Fifteen GIGS have been established including five river groups for the regions Mediterranean, Central, Alpine, Eastern Continental and Northern. A defined number of countries comprise each GIG but individual countries may belong to more than one GIG if the variation in the river types within its borders qualifies it to do so.

Supportive European Commission research projects

AQEM

In support of the technical activities associated with the implementation of the WFD, the European Union has commissioned a series of research projects designed to provide scientific support for the technical processes. The first of these projects specifically concerned with the assessment of Ecological Status was the AQEM project (EVK1-CT1999-00027). The structure and objectives of the project and the main scientific findings and applied outputs are described in a special issue of *Hydrobiologia* (Hering et al., 2004b).

The AQEM project established a standard macroinvertebrate sampling protocol, the AQEM method, and a common field protocol for recording hydromorphological, physical, chemical and geographical information concerning the study sites and their upstream, downstream and riparian environs (Hering et al., 2004a). Outputs of the project include a database (AQEMDip) for the orderly storage and retrieval of macroinvertebrate and environmental data and a river assessment program (now termed ASTERICS) for calculating the values of almost 200 biological metrics and selected national multi-metric systems.

Whilst the AQEM project addressed many of the key questions associated with the use of macroinvertebrate data for assessing the Ecological Status of surface waters, the directive also required the integration of other biological quality elements together with the hydromorphological, chemical and physical elements that support the biological elements.

STAR

STAR is a European Commission Framework V project (EVK1-CT-2001-00089) with the full title of 'Standardisation of river classifications: Framework method for calibrating different biological survey results against ecological quality classifications to be developed for the Water Framework Directive'. The project is categorised as 'Pre-normative, co-normative research and standardisation'. It therefore seeks to provide practical solutions to some of the additional problems associated with the implementation of the directive. Issues addressed include comparison of macroinvertebrate sampling methods, the effectiveness of the use of different organism groups in different stream types and for different stressors, variation and uncertainty in the collection and interpretation of biological data, the inter-calibration of assessment methods for the allocation of Ecological Status, the formulation of drafts for the relevant CEN bodies, and the development of a decision support system to assist water managers in applying the project findings. In this paper the objectives of the STAR project and the methodological approach adopted to achieve these aims will be described. It will provide the background for the remaining papers that comprise this special issue of Hydrobiologia.

FAME and REBECCA

Clustered with the STAR project and collaborating closely with it has been another EC Framework V project, FAME (EVK1-CT-2001-00094). This project has developed a specific system for the assessment of the Ecological

Status of surface waters based on the fish communities that they support (Noble & Cowx, 2002). The STAR project is also working collaboratively with the EC Framework VI project, REBECCA (SSP1-CT-2003-502158), that aims to provide new interpretations of the relationships between Chemical and Ecological Status of surface waters in order to support the implementation of the WFD.

Objectives

The central objectives of STAR and the papers in this volume that address them are:

- *Which methods or biological quality elements are best able to indicate certain stressors?*

The varying responses to stressors of different biological quality elements will allow WFD monitoring data to be interpreted in a diagnostic manner in order to identify the pressures operating on aquatic systems. Advice on the selection of the most appropriate BQEs for specific objectives and in specific regions is provided by Johnson et al. (2006a, b) and Pinto et al. (2006). Other authors consider specific techniques (Kokeš et al., 2006 – PERLA) or taxonomic groups and stressors (Szoszkiewicz et al., 2006b – macrophytes and organic pollution; O'Hare et al., 2006 – macrophytes and habitat alteration). In addition the application of River Habitat Survey Techniques (Raven et al., 1998) to the evaluation of the hydromorphological condition of watercourses is evaluated by Erba et al. (2006) and Szoszkiewicz et al. (2006a).

- *Which method can be used on which scale?*

The organism groups that the WFD require to be considered in assessing the Ecological Status of waterbodies indicate environmental change on different scales. The issue of scale has been considered by Springe et al. (2006) and Verdonschot (2006a).

- *Which methods are suited for early and late warnings?*

Besides the spatial dimension, different organism groups indicate change on different temporal dimensions, thus providing different signals of early or late warning. Johnson et al. (2006b) address this issue for all BQEs except phytoplankton.

- *How are different assessment methods affected by errors and how can 'signal' be distinguished from 'noise'?*

STAR has investigated a range of factors that confound the ability of bioassessment procedures to detect change and many papers in this volume address the issue of uncertainty. These include Besse-Lotoskaya et al. (2006) who investigate uncertainty associated with diatom sampling and interpretation, Clarke et al. (2006a, b) and Lorenz & Clarke (2006) who look at the impact of sampling variation on macro-invertebrate assessments, Haase et al. (2006) who consider the effects of macro-invertebrate sorting and identification errors; Staniszewski et al. (2006) and Baattrup-Pedersen et al. (2006) who examine uncertainty associated with macrophyte surveys and Johnson et al. (2006a) who explore the incidence and effects of Type I and Type II errors for most BQEs. One factor that may influence the evaluation of sites is the method used to define reference conditions. Davy-Bowker et al. (2006) consider the implications of using type specific conditions based on a physical/chemical typology with those site specific reference conditions produced by predictive systems such as RIVPACS and PERLA (Kokeš et al., 2006).

- *How can data from different assessment methods and taxonomic groups be compared and intercalibrated and how can the results of the STAR programme be used to assist the WFD intercalibration exercise?*

A central problem, for the implementation of the WFD, is how biological data collected using different national protocols and biological quality elements (BQEs) can be compared and integrated in order to derive comparable allocations of sites to standard European classes of environmental degradation. Friberg et al. (2006) compare the main macroinvertebrate sampling procedures used in Europe, whilst alternative mechanisms for inter-calibration are discussed by Birk & Hering (2006 – macroinvertebrates), Birk et al. (2006 – macrophytes) and Buffagni et al. (2006 – general but principally macroinvertebrates).

- *How can the cost-effectiveness of field and laboratory protocols for the collection and processing of macroinvertebrate samples be optimised?*

Methodologically, standardisation must also take a balanced account of the relative costs and ecological effectiveness of different field and laboratory procedures. Experimental field studies were devised to consider a spectrum of relevant issues (Šporka et al., 2006; Vlek et al., 2006) whilst Verdonschot (2006b) examined the significance of varying levels taxonomic precision on the biological typology of European streams and rivers.

- *Can species trait analysis provide a unifying procedure for the establishment of reference conditions and the assessment of Ecological Status?*

An aim of STAR was to test the applicability of a species trait analysis as a unifying theme for the derivation of functionally based reference conditions and, as a result, for the assessment of Ecological Status. These results are presented elsewhere including as Deliverable N2 (Bis & Usseglio-Polatera, 2004) on the STAR website – www.eu-star.at

- *How can boundaries of the five classes of Ecological Status recognised by the WFD be best set?*

On the basis of the field and laboratory protocols and metrics that will be tested in STAR, an aim of the project is investigate and to elaborate standard procedures for the determination of European class boundaries of Ecological Status. Mechanisms for setting and inter-calibrating class boundaries are considered by Birk & Hering (2006), Birk et al. (2006) and Buffagni et al. (2006).

- *How can the results of the STAR programme be used to make recommendations for common European standards?*

The STAR consortium have suggested outline standards, on methodological issues related to the implementation of the WFD that are being considered by CEN (Comité Européen de Normalisation) for adoption as full standards. These include multi-habitat sampling for invertebrates, the construction of multi-metric assessment systems and the selection of the best suited organism groups for specific monitoring purposes. Methodologies for developing multi-metric indices are elaborated in this volume by Hering et al. (2006). An additional standard tool for the use of the European water industry and academia is a pan-European macro-invertebrate ecological database and taxa inventory described here by Schmidt-Kloiber et al. (2006).

Approaches: site selection

Research framework

The STAR consortium comprised 22 partners from 14 countries including four countries who were candidate states, the Czech Republic, Slovakia, Poland and Latvia, that acceded to the European Union during the course of the project on 1st May 2004.

The project was divided into 19 discrete but inter-linked workpackages (Table 1). Most workpackages (WPs) could be allocated to one or other of two loose groupings. There were ten core WPs in which most partners worked collaboratively on a common activity and nine that were specific research programmes contributed to by a small minority of the partners and predominantly engaged in by a dominant leading institute (Table 1).

Stream types studied

The central components of the STAR project were the two WPs devoted to the collection of new biological, hydromorphological and other environmental data (WP7 and WP8).

WP7 (Table 1) involved the selection and monitoring of sites in two core groups of stream types (Table 2). The variables and their ranges used to define each group were those involved in the system A approach to surface water body typology given in the WFD (European Commission, 2000). Sites in core group 1 were defined as 'Small, shallow, upland streams' in early STAR project documentation. In WFD system A terms they are sites with a 'small' catchment situated in the lower 60% of the 'mid-altitude' range. Core group 2 sites were defined in early STAR

Table 1. The 19 STAR project workpackages (WPs)

No.	Theme
1	*Project co-ordination*
2	*Project homepage*
3	*Review of data on reference conditions and existing assessment methods using benthic invertebrates, fish, phytobenthos, macrophytes and river habitat surveys, national standards on sampling, analysis and quality evaluation, related national projects and existing databases*
4	*Acquisition of existing data*
5	Selection of sampling sites
6	Sampling workshops to standardise the understanding and application of sampling protocols between participants and to undertake replicate sampling programmes for diatoms and macroinvertebrates
7	Investigation of core stream types 1 (small, shallow, mountain streams) and 2 (medium-sized, deeper, lowland streams)
8	Investigation of additional stream types
9	Audit of performance in the processing and identification of macroinvertebrate and diatom samples
10	Generation and hosting of the project database
11	Comparison and linking assessment systems based on invertebrates
12	Linking of assessment systems working with different organism groups
13	Linking of the project database and the database of existing data
14	Recommendations for standardisation to support CEN in its development of appropriate standard methods for the WFD
15	*Elaboration of a decision support system, implemented through a DSS computer program, to provide practical guidance in the application of monitoring programmes necessary to meet the terms and objectives of the Water Framework Directive*
16	*Examination of the effectiveness of and relative cost-efficiency of different field and laboratory protocols for the collection and processing of macroinvertebrate samples*
17	*Examination of the value of species trait analysis as a unifying system for the establishment of functionally based reference conditions and the assessment of Ecological Status*
18	*Spatial scale analyses*
19	*Study of errors and variation associated with field protocols for the collection and application of macrophyte and hydro-morphological data in the implementation of the WFD*

The 10 core collaborative WPs are shown in regular font.

documentation as 'Medium-sized, deeper lowland streams'. In WFD system A terms they have 'medium' catchment sizes and are situated at 'lowland' altitudes.

WP8 (Table 1) involved the selection and sampling of a group of 'additional' stream types. Additional streams types were not prescriptively allocated to any WFD system A typology and could include sites whose combination of altitude and catchment size characteristics might or might not fit the definition of either core stream type groups 1 or 2. In general terms they were confined to either the system A mid-altitude or lowland categories and to the system A small, medium or, very occasionally, large catchment size categories. Initially the additional stream types were selected to fulfil four specific roles. These were to:

- allow new, characteristic sites of individual states to be included in the analysis;

Table 2. Definitions of the two STAR core stream type groups

Core stream type		Theoretical value range of typological variables		
No.	Description	Altitude	Catchment size	Geology
1	Small, shallow, upland streams	200–500 m	10–100 km^2	Calcareous or siliceous
2	Medium-sized, deeper lowland streams	<200 m	>100–1000 km^2	Calcareous, siliceous or organic

- provide an opportunity to extend the range of sites in existing European assessment systems;
- extend the range of sites at which the specific field methods are compared;
- provide an opportunity to test alternative sampling/assessment methods of specific importance to individual consortium Member States.

However, the data for core and additional stream types were used jointly in most analyses. Core and additional stream types could also be defined as either calcareous, siliceous or, occasionally, organic but, with a few exceptions, sites within specific site sets (see the following section) were all in the same geological category.

Selection of site sets

Each participating partner in WP7 and/or WP8 selected a minimum of one and a maximum of three sets of sites to sample. Each set of sites was in one of the three basic stream type groups (core 1, core 2 or additional) described in the previous section. Partners with two or more site sets selected these sets to be either in the same or different stream type groups. Sets of sites defined by their stream type group, eco-region or sub-eco-region and, optionally, other geographical criteria are termed 'stream types'. The definition of stream types used here is that established by the AQEM project and is "an artificially delineated but potentially ecologically meaningful entity with limited internal biotic (taxa composition) and abiotic (chemical and hydromorphological) variation and a biotic and abiotic discontinuity toward other types" (Hering et al., 2004a). Selection of specific stream types within the three stream type groups defined in the previous section took account of many of the criteria for stream typology in System B of the WFD.

In total, 22 stream types were selected for study as part of either WP7 or WP8 (Table 3). In addition, two other stream types in Italy (small-sized calcareous streams in the Southern Apennines and medium-sized calcareous streams in the Northern Apennines) and three other stream types in Greece (small-sized siliceous streams in Northern Greece, medium-sized calcareous streams in Southern Greece and small-sized siliceous streams on the Aegean Islands) were sampled for other national purposes connected with the STAR project.

For each stream type, a minimum of ten and a maximum of 24 sites were sampled (Table 4). For each stream type, sites were selected to represent a gradient of degradation usually due to a pre-identified dominant stressor (Table 4). For the purpose of site selection, these dominant stressors were divided into three broad categories: organic pollution (including eutrophication), toxic pollution (including acidification) and habitat degradation. In one case, (stream type I06 – Italy) a single dominant stressor could not be identified and the category 'general' stressors was applied. In a few other cases (see Table 4) different dominant stresses applied to specific sites within a stream type and some of these only became apparent during the sampling programme.

In general, approximately 25% of sites in each site set were selected to be likely to be of 'high', 25% of 'good', 25% of 'moderate' and 25% of 'poor'/'bad' Ecological Status. The 'high' status sites were selected to represent the reference condition for their particular stream type. Reference condition sites were selected through a combination of site visits, cartographic information and information derived from new biological sampling or existing sample data held by internal (i.e. partner's own) or external (e.g. national monitoring organisations) sources. Where adequate data were available, all biological quality elements and hydromorphological and chemical quality elements were considered. However, in many cases the most important elements considered were macroinvertebrates, hydromorphology and nutrient status. In order to aid the process of reference site selection a list of criteria was developed (Table 5) based on Hering et al. (2003) but modified in response to the ongoing discussions of the REFCOND group.

In many cases, e.g. some lowland stream types or larger streams, no reference sites meeting all of the criteria above were available. For these stream types the 'best available' existing sites were selected. However, where possible, the description of reference communities of these types could be supplemented by evaluation of historical data and possibly the biotic composition of comparable stream types, e.g. streams of a similar size but located in different ecoregions.

The remaining sites, other than reference sites, were pre-classified using the same sources of information but with particular attention to

Table 3. The 22 stream types sampled as part of WP7 (core streams) or WP8 (additional streams)

Country	Core stream type group 1		Core stream type group 2		Additional stream type	
	No.	Name	No.	Name	No.	Name
Austria	A05	Small-sized, shallow mountain streams			A06	Small-sized, crystalline streams of the ridges of the Central Alps
Czech Republic	C04	Small-sized, shallow mountain streams			C05	Small-sized streams in the Central sub-alpine Mountains
Denmark			K02	Medium-sized lowland streams		
France					F08	Small-sized, shallow headwater streams in Eastern France
Germany	D04	Small-sized, shallow mountain streams	D03	Medium-sized lowland streams	D06	Small-sized, Buntsandstein streams
Greece					H04	Small-sized, calcareous mountain streams in Western, Central and Southern Greece
Italy					I05	Small-sized streams in the southern calcareous Alps
					I06	Small-sized, calcareous streams in the Central Apennines
Latvia			L02	Medium-sized lowland streams		
Poland			O02	Medium-sized lowland streams (Ecoregion 14)		
			O03	Medium-sized lowland streams (Ecoregion 16)		
Portugal					P04	Medium-sized streams in lower mountainous areas of Southern Portugal
Slovakia	V01	Small-sized, calcareous mountain streams in the Eastern Carpathians				
	V02	Small-sized, siliceous mountain streams in the Western Carpathians				
Sweden			S05	Medium-sized lowland streams	S06	Medium-sized streams on calcareous soils
United Kingdom			U23	Medium-sized lowland streams	U15	Small-sized, shallow, lowland streams

Table 4. The number of WP7 and WP8 sites sampled for each biological quality element, where the qualifying criterion is that sites must have been fully sampled for macroinvertebrates (except Slovakia, where regular 'national' sampling was not undertaken)

Country	Site type	Number of sites	Geology	Degradation	Number of sites sampled for each quality element				
					Diatoms	Macrophytes	Macroinvertebrates	Fish	Hydromorphology
Austria	A05	15	Siliceous	Stream morphology	11	11	13	12	13
	A06	21	Siliceous	Stream morphology	7	8	8	0	8
Czech Republic	C04	14	11 siliceous and 3 calcareous	Organic pollution	14	14	14	14	14
	C05	10	Siliceous	Stream morphology	10	10	10	10	10
Denmark	K02	12	9 calcareous and 3 siliceous	Stream morphology	11	11	11	11	11
France	F08	12	Calcareous	Organic pollution	12	12	12	12	12
Germany	D03	13	Siliceous	Stream morphology	10	12	12	9	12
	D04	12	Siliceous	Stream morphology	10	12	12	10	12
	D06	10	Siliceous	Stream morphology	6	6	6	6	6
Greece	H04	10	Calcareous	Organic pollution	10	10	10	10	10
Italy	I05	10	Calcareous	Stream morphology	10	0	10	10	10
	I06	11	Calcareous	General	11	11	11	11	11
Latvia	L02	24	16 siliceous, 6 calcareous and 2 siliceous/calcareous	Organic pollution	24	24	24	24	24
Poland	O02	13	12 siliceous and 1 siliceous/calcareous	Organic pollution	13	13	13	13	13
	O03	12	8 organic, 2 siliceous and 2 organic/siliceous	Organic pollution	11	11	11	11	11
Portugal	P04	10	Siliceous	Organic pollution	10	10	10	10	10
Slovakia	V01	12	Calcareous	Organic pollution	12	12	12	12	12
	V02	12	Siliceous	Organic pollution	12	12	12	12	12
Sweden	S05	16	Siliceous	Acidification/toxic (5 sites), organic (4 sites), reference (3 sites), undefined (4 sites)	14	16	16	16	16
	S06	11	Calcareous	Organic pollution	9	11	11	11	11
United Kingdom	U15	13	Calcareous	Organic pollution	13	13	13	13	13
	U23	12	11 calcareous and 1 siliceous	Organic pollution	12	12	12	12	12
Totals		285			252	251	263	249	263

Table 5. Criteria for reference site selection

General

- The reference condition must be politically palatable and reasonable
- A reference site, or process for determining it, must hold or consider important aspects of 'natural' conditions
- The reference conditions must reflect only minimal anthropogenic disturbance

Land use practices in the catchment area

- The degree of urbanisation, agriculture and silviculture should be as low as possible for a site to serve as a reference site
- Least-influenced sites with the most natural vegetation are to be chosen

River channel and habitats

- The reference site floodplain should not be cultivated. If possible, it should be covered with natural climax vegetation and/or unmanaged forest
- Coarse woody debris should not be removed (minimum demand: presence of coarse woody debris)
- Stream bottoms and stream margins must not be fixed
- Spawning habitats for the natural fish population (e.g. gravel bars, floodplain ponds connected to the stream) should be present
- Preferably, there should be no migration barriers (affecting the bed load transport and/or the biota of the sampling site)
- In stream types in which naturally anadromous fish species would occur, the accessibility of the reference site from downstream is an important aspect for the site selection
- Only moderate influence due to flood protection measures can be accepted

Riparian vegetation and floodplain

- Natural riparian vegetation and floodplain conditions must still exist
- Lateral connectivity between the stream and its floodplain should be possible
- The riparian buffer zone should be greater or equal to 3× channel width

Hydrologic conditions and regulation

- No alterations of the natural hydrograph and discharge regime should occur
- There should be no or only minor upstream impoundments, reservoirs, weirs and reservoirs retaining sediment; no effect on the biota of the sampling site should be recognisable
- There should be no effective hydrological alterations such as water diversion, abstraction or pulse releases

Physical and chemical conditions

- No point sources of pollution or nutrient input affecting the site
- No point sources of eutrophication affecting the site
- No sign of diffuse inputs or factors which suggest that diffuse inputs are to be expected
- 'Normal' background levels of nutrient and chemical base load, which reflect a specific catchment area
- No sign of acidification
- No liming activities
- No impairments due to physical conditions
- Thermal conditions must be close to natural
- No local impairments due to chemical conditions; especially no known point-sources of significant pollution, all the while considering near-natural pollution capacity of the water body
- No sign of salinity

Biological conditions

- No significant impairment of the indigenous biota by introduction of fish, crustaceans, mussels or any other kind of plants and animals
- No significant impairment of the indigenous biota by fish farming
- No intensive management, e.g. of the fish population

Underlined criteria were mandatory and as many of the other criteria as possible were met.

reported known sources of stress as supplied by national and regional agencies with responsibility for monitoring water quality and hydromorphology. The relative emphasis placed on individual quality elements varied from partner to partner with some placing more emphasis on biological data and less on hydromorphological and chemical elements than others in establishing their site pre-classification. However, each partner used their specific approach to establish a set of sites with a marked degradation gradient according to their chosen dominant stressor.

In total, excluding the Italian and Greek sites sampled for other purposes, 285 sites were selected for possible sampling for WP7 or WP8. Of these, 263 (Fig. 1) were sampled for macroinvertebrates in each sampling season using both the AQEM and, with the single exception of Slovakia, a second, mainly 'national' sampling protocol in each sampling season. These were the sites included in most of the central analyses undertaken in the project. All of these 263 sites were also subject to hydromorphological surveys and 252 were sampled for phytobenthos, 251 for macrophytes and 249 for fish. A final total of 233 were fully sampled for all biological quality elements.

Selection of quality elements

Three biological quality elements were sampled in all or almost all of the sites contributing to the central project analyses. These were 'aquatic flora', 'benthic invertebrate fauna' and 'fish fauna'. The aquatic flora was subdivided into phytobenthos and macrophytes for the purposes of this project. The only component of the flora not sampled was phytoplankton because this element was considered not to be a significant component of the biota of the small to medium-sized, often fast-flowing streams that predominated in the STAR sampling programme.

At least two survey protocols were used to record components of the hydromorphological quality element supporting the biological elements. Information on chemical and physico-chemical quality elements supporting the biological elements was, in all cases collected both from direct field sampling and surveys and also, in some cases, from data collected by the national water quality monitoring agencies.

Selection of sampling reach

Prior to starting field sampling and surveying, the study reach at each site was selected. The reach was 500 m long and was selected as representative of the hydromorphological conditions of the stretch of river under investigation. A stretch of river is a continuous section of river without any significant tributaries or point sources of pollution likely to modify its Chemical Status (equivalent to a 'water body' as defined by the WFD). The

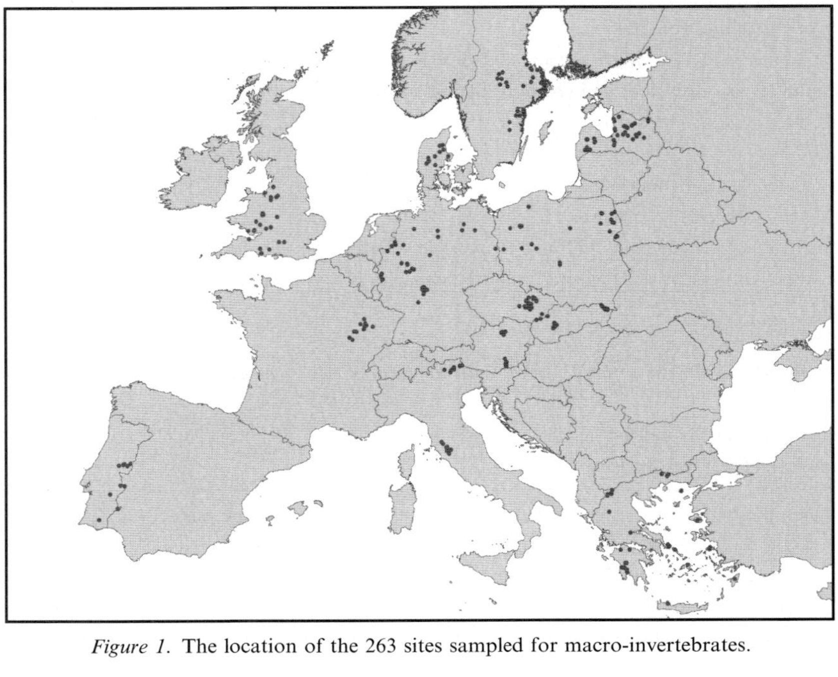

Figure 1. The location of the 263 sites sampled for macro-invertebrates.

selection of the sampling and survey reach normally followed the completion of the AQEM site protocol (see below).

Wherever possible, at each site, a common monitoring strategy was adopted in relation to the relative positions of the different sampling/surveying points. Field surveyors were provided with a conceptual diagram of this strategy (Fig. 2).

A preliminary reach was first located and, within this, the STAR-AQEM invertebrate sampling area was selected. This was a length of river of up to 100 m, depending on stream width (see the STAR-AQEM section below) at which all of the common representative habitat types of that river stretch were present, including both erosional ('riffle') and depositing ('pool') areas if possible. The centre of this sampling length was taken to be River Habitat Survey (RHS) spot check 9 (Raven et al., 1998) and was used to define the exact position of the whole 500 m RHS survey reach.

The 'national' invertebrate sampling and the phytobenthos sampling were undertaken in the same 100 m section as the STAR-AQEM method. Care was taken to minimise the disturbance to the river by each sampling method and overlap between the different precise sampling locations.

The macrophyte survey was undertaken in the 100 m reach immediately upstream of the invertebrate and diatom sampling reach and after these elements had been sampled. Where all three elements were sampled on the same day this spatial separation and sequence of sampling was designed to minimise any trampling of plants resulting from the sampling of the other two elements.

Fish sampling took place over a ≥100 m section immediately upstream of the macrophyte survey area. Fish sampling was normally undertaken on a

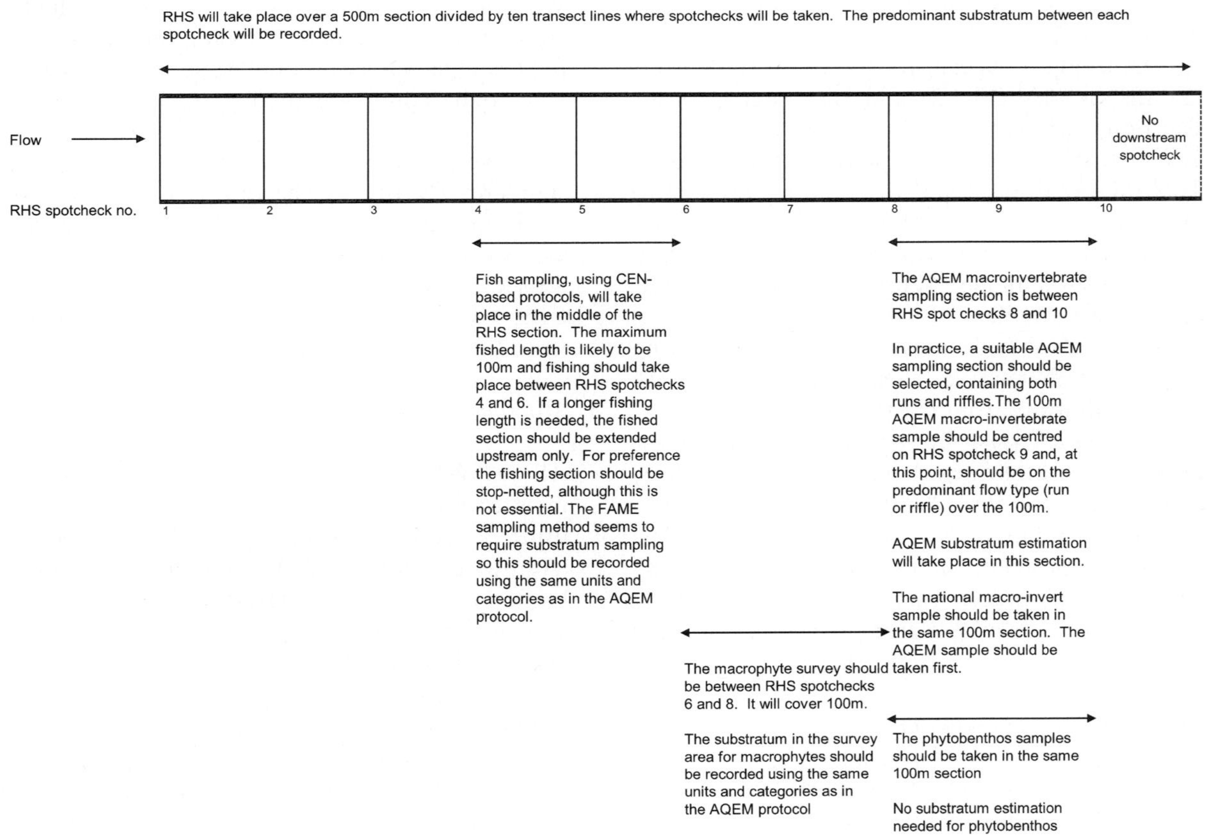

Figure 2. The conceptual locations of the STAR sampling areas for each of the five recorded quality elements at each site, as provided to field surveyors.

separate date to the other sampling. Chemical sampling was from within this 500 m reach of river and avoided any disturbance to the sediment caused by biological sampling.

Whilst this strategy represented the ideal, on occasions local conditions required variations in the general pattern of sampling. Such departures from the optimal were kept to the minimum.

Approaches: field and laboratory protocols

Phytobenthos – diatoms

Diatoms were sampled once only at the WP7 and WP8 sites. Samples were collected during periods of stable stream flow and at least four weeks after a period of extreme conditions like a major storm or drought. The time of these stable conditions varied from region to region. Subject to this criterion, spring was the preferred season for sampling as diatoms dominate the phytobenthos during this season (Moore, 1977).

The location selected for STAR-AQEM invertebrate samples (Fig. 2) and the criteria for its selection, ensured that it also had the most suitable available substrata for sampling benthic diatoms. This ranged from stones to macrophytes and to mineral sediments, depending on the type of river. Selection criteria also ensured that the sampled section also combined riffles and pools and thus enabled the sampling of a good variety of natural substrata. Bank side areas were avoided during sampling with samples taken at least 10% of the river width away from the river edge.

In general terms, the sampling and processing protocols used followed those of Kelly et al. (1998) and Winter & Duthie (2000). Methods conform to the CEN standards EN 13946 and EN 14407. The full STAR protocol for the sampling, processing and audit of diatom samples was prepared by Alterra and is available from the STAR website (www.eu-star.at).

Phytobenthos – non-diatoms

Collection of non-diatom phytobenthos was voluntary and not all partners collected information on this taxonomic group. Where partners did collect and process material they adopted the methods described in the project protocol for sampling, processing and audit of non-diatom benthic algal samples, which was also prepared by Alterra and is available from the STAR website.

Macrophytes

Macrophytes were surveyed once only at the WP7 and WP8 sites. Surveys were undertaken using a slightly adapted form of the Mean Trophic Rank (MTR) field protocol developed in the United Kingdom (Holmes et al., 1999). Most surveys were carried out between mid-June and mid-September after several days of low flow or low-normal flow as opposed to high flow/spate.

The MTR survey procedure is based on the presence and abundance of species of aquatic macrophytes, where a macrophyte is defined as 'any plant observable with the naked eye and nearly always identifiable when observed' (Holmes & Whitton, 1977). This definition includes all higher aquatic plants, vascular cryptograms and bryophytes, together with groups of algae which can be seen to be composed predominantly of a single species.

Survey techniques conformed to the CEN standard EN 14184. The full STAR survey protocol for macrophytes (Guidance for the field assessment of macrophytes of rivers within the STAR project) was prepared for STAR by the Centre for Ecology and Hydrology and is available from the STAR website.

Macroinvertebrates

All 263 WP7 and WP8 sites listed in the macroinvertebrates column of Table 4 were sampled using a modified form of the AQEM method (AQEM consortium, 2002; Hering et al., 2004a), known as the STAR-AQEM method. With the exception of all Slovakian sites in stream type V01 and six sites in V02, all sites were also sampled using a current national method of the country (Table 6). Where no consistent national sampling method existed for the country, either the RIVPACS (Austria, Germany and Greece) or PERLA (Slovakia) methods were used instead (Table 6). With the exception of the three non-UK RIVPACS users, the national methods used were

Table 6. 'National' sampling methods applied in each STAR country participating in project workpackages 7 and 8

Country	Methods applied	Reference
Denmark	Danish Stream Fauna Index (DSFI)	Danish Environmental Protection Agency (1998)
Italy	Indice Biotico Esteso	Ghetti (1997)
France	Indice Biologique Global Normalisé (IBGN)	GAY, Cabinet en Environnement (1994)
Latvia	LVS 240:1999	Unpublished (see 'Protocols' on www.eu-star.at)
Czech Republic	PERLA	Kokeš et al. (2006)
Slovakia	PERLA	
Poland	Polish national method	Unpublished (see 'Protocols' on www.eu-star.at)
Portugal	Portuguese national method (PMP)	Unpublished (see 'Protocols' on www.eu-star.at)
Austria	River In-Vertebrate Prediction And Classification System (RIVPACS)	Murray-Bligh et al. (1997)
Germany	River In-Vertebrate Prediction And Classification System (RIVPACS)	
Greece	River In-Vertebrate Prediction And Classification System (RIVPACS)	
United Kingdom	River In-Vertebrate Prediction And Classification System (RIVPACS)	
Sweden	Swedish national method	Swedish Environmental Protection Agency (1996)

assumed to be the methods likely to be adopted by their countries for implementing the WFD.

Immediately prior to sampling, the length of river to be sampled was surveyed as part of the AQEM field protocol and the proportions of the different habitats present at the river bottom were estimated (Hering et al., 2004a). This knowledge was used to establish the precise STAR-AQEM sampling area and the proportions of micro-habitats to be sampled (Hering et al., 2004a). Normally, the STAR-AQEM sample was the first to be collected followed by the 'national' sample. For the national sample, care was taken to avoid the specific locations at which the STAR-AQEM sample was collected.

Most samples were fixed and/or preserved in the field using a fixative/preservative of the partner's choice which was normally either formaldehyde solution or ethanol of varying strength. Exceptions to this generalisation were the Italian IBE and most Latvian LVS 240:1999 samples that were sorted at the bankside, and some Portuguese 'national' samples that were sorted live in the laboratory within 48 h of collection. Prior to preservation and/or transport to the laboratory, large and easily identified specimens and identifiable specimens of taxa of known conservation importance or particular fragility to damage were recorded and returned live to the river.

The laboratory sample processing techniques were specific to the particular field protocols and differed between the STAR-AQEM and 'national' samples and between the different 'national' field protocols. However, all partners were trained to collect and process STAR-AQEM samples in a consistent and prescriptive manner.

In all cases taxa were identified to the best achievable level, according to the expertise of the partner and the availability of adequate national keys. Most partners achieved species level identification for most groups but this was not possible in Latvia, where only some groups could be identified to this level, nor in Greece, Italy or Portugal where most identifications were to family level.

It is not possible to describe each field and laboratory protocol here but the key features of each method are provided Friberg et al. (2006). For further details the reader is directed to the references given in Table 6, to the 'Protocols' section of the STAR website (www.eu-star.at), the AQEM website (www.aqem.de) for the key principles of the AQEM method that formed the basis of the STAR-AQEM procedure and to the 'Waterview' database developed during the STAR project (Birk & Hering, 2002) and accessible via the 'Review' section of the STAR website. LVS 240:1999, the Latvian national sampling protocol (Latvian Standard Ltd., 1999) and can also be accessed via www.lvs.lv/en/services/services_EP.html.

In all cases where hand-net sampling was employed sampling and equipment specifications were consistent with CEN standard EN 27828. Where Surber sampling was used, as in the case of the French IBGN method and, occasionally, in the STARAQEM method, sampling and equipment specifications were consistent with CEN standard EN 28265.

Fish

The fishing strategy used conformed to CEN standard EN 14011 and was developed following discussions with STAR's cluster project FAME (http://fame.boku.ac.at). Almost all STAR sites were, on average, less than one metre deep and, in these circumstances, the STAR protocol was based on the section of EN 14011 relating to electric fishing of wadeable rivers.

Where possible, fishing was carried out using direct current (dc) fields. However where this was not possible, due to high conductivity water, variable electrical characteristics of stream topography or poor fish response to dc field pulsed, direct current (pdc) fields were used. In all cases, fields were adjusted to the minimum voltage gradient and current density concomitant with efficient fish capture.

Optimally, the length of river fished was a minimum of 100 m and located in the centre of the RHS survey area (Fig. 2). Normally the full width of the river was surveyed over this length. However, in a small minority of cases the fishing reach was slightly shorter than 100 m for logistical reasons (Table 7). The relative position of the fishing area within the survey area was also sometimes varied for practical reasons. Wherever possible the fishing area was demarcated by upstream and downstream stop nets (Table 7). Net mesh sizes were suitable for preventing fish >5 cm from escaping. A minimum of two fishing runs was undertaken at most sites.

In the small number of cases where sites were not wadeable, fishing was undertaken from a boat (Table 7) and at a series of spot locations within the RHS survey area. In such circumstances, stop nets were not used, the sites were normally >10 m wide and the length of river sampled was often less than 100 m. Some wide, wadeable sites in Sweden were also sampled discontinuously without stop nets.

All or most of the following elements of the fish population in the sample area were recorded:

- Number of species
- Species composition (percentage of each species by number)
- Fish density by species (number of fish per m^2) of individuals other than young of the year. There was no requirement to measure or age fish
- Young of the year per species (qualitative assessment by class, e.g. abundant, common or rare)
- Ratio between number of phytophils and limnophils (fish species grouped by reproductive guild (Balon, 1975; Mann, 1996)
- Number of intolerant or sensitive species in terms of functionally descriptive fish species (i.e., salmonids for water quality, migratory species for connectivity, etc.)
- Number of endemic species (species which are only present in the river basin under study)
- Number of native species (species known to be present in the watercourses of the country for a long period of time i.e. >200 years)
- Subjective assessment of degree of infestation of external parasites or other diseases

Final population estimates, capture efficiency and standard errors of population numbers were also determined. Two catch estimates were based on the Seber & LeCren (1967) method but where more than two capture runs were undertaken values were calculated using the Exact Maximum Likelihood methodology.

Hydromorphology

At least two standard site assessment protocols, River Habitat Survey and the AQEM site protocol, were conducted at each STAR WP7 and WP8 site. Only one RHS survey was undertaken in each

Table 7. Fish sample procedures operated by STAR partners in each country participating in WP7 and WP8

Parameter	Units /categories	United Kingdom	Germany	Austria	Sweden	Czech Republic	Greece	Italy (LABBIO)	Portugal	Denmark	France	Poland	Latvia	Slovakia (Eastern)	Slovakia (Central)
Normal length river fished at each site	Metres	100	280	100	50	100	100	100	100	100	100	100	70	100	100
Maximum length of river fished at each site	Metres	100	810	130	50	123	100	100	100	100	100	100	80	120	100
Minimum length of river fished at each site	Metres	80	100	30	50	63	100	100	100	100	80	100	50	50	30
Most common depth category at sites	A = <25 cm B = 25–<50 cm C = 50–<100 cm D = ≥1 m	B	B	B	B	A	B	A	B/C	C	B/C	C	B	B	A
Use of boat to undertake fishing	A = Always B = Sometimes C = Never	C	B	C	C	C	C	C	C	B	C	C	C	C	C
Use of stop nets to enclose the fishing area	A = Always B = Sometimes C = Never	A	B	A	C	B	A	C	A	C	A	A	C	B	B
Multiple fishing runs at each sampling site	A = Always B = Sometimes C = Never	A	B	A	A	B	B	A	A	A	A	A	A	A	A
Maximum number of fishing runs at a single site	Number	3	5	3	3	2	2	2	3	3	2	3	3	2	2
Season most commonly sampled	A = Spring only B = Summer only C = Summer & autumn D = Autumn only	D	C	C	D	C	C	D	A	B	C	D	C	B	B

reach and normally in the period July to September. However, the AQEM site protocol comprised both time variant and time invariant variables. Time invariant variables were recorded during the first site visit only but time variant variables were recorded at the time of each macro-invertebrate sampling.

RHS is a system for assessing the quality of rivers based on their physical structure. The technique and associated interpretation comprise a standard field survey, a bespoke database and indexation systems for Habitat Quality Assessment (HQA) and Habitat Modification Score (HMS). River Habitat Survey (RHS) was developed and applied in the United Kingdom (Raven et al., 1998) but has also been applied in other European countries and has been specially adapted for use in southern Europe (Buffagni & Kemp, 2002). The southern European modifications to the system were principally, but not exclusively, to cater for the braided channels that commonly occur there (Buffagni & Kemp, 2002). The procedures conform to the evolving CEN standard prEN 14614.

The AQEM site protocol, as its name implies, was developed for the AQEM project and subsequently modified and simplified for the STAR project. A brief outline of the method is provided by Hering et al. (2004a) and fuller details of the original AQEM system and the modified STAR version are given on the two project websites (www.aqem.de and www.eu-star.at).

In addition separate site assessment protocols for phytobenthos, MTR and 'national' macroinvertebrate sampling were completed at most sites. Field measurements were complemented by cartographic information assembled for most of the protocols. The fish surveys were also complemented by a standard suite of site information required for the Fides database and site indexation system developed by the FAME project. These including field measurements, a broad suite of cartographic information and information on the biological, hydromorphological and chemical condition of the site collated from published data and from data supplied by national monitoring agencies.

The various site protocols adopted at STAR sites are each too complex to document in detail here and the reader is generally referred to the cited literature for a more complete understanding of the components and implementation of the RHS and AQEM site protocol techniques.

Approaches: quality control and uncertainty

A specific requirement of the WFD is that Member States support their assessments of Ecological Status of water bodies with estimates of the level of confidence and precision of the results of the monitoring programmes. The sources of uncertainty associated with results will include components resulting from each of the sampling and surveying process, the sorting of samples, identification of the sorted material, data logging and the precision of the models, hind casting or other procedures used to set the reference condition to calculate EQR values.

The assessment of many of these sources of uncertainty has been researched by Clarke (2000) and Clarke et al. (2002, 2003). In the STAR project, focus was on the uncertainty associated with sampling, sample sorting and the identification of specimens, where errors of data logging were considered together with identification errors.

A presumption of a well-implemented monitoring programme is that the persons responsible for carrying out each stage in the process are well trained and competent in the tasks that they are undertaking. In the STAR project partners carrying out the sampling process initially had variable experience of the tasks that they were required to perform. Even where they were experienced and proficient in parts of some tasks, such as collecting macroinvertebrates using their specific national method, they were often less well trained and experienced in collecting STAR-AQEM samples.

Sampling and survey training and identification courses

Prior to any sampling, extensive training course were arranged in the field and laboratory procedures to be used. Representatives of all partners were trained in diatom sampling and preservation, MTR procedures and River Habitat Survey. Particular emphasis was based on consistent application of the STAR-AQEM site and sampling procedures since these were the common standards against which other sampling protocols were to be compared. An initial week's training course in France included training sessions in sampling of diatoms, macrophytes and macroinvertebrates, including RIVPACS sampling training for the Austrian, German and Greek partners, who were using it as their 'national' method. The specialist diatom and macrophyte trainers were international experts Martyn Kelly and Nigel Holmes respectively. Macroinvertebrate sampling training was provided by highly experienced STAR internal partners.

The French training course also included the requisite three day RHS accreditation course that required all participants to pass a rigorous exam in the application of the method before they were able to undertake it in STAR. The training and accreditation was led by Helena Parsons of the UK Environment Agency.

The French course was supplemented by additional macroinvertebrate sampling training in Denmark and Poland. A separate diatom and macrophyte training course also took place in Poland and included the three day RHS training course. Additionally specialist training courses in the processing and identification of diatoms and identification courses on Oligochaeta, Plecoptera and Trichoptera were organised by expert STAR partners for other STAR partners and external scientists.

Diatom ring test, replicate sampling programme and audit

A diatom ring test was undertaken during the French training course. It was used to compare the results of simultaneous sampling of diatoms by the STAR personnel responsible for sampling this biological quality element during the main sampling programme. Parameters compared included intra- and inter-substratum variability and intra- and inter-operator variability in the type and relative proportions of taxa collected and identified and the indices derived from the results of sampling.

For the ring test, samples were collected from two locations on the Plaine River, in the Vosges region of France. Samples were collected from three different habitat types; stones, macrophytes and sediments. Sampling methods followed the STAR sampling protocol (see above). Each partner collected three samples from each of two substrata at each of the two test sites. The participants who collected the samples also prepared the samples in their respective laboratories and identified and counted a minimum of 300 valves. The results of the ring test, including identification checks were evaluated by specialists at STAR partner Alterra (Besse-Lotoskaya et al., 2006).

An additional replicate sampling programme was also carried out by partners in the Czech Republic, France, Greece, Portugal, Sweden and the United Kingdom (Table 8). All samples were prepared and identified by the organisation that collected them.

All partners collecting and processing diatom samples for WP7 and/or WP8 were subject to auditing of their taxon counts and identifications. Thirty-eight percent of all core and additional stream samples were subject to audit by experts at Alterra. The samples to be re-analysed were selected randomly from all the samples taken by each partner. Therefore, all samples were numbered and the numbers of samples to be audited were selected using a list of random digits.

The identification of taxa was initially to the most precise taxonomic level that was achievable (species or variety/forma). Subsequently, following discussions amongst the STAR partners, the level of identification of some difficult taxa was made less rigorous. This provided a more consistent level of achievable identification but one that remained compatible with the metrics to be used for Ecological Status assessments. After resolving nomenclatural differences, the results of taxa and counts obtained by the primary analysts and the auditor were compared to determine the error rates (Besse-Lotoskaya et al., 2006).

Macroinvertebrate replicate sampling programme and audit

A replicate macroinvertebrate sampling programme was undertaken by all partners involved in WP7 and/or WP8 except in Slovakia. Replicate sampling was undertaken at 80 sites (Table 7). At each replicate site two STAR-AQEM samples and two national samples were collected in the same 100 m section of the river on the same visit in each of the two macroinvertebrate sampling

Table 8. The number of sites subject to replicate diatom and/or macroinvertebrate sampling in the main WP7 and WP8 programmes

Country	Stream type	Number of replicate sites	
		Diatoms: one sample per site in one sampling season	Macroinvertebrates: two samples per method per site in each of two sampling seasons
Austria	A05	0	3
	A06	0	3
Czech	C04	3	3
Republic	C05	3	3
Denmark	K02	0	6
France	F08	3	6
Germany	D03	0	2
	D04	0	2
	D06	0	4
Greece	H04	6	6
Italy	I05	0	0
	I06	0	6
Latvia	L02	0	6
Poland	O02	0	3
	O03	0	3
Portugal	P04	6	6
Slovakia	V01	0	0
	V02	0	6
Sweden	S05	1	3
	S06	5	3
United	U15	3	3
Kingdom	U23	3	3
Totals		33 (33 samples)	80 (320 samples)

seasons. In the case of STAR-AQEM samples, two separate blind estimates of the proportion of the habitats present were normally made and the number of sample units on each habitat in each sample was based on their respective substratum recording forms. Additional replicate sampling was undertaken at some sites as itemised by Clarke et al. (2006a, 2006b).

Sorting and processing of all main and replicate samples were undertaken by the partner collecting the samples. Each partner also conducted a separate replicate sub-sampling programme for their STAR-AQEM samples where feasible. Replicate sub-sampling was attempted on all of the 160 replicate STAR-AQEM samples. In this programme a standard STAR-AQEM sub-sample was processed for the replicate sample, involving the sorting of either five cells or the number of cells needed to obtain the 700+ specimens required. On completion of this sub-sample a second sub-sample was processed using the material from five, or more if necessary, of the remaining cells from the sample tray grid. Replicate sub-sampling could not be undertaken in specimen-poor sites, particularly in Greece, where two sub-samples of 700+ specimens could not be achieved.

An audit programme was undertaken involving re-sorting and re-identification of samples collected and first processed by partners. Audits were undertaken on replicate samples collected as part of the replicate sampling programme. A single replicate sample was audited for each method at each site with half the audited samples being collected in spring and half in the second sampling season. The single exception was Italy where it was only possible to audit spring samples for operational reasons. For STAR-AQEM samples the first sub-sample of the replicate sub-sampling programme was normally selected for auditing but occasionally the second sub-sample was audited instead.

Partners were not informed of which samples had been selected for auditing, nor that only replicate samples would be chosen, until all project samples had been processed. Prior to notification of which samples were to be audited, partners were required to provide the auditors with copies of their full taxon lists for all samples. This prevented modification of results prior to dispatching the notified samples for audit.

The audit programme was in two parts. The first part was the sorting audit to record any families of macroinvertebrates that had not been removed by the partner of origin of the sample. Representative specimens of all taxa present, including those not found by the original partner were removed from the sample and retained for further audit. Sorting audits were all undertaken by the Centre for Ecology and Hydrology except for their own samples, which were audited by the University of Duisburg-Essen.

The second stage of the audit process was the identification audit. This involved re-identification of all the taxa removed from the sample and identified by the partner of origin plus the identification of all of the additional specimens, including any new families, removed from the sample during the sorting audit process. Identification audits were shared amongst most STAR partners with the auditing partner being in the same, adjacent or similar eco-region to that of the audited partner and therefore familiar with the majority of taxa in the audited partner's region.

The output of the auditing process was a list of the families gained to samples by the sorting audit, which were genuine errors, plus a comparison of the two different lists of taxa separately identified by the audited and auditing partner. The latter differences were perceived to be differences of opinion since no arbitration or consensus of identifications were attempted. The audit therefore compared the uncertainty involved in two different experts both identifying the same set of specimens.

Approaches: software development

The project data were curated, managed and analysed using a series of bespoke database and software products developed by STAR or modified from existing software created within the AQEM project or by the Centre for Ecology and Hydrology (Table 9). The STAR website (www.eu-star.at) provided an internal discussion forum, data repository and portal to the outside world for information, ideas and reports developed during the project and the protocols used in the project for data collection. Access routes to the databases and software products in the public domain, including the Decision Support System, MONSTAR, developed as a project deliverable, are provided in Table 9.

In addition to the software developed in the STAR, AQEM and Euro-limpacs projects, phytobenthos metric values were calculated using the version 3.2 of the Omidia software (Lecointe et al., 1993). Fish data collected in the STAR project were stored and retrieved using the Fides software developed by the FAME project. Fides software was used to calculate metric values including the new European Fish Index and classification system (EFI) developed by FAME. FAME software may be accessed from http://fame.boku.ac.at/downloads.htm.

Approaches: output

STAR has been one of the central research projects contributing to the implementation of the Water Framework Directive. The STAR research programme has made significant contributions to this process through membership of, or formal advice to the Common Implementation Strategy (CIS) Working Groups, including ECOSTAT and the Geographical Inter-calibration Groups and to the Comité Européen de Normalisation (CEN) responsible for producing the methodological standards for use in conjunction with the WFD monitoring programmes.

The detailed programme of work undertaken by the group, the project deliverables, in the form of reports, data and software and much other information, including the Waterview database (Birk & Hering, 2002) are available from the project website (www.eu-star.at). The objective of this special issue of Hydrobiologia has been to make the major findings of the project available to a wider audience via a series of individual papers.

Table 9. STAR bespoke databases and software

Product	Brief product description	Access route
AQEMDip	An Access®-based program for storage and retrieval of macroinvertebrate and phytobenthos data in standard input and output formats. Output macroinvertebrate data are supplied in the input format structure for ASTERICS. AQEMDip was developed using the platform of the Austrian assessment program ECOPROF. AQEMDip and ECOPROF were developed by Softwarehaus Graf and Partner	AQEMDip software and manual are available for download from the STAR website (www.eu-star.at). ECOPROF is available from www.ecoprof.at
STARMTR1	A database for the storage and retrieval of macrophyte data collected using Mean Trophic Rank survey and similar procedures. The Access®-based program has functions for standardising (adjusting) taxonomic levels of surveys prior to analysis. Standard taxonomic adjustments are provided but user defined adjustments may also be created. Mean Trophic Rank scores may be calculated in order to assess the degree to which a stream is impacted by organic pollution or enrichment. The STAR software is an extension of a program first developed by the Centre for Ecology and Hydrology	STARMTR1 software, with a detailed internal help system, is available for download from the STAR website (www.eu-star.at).
STARRHS1	An Access®-based program for storage and retrieval of survey data collected using the River Habitat Survey (RHS) methodology. The software is compatible with the 1997, 2001 and 2003 versions of the British RHS system and also the 2001 southern European modification (Buffagni & Kemp, 2002). STARRHS1 can be used to calculate the Habitat Quality Assessment (HQA) and Habitat Modification Score (HMS) indices. The STAR software is an extension of a program first developed by the Centre for Ecology and Hydrology	STARRHS1 software, with a detailed internal help system, is available for download from the STAR website (www.eu-star.at)
ASTERICS	A program for calculating the values of approximately 200 macroinvertebrate metrics used to assess the Ecological Status of running waters. The program, referred to as the AQEM River Assessment Program can also be used to calculate the values of multi-metric systems developed for specific stream, stressor and/or countries during the AQEM project (Hering et al., 2004a) and several national projects. The program was developed by Alterra Green World Research and by the University of Duisburg-Essen, using funds from the European Union (AQEM and STAR) and national funding from Germany (Umweltbundesamt, Länderarbeitsgemeinschaft Wasser)	Software and manual are available from: www.eu-star.at, www.aqem.de and www.fliessgewasserbewertung.de

Macroinvertebrate Taxa and Autecology Database	A database of approximately 10 000 macroinvertebrate taxa occurring in Europe and the associated taxon codes applied to each taxon in each of six different European systems. The database was principally developed during the AQEM project, extended in the STAR project and is now being further developed and enhanced with comprehensive autecological data in the EC PPVI project Euro-limpacs (GOCE-CT-2003-505540). The database has been developed under the management of the University of Duisburg-Essen and BOKU, the University of Natural Resources and Applied Life Sciences, Vienna	The database taxon lists and codes can be accessed on the Euro-limpacs website at www.freshwaterecology.info
STARBUGS	STARBUGS is an Excel®-based system for estimating the impact of known sources and levels of uncertainty on the values of single and multi-metrics and on the confidence and precision with which metric values can be assigned to individual classes of Ecological Status. STARBUGS may be used to test the development of multi-metrics. The system was designed for running waters but is portable to other surface waters or even to the terrestrial environment. STARBUGS was developed by the Centre for Ecology and Hydrology	STARBUGS and its associated manual are available for download from the STAR website (www.eu-star.at)
MONSTAR	MONSTAR is a software product that provides guidance in the design of monitoring programmes necessary to meet the terms and objectives of the Water Framework Directive (WFD). MONSTAR gives insight in the consequences of different choices in terms of costs and results that enables users to optimize their monitoring programme. MONSTAR specifically focus on running waters and addresses stream types and stressors studied in the STAR project with a strong focus on macro-invertebrates. However, MONSTAR provides a framework for including additional information related to other stream types, stressors and quality elements. MONSTAR was developed by Alterra Green World Research	

Each of these papers directly or indirectly addresses particular practical issues faced by the CIS and by those charged with implementing the WFD. For this purpose, the volume has been divided into a series of sections devoted to specific generic issue and each comprising a set of two or more papers. The seven component sections are (1) typology, (2) organism groups, (3) macrophytes and diatoms, (4) hydromorphology, (5) tools for assessing European streams with macroinvertebrates, (6) intercalibration and comparison and (7) errors and uncertainty. This structure mirrors the sequence of practical considerations that need to be addressed in delivering a coherent monitoring programme for evaluating the Ecological Status of streams and rivers within the European Union.

The objective of this paper, as its title implies, has been to introduce the project and to set out the context, objectives and approaches taken to provide some of the scientific information needed to best implement the WFD. In doing so it provides a reference point for many of the individual papers contained within this volume and obviates the necessity to repeat this information elsewhere. In order to assimilate the key findings of the research programme, as presented here, each of the seven separate sections of this volume is prefaced by a summary paper drawing out the key results of the component papers and highlighting the recommendations of these papers and their practical contribution to the implementation process.

Acknowledgements

STAR was funded by the European Commission, 5th Framework Program, Energy, Environment and Sustainable Development, Key Action Water, contract no. EVK1-CT-2001-00089. AQEMDip and ASTERICS (formerly the AQEM River Assessment Program) were initially developed during the European Commission, 5th Framework Project AQEM, contract No. EVK1-CT1999-00027. ASTERICS was co-funded by Umweltbundesamt, Länderarbeitsgemeinschaft Wasser. The authors acknowledge the support of all their colleagues, too numerous to mention, who have contributed to the work described here. We are grateful to Stefan Schmutz and his colleagues in the FAME project for the use of the Fides database and their EFI software. We also wish to thank Hartmut Barth, Mogens Gadeberg and Elena Domínguez, EC Scientific Officers for STAR for their constant help and encouragement.

References

AQEM Consortium, 2002. Manual for the application of the AQEM system. A comprehensive method to assess European streams using benthic macroinvertebrates, developed for the purpose of the Water Framework Directive. Version 1.0, February 2002.

Balon, E. K., 1975. Reproductive guilds of fishes: a proposal and definition. Journal of the Fisheries Research Board of Canada 32: 821–864.

Baattrup-Pedersen, A., K. Szoszkiewicz, R. Nijboer, M. O'Hare & T. Ferreira, 2006. Macrophyte communities in unimpacted European streams: variability in assemblage patterns, abundance and diversity. Hydrobiologia 566: 179–196.

Besse-Lotoskaya, A., P. F. M. Verdonschot & J. A. Sinkeldam, 2006. Uncertainty in diatom assessment: Sampling, identification and counting variation. Hydrobiologia 566: 247–260.

Birk, S. & D. Hering, 2002. Waterview Web-Database: a comprehensive review of European assessment methods for rivers. FBA News 20: 4.

Buffagni, A., S. Erba, M. Cazzola, J. Murray-Bligh, H. Soszka & P. Genoni, 2006. The STAR common metrics approach to the WFD intercalibration process: Full application for small, lowland rivers in three European countries. Hydrobiologia 566: 379–399.

Buffagni, A. & L. J. Kemp, 2002. Looking beyond the shores of the United Kingdom: addenda for the application of River Habitat Survey in southern European rivers. Journal of Limnology 61: 199–214.

Birk, S. & D. Hering, 2006. Direct comparison of assessment methods using benthic macroinvertebrates: a contribution to the EU Water Framework Directive intercalibration exercise. Hydrobiologia 566: 401–415.

Birk, S., T. Korte & D. Hering, 2006. Intercalibration of assessment methods for macrophytes in lowland streams: direct comparison and analysis of common metrics. Hydrobiologia 566: 417–430.

Bis, B. & P. Usseglio-Polatera, 2004. Species traits analysis. STAR deliverable N2 to the European Commission, 148 pp.

Clarke, R. T., 2000. Uncertainty in estimates of river quality based on RIVPACS. In Wright, J. F., D. W. Sutcliffe & M. T. Furse (eds), Assessing the Biological Quality of Freshwaters: RIVPACS and Similar Techniques. Freshwater Biological Association, Ambleside, 39–54.

Clarke, R. T., J. Davy-Bowker, L. Sandin, N. Friberg, R. K. Johnson & B. Bis, 2006a. Estimates and comparisons of the effects of sampling variation using 'national' macroinvertebrate sampling protocols on the precision of metrics used to assess ecological status. Hydrobiologia 566: 477–503.

Clarke, R. T., M. T. Furse, R. J. M. Gunn, J. M. Winder & J. F. Wright, 2002. Sampling variation in macroinvertebrate data and implications for river quality indices. Freshwater Biology 47: 1735–1751.

Clarke, R. T., A. Lorenz, L. Sandin, A. Schmidt-Kloiber, J. Strackbein, N. T. Kneebone & P. Haase, 2006b. Effects of sampling and sub-sampling variation using the STAR-AQEM sampling protocol on the precision of macroinvertebrate metrics. Hydrobiologia 566: 441–459.

Clarke, R. T., J. F. Wright & M. T. Furse, 2003. RIVPACS models for predicting the expected macroinvertebrate fauna and assessing the ecological quality of rivers. Ecological Modelling 160: 219–233.

Danish Environmental Protection Agency, 1998. Biological Assessment of Watercourse Quality. Guidelines, No. 5. – Danish Environmental Protection Agency, Ministry of Environment and Energy, Copenhagen.

Davies, P. E., 2003. Development of a national river bioassessment system (AUSRIVAS) in Australia. In Wright, J. F., D. W. Sutcliffe & M. T. Furse (eds), Assessing the Biological Quality of Freshwaters: RIVPACS and Similar Techniques. Freshwater Biological Association, Ambleside, 113–124.

Davy-Bowker, J., R. T. Clarke, R. K. Johnson, J. Kokes, J. F. Murphy & S. Zahrádková, 2006. A comparison of the European Water Framework Directive physical typology and RIVPACS-type models as alternative methods of establishing reference conditions for benthic macroinvertebrates. Hydrobiologia 566: 91–105.

Erba, S., A. Buffagni, N. Holmes, M. O'Hare, P. Scarlett & A. Stenico, 2006. Preliminary testing of River Habitat Survey features for the aims of the WFD hydro-morphological assessment: an overview from the STAR Project. Hydrobiologia 566: 281–296.

European Commission, 2000. Directive of the European Parliament and of the Council 2000/60/EC establishing a framework for Community action in the field of water policy. European Commission PE-CONS 3639/1/00 REV 1, Luxembourg.

European Commission, 2001. Common implementation strategy for the Water Framework Directive (2000/60/EC). Strategic document as agreed by the Water Directors under Swedish presidency, 2 May 2001. European Commission, 81 pp.

European Commission, 2002. Water Framework Directive (WFD) Common Implementation Strategy Working Group 2.5. Intercalibration: Towards a guidance on establishment of the intercalibration network and on the process of the intercalibration exercise. European Commission, 50 pp.

European Commission, 2003a. Water Framework Directive (WFD) Common Implementation Strategy Working Group 2.3 Reference conditions for inland surface waters (REFCOND). Guidance on establishing reference conditions and Ecological Status class boundaries for inland surface waters. Final version, 30 April 2003. European Commission, 86 pp.

European Commission, 2003b. Water Framework Directive Common Implementation Strategy Working Group 2.7 Monitoring. Guidance on monitoring for the Water Framework Directive. Final version. 23 January 2003. European Commission, 170 pp.

European Commission, 2003c. Water Framework Directive Common Implementation Strategy Working Group 2.A Ecological Status (ECOSTAT). Overall approach to the classification of Ecological Status and Ecological Potential. European Commission, 86 pp.

GAY, Cabinet en Environnement, 1994. Indice Biologique Global Normalisé. NF T 90-350. Guide Technique. Agences de l'eau et Ministère de l'Environnement, Paris.

Friberg, N., L. Sandin, M. T. Furse, S. E. Larsen, R. T. Clarke & P. Haase, 2006. Comparison of macroinvertebrate sampling methods in Europe. Hydrobiologia 566: 365–378.

Ghetti, P. E., 1997. Manuale di Applicazione. Indice Biotico Esteso (I. B. E.). I Macroinvertebrati Nell Contro uo Della Qualita Degli Ambienti di Acquae Correnti. – Provinzia Autonoma di Trento, Agenzia Provinciale per la Protezione dell'Ambiente, Trento.

Haase, P., J. Murray-Bligh, S. Lohse, S. Pauls, A. Sundermann, R. Gunn & R. Clarke, 2006. Assessing the impact errors in sorting and identifying macroinvertebrate samples. Hydrobiologia 566: 505–521.

Hellawell, J. M., 1978. The Biological Surveillance of Rivers: A Biological Monitoring Handbook. Water Research Centre, Stevenage.

Hellawell, J. M., 1986. Biological Indicators of Freshwater Pollution and Environmental Management. Pollution Monitoring Series. Elsevier Applied Science, London, New York.

Hering, D., A. Buffagni, O. Moog, L. Sandin, M. Sommerhäuser, I. Stubauer, C. Feld, R. K. Johnson, P. Pinto, N. Skoulikidis, P. F. M. Verdonschot & S. Zahrádková, 2003. The development of a system to assess the ecological quality of streams based on macroinvertebrates – design of the sampling programme within the AQEM project. Internationale Revue der gesamten Hydrobiologie 88: 345–361.

Hering D., C. K. Feld, O. Moog & T. Ofenböck, 2006. Cook book for the development of a Multimetric-Index for biological condition of aquatic ecosystems: experiences from the European AQEM and STAR projects and related initiatives. Hydrobiologia 566: 311–324.

Hering, D., O. Moog, L. Sandin & P. F. M. Verdonschot, 2004a. Overview and application of the AQEM assessment system. Hydrobiologia 516: 1–21.

Hering, D., P. F. M. Verdonschot, O. Moog & L. Sandin (eds), 2004b. Integrated assessment of running waters in Europe. Hydrobiologia. 516 pp.

Holmes, N. T. H., J. R. Newman, S. Chadd, K. J. Rouen, L Sharp & F. H. Dawson, 1999. Mean Trophic Rank: A Users' Manual. R&D Technical Report No. E38, Environment Agency, Bristol.

Holmes, N. T. H. & B. A. Whitton, 1977. Macrophytic vegetation of the River Swale, Yorkshire. Freshwater Biology 7: 545–558.

Hughes, R. M., 1995. Defining acceptable status by comparing with reference conditions. In Davis, W. S. & T. P. Simon (eds), Biological Assessment and Criteria. Tools for Water Resource Planning and Decision Making. Lewis Publishers, Boca Raton, FL, 31–47.

Illies, J. (ed.), 1978. Limnofauna Europaea, 2nd edn. Gustav Fischer Verlag, Stuttgart, New York; Swets and Zeitlinger B. V., Amsterdam.

Johnson, R. K., D. Hering, M. T. Furse & R. T. Clarke, 2006a. Detection of ecological change using multiple organism groups: metrics and uncertainty. Hydrobiologia 566: 115–137.

Johnson, R. K., D. Hering, M. T. Furse & P. F. M. Verdonschot, 2006b. Indicators of ecological change: comparison of the early response of four organism groups to stress gradients. Hydrobiologia 566: 139–152.

Kelly, M. G., A. Cazaubon, E. Coring, A. Dell'Uomo, L. Ector, B. Goldsmith, H. Guasch, J. Hürlimann, A. Jarlman, B. Kawecka, J. Kwandrans, R. Laugaste, E.-A. Lindstrøm, M. Leitao, P. Marvan, J. Padisák, E. Pipp, J. Prygiel, E. Rott, S. Sabater, H. van Dam & J. Vizinet, 1998. Recommendations for the routine sampling of diatoms for water quality assessments in Europe. Journal of Applied Phycology 10: 215–224.

Knoben, R. A. E., C. Roos & M. C. van Oirschot, 1995. Biological assessment methods for watercourses. UN/ECE Task Force on Monitoring and Assessment, 86 pp.

Kokeš, J., S. Zahrádková, D. Němejcová, J. Hodovský, J. Jarkovský & T. Soldán, 2006. The PERLA system in the Czech Republic: a multivariate approach for assessing the ecological status of running waters. Hydrobiologia 566: 343–354.

Latvian Standard Ltd., 1999 LVS 240:1999 Water quality – operative evaluation biological quality of small stream by saprobity index of macrozoobenthos community. In Catalogue of Latvian standards, Riga, Latvian Standard Ltd, 1999: Group 13.060, 1(11)–11.

Lecointe, C., M. Coste & J. Prygiel, 1993. "OMNIDIA" software for taxonomy, calculation of diatom indices and inventories management. Hydrobiologia 269/270: 509–513.

Lorenz, A. & R. T. Clarke, 2006. Sample coherence – a field study approach to assess similarity of macroinvertebrate samples. Hydrobiologia 566: 461–476.

Mandl, V., 1992. Draft EC directive on ecological quality of surface waters. In Newman, P. J., M. A. Piavaux & R. A. Sweeting (eds), River Water Quality. Ecological Assessment and Control. Publication EUR 14606 EN-FR. Commission of the European Communities, Luxembourg, 18.

Mann, R. H. K., 1996. Environmental requirements of European non-salmonid fish in rivers. Hydrobiologia 323: 223–235.

Metcalfe, J. L., 1989. Biological water quality assessment of running waters based on macroinvertebrate communities: history and present status in Europe. Environmental Pollution 60: 101–139.

Metcalfe-Smith, J. L., 1994. Biological water quality assessment of rivers: use of macroinvertebrate communities. In Calow, P. & G. E. Petts (eds), The Rivers Handbook, Vol. II. Blackwell Scientific Publications, London, 144–170.

Moore, W. W., 1977. Seasonal succession of algae in a eutrophic stream in southern England. Hydrobiologia 53: 181–192.

Murray-Bligh, J. A. D., M. T. Furse, F. H. Jones, R. J. M. Gunn, R. A. Dines & J. F. Wright, 1997. Procedure for collecting and analysing macroinvertebrate samples for RIVPACS. Joint publication by the Institute of Freshwater Ecology and the Environment Agency, 162 pp.

Noble, R. & I. Cowx, 2002. Development, evaluation & implementation of a standardised fish-based assessment method for the Ecological Status of European rivers – a contribution to the water framework directive (FAME). A report to the European Commission, 100 pp.

Norris, R. H., 1994. Rapid biological assessment, natural variability and selecting reference sites. Classification of rivers and environmental health indicators. In Uys, M. C. (ed.), Proceedings of a Joint South African/Australian Workshop, Cape Town, South Africa. Water Research Commission, Report No. TT/63/94: 129–166.

O'Hare, M. T., A. Baattrup-Pedersen, R. Nijboer, K. Szoszkiewicz & T. Ferreira, 2006. Macrophyte communities of European streams with altered physical habitat. Hydrobiologia 566: 197–210.

Pinto, P., M. Morais, M. Ilhéu & L. Sandin, 2006. Relationships among biological elements (macrophytes, macroinvertebrates and ichyofauna) for different river types across Europe at two different spatial scales. Hydrobiologia 566: 75–90.

Raven, P. J., N. T. H. Holmes, F. H. Dawson, P. J. A. Fox, M. Everard, I. R. Fozzard & K. J. Rouen, 1998. River Habitat Quality – The Physical Character of Rivers and Streams in the UK and Isle of Man. River Habitat Survey Report Number 2. Environment Agency, Bristol: Scottish Environment Protection Agency, Stirling: Environment and Heritage Service, Belfast, 1–84.

Reynoldson, T. B., R. C. Bailey, K. E. Day & R. H. Norris, 1995. Biological guidelines for freshwater sediment based on BEnthic Assessment of SedimenT (the BEAST) using a multivariate approach for predicting biological state. Australian Journal of Ecology 20: 198–219.

Reynoldson, T. B., K. E. Day & T. Pascoe, 2000. The development of the BEAST: a predictive approach for assessing sediment quality in the Great Lakes. In Wright, J. F., D. W. Sutcliffe & M. T. Furse (eds), Assessing the Biological Quality of Freshwaters: RIVPACS and Similar Techniques. Freshwater Biological Association, Ambleside, 165–180.

Rosenberg, D. M., T. B. Reynoldson & V. H. Resh, 2000. Establishing reference conditions in the Fraser River catchment, British Colombia, Canada, using the BEAST (Benthic Assessment of SedimenT) predictive model. In Wright, J. F., D. W. Sutcliffe & M. T. Furse (eds), Assessing the Biological Quality of Freshwaters: RIVPACS and Similar Techniques. Freshwater Biological Association, Ambleside, 181–194.

Schmidt-Kloiber, A., W. Graf, A. Lorenz & O. Moog, 2006. The AQEM/STAR taxalist – a pan-European macro-invertebrate ecological database and taxa inventory. Hydrobiologia 566: 325–342.

Seber, G. A. F. & E. D. Le Cren, 1967. Estimating population parameters from catches large relative to the population. Journal of Animal Ecology 36: 631–643.

Sladecek, V., 1973. System of water quality from the biological point of view. Archiv für Hydrobiologie Ergebnisse der Limnologie 7: 1–218.

Šporka, F., H. E. Vlek, E. Bulánková & I. Krno, 2006. Influence of seasonal variation on bioassessment of streams using macroinvertebrates. Hydrobiologia 566: 543–555.

Springe, G., L. Sandin, A. Briede & A. Skuja, 2006. Biological quality metrics: their variability and appropriate scale for assessing streams. Hydrobiologia 566: 153–172.

Staniszewski, R., K. Szoszkiewicz, J. Zbierska, J. Lesny, S. Jusik & R. T. Clarke, 2006. Assessment of sources of uncertainty in macrophyte surveys and the consequences for river classification. Hydrobiologia 566: 235–246.

Swedish Environmental Protection Agency, 1996. Bottenfauna i sjöars litoral och I vattendrag – tidsserier. [In Swedish: Benthic fauna in lake litoral and running waters – time series]. Swedish EPA monitoring handbook, Fresh waters. [Published digitally at: www.naturvardsverket.se].

Szoszkiewicz, K., A. Buffagni, J. Davy-Bowker, J. Lesny, B. H. Chojnicki, J. Zbierska, R. Staniszewski & T. Zgola, 2006a. Occurrence and variability of River Habitat Survey features across Europe and the consequences for data collection and evaluation. Hydrobiologia 566: 267–280.

Szoszkiewicz, K., T. Ferreira, T. Korte, A. Baattrup-Pedersen, J. Davy-Bowker & M. O'Hare, 2006b. European river plant communities: the importance of organic pollution and the usefulness of existing macrophyte metrics. Hydrobiologia 566: 211–234.

Verdonschot, P. F. M., 2006a. Evaluation of the use of Water Framework Directive typology descriptors, reference sites and spatial scale in macroinvertebrate stream typology. Hydrobiologia 566: 39–58.

Verdonschot, P. F. M., 2006b. Data composition and taxonomic resolution in macroinvertebrate stream typology. Hydrobiologia 566: 59–74.

Vlek, H. E., F. Šporka & I. Krno, 2006. Influence of macroinvertebrate sample size on bioassessment of streams. Hydrobiologia 566: 523–542.

Winter, J. G. & H. C. Duthie, 2000. Stream epilithic, epipelic and epiphytic diatoms: habitat fidelity and use in biomonitoring. Aquatic Ecology 34: 345–353.

Wright, J. F., P. D. Armitage & M. T. Furse, 1989. Prediction of invertebrate communities using stream measurements. Regulated Rivers: Research and Management 4: 147–155.

Wright, J. F., D. W. Sutcliffe & M. T. Furse (eds), 2000. Assessing the Biological Quality of Freshwaters: RIVPACS and Similar Techniques. Freshwater Biological Association, Ambleside.

Stream and River Typologies

Hydrobiologia (2006) 566:33–37
M.T. Furse, D. Hering, K. Brabec, A. Buffagni, L. Sandin & P.F.M. Verdonschot (eds), The Ecological Status of European Rivers: Evaluation and Intercalibration of Assessment Methods
DOI 10.1007/s10750-006-0072-9

Stream and river typologies – major results and conclusions from the STAR project

Leonard Sandin[1,*] & Piet F. M. Verdonschot[2]
[1]*Department of Environmental Assessment, Swedish University of Agricultural Sciences, P.O. Box 7050, SE-750 07 Uppsala, Sweden*
[2]*Alterra, Green World Research, P.O. Box 47, 6700 AA Wageningen, the Netherlands*
*(*Author for correspondence: E-mail: Leonard.sandin@ma.slu.se)*

Key words: environmental assessment, river, stream, typology, Europe, water framework directive

Abstract

The EU Water Framework Directive uses abiotic variables for classifying streams and rivers into types. For rivers, the EU Water Framework Directive fixed typology i.e. ‘System A’ typology are defined by ecoregions, size based on the catchment area, catchment geology and altitude. Within any given part of the WFD typology, it is assumed that biological communities at undisturbed sites will be broadly similar and will therefore constitute a type-specific biological target and a way to stratify the spatial variability in stream and river monitoring and assessment. The data collected for the STAR project cover 13 countries and include 22 stream types. A total of 233 sites were fully sampled for all biological quality elements (fish, macrophytes, benthic macroinvertebrates, and diatoms) in the study. Analysing the STAR macroinvertebrate dataset in relation to environmental and biogeographical variables resulted in three major groups of stream types that correspond to three major landscape types in Europe: Mountains, Lowlands and Mediterranean. Similar results were found when analysing all four biological quality elements (fish, macrophytes, benthic macroinvertebrates, and diatoms) sampled in the STAR project. The studies also showed that the stream types using the WFD ‘System A’ descriptors are probably less useful at finer scales and it is suggested that a stream typology should take three main parameters as a starting point, i.e., climate (temperature), slope (current velocity) and stream size. Existing site-specific multivariate RIVPACS-type predictive models were also compared to both null models and the WFD ‘System A’ physical typology as methods of predicting macroinvertebrate reference conditions. It was concluded that the multivariate models are more effective in predicting reference conditions primarily because they make use of continuous rather than categorical predictor variables and because the multivariate RIVPACS-type models are not constrained by the use of a limited number of variables.

Introduction

In environmental assessment studies of environmental impact, the objective is to separate the change generated by anthropogenic stress from the natural spatial and temporal variability (Johnson, 1998). If the natural variability is large and the anthropogenic induced change is small it will be difficult to detect a real change in the measured variable(s) caused by the pollutant (Johnson, 1998). Geographical classifications (e.g., by ecoregions) can be a useful tool in partitioning natural spatial variability, thereby optimizing monitoring, assessment, and conservation programs. Simply put, by using geographical classification sampling is more cost-effective (less samples are needed to detect anthropogenic stress) and a water quality baseline for each geographic area can be defined.

A renewed interest in regionalization in the environmental assessment of fresh waters (mainly in the USA and Europe) is largely a result of government agencies wanting to shift their efforts in resource management from single issues (a particular stream or lake) to a more holistic approach (e.g., aquatic systems nested in catchments or ecoregions) (Omernik & Bailey, 1997). The use of a typology to classify streams has also become an accepted part of ecological assessment (Wright et al., 1999; Hering et al., 2004), but the underlying factors (variables) determining the typology differs strongly among approaches.

The EU Water Framework Directive (European Commission, 2000) uses only abiotic variables for classifying streams and rivers into types (Annex II, section 1 of the WFD), whereas other approaches such as the RIVPACS predictive model (e.g., Wright et al., 1984) uses biotic variables, and others uses combinations of biotic and abiotic variables (e.g., Reynoldson et al., 1997). For rivers, the EU Water Framework Directive fixed typology i.e., 'System A' typology is defined by ecoregions (according to Illies, 1978), size based on the catchment area (small 10–100, medium 100–1000, large 1000–10,000 and very large >10,000 km^2), catchment geology (calcareous, silicious, and organic), and altitude (lowland, <200, mid-altitude 200–800 and high altitude >800 m.a.s.l.). Within any given part of the WFD typology, it is assumed that biological communities at undisturbed sites will be broadly similar, and will therefore constitute a type-specific biological target and the typology is thus a way to stratify the spatial variability in stream and river monitoring and assessment.

The EU Water Framework Directive also allows each member state to adopt an alternative characterization 'System B' with five obligatory factors (altitude, latitude, longitude, geology and size), and an additional 15 optional factors (e.g., distance from river source, mean water depth, and mean substratum composition). No specific categories of value ranges are suggested for each factor in 'System B' and the member state is left to decide how many of the optional factors they wish to use. In consequence a very extensive set of stream types could be defined by individual Member States for each ecoregion within their territorial limits, since with e.g., 'System A' each ecoregion has a theoretical maximum of 4 (size)×3 (altitude)×3 (geology) = 36 types, which means that within a country such as Sweden containing three (Illies) ecoregions, the maximum number of theoretical stream and river types is 108, but in reality only some of these types do exist, mainly because altitude is strongly related to ecoregion.

River and stream types

The data collected for the STAR project cover 13 countries (Austria, Czech Republic, Denmark, France, Germany, Greece, Italy, Latvia, Poland, Portugal, Slovakia, Sweden and UK). The sampling included 22 stream types, where five were defined as being of the STAR project type "Core stream type 1" (mid altitude, 200–500 m.a.s.l. and with a "small" catchment area 10–100 km^2), seven were of the STAR project type "Core stream type 2" (lowland, <200 m.a.s.l. and "medium" catchment areas 100–1000 km^2), whereas ten other stream types were defined as STAR project type "Additional stream type" (having a different characterisation). Core and additional stream types could also be defined as either calcareous, siliceous or, occasionally, organic but, with a few exceptions, sites within a type sampled by a partner were all in the same geological category. These stream types sampled in the project are situated in 11 Ecoregions according to Illies definition (Illies, 1978; as used in the Water Framework Directive), these were regions 3, 4, 6, 7, 8, 9, 10, 14, 15, 16 and 18. Within these 22 types, 263 sites (streams or rivers) were sampled for macro-invertebrates in each sampling season (see Furse et al., 2006). All of these 263 sites were also subject to hydromorphological surveys and 252 were sampled for phytobenthos, 251 for macrophytes and 249 for fish. A final total of 233 sites were fully sampled for all biological quality elements.

Testing of the WFD typology

As part of the STAR project, the amount of variation in benthic macroinvertebrate composition explained by differences in stream types was tested in the six countries where at least two types were sampled. The amount of variation in macroinvertebrate community composition explained by type differed between 16.0% in the Czech Republic and 67.9% of the total explained variation in Greece (Sandin, Friberg, Furse, Clarke & Larsen unpublished). In comparison, the difference between two sampled seasons explained between 11.6% of the total explained variation (in Greece) and 56.0% of the total explained variation in Latvia. The pre-defined stress gradient (here divided into sites pre-defined as having a high or good ecological status versus those pre-defined as having a moderate, poor or bad ecological status [see Furse et al., 2006]) explained between 15.3% (in Greece) and 55.3% of the total explained variation in France (Sandin, Friberg, Furse, Clarke & Larsen unpublished). Stream type, differences between seasons, and the pre-defined stress gradient were always statistically significant explanatory variables. When looking at these comparisons one must of course take into account the fact that e.g., differences in how much variation is explained by type in relation to the other factors in dependent on how large differences there are in types analysed.

Analysing the STAR macroinvertebrate dataset in relation to environmental and biogeographical variables resulted in three major groups of stream types that correspond to three major landscape types in Europe: Mountains, Lowlands, and Mediterranean (Verdonschot, 2006a). This author suggests that the three major groups probably represent the major combination of geomorphological and/or climatological conditions of the sampled sites and that the driving forces behind these differences are most probably climate (temperature), slope (current velocity) and stream size, where benthic macroinvertebrates respond to the driving forces of these three major factors. Similarly, Pinto et al. (2006) also concluded that the biotic data (in this case all four sampled biological quality element) could be divided into three main groups, i.e., Mediterranean, mountain, and lowland streams or rivers.

The study by Verdonschot (2006a) also showed that the stream types using the WFD 'System A' descriptors are probably less useful at finer scales and he also suggests that a stream typology should take the three main parameters as a starting point. Next streams with comparable major environmental conditions can be mapped and reference conditions can be defined as such. Verdonschot (2006a) also concluded that, the geographic descriptors (e.g., ecoregions) did not fit well within the benthic macroinvertebrate typology testing.

Earlier, Verdonschot & Nijboer (2004) tested if the typology suggested in the WFD was useful for developing an assessment system for macroinvertebrates in streams. They concluded that the major macroinvertebrate distribution patterns in European streams follow climatological and geomorphological conditions and are well distinguished in terms of stream types. Thus, the WFD typology was useful for the development of type-specific assessment systems for streams using macroinvertebrates. Furthermore, it was shown that large-scale factors affected the macroinvertebrate distribution even on a very fine scale. On the other hand Sandin & Johnson (2000) concluded that large-scale variables such as an ecoregional delineation is not sufficient for stratifying streams or rivers for monitoring and assessment based on benthic macroinvertebrates. Testing differences in taxonomic composition among sites or streams between types (based on abiotic data) is clearly also dependent on taxonomic resolution used in the study (Verdonschot, 2006b), the finer the taxonomic resolution used, the more distinctive the types become. In this study it was also shown that species (or 'best available') taxonomic level performed better at a practical (fine) scale in comparison to family-level taxonomy. Another complication is that human stress diminishes the natural differences between stream communities and typologies should therefore be based on reference conditions (Verdonschot, 2006b). If the reference condition criteria used within different stream or river types differ, when defining references used for comparisons, types that are in reality distinct might seem to be similar enough to merit them to be joined into a common type.

Finally, a different approach was taken by Davy-Bowker et al. (2006). These authors used existing site-specific multivariate RIVPACS-type predictive models already in place in Great Britain (RIVPACS), Sweden ($SWEPAC_{SRI}$) and the Czech Republic (PERLA) and compared them to both null models and the WFD 'System A' physical typology as methods of predicting macro invertebrate reference conditions. They conclude that the multivariate models are more effective in predicting reference conditions primarily because they make use of continuous rather than categorical predictor variables (that have been selected for their value as good correlates of macroinvertebrate community composition) and because the multivariate RIVPACS-type models are not constrained by the use of a limited number of variables (Davy-Bowker et al., 2006). A further problem with a priori typological approaches such as the WFD 'System A' as demonstrated by Davy Bowker et al. (2006) is that they usually utilise variables gathered solely at large geographical scales, whereas their analysis shows that substratum composition, width and depth, all of which are local scale variables measured at the time of sampling, can also be strong correlates of macroinvertebrate community composition. The importance of both large-scale and local factors as determinants of macroinvertebrate communities should therefore not be overlooked when setting up montoring and environmental assessment systems in streams and rivers. Similar conclusions were reached by Heino et al. (2003) based on a study of macroinvertebrate diversity in headwater streams.

Conclusions

- Stream type, differences between seasons, and the pre-defined stress gradient were always statistically significant explanatory variables when testing their relation to the benthic macroinvertebrate community composition.
- Analysing the STAR macroinvertebrate dataset in relation to environmental and biogeographical variables resulted in three major groups of stream types that correspond to three major landscape types in Europe: Mountains, Lowlands, and Mediterranean.
- Similarly, it was concluded that the biotic data (in this case all four sampled biological quality element) could be divided into three main groups, i.e. Mediterranean, mountain, and lowland streams or rivers.
- The driving forces for benthic macroinvertebrates are most probably climate (temperature), slope (current velocity), and stream size.
- Testing differences in taxonomic composition among sites or streams between types (based on abiotic data) is clearly also dependent on taxonomic resolution used in the study, the finer the taxonomic resolution used, the more distinctive the types become.
- It was also concluded that multivariate (RIVPACS type) models are more effective in predicting reference conditions than either null models or WFD 'System A' typology primarily because they make use of continuous rather than categorical predictor variables.

Acknowledgements

STAR was funded by the European Commission, 5th Framework Program, Energy, Environment and Sustainable Development, Key Action Water, Contract no. EVK1-CT-2001–00089.

References

Davy-Bowker, J., R. T. Clarke, R. K. Johnson, J. Kokes, J. F. Murphy & S. Zahrádková, 2006. A comparison of the European Water Framework Directive physical typology and RIVPACS-type models as alternative methods of establishing reference conditions for benthic macroinvertebrates. Hydrobiologia 566: 91–105.

European Commission, 2000. Directive 2000/60/EC. Establishing a framework for community action in the field of water policy. European Commission PE-CONS 3639/1/100 Rev 1, Luxembourg.

Furse, M., D. Hering, O. Moog, P. Verdonschot, R. K. Johnson, K. Brabec, K. Gritzalis, A. Buffagni, P. Pinto, N. Friberg, J. Murray-Bligh, J. Kokes, R. Alber, P. Usseglio-Polatera, P. Haase, R. Sweeting, B. Bis, K. Szoszkiewicz, H. Soszka, G. Springe, F. Sporka & I. Krno, 2006. The STAR project: context, objectives and approaches. Hydrobiologia 566: 3–29.

Heino, J., T. Muotka & R. Paavola, 2003. Determinants of macroinvertebrate diversity in headwater streams: regional and local influences. Journal of Animal Ecology 72: 425–434.

Hering, D., O. Moog, L. Sandin & P. F. M. Verdonschot, 2004. Overview and application of the AQEM assessment system. Hydrobiologia 516: 1–20.

Illies, J., 1978. Limnofauna Europaea. Gustav Fischer Verlag, Stuttgart.

Johnson, R. K., 1998. Spatiotemporal variability of temperate lake macroinvertebrate communities: detection of impact. Ecological Applications 8: 61–70.

Omernik, J. M. & R. G. Bailey, 1997. Distinguishing between watersheds and ecoregions. Journal of the American Water Resources Association 33: 935–949.

Pinto, P., M. Morais, M. Ilhéu & L. Sandin, 2006. Relationships among biological elements (macrophytes, macroinvertebrates and ichthyofauna) for different core river types across Europe at two different spatial scales. Hydrobiologia 566: 75–90.

Reynoldson, T. B., R. H. Norris, V. H. Resh, K. E. Day & D. M. Rosenberg, 1997. The reference condition: a comparison of multimetric and multivariate approaches to assess water-quality impairment using benthic macroinvertebrates. Journal of the North American Benthological Society 16: 833–852.

Sandin, L. & R. K. Johnson, 2000. Ecoregions and benthic macroinvertebrate assemblages of Swedish streams. Journal of the North American Benthological Society 19: 462–474.

Verdonschot, P. F. M. & R. C. Nijboer, 2004. Testing the European stream typology of the water Framework Directive for macroinvertebrates. Hydrobiologia 175: 35–54.

Verdonschot, P. F. M., 2006a. Evaluation of the use of Water Framework Directive typology descriptors, reference sites, and spatial scale in macroinvertebrate stream typology. Hydrobiologia 566: 39–58.

Verdonschot, P. F. M., 2006b. M. Data composition and taxonomic resolution in macroinvertebrate stream typology. Hydrobiologia 566: 59–74.

Wright, J. F., D. Moss, P. D. Armitage & M. T. Furse, 1984. A prelimnary classification of running-water sites in Great Britain based on macroinvertebrate species and the prediction of community type using environmental data. Freshwater Biology 14: 221–256.

Wright, J. F., D. W. Sutcliffe & M. T. Furse (eds), 1999. Assessing the biological quality of fresh waters: RIVPACS and other techniques. Freshwater Biological Association, Ambleside, Cumbria, UK. The RIVPACS International Workshop, 16–18 September 1997, Oxford, UK.

Hydrobiologia (2006) 566:39–58
M.T. Furse, D. Hering, K. Brabec, A. Buffagni, L. Sandin & P.F.M. Verdonschot (eds), The Ecological Status of European Rivers: Evaluation and Intercalibration of Assessment Methods
DOI 10.1007/s10750-006-0071-x

Evaluation of the use of Water Framework Directive typology descriptors, reference sites and spatial scale in macroinvertebrate stream typology

Piet F.M. Verdonschot
Alterra, Green World Research, P.O. Box 47, 6700 AA Wageningen, The Netherlands
(Fax: +31-317-424988; E-mail: piet.verdonschot@wur.nl)

Key words: Europe, stream typology, macroinvertebrates, ordination, reference condition, ecological quality

Abstract

The aim of this study was to test the effect of the Water Framework Directive typology descriptors on a macroinvertebrate-based stream typology, the use of reference sites in comparison to the use of degraded sites, and both degraded and reference sites. The EU research projects AQEM and STAR provided 1660 samples of 48 stream types sampled all over the major geographical gradients in Europe. The samples included gradients from reference conditions to samples with bad ecological quality. These stream types fit the WFD typological demands. The macroinvertebrate data were analysed by using Detrended Correspondence Analysis (DCA). The observed macroinvertebrate distribution was tested against the WFD river typology by a graphical interpretation of the ordination diagrams. The major macroinvertebrate distribution patterns in European streams were based on climate (temperature), slope (current velocity), and stream size. The WFD descriptors 'System A' for stream types are too rigid and should be replaced by temperature, current, and size. The differences in average numbers of taxa between the 1660 sites distributed over Europe were either caused by differences between local environmental factors or by sampling effort, not by temperature, elevation, stream order or latitudinal position. The distribution patterns using all samples, only reference samples, and only degraded samples showed that human stress diminished the natural differences between stream communities and typologies should therefore be based on reference conditions.

Introduction

It is of practical value to use stream types because numbers of comparable streams can be treated with the same method. But the finer the spatial scale the less clear type boundaries become and the more the applicability of a typology decreases. An ideal typology fulfils requirements of different objectives, is robust and easy to understand by non-specialists. Such an ideal typology remains a utopy. A typology will always be subjective and based on the objectives it is designed for (Verdonschot, 1990; Nijboer, in prep.).

Differences between climate, hydrology, geomorphology, geology, soil composition, land-use, vegetation and ecology make comparison of communities in running waters difficult or even impossible (Macan, 1961; Maitland, 1966). On the other hand, a typology generalises knowledge that can be applied on a wider scale (Pennak, 1971), and improves the comparability of running waters in management, assessment, and prediction (Hawkes, 1975). A typology thus adds to the intercalibration of the 10.

The use of a typology to classify streams has become an accepted part of ecological assessment (Wright et al., 1999; Hering et al., 2004). Stream types serve as 'classes' for which assessment systems can be developed and applied (Verdonschot, 1990). The comparison of conditions at a current site with those of a reference site belonging to the same stream type allows a type-specific evaluation (Hering et al., 2004). Reference conditions are best described at the scale of a type (Nijboer et al., 2004).

The underlying descriptors of typologies differ strongly. Three major approaches can be distinguished:

1. Biotic descriptors (such as the WFD predescribes; European Commission, 2000)
2. Biotic descriptors (e.g., Wright et al., 1984)
3. A combination of abiotic and biotic descriptors (e.g., Reynoldson et al., 1997).

The EU Water Framework Directive defined abiotic descriptors to classify stream types. This typology is an essential building block of the implementation of the WFD and offers a framework for assessment. For rivers, the Directive defined abiotic descriptors to establish the 'System A' typology. 'System A' descriptors are defined by ecoregions (according to Illies, 1978), size based on the catchment area classes (small 10–100, medium 100–1000, large 1000–10 000, very large >10 000 km^2), catchment geology (siliceous, calcareous, organic), and altitude (lowland <200, medium-altitude 200–800, high >800 m). Using these abiotic descriptors, the participating countries in the EU research projects AQEM and STAR (Austria, Czech Republic, Germany, Greece, Italy, Latvia, Netherlands, Portugal, Sweden and United Kingdom) selected sites in 48 stream types to construct a standardised European stream classification. But do the abiotic descriptors based on the 'System A' of the WFD fit the distribution patterns of the organisms or communities present in the European streams?

The European Commission further recognised that the ecological status of water bodies should be determined by comparing these to near-natural or reference conditions. The WFD approach of using reference conditions in assessment is in agreement with the assessment approaches adopted in the USA (e.g., USEPA, 1996) and Australia (Davies, 2000). Communities are optimally developed under reference conditions (e.g., Karr & Chu, 1999). It is commonly accepted that human disturbance affects a stream ecosystem in such a way that communities become poor and look more alike (e.g., Wright et al., 1984; Verdonschot, 1990). Yet, would a stream typology become most explicit using only reference sites?

Verdonschot & Nijboer (2004) tested if the typology suggested in the WFD was useful for developing an assessment system for macroinvertebrates in streams. They concluded that the major macroinvertebrate distribution patterns in European streams follow climatological and geomorphological conditions and are well distinguished in terms of stream types. Thus, the WFD typology was useful for the development of type-specific assessment systems for streams using macroinvertebrates. Furthermore, it was shown that large-scale factors affected the macroinvertebrate distribution even on a very fine scale. The large-scale factors were indeed the variables that explained most of the variation in species composition. But as these factors even strongly act at the scale of stream types, a further refinement is most probably necessary to disentangle typological actors from water quality ones.

In this follow up study, additional data became available, which implied that enough data of reference sites provided the opportunity to do analyses with such sites solely, and the question of scale could be tackled. The objectives of this study were:

- To test whether the WFD abiotic descriptors for rivers are valid and fit the biotic ones when using a larger data set.
- To explore whether the stream typology should be based on reference sites only or also can include or solely use degraded sites.
- To explore the effect of scale in typology.

Methods

Data collection

In the EU research project AQEM, in total 889 macroinvertebrate samples representing 29 stream types were taken in 8 countries in 2000 and 2001. In the EU research project STAR, an additional 771 samples were taken in 13 countries in 2002 and 2003. The combined AQEM–STAR database composed 1660 samples representing 48 stream types (see Verdonschot, 2006). All samples together cover the major geographical gradients in

Europe. The AQEM site selection, sampling, sorting, and identification procedure was explained by Hering et al. (2004). The STAR samples were either processed according to a slightly adapted AQEM protocol (Furse et al., 2004) or according to several national sampling protocols: RIVPACS (Germany, Austria, Greece and United Kingdom), IBE (Italy), IBGN (France), DSFI (Denmark), LVS 240 (Latvia), PERLA (Czech Republic) and the national protocols of Poland, Sweden, and Portugal.

Handnets were used in all methods. All samples were taken within a stream stretch of <500 m of the respective stream. All samples were collected in at least two seasons, of which one was spring. The second sample was collected in summer or autumn, depending on the regional, geographical and climatological conditions. At the STAR-related sites replicate samples were taken. All samples were further processed in the same standardised way. Finally, different samples from the same site, either being replicates or taken using a different method, and samples taken in different seasons from the same site were kept in the analyses and treated as separate samples. Hereby, the variation caused by the different methods is accepted.

Identification took place to species-level when possible. In some areas, identification was limited to higher taxonomic levels due to a lack of taxonomic knowledge. Finally, all samples were combined into one European database.

Data analyses

For several reasons taxonomic levels within and between samples differed. This can be because of damaged specimens, lack of taxonomic knowledge in certain areas of Europe, lack of certain life stages, or lack of certain taxonomic groups in general. Therefore, taxonomic adjustment was needed to assure unambiguous data processing. Differences in taxonomic level could otherwise later prove to be the cause of differences between sample groupings. In this study a weighed taxonomic adjustment was applied according to the criteria described by Vlek et al. (2004). These criteria were applied to the total database. After taxonomic adjustment the macroinvertebrate abundances of each sample were transformed ($^2\log(x+1)$) (Preston, 1962; Verdonschot, 1990).

The multimetric AQEM assessment system (Hering et al., 2004) was used to classify all AQEM samples into an Ecological Quality Class ranging from 5 (high quality) to 1 (bad quality). For all STAR samples only a pre-classification was available assigning the samples to the same quality classes based on the expert knowledge and abiotic field measurements. For data analysis three datasets were compiled: (i) 'all samples'; 1660 samples, (ii) 'reference samples'; 876 samples including only samples with an ecological quality classification good (class 4) and high (class 5), and (iii) 'degraded samples'; 784 samples including only samples with an ecological quality classification moderate (class 3), poor (class 2), and bad (class 1). The inclusion of class 4 (good) in the group of reference samples was done because (i) the quality deviation from the reference is only slight and (ii) to obtain enough samples for reliable analyses.

Ordination was designed for data analysis in community ecology. Used in an explorative way it shows an ordination diagram that optimally displays how community composition varies (ter Braak & Šmilauer, 2002). In order to analyse the macroinvertebrate species composition in relation to stream type, detrended correspondence analysis (DCA) was used. DCA is an indirect ordination technique and part of the program CANOCO for Windows, version 4.2 (ter Braak & Šmilauer, 2002). In DCA, the samples are patterned in a multidimensional space based on their taxonomic composition.

The options chosen in CANOCO will influence the result of the DCA ordination. In this study the following options were selected (ter Braak & Šmilauer, 2002):

- Detrending by 2nd order polynomials to reduce the 'arch' effect;
- Downweighting of rare species which reduces the influence of rare species and stresses the importance of more common ones in the analysis;
- Inter-sample distance that optimises the position of the samples in the ordination diagram;
- Hill's scaling to allow for long gradients the sample distances to represent turn over distances.

To establish the percentage of overlap between groups of samples a graphical approach was used. Using more 'classical' clustering techniques, such as hierarchical agglomerative clustering, a number of reproducible but more or less subjective choices within the program must be made by the user and decide the results of the classification. The technique chosen in this study is based on the interpretation of the DCA ordination diagram by counting the number of samples present in adjacent groups. Therefore, within the resulting ordination diagram, which included the first and second ordination axes, the samples were labelled *a priori* according to stream type as defined in 'System A' of the WFD. The overlap between stream types was established by drawing contour lines, straight lines between adjacent sites of the same type, around each of the types and summing up all of the overlapping samples. The position of the contour line was the result of an iterative process of repositioning the contour line and re-counting the overlap until a minimum overlap was reached. Overlapping stream types were grouped into larger groups if more than 25% of the samples were positioned within the other type or group, and next the overlap between these new established groups was calculated again by summing up all the remaining overlapping samples. The groups with an overlap <25% of the samples were identified as an identifiable group. Each group was considered to represent a recognisable typological unit, and a next DCA run was performed for this respective group to identify groups within. This process was repeated until no groups could further be disentangled or the level of stream type as recognisable group was reached. Starting with the whole database, the groups recognised in the first ordination were considered to represent the highest hierarchical units and are considered the major groups in Europe, the further the 'pealing off' was done the lower hierarchical position a group represented: groups, sub-groups, and stream types, respectively.

The calculation of the overlap was restricted to axes one and two, as in each run only two to three major groups were separated. DCA plots the major grouping of samples along the first and the second major grouping along the second axis (ter Braak & Šmilauer, 2002). The DCA analyses were repeated for all six datasets; 'all samples', 'reference samples' and 'degraded samples'.

Based on the results a schematic overview was made including the hierarchy and clustering of the European stream types.

For a selected number of environmental variables the average value per stream type was calculated to support the interpretation of the group identification.

Results

Hierarchical grouping of stream types

In general, the loss of species due to taxonomic adjustment was very high (Verdonschot, 2006). Many species and combinations of species were assigned to genus-level and family-level. All major taxonomic groups were strongly reduced in number of taxa after adjustment. Major losses occurred in the Chironomidae, but also the numbers of Hydrachnidia, Megaloptera, Plecoptera, and Coleoptera taxa were strongly reduced. Gastropoda seemed to be best known throughout Europe and the decrease of the number of taxa was restricted to 61%.

The hierarchical position and number of samples per major group as well as all other groupings discussed further on and resulting from the DCA analyses are listed in Table 1. The most important environmental variables were averaged per stream type (Table 2).

The first DCA analysis of the 876 reference samples using species data, and stream types as labels, resulted in three major groups of stream types that correspond to three major landscape types in Europe (Fig. 1): Mountains, Lowlands and Mediterranean. These major groups have an average altitude of 481, 130, and 313 m, respectively. The ordination diagram (Fig. 1) shows that the widest spread of data points occurred within the samples of the Lowlands, samples belonging to the Mountains are more similar and, thus, less widely spread over the diagram, and finally, samples of the Mediterranean were projected along both Mountains and Lowlands groups. The Mediterranean dataset had a lower number of samples, while these samples originated

Table 1. Hierarchical grouping of European (groups of) stream types for Europe. Number of samples per group/type indicated for all samples, reference samples and degraded samples, respectively

	major group	group	sub-group	stream type	all samples	reference samples	degraded samples
Europe: all samples					1660	876	784
Mountains	x				645	339	306
Northern European Mountains		x			120	68	52
Northern Sweden			x	S01, S02	60	40	20
Boreal Highlands			x		60	28	32
				S03	30	22	8
				S04	30	6	24
Central European Mountains		x			441	224	217
Central European Mountains (medium-sized)			x		64	29	35
				C14	24	14	10
				D05	40	15	25
Central European Mountains (small)			x	A04, A05, A06, C04, C05, C15, C16, D04, D06, V01	377	195	182
Central Alps		x			84	47	37
				A02	26	14	12
				A03	26	13	13
				I05	32	20	12
Mediterranean	x				217	128	89
Central and Eastern Mediterranean		x			159	92	67
Greece (Medit.)			x	H01, H02, H03	80	53	27
Central Apennines			x	I06	34	18	16
Northern Apennines			x		45	21	24
				I23	23	11	12
				I24	22	10	12
Western Mediterranean		x			58	36	22
S-Portugal (medium-sized)			x	P04	32	20	12
S-Portugal (small)			x	P01, P02	26	16	10
Lowlands	x				798	409	389
Central and Southern Lowlands		x			572	257	315
Hungarian Plain			x	A01	24	11	13
Hellenic Balkans			x	H04, H05, H06, H07	72	43	29
Central European Lowlands			x	D01, D02, D03, I22, K02, N13, N14, O02, P03, U15, U23	476	203	273
Northern Lowlands		x			226	152	74
Southern Sweden			x	S05, S06	97	61	36
Baltic Province			x	L02	93	71	22
Western sub-alpine Mountains			x	F08	36	20	16

Table 2. Average values of selected major environmental variables per stream type for 48 stream types sampled in the AQEM and STAR research projects

Stream type	A01	A02	A03	A04	A05	A06	C04	C05	C14	C15	C16	D01	D02	D03	D04	D05
Altitude (m)	249	617	1060	433	478	458	348	367	419	303	322	56	41	41	363	287
S.D.	56	144	166	114	27	72	54	65	51	75	135	29	6	13	79	52
Catchment area (km^2)	199	255	32	262	45	32	28	32	587	29	261	55	3	538	18	382
S.D.	101	122	14	3	8	17	7	10	306	18	128	36	3	1139	10	244
Distance to source (km)	38.5	29.6	9.5	38.0	14.8	11.8	9.9	10.9	50.4	9.4	29.0	11.4	1.9	47.9	7.2	44.4
S.D.	10.7	8.9	3.1	3.9	2.4	4.0	2.1	2.8	20.2	2.7	7.6	7.3	1.6	60.0	2.5	18.3
Slope (%)	0.42	0.51	5.74	0.68	2.53	4.05	0.99	1.02	0.34	1.30	21.66	0.00	0.01	0.10	1.74	0.42
S.D.	0.30	0.24	5.14	0.17	1.68	3.88	0.41	0.58	0.14	0.75	89.75	0.00	0.01	0.06	0.67	0.24
Stream order	4.3	4.7	3.6	4.6	3.8	3.4	3.3	3.0	5.8	3.9	5.2	2.7	1.2	3.4	2.3	3.1
S.D.	0.9	0.5	0.6	0.5	0.4	0.8	0.5	0.4	0.6	0.7	0.7	0.6	0.4	0.7	0.7	0.3
Intermittent % sites	0	0	0	0	0	0	5	0	0	7	0	0	47	0	0	0
S.D.	0	0	0	0	0	0	22	0	0	26	0	0	52	0	0	0
Current velocity (m/s)	0.49	0.59	0.71	0.37	0.48	0.45	0.24	0.30	0.40	0.31	0.45	0.14	0.11	0.23	0.42	0.61
S.D.	0.18	0.25	0.26	0.18	0.40	0.26	0.13	0.13	0.25	0.21	0.23	0.16	0.20	0.12	0.30	0.25
Width (m)	5.9	22.7	6.7	11.4	4.0	5.1	2.8	3.2	17.3	4.0	12.4	6.5	3.6	10.2	3.6	17.6
S.D.	2.9	7.2	3.4	2.5	1.2	2.0	0.5	1.5	7.8	1.3	5.7	2.5	3.8	4.5	1.3	8.8
Depth (cm)	25	29	17	38	27	25	24	16	44	19	29	26	7	63	21	39
S.D.	6	11	6	13	10	19	13	4	11	9	8	16	5	35	13	13
EC (μS/cm)	415	331	113	112	100	90	454	289	235	608	480	695	340	553	192	240
S.D.	191	37	54	10	19	54	191	171	101	196	212	341	123	131	109	93
Cl (mg/l)	22.4	2.8	2.2	6.3	3.2	2.8	18.9	10.6	19.7	17.0	22.1	55.9	37.9	37.9	20.7	25.3
S.D.	12.3	0.3	0.3	1.7	1.8	1.1	17.8	6.1	13.5	14.5	20.5	29.3	10.0	13.7	19.1	6.2
pH	7.9	8.3	7.9	7.6	7.5	7.5	8.4	7.8	7.7	8.1	8.3	7.7	6.1	7.8	7.6	7.7
S.D.	0.2	0.3	0.2	0.3	0.2	0.3	0.4	0.4	0.7	0.5	0.6	0.4	0.9	0.3	0.5	0.7
Total hardness (mmol/l)	1.89	1.73	0.53	0.36	0.59	0.67	1.86	1.19	0.88	3.10	2.27	2.73	1.31	2.98	0.75	0.98
S.D.	1.24	0.19	0.24	0.04	0.13	0.51	0.78	0.86	0.39	1.09	1.02	0.88	0.46	2.10	0.25	0.39
O_2 (mg/l)	9.9	12.0	10.5	10.4	9.6	9.8	10.8	10.9	9.5	10.1	9.7	10.2	10.2	10.8	11.6	11.5
S.D.	2.6	1.3	0.9	1.4	0.4	0.4	3.2	2.0	2.5	2.7	3.1	2.8	0.8	1.6	1.9	1.3
Ammonium (mg/l)	0.35	0.01	0.01	0.01	0.01	0.03	0.34	0.11	0.09	0.21	0.41	0.41	0.14	0.19	0.13	0.20
S.D.	0.63	0.01	0.00	0.00	0.00	0.03	0.70	0.23	0.07	0.33	0.61	0.88	0.25	0.31	0.37	0.38
Ortho-phosphate (μg/l)	116	2	2	19	22	6	227	128	439	440	716	222	44	168	166	174
S.D.	161	1	2	5	15	4	195	116	260	418	732	210	44	185	330	188
Total phosphate (μg/l)	219	7	46	35	38	22	595	215	961	653	1211	413	91	390	340	244
S.D.	195	2	112	3	20	12	564	235	498	554	1171	351	76	209	698	193
Alkalinity (mmol/l)	1.62	1.61	0.54	0.29	0.47	0.62	2.84	1.53	1.14	4.86	3.63	2.42	0.40	3.41	1.02	0.56
S.D.	0.91	0.15	0.24	0.03	0.09	0.43	1.60	1.21	0.50	1.31	1.37	1.35	0.39	1.42	0.61	0.57

Stream type	D06	F08	H01	H02	H03	H04	H05	H06	H07	I05	I06	I22	I23	I24	K02	L02
Altitude (m)	205	326	243	678	365	310	626	166	75	1395	381	133	374	478	22	64
S.D.	49	41	269	343	395	274	247	69	57	215	116	14	96	111	19	50
Catchment area (km^2)	42	189	103	167	156	48	114	17	296	28	47	6	355	60	106	159
S.D.	16	111	155	181	228	56	137	11	319	35	0	5	214	75	48	110
Distance to source (km)	9.3	25.9	19.0	16.9	20.6	10.6	17.0	5.8	23.0	5.9	9.0	1.7	42.7	13.1	18.8	31.0
S.D.	2.4	10.5	18.3	12.0	16.5	5.6	9.6	2.9	19.2	4.1	0.0	1.1	18.0	9.1	5.4	18.8
Slope (%)	0.91	1.98	0.92	1.83	1.18	1.19	1.23	1.41	0.58	9.62	1.53	0.23	0.79	2.13	1.38	0.35
S.D.	0.30	0.95	1.38	2.24	1.81	1.23	1.50	1.65	0.71	4.32	0.85	0.08	0.33	1.24	0.93	0.23
Stream order	2.3	2.8	4.6	5.1	5.0	3.2	4.8	2.6	1.0	2.7		1.0	5.7	4.3	3.3	2.0
S.D.	0.4	0.5	1.8	1.2	1.1	1.5	1.2	0.5	0.0	0.9		0.0	0.5	0.6	0.5	0.5
Intermittent % sites	0	0	8	4	20	0	0	0	0	0	0	0	0	18	0	1
S.D.	0	0	28	20	41	0	0	0	0	0	0	0	0	39	0	10
Current velocity (m/s)			0.46	0.45	0.51				0.24	0.54		0.15	0.28	0.24		0.48
S.D.			0.33	0.36	0.36				0.21	0.18		0.22	0.10	0.16		0.24
Width (m)	4.4	8.4	6.2	5.0	7.0	9.3	5.3	2.4	5.6	5.1		2.9	14.4	3.1	5.8	6.8
S.D.	1.6	2.5	11.0	8.9	5.0	4.9	4.4	1.4	2.8	5.5		0.9	9.5	2.3	1.4	2.8
Depth (cm)	24	28	18	15	19	24	19	18	79	24	53	26	22	15	72	35
S.D.	6	11	8	7	7	18	10	11	62	10	14	9	5	4	23	13
EC (μS/cm)	128	660	400	303	488	432	191	466	584	327	782	247	298	469	401	419
S.D.	58	322	577	202	239	151	113	258	236	57	443	99	56	137	167	99
Cl (mg/l)	14.1		11.1	7.3	15.9	13.3	6.4	39.5	50.8	1.3	19.1	11.3	8.6	20.0	32.5	7.0
S.D.	14.1		13.6	7.4	14.8	11.8	5.6	33.0	46.4	1.4	5.6	6.6	7.4	12.8	7.6	2.7
pH	7.5	8.2	8.2	8.1	8.1	8.5	8.6	8.3	8.0	8.3	7.8	7.7	8.4	7.8	7.5	8.0
S.D.	0.6	0.1	0.5	0.7	0.4	0.4	0.5	0.6	0.4	0.2	0.2	0.2	0.2	0.3	0.3	0.2
Total hardness (mmol/l)	0.73		0.04	0.21	0.14	1.97	0.73	1.83	3.21	1.76	3.46	1.15	1.64	2.46		9.08
S.D.	0.78		0.08	0.41	0.19	0.83	0.47	1.20	1.02	0.29	2.50	0.18	0.29	0.78		2.16
O_2 (mg/l)	11.0	9.4	9.4	9.9	9.7	10.3	9.1	9.5	8.1	11.0	7.7	6.6	9.4	8.2		9.5
S.D.	1.3	0.2	2.1	2.0	2.0	2.1	1.4	1.6	2.8	2.3	0.9	1.4	1.0	3.0		2.2
Ammonium (mg/l)	0.22		0.09	0.34	0.14	0.02	0.23	0.01	0.03	0.00	0.08	0.16	0.40	0.62	0.10	0.27
S.D.	0.37		0.16	0.91	0.37	0.06	0.60	0.01	0.03	0.00	0.14	0.34	1.70	1.71	0.08	0.08
Ortho-phosphate (μg/l)	112		371	675	236	32	272	31	39	698	157	95	3	313	17	34
S.D.	172		623	674	371	14	525	3	19	835	218	174	9	466	11	17
Total phosphate (μg/l)			158	256	91	71	411	59	45	7	133	155	8	381	46	200
S.D.			232	227	134	96	619	33	24	4	213	237	17	589	31	49
Alkalinity (mmol/l)	0.45		2.33	2.57	3.52	3.59	1.53	3.43	4.37	2.90	3.61	0.79	2.93	2.33	1.99	4.07
S.D.	0.31		1.35	1.90	1.08	1.36	1.03	2.40	0.50	0.44	0.75	0.16	0.63	0.55	1.22	0.91

Continued on p. 46

Table 2. (Continued)

Stream type	N13	N14	O02	P01	P02	P03	P04	S01	S02	S03	S04	S05	S06	U15	U23	V01
Altitude (m)	47	19	128	294	87	94	263	134	315	501	790	107	10	27	66	351
S.D.	50	14	37	91	35	47	43	53	74	76	70	81	8	22	45	88
Catchment area (km^2)	26	33	321	31	44	249	190	121	63	72	37	193	234	49	168	32
S.D.	26	49	196	18	32	96	127	151	55	76	34	213	275	96	111	24
Distance to source (km)	8.6	6.6	29.8	14.3	15.3	46.4	32.9	20.7	13.9	15.8	9.7	28.5	30.7	7.3	27.5	9.2
S.D.	11.1	7.0	13.6	5.7	5.5	20.2	9.7	13.4	5.8	8.8	4.9	15.4	20.2	2.2	7.3	4.1
Slope (%)	2.85	1.02	0.11	1.13	0.70	0.39	1.14					1.50	0.21	3.60	3.75	5.45
S.D.	4.98	1.96	0.10	0.97	0.61	0.35	0.50					3.94	0.22	6.63	4.49	2.60
Stream order	2.0	1.9	3.1	1.7	1.6	2.9	2.8	3.2	3.0	3.5	3.3	4.1	4.6	2.8	4.3	3.5
S.D.	0.9	0.8	0.7	0.6	0.7	0.6	0.4	1.1	0.5	0.7	0.6	0.9	0.7	1.0	0.8	0.7
Intermittent % sites	6	16	0	93	55	62	100	0	0	0	0	6	27	0	0	0
S.D.	25	37	0	26	52	51	0	0	0	0	0	24	45	0	0	0
Current velocity (m/s)	0.27	0.24	0.09	0.31	0.22	0.33	0.25	0.00	0.00	0.00	0.00	0.00		0.44		0.41
S.D.	0.23	0.22	0.07	0.16	0.11	0.21	0.12	0.00	0.00	0.00	0.00	0.00		0.23		0.14
Width (m)	2.5	3.7	7.6	4.1	5.7	6.3	14.1	6.4	7.3	9.5	19.8	9.0	6.3	2.3	9.6	4.6
S.D.	1.8	3.2	3.2	2.1	3.2	4.0	5.5	2.1	5.7	3.5	9.7	6.0	3.8	0.8	4.6	2.1
Depth (cm)	20	26	44	20	31	26	68	42	35	37	37	38	31	43	41	16
S.D.	16	27	14	8	18	10	19	21	14	18	11	18	14	17	19	3
EC (μS/cm)	486	402	458	228	436	671		39	25	26	21	177	251	592	466	260
S.D.	169	174	233	223	210	585		22	4	21	16	249	99	110	97	132
Cl (mg/l)	27.7	31.5	21.9	25.8	56.1	166.3	10.8	1.7	0.8	0.8	0.7	15.7	11.6			3.2
S.D.	15.4	11.7	25.3	20.5	29.4	180.8	14.7	1.7	0.3	0.4	0.2	37.6	6.6			2.1
pH	7.0	7.2	7.5	7.5	7.7	7.8		6.7	6.8	7.0	6.8	7.0	7.5			7.8
S.D.	0.9	0.4	0.5	0.4	0.3	0.5		0.3	0.3	0.4	0.5	0.4	0.3			0.4
Total hardness (mmol/l)	1.83	1.54	2.22	0.74	1.06	1.89	2.42	0.14	0.09	0.11	0.08	0.51	1.16			2.29
S.D.	1.01	0.84	0.60	0.74	0.53	1.68	2.83	0.08	0.02	0.11	0.08	0.63	0.47			2.18
O2 (mg/l)	8.7	9.3	8.6	7.7	8.4	7.6	9.1	10.3	10.6	11.2	11.6	10.5	10.2			12.0
S.D.	1.7	2.2	2.8	2.3	0.6	1.5	1.2	1.1	1.1	1.2	0.9	1.8	1.8			1.0
Ammonium (mg/l)	0.29	0.21	1.06	0.20	0.03	0.09	0.04	0.01	0.01	0.00	0.00	0.28	0.07			0.12
S.D.	0.36	0.19	2.77	0.39	0.05	0.07	0.05	0.01	0.01	0.00	0.00	0.84	0.15			0.14
Ortho-phosphate (μg/l)	86	45	615	25				8	5	4	4	31	16			97
S.D.	187	39	1743					6	4	2	2	77	21			284
Total phosphate (μg/l)	177	124	1051	314	225	377	0	43	22	14	10	196	40			73
S.D.	240	90	2588	684	433	1089	0	29	12	11	12	825	31			131
Alkalinity (mmol/l)	2.06	1.74		1.00	0.97	2.04	1.71	0.08	0.06	0.08	0.06	1.95	1.74			2.20
S.D.	1.60	1.23		1.19	0.47	1.67	1.16	0.06	0.02	0.10	0.08	12.30	0.78			1.17

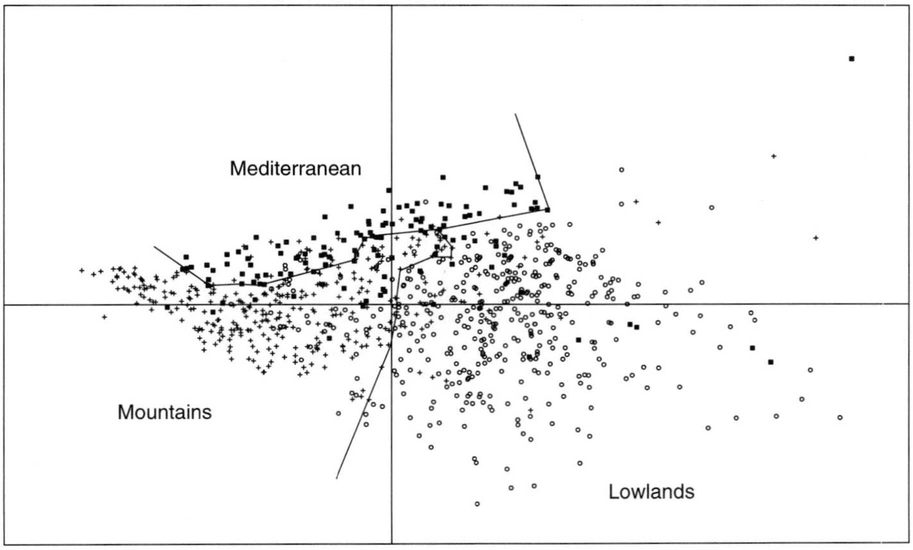

Figure 1. DCA ordination diagram of the axis 1 (horizontal; eigenvalues: 0.23) and 2 (vertical: eigenvalues: 0.13) of the (groups of) stream types within Europe based on species data of reference samples.

from a wider variety of landscapes of both high and low altitude. The dissimilarity of samples within Mountains and Lowlands suggests a wider variety of species combinations in the Lowlands.

The major group Mountains was divided into three groups (Fig. 2a; Table 1): Central Alps, Northern European Mountains and Central European Mountains.

In the diagram, the Central Alps are positioned more or less as an extension of the Central European Mountains (Fig. 2a). The Central Alps were further divided into stream: A02 and I05 types (Fig. 4c), both of which are calcareous streams (see Verdonschot, 2006), the former being medium-sized and the latter small, and stream type A03, small siliceous streams. The samples of the three stream types within the group Central Alps (average altitude 1024 m) were situated along a strong altitudinal gradient: two stream types were situated at altitudes higher than 800 m (stream type I05; average altitude 1395 m; stream type A03; average altitude 1060 m) and one just below the 800 m (stream type A02; average altitude 617 m). The catchment size of the latter is also much larger in comparison to the first two stream types.

The groups Northern vs. the Central European Mountains do not differ in average altitude; 435 m vs. 370 m. The difference in conductivity, alkalinity, and total hardness is evident (Table 2). These differences in ionic composition are most probably caused the very different geology of both groups.

Within the group Central European Mountains two sub-groups were identified (Fig. 3b, Table 1): the medium-sized stream types (with also a larger catchment) of the Central European Mountains (in the right upper corner of the diagram) and the small streams (with a smaller catchment) of the Central European Mountains (composed of 10 stream types). The samples show a gradual transition between both groups of stream types, whereby the samples of the medium-sized stream types of the Central European Mountains constituted one end of the gradient (right upper corner in the diagram: Fig. 3b) and overlap with the small stream types of the Central European Mountains which as a group could not be disentangled further. The medium-sized streams in the Central European Mountains were further divided into two stream types (Fig. 4b): C14 and D05, situated in the Czech Republic and Germany, respectively. The stream type C14 refers to acid-silicate geology (see Verdonschot, 2006), while the stream type D05 refers to a calcareous one.

The group Northern European Mountains was divided into two sub-groups (Fig. 3a) again caused by altitudinal differences: Northern Sweden with an average altitude of 224 m vs. the Boreal Highlands with an average altitude of 645 m. In

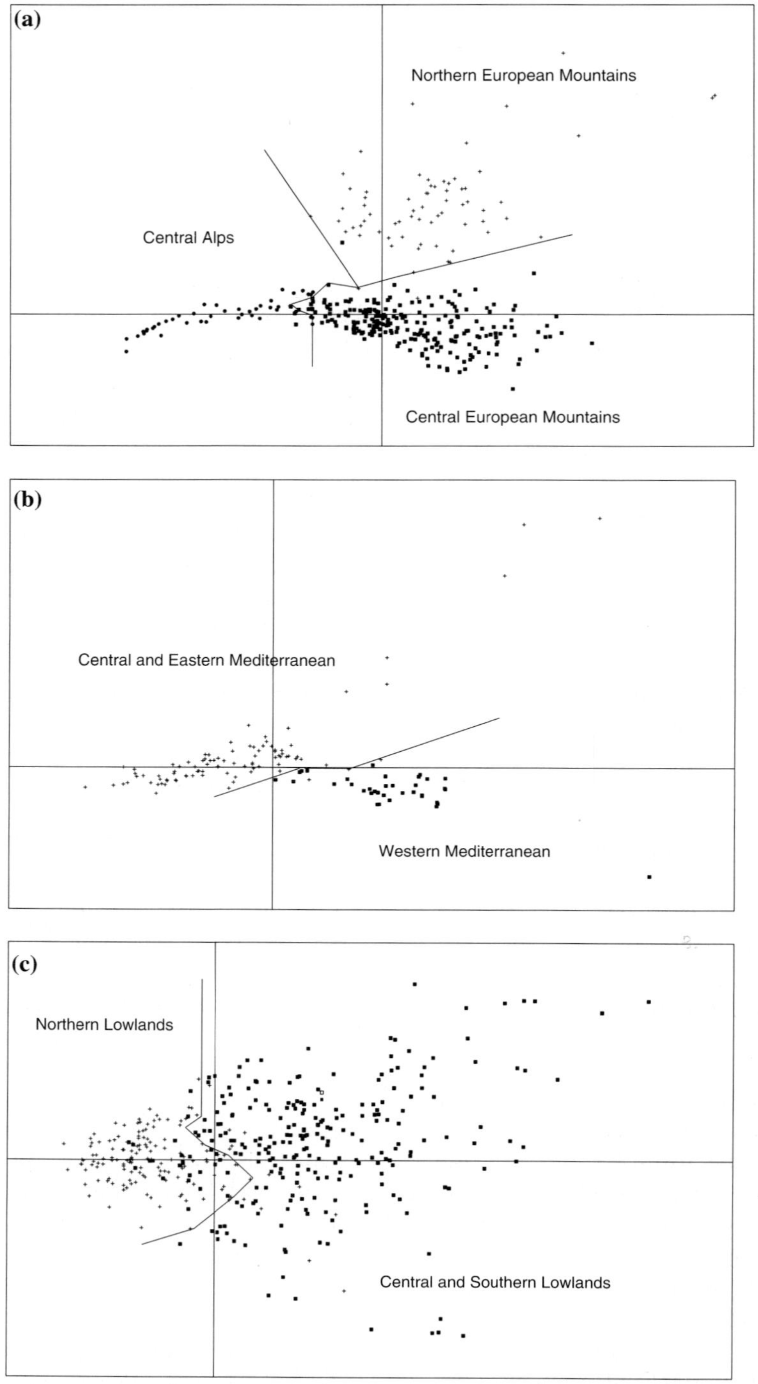

Figure 2. DCA ordination diagrams of the axis 1 (horizontal) and 2 (vertical) of the groups of stream types within the major regions in Europe based on species data of reference samples. (a) Mountains (eigenvalues axis 1: 0.17, axis 2: 0.14), (b) Mediterranean (eigenvalues axis 1: 0.31, axis 2: 0.23), (c) Lowlands (eigenvalues axis 1: 0.16, axis 2: 0.13).

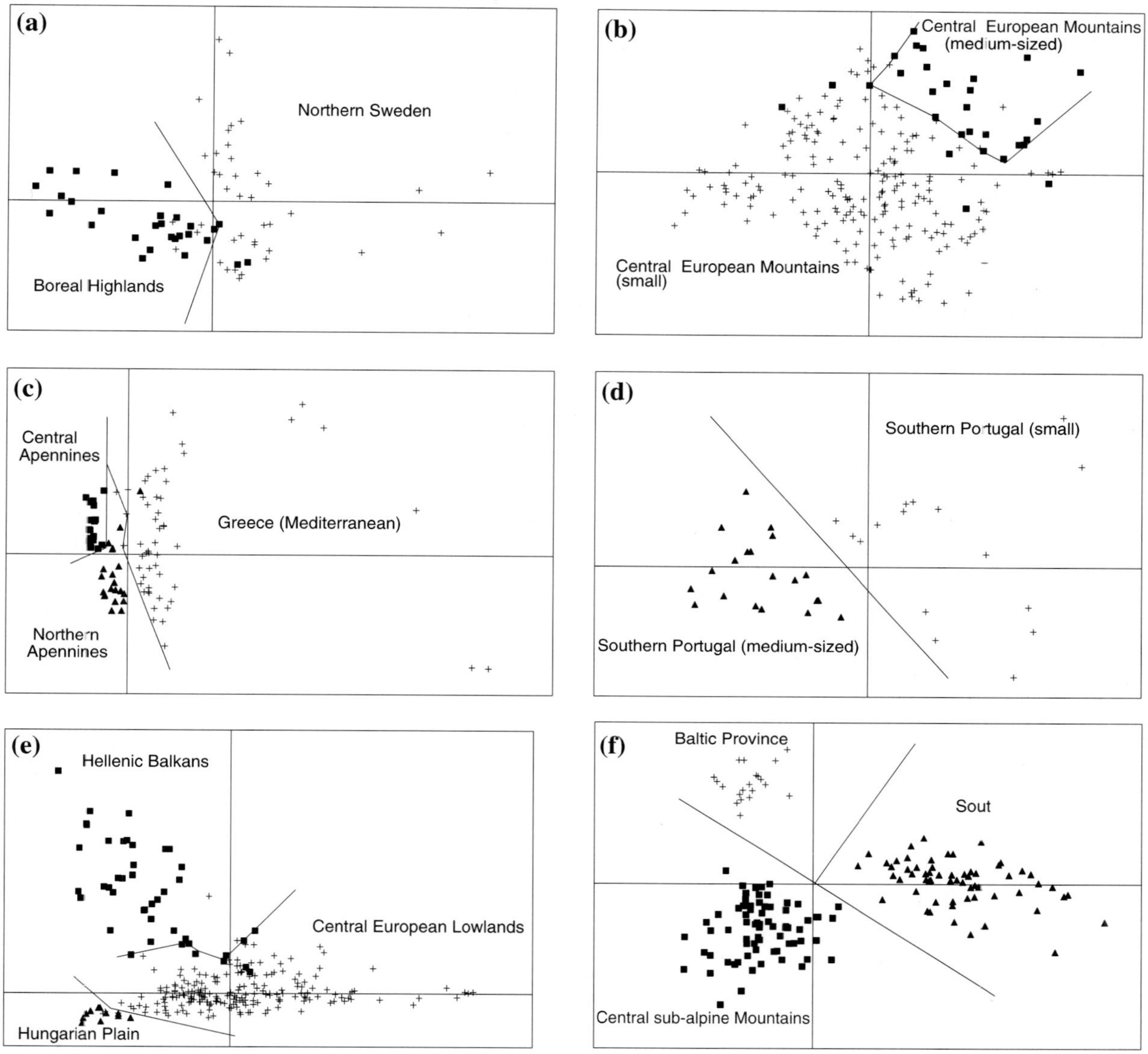

Figure 3. DCA ordination diagrams of the axis 1 (horizontal) and 2 (vertical) of the groups of stream types within the regions in Europe based on species data of reference samples. (a) Northern European Mountains (eigenvalues axis 1: 0.24, axis 2: 0.14), (b) Central European Mountains (eigenvalues axis 1: 0.15, axis 2: 0.11), (c) Central and Eastern Mediterranean (eigenvalues axis 1: 0.36, axis 2: 0.19), (d) Western Mediterranean (eigenvalues axis 1: 0.28, axis 2: 0.16), (e) Central and Southern Lowlands (eigenvalues axis 1: 0.19, axis 2: 0.16), (f) Northern Lowlands (eigenvalues axis 1: 0.16, axis 2: 0.12).

the diagram the samples of sub-group Northern Sweden (two stream types) were much more diverse in comparison to the samples of the sub-group Boreal Highlands. The sub-group Boreal Highlands contained two stream types (Fig. 4a): S03 and S04, with a medium-altitude (average altitude 501 m) vs. the high-altitude (average altitude 790 m) samples (see Verdonschot, 2006). The two stream types in Northern Sweden could not be disentangled further.

The major group Mediterranean divided into two groups (Fig. 2b; Table 1): the group Western Mediterranean refers to the Portuguese samples at an average altitude of 215 m, mostly intermittent streams, and the group Central and Eastern Mediterranean, that refers to the Italian and Greek samples situated at an average altitude of 362 m, mostly permanent streams. The Italian streams showed higher hardness and phosphate concentrations in comparison to the Greek ones (Table 2). The group Central and Eastern Mediterranean clearly included some outliers in the right upper corner of the diagram (Fig. 2c), while the group Western Mediterranean had one outlier in the lower right corner. The group Central and Eastern Mediterranean was divided further, especially along the first axis, into the three local regions (Fig. 3c): both Italian sub-groups of the Northern and Central Apennines (one stream type), respectively, on the left of the diagram and the Greek (three stream types) samples on the right. All three groups of samples were situated along a vertical gradient parallel to the vertical axis in the diagram (Fig. 3c). The sub-group Northern Apennines was further divided into two stream types (Fig. 4d): I23 and I24. The group Western Mediterranean was divided into the sub-groups of the Southern Portugese medium-sized (one stream type) and small streams (two stream types) (Fig. 3d).

The major group Lowlands was divided along the first axis into two groups (Fig. 2c; Table 1): the group Central and Southern Lowlands, a heterogenous group of samples that is quite widely scattered over the right side of the diagram (average altitude of 131 m) and the group Northern Lowlands that is positioned as a more homogeneous group of samples at the left of the diagram (average altitude of 127 m). Only differences in chloride and conductivity are clear (Table 2). The group Central and Southern Lowlands was further divided into three sub-groups (Fig. 3e): the sub-group Hellenic Balkans (four stream types), and the sub-group Hungarian Plains (one stream type), both situated at the left side of the diagram along the second axis, and the

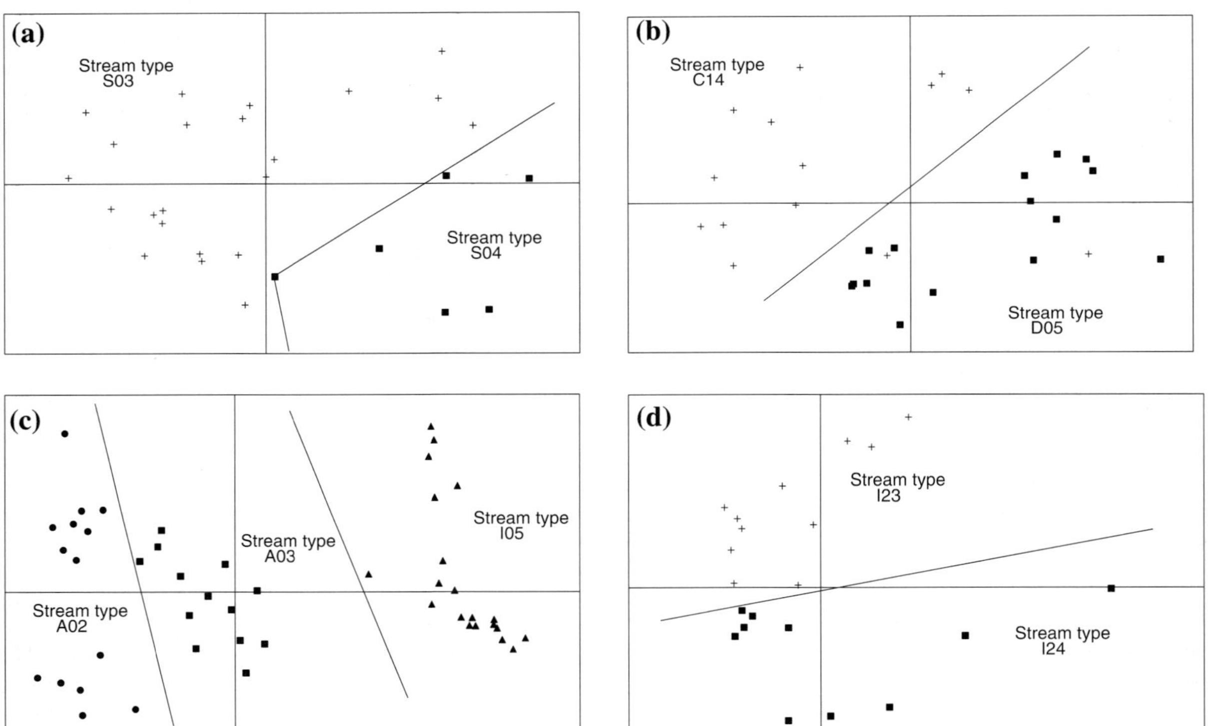

Figure 4. DCA ordination diagrams of the axis 1 (horizontal) and 2 (vertical) of the (groups of) stream types within the local regions in Europe based on species data of reference samples. (a) Boreal Highlands (eigenvalues axis 1: 0.20, axis 2: 0.11), (b) Central European Mountains (medium-sized; eigenvalues axis 1: 0.14, axis 2: 0.10), (c) Central Alps (eigenvalues axis 1: 0.17, axis 2: 0.13), (d) Northern Apennines (eigenvalues axis 1: 0.24, axis 2: 0.16). For explanation of stream type codes see Verdonschot (2006).

sub-group Central European Lowlands (11 stream types), a diverse and widely spread group of samples along the first axis. The sub-group Hellenic Balkans is situated at a higher altitude (average of 294 m) and had a steeper slope. The sub-group Central European Lowlands could not be disentangled further. The group Northern Lowlands was very clearly divided into three sub-groups (Fig. 3f): in the left upper corner the sub-group Baltic Province (one stream type), in the left lower corner the sub-group Western sub-alpine Mountains (one stream type), and to the right along the first axis the sub-group Southern Sweden (two stream types).

Diversity along European environmental gradients

To explore further the drivers of differences in data composition, changes of taxon diversity along major environmental gradients that can be linked to ecoregions were explored. Macroinvertebrates distribute along temperature gradients which are best expressed in either latitudinal, elevational and stream order gradients (Ward, 1985). By plotting the average number of taxa per sample along the latitudinal gradient from Sweden down to Portugal for the reference samples, no relation at all became evident (Fig. 5). There was even a decrease in the average number at the lower latitudes (more southern samples) indicated, which contradicts the findings of Vannote & Sweeney (1980) and Jacobsen et al. (1997). The R^2 value indicates that a correlation is completely absent. The results were similar for the altitudinal gradient. At higher altitude, temperature decreases and the numbers of taxa would be expected to decrease as well (Ward, 1982; Furse et al., 1984; Quinn & Hickey, 1990). The relation between the numbers of taxa and altitude in Europe is shown in Fig. 6. Although the regression line goes somewhat down, the R^2 value shows that there was no trend between altitude and number of taxa.

Going down from a first to a seventh order stream, along the river continuum, temperature again should rise (Hawkes, 1975; Vannote et al., 1980). The relation between the number of macro-invertebrate taxa and stream order in the studied European dataset showed no relationship (Fig. 7).

The average number of individuals per sample showed huge variation between countries. Densities of macroinvertebrates can differ due to the stream and the habitat. Fast and varying current velocities (e.g., Townsend et al., 1997) as well as presence/absence of shelter often relate to lower numbers of specimens (Hynes, 1970). Other authors indicated additional factors being responsible for density differences, such as substrate type (Gore & Judy, 1981), presence of (bank)vegetation, alkalinity (Armitage, 1958),

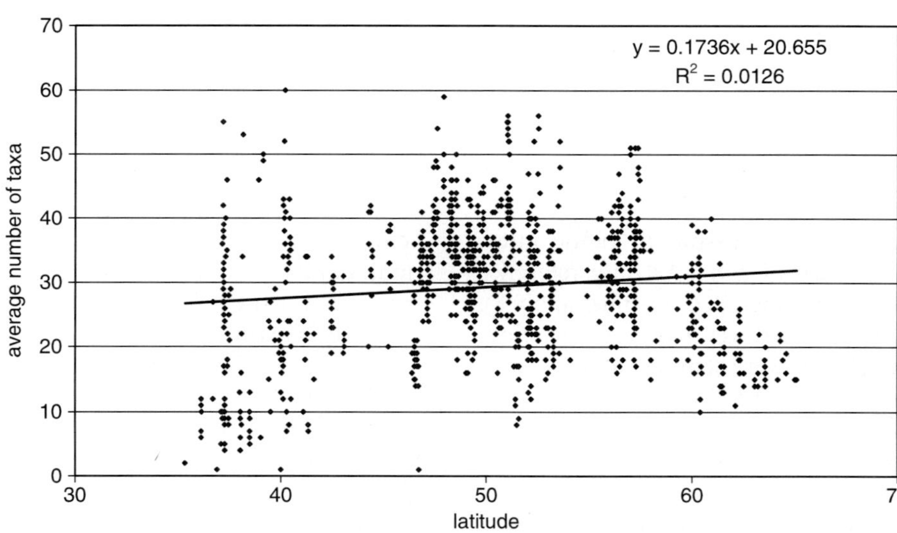

Figure 5. The average number of taxa per sample of the reference sites plotted against latitude.

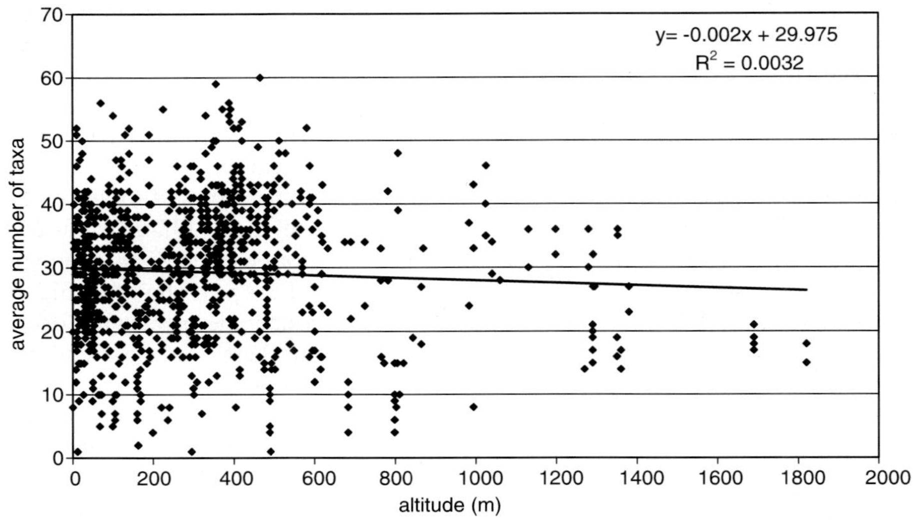

Figure 6. The average number of taxa per sample of the reference sites plotted against altitude.

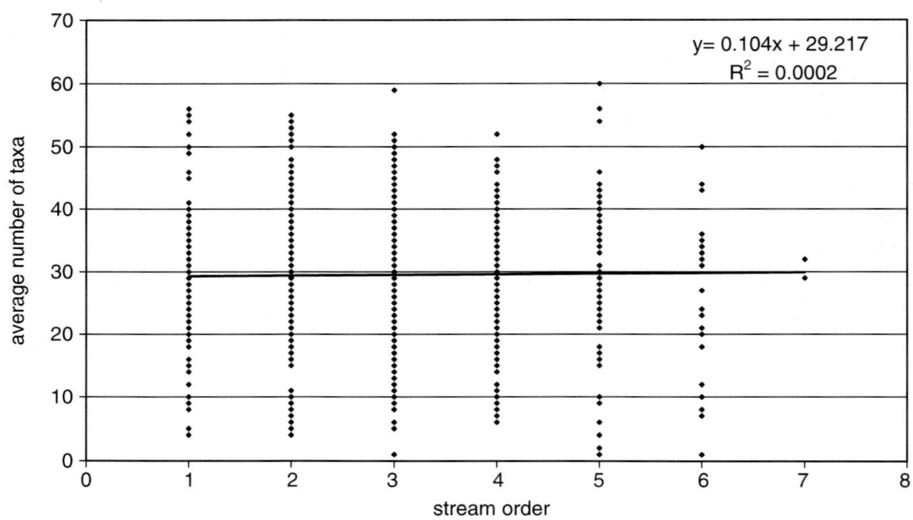

Figure 7. The average number of taxa per sample of the reference sites plotted against stream order.

pollution (Hynes, 1960), season, or biotic interactions (e.g., presence of fish). As only reference and good sites were included in this analysis, pollution can be excluded as a cause of variation. A high number of sites were sampled at least twice which excludes season. Plotting the average current velocity versus the average number of taxa (adjusted) no relation was shown ($R^2=0.05$) (Fig. 8). Similar results were observed for valley slope, a more general timeless parameter for potential current velocity ($R^2=0.01$; figure not shown). As all samples were taken by using a multihabitat sampling approach, all habitats present at a site were sampled. But the number of habitats present per stream can differ between types and as the specimens of most populations show irregular distributions, density estimates are always difficult (Statzner et al., 1998). As Hynes (1970) stated "by their very nature river beds are difficult to sample accurately", most probably this is also one of the major causes for the density differences found in this study.

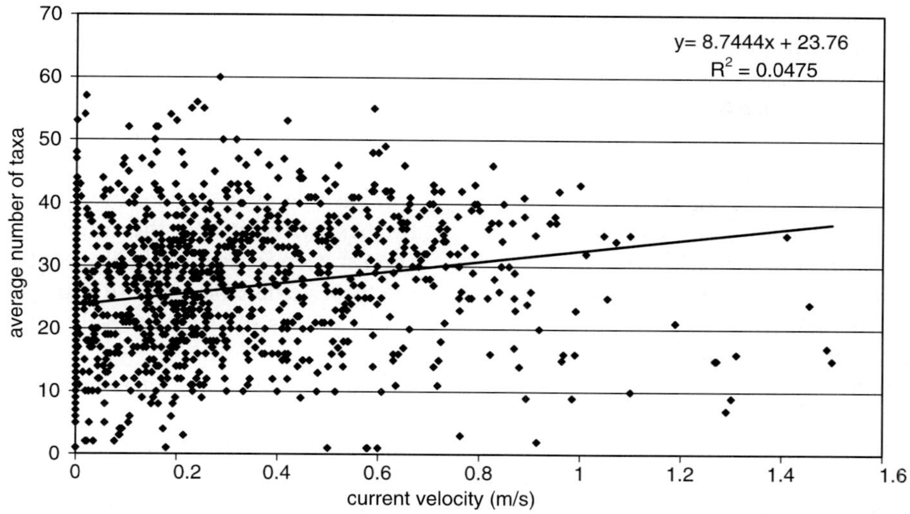

Figure 8. The average number of taxa per sample of the reference sites plotted against current velocity.

Reference or degraded samples in relation to taxonomical level

At the European level, the separation of the three major groups performed best for species-level data (±15%) (Table 3) in comparison with family-level data (±20%) (Table 3; see Verdonschot, this issue). Within the species data the datasets of reference samples and all samples scored even.

Within the major groups the overlap of species-level data was much smaller than that of family-level data: 2–4% vs. 7–9%, respectively. In both datasets the separation between groups of stream types was best using samples of reference sites:

Table 3. Percentage overlap of species-level (before /) and family-level data (after /) of samples from all sites, only reference sites, and only degraded sites for major groups, groups, sub-groups, and stream types (see Table 1)

Overlap for	Species data	All samples	Reference	Degraded
Major groups	Europe	15.1/20.2	15.1/20.1	15.6/17.0
Groups	Mountains	2.0/6.0	1.8/2.4	6.9/7.5
	Lowlands	8.0/12.3	3.8/11.7	4.6/12.6
	Mediterranean	0.5/8.3	0.1/7.8	0.2/5.6
	Average	3.5/8.9	1.9/7.3	3.9/8.6
Sub-groups	Northern European Mountains	6.7/7.5	8.8/7.4	13.5/13.5
	Central European Mountains	3.4/3.9	2.7/3.1	7.8/7.4
	Central and Eastern Mediterranean	5.0/20.8	1.1/10.9	4.5/16.4
	Western Mediterranean	0.0/0.0	0.0/2.8	0.0/4.5
	Northern Lowlands	3.1/8.8	0.0/2.0	0.0/4.1
	Central and Southern Lowlands	8.0/7.5	1.9/3.5	11.4/8.6
	Average	4.4/8.1	2.4/4.9	6.2/9.1
Stream types	Central Alps	1.2/2.4	0.0/2.1	5.4/2.7
	Boreal Highlands	5.0/3.3	0.0/3.6	0.0/3.1
	Central European Mountains (medium-sized)	9.4/3.1	6.9/0.0	5.7/0.0
	Northern Apennines	4.4/0.0	0.0/0.0	8.3/0.0
	Average	5.0/2.2	1.7/1.4	4.9/1.5
	Overall average	27.9/39.4	21.1/33.8	30.5/36.1

1.9% for species data vs. 7.3% for family data (Table 3). The separation between groups of samples was best in the major group Mediterranean based on species-level data but based on family-level data the major group Mountains showed the least overlap. In both species and family-level datasets, the major group Lowlands showed the greatest overlap. This is in concordance with the diverse distribution of samples in the ordination diagrams.

Within the groups the overlap of species-level data is much smaller than that of family-level data: 2–6% vs. 5–9%, respectively. Again, in both datasets the separation between groups of stream types was best in the reference samples: 2.4% for species-level data vs. 4.9% for family-level data (Table 3). For species-level data the group Northern Lowlands showed little overlap. The separation between groups of samples was best in the group Western Mediterranean, a group that even did not show any overlap in the species-level data. Most overlap was seen in the group Northern European Mountains for species-level data and in the group Central and Eastern Mediterranean for family-level data.

Within the sub-groups the overlap of family-level data was somewhat smaller than that of species-level data: 1–2% vs. 2–5%, respectively. Again, in both datasets the separation between groups of stream types was best in the reference sites: 1.7% for species-level data vs. 1.4% for family-level data (Table 3). The smallest overlap was shown in the sub-group Boreal Highlands for species-level data and in the sub-group Northern Apennines for family-level data.

In general, the degraded samples showed largest variation in overlap, whereby large overlap indicated a higher number of degraded samples.

Discussion

In this analysis taxonomic adjustment was done as well as downweighting of rare species. Both choices reduce the variation within the dataset. This was necessary to make all data mutual comparable but at the same time information got lost. One option would have been to redo the taxonomical adjustment after each DCA run per resulting group, e.g., after the first run the three major regions in Europe could be re-adjusted. The advantage is that each finer grouping would be based on more information. The disadvantage is that results would become uncomparable between different groupings within each major region as well as within all other groupings. The objective of this analysis was to compare groupings within Europe in a defined and comparable way. Therefore, all items discussed further on relate to the adjusted data and one should keep in mind that these were data were on the 'best achievable' overall European level, which is not always the species level.

Hierarchical grouping of stream types

Based on the AQEM data, Verdonschot & Nijboer (2004) concluded that the macroinvertebrate distribution over Europe appeared to be strongly related to geographical position. Stream types were hierarchically grouped over major regions, regions and local regions. The addition of the STAR research project data almost doubled the number of macroinvertebrate samples of European rivers. The analyses of this study showed that again three major groups were distinguished. This is in accordance with the AQEM results (Verdonschot & Nijboer, 2004), although each group was less restricted to specific geographical regions, for example the major group Lowlands which now included the Northern Lowlands, composed of the Scandinavian Lowlands and the more continental situated Baltic Province (Latvia), along with the sub-mountainous (atlantic) French area, the Po valley and the Hellenic Balkans. Thus, the term Lowlands with an average altitude of 130 m covered a wide and discontinuous area over Europe and can be better referred to as a low slope landscape, then as a geographical area of (North-Western) Europe. The Mountains, with an average altitude 481 m, included the sub-mountainous to alpine areas of Central and Northern Europe. This major group was less geographically restricted and more related to a steep slope landscape. The major group Mediterranean was solely restricted to the area with a Mediterranean sea climate and situated at lower altitudes with an average altitude of 313 m. The three major groups probably represent the major combination of geomorphological and/or climatological conditions of the sampled sites. The driving forces behind are most probably current (slope) and temperature.

Verdonschot & Nijboer (2004) divided the Mountains into Northern Scandinavia, and the high and low alpine regions. These three regions are much alike to the present groups of Northern European Mountains, Central Alps and Central European Mountains, respectively. These names better define the groups distinguished. The division between the Northern and both the Central European Mountains are most probably due to differences in climatological conditions. The Northern European Mountains and the Central Alps differ in altitude, which can be seen as differences in climatological and geomorphological or slope conditions. The Central European Mountains were separated into the small and the medium-sized streams; size or dimensions was probably the dividing factor.

The major group Mediterranean was divided according to the same scheme as presented by Verdonschot & Nijboer (2004). It could be taken into consideration to name the Western Mediterranean as the Mediterranean Lowlands or Atlantic Mediterranean due to the influence of the Atlantic climate and as it only refers to Portuguese sites not Spanish ones. The Central and Eastern Mediterranean could also be indicated as the Mediterranean Mountains, as the sites were all situated at higher altitudes. The differences with the Hellenic Balkans are the sub-continental climatological influences in the latter. The Western Mediterranean streams were divided in small- and medium-sized streams.

The major group Lowlands was separated into the Central and Southern Lowlands and the Northern Lowlands. This clearly deviates from the former Western and North-Eastern Lowlands (Verdonschot & Nijboer, 2004). This new grouping is probably due to differences in climatological conditions, caused by the inclusion of newly sampled lowland stream types all over Europe in the more flat or low slope areas in Europe. The stream type Western sub-alpine Mountains (F08) was classified among the Northern Lowlands, possibly due to the atlantic climatological conditions at somewhat higher altitude in this mountain area. The conditions are probably comparable to the climatological colder lowland areas of Southern Sweden and of the Baltic Province (Latvia), in combination with a lower slope that could cause comparable environmental circumstances. The latter two can be distinguished based on substrate composition. Another explanation could be a taxonomical composition or the identification level used of the Western sub-alpine Mountains sites that differ from the other Central European Mountains, as the French data did only lose 47% of their taxa due to species-level data adjustment. This means that these data more often were identified to higher taxonomical levels (genus or family).

The WFD stream typology descriptors were linked to ecoregion, catchment size class, geology of the catchment and altitude class. Ecoregion and altitude are both related to climate (temperature, precipitation) and geomorphology. Precipitation and geomorphology (especially slope) set the conditions for the streams current velocity and size. The latter is also directly linked to the catchment size. Finally, the geology is related to geomorphology, hydrology and chemistry of the stream. It is a question whether chemistry is of importance at the scales of this study with only using reference sites. But geology affects hydrology, e.g., calcareous mountains will be much drier then siliceous ones. This in its turn affects current velocity, permanency (not included in this study), and water temperature. All together the driving forces behind these descriptors are temperature, current velocity and stream size.

Several larger groups of stream types could not be further separated, e.g., the Central European Lowlands and the Central European Mountains with 11 and 10 types, respectively. This especially occurred in geographical areas where stream types that are situated close to each other were sampled. This is conform the River Continuum Concept (RCC) that states that stream communities can be viewed as continua consisting of mosaics of population aggregations responding to the gradient of physical factors formed by the drainage network (Vannote et al., 1980). The thought that communities gradually change along environmental gradients is not only true for gradients along one river, but this is also true along landscape gradients that run over different catchments. These gradients will not always change gradually and some gradients can be quite short and then even look abrupt. Where such changes occur communities will overlap and some species are found in neighbouring communities. In such situations these species produce transitional zones or ecotones (Sobolev & Utekhin, 1979; Park, 1948). Short gradients can also be found going uphill where the slope increases and climatological conditions become more and more extreme. Species turn-over along such a gradient will increase and transitions in species composition will occur.

The European landscape is a mixture of mountains and lowlands across two climatological gradients; one north–south from the tundra down to the Mediterranean climate, and one west–east gradient from the atlantic to the continental climate. Over this macro-mosaic the WFD stream type system is set as the basis for stream typology and the starting point for intercalibration. The study showed that the stream types using the WFD 'System A' descriptors are probably less useful at finer scales. Macroinvertebrates responded to the driving forces of the three major factors of temperature, slope and size. Thus, the stream typology should take these three parameters as a starting point. Next streams with comparable major environmental conditions can be mapped and reference conditions can be defined as such. These groups of streams will cross boundaries of stream types, as can be seen in Central European Mountains as well as in the Central European Lowlands, and will also cross boundaries of individual countries. For intercalibration refined analyses are needed, especially for large areas with comparable environmental conditions, to reach a more ecologically relevant typology. This will go beyond the current WFD descriptors of 'System A' for stream types.

Environmental variables and gradients

Despite a standardised protocol the environmental variables measured showed a scattered result. A number of variables was only measured in a restricted number of streams. This affected the interpretation of the data and made a direct gradient analysis approach less effective. Some stream types could clearly be distinguished and identified by their abiotic description while others were much harder to interpret. The results showed that more attention should be given to not only keep with the protocol but also include in a protocol only the relevant variables.

Going along some major European environmental gradients, i.e., latitude, elevation and stream order, each one of these did not cause large differences in taxon richness. This means that the problems of standardising the sampling protocol still can be a major cause of differences in data composition. On the other hand, such suggested gradients may not be existing?

Reference or degraded samples

One of the criticisms on the European stream typology of Verdonschot & Nijboer (2004) was the use of samples from reference as well as degraded sites to construct the typology. It is commonly accepted that stress will degrade a community and degraded communities of different stream types become more similar (e.g., Karr & Chu, 1999). Therefore, it was tested whether the use of reference sites would give better results. Indeed, the reference samples performed best which supports the hypothesis that human stress diminishes the natural differences between stream communities. The higher overlap in the degraded samples also indicates the higher number of degraded samples taken into account, e.g., Northern European Mountains, Central European Mountains and Central and Southern Lowlands (Table 3). This does not mean that all our samples consisted of completely undisturbed conditions (e.g., Nijboer et al., 2004). Human impact in Europe, especially in accessible areas such as the lowlands, goes back to far before medieval times. Still, samples of these recent reference conditions performed best and were most optimally separated. This underlines the basic principle of the WFD that European Member States are required to identify reference conditions for defining the reference community, setting the upper anchor for quality classification and expressing degradation as deviation from this upper anchor (Wallin et al., 2003).

Conclusions

The conclusion of this study were:

- Not all WFD abiotic descriptors for rivers appeared to be valid and fit biotic ones. Three major parameters further divided the three major groups of stream types in Europe; climate (temperature), slope (current velocity) and stream size. Especially, the geographic descriptors (e.g., ecoregion) did not fit well. Thus, the WFD descriptors for stream types should be interpreted in such way that temperature, slope and stream size constitute the basic parameters to define stream types.

- Human stress diminishes the natural differences between stream communities and typologies should therefore be based on reference conditions.
- Neither temperature, nor elevation, stream order or latitudinal position is solely causes the differences in average numbers of taxa between the 1660 sites distributed over Europe.

Acknowledgements

The author would like to thank all AQEM and STAR partners for the use of the data. The EU research projects AQEM and STAR were funded by the European Commission, 5th Framework Program, Energy, Environment and Sustainable Development, Key Action Water, Contract no. EVK1-CT1999-00027 and Contract no. EVK1-CT2001-00089, respectively.

References

Armitage, P. D., 1958. Ecology of riffle insects of the Firehole River, Wyoming. Ecology 39: 571–580.

Davies, P. E., 2000. Development of a national river bioassessment system, AUSRIVAS in Australia. In Wright, J. F., D. W. Sutcliffe & M. T. Furse (eds), Assessing the Biological Quality of Fresh Waters – RIVPACS and Other Techniques. Freshwater Biology 113–124.

European Commission, 2000. Directive 2000/60/EC. Establishing a framework for community action in the field of water policy. European Commission PE-CONS 3639/1/100 Rev 1. Luxembourg.

Furse, M. T., D. Moss, J. F. Wright & P. D. Armitage, 1984. The influence of seasonal and taxonomic factors on the ordination and classification of running-water sites in Great Britain and on the prediction of their macro-invertebrate communities. Freshwater Biology 14: 257–280.

Gore, J. A. & R. D. Judy, 1981. Predictive models of benthic macroinvertebrate density for use in instream flow studies and regulated flow management. Canadian Journal of Fisheries and Aquatic Sciences 38: 1363–1370.

Hawkes, H. A., 1975. River zonation and classification. In Whitton, B.A. (ed.), River Ecology. Studies in Ecology (Vol. 2). University of California Press, 312-374.

Hering, D., O. Moog, L. Sandin & P. F. M. Verdonschot, 2004. Overview and application of the AQEM assessment system. Hydrobiologia 516: 1–20.

Hynes, H. B. N., 1960. Biology of Polluted Waters. Liverpool Univ. Press, Liverpool, 202 pp.

Hynes, H. B. N., 1970. The Ecology of Running Waters. Liverpool Univ. Press, Liverpool, 1 202 pp.

Illies, J., 1978. Limnofauna Europaea. Gustav Fischer Verlag, Stuttgart 532 pp.

Jacobsen, D., R. Schultz & A. Encalada, 1997. Structure and diversity of stream invertebrate assemblages: the influence of temperature with altitude and latitude. Freshwater Biology 38: 247–261.

Karr, J. R. & E. W. Chu, 1999. Restoring Life in Running Waters: Better Biological Monitoring. Island Press, Washington, DC.

Macan, T. T., 1961. A review of running water studies. Verhandlungen Internationale Verein für Limnology 14: 587–602.

Maitland, P. S., 1966. The Fauna of the River Endrick. Studies on Loch Lomond. 2 Publ Univ, Glasgow, 194 pp.

Nijboer, R. C., R. K. Johnson, M. Sommerhäuser, A. Buffagni & P. F. M. Verdonschot, 2004. Reference conditions for European streams. Hydrobiologia 516: 91–105.

Park, T., 1948. Population Ecology. Encyclopedia Brittanica.

Pennak, R. W., 1971. Towards a classification of lotic habitats. Hydrobiologia 38: 321–324.

Preston, F. W., 1962. The canonical distribution of commonness and rarity: part 1. Ecology 43: 185–215.

Quinn, J. M. & C. W. Hickey, 1990. Characterisation and classification of benthic invertebrate communities in 88 New Zealand river in relation to environmental factors. New Zealand Journal of Marine and Freshwater Research 24: 387–409.

Reynoldson, T. B., R. H. Norris, V. H. Resh, K. E. Day & D. M. Rosenberg, 1997. The reference condition: a comparison of multimetric and multivariate approaches to assess water-quality impairment using benthic macroinvertebrates. Journal of North American Benthological Society 16: 833–852.

Sobolev, L. N. & V. D. Utekhin, 1979. Russian (Ramensky) approaches to community systematization. In Whittaker, R. H. (ed.), Ordination of Plant Communities. Junk, The Hague, 71–98.

Statzner, B., J. A. Gore & V. H. Resh, 1998. Monte Carlo simulation of benthic macroinvertebrate populations: estimates using random stratified and gradient sampling. Journal of the North American Benthological Society 17: 324–337.

ter Braak, C. J. F. & P. Ŝmilauer, 2002. CANOCO Reference Manual and Users Guide to Canoco for Windows. Software for Canonical Community Ordination (version 4.5). Centre for Biometry, Wageningen, The Netherlands.

Townsend, C. R., M. R. Scarsbrook & S. Doledec, 1997. Quantifying disturbance in streams: alternative measures of disturbance in relation to macroinvertebrate species traits and species richness. Journal of the North American Benthological Society 16: 531–544.

USEPA (U.S. Environmental Protection Agency), 1996. Biological Criteria: Technical Guidance for Streams and Small Rivers. U.S. Environmental Protection Agency, Office of Water, Washington, DC. EPA-822-B96-001.

Vannote, R. L. & B. W. Sweeney, 1980. Geographic analysis of thermal equilibria: a conceptual model for evaluating the effect of natural and modified thermal regimes on aquatic insects. American Naturalist 115: 667–695.

Vannote, R. L., G. W. Minshall, K. W. Cummins, J. R. Sedell & C. E. Cushing, 1980. The River Continuum Concept. Canadian Journal of Fisheries and Aquatic Sciences 37: 130–137.

Verdonschot, P. F. M. & R. C. Nijboer, 2004. Testing the European stream typology of the water Framework Directive for macroinvertebrates. Hydrobiologia, 175: 35–54.

Verdonschot, P. F. M., 1990. Ecological characterization of surface waters in the province of Overijssel. Thesis, Agricultural University Wageningen, The Netherlands 255 pp.

Verdonschot, P. F. M., 2006. Data composition and taxonomic resolution in macroinvertebrate stream typology. Hydrobiologia 566: 59–74.

Vlek, H., P. F. M. Verdonschot & R. C. Nijboer, 2004. Towards a multimetric index for the assessment of Dutch streams using benthic macroinvertebrates. Hydrobiologia 516: 173–189.

Wallin, M., T. Wiederholm & R. K. Johnson, 2003. Guidance on Establishing Reference Conditions and Ecological Status Class Boundaries for Inland Surface Waters. CIS Working Group 2.3 – REFCOND. 7th Version.

Ward, J. V., 1982. Altitudinal zonation of Plecoptera in a Rocky Mountain stream. Aquatic Insects 2: 105–110.

Ward, J. V., 1985. Thermal characteristics of running waters. Hydrobiologia 25: 31–46.

Wright, J. F., D. Moss, P. D. Armitage & M. T. Furse, 1984. A preliminary classification of running-water sites in Great Britain based on macroinvertebrate species and the prediction of community type using environmental data. Freshwater Biology 14: 221–256.

Wright, J. F., D. W. Sutcliffe & M. T. Furse (eds), 1999. Assessing the biological quality of fresh waters: RIVPACS and other techniques. Freshwater Biological Association, Ambleside, Cumbria, UK. The RIVPACS International Workshop, 16–18 September 1997, Oxford, UK.

Hydrobiologia (2006) 566:59–74

M.T. Furse, D. Hering, K. Brabec, A. Buffagni, L. Sandin & P.F.M. Verdonschot (eds), The Ecological Status of European Rivers: Evaluation and Intercalibration of Assessment Methods
DOI 10.1007/s10750-006-0070-y

Data composition and taxonomic resolution in macroinvertebrate stream typology

Piet F.M. Verdonschot
Alterra, Green World Research, P.O. Box 47, 6700 AA Wageningen, The Netherlands
(Fax: +31-0-317-424988; E-mail: piet.verdonschot@wur.nl)

Key words: Europe, taxonomic resolution, data composition, macroinvertebrates, stream typology, ordination, reference condition, ecological quality

Abstract

In the EU water framework directive (WFD) a typological framework is defined for assessing the ecological quality of water bodies in the future. The aim of this study was to test the effect of data composition and taxonomic resolution on this typology. The EU research projects AQEM and STAR provided 1660 samples of 48 stream types sampled all over the major geographical gradients in Europe. These stream types fit the WFD typological demands and fit to the major European geographic regions (ecoregions). The samples included gradients from reference conditions to samples with bad ecological quality. Despite standardisation, there were large differences between the participating countries concerning the number of taxa, the number of specimens, and the taxonomic resolution. The macroinvertebrate data were analysed by using detrended correspondence analysis (DCA). The distribution patterns using all samples, only reference samples, and only degraded samples showed that the use of species-level (or 'best available taxonomic' level) performed better at a practical (fine) scale in comparison to family-level. The analyses further showed that even the use of a standardised protocol can not easily overcome (i) differences in site conditions that force the researcher to deviate from the protocol as well as (ii) the experiences of the researcher(s) and (iii) the available taxonomic knowledge.

Introduction

Can European stream types be based on orders or families while local stream types must be based on species-level identifications? Moog et al. (2004) concluded that a finer spatial resolution required a finer taxonomic resolution which is in concordance with the hierarchical approach described by Frissell et al. (1986). The strength and amount of detailed information that can be extracted from species-level data was already shown by several authors (e.g., Resh & Unzicker, 1975; Moog et al., 1997; Lenat & Resh, 2001). In stream or river assessment different taxonomic resolutions were used on different scales (e.g., Resh & McElravy, 1993; Verdonschot, 2000).

The European Commission recognised that the ecological status of water bodies should be determined compared to near-natural or reference conditions (European Commission, 2000). The water framework directive (WFD) approach of using reference conditions in assessment is in agreement with assessment approaches adopted in the USA (e.g., USEPA, 1996) and Australia (Davies, 2000). Communities are optimally developed under reference conditions (e.g., Karr & Chu, 1999). It is commonly accepted that human disturbance affects a stream ecosystem in such a way that communities become poor and look more alike (e.g., Wright et al., 1984; Verdonschot, 1990). Yet, would a stream typology become most explicit using only reference sites and species-level identifications? In this study the amount of data of reference sites provided the opportunity to do analyses with such sites solely, and the question of taxonomic resolution could therefore be tackled.

Furthermore, the data composition was analysed to explore the variation to be expected in future assessment. The objectives of this study were:

- To explore the effects of variation in data composition on analyses results;
- To explore whether the stream typology depends on taxonomic resolution, whereby species-level (or 'best available taxonomic level') is compared to family-level at different scales;
- To explore whether the stream typology should be based on reference sites only or also on degraded sites.

Methods

Data collection

In the EU research project AQEM, in total 889 macroinvertebrate samples representing 29 stream types were taken in eight countries in 2000 and 2001. In the EU research project STAR, an additional 771 samples were taken in 13 countries in 2002 and 2003. The combined AQEM-STAR database composed 1660 samples representing 16 countries and 48 stream types (Table 1). All samples together cover the major geographical gradients in Europe. The AQEM site selection, sampling, sorting, and identification procedure was explained by Hering et al. (2004). The STAR samples were either processed according to a slightly adapted AQEM protocol (Furse et al., 2004) or according to several national sampling protocols: RIVPACS (Germany, Austria, Greece, and United Kingdom), IBE (Italy), IBGN (France), DSFI (Denmark), LVS 240 (Latvia), PERLA (Czech Republic), and the national protocols of Poland, Sweden, and Portugal.

Handnets were used in all methods. All samples were taken within a stream stretch of <500 m of the respective stream site. All samples were collected in at least two seasons, of which one was spring. The second sample was collected in summer or autumn, depending on the regional, geographical and climatological conditions. At the STAR related sites replicate samples were taken. All samples were further processed in the same standardised way. Finally, different samples from the same site, either being replicates or taken using a different method, and samples taken in different seasons from the same site were kept in the analyses and treated as separate samples. Hereby, the variation caused by the different methods and seasons is accepted, because these differences will also be present when applying assessment in practical water management.

Identification took place to species-level when possible. In some areas, identification was limited to higher taxonomic levels due to a lack of taxonomic knowledge. Finally, all samples were combined into one European database.

Data composition

Despite the sampling, sorting, and identification protocols agreed upon within the consortia, differences in sample size, sorted number of specimens, and levels of identification occurred. Therefore, for all data per country the number of samples, the total number of taxa, and the total number of individuals were compared to get an overview of this source of variation.

Taxonomic adjustment

For several reasons taxonomic resolution within and between samples differed. This can be because of damaged specimens, lack of taxonomic knowledge in certain areas of Europe, lack of certain life stages, or lack of certain taxonomic groups in general. Therefore, taxonomic adjustment was needed to assure unambiguous data processing. Differences in taxonomic resolution could otherwise later prove to be the cause of differences between sample groupings in typology. To study the influence of taxonomic resolution on the analyses results, two datasets were extracted. The first dataset is based on the best available taxonomic level possible. To reach the best available taxonomic resolution a weighed taxonomic adjustment was applied according to the criteria described by Vlek et al. (2004). This dataset is indicated as 'species data'. The second dataset is composed of the family-level as best achievable level, and is indicated as 'family data'. Therefore, all taxa

Table 1. Stream type code and name

Code	Name
A01	Small to medium-sized streams in the Hungerian Plains
A02	Medium-sized, calcareous streams in the Alps
A03	Small, siliceous streams in the Alps
A04	Medium-sized, siliceous streams in the Bohemian Massif
A05	Small, shallow streams in the Central sub-alpine Mountains
A06	Small, crystalline streams of the ridges of the Central Alps
C04	Small, shallow, siliceous, mountain streams in the Carpathians
C05	Small streams in the Central sub-alpine Mountains
C14	Medium-sized, siliceous streams in the Central sub-alpine Mountains
C15	Small, calcareous streams in the Carpathians
C16	Small to medium-sized, calcareous streams in Carpathians
D01	Small, sand-bottom streams in the German Lowlands
D02	Small, organic type brooks in the German Lowlands
D03	Medium-sized, sand-bottom streams in the German Lowlands
D04	Small streams in the Central and Western Mountains (Germany)
D05	Medium-sized streams in the Central Mountains (Germany)
D06	Small, Buntsandstein streams in the Central Mountains (Germany)
F08	Small, shallow to medium-sized, headwater streams in the Western sub-alpine Mountains (Eastern France)
H01	Small to large, siliceous streams in North-Eastern Greece
H02	Small to large streams in Central and North Greece
H03	Small to large, calcareous streams in Western Greece
H04	Small, calcareous streams in the Hellenic Western Balkans (Greece)
H05	Small, siliceous streams in the Eastern Balkans and Hellenic Western Balkans (Greece)
H06	Small, siliceous streams on the Aegean Islands
H07	Medium-sized, calcareous streams in Hellenic Western Balkans (Southern Greece)
I05	Small streams in the southern calcareous Alps
I06	Small, calcareous streams in the Central Apennines
I22	Small, siliceous, source streams in the Po valley
I23	Small to medium-sized, lower mountain, siliceous streams in the Northern Apennines
I24	Small to medium-sized, lower mountain, siliceous streams in the Apennines (Southern Italy)
K02	Small to medium-sized, siliceous, lowland streams in the Central Lowlands
L02	Medium-sized, siliceous, lowland rivers in the Baltic Province
N13	Small, siliceous, sand-bottom streams in the Dutch Lowlands
N14	Small to medium-sized, organic and siliceous, sand-bottom streams in the Dutch Lowlands
O02	Medium-sized, siliceous, lowland streams in the Central and Eastern Lowlands
P01	Small, lower mountain streams in Southern Portugal
P02	Small to medium-sized, lowland streams in Southern Portugal
P03	Medium-sized, lowland streams in Southern Portugal
P04	Medium-sized, lower mountain streams in Southern Portugal
S01	Small to medium-sized, lowland streams in Northern Sweden
S02	Small to medium-sized, medium-altitude streams in Northern Sweden
S03	Small to medium-sized, medium-altitude streams in the Boreal Highlands
S04	Small, high-altitude streams in the Boreal Highlands
S05	Small and medium-sized, lowland streams in the Swedish Lowlands and Northern Sweden

Continued on p. 62

Table 1. (Continued)

Code	Name
S06	Small to large, lowland and calcareous streams in the Swedish Lowlands and Northern Sweden
U15	Small to medium-sized, shallow, lowland streams in England
U23	Small to medium-sized. lowland streams in England
V01	Small, mountain streams in the East and West Carpathians

below family-level (e.g., species or genus level) were adjusted to their respective family, and all familial and higher taxonomic units (e.g., suborder or order) were kept as such. Both adjustments were done for the total database.

Data analyses

After taxonomic adjustment the macroinvertebrate abundances of each sample were transformed [$^2\log(x+1)$] (Preston, 1962; Verdonschot, 1990).

The multimetric AQEM assessment system (Hering et al., 2004) was used to classify all AQEM samples into an Ecological Quality Class ranging from 5 (high quality) to 1 (bad quality). For all STAR samples only a pre-classification was available assigning the samples to the same quality classes based on the expert knowledge and abiotic field measurements. For data analysis three datasets were compiled: (i) 'all samples'; 1660 samples, (ii) 'reference samples'; 876 samples including only samples with an ecological quality classification good (class 4) and high (class 5), and (iii) 'degraded samples'; 784 samples including only samples with an ecological quality classification moderate (class 3), poor (class 2), and bad (class 1). The inclusion of class 4 (good) in the group of reference samples was done because (i) the quality deviation from the reference is only slight, and (ii) to obtain enough samples for reliable analyses.

Ordination was designed for data analysis in community ecology. Used in an explorative way it shows an ordination diagram that optimally displays how community composition varies (ter Braak & Šmilauer, 2002). In order to analyse the macroinvertebrate species composition in relation to stream type, detrended correspondence analysis (DCA) was used. DCA is an indirect ordination technique and part of the program CANOCO for Windows, version 4.2 (ter Braak & Šmilauer, 2002). In DCA, the samples are patterned in a multidimensional space based on their taxonomic composition.

The options chosen in CANOCO will influence the result of the DCA ordination. In this study the following options were selected (ter Braak & Šmilauer, 2002):

- detrending by 2nd order polynomials to reduce the 'arch' effect;
- downweighting of rare species which reduces the influence of rare species and stresses the importance of more common ones in the analysis;
- inter-sample distance that optimises the position of the samples in the ordination diagram;
- Hill's scaling to allow for long gradients the sample distances to represent turn over distances.

To establish the percentage of overlap between groups of samples a new approach was used. Using more 'classical' clustering techniques such as hierarchical agglomerative clustering a number of reproducible but more or less subjective choices within the program to be made by the user decide the results of the classification. The technique chosen is based on the interpretation of the DCA ordination diagram by counting the number of samples present in adjacent groups. Therefore, within the resulting ordination diagram, which included the first and second ordination axes, the samples were labelled according to stream type. The overlap between stream types was established by drawing contour lines around the types and summing up all the overlapping samples. The position of the contour line was the result of an iterative process of repositioning the line and counting the overlap until a minimum overlap was reached. Overlapping stream types were grouped into larger groups, if more than 25% of the samples were positioned within the other type or group, and next the overlap between these new established groups was calculated again by summing up all the remaining overlapping samples. The groups with an overlap <25% of the samples were identified as an identifiable group. Each group was considered to represent a recognisable typological unit, and a next DCA run was performed for this respective group to identify groups within. This process was repeated until no groups could further be disentangled or the level of stream type as recognisable group was reached. Starting with the whole database, the groups recognised in the first ordination were considered to represent the highest hierarchical units and are considered the major groups in Europe, the further the 'pealing off' was done the lower hierarchical position a group represented: groups, sub-groups, and stream types, respectively.

The calculation of the overlap was restricted to axes one and two, as in each run only two to three major groups were separated. DCA plots the major grouping of samples along the first and the second major grouping along the second axis (ter Braak & Šmilauer, 2002). The DCA analyses were repeated for all six datasets; 'all samples', 'reference samples' and 'degraded samples'; each as 'species data' and 'family data'.

Results

Data composition

The number of samples, taxa, and adjusted taxa, and percentage of adjusted taxa differed strongly between countries (Table 2), partly due to the fact that some partners were in both the AQEM and STAR project and some partners were not. Denmark and France had the lowest numbers of samples (34 and 36, respectively) and Germany had the highest number of samples (279) taken. Germany also had the highest total number of taxa (912) collected, and both Austria and the Netherlands also collected high numbers of taxa. The overall lowest numbers of taxa were collected by France (224) and Denmark (237), though this is related to their low number of samples.

Differences in number of samples taken by each country will affect the distribution of samples over stream types and thus influences the analyses. For example, only 24 samples were taken in the Hungarian plains vs. 93 in the Baltic province. Furthermore, the lower the number of samples taken in a geographical area the lower the number of taxa collected is. This skew distribution of data must be taken into account interpreting the results.

Table 2. The number of samples, taxa, 'species' adjusted taxa, and percentage of 'species' adjusted taxa per country

Country	Number of samples	Number of taxa	Number of taxa left after adjustment	% Taxa left after adjustment
Austria	169	868	160	18
Czech Republic	146	717	172	24
Denmark	34	237	97	41
France	36	224	118	53
Germany	279	912	231	25
Greece	152	595	198	33
Italy	133	422	152	36
Latvia	93	450	132	29
Netherlands	156	885	215	24
Poland	64	515	158	31
Portugal	71	416	160	38
Slovak Republic	48	375	97	26
Sweden	217	352	130	37
United Kingdom	62	388	136	35

Taxonomic adjustment

After taxonomic adjustment, Germany still showed the highest number of taxa (231) together with the Netherlands (215 taxa). The Slovak Republic and Denmark collected least numbers of taxa (both 97 taxa). On average per sample Austria collected most taxa before adjustment and Latvia after, while Greece collected lowest average number of taxa before as well as after adjustment.

The loss of taxa due to taxonomic adjustment was lowest in France and highest in Austria (Fig. 1). Also the Czech Republic and the Netherlands lost more than 50% of the average number of taxa per sample after adjustment.

The loss of individuals through adjustment was negligible (Table 3). Only Greece and Italy lost

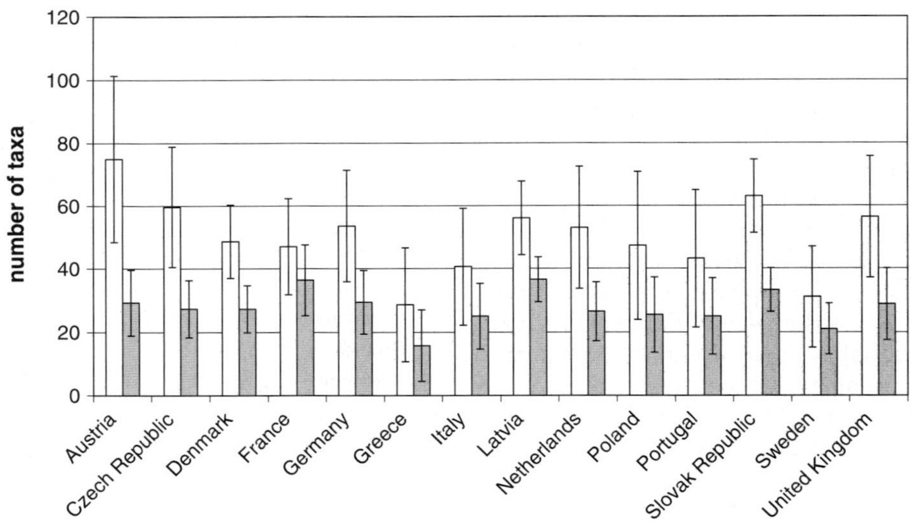

Figure 1. Average number of taxa (white bar) and adjusted taxa (grey bar) per sample per country (standard deviation indicated).

38% and 11%, respectively, of their numbers of individuals due to specimens only identified to very high taxonomic levels which were deleted in taxonomic adjustment. The differences in average number of individuals per sample between countries were very large (Fig. 2). The lowest number was less then 2% of the highest. Furthermore, the standard deviation was large in all countries. So, the one country on average per sample collected 50 times more specimens in comparison to the other.

Table 3. The total number of individuals for raw and 'species' adjusted taxa data per country

Country	Raw	Adjusted	% Loss
Austria	1520784	1503759	1.12
Czech Republic	621715	618294	0.55
Denmark	183711	182466	0.68
France	360518	358886	0.45
Germany	546684	532279	2.63
Greece	38984	24316	37.63
Italy	568866	507013	10.87
Latvia	414679	404294	2.50
Netherlands	450035	441351	1.93
Poland	227760	220340	3.26
Portugal	294708	288746	2.02
Slovak Republic	149289	144296	3.34
Sweden	220496	219266	0.56
United Kingdom	570897	569158	0.30

In general, the loss of species due to taxonomic adjustment was very high (Table 4). Many species and combinations of species were assigned to the genus-level and family-level.

All major taxonomic groups were strongly reduced in number of taxa after adjustment (Table 5). Major losses occurred in the Chironomidae, but also the numbers of Hydrachnidia, Megaloptera, Plecoptera, and Coleoptera taxa were strongly reduced. Gastropoda seemed to be best known throughout Europe and the decrease of the number of taxa was restricted to 61%.

Reference or degraded samples and taxonomic resolution

The full typological analyses are described by Verdonschot (2006). This manuscript focuses on the importance of taxonomic resolution. The hierarchical grouping of European (groups of) stream types is listed in Table 6 and reflects the full typological analyses. The number of samples per major group, group, sub-group of stream types, and stream types is indicated for all samples, reference samples and degraded samples, respectively (Table 6). The differences of number of samples for all groups must be taken into account interpreting the further analyses. The overlap between

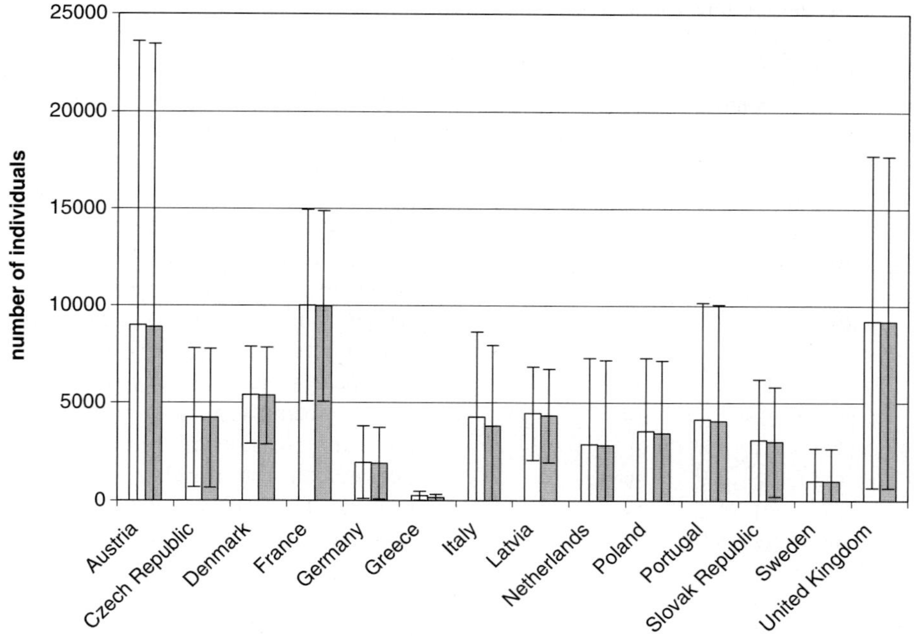

Figure 2. Average number of individuals for raw data (white bar) and adjusted data (grey bar) per country (positive standard deviation indicated).

the groups of stream types was analysed and calculated for the reference samples based on family-level data as well as for the datasets of all samples and degraded samples using species-level or family-level data. Because the number of samples per group differed, the percentage overlap was calculated for each dataset and both species-level and for family-level data (Table 7). Certain groups do show a much higher percentage of overlap then others. For example, the highest percentage of overlap is between Mediterranean and Lowlands for family-level data of degraded sites (64%). Degraded sites tend to have a poorer taxa composition, and the slope, current and substrate composition of sites within the Lowlands and the Mediterranean is mutual more alike and differs from the Mountains. On the other hand, the overlap within sub-groups and stream types is often 0%.

As an example the DCA ordination diagrams for species-level and family-level data of reference samples of all stream types at the European level are shown in Figures 3 and 4. The overlap between the major groups Mountains, Mediterranean and Lowlands for species-level data of reference samples differs strongly. Only 0.8% of the samples belonging to the Mountains or the Lowlands is projected within the Mediterranean, while on the other hand 15.2% and 26.3% of the Mediterranean samples is situated within the Mountains and Lowlands, respectively. On average there is an overlap of 15.1% (Table 7). For family-level data this overlap is larger (20.2%). This

Table 4. The number of taxa per taxonomic level before (raw taxa data) and after taxonomic adjustment ('species' adjusted)

Taxonomic level	Raw taxa data	'Species' adjusted taxa data
Phylum	2	3
Class	9	153
Order/suborder	38	38
Family	270	1170
Subfamily/tribus	56	10
Genus/subgenus	707	1377
Species group/aggregate /combination	214	3
Species	1851	188
Subspecies	72	6
Deleted taxa	0	271
Total	3219	3219

Table 5. The number of taxa before and after taxonomic 'species' adjustment, and percentage left after 'species' adjustment per major macroinvertebrate taxonomic group

Group	Number of taxa		% Taxa left after adjustment
	Raw	Adjusted	
Aranea	1	1	100
Bivalvia	37	8	22
Chironomidae	478	1	0.2
Coelenterata	2	0	0
Coleoptera	407	79	19
Collembola	1	0	0
Crustacea	53	17	32
Diptera	296	58	20
Ephemeroptera	215	34	16
Gastropoda	97	38	39
Heteroptera	87	24	28
Hirudinea	49	15	31
Hydrachnidia	8	1	13
Lepidoptera	10	0	0
Megaloptera	8	1	13
Mermithidae	1	0	0
Nematoda	1	0	0
Nematomorpha	3	1	33
Odonata	92	27	29
Oligochaeta	117	1	1
Planipennia	5	0	0
Plecoptera	142	26	18
Trichoptera	352	70	20
Turbellaria	23	1	4

larger overlap can be seen comparing Figures 3 and 4.

Discussion

Data composition

Although, both the AQEM and the STAR project used a standardised protocol for sampling, sorting and identification of samples (Hering et al., 2004; Furse et al., 2006) the analyses of the composition of the data used in this study showed a wide spread in number of samples, number of taxa, number of specimens, and taxonomic level achieved through identification amongst participating countries. Verdonschot & Nijboer (2004) already listed some arguments that could explain these differences: differences in taxonomic knowledge, differences in natural population densities, changes of taxon diversity along geographical and altitudinal gradients, and local environmental differences that affected sampling efficiency or forced changes to the protocol. The data showed that all these arguments could also be valid for the STAR and national sampling procedures. The experience of the AQEM project results on the level of standardisation achievable was the same in the follow-up project STAR. The local environmental conditions and the researchers training and experience set the limits for a European wide standardisation.

As ordination is a robust technique, major patterns are shown and such patterns strongly depend on the number of comparable samples present. The differences in number of samples per country and per (groups of) stream type(s) affected the ordination results. Some major groups were represented by much more samples than others, like Mountains (645) and Lowlands (798) vs. Mediterranean (217). Furthermore, as standardisation showed its limitations, the average number of taxa and individuals strongly differed between countries and (groups of) stream type(s) and this affects ordination. As a consequence, some groupings or divisions of stream types will have been influenced by these differences in data composition.

Taxonomic adjustment

The species is the basic unit that carries features related to its ecological requirements (Resh & Unzicker, 1975; Stubauer & Moog, 2000). Higher taxonomical units, like genus, family or order are aggregates of different species. This aggregation is commonly based on morphological characteristics (especially of the reproductive organs). This implies that species within a higher taxonomical unit can carry different ecological features (Moog, 1995). The consequence is that higher taxonomical units will show wider varieties in ecological response and thus have wider distribution ranges. Taxonomical adjustment, especially in this data set, led to a number of groupings of taxa to higher taxonomical levels. Adjustment to higher taxonomical level automatically implies loss of ecological information (Nijboer & Verdonschot, 2000; Nijboer & Schmidt-Kloiber, 2004; Schmidt-Kloiber & Nijboer, 2004), because the most refined

Table 6. Hierarchical grouping of European (groups of) stream types for Europe. Number of samples per group/type indicated for all samples, reference samples and degraded samples, respectively

Europe				**number of samples**		
major group	**group**	**sub-group**	**stream type**	**all**	**reference**	**degraded**
all samples				1660	876	784
Mountains				645	339	306
	Northern European Mountains			120	68	52
		Northern Sweden	no further grouping of S01 and S02	60	40	20
		Boreal Highlands		60	28	32
			S03	30	22	8
			S04	30	6	24
	Central European Mountains			441	224	217
		Central European Mountains (medium-sized)		64	29	35
			C14	24	14	10
			D05	40	15	25
		Central European Mountains (small)	no further grouping of A04, A05, A06, C04, C05, C15, C16, D04, D06, V01	377	195	182
	Central Alps			84	47	37
			A02	26	14	12
			A03	26	13	13
			I05	32	20	12
Mediterranean				217	128	89
	Central and Eastern Mediterranean			159	92	67
		Greece (Medit.)	no further grouping of H01, H02 and H03	80	53	27
		Central Apennines	I06	34	18	16
		Northern Apennines		45	21	24
			I23	23	11	12
			I24	22	10	12
	Western Mediterranean			58	36	22
		S-Portugal (medium-sized)	P04	32	20	12
		S-Portugal (small)	no further grouping of P01 and P02	26	16	10
Lowlands				798	409	389
	Central and Southern Lowlands			572	257	315
		Hungarian Plain	A01	24	11	13
		Hellenic Balkans	no further grouping of H04, H05, H06 and H07	72	43	29
		Central European Lowlands	no further grouping of D01, D02, D03, I22, K02, N13, N14, O02, P03, U15, U23	476	203	273
	Northern Lowlands			226	152	74
		Southern Sweden	no further grouping of S05 and S06	97	61	36
		Baltic Province	L02	93	71	22
		Western sub-alpine Mountains	F08	36	20	16

Table 7. Percentage of overlap between (groups of) stream types using all samples, reference samples and degraded samples at both species and family-level

Source group	Overlap with	All samples		Reference		Degraded	
		Species data	Family data	Species data	Family data	Species data	Family data
Major groups							
Mediterranean	Mountains	15.2	8.3	15.6	18.0	3.4	2.2
Mediterranean	Lowlands	26.3	31.3	13.3	8.6	60.7	64.0
Lowlands	Mountains	8.0	8.6	10.0	15.4	2.3	3.3
Lowlands	Mediterranean	0.8	3.5	1.2	7.1	1.0	2.8
Mountains	Mediterranean	0.8	1.4	5.3	5.3	8.8	7.5
Mountains	Lowlands	13.2	22.3	9.1	9.4	8.2	8.8
	Overall overlap	15.1	20.2	15.1	20.1	15.6	17.0
Groups							
Mountains							
Northern European Mountains	Central European Mountains	0.8	18.3	1.5	5.9	17.3	26.9
Northern European Mountains	Central Alps	0.8	1.7	0.0	0.0	3.8	9.6
Central European Mountains	Central Alps	0.2	0.0	0.9	0.9	0.0	0.0
Central European Mountains	Northern European Mountains	0.5	1.4	0.4	0.4	0.9	0.0
Central Alps	Northern European Mountains	1.2	8.3	0.0	0.0	10.8	5.4
Central Alps	Central European Mountains	8.3	2.4	4.3	2.1	10.8	5.4
	Overall overlap	2.0	6.0	1.8	2.4	6.9	7.5
Lowlands							
Lowlands	Northern Lowlands	5.2	8.2	8.2	12.1	2.9	4.4
Northern Lowlands	Lowlands	15.0	22.6	7.9	11.2	41.9	47.3
	Overall overlap	8.0	12.3	3.8	11.7	4.6	12.6
Mediterranean							
Central and Eastern Mediterranean	Western Mediterranean	0.6	6.3	1.1	3.3	3.0	6.0
Western Mediterranean	Central and Eastern Mediterranean	0.0	13.8	0.0	19.4	0.0	4.5
	Overall overlap	0.5	8.3	0.1	7.8	0.2	5.6
Sub-groups							
Northern European Mountains							
N-Sweden	Boreal Highlands	3.3	5.0	10.0	7.5	25.0	25.0
Boreal Highlands	N-Sweden	10.0	10.0	7.1	7.1	6.3	6.3
	Overall overlap	6.7	7.5	8.8	7.4	13.5	13.5

Central European Mountains							
Central European Mountains (medium-sized)	Central European Mountains (small)	14.1	12.5	17.2	17.2	25.7	25.7
Central European Mountains (small)	Central European Mountains (medium-sized)	1.6	2.4	0.5	1.0	4.4	3.8
	Overall overlap	3.4	3.9	2.7	3.1	7.8	7.4
Central and Eastern Mediterranean							
Greece (Medit.)	Central Apennines	0.0	3.8	0.0	0.0	0.0	3.7
Greece (Medit.)	Northern Apennines	1.3	12.5	0.0	15.1	3.7	7.4
Central Apennines	Greece (Medit.)	0.0	11.8	0.0	0.0	0.0	6.3
Central Apennines	Northern Apennines	2.9	20.6	0.0	0.0	0.0	31.3
Northern Apennines	Greece (Medit.)	4.4	0.0	4.8	9.5	0.0	4.2
Northern Apennines	Central Apennines	8.9	20.0	0.0	0.0	8.3	4.2
	Overall overlap	5.0	20.8	1.1	10.9	4.5	16.4
Western Mediterranean							
S-Portugal (medium-sized)	S-Portugal (small)	0.0	0.0	0.0	0.0	0.0	8.3
S-Portugal (small)	S-Portugal (medium-sized)	0.0	0.0	0.0	6.3	0.0	0.0
	Overall overlap	0.0	0.0	0.0	2.8	0.0	4.5
Northern Lowlands							
S-Sweden	Western sub-alpine Mountains	0.0	1.0	0.0	0.0	0.0	5.6
S-Sweden	Baltic Province	0.0	0.0	0.0	0.0	0.0	0.0
Baltic Province	S-Sweden	0.0	1.1	0.0	1.4	0.0	0.0
Baltic Province	Western sub-alpine Mountains	3.2	5.4	0.0	2.8	0.0	0.0
Western sub-alpine Mountains	S-Sweden	0.0	0.0	0.0	0.0	0.0	6.3
Western sub-alpine Mountains	Baltic Province	11.1	36.1	0.0	0.0	0.0	0.0
	Overall overlap	3.1	8.8	0.0	2.0	0.0	4.1
Lowlands							
Hungarian Plain	Hellenic Balkans	4.2	0.0	0.0	0.0	0.0	0.0
Hungarian Plain	Plains	54.2	58.3	0.0	0.0	38.5	38.5
Hellenic Balkans	Plains	22.2	13.9	4.7	16.3	34.5	27.6
Hellenic Balkans	Hungarian Plain	0.0	1.4	0.0	2.3	0.0	0.0

Continued on p. 70

Table 7. (Continued)

Source group	Overlap with	All samples		Reference		Degraded	
		Species data	Family data	Species data	Family data	Species data	Family data
Plains	Hungarian Plain	0.2	0.2	0.5	0.0	1.1	2.9
Plains	Hellenic Balkans	3.2	3.6	1.0	0.5	6.6	2.2
	Overall overlap	8.0	7.5	1.9	3.5	11.4	8.6
Stream types							
Central Alps							
A02	A03	3.8	3.8	0.0	0.0	8.3	0.0
A02	I05	0.0	0.0	0.0	0.0	0.0	0.0
A03	A02	0.0	3.8	0.0	7.7	7.7	7.7
A03	I05	0.0	0.0	0.0	0.0	0.0	0.0
I05	A02	0.0	0.0	0.0	0.0	0.0	0.0
I05	A03	0.0	0.0	0.0	0.0	0.0	0.0
	Overall overlap	1.2	2.4	0.0	2.1	5.4	2.7
Boreal Highlands							
S03	S04	6.7	6.7	0.0	4.5	0.0	12.5
S04	S03	3.3	0.0	0.0	0.0	0.0	0.0
	Overall overlap	5.0		0.0	3.6	0.0	3.1
Central European Mountains (medium-sized)							
C14	D05	8.3	8.3	14.3	0.0	0.0	0.0
D05	C14	10.0	0.0	0.0	0.0	8.0	0.0
	Overall overlap	9.4	3.1	6.9	0.0	5.7	0.0
Northern Apennines							
I23	I24	4.3	0.0	0.0	0.0	0.0	0.0
I24	I23	4.5	0.0	0.0	0.0	16.7	0.0
	Overall overlap	4.4	0.0	0.0	0.0	8.3	0.0

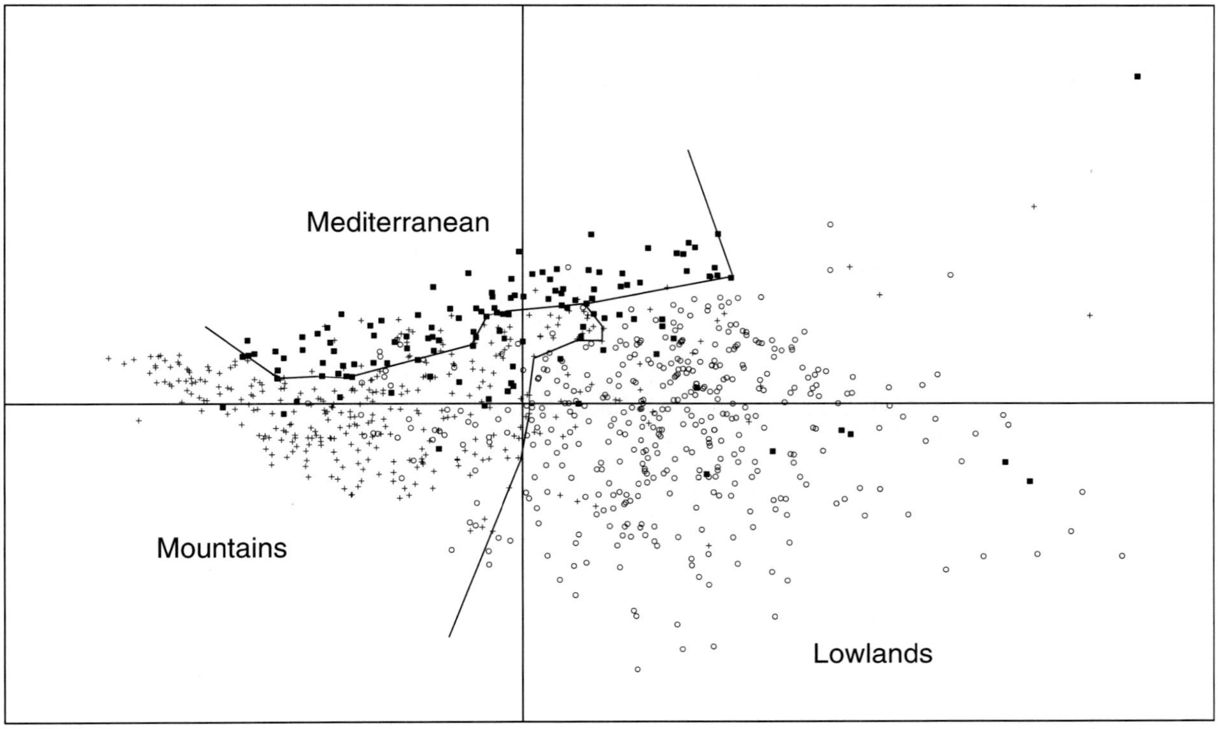

Figure 3. DCA ordination diagram of the axis 1 (horizontal; eigenvalues: 0.23) and 2 (vertical: eigenvalues: 0.13) of the (groups of) stream types within Europe based on species level data of reference samples.

level of information is the ecological response of the species. Even in the 'species data' part of this study, quite a large amount of information got lost since adjustment forced in a number of cases an up-scaling of the respective taxonomical level. On the other hand, to keep samples comparable all over Europe and to perform a balanced analysis, adjustment was needed. In conclusion, the need for standardising taxonomic levels to be achieved in European assessment projects is crucial. Areas in Europe where taxonomy is less developed should get and give more attention to improve taxonomic knowledge. In general, taxonomy is the fundament of ecological research and in practical applications. Therefore, also water management and water policy makers should become aware of the importance of a well established knowledge of taxonomy all over Europe.

Taxonomic resolution

First, it was tested whether the use of reference sites would give better results. Indeed, the reference samples performed best and were most optimally separated. This supports the hypothesis that human stress diminishes the natural differences between stream communities.

The need to establish reference conditions for typology and classification purposes is one, but the biotic parameters to express these conditions are as important. This study demonstrated that the use of species vs. family-level data changed the results. The use of the family-level data led to a less distinct separation of reference sites. This implies that the description of reference conditions must be based on species-level data ('best achievable taxonomic level'). But this conclusion reaches further. It also demands two other improvements in the current approaches, one that deals with the use of metrics and the autecological information within, and one with the question what is 'best achievable' in identification. The second is related to the former plea of improvement of taxonomy in European research. The first touches the multimetrics approach in its fundament. In Europe, at the moment there is a very strong tendency to use

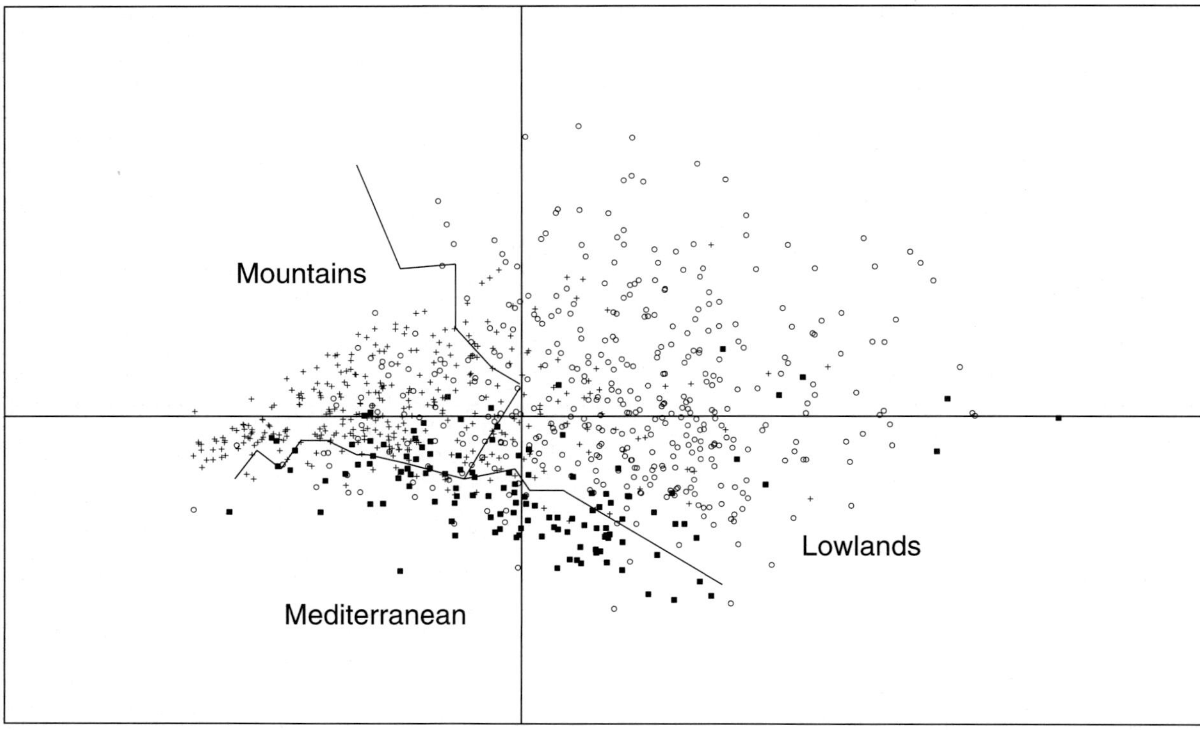

Figure 4. DCA ordination diagram of the axis 1 (horizontal; eigenvalues: 0.18) and 2 (vertical: eigenvalues: 0.10) of the (groups of) stream types within Europe based on family level data of reference samples.

multimetrics in assessment (Hering et al., 2004). The AQEM research project tested over one hundred metrics and came up with a list of 18 suitable core metrics (Hering et al., 2004). Most of these metrics are based on the use of autecological information, often at high taxonomic (family) level. It was shown that the use of family-level data resulted in a lower resolution thus smaller differences between stream types. Families aggregate information of individual species and thus generalise information. Testing the use of family-level data already showed poorer results (Lenat & Resh, 2001; Schmidt-Kloiber & Nijboer, 2004). Using multimetrics means approaching communities by their individual taxonomic features. Each taxon that is included in the respective metric is used as an indicator. But most metrics are (i) dependent on the autecological, often family-level, information, and (ii) restricted to inclusion of a limited number of taxa. A metric extracts only part of the information of the community and expresses it into a value, the ecological quality class. The alternative is the community approach. The classification is herein based on the use of the community as a whole (e.g., Wright et al., 1999). From community descriptions further information on the ecology of the composing species can be extracted and used in metrics. This improves the metrics as well as the supporting autecological information. The multimetric and community approach are both complementary ecological tools and can strengthen each other.

Conclusions

The conclusion of this study were:

- Human stress diminishes the natural differences between stream communities and typologies should therefore be based on reference conditions.
- Stream typology depends on the taxonomic resolution, the finer the resolution the more distinctive the types become.

- Species (or 'best available') taxonomic level performed better at a practical (fine) scale in comparison to family-level.
- Even a standardised protocol can not easily overcome (i) local differences in site conditions that cause deviations from the protocol as well as (ii) the experience of the researcher(s).

Acknowledgements

The author would like to thank all AQEM and STAR partners for the use of the data. The EU research projects AQEM and STAR were funded by the European Commission, 5th Framework Program, Energy, Environment and Sustainable Development, Key Action Water, Contract no. EVK1-CT1999-00027 and Contract no. EVK1-CT2001-00089, respectively.

References

Davies, P. E., 2000. Development of a national river bioassessment system, AUSRIVAS in Australia. In Wright, J. F., D. W. Sutcliffe & M. T. Furse (eds), Assessing the Biological Quality of Fresh Waters – RIVPACS and Other Techniques. Freshwater Biology, 113–124.

European Commission, 2000. Directive 2000/60/EC. Establishing a framework for community action in the field of water policy. European Commission PE-CONS 3639/1/100 Rev 1, Luxembourg.

Frissell, C. A., W. J. Liss, C. E. Warren & M. D. Hurley, 1986. A hierarchical approach to classifying stream habitat features: viewing streams in a watershed context. Environmental Management 10: 199–214.

Furse, M. T., A. Schmidt-Kloiber, J. Strackbein, J. Davy-Bowker, A. Lorenz, J. van der Molen & P. Scarlett, 2004. Standardisation of river classifications. Results of the sampling programme. European 5th Framework Porgramme, research project STAR, 6th Deliverable, 31/07/04, 130.

Furse, M., D. Hering, O. Moog, P. Verdonschot, R. K. Johnson, K. Brabec, K. Gritzalis, A. Buffagni, P. Pinto, N. Friberg, J. Murray-Bligh, J. Kokes, R. Alber, P. Usseglio-Polatera, P. Haase, R. Sweeting, B. Bis, K. Szoszkiewicz, H. Soszka, G. Springe, F. Sporka & I. Krno, 2006. The STAR project: context, objectives and approaches. Hydrobiologia 566: 3–29.

Hering, D., O. Moog, L. Sandin & P. F. M. Verdonschot, 2004. Overview and application of the AQEM assessment system. Hydrobiologia 516: 1–20.

Karr, J. R. & E. W. Chu, 1999. Restoring Life in Running Waters: Better Biological Monitoring. Island Press, Washington, DC.

Lenat, D. R. & V. H. Resh, 2001. Taxonomy and stream ecology – the benefits of genus- and species-level identifications. Journal of the North American Benthological Society 20: 287–298.

Moog, O., 1995. Fauna Aquatica Austriaca. A Comprehensive Species Inventory of Austrian Aquatic Organisms with Ecological Notes. Federal Ministry for Agriculture and Forestry, Wasserwirtschaftskataster Vienna.

Moog, O., E. Bauernfeind & P. Weichselbaumer, 1997. The use of Ephemeroptera as saprobic indicators in Austria. In Landolt, P. & M. Sartori (eds), Ephemeroptera & Plecoptera. Biology- Ecology-Systematics, Fribourg, 254–260.

Moog, O., A. Schmidt-Kloiber, T. Ofenböck & J. Gerritsen, 2004. Does the ecoregion approach support the typological demands of the EU "Water Framework Directive"? Hydrobiologia 516: 21–33.

Nijboer, R. C. & A. Schmidt-Kloiber, 2004. The effect of excluding rare taxa on the ecological quality assessment of running waters. Hydrobiologia 516: 347–363.

Nijboer, R. C. & P. F. M. Verdonschot, 2000. Taxonomic adjustment affects data analysis: an often forgotten error. Verhandlungen Internationale Verein für Limnology 27: 1–4.

Preston, F. W., 1962. The canonical distribution of commonness and rarity: part 1. Ecology 43: 185–215.

Resh, V. H. & E. P. McElravy, 1993. Contemporary quantitative approaches to biomonitoring using benthic macroinvertebrates. In Rosenberg, D. M. & V. H. Resh (eds), Freshwater Biomonitoring and Benthic Macroinvertebrates. Chapman & Hall, New York, 159–194.

Resh, V. H. & J. D. Unzicker, 1975. Water quality monitoring and aquatic organisms: the importance of species identification. Journal of Water Pollution Federation 47: 9–19.

Schmidt-Kloiber, A. & R. C. Nijboer, 2004. The effect of taxonomic resolution on the assessment of ecological water quality classes. Hydrobiologia 516: 269–283.

Stubauer, I. & O. Moog, 2000. Taxonomic sufficiency versus need for information – comments on Austrian experience in biological water quality monitoring. Verhandlungen Internationale Verein für Limnology 27: 2562–2566.

ter Braak, C. J. F. & P. Ŝmilauer, 2002. CANOCO Reference Manual and Users Guide to Canoco for Windows. Software for Canonical Community Ordination (version 4.5). Centre for Biometry, Wageningen, The Netherlands.

USEPA (U.S. Environmental Protection Agency), 1996. Biological Criteria: Technical Guidance for Streams and Small Rivers. U.S. Environmental Protection Agency, Office of Water, Washington, DC EPA-822-B96–001.

Verdonschot, P. F. M., 1990. Ecological characterization of surface waters in the province of Overijssel (The Netherlands). Thesis, Agricultural University Wageningen, 255.

Verdonschot, P. F. M., 2000. Integrated ecological assessment methods as a basis for sustainable catchment management. In Jungwirth, M., S. Muhar & S. Schmutz (eds), Assessing the Ecological Integrity of Running Waters. Proc. Int. Conf., Vienna, Austria. Developments in Hydrobiology 149. Hydrobiologia 422/423: 389–412.

Verdonschot, P. F. M., 2006. Evaluation of the use of Water Framework Directive typology descriptors, reference sites, and spatial scale in macroinvertebrate stream typology. Hydrobiologia 566: 39–58.

Verdonschot, P. F. M. & R. C. Nijboer, 2004. Testing the European stream typology of the water Framework Directive for macroinvertebrates. Hydrobiologia 175: 35–54.

Vlek, H., P. F. M. Verdonschot & R. C. Nijboer, 2004. Towards a multimetric index for the assessment of Dutch streams using benthic macroinvertebrates. Hydrobiologia 516: 173–189.

Wright, J. F., D. Moss, P. D. Armitage & M. T. Furse, 1984. A preliminary classification of running-water sites in Great Britain based on macroinvertebrate species and the prediction of community type using environmental data. Freshwater Biology 14: 221–256.

Wright, J. F., D. W. Sutcliffe & M. T. Furse (eds), 1999. Assessing the biological quality of fresh waters: RIVPACS and other techniques. Freshwater Biological Association, Ambleside, Cumbria, UK. The RIVPACS International Workshop, 16–18 September 1997, Oxford, UK.

Hydrobiologia (2006) 566:75–90

M.T. Furse, D. Hering, K. Brabec, A. Buffagni, L. Sandin & P.F.M. Verdonschot (eds), The Ecological Status of European Rivers: Evaluation and Intercalibration of Assessment Methods
DOI 10.1007/s10750-006-0069-4

Relationships among biological elements (macrophytes, macroinvertebrates and ichthyofauna) for different core river types across Europe at two different spatial scales

Paulo Pinto[1,*], Manuela Morais[1], Maria Ilhéu[1] & Leonard Sandin[2]
[1]*Centre of Applied Ecology, Water Laboratory, University of Évora, Largo dos Colegiais, 7001 Évora codex, Portugal*
[2]*Department of Environmental Assessment, Swedish University of Agriculture Sciences, Valvagen 3, PO Box 7050, S-75007 Uppsala, Sweden*
*(*Author for correspondence: Fax: +35-1-1847 35 71; E-mail: ppinto@uevora.pt)*

Key words: macrophytes, macroinvertebrates, fishes, mantel correlations, lotic ecosystems, spatial scale, linkages

Abstract

The objective of this study was to evaluate differences in correlations among Biological Elements and environmental parameters for different river types, analysed at two different spatial scales. A total of 82 sites, with at least good ecological status, were sampled across Europe, representing three core river types: Mountain rivers (26 sites); Lowland rivers (29 sites) and Mediterranean rivers (17 sites). At each site samples of macrophytes, macroinvertebrates and fishes were taken during spring, following the methodological procedures established by the European STAR project. Environmental parameters were also recorded, based on a site protocol developed by the European projects AQEM and STAR. Environmental parameters were divided into three categories: aquatic habitats (mesohabitat scale), global features (reach scale) and obligatory typology parameters of Water Framework Directive (WFD) (geographical scale). Data were analysed to evaluate at the two scales, first, relationships among biological elements, and second, relationships between biological elements and environmental parameters. Within each river type, correlation matrices (Bray–Curtis distance) were calculated separately for each biological element and for each category of environmental parameters. All biological elements were correlated ($p < 0.01$) to the larger spatial scale: macrophytes and macroinvertebrates are more correlated in lowland and mountain rivers, while in Mediterranean rivers, fish and macrophytes presented higher correlations. These links tend to be consistent for different spatial scales, except if they are weak on a larger regional scale, obligatory parameters of WFD were, in most cases, significantly correlated with the three biological communities ($p < 0.05$). Results at different spatial scales supported the hierarchical theory of river formation. Reach and mesohabitat environmental parameters tend to explain aquatic communities at a lower spatial scale, while geographical parameters tend to explain the communities at a major spatial scale.

Introduction

Rivers and streams are composed of a hierarchical system of patches of different ages, sizes and environmental conditions, thus creating a multiplicity of ecological niches (Beisel et al., 1998; Crook et al., 2001; Li et al., 2001). These niches are occupied by communities of organisms with different biological and ecological characteristics (algae, macrophytes, invertebrates and fishes), permitting the establishment of a complex net of relationships among organisms and communities.

Biological interactions like predation and competition are generally recognized as biological interactions with a direct influence on aquatic biodiversity (Rosenfeld, 1997; Dahl & Greenberg, 1998; Warfe & Barmuta, 2004). However, other indirect relationships may occur, like those related to macrophyte growth. Macrophyte abundances tend to increase habitat diversity, thus creating refuges for invertebrates and young fishes (Dahl & Greenberg, 1998; Cheruvelil et al., 2000; Allouche, 2002; Wright et al., 2002; Zrum & Hann, 2002; Balci & Kennedy, 2003), providing surface areas for periphyton development and also influencing current velocity (Armitage, 1995; Armitage & Gunn, 1996).

Aquatic communities also depend on different environmental scales: mesohabitat scale; reach scale; catchment scale and regional scale (see Jensen et al., 1996; Beisel et al., 1998; Verdonschot, 2000; Crook et al., 2001; Verdonschot & Nijboer, 2004). The influence of these different scales on aquatic communities is dependent on organism's characteristics, namely ecological sensitivity, life cycles, mobility and size (Tolonen et al., 2003). Fishes that can move along the river are expected to be more dependent on catchment scale than other communities with lower mobility, and for this reason they are more dependent on reach or mesohabitat parameters.

The intensity and frequency of disturbance (Townsend, 1989; Voelz & McArthur, 2000; Ward & Tockner, 2001; Vieira et al., 2004) are clearly related to catchment and regional scales (Beisel et al., 1998; Crook et al., 2001; Li et al., 2001; Reyjol et al., 2003), and may be determinant factors for the strength of interaction among organisms and communities. Flood events may induce modifications on riverbed shape (Armitage & Cannan, 2000; Petts, 2000; Bio et al., 2002) and influence the shift in species composition, particularly in streams presenting high inter-annual flow variability (Bernardo et al., 2003). These interactions are also influenced by the longitudinal dimension of lotic ecosystems (Vannote et al., 1980; Ward, 1989), being expected higher dependency from the surrounding terrestrial ecosystem to upstream reaches (Mountain rivers) than to downstream reaches (Lowland rivers).

Many studies have theorized about the implications of all these relationships on aquatic ecosystem functioning, with the main focus at a very local scale (Casas, 1997; Cheruvelil et al., 2000; Zrum & Hann, 2002; Wagner & Bretschko, 2003; Zimmer et al., 2003; Warfe & Barmuta, 2004; Willis et al., 2005). However, few studies have addressed the hierarchy of these relationships and their patterns at a wider regional scale.

In this study, three different core river types (mountain, lowland and Mediterranean rivers) with low-human impacts were investigated at a European regional level, during spring. The objective is to answer to the following questions: (1) are links among biological elements different for different core river types?; (2) if yes, are they consistent at different spatial scales?; (3) are biological elements of each core river type explained for specific environmental parameters?; (4) if yes, are they specific to precise spatial scales?

Study area

Within the framework of the European project STAR (EVK1-CT-2001-00089), 82 sites were sampled across Europe during spring 2003, covering nine different countries. Sites were grouped into three core river types, according to the results of previous studies across Europe (Verdonschot & Nijboer, 2004; Verdonschot, 2006) based on the hierarchical approach to stream formation (Jensen et al., 1996; Verdonschot, 2000): (1) mountain rivers (26 sites) including an Austrian river (A05), two Czeck rivers (C04, C05) and two German rivers (D04, D06); (2) lowland rivers (29 sites) including a German river (D03), a Danish river (K02), a British river (U23) and two Swedish rivers (S05, S06); (3) Mediterranean rivers (17 sites), including a Portuguese river (P04), an Italian river (I06) and two Greek rivers (H04, H06). These groups were established in order to understand differences in river functioning at a large geographical scale. Since river groups include sites with contrasting geographical locations, a set of subgroups with relatively homogenous features were used in data analysis. These groups and subgroups correspond to regional and local scales of the hierarchical approach (Jensen et al., 1996; Verdonschot, 2000).

Methodology

Environmental parameters

Each site was described by a protocol developed within the European projects AQEM and STAR. This site protocol covers a set of environmental parameters related to different spatial scales (AQEM consortium, 2002). Some of these parameters were selected and grouped into three categories, based on different spatial scales (Table 1): aquatic habitats (mesohabitat scale); global features of the site (reach scale) and obligatory variables of Water Framework Directive (WFD) System B (larger regional scale). Aquatic habitats and global features were evaluated in the field, while the WFD parameters were evaluated using Geographical Information Systems.

Biotic parameters

Fish, macroinvertebrates and macrophytes were sampled at all sites during the spring of 2003. Macrophytes were recorded and percentage cover of each species was estimated along a reach of 100 m. The final results were expressed by nine abundance classes: $1 \leq 0.1\%$ cover; $2 = 0.1$–1%; $3 = 1$–2.5%; $4 = 2.5$–5%; $5 = 5$–10%; $6 = 10$–25%; $7 = 25$–50%; $8 = 50$–75% and $9 \geq 75\%$ cover.

A multihabitat procedure developed by AQEM consortium (2002) was adopted to sample benthic macroinvertebrates. A total of 20 Surber samples (25 cm square side with a mesh size of 0.5 mm)

Table 1. Evaluated environmental parameters included in each category. In bold are mentioned the abbreviations of some environmental parameters used in the subsequent tables

Aquatic habitats (% of coverage)	Global features	Water framework directive system B
(Mesohabitat scale)	(Reach scale)	(Geographical scale)
Mineral substrates	Mean depth water body (m)	Longitude
Hygropetric sites	Maximum depth water body (m) **max. depth**	Latitude
Megalithal > 40 cm	Mean slope of the valley floor (%) **mean slope val.**	Altitude (m)
Macrolithal > 20–40 cm	Shading at zenith (foliage cover) **shading**	Catchments area (km^2) **catch. Area**
Mesolithal > 6–20 cm	Average width of woody riparian vegetation (m) left **width rip**.	
Microlithal > 2–6 cm	Average width of woody riparian vegetation (m) right **width rip**.	
Akal > 0.2–2 cm	Shoreline covered with woody riparian vegetation left **length rip**.	
Psammal/Psammopelal	Shoreline covered with woody riparian vegetation right **length rip**.	
Argyllal < 6 μm		
Biotic microhabitats		
Macro-algae		
Micro-algae		
Submerged macrophytes		
Emergent macrophytes		
Living parts of terrestrial plants		
Ter. plants		
Xylal		
CPOM		
FPOM		

were taken, in a reach of 100 m, covering the different habitats. The proportion of Surber samples from each habitat was determined on the basis of the proportion of total reach area occupied by each habitat. Habitats represented by less than 5% of the total area were excluded. The samples were fixed *in situ* with 96% alcohol or with a 40% formalin solution. In the laboratory, samples were sieved (0.5 mm mesh size) and the organisms sorted by naked eye. Sorted organisms were identified to the lowest taxonomic level possible.

Fish sampling took place in wadeable reaches with high-habitat diversity. Reach length was 10 times stream width (minimum 100 m), with special exceptions (minimum 50 m). Stop nets to enclose the fishing area and multiple fishing runs (minimum 2) were recommended. Fishing was conducted in a discontinuous way, always in a downstream–upstream direction. Equipment disinfection between watercourses was recommended to prevent disease spread. All captured fish were identified to the species level and measured in the field. All specimens were returned to water (with exceptions to confirm identification). Mesh-cages and an oxygen diffuser were used whenever necessary. For additional information on sampling methods of all biological elements see Furse et al. (2006).

Data analysis

A taxonomic adjustment was made to avoid the inclusion of different taxonomic levels. For this reason, Mediterranean macroinvertebrates data were treated at the family level, because one of the countries only attained this level of identification.

Taxa with a frequency occurrence lower than 10% and present in samples with less than two individuals were excluded. To prevent distortions caused by the most abundant taxa, species abundances were log $(x+1)$ transformed. Environmental data were standardized to centre and reduce variation:

$$\mathrm{ST} = \frac{x - \text{mean}}{\text{medium deviation}},$$

where x is raw data and ST is standardized data.

Indirect gradient analysis was carried out to detect, inside each core river type, ecological gradients and subgroups of sites ecologically consistent to a lower regional scale. Sites from countries that, in the ordinations of the biological elements, are consistently close together were assumed as subgroups of the respective core river type. Principal Component Analysis (PCA) or Correspondence Analysis (CA) were carried out if the gradient lengths of a preliminary Detrended Correspondence Analysis (DCA) were, respectively, lower or higher than three (ter Braak & Smilauer, 1998).

For each core river type, similarity matrices (site × site) were calculated, using Bray–Curtis distance, to the three biological elements (taxa abundances for macroinvertebrate and fishes, and percentage of cover for macrophytes) and to the three categories of environmental parameters. Mantel correlations were first calculated among the three biological elements, and with these results a new similarity matrix was calculated to build a cluster showing the hierarchical linking among the biological elements. Second, new mantel correlations were calculated between each environmental data category and each biological element. The objective of this second step was to evaluate global dependencies between biological elements and environmental parameters. To go into further detail, it was necessary to evaluate the environmental parameters that account more for the global mantel correlations carried out in the second step. This evaluation was obtained by performing, within each environmental data category, a set of mantel correlations between the single parameters similarity matrices and the biological similarity matrices. The single parameters that got higher correlation coefficients were considered to be the most important to explain the global mantel correlations. This more detailed evaluation was done to mantel correlations whose percentage of significance was lower than 0.1. Mantel correlations among matrices were calculated by Pearson correlations and the respective percentage of significance evaluated by a Monte–Carlo permutation test (999 permutations). This same procedure was carried out on the subgroups of sites established within each core river type, to evaluate if the observed patterns are similar within a lower spatial scale. Subgroups composed by only one country were excluded from this new analysis, because the number of observations was too small to get robust statistical tests. A flow diagram summarizing all these steps can be seen in Fig. 1. Mantel correlations and clusters were done by the software PRIMER 5 for windows, Version 5.2.2 (Clark & Warwick, 1980).

Results

Differences in mean abundance and mean richness of the three aquatic communities in each core river type were evident (Table 2). Macrophyte and fish richness were higher in lowland rivers, while the highest macroinvertebrate richness was observed in Mountain Rivers, although presenting a high variability. Fish communities were at their lowest abundances in Mountain Rivers. Macrophytes and fishes presented low richness for Mediterranean river although slightly higher than Mountain Rivers.

Table 2. Mean abundance and richness of the three aquatic communities in each core river type. Standard deviation in parentheses. Mean abundances of macrophytes were not calculated because data are expressed in abundance classes

	Mean abundance	Mean richness
Lowland rivers		
Macrophytes	–	7.1 (4.6)
Invertebrates	2817 (2864)	38.7 (8.9)
Fishes	10,771 (24,531)	6.1 (2.5)
Mountain rivers		
Macrophytes	–	3.1 (1.4)
Invertebrates	3739 (2824)	49.1 (8.9)
Fishes	3792 (2362)	2.2 (1.1)
Mediterranean rivers		
Macrophytes	–	4.4 (2.1)
Invertebrates	2060 (2822)	20.2 (8.1)
Fishes	10,727 (12,545)	3.8 (2.2)

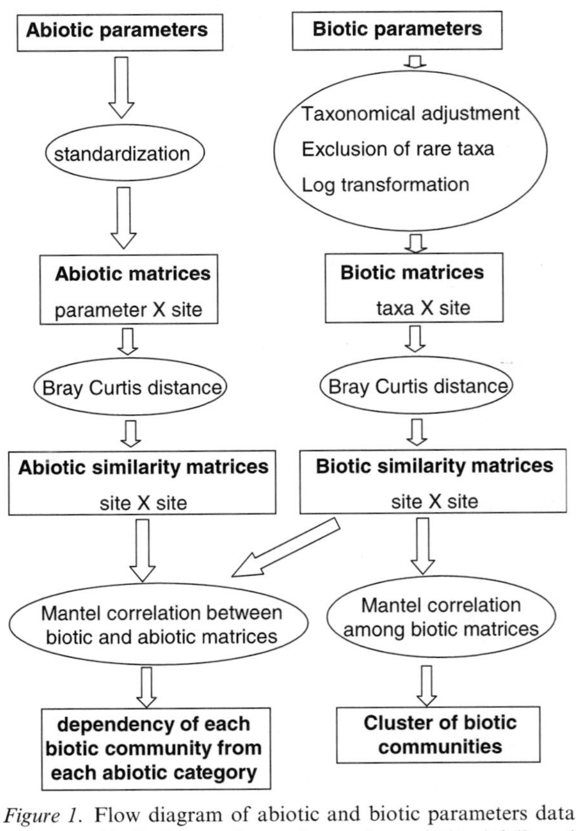

Figure 1. Flow diagram of abiotic and biotic parameters data treatment (similarity matrices and mantel correlations) followed for each core river type and to its respective subgroups. Inside the rectangle are the partial and the final results obtained during the treatment. Inside the elliptical shapes are mentioned the actions carried out.

The ordinations plotted in Figures 2–4 showed that, although the different spatial patterns observed, within each core river type, for each biological element, it is possible to extract consistent groups of countries representing lower spatial scales. For Mountain Rivers (Fig. 2), German (D05, D06) and Czeck rivers (C04, C05) tend to form two distinct groups in both invertebrate and fish ordinations, as opposed to the macrophyte ordination where no consistent group is detectable. The Austrian river (A05) is grouped with German rivers concerning fish ordination, while the invertebrate ordination, despite the proximity to the German group, tends to be separate. A similar tendency of the Austrian river to be an independent group was observed in the macrophyte ordination. In this ordination, Austrian sites tend to be located on the extreme negative of the first axis. Thus, two subgroups of rivers, mainly defined by invertebrate and fish communities, can be established within Mountain Rivers, D04/D06 and C04/C05, respectively.

Concerning Lowland rivers (Fig. 3), as detected for Mountain Rivers, it was only possible to establish different groups in relation to

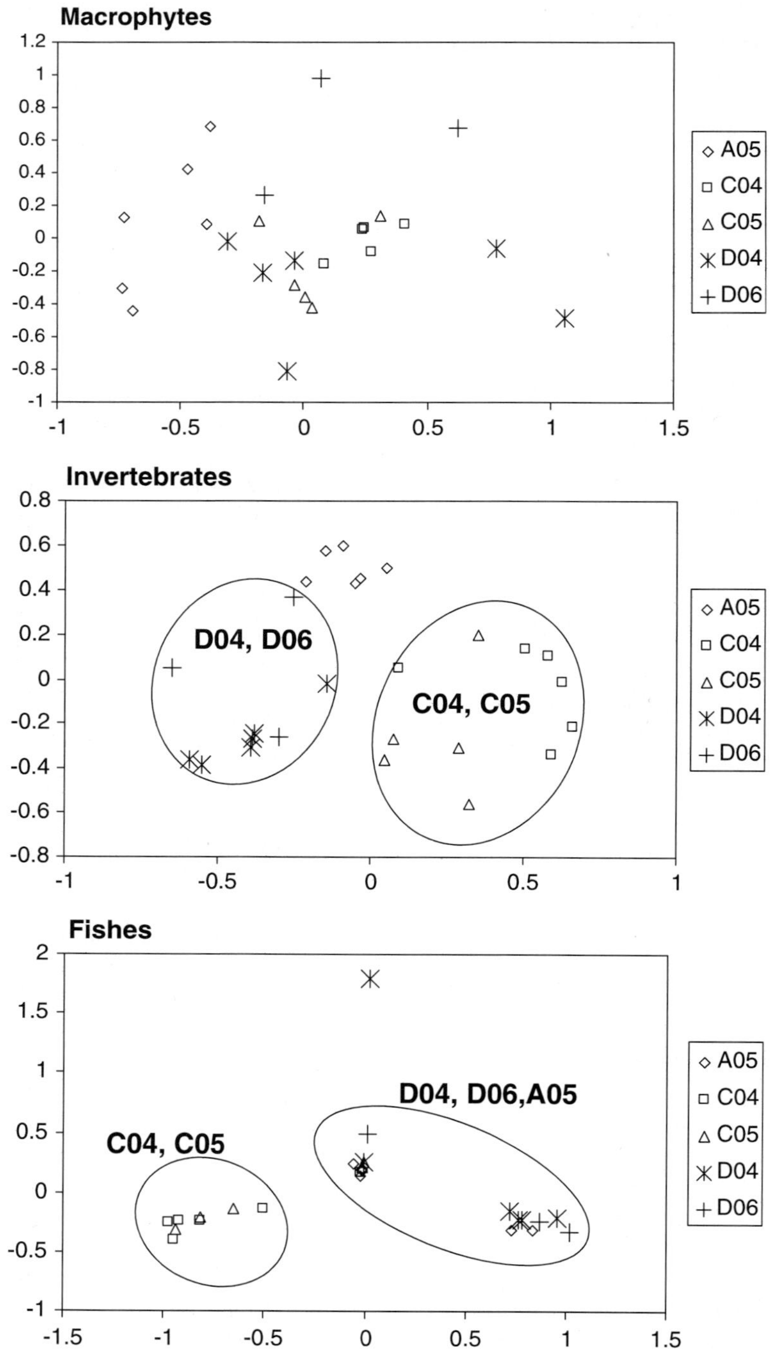

Figure 2. Ordination of biological elements to Mountain Rivers. The rounded shapes indicate the subgroups pointed out by the plotted ordinations.

invertebrate and fish ordinations. However, the established groups are no longer the same. Swedish rivers (S05, S06) as well as Danish and British rivers (K02, U23) tend to form two different groups. However, a lack of consistency was observed in relation to the German river (D03). It

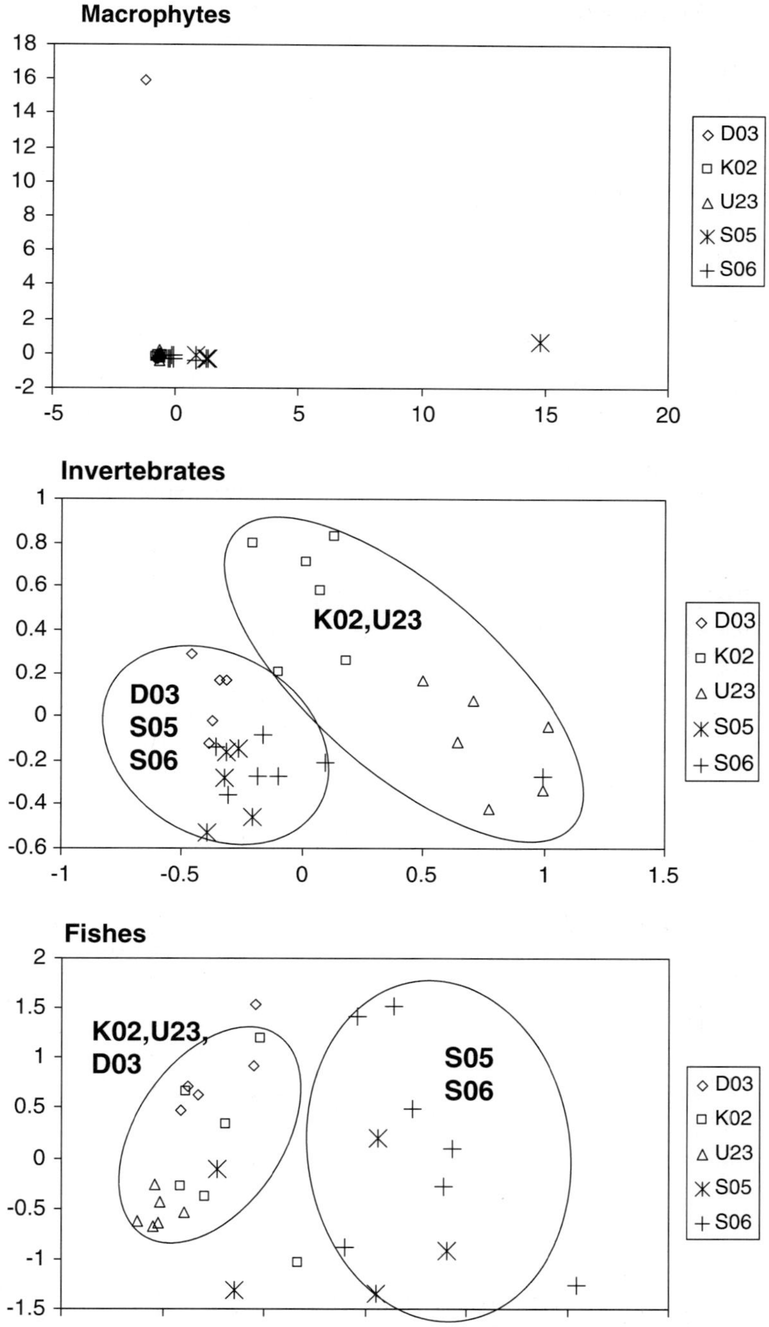

Figure 3. Ordination of biological elements to Lowland rivers. The rounded shapes indicate the subgroups pointed out by the plotted ordinations.

is grouped with Swedish rivers for the invertebrate ordination, being included in the other group for the fish ordination. As a result of this lack of consistency, this German river was excluded and two subgroups were established for Lowland rivers S05/S06 and K02/U23, respectively.

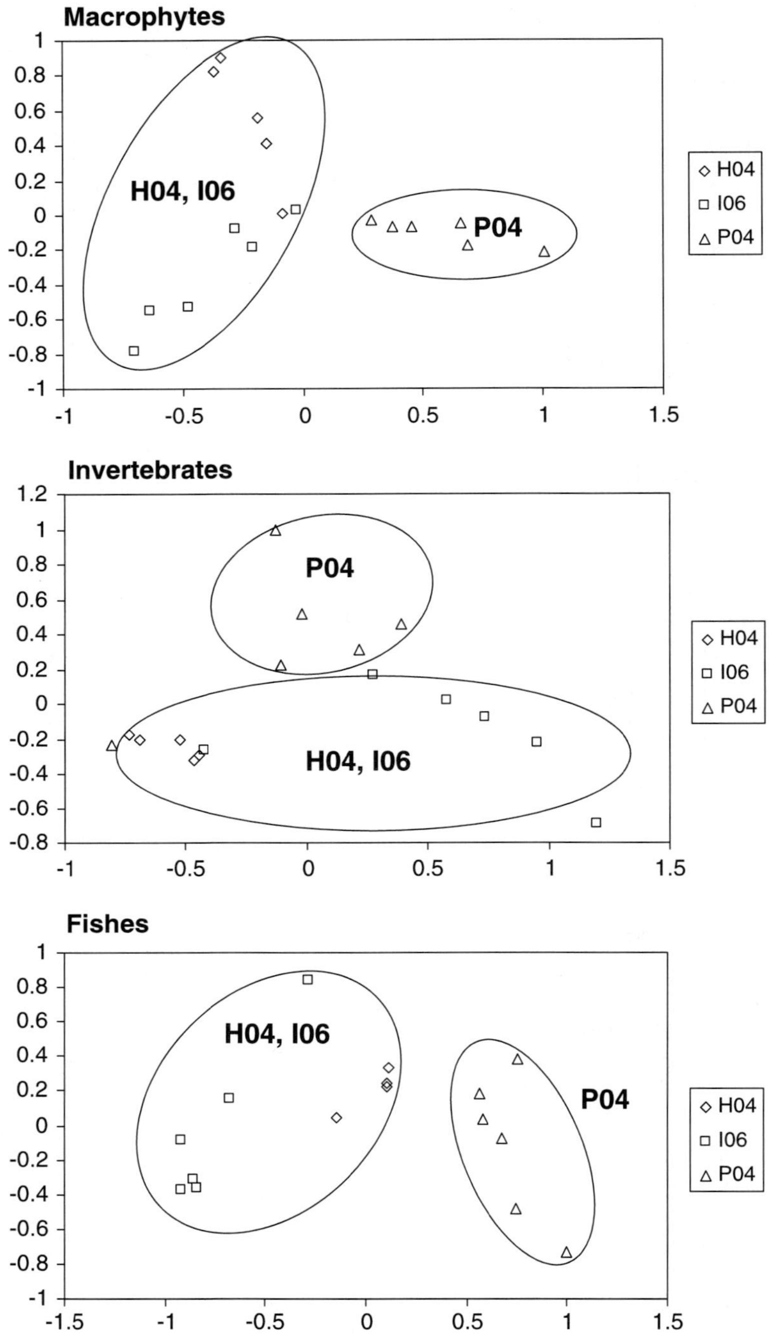

Figure 4. Ordination of biological elements to Mediterranean rivers. The rounded shapes indicate the subgroups pointed out by the plotted ordinations.

The ordinations plotted for Mediterranean rivers (Fig. 4) permitted the detection of the two same groups for the three biological elements. Portuguese rivers tend to be consistently separated from Greek and Italian rivers and, for this reason, only one subgroup was established for Mediterranean rivers H04/I06.

Cluster analysis and Mantel tests, carried out to a larger spatial scale, showed significant correlations among the three aquatic communities ($p < 0.01$) for the three core river types (Fig. 5). Lowland and Mountain rivers displayed similar patterns. For those core river types, the clusters first agglomerated macrophytes and invertebrates. In contrast, the first agglomeration for Mediterranean rivers was between macrophytes and fishes (Fig. 5).

The mantel correlations obtained between environmental parameter matrices and aquatic community matrices, as well as the environmental parameters that account more for the mantel

Table 3. Percentage of significance obtained by mantel correlations between environmental categories and each biological element, for the three core river types (* $p < 0.05$; ** $p < 0.01$). Down mantel correlations whose $p < 0.1$ are mentioned the most important single parameters that account for the global correlation

	Aquatic communities		
	Macrophytes	Invertebrates	Fishes
Mountain			
Habitat	22.8%	5.6%	16.6%
Global features	4.8% (*)	5%	8.1%
	Stream width (0.319)	Length rip. vegt. (0.295)	Length rip. vegt. (0.207)
	Length rip. vegt. (0.284)	Stream width (0.182)	Shading (0.139)
	Width rip. veget.(0.165)	Shading (0.128)	
WFD	6.3%	0.1% (**)	0.3% (**)
	Longitude (0.444)	Latitude (0.550)	Latitude (0.415)
	Latitude (0.214)	Altitude (0.339)	Catch. area (0.377)
			Altitude (0.377)
Lowland			
Habitat	0.1% (**)	2.9% (*)	0.1% (**)
	Psammal (0.268)	Psammal (0.344)	FPOM (0.149)
	Microlithal (0.256)	Microlithal (0.318)	Xylal (0.141)
	Subm. macroph. (0.249)	CPOM (0.169)	Microlithal (0.118)
	FPOM (0.202)	Akal (0.162)	Megalithal (0.117)
		FPOM (0.152)	
Global features	0.6% (**)	21.3%	27.3%
	Shading (0.241)		
	Max. depth (0.225)		
	Length rip. vegt. (0.156)		
WFD	4.0% (*)	0.1% (**)	0.1% (**)
	Longitude (0.128)	Altitude (0.302)	Longitude (0.124)
		Longitude (0.295)	
Mediterranean			
Habitat	32.7%	38.5%	0.5%(**)
			Mesolithal (0.310)
			CPOM (0.288)
			Megalithal (0.248)
			Akal (0.222)
Global features	–	–	–
WFD	0.1% (**)	1.8% (*)	0.2 (**)
	Longitude (0.588)	Catch. area (0.230)	Longitude (0.694)
	Catch. area (0.528)	Latitude (0.216)	Catch. area (0.614)
	Latitude (0.456)	Longitude (0.216)	Latitude (0.511)

correlations are shown in Table 3. Habitat data category was only significantly correlated with all biological elements on Lowland rivers, FPOM being an important parameter in all correlations. Concerning the inorganic habitats, fine sediments (psammal) tend to be more important to macrophytes and invertebrates than to fishes, where sediment with larger granulometry (megalithal) tends to increase its importance. The same importance of larger inorganic sediments to fish communities was also observed to Mediterranean rivers (mesolithal and megalithal).

Macrophytes were the only biological element to show significant correlations ($p < 0.05$) with the habitat data category, the importance of the shoreline covered with woody riparian vegetation (length rip. veget.) being noticeable for both Mountain and Lowland rivers.

Finally, and also at the same larger spatial scale, for almost all cases, biological elements were significantly correlated with WFD data category. Latitude and longitude were consistently important in the great majority of the situations, denoting the biogeographical distribution of the taxa across Europe. In the specific case of Mediterranean rivers, catchment area, at this spatial scale, seems to be a key factor in explaining the establishment of aquatic communities.

Going into further detail to a lower spatial scale (subgroups within each core river type), the same data treatment was carried out in relation to the subgroups, and their patterns compared with the respective core river type.

Cluster analysis for subgroups of each core river type (Figs. 5–7) showed similar results for Lowland and Mediterranean rivers (Figs. 6 and 7). The Mountain rivers clusters were different for each subgroup (Fig. 5), and in contrast with the two other core river types, no significant correlations ($p > 0.05$) were detected among the three aquatic communities (Fig. 6).

Biological elements of Mountain Rivers (two subgroups, Table 4) showed the least number of significant correlations with the environmental parameter categories (three significant correlations at $p < 0.05$ for the 18 correlations carried out). In contrast, Lowland rivers (two subgroups, Table 5) presented the highest number of significant correlations (13 significant correlations at $p < 0.05$ for

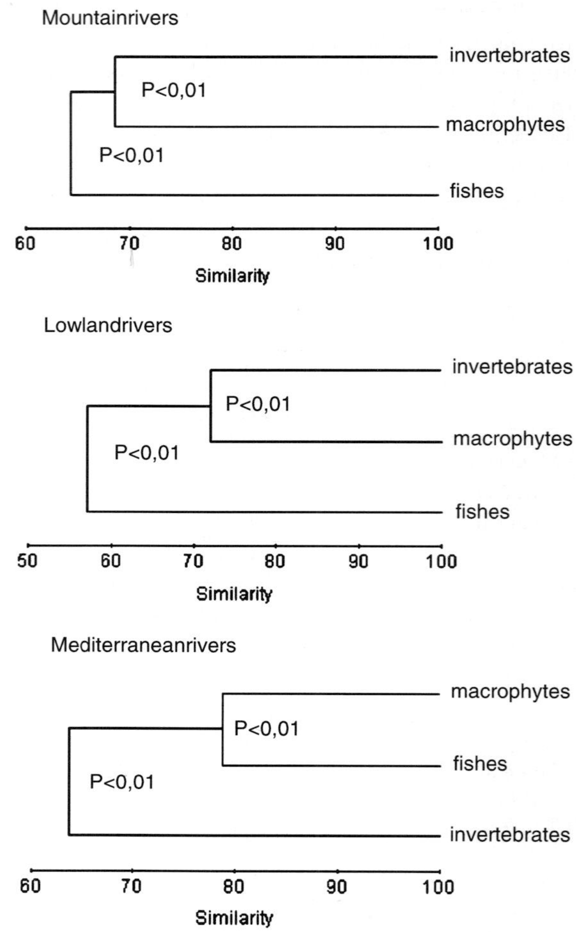

Figure 5. Cluster analyses of all biological elements for Mountain, Lowland and Mediterranean rivers. Critical level of significance (p) mentioned for all the partial agglomerations.

the 18 correlations carried out). Aquatic communities of the Mediterranean subgroup were also significantly correlated ($p < 0.05$) to both habitat data category and WFD data category parameters (Table 6), with the exception of macrophytes (five significant correlations at $p < 0.05$ for the six correlations carried out).

Macrophyte and fish communities from subgroups of Mountain rivers (biological elements with the lowest observed richness, see Table 2), contrasting with the other core river types, are the only biological element that did not present any significant correlation with the environmental parameters. Concerning invertebrates, the influence of latitude and longitude on the global

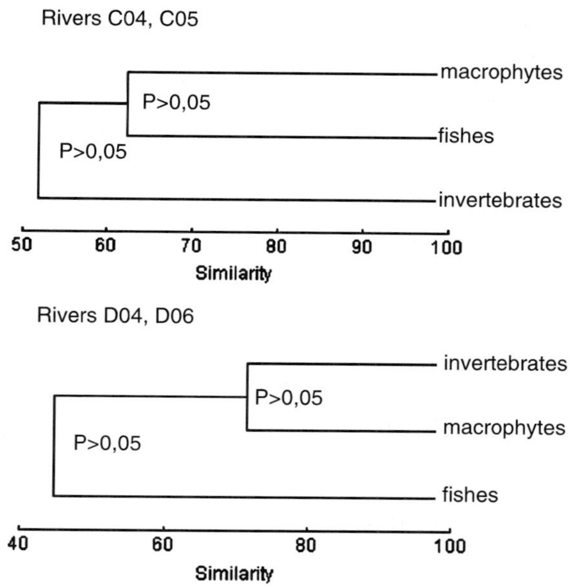

Figure 6. Cluster analyses of all biological elements for subgroups of Mountain Rivers. Critical level of significance (*p*) mentioned for all the partial aglomerations.

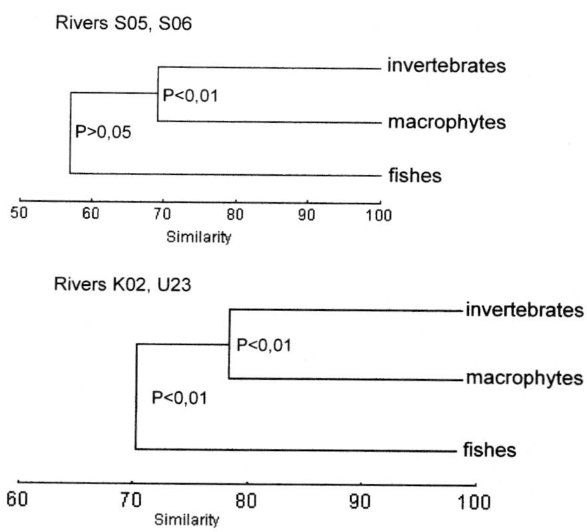

Figure 7. Cluster analyses of all biological elements for subgroups of Lowland, rivers. Critical level of significance (*p*) mentioned for all the partial aglomerations.

correlations is remarkable, denoting that biogeographical aspects of taxa distributions are still important at this lower spatial scale.

Two quite different patterns were observed for the subgroups of Lowland rivers. Swedish rivers (S05, S06), in contrast to Danish and British rivers (K02, U23), presented a low number of significant correlations between biological elements and environmental parameters categories. Analysing the WFD data category parameters in more detail, it was observed that longitude is very important to the K02/U23 subgroup, a fact that could be expected as a result of the accentuated geographical isolation from Denmark and the UK. In any case, habitat and global features data categories are all significantly correlated with all biological elements ($p < 0.01$), pointing out that, despite the geographical isolation, rivers of this subgroup are clearly consistent and the aquatic communities are clearly predicted by the environmental parameters. Also, for these rivers, fine inorganic sediments are important to all biological elements, an expected feature due to the importance of the deposition processes in Lowland rivers. Concerning Swedish rivers, catchment area is an important WFD parameter, denoting variability in discharge and hydrodynamics inside this subgroup, a fact that agrees with the importance of inorganic habitats of very different granulometries for invertebrate communities. The mean slope of the valley (important to fish communities) is also a parameter that can be related to the catchment area.

In relation to Mediterranean rivers, as occurred in the larger spatial scale, catchment area continues to be an in important factor in this lower spatial scale, emphasizing the dependence of all biological elements on the water availability throughout the year. At this lower scale, the fine inorganic sediments are important to invertebrate and fish communities, but larger inorganic sediments also account for the correlation with fishes (megalithal). Due to the geographical isolation from Italy and Greece, longitude is an important factor, still at this scale (Fig. 8).

Discussion

Mountain Rivers, due to their higher hydrodynamics, are less suitable for macrophyte development, which also may affect fish communities. Vegetated habitats are important refuge areas against high currents and aquatic predators, particularly for small fish (e.g., Allouche, 2002; Shoup et al., 2003). For invertebrates, the pattern of greater richness may result from higher hydrodynamic disturbance in mountain rivers,

Table 4. Percentage of significance obtained by mantel correlations between environmental categories and each biological element, for Mountain Rivers (*$p < 0.05$; **$p < 0.01$). Down mantel correlations whose $p < 0.1$ are mentioned the most important single parameters that account for the global correlation

	Aquatic communities		
	Macrophytes	Invertebrates	Fishes
Rivers C04, C05			
Habitat	69.8%	19.6%	45.8%
Global features	56.4%	71.0%	93.9%
WFD	23.0%	2.0% (*)	78.2%
		Longitude (0.630)	
		Latitude (0.530)	
Rivers D04, D06			
Habitat	54.0%	0.1% (**)	12.3%
		Ter. plants (0.610)	
		Xylal (0.436)	
		Microlithal (0.369)	
		CPOM (0.364)	
Global features	25.0%	9.2%	24.6%
		Stream width (0.540)	
		Mean solpe val. (0.320)	
		Max. depth (0.285)	
		Width rip. veget. (0.258)	
WFD	29.0%	0.1% (**)	19.4%
		Catch. area (0.634)	
		Longitude (0.527)	
		Altitude (0.512)	

thus maintaining invertebrate communities under a constant level of intermediate disturbance, diminishing competition and increasing diversity (Townsend, 1989; Voelz & McArthur, 2000; Wright & Li, 2002; Willis et al., 2005). Despite those strong differences observed between the richness of aquatic communities between Mountain and Lowland rivers, similar clusters were obtained with respect to the correlations among the three communities. Higher correlations were observed between macrophytes and invertebrates, suggesting the importance of macrophytes as refuges for invertebrates (Rosenfeld, 1997; Voelz & McArthur, 2000; Zrum & Hann, 2002; Balci & Kennedy, 2003; Strayer et al., 2003; Warfe & Barmuta, 2004) from predation and hydrodynamic peaks. Fishes that can easily move along rivers are less dependent on local variables than invertebrates and macrophytes. Despite the high-correlation observed between macrophytes and fishes in Mediterranean rivers, no direct relationship occurs between these two communities. This relationship may result from water availability under discharge fluctuations, as both fish and macrophytes are prone to dessication and require aquatic habitat persistence during the dry-season. Invertebrates, due to several well-known strategies to resist to water level fluctuations (Stanley et al., 1994; Vieira et al., 2004), are less dependent on water availability. This idea can be supported by the observed importance of the catchment area (clearly related to water availability) in the correlations between WFD data and all biological elements. Catchment area was also important to the correlation with fishes in Mountain Rivers, were the water availability is a key factor in supporting fish communities.

Lowland rivers showed a high number of significant correlations between aquatic communities and environmental parameters categories, which may also be related to more stable conditions and to the greater water availability in Lowland rivers,

Table 5. Percentage of significance obtained by mantel correlations between environmental categories and each biological element, for Lowland rivers (*$p < 0.05$; **$p < 0.01$). Down mantel correlations whose $p < 0.1$ are mentioned the most important single parameters that account for the global correlation

	Aquatic communities		
	Macrophytes	Invertebrates	Fishes
Rivers S05, S06			
Habitat	6.9%	0.1% (**)	56.0%
		Mesolithal (0.533)	
		Akal (0.417)	
		Microlithal (0.391)	
		Macrolithal (0.357)	
		Psammal (0.259)	
Global features	29.3%	92.0%	3.6% (*)
			Shading (0.306)
			Mean slope val. (0.273)
			Length rip. vegt. (0.236)
			Width rip. vegt. (0.199)
WFD	0.6% (**)	12.8%	3.5% (*)
	Catch. area (0.298)		Catch. area (0.268)
	Altitude (0.118)		Altitude (0.121)
Rivers K02, U23			
Habitat	0.1% (**)	0.2% (**)	0.7% (**)
	Psammal (0.593)	Microlithal (0.338)	Microlithal (0.409)
	Microlithal (0.304)	Psammal (0.294)	Xylal (0.243)
		Xylal (0.104)	Mesolithal (0.224)
Global features	0.7% (**)	0.1% (**)	0.1% (**)
	Shading (0.425)	Shading (0.518)	Riffles/pools rel. (0.560)
	Mean slope val. (0.354)	Max. depth (0.373)	Width rip. vegt. (0.465)
		Riffles/pools rel. (0.359)	Max. depth (0.387)
WFD	25.6%	1.0% (*)	4.6% (*)
		Longitude (0.555)	Longitude (0.412)

than in Mountain rivers (Reyjol et al., 2003). Under these conditions, more stable communities can establish themselves in aquatic ecosystems and maintain a greater dependency on environmental parameters. For Mediterranean rivers, apart from WFD parameters, only fishes had a significant correlation with habitat parameters. This may be due to the lower stability of Mediterranean aquatic communities. Fish habitat relationships in Mediterranean rivers, particularly in intermittent ones, are highly dynamics due to seasonal habitats patchiness. Thus although fish can present a quite plastic habitat use along the year, in certain periods there are a strong habitat selectivity, namely for reproduction in spring and for individual survival during the dry-season (Ilhéu, 2004).

In Mountain rivers, all biological elements were correlated with the global features data, contrasting with Lowland rivers were only a significant correlation was detected for macrophytes. This fact confirms the assumptions of the river continuum concept (Vannote et al., 1980), predicting higher dependency of headwaters (Mountain rivers) from the allochtoneous inputs, as supported by the importance assumed by the shoreline covered with woody riparian vegetation (length rip.).

At the same major spatial scale the importance of latitude and longitude on the correlations observed between WFD data and biological elements was evident. This fact can result from climatic aspects dependent on the geographical localization or on the biogeographic distribution

Table 6. Percentage of significance obtained by mantel correlations between environmental categories and each biological element, for Mediterranean rivers (* $p < 0.05$; ** $p < 0.01$). Down mantel correlations whose $p < 0.1$ are mentioned the most important single parameters that account for the global correlation

	Aquatic communities		
	Macrophytes	Invertebrates	Fishes
Rivers H04, I06			
Habitat	50.9%	1.4% (*)	1.7% (*)
		Akal (0.393)	Mesolithal (0.520)
		Microlithal (0.383)	Microlithal (0.502)
		Psammal (0.345)	Akal (0.453)
		Mesolithal (0.271)	CPOM (0.391)
		Ter. plants (0.217)	
Global features	–	–	–
WFD	0.2% (**)	0.2% (**)	0.9% (**)
	Altitude (0.528)	Catch. area (0.777)	Catch. area (0.649)
	Catch. area (0.496)	Altitude (0.624)	Latitude (0.533)
		Longitude (0.622)	Longitude (0.518)

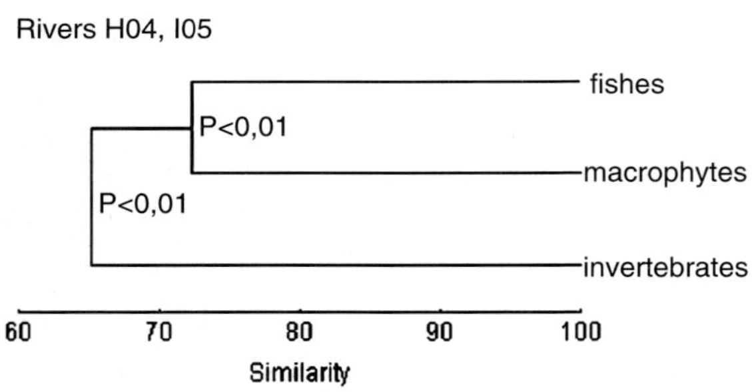

Figure 8. Cluster analyses of all biological elements for the subgroup of Mediterranean, rivers. Critical level of significance (*p*) mentioned for all the partial aglomerations.

of taxa across Europe. Due to the general similarities of climatic aspects of the regions covered by each river type, the second hypothesis seems to be a better explanation.

Separate ordinations for each aquatic community within each core river type denoted a high dependence on the geographical location. However, a more detailed analysis of subgroups of major river types (i.e., regional and local scale) showed similar results. The only exception was the two subgroups of Mountain Rivers, where no significant correlations were obtained among the aquatic communities. In contrast to Lowland rivers, only two significant correlations were obtained between environmental parameters categories and aquatic communities. Greater hydraulic disturbance, leading to less stable communities (Townsend et al., 1983; Reichard et al., 2002; Johnson et al., 2003; Smith et al., 2003) can decrease the strength of linkages between aquatic communities and environmental parameters. These results emphasize the importance of physical factors on aquatic ecosystems, causing different linkages between aquatic communities and different dependencies on environmental parameters (Townsend et al., 1983; Townsend, 1989; Voelz & McArthur, 2000; Wright & Li, 2002). Concerning Mountain Rivers, the low-richness observed of macrophytes and fishes can also contribute to the weakness of the links among the biological elements, as well as to the absence of significant correlations between all environmental data categories and those two biological elements.

Biological elements of the two subgroups of Lowland rivers (S05/S06 and K02/U23) showed similar clusters, but with completely different correlations with environmental category data. This fact suggests that links among biological elements are more dependent on relationships established among aquatic communities than on the influence of environmental parameters.

Concerning Mediterranean rivers, at a lower spatial scale, catchment area continued to be an important parameter controlling the establishment of aquatic communities, thus suggesting the importance of water availability at different scales for Mediterranean rivers.

Generally, when mantel correlations were carried out at the larger regional level, WFD parameters were significantly correlated with almost all aquatic communities. However, when this analysis was carried out at a smaller spatial scale (regional or local), the number of significant correlations to WFD parameters diminished, while significant correlations to habitats and global features increased. This tendency partially agrees with the hierarchical theory of river formation (Jensen et al., 1996; Verdonschot, 2000; Crook et al., 2001; Li et al., 2001; Weigel et al., 2003) because there are no environmental parameters specific to any spatial scale.

The present study confirmed that links among biological elements are different for different core river types, and are influenced by water availability. These links tend to be consistent for different spatial scales, except if they are weak. The influence of environmental parameters is dependent on the spatial scale. Reach and mesohabitat environmental parameters tend to explain aquatic communities at a lower spatial scale, while geographical parameters tend to explain the communities at a major spatial scale.

Acknowledgements

This study was funded by the EU research project STAR from the European Commission, 5th Framework Program, Energy, Environment and Sustainable Development, Key Action Water, Contract No. EVK1-CT2001-00089. We would like to thank to STAR partners for the data availability and to two unknown reviewers for their comments that largely contributed to the scientific improvement of this paper.

References

Allouche, S., 2002. Nature and function of cover for riverine fish. Bulletin Francais de La Peche et la Pisciculture 356(66): 297–324.

AQEM consortium, 2002. Manual for the application of the AQEM system. A comprehensive method to assess European streams using benthic macroinvertebrates, developed for the purpose of the Water Framework Directive. Version 1.0 February 2002.

Armitage, P. D., 1995. Faunal Community change in response to flow manipulation. In Harper, D. M. & J. D. Ferguson, (eds), The Ecological Basis of River Management. John Wiley & Sons, New York, 59–78.

Armitage, P. D. & C. E. Cannan, 2000. Annual changes in summer patterns of mesohabitat distribution and associated macroinvertebrate assemblages. Hydrological Processes 14: 3161–3179.

Armitage, P. D. & R. J. M. Gunn, 1996. Differential response of benthos to natural and anthropogenic disturbances in 3 lowland streams. Internationale Revue der Gesamten Hydrobiologia 81: 161–181.

Balci, P. & J. H. Kennedy, 2003. Comparison of chironomids and other macroinvertebrates associated with Myriophyllum spicatum and Heteranthera dubia. Journal of Freshwater Ecology 18: 235–247.

Beisel, J. N., P. Usseglio-Polaterra, S. Thomas & J. C. Moretou, 1998. Stream community structure in relation to spatial variation: the influence of mesohabitat characteristics. Hydrobiologia 389: 73–88.

Bernardo, J. M., M. Ilhéu, P. Matono & A.M. Costa, 2003. Interannual variation of fish assemblage structure in a Mediterranean river: implications of streamflow on the dominance of native or exotic species. River Research and Application 19: 1–12.

Bio, A. M. F., P. De Becker, E. De Bie, W. Huybrechts & M. Wassen, 2002. Prediction of plant species distribution in lowland river valleys in Belgium: modelling species response to site conditions. Biodiversity and Conservation 11: 2189–2216.

Clark, K. R. & R. M. Warwick, 1980. Changes in Marine Communities: An Approach to Statistical Analysis and Interpretation. Natural Environmental Council, UK.

Casas, J. J., 1997. Invertebrate assemblages associated with plant debris in a backwater of a mountain stream: natural leaf packs vs. debris dam. Journal of Freshwater Ecology 12: 39–49.

Cheruvelil, K. S., P. A. Soranno & R. D. Serbin, 2000. Macroinvertebrates associated with submerged macrophytes: sample size and power to detect effects. Hydrobiologia 441: 133–139.

Crook, D. A., A. I. Robertson, A. J. King & P. Humphries, 2001. The influence of spatial scale and habitat arrangement on diel patterns of habitat use by two lowland river fishes. Oecologia 129: 525–533.

Dahl, J. & L. A. Greenberg, 1998. Effects of fish predation and habitat type on stream benthic communities. Hydrobiologia 361: 67–76.

Furse, M., D. Hering, O. Moog, P. Verdonschot, R. K. Johnson, K. Brabec, K. Gritzalis, A. Buffagni, P. Pinto, N. Friberg, J. Murray-Bligh, J. Kokes, R. Alber, P. Usseglio-Polatera, P. Haase, R. Sweeting, B. Bis, K. Szoszkiewicz, H. Soszka, G. Springe, F. Sporka, I. Krno, 2006. The STAR project: context, objectives and approaches. Hydrobiologia 566: 3–29.

Ilhéu, M., 2004. Patterns of habitat use by freshwater fishes in Mediterranean rivers. PhD Thesis, University of Évora, Évora.

Jensen, M. E., P. Bourgeron, R. Everett & I. Goodman, 1996. Ecosystem management: a landscape ecology perspective. Water Research 32: 203–216.

Johnson, L. B., D. H. Breneman & C. Richards, 2003. Macroinvertebrate community structure and function associated with large wood in low gradient streams. River Research and Applications 19: 199–218.

Li, J. L., A. Herlihy, W. Gerth, P. Kaufmann, S. Gregory, S. Urquhart & D. P. Larsen, 2001. Variability in stream macroinvertebrates at multiple spatial scales. Freshwater Biology 46: 1–87.

Petts, G. E., 2000. A perspective on the abiotic process sustaining the ecological integrity of running waters. Hydrobiologia 422/423: 15–27.

Reichard, M., P. Jurajda & M. Ondrackova, 2002. Interannual variability in seasonal dynamics and species composition of drifting young-of-the-year fishes in two European lowland rivers. Journal of Fish Biology 60: 87–101.

Reyjol, Y., A. Compin, A. Ibarra & P. Lim, 2003. Longitudinal diversity patterns in streams: comparing invertebrates and fish communities. Archiv fur Hydrobiologie 157: 525–533.

Rosenfeld, J. S., 1997. The effect of large macroinvertebrate herbivores on sessile epibenthos in a mountain stream. Hydrobiologia 344: 75–79.

Smith, H., P. J. Wood & J. Gunn, 2003. The influence of habitat structure and flow permanence on invertebrate communities in karst spring systems. Hydrobiologia 510: 53–66.

Stanley, E. H., D. L. Buschman, A. J. Boulton, N. B. Grimm & S. G. Fisher, 1994. Invertebrate resistance and resilience to intermittency in a desert stream. American Midland Naturalist 131: 288–300.

Strayer, D. L., C. Lutz, H. M. Malcom, K. Munger & W. H. Shaw, 2003. Invertebrate communities associated with a native (*Vallisneria americana*) and an alien (*Trapa natans*) macrophyte in a large river. Freshwater Biology 48: 1938–1949.

Shoup, D.E., R. E. Carlson & R. T. Heath, 2003. Effects of predation risk and foraging return on the diel use of vegetated habitat by two size-classes of bluegills. Transactions of the American Fisheries Society 132: 590–597.

ter Braak, C. J. F. & P. Smilauer, 1998. 'CANOCO Reference Manual and User's Guide to Canoco for Windows'. Microcomputer Power, Ithaca, New York, 351 pp.

Tolonen, K. T., H. Hamalainen, I. J. Holopainen, K. Mikkonen & J. Karjalainen, 2003. Body size and substrate association of littoral insects in relation to vegetation structure. Hydrobiologia 499: 179–190.

Townsend, C. R., 1989. The patch dynamic concept of stream community ecology. Journal of North American Benthological Society 8: 36–50.

Townsend, C. R., A. G. Hidrew & J. Francis, 1983. Community structure in some southern English streams: the influence of physicochemical factors. Freshwater Biology 13: 521–544.

Vannote, R. L., G. W. Minshall, K. W. Cummins, J. R. Sedell & E. Cushing, 1980. The river continuum concept. Canadian Journal of Fisheries and Aquatic Sciences 37: 130–137.

Verdonschot, P. F., 2000. Integrated ecological assessment methods as a basis for sustainable catchment management. Hydrobiologia 422/423: 389–412.

Verdonschot, P. F. M., 2006. Evaluation of the use of Water Framework Directive typology descriptors, reference sites and spatial scale in macroinvertebrate stream typology. Hydrobiologia 566: 39–58.

Verdonschot, P. F. & R. C. Nijboer, 2004. Testing the European stream typology of the water framework directive for macroinvertebrates. Hydrobiologia 516: 35–54.

Vieira, N. K. M., W. H. Clements, L. S. Guevara & B. F. Jacobs, 2004. Resistance and resilience of stream insect communities to repeated hydrologic disturbances after a wildfire. Freshwater Biology 49: 1243–1259.

Voelz, N. J. & J. V. McArthur, 2000. An exploration of factors influencing lotic insect species richness. Biodiversity and Conservation 9: 1543–1570.

Wagner, F. H. & G. Bretschko, 2003. Riparian trees and flow paths between the hyporheic zone and groundwater in the Oberer Seebach, Austria. International Review of Hydrobiology 88: 129–138.

Ward, J. V., 1989. The four dimensional nature of lotic ecosystems. Journal of North American Benthological Society 8: 2–8.

Ward, J. V. & K. Tockner, 2001. Biodiversity: towards a unifying theme for river ecology. Freshwater Biology 46: 807–819.

Warfe, D. M. & L. A. Barmuta, 2004. Habitat structural complexity mediates the foraging success of multiple predator species. Oecologia 141: 171–178.

Weigel, B. M., L. Z. Wang, P. W. Rasmussen, J. T. Butcher, P. M. Stewart, T. P. Simon & M. J. Wiley, 2003. Relative influence of variables at multiple spatial scales on stream macroinvertebrates in the Northern Lakes and Forest ecoregion, USA. Freshwater Biology 48: 1440–1461.

Willis, S. C., K. O. Winemiller & H. Lopez-Fernandez, 2005. Habitat structural complexity and morphological diversity of fish assemblages in a Neotropical floodplain river. Oecologia 142: 284–295.

Wright, J. F., R. J. M. Gunn, J. M. Winder, R. Wiggers, K. Vowles, R. T. Clarke & I. Harris, 2002. A comparison of the macrophyte cover and macroinvertebrate fauna at three sites on the River Kennet in the mid 1970s and late 1990s. Science of the Total Environment 282: 121–142.

Wright, K. K. & J. L. Li, 2002. From continua to patches: examining stream community structure over large environmental gradients. Canadian Journal of Fisheries and Aquatic Sciences 59: 1404–1417.

Zimmer, K. K. D., M. A. Hanson & M. G. Butler, 2003. Relationships among nutrients, phytoplankton, macrophytes, and fish in prairie wetlands. Canadian Journal of Fisheries and Aquatic Sciences 60: 721–730.

Zrum, L. & B. J. Hann, 2002. Invertebrates associated with submersed macrophytes in a prairie wetland: effects of organophosphorus insecticide and inorganic nutrients. Archiv fur Hydrobiologie 154: 413–445.

Hydrobiologia (2006) 566:91–105
M.T. Furse, D. Hering, K. Brabec, A. Buffagni, L. Sandin & P.F.M. Verdonschot (eds), The Ecological Status of European Rivers: Evaluation and Intercalibration of Assessment Methods
DOI 10.1007/s10750-006-0068-5

A comparison of the European Water Framework Directive physical typology and RIVPACS-type models as alternative methods of establishing reference conditions for benthic macroinvertebrates

John Davy-Bowker[1,*], Ralph T. Clarke[1], Richard K. Johnson[2], Jiri Kokes[3], John F. Murphy[1] & Svetlana Zahrádková[4]

[1]*Centre for Ecology & Hydrology, Winfrith Technology Centre, Dorchester, Dorset DT2 8ZD, United Kingdom*
[2]*Department of Environmental Assessment, Swedish University of Agricultural Sciences, P.O. Box 7050 SE-750 07 Uppsala, Sweden*
[3]*T.G.M. Water Research Institute, Drevarska 12, 657 57 Brno, Czech Republic*
[4]*Department of Zoology and Ecology, Faculty of Science, Masaryk University Brno, Kotlářská 2, 611 37 Brno, Czech Republic*
*(*Author for correspondence: E-mail: jobo@ceh.ac.uk)*

Key words: reference condition, physical typology, RIVPACS, $SWEPAC_{SRI}$, PERLA

Abstract

The EU Water Framework Directive requires European Union Member States to establish 'type-specific biological reference conditions' for streams and rivers. Types can be defined by using either a fixed typology (System-A), defined by ecoregions and categories of altitude, catchment area and geology, or by means of an alternative characterisation (System-B) that can use a variety of physical and chemical factors. Several European countries also have existing RIVPACS-type models that give site (rather than stream type) specific predictions of benthic macroinvertebrate communities. In this paper we compare the Water Framework Directive (WFD) System-A physical typology and three existing European multivariate RIVPACS-type models as alternative methods of establishing reference conditions. This work is carried out in Great Britain – using RIVPACS, Sweden – using $SWEPAC_{SRI}$ and the Czech Republic – using PERLA. We found that in all three countries, all seasons and season combinations, and for all biotic indices tested, RIVPACS-type models were more effective (lower standard deviations of O/E ratios) than models based solely on the WFD System-A variables or null models (based on a single expectation for all sites). We also investigated the explanatory power of whole groups of WFD System-A variables and RIVPACS-type model variables, and the explanatory power of individual variables. We found that variables used in the RIVPACS-type models were often better correlates of macroinvertebrate community variation than the WFD System-A variables. We conclude that this is primarily because while the latter use very broad categories of map-derived variables, the former are based on continuous variables selected for their ecological significance.

Introduction

The EU Water Framework Directive (Council of the European Communities, 2000), hereafter referred to as the WFD, requires Member States of the European Union (EU) to assess, monitor and, where necessary, improve the ecological quality status of its surface waters. This landmark piece of environmental legislation seeks to achieve at least 'good ecological status' for all surface water by 2015 and, for the first time, recognises the importance of the aquatic biota in determining the quality of fresh and marine waters (Sweeting, 2001; Logan & Furse, 2002). In placing the aquatic biota at the forefront of European environmental assessment the WFD recognises the importance of biogeographical drivers of species distribution patterns in setting targets for the biota (e.g., Illies, 1978). To achieve this, the WFD has established a hierarchical water body typology. Within any given part of this typology, it is assumed that biological communities at undisturbed sites will be broadly similar and will therefore constitute a type-specific biological target. The WFD typology (WFD, Annex II, Section 1) is organised by firstly placing surface water bodies into broad categories (rivers, lakes, transitional/coastal waters, artificial water bodies or heavily modified water bodies), and secondly, within these categories, by differentiating water bodies into types. This is achieved by using either a fixed typology, 'System-A', which in the case of rivers categorises sites based on ecoregion, altitude, catchment area and geology (Table 1), or an alternative typology, 'System-B', comprising a mixture of obligatory and optional factors.

In contrast to the *a priori* stream typologies set out in the WFD, RIVPACS-type predictive models are not based on predefined physical categories. Indeed, the site classification step within the development of a RIVPACS-type model makes no reference to physical variables (Clarke et al., 2003). Reference sites (sites considered to be of high ecological and physicochemical quality) that have been selected to encompass the full range of river types within a geographical area, are first classified into groups based solely on their macroinvertebrate fauna. Secondly, discriminant analysis is used to derive predictive equations that relate a range of recorded environmental variables to the biological classification. New sites are tested by applying the discriminant equations to the environmental variables recorded at the test sites. The macroinvertebrate communities expected to occur in the absence of environmental stress are then predicted (the expected fauna, E). The observed macroinvertebrate fauna (O) from test sites can then be compared to the expected fauna (E) by calculating observed/expected (O/E) ratios for a biotic index (e.g., O/E Number of Taxa). These are equivalent to the Ecological Quality Ratios (EQRs) described in the Water Framework Directive (WFD, Annex V, Section 1.4.1.ii). RIVPACS-type models therefore differ from *a priori* typologies in the use they make of physical data. Firstly, and perhaps most fundamentally, they use only biological data for site classification. Secondly they do not make *a priori* judgements about which variables are good correlates of community composition. And thirdly, RIVPACS-type models utilise multiple environmental variables (including local site characteristics such as substrate composition) to reveal correlations with macroinvertebrate communities rather than a restricted set of large-scale bio-geographical or physical factors.

The WFD requires EU Member States to establish 'type-specific biological reference conditions' for each water body type (WFD, Annex II, Section 1.1.iv), where reference condition equates to the definitions of high status in Annex V of the Directive. The choice of using either a System-A or System-B typology is left for individual Member States to decide. However, if choosing System-B, 'Member States must achieve at least the same degree of differentiation as would be achieved using System-A' (WFD, Annex II, Section 1.1.iv). In this paper we compare the relative effectiveness of the WFD System-A and RIVPACS-type multivariate models as alternative approaches for setting biological reference states. This comparison is made in three EU Member States, namely, Great Britain – using RIVPACS (River InVertebrate Prediction And Classification System), Sweden – using $SWEPAC_{SRI}$ (Swedish Prediction And Classification system for Stream Riffle Invertebrates), and the Czech Republic – using PERLA (after the Plecoptera genus).

Materials and methods

An assessment of biotic community variance within the WFD system-A typology and existing RIVPACS-type models was performed by assessing their ability to predict the observed values of biotic indices for the reference sites (Fig. 1) in all separate seasons and all combinations of seasons. We also sought to explore the relative explanatory power of the environmental variables (as correlates of community composition) in the WFD

Table 1. Distribution of RIVPACS-type model sites within WFD System-A stream types (as a percentage within each Country)

	Country	Great Britain RIVPACS III+			Sweden SWEPAC$_{SRI}$			Czech Republic PERLA											
	Ecoregion	18			14, 20, 22			9			10			11			14		
	# Sites in model	$n=614$			$n=389$			$n=238$			$n=33$			$n=8$			$n=21$		
Site altitude (m)	Catchment size (km^2)	*Geology*																	
		Siliceous	Calcareous	Organic	Siliceous	Calcareous	Organic	Siliceous	Calcareous	Organic	Siliceous	Calcareous	Organic	Siliceous	Calcareous	Organic	Siliceous	Calcareous	Organic
< 200	10–100	17.3	22.0	0.8		1.0	41.4	0.3											
	> 100–1000	14.5	26.2	1.8		0.5	6.0	0.3											
	> 1000–10,000	2.4	6.0					0.3											
	> 10 000							0.3						0.7					
200–800	10–100	2.6	3.9	1.0	11.1	5.1	24.9												
	> 100–1000	1.0	0.3	0.2	3.3	1.3	5.1	26.9	3.0		0.7	6.0			0.3		3.7	0.7	
	> 1000–10,000							20.5	1.3		0.3	2.3		1.0	0.3		1.7		
	> 10 000							20.9	0.3		1.0			0.3			0.7		
> 800	10–100					0.3		4.3	0.3		0.7						0.3		
	> 100–1000																		
	> 1000–10,000							0.3											
	> 10 000							0.3											

Ecoregion 9, Central highlands; 10, The Carpathians; 11, Hungarian lowlands; 14, Central plains; 18, Great Britain; 20, Borealic uplands and; 22, Fenno-Scandian shield.

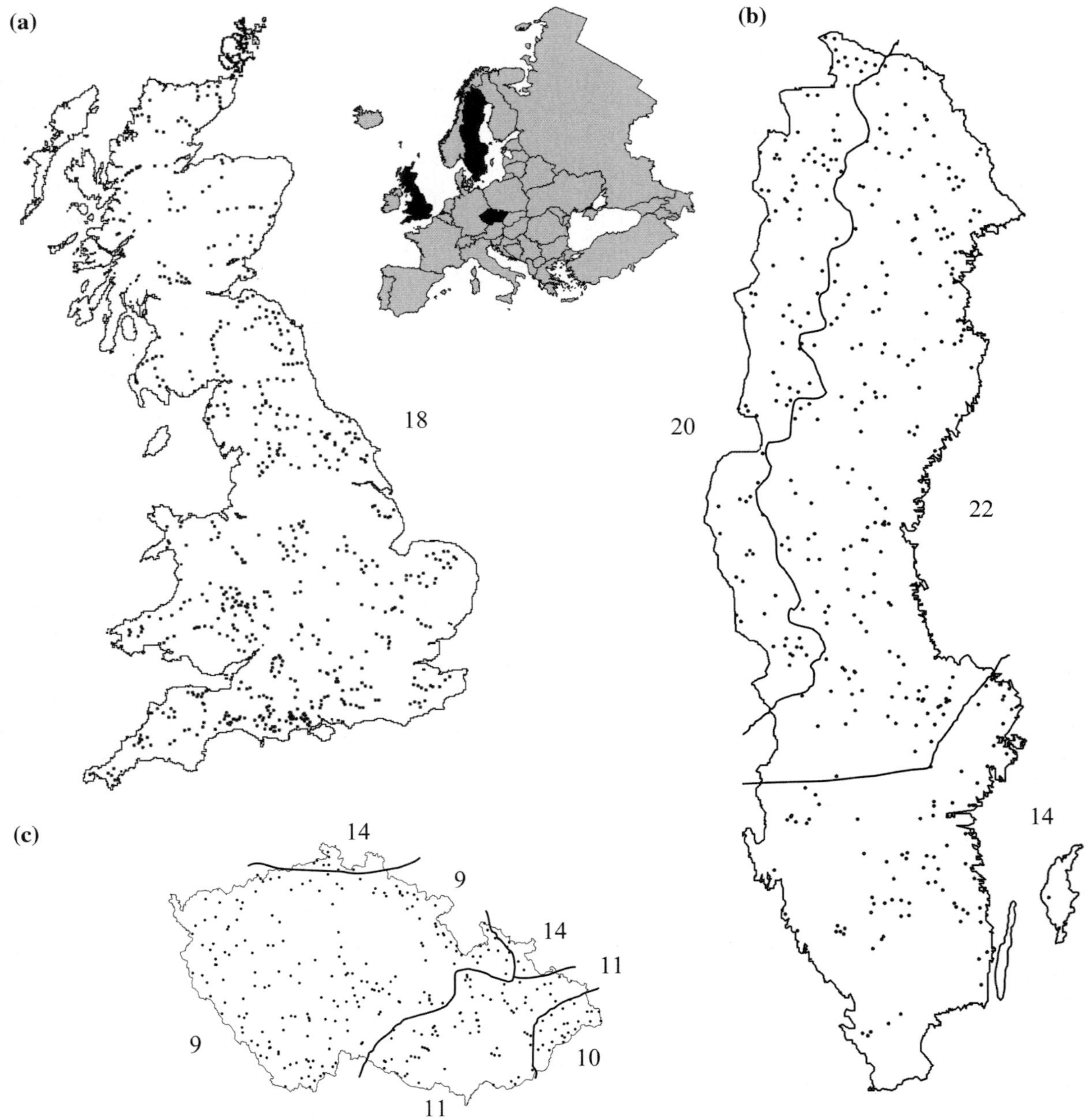

Figure 1. Geographical distribution of (a) the 614 RIVPACS III+ reference sites in Great Britain (ecoregion 18), (b) the 389 $SWEPAC_{SRI}$ reference sites in Sweden (ecoregion 14, Central plains; 20, Borealic uplands; 22, Fenno-Scandian shield), and (c) the 300 PERLA reference sites in the Czech Republic (ecoregion 9, Central highlands; 10, The Carpathians; 11, Hungarian lowlands; 14, Central plains).

System-A and RIVPACS-type models. Our analysis was done in four stages:

1. Calculation of expected biotic index values (WFD System-A typology)
2. Calculation of expected biotic index values (RIVPACS-type models)
3. Assessment of relative prediction accuracy
4. Analysis of correlates of community composition.

These stages are described below.

Expected biotic index values (WFD system-A typology)

The WFD system-A typology classifies rivers across Europe into 25 ecoregions, three altitude categories (<200 m, 200–800 m and >800 m), four catchment size categories (10–100 km^2, >100–1000 km^2, >1000–10,000 km^2 and >10,000 km^2) and three geology categories (siliceous, calcareous and organic). There are some uncertainties in the application of the System-A typology to streams and rivers. Firstly, the ecoregions set out in the WFD are defined at a very crude scale making the interpretation of local boundaries between ecoregions difficult. Secondly, it is unclear whether the geological class should be based on the geology underlying the biological sampling site itself or the geology of the upstream catchment, and thirdly there appears to be no legislative provision for streams with catchment areas less than 10 km^2; an important source of biodiversity (Furse, 2000).

In Great Britain all sites were assigned to the WFD ecoregion 18 'Great Britain.' WFD System-A altitude categories were taken as the altitude at each RIVPACS reference site. Under licence from the Centre for Ecology and Hydrology (CEH), the Environment Agency and the Scottish Environmental Protection Agency used the CEH Intelligent River Network (Dawson et al., 2002) to determine WFD catchment size categories. The agencies derived WFD geology categories by overlaying the RIVPACS reference sites on the British Geological Survey 1:625,000-scale solid geology GIS map, defining geology categories from the geology in the immediate vicinity of each site. The WFD geology classes where categorised as 'calcareous' where the bedrock was wholly or partially composed of calcium carbonate, 'siliceous' where the bedrock was acid igneous or there was other bedrock that did not contain calcium carbonate, or 'organic' where surface deposits were composed primarily of peat. In Sweden, the Swedish Environmental Protection Agency is responsible for implementing the WFD, although much of the work has been subcontracted to academics and consultants. A Nordic-funded project is also responsible for harmonising work among the Nordic countries (Johnson et al., 2001). A System-A typology-based classification is presently being used, consisting of ecoregion delineation, altitude, catchment size and geology classification. Following the recommendations of the WFD, Illies ecoregions (Illies, 1978) are used (regions 14, 20 and 22) and site altitude and catchment size are classified into three and four WFD classes, respectively. In Sweden, geological categories are defined as siliceous (alkalinity <0.2 meq/l), calcareous (alkalinity 0.2–1.0 meq/l) and organic (absorbance 420 nm of filtered water $>0.06 \approx 30$ mgPt/l). In the Czech Republic, individual PERLA sites were categorised as either calcareous or siliceous depending upon the calcareous/siliceous category of the overall water body (designated by the Czech Geological Institute, where calcareous rock types were defined as those with an equivalent content of alkaline elements (Na, K, Mg, Ca) >5.2, and water bodies with >40% calcareous rock types in their basin were categorised as calcareous).

Expected biotic index values based on the WFD System-A typology was derived by averaging index values for all references sites in each WFD System-A group. Using this method, the biotic indices Number of Taxa (number of BMWP scoring families) and ASPT (average score per taxon of BMWP families) (National Water Council, 1981) were derived for all reference sites. Number of Taxa and ASPT were chosen because while these are the indices currently used to report the biological quality of streams and rivers in Great Britain, they can be easily applied to all three countries. Additionally, in Great Britain, the Lotic Invertebrate Index for Flow Evaluation (LIFE), an index describing macroinvertebrate sensitivity to low flow (Extence et al., 1999), was predicted for spring summer and autumn samples, and in the Czech Republic, a Saprobic index was predicted (based on an index originally developed by Pantle & Buck (1955)) with the addition of taxon weightings (Marvan, 1969) and using standard Saprobic values and weights (CSN 75 7716, 1998). These indices were included to extend the generality of our analyses beyond the indices typically used when comparing the relative effectiveness of different models. Indices were calculated separately for spring, summer and autumn samples and (where models permitted) for all seasonal combinations of samples (spring and summer;

spring and autumn; summer and autumn; and spring, summer and autumn combined).

Expected biotic index values (RIVPACS-type models)

The current version of RIVPACS (RIVPACS III+) is based 614 reference sites. For a full review of RIVPACS see (Wright et al., 1984; Moss et al., 1987; Wright, 2000; Clarke et al., 2003). RIVPACS predicted biotic index values were obtained by running RIVPACS III+ on the 614-reference site dataset (generating predicted faunal lists and biotic index values for each reference site). Number of Taxa and ASPT were predicted for spring, summer and autumn and for all season combinations (spring and summer; spring and autumn; summer and autumn; and spring, summer and autumn combined). The LIFE index was predicted for spring, summer and autumn samples.

$SWEPAC_{SRI}$ models (Johnson, unpublished) were calibrated following the procedure outlined in Johnson & Sandin (2001), but combining data from all three ecoregions (14 – Central Plains; 20 – Borealic Uplands and 22 – Fenno-Scandian Shield). Data consisted of benthic macroinvertebrate samples collected in the year 2000 national stream survey (Wilander et al., 2003). Sites deemed to be affected by liming, agriculture (> 10% agricultural land use and TP > 8 or 10 μg/l – compensated for humic P) or acidification (pH ≤ 6.0 – compensated for natural acidity) were removed from the dataset resulting in 389 'unperturbed' sites distributed across the country. Predicted faunal lists and biotic indices were calculated using the combined $SWEPAC_{SRI}$ model. $SWEPAC_{SRI}$ predicted Number of Taxa and ASPT values were calculated for autumn samples ($SWEPAC_{SRI}$ is currently based on autumn samples).

The PERLA predictive system (Kokes et al., 2006) is based on 300 reference sites throughout the Czech Republic and is programmed into the HOBENT software package. Chironomidae identifications were not available for all PERLA reference sites in summer and autumn. Chironomidae were therefore excluded from the PERLA predictive model in summer and autumn, although they were included in the spring model used in our analyses. Number of Taxa, ASPT and the Saprobic index were predicted for the 238 reference sites in Ecoregion 9 (Central highlands). These were predicted separately for spring, summer and autumn samples.

Assessment of relative prediction accuracy

The effectiveness of the WFD System-A typology was then compared directly with the RIVPACS-type model predictions of expected index values. Prediction accuracy was measured as the standard deviation (SD) of the ratios of the observed (O) to expected (E) values of each biotic index for the reference sites. Van Sickle et al. (2005) introduced the idea of a null model in which the predicted reference condition index value for a site is the average observed value of the index for all reference sites. The SD of O/E, hereafter denoted SD(O/E), based on such as null model provides a useful upper limit to the SD(O/E) based on any model (an effective model should achieve a lower SD(O/E) than that of the null model). The relative sizes of the SD(O/E) enable a comparison to be made of the relative effectiveness of the WFD System-A typology and the RIVPACS-type models in terms of their ability to predict the observed values of the biotic indices and hence define reference conditions.

Analysis of correlates of community composition

After assessing the relative performance of the WFD System-A and RIVPACS-type models, we also sought to identify which of the environmental variables (both collectively and individually) used in the WFD System-A and RIVPACS-type models were the best correlates of macroinvertebrate community composition across the reference sites. This was investigated with canonical correspondence analysis (CCA) using the CANOCO 4.5 software package (ter Braak & Smilauer, 2002). The biological data sets used in our analyses were prepared in the same way as those used in the biological classifications underpinning each of the RIVPACS-type models. For RIVPACS III+, spring, summer and autumn data were combined into a 3 seasons combined dataset where family ($\log_{10}$ categories) and species (presence/absence) records included any of the species and families occurring in any separate season. Family $\log_{10}$ abundances were taken as the maximum $\log_{10}$

abundance (except where all three $\log_{10}$ abundances were the same, in which case the $\log_{10}$ abundance was increased by 1 category). The $SWEPAC_{SRI}$ biological classification was based on autumn sample, species level, presence/absence data while that for PERLA was based spring sample, species level, abundance data. Preliminary detrended correspondence analyses (DCA) revealed that the rate of turnover of macroinvertebrate taxa across the sites on the first axis of variation was >3 (DCA axis 1 length 3.01, 3.16 and 4.47 for RIVPACS, $SWEPAC_{SRI}$ and PERLA, respectively). The unimodal model within CCA was therefore considered to be appropriate for use with these datasets (ter Braak & Prentice, 1988). Within each country we determined the proportion of variation in the biotic data that could be accounted for by:

(1) The WFD System-A environmental variables as a whole
(2) The RIVPACS-type model environmental variables as a whole
(3) Both sets of variables combined.

This was calculated as the sum of all conditional effect eigenvalues (the collective contribution of the first variable and each successive variable to a forward selection model) as a proportion of total inertia (the total extent of variation in the macroinvertebrate communities across the reference sites). Three further CCAs (using both sets of variables combined) were used to determine the individual explanatory power of each environmental variable. These were the marginal effect eigenvalues, i.e., the variance explained when a particular variable is the only explanatory variable (ter Braak & Smilauer, 2002).

Results

In all three countries the reference sites were unevenly distributed throughout the WFD System-A stream types (Table 1). In Great Britain 91% of the reference sites were below 200 m altitude and 92% had catchment areas of 1000 km^2 or less. In terms of geology, 58% were calcareous and 38% were siliceous while only 4% were organic. In Sweden the overall percentage of geologically organic sites was much higher (77%) and calcareous sites were much less common (8%). The Swedish reference sites were distributed evenly between WFD altitude categories <200 m (49%) and 200–800 m (51%) and mainly drained 10–100 km^2 catchments (83%), with no sites from catchments above 1000 km^2. In the Czech Republic the reference sites were distributed across four ecoregions with 79% in ecoregion 9 (the central highlands) and only 21% in the other three ecoregions combined. While the Czech reference sites were distributed over a fairly broad range of catchment area categories (only catchment area 10–100 km^2 having a low number of sites), the Czech sites had altitudes predominantly between 200 and 800 m (92%), and were mainly siliceous (85%). In general, the matrix of possible WFD stream types (Table 1) contained many blank cells where combinations of ecoregion, altitude, catchment size and geology were not represented in each country. Where WFD System-A stream types were represented, the proportion of sites in each type was often imbalanced towards a few predominant types.

For all countries (Great Britain, Sweden and the Czech Republic), all indices (Number of Taxa, ASPT, LIFE and Saprobic) and for all seasons and season combinations, the SD(O/E) ratios were consistently highest for the null models, indicating relatively high uncertainty, and consistently lowest for the RIVPACS-type models, indicating relatively lower uncertainty (Table 2). In Great Britain, SD(O/E) ratios were lowest in the combined season RIVPACS models compared to the separate season models. For all three-model types (null, WFD System-A and RIVPACS-type models) the SD(O/E) ratios were lower for ASPT than for Number of Taxa (Fig. 2). In Great Britain, the percent reductions in SD(O/E) for the LIFE index were also greater than (or in one case equal to) those for ASPT. In Sweden there was a greater reduction in SD(O/E) for ASPT than for Number of Taxa. In the Czech Republic, while the percentage reductions in SD(O/E) were always greatest for PERLA, the reduction in ASPT was greatest in the spring (22%), but not so great in summer (4%) or autumn (7%). This could be because Chironomidae, which make up 108 of the 564 taxa in the spring dataset, were included in the spring model but excluded from the summer and autumn models. Although all the species in the

Table 2. Standard deviations of the observed/expected ratios of biotic indices for RIVPACS-type model reference sites based on null models, WFD System-A models and the RIVPACS-type models used in Great Britain, Sweden and the Czech Republic (% reduction in SD(O/E) compared to null models in parentheses)

Country ecoregion index	Season	Prediction method		
		Null model	WFD System-A	RIVPACS type model
Great Britain				
Ecoregion 18 (Great Britain)				RIVPACS III +
No. of Taxa	Spring	0.244	0.231 (5)	0.198 (19)
	Summer	0.255	0.243 (5)	0.196 (23)
	Autumn	0.263	0.251 (5)	0.218 (17)
	Spr + Sum	0.205	0.193 (6)	0.156 (24)
	Spr + Aut	0.206	0.193 (6)	0.156 (24)
	Sum + Aut	0.211	0.200 (5)	0.161 (24)
	Spr + Sum + Aut	0.188	0.176 (6)	0.138 (27)
ASPT	Spring	0.125	0.107 (14)	0.075 (40)
	Summer	0.122	0.109 (11)	0.081 (34)
	Autumn	0.132	0.113 (14)	0.086 (35)
	Spr + Sum	0.109	0.094 (14)	0.062 (43)
	Spr + Aut	0.109	0.096 (12)	0.064 (41)
	Sum + Aut	0.112	0.096 (14)	0.067 (40)
	Spr + Sum + Aut	0.103	0.088 (15)	0.057 (45)
LIFE	Spring	0.081	0.072 (11)	0.048 (41)
	Summer	0.091	0.085 (7)	0.059 (35)
	Autumn	0.085	0.077 (9)	0.055 (35)
Sweden				
Whole-country model				$SWEPAC_{SRI}$
No. of Taxa	Autumn	0.355	0.345 (3)	0.312 (12)
ASPT	Autumn	0.139	0.122 (12)	0.102 (27)
Czech Republic				
Ecoregion 9 (Central highlands)				PERLA
No. of Taxa	Spring	0.337	0.295 (12)	0.221 (34)
	Summer	0.335	0.289 (14)	0.265 (21)
	Autumn	0.383	0.317 (17)	0.276 (28)
ASPT	Spring	0.096	0.093 (3)	0.075 (22)
	Summer	0.081	0.079 (2)	0.078 (4)
	Autumn	0.091	0.089 (2)	0.085 (7)
Czech Saprobic	Spring	0.407	0.336 (17)	0.225 (45)
	Summer	0.461	0.345 (25)	0.266 (42)
	Autumn	0.452	0.382 (15)	0.279 (38)

family Chironomidae have the same BMWP score (2), the Chironomidae species differ in their sensitivity to organic pollution (Armitage & Blackburn, 1985), and hence models including the Chironomidae species in their biological classification may be better able to distinguish sites that differ in their natural nutrient levels.

While the environmental variables used by the three RIVPACS-type predictive models differ (Table 3), several variables (or types of variables), either in their $\log_{10}$ transformed or untransformed form, are used in either two or all three of the models (latitude, longitude, altitude, air temperature, distance from source, slope, depth, width,

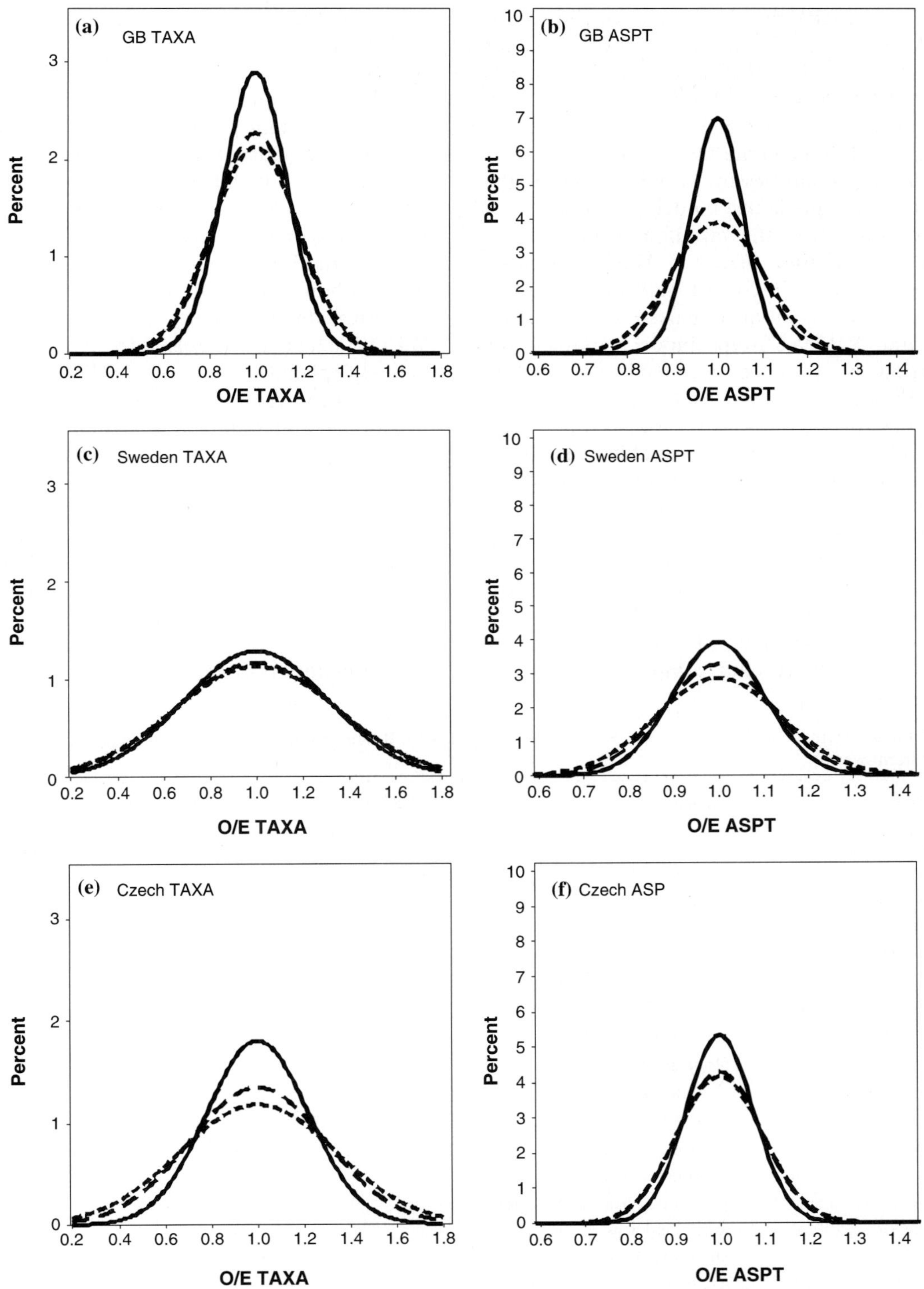

Figure 2. Observed/expected ratios for Number of Taxa (a, c, e) and ASPT (b, d, f) based on the null model (·····), WFD System-A typology (– – – –) and RIVPACS-type models (——). O/E Ratios based on spring summer and autumn data combined data for Great Britain, autumn data for Sweden and spring data for the Czech Republic.

substratum, discharge/velocity and alkalinity). In RIVPACS (and in the case of $\bar{x}$ catchment air temperature in $\text{SWEPAC}_{\text{SRI}}$) several variables were used in their $\log_{10}$ form as this has been found to improve their strength as correlates of macroinvertebrate community composition.

In all three countries the variables used by the RIVPACS-type predictive models could explain a larger proportion of the variation in macroinvertebrate communities than the WFD System-A variables (Table 4). In Great Britain and the Czech Republic, in the combined analyses of WFD System-A and RIVPACS-type model variables, the proportion of variance explained was slightly higher than that achieved by the RIVPACS-type model variables alone. This suggests that while the RIVPACS-type model variables are more effective as environmental predictors, the WFD System-A variables may be contributing a small amount of unique explanatory power not already encapsulated within the variables used by these RIVPACS-type models. In Sweden, the combined analyses of WFD System-A and RIVPACS-type model variables explained 14.2% of the total variation in community data compared to 8.0 and 9.9% for the WFD System-A and $\text{SWEPAC}_{\text{SRI}}$ model variables, respectively. This suggests that the System-A variables and the $\text{SWEPAC}_{\text{SRI}}$ variables are somewhat more distinct in the aspects of community variation they describe.

The individual explanatory power of each environmental variable (from a pool of RIVPACS-type model variables and WFD System-A variables within a given country) is shown in Table 5. In Great Britain, the RIVPACS variables were without exception better correlates of macroinvertebrate community composition than the WFD System-A variables. In several cases the greater explanatory power of a RIVPACS variable versus its equivalent WFD System-A variable was probably because that variable (or a closely related variable) was used in continuous rather than categorical form (e.g., the RIVPACS continuous variable $\log_{10}$ altitude versus WFD System-A altitude category, and RIVPACS $\log_{10}$ distance from source (which is highly correlated with catchment size, r_s 0.806, $p < 0.001$) versus WFD System-A catchment size category).

In Sweden the System-A variable ecoregion (which divides northern Sweden into the Borealic

Table 3. Environmental variables used in RIVPACS-type models in Great Britain, Sweden and the Czech Republic (•, untransformed; □, $\log_{10}$ transformed)

Variable type	Environmental variable	PERLA Czech Republic	RIVPACS III+ Great Britain	$\text{SWEPAC}_{\text{SRI}}$ Sweden
Geographical	Latitude	•	•	
	Longitude	•	•	
Altitude	Altitude	•	□	
Meteorological	$\bar{x}$ Air temperature		•	
	$\bar{x}$ Catchment air temperature			□
Catchment dimensions	Distance from source	•	□	
	Catchment area size			•
Gradient	Slope at site	•	□	
Channel dimensions	$\bar{x}$ Water depth	•	□	
	$\bar{x}$ Water width	•	□	•
Substratum	$\bar{x}$ Substratum composition	•	•	
	% Fine sediment			•
	% Floating-leaved vegetation			•
Hydrological	River discharge category		•	
	Stream velocity			•
Alkalinity	Alkalinity		•	•
	Alkalinity		□	
Catchment vegetation	% Forest in catchment			•

Table 4. Percentage of the total variation across macroinvertebrate communities explained collectively (conditional effects) by the WFD System-A variables, the RIVPACS-type model variables, and both groups of variables combined

	WFD System-A variables	RIVPACS-type model variables	WFD System-A and RIVPACS-type model variables combined
Great Britain	7.2	18.9	20.0
Sweden	8.0	9.9	14.2
Czech Republic	4.3	11.3	12.6

Uplands (20) in the west and the Fenno-Scandian Shield (22) in the east, and separates southern Sweden as the Central Plains (14) – Fig. 1) was a particularly strong correlate of macroinvertebrate community variation. While our analyses of reference sites in Great Britain and the Czech Republic did not traverse ecoregion boundaries, in Sweden these boundaries appear to have useful ecological meaning. The WFD System-A variable altitude was also a good descriptor. However, unlike the RIVPACS-type models in Great Britain and the Czech Republic, the continuous variables latitude, longitude and altitude are not part of the combined $SWEPAC_{SRI}$ model. The high descriptive power of the categorised ecoregion and altitude variables suggest that ecoregion and altitude

Table 5. Percentage of the total variation across macroinvertebrate communities explained (% VE) when a particular variable is the only explanatory variable (marginal effect). The % VE for all WFD System-A and RIVPACS-type predictive model variables are presented

Great Britain – Ecoregion 18	% VE	Sweden – Ecoregions 14, 20 & 22	% VE	Czech Republic – Ecoregion 9	% VE
Alkalinity	7.0	WFD Ecoregion	8.4	Distance from source	3.4
$\bar{x}$ Substratum composition	6.4	WFD Altitude category	3.9	WFD Catchment size category	3.2
Log_{10} alkalinity	5.9	Percent fine sediment	3.2	$\bar{x}$ Water width	2.8
Log_{10} Slope	5.9	WFD Geology – organic category	2.3	$\bar{x}$ Water depth	2.7
Longitude	5.4	$\bar{x}$ Water width	2.0	Slope	2.6
Log_{10} distance from source	4.3	Stream velocity	1.8	Altitude	2.1
Log_{10} Altitude	3.7	Percent floating-leaved vegetation	1.6	$\bar{x}$ Substratum composition	1.4
Log_{10} $\bar{x}$ water depth	3.7	WFD Geology – siliceous category	1.6	Longitude	1.4
Latitude	3.7	Alkalinity	1.4	Latitude	1.3
Log_{10} $\bar{x}$ water width	3.2	Percent forest in catchment	1.1	WFD Altitude category	0.6
River discharge (flow) category	3.2	Log_{10} $\bar{x}$ catchment air temperature	1.0	WFD Geology – calcareous category	0.4
$\bar{x}$ Air temperature	3.2	WFD Geology – calcareous category	0.7	WFD Geology – siliceous category	0.4
WFD Catchment size category	2.7	Catchment size	0.4	WFD Geology – organic category	0.0
WFD Geology – calcareous category	2.7	WFD Catchment size category	0.4		
WFD Geology - siliceous category	2.1				
WFD Altitude category	1.6				
WFD Geology – organic category	0.5				

as continuous variables (latitude, longitude and altitude) could be even stronger correlates. The way in which the boundary between ecoregions 20 and 22 also represents an altitude delineation (ecoregion 20 generally being at higher altitude than 22, and both being higher than 14) perhaps accounts for why both these variables are good predictors in Sweden. The percentage of fine sand was the third strongest variable. This is (to some extent) equivalent to $\bar{x}$ substratum size, the second strongest variable in Great Britain. However, in contrast to Great Britain and the Czech Republic, in Sweden the WFD System-A variable geology (organic category) was the fourth strongest variable. Sweden is also the country with the highest proportion of organic sites (77% versus 4% in Great Britain and 0% in the Czech Republic).

In the Czech Republic the WFD System-A variables were generally relatively weak descriptors of macroinvertebrate community variation, with the notable exception of the WFD System-A variable catchment size category. The high explanatory power of catchment size category is probably due to the relatively even spread of PERLA reference sites across the catchment size categories 100–1000 km^2, 1000–10,000 km^2 and >10,000 km^2 in comparison to the spread of sites across other WFD System-A variable categories. It is also interesting to note that distance from source (highly correlated with System-A catchment size category, r_s 0.875, $p < 0.001$) was the strongest explanatory variable. The continuous variable altitude was also a stronger descriptor than the WFD System-A variable altitude category.

While the RIVPACS-type model variables were collectively always better descriptors of community variation, the usefulness of individual variables differed between countries. In general the RIVPACS-type model variables were individually better descriptors, although some WFD System-A variables in particular countries were also relatively strong correlates of macroinvertebrate community variation. In all cases where continuous and categorical variables (or closely correlated variables such as catchment size and distance from source) were present in the CCA analyses within in country, the continuous variables were better descriptors of macroinvertebrate community variation.

Discussion

Attempts to use fixed *a priori* stream typologies (especially ecoregions) to define biotic communities have had mixed results. For example, Verdonschot & Nijboer (2004) in their analysis of 889 streams across eight European countries concluded that large-scale typological factors explained most of the variation in macroinvertebrate assemblages. Similarly, Rabeni & Doisy (2000) and Feminella (2000) found that benthic macroinvertebrate assemblages in Missouri and parts of the southeastern USA coincided well with existing ecoregions. However, Sandin & Johnson (2000) in their study of Swedish streams concluded that ecoregion classifications need to be augmented with other factors such as altitude, stream size and catchment characteristics to discriminate macroinvertebrate communities and Waite et al. (2000) found that there was a large variation in macroinvertebrate community composition across the Mid-Atlantic Highlands of the USA both within and between ecoregions and that ordination did not reveal a distinct clustering of sites by ecoregion. Further, Van Sickle and Hughes (2000), in their study in Western Oregon, USA argue that geographic partitions can be expected to account for only a minor proportion of the total variation seen in macroinvertebrate communities across a large region and based on their study of five *a priori* landscape classifications in several regions of the USA, Hawkins & Vinson (2000) conclude that benthic macroinvertebrates vary continuously along environmental gradients so that methods of bioassessment that seek to place sites into discrete categories are fundamentally limited compared to approaches that recognise biological continua. These studies indicate that while fixed typologies provide a useful large-scale framework for setting ecological targets, they do not necessarily account for all sources of observed biological variation. Ecological targets set solely in terms of fixed typologies may not be precise enough to accurately define target communities.

The hierarchical water body typology set out in the EU Water Framework Directive defines streams and rivers across Europe in terms of ecoregions (Illies, 1978) and broad categories of altitude, catchment area and geology. It is implicit within this typology that macroinvertebrate communities at undisturbed sites should be broadly similar and therefore predictable. Our results from three European countries suggest that while the WFD System-A typology is more effective than a null model as a means of predicting reference values for macroinvertebrate biotic indices, it is considerably less effective than the site-specific multivariate RIVPACS-type models already in place in Great Britain (RIVPACS), Sweden ($SWEPAC_{SRI}$) and the Czech Republic (PERLA). This is probably due to the inclusion of a wider range of continuous (rather than categorical) variables, which are both map derived and site/sampling date specific. While there are some exceptions in the case of individual variables (particularly in Sweden), as a group the RIVPACS-type model variables have greater ecological significance than the WFD System-A variables and therefore enable more effective predictive models to be built.

Of the environmental variables (or types of variables) used in either two or all three of the RIVPACS-type models developed in Europe, many of these are also used in RIVPACS-type models developed for other geographical areas. For example, many of the Australian AUSRIVAS models utilise latitude, longitude, alkalinity, altitude, distance from source, slope, width, substratum, and discharge (Simpson & Norris, 2000) and eight of the 10 variables used in the Canadian BEAST model are altitude, longitude, substratum composition (three separate measures), depth, velocity and alkalinity (Rosenberg et al., 2000). Similarly in predictive models developed in California (Hawkins et al., 2000), seven of the 11 variables included in a species level model (longitude, altitude, depth, latitude, distance from source, width and slope) and six of the nine variables in a family level model (depth, longitude, altitude, slope, distance from source and width) are the same or similar to the variables used in two or all three of the European RIVPACS-type models. The variables used in each of the models above have themselves been selected from longer lists of candidate variables, and it is interesting to consider how the same or similar variables tend to emerge as good predictors of macroinvertebrate communities within models developed to serve such geographically widespread areas.

Within Europe, while the variables used by the three RIVPACS-type models may be broadly similar, they are not exactly the same. It is the use of multiple variables and the selection of these variables for their ecological significance in a given region that is key to the success of these models. In contrast, the WFD System-A typology is constrained in its use of a limited number of categorical variables. While in some cases these variables are good predictors, in many cases they are relatively weak correlates of variation in macroinvertebrate community composition. The variables WFD Ecoregion and WFD Altitude (in Sweden) and WFD Catchment size category (in the Czech Republic, Ecoregion 9 – Central Highlands) are examples of WFD System-A variables identified in this study as strong correlates of macroinvertebrate community variation. However, the usefulness of the WFD System-A typology is limited to certain variable type and geographical area combinations. This highlights the problem of the Europe-wide application of the same category boundaries in the WFD System-A typology. While the 238 PERLA reference sites in Ecoregion 9 of the Czech Republic are quite evenly spread among the three largest WFD System-A catchment size categories, the RIVPACS and $SWEPAC_{SRI}$ reference sites are distributed almost exclusively across the two smallest catchment size categories. The Europe-wide application of the same intervals of catchment size category appears to be useful in the Czech Republic, but has little ecological significance in Great Britain or Sweden.

Another problem with the WFD System-A typology is the use of categorical rather than continuous variables. This is exemplified by the case of altitude. In Great Britain and the Czech Republic the continuous variables $\log_{10}$ altitude and altitude, respectively, are both better predictor variables than the WFD System-A variable altitude category. The loss of predictive power by summarising the continuous variable altitude into broad categories is considerable. The same problem almost certainly contributes to the low percentage variance explained by many of the other WFD System-A variables.

A further problem with *a priori* typological approaches such as the WFD System-A is that they usually utilise variables gathered solely at large geographical scales. In Europe this can be addressed by opting for a System-B approach (which can incorporate a range of additional variables gathered at a variety of scales). Our analysis shows that substratum composition, width and depth, all of which are local scale variables measured at the time of sampling, can also be strong correlates of macroinvertebrate community composition. The importance of both large-scale and local factors as determinants of macroinvertebrate communities should therefore not be overlooked. Similar conclusions were reached by Heino et al. (2003) based on a study of macroinvertebrate diversity in headwater streams.

The percentage reduction in SD(O/E) achieved by the three European RIVPACS-type models compared to null models (based on Number of Taxa at BMWP family level) varied between 12 and 34% depending on the season model used. These generally exceeded the percent reductions in SD(O/E) achieved by a predictive model built from 86 reference sites in the Mid-Atlantic Highlands region of the USA (Van Sickle et al., 2005) where SD(O/E), based on Number of Taxa at species and genus level, was reduced by 13.7%. Another predictive model built from 209 sites in North Carolina, USA (Van Sickle et al., 2005) again based on Number of Taxa at species and genus level, achieved an impressive reduction in SD(O/E) compared to a null model of 52.5%.

The SD(O/E) of a null model is equivalent to the coefficient of variation (cv) in the observed metric values of the reference sites and reflects the natural variability in the values of the metric within a region. For example, null model Number of Taxa SD(O/E) values in Great Britain are lower than those in Sweden and the Czech Republic indicating that Number of Taxa is inherently more variable in the latter two countries. Also, different metrics have different null model SD(O/E) values within countries. For example, null model ASPT SD(O/E) values in the Great Britain are lower than null model Number of Taxa SD(O/E) values. In contrast, in relation to a null model the percent reduction in SD(O/E) obtained by using a predictive model (of any type) indicates the predictive model's effectiveness. By making reference to null models, the statistic, percent reduction in SD(O/E), allows us to compare the performance of models both between indices and between regions. The null model approach proposed by Van Sickle et al. (2005) therefore appears to provide an objective test of model performance and as such is likely to be extremely useful.

Conclusion

This study has shown that the site-specific multivariate RIVPACS-type predictive models already in place in Great Britain (RIVPACS), Sweden ($SWEPAC_{SRI}$) and the Czech Republic (PERLA) are more effective than both null models and the WFD System-A physical typology as methods of predicting macroinvertebrate reference conditions. The multivariate models are more effective primarily because they make use of continuous rather than categorical predictor variables (that have been selected for their value as good correlates of macroinvertebrate community composition) and because the multivariate RIVPACS-type models are not constrained by the use of a limited number of variables.

Acknowledgements

We would like to acknowledge Robin Guthrie (Scottish Environment Protection Agency) and John Murray-Bligh (Environment Agency) for their help defining WFD System-A types for the RIVPACS reference sites in Great Britain.

References

Armitage, P. & J. Blackburn, 1985. Chironomidae in the Pennine stream system receiving mine drainage and organic enrichment. Hydrobiologia 121: 165–172.

Clarke, R. T., J. F. Wright & M. T. Furse, 2003. RIVPACS models for predicting the expected macroinvertebrate fauna and assessing the ecological quality of rivers. Ecological Modelling 160: 219–233.

Council of the European Communities., 2000. Directive 2000/60/EC, Establishing a Framework for Community Action in the Field of Water Policy. European Commission PE-CONS 3639/1/100 Rev 1, Luxembourg.

CSN 75 7716, 1998. Water quality, biological analysis, determination of saprobic index. Czech Technical State Standard. Czech Standards Institute, Prague, 174 pp.

Dawson, F. H., D. D. Hornby & J. H. Hilton, 2002. A method for the automated extraction of environmental variables to help the classification of rivers in Britain. Aquatic Conservation: Marine and Freshwater Ecosystems 12: 391–403.

Extence, C. A., D. M. Balbi & R. P. Chadd, 1999. River flow indexing using British benthic macroinvertebrates: a framework for setting hydroecological objectives. Regulated Rivers Research and Management 15: 543–574.

Feminella, J. W., 2000. Correspondence between stream macroinvertebrate assemblages and 4 ecoregions of the southeastern USA. Journal of the North American Benthological Society 19: 442–461.

Furse, M. T., 2000. The application of RIVPACS procedures in headwater streams – an extensive and important national resource. In Wright, J. F., D. W. Sutcliffe & M. T. Furse (eds), Assessing the Biological Quality of Fresh Waters. Freshwater Biological Association, Ambleside, 79–91.

Hawkins, C. P., R. H. Norris, J. N. Hogue & J. W. Feminella, 2000. Development and evaluation of predictive models for measuring the biological integrity of streams. Ecological Applications 10: 1456–1477.

Hawkins, C. P. & M. R. Vinson, 2000. Weak correspondence between landscape classifications and stream macroinvertebrate assemblages: implications for bioassessment. Journal of the North American Benthological Society 19: 501–517.

Heino, J., T. Muotka & R. Paavola, 2003. Determinants of macroinvertebrate diversity in headwater streams: regional and local influences. Journal of Animal Ecology 72: 425–434.

Illies, J., 1978. Limnofauna Europaea. Gustav Fisher Verlag, Stuttgart.

Johnson, R. K. & L. Sandin, 2001. Development of a Prediction and Classification System for Lake (Littoral) and Stream (Riffle) Macroinvertebrate Communities. Stencil. Department of Environmental Assessment, SLU, Uppsala.

Johnson, R. K., K. Aagaard, K. J. Aanes, N. Friberg, G. M. Gislason, H. Lax & L. Sandin, 2001. Macroinvertebrates. In Skriver, J. (ed), Biological Monitoring in Nordic Rivers and Lakes. TemaNord Environment 513: 43–52.

Kokeš, J., S. Zahrádková, D. Němejcová, J. Hodovský, J. Jarkovský & T. Soldán, 2006. The PERLA system in the Czech Republic: a multivariate approach for assessing the ecological status of running waters. Hydrobiologia 566: 343–354.

Logan, P. & M. Furse, 2002. Preparing for the European Water Framework Directive – making the links between habitat and aquatic biota. Aquatic Conservation: Marine and Freshwater Ecosystems 12: 425–437.

Marvan, P., 1969. Primechania k primeneniu statisticheskich metodov po opredeleniu saprobnosti. Symposium SEV Voprosy saprobnosti, Zivohost, 19–43.

Moss, D., M. T. Furse, J. F. Wright & P. D. Armitage, 1987. The prediction of the macro-invertebrate fauna of unpolluted running-water sites in Great Britain using environmental data. Freshwater Biology 17: 41–52.

National Water Council., 1981. River Quality: The 1980 Survey and Future Outlook. National Water Council, London.

Pantle, E. & H. Buck, 1955. Die biologische Uberwachung der Gewasser und die Darstellung der Ergebnisse. Gas und Wasserfach. 96: 604.

Rabeni, C. F. & K. E. Doisy, 2000. Correspondence of stream benthic invertebrate assemblages to regional classification schemes in Missouri. Journal of the North American Benthological Society 19: 419–428.

Rosenberg, D. M., T. B. Reynoldson & V. H. Resh, 2000. Establishing reference conditions in the Fraser River catchment, British Columbia, Canada, using the BEAST (BEnthic Assessment of SedimenT) predictive model. In Wright, J. F., D. W. Sutcliffe & M. T. Furse (eds), Assessing the Biological Quality of Fresh Waters. Freshwater Biological Association, Ambleside, 181–194.

Sandin, L. & R. K. Johnson, 2000. Ecoregions and benthic macroinvertebrate assemblages of Swedish streams. Journal of the North American Benthological Society 19: 462–474.

Simpson, J. C. & R. H. Norris, 2000. Biological assessment of river quality: development of AusRivAS models and outputs. In Wright, J. F., D. W. Sutcliffe & M. T. Furse (eds), Assessing the Biological Quality of Fresh Waters. Freshwater Biological Association, Ambleside, 125–142.

Sweeting, R., 2001. Classification of ecological status of lakes and rivers – biological elements in the classification. In Back, S. & K. Karttunnen (eds), Classification of Ecological Status of Lakes and Rivers. TemaNord Environment 2001:584, Nordic Council of Ministers, Copenhagen, 9.

ter Braak, C. J. F. & I. C. Prentice, 1988. A theory of gradient analysis. Advances in Ecological Research 18: 271–317.

ter Braak, C. J. F. & P. Smilauer, 2002. CANOCO Reference Manual and CanoDraw for User's Guide: Software for Canonical Community Ordination (version 4.5). Microcomputer Power (Ithaca NY, USA), 500 pp.

Van Sickle, J. & R. M. Hughes, 2000. Classification strengths of ecoregions, catchments, and geographical clusters for aquatic vertebrates in Oregon. Journal of the North American Benthological Society 19: 370–384.

Van Sickle, J., C. P. Hawkins, D. P. Larsen & A. H. Herlihy, 2005. A null model for the expected macroinvertebrate assemblage in streams. Journal of the North American Benthological Society 24: 178–191.

Verdonschot, P. F. M. & R. C. Nijboer, 2004. Testing the European stream typology of the Water Framework Directive for macroinvertebrates. Hydrobiologia 516: 35–54.

Waite, I. R., A. T. Herlihy, D. P. Larsen & D. J. Klemm, 2000. Comparing strengths of geographic and non-geographic classifications of stream benthic macroinvertebrates in the Mid-Atlantic Highlands, USA. Journal of the North American Benthological Society 19: 429–441.

Wilander, A., R. K. Johnson & W. Goedkoop, 2003. Riksinventering 2000: En synoptisk studie av vattenkemi och bottenfauna i svenksa sjöar och vattendrag. Department of Environmental Assessment, Swedish University of Agricultural Sciences, Uppsala, Report 2003, 1 pp.

Wright, J. F., D. Moss, P. D. Armitage & M. T. Furse, 1984. A preliminary classification of running water sites in Great Britain based on macro-invertebrate species and prediction of community type using environmental data. Freshwater Biology 14: 221–256.

Wright, J. F., 2000. An introduction to RIVPACS. In Wright, J. F., D. W. Sutcliffe & M. T. Furse (eds), Assessing the Biological Quality of Fresh Waters. Freshwater Biological Association, Ambleside, 1–24.

Linking Organism Groups

Hydrobiologia (2006) 566:109–113

M.T. Furse, D. Hering, K. Brabec, A. Buffagni, L. Sandin & P.F.M. Verdonschot (eds), The Ecological Status of European Rivers: Evaluation and Intercalibration of Assessment Methods
DOI 10.1007/s10750-006-0098-z

Linking organism groups – major results and conclusions from the STAR project

Daniel Hering[1,*], Richard K. Johnson[2] & Andrea Buffagni[3]
[1]*Department of Hydrobiology, University of Duisburg-Essen, D-45117 Essen, Germany*
[2]*Department of Environmental Assessment, Swedish University of Agricultural Sciences, 7050, SE-750 07 Uppsala, Sweden*
[3]*CNR – IRSA, Water Research Institute, Via della Mornera, 25, 20047, Brugherio (MI), Italy*
*(*Author for correspondence: E-mail: daniel.hering@uni-essen.de)*

Key words: fish, benthic invertebrates, macrophytes, diatoms, assessment, rivers, Europe

Abstract

Here we summarize results of the EU funded research project STAR concerning the suitability of different organism groups (fish, benthic invertebrates, macrophytes, diatoms) for monitoring European rivers. In a general way, the suitability of the organism groups is classified by monitoring type, stress type, river type, temporal scale and taxonomic resolution. For example, although all organism groups are affected by acidification, the relatively low species richness of fish and macrophytes in small mountain streams makes these two groups less suitable, and, hence, we argue that benthic diatoms and/or invertebrates may be considered as more robust indicators. Similar, lines of reasoning are given for a number of stressor and stream types.

Introduction

Biological response variables are often selected over physical–chemical variables because they represent valued ecosystem attributes such as diversity or productivity. The use of complementary indicators, as stipulated by the European Water Framework Directive, is based on the premise that using multiple organism groups/assemblages can help to distinguish the effects of human-induced stress more efficiently (with less uncertainty) and more effectively (by detecting the effects of multiple stressors). A number of factors lend support to this conjecture. For example, different organism groups (or assemblages) supposedly respond differently to stress depending on inherent life history attributes: 1. *Physiological* constraints; e.g., (i) Complex, multicellular, organisms such as fish may be better indicators of changes in ambient temperature than single-celled organisms like algae. (ii) Organisms with short generation times, from weeks to months (e.g., algae and invertebrates), may respond more rapidly to environmental changes than organisms with relatively long generation times, from months to years (e.g., fish and macrophytes). 2. *Behavioural* constraints; e.g., (i) Organisms that are acquire nutrients directly from their surroundings (e.g., algae) may be better indicators of nutrient enrichment, in systems where nutrients are a limiting, than organisms (e.g., fish) that acquire their nutrients 'indirectly' (e.g., through a benthic pathway such as nutrients – diatoms – invertebrates – fish). (ii) Relatively large and mobile organisms that use a wide range of habitats [e.g., fish habitats range from small ($<m^2$) to large ($>km^2$)], may be more influenced by factors acting on large spatial scales (e.g., reach and catchment-level variables), than relatively small and sessile organisms (e.g., benthic algae or invertebrates) that are probably more influenced by their immediate surroundings or microhabitat quality. Hence, in theory differences among organism groups can be used to select complementary indicators.

To tests these conjectures, comparative investigations on the response of different organism groups to stress are needed, which have, for example, been performed in the EU funded project STAR and are described in detail by Johnson et al. (2006a, b), Springe et al. (2006) and in addition by Hering et al. (in press) and Johnson et al. (in press). With this paper, we try to transform these results into a more general guidance, with focus on which organism group/groups can be used in biomonitoring.

Parameters for indicator selection

In designing biomonitoring programs, consideration should be given to the river type being addressed, the type of stress(ors) potentially affecting the integrity of the river ecosystem, and the time frame of the study, including knowledge of interannual variability and potential lag-phase responses of degradation and recovery (e.g., Stevenson et al., 2004). By combining conceptual models (expert opinion) and empirical data, more cost-effective monitoring programs, incorporating knowledge of how different organism groups react to different human-generated stressors, can be designed (e.g., USEPA, 2000). For example, since the response of the four organism groups addressed by the Water Framework Directive for river monitoring (fish, benthic invertebrates, benthic diatoms and macrophytes) are often correlated, Hering et al. (in press) and Johnson et al. (in press) argue that it is not necessary to monitor all groups simultaneously.

Based on the STAR results we suggest which organism group(s) is/are most suited for monitoring different types of human-induced stress (Table 1). This is a general guidance and the response of specific metrics should always be taken into consideration.

Type of monitoring

For monitoring European rivers formal criteria for indicator selection are set by the Water Framework Directive. *Surveillance monitoring for the Water Framework Directive* requires the use of all organism groups (fish, benthic invertebrates, benthic diatoms, macrophytes). However, all four organism groups are not necessarily required for covering different types of stressors and different scales; thus, *surveillance monitoring for purposes other than the Water Framework Directive* should mainly ensure that all relevant stressors potentially affecting the monitored rivers and the relevant spatial and temporal scales are covered. This can either be achieved by monitoring benthic invertebrates, which respond to many stressors, or by the combination of diatoms (mainly reacting on eutrophication and land use pressures) and fish (mainly responding to large scale hydromorphological degradation) (Hering et al., in press; Johnson et al., in press). For *operational monitoring* indicators for assessing the main stressor affecting the integrity of the river being monitored should be selected (see below). For *assessing the success of restoration measures*, an indicator group mainly addressing the stress type which effect is restored should be selected. Early warning indicators should be used with caution, since their signal may be subject to high natural variability.

River type

In *small mountain streams in Central and Northern Europe* benthic diatoms and invertebrates are the most diverse organism groups and, consequently, most suited for monitoring. Fish assemblages are usually species-poor and, with the exception of down-stream weir effects, this organism group is not recommended for monitoring many stressors. Further, macrophytes are often patchily distributed and, thus, less suited for monitoring purposes. In *medium-sized mountain streams in Central and Northern Europe*, in *small and medium-sized lowland streams in Central and Northern Europe* and in *large rivers in Central and Northern Europe* all organism groups are, in principal, suited for monitoring. The selection of indicator(s) depends on the stressor-type being assessed and the monitoring type. Due to poor taxonomical knowledge, benthic invertebrates are less suited for monitoring the effects of hydromorphological degradation in southern European rivers. For the effects of land use, eutrophication and other anthropogenic effects all organism groups (fish, benthic invertebrates, benthic diatoms, macrophytes) can be used.

Table 1. Recommended (combination of) indicator(s) for biomonitoring of European rivers

Monitoring type	River types	Diatoms	Macrophytes	Invertebrates	Fish	Comments
Surveillance monitoring WFD	All	+	+	+	+	
Surveillance monitoring other purposes	All			+		cost-effective option
Surveillance monitoring other purposes	All, except small mountain streams	+			+	
Surveillance monitoring other purposes	Small mountain streams	+		+		
Operational monitoring, eutrophication/pollution	Small mountain streams	+				cost-effective option if no other stressors are present
Operational monitoring, eutrophication/pollution	All, except small mountain streams	+	+	+	+	
Operational monitoring, hydromorphological degradation	Small mountain streams			+		
Operational monitoring, hydromorphological degradation	All, except small mountain streams		+	+	+	
Operational monitoring, catchment land use effects	Small mountain streams	+		+		
Operational monitoring, catchment land use effects	All, except small mountain streams	+		+	+	
Operational monitoring, acidification	Small mountain streams	+		+		
Operational monitoring, different stressors	Small mountain streams	+		+		
Operational monitoring, different stressors	Medium-sized mountain streams, lowland streams	+			+	
Operational monitoring, different stressors	Medium-sized mountain streams, lowland streams		+		+	
Operational monitoring, different stressors	Medium-sized mountain streams, lowland streams	+		+		

Type of anthropogenic stress

Although the effects of *eutrophication* (nutrient enrichment) and *organic pollution* (e.g., increased BOD) are of different origin, they are often correlated and, thus, similar indicators can be used in most cases to detect both types of stressors. All organism groups (fish, benthic invertebrates, benthic diatoms, macrophytes) respond to eutrophication/organic pollution and are thus, in principal, suited as indicators (Hering et al., submitted; Johnson et al., manuscript). However, the rates and trajectories of change may vary among the organism groups. For example, benthic diatoms often show a stronger response (high sensitivity) compared to the other three organism groups (Johnson et al., 2006a). Hence, benthic diatoms may be best suited for situations in which only pollution/eutrophication is assessed. If multiple stressors are being assessed then benthic invertebrates and/or macrophytes should be considered, which also respond to other stress types (Hering et al., submitted; Johnson et al., 2006a). If the focus of the study is on nutrient enrichment, benthic diatoms and/or macrophytes should be considered, since nutrient enrichment may be the main factor directly affecting both groups. If the focus of the study is on organic pollution, benthic invertebrates and/or fish should be considered, since these groups are more directly affected by oxygen condition (Figure 1).

Fish, benthic invertebrates and macrophytes respond to a varying degree to *hydromorphological degradation* (Hering et al., submitted; Johnson et al., manuscript; Johnson et al., 2006a). The selection of the most appropriate organism group for monitoring this stressor is dependent on stream type: In lowland streams and in medium-sized to large rivers all three groups can be considered. The relatively species-poor fish and macrophyte assemblages in small streams may limit the use of these two organism groups, and hence benthic invertebrates should be considered for monitoring the effects of hydromorphological degradation on the reach scale. For hydromorphological effects on smaller spatial scales (microhabitat scales) benthic invertebrates should be considered.

Land-use affects river communities by altering, for example, nutrient levels (eutrophication), habitat quality (sedimentation) and toxicity (e.g., pesticides). These effects are most strongly reflected by fish, benthic invertebrates and benthic diatoms (Hering et al., submitted). This contradicts to some degree with the results of Springe et al. (2006), who found macrophytes and fish being more suitable for assessing ecological quality at the river basin scale, whereas metrics of macroinvertebrates and benthic diatoms were more appropriate at smaller scales.

All organism groups (fish, benthic invertebrates, benthic diatoms, macrophytes) are affected by *acidification* (Stokes et al., 1989; Brodin, 1995).

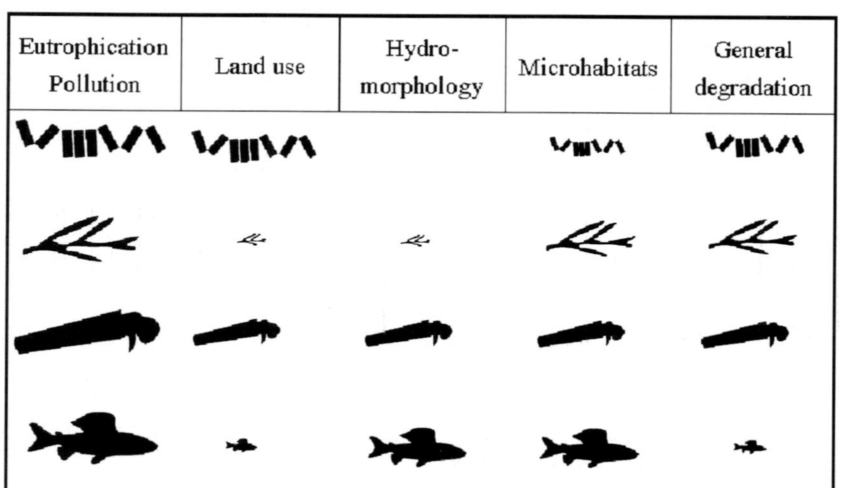

Figure 1. Response of diatoms (first line), macrophytes (second line), benthic invertebrates (third line) and fish (fourth line) to different stress types. The size of the symbols reflects the strength of the response (based on unpublished data and Johnson et al. (manuscript)).

The most profound effects are found, however, in small mountain streams with low buffering capacity. The relatively species-poor fish and macrophyte assemblages in small streams may limit the use of these two organism groups, and hence benthic diatoms and/or benthic invertebrates should be considered for monitoring the effects of acidification stress.

In cases of *different stressors* affecting a river or of *unknown stress type(s)* no general guidance is possible. If only one organism group can be investigated, then benthic invertebrates should be considered since they respond to most stressor types in all river types. If multiple organism groups can be monitored, the following alternatives may be useful: (1) benthic diatoms (for eutrophication and acidification effects) and benthic invertebrates (for various stressors) in small mountain streams; (2) benthic diatoms or macrophytes (for eutrophication and land use effects) and benthic invertebrates or fish (for hydromorphological and land use effects) in medium-sized mountain streams and lowland streams.

Temporal scale

Diatoms, with relatively short generation times, are often considered as early warning indicators, detecting short-term pollution events. Fish, with their relatively long generation times, might be considered for monitoring long-term changes (late-warning indicators). Benthic invertebrates, a taxonomically diverse organism group, have generation times ranging from weeks to years and hence may be considered as both early- and late-warning indicators. In contrast to these considerations, Johnson et al. (2006b) found macrophytes and fish to be superior to diatoms as early warning indicators in European mountain streams.

Taxonomic resolution

At present, most fish-, diatom-, and macrophyte-metrics commonly used in biomonitoring require species-level data. Similarly, for assessing the effects of hydromorphological degradation and land use using benthic invertebrates most metrics require species-level taxonomic resolution. If only family-data are available, invertebrates can only be used for assessing the effects of general degradation.

Acknowledgements

STAR was funded by the European Commission, 5th Framework Program, Energy, Environment and Sustainable Development, Key Action Water, Contract no. EVK1-CT-2001-00089.

References

Brodin, Y. W., 1995. Acidification of Swedish freshwaters. In Henrikson, L. & Y. W. Brodin (eds), Liming of Acidified Surface Waters. Springer-Verlag, Berlin, 63–76.

Johnson, R. K., D. Hering, M. T. Furse & R. T. Clarke, 2006a. Detection of ecological change using multiple organism groups: metrics and uncertainty. Hydrobiologia 566: 115–137.

Johnson, R. K., D. Hering, M. T. Furse & P. F. M. Verdonschot, 2006b. Indicators of ecological change: comparison of the early response of four organism groups to stress gradients. Hydrobiologia 566: 139–152.

Johnson, R. K. Hering, D. & M. T. Furse. Comparing the response of fish, macroinvertebrate, macrophyte and diatom assemblages in European streams to human-generated stress. Manuscript.

Stokes, P. M., E. T. Howell & G. Krantzberg, 1989. Effects of acidic precipitation on the biota of freshwater lakes. In Adriano, D. C. & A. H. Johnson (eds), Acidic Precipitation, 2: Biological and Ecological Effects. Springer-Verlag, New York: 273–304.

Stevenson, R. J., R. C. Bailey, M. C. Harrass, C. P. Hawkins, J. Alba-Tercedor, C. Couch, S. Dyer, F. A. Fulk, J. M. Harrington, C. T. Hunsaker & R. K. Johnson, 2004. Designing data collection for ecological assessments. In Barbour, M. T., S. B. Norton, H. R. Preston & K. W. Thornton (eds), Ecological Assessment of Aquatic Resources: Linking science to decision making. SETAC, Pensacola, Florida, USA, 55–84.

Springe, G., L. Sandin, A. Briede & A. Skuja, 2006. Biological quality metrics: their variability and appropriate scale for assessing streams. Hydrobiologia 566: 153–172.

USEPA, 2000. Stressor Identification Guidance Document. U.S. Environmental Protection Agency, Washington, USA 228 pp, EPA-822-B-00-025.

Hydrobiologia (2006) 566:115–137

M.T. Furse, D. Hering, K. Brabec, A. Buffagni, L. Sandin & P.F.M. Verdonschot (eds), The Ecological Status of European Rivers: Evaluation and Intercalibration of Assessment Methods
DOI 10.1007/s10750-006-0101-8

Detection of ecological change using multiple organism groups: metrics and uncertainty

Richard K. Johnson[1,*], Daniel Hering[2], Mike T. Furse[3] & Ralph T. Clarke[3]
[1]*Department of Environmental Assessment, Swedish University of Agricultural Sciences, P.O. Box. 7050, SE-750 07 Uppsala, Sweden*
[2]*Department of Hydrobiology, University of Duisburg-Essen, D-45117 Essen, Germany*
[3]*Centre for Ecology and Hydrology, Winfrith Technology Centre, Winfrith Newburgh, Dorchester, Dorset DT2 8ZD, UK*
*(*Author for correspondence: E-mail: richard.johnson@ma.slu.se)*

Key words: metrics, streams, bioassessment, uncertainty, monitoring

Abstract

A number of biological approaches are commonly used to assess the ecological integrity of stream ecosystems. Recently, it is becoming increasingly common to use multiple organism groups in bioassessment. Advocates of the multiple organism approach argue that the use of different organism groups should strengthen inference-based models and ultimately result in lower assessment error, while opponents argue that organism groups often respond similarly to stress implying a high degree of redundancy. Using fish, macroinvertebrate, macrophyte and benthic diatom data, site-specific parameters (e.g., water chemistry and substratum) and catchment variables from European mountain ($n=77$) and lowland ($n=85$) streams we evaluated the discriminatory power and uncertainty associated with the use of a number of biological metrics commonly used in stream assessment. The primary environmental gradient for both streams types was land use and nutrient enrichment. Secondary and tertiary gradients were related to habitat quality or alterations in hydromorphology. Benthic diatom and macroinvertebrate metrics showed high discriminatory power (R^2 values often >0.50) and low error ($<30\%$) with the primary (nutrient) gradient, while both fish and macrophyte metrics performed relatively poorly. Conversely, both fish and macrophyte metrics showed higher response (high coefficients of determination) than either benthic diatom or macroinvertebrate metrics to the second (e.g., alteration in habitat/hydromorphology) gradient. However, the discriminatory power and error associated with individual metrics varied markedly, indicating that caution should be exercised when selecting the 'best' organism group or metric to monitor stress.

Introduction

A principle challenge in studies of ecological assessment is the ability to isolate human-induced effects (signal) from the natural, inherent variability (noise) associated with ecosystem structure and function. Stressed systems often show a reduction in species richness, with a change in the number of individuals within a species and a predominance of stress-tolerant species. This knowledge was used early in the 1900s by German aquatic ecologists in the development of the Saprobien system to assess the effects of organic pollution on stream systems (Kolkwitz & Marsson, 1902), and in the past century, in particular in the last two to three decades, a number of approaches have been developed to evaluate the ecological effects of stress on stream ecosystems (e.g., Metcalfe, 1989; Johnson et al., 1993; Knoben et al., 1995). Despite the large number and widespread use of biological approaches in monitoring and assessment programmes, surprisingly little is known of their inherent errors or weaknesses. Indeed, few studies have compared the discriminatory power of different approaches to detect change (e.g., Fore et al., 1996; Reynoldson et al., 1997), and fewer still have compared the precision and sensitivity of different organism groups to detect ecological change (e.g., Hirst et al., 2002; Paavola et al., 2003).

The EU recently passed legislation, the Water Framework Directive (WFD), focused on hindering deterioration and improving the ecological quality of inland and coastal waters (European Commission, 2000). One of the innovative aspects of this aquatic Directive is the underpinning role of the use of multiple groups of organisms to detect ecological change. Accordingly, Member States of the European Union are required to use a suit of indicator metrics in their monitoring programmes. For example, in the surveillance monitoring of streams it is recommended that biological parameters such as fish, macroinvertebrates, macrophytes and periphyton are combined with physico-chemical parameters to assess the ecological status. Although the simultaneous use of multiple organism groups might be recommended in areas where the primary stressor is unknown, the costs associated with sampling and processing (e.g., taxonomic identification) are not trivial and, ideally, selection of (an) organism group(s) should be based on the response of the organism group/metric to the stressor of interest as well as the errors or levels of uncertainty associated with the selected metric. For example, as the landscape of Europe has been altered for centuries, much is known regarding the effects of certain stressors (e.g., nutrient enrichment/organic pollution, acidification), and using this knowledge, cost-effective assessment programmes should focus on the selection of the most appropriate indicators (high discriminatory power and low error) to monitor the stressor(s) of interest.

The use of biological variables for monitoring aquatic integrity has a long history in Europe (e.g., Metcalfe, 1989). For example, benthic diatoms have been used for assessing the effects of acidification and eutrophication (e.g., Battarbee et al., 1997; Coring, 1999), macrophytes and fish are considered as reliable indicators of alterations to flow and habitat quality (e.g., Gorman & Karr, 1978; Bain et al., 1988; Tremp & Kohler, 1995), and benthic invertebrates are commonly used for monitoring the effects of organic pollution, acidification and alterations in hydromorphology (e.g., Armitage et al., 1983; Buffagni et al., 2004; Sandin et al., 2004). Although different organisms and metrics are frequently used in the assessment of aquatic ecosystems, the decision of what organism group(s)/metric(s) is probably based more on local taxonomic expertise than knowledge of discriminatory power or precision for the stressor of interest. In this study, we compare and contrast the response of four ecologically different organism groups (fish, macrophytes, benthic diatoms and macroinvertebrates) and metrics (ca. 10 metrics per organism group) to putative stress gradients. Two endpoints are considered in evaluating the organism groups/metrics to stress; namely, the response of the indicator to stress and the uncertainty associated with the response.

Methods

Study sites

Some 162 streams, sampled as part of the European funded STAR project (Hering & Strackbein, 2002; Furse et al., 2004), were used here to compare and contrast the response of different organism groups and metrics to stress. A large number of physical, chemical and geographical variables were sampled or obtained (e.g., using GIS) for each of the streams sampled (Furse et al., 2004). The substratum of each of the sampling sites was classified (in percent) according to seven inorganic size classes (i.e., from silt/clay <6 μm to large cobbles, boulders and blocks >40 cm) and 10 organic classes/fractions such as the amount of algae (macro and micro), vegetation (aquatic submerged and emergent and living parts of terrestrial plants) and detritus (e.g., woody debris, coarse and fine particulate organic matter, CPOM and FPOM). A number of physical-chemical metrics representative of nutrient (nitrogen and phosphorus fractions), acidity (pH) status as well as oxygen conditions (BOD5) were measured for each site. Catchment and riparian land use/type was classified according to 16 classes (e.g., forest type, cropland, pasture, urbanization). A number of measures of stream hydrology and morphology were recorded such as mean annual discharge, valley and channel form (six classes), stream width and depth using the RHS survey technique (Raven et al., 1998). Also included in hydromorphology were measures of the number of debris dams and woody debris in and long the stream channel, bank and bed fixation.

Biological samples

Four organism groups were sampled at each stream site; namely, fish, macroinvertebrates, macrophytes and periphytic diatoms. A brief description of the sampling method used is given here, for more detailed information refer to the STAR website (www.eu-star.at).

Fish were normally sampled by electric fishing in accordance with the procedures set out in CEN prestandard PrEN 14011. Fishing was undertaken along two runs of a stop-netted area on a single occasion in late summer or early autumn. The recommended sampling length was 10× the stream width, with a minimum of a 100 m stream length sampled. The fish variables recorded were number of species, life history stage (young of the year per species), density (number of fish per m^2) and assessment of degree of infestation of external parasites or other diseases.

Benthic macroinvertebrates were sampled in spring and either summer or autumn using a Surber sampler or by standardised kick-sampling with a handnet (area 625 cm^2, mesh size 500 μm). Generally the sampling section consisted of 20–50 m in length in small (1–100 km^2) and 50–100 m in length in medium (100–1000 km^2) sized streams. Each sampling site encompassed the whole width of the stream and was deemed to be representative of a minimum area surveyed (i.e., 500 m of length or 100× average width). Before sampling, the site was first classified according to the coverage of all microhabitats with at least 5% cover. A multi-habitat sampling strategy was then adopted that reflected the proportion of different habitat types present at each stream site. Each complete sample consisted of 20 sample units of dimensions 25 cm×25 cm. These sampling units were proportionally situated in all microhabitats with >5% coverage. The 20 sample units resulted in ca. 1.25 m^2 of stream bottom being sampled. Each composite sample was preserved with formalin (4% final concentration) or 95% ethanol to a final concentration of ca. 70%. Macroinvertebrates were sorted (subsampling with the target of 700 individuals) and identified (usually to genus/species level).

Macrophytes were sampled using a single survey in late summer or early autumn. Macrophytes included higher aquatic plants, vascular cryptograms, bryophytes as well as groups of algae. A 100 m stream length was surveyed in each stream by wading, walking along the bank or by boat according to the MTR method described by Holmes et al. (1999). All macrophytes species were recorded as well as the percent cover of the overall macrophyte growth. Submerged vegetation was observed using a glass-bottom bucket. If identification was uncertain, representative samples were collected and later identified.

Periphytic diatoms were sampled from hard (usually cobbles or pebbles) or soft (sand/silt) substratum or macrophytes. Wherever possible periphyton samples were collected within the same sampling area as benthic macroinvertebrates. In brief, a minimum of five cobbles were arbitrarily selected at each site (the combined exposed surface area comprised ca. 100 cm^2). The stones were individually placed in a plastic tray and 100–200 ml of distilled or filtered water added to the tray. The upper part of the stone substratum was washed using a toothbrush, and the dislodged material was decanted into a sample bottle and a composite sample was preserved (e.g., using formalin or Lugol's iodine solution) if the sample could not be processed within 24 h. Submerged macrophytes and parts of emergent ones were collected, placed in a wide-mouth 1-l container, ca. 100–200 ml of distilled/filtered water added and the container was shaken vigorously for about 60 s. A 250 ml aliquot of the sample was decanted to a sample bottle as preserved as above if not analysed within 14 h. Mineral sediments were sampled using a glass tube submerged in the sediment and extracting sediment and interstitial water. Replicate samples were collected until volume of ca. 200 ml was obtained. Light microscopy was used to identify the living and dead diatom cells. The diatom taxa were counted (a minimum of 300 diatom valves) and identified to species at 400 and 1000× magnification.

Analyses

Stream types
Two stream types were studied here; namely, small, shallow mountain streams and medium-sized, lowland streams (Fig. 1). To ensure adequate sampling of stressor gradients, prior to sampling, all sites were pre-classified into five classes of ecological status using physico-chemical and in some instances biological information and/or expert opinion: (i) high (no or only minimal disturbance), (ii) good (slight deviation from high status), (iii) moderate (moderate deviation from high status and significantly more disturbed than good), (iv) poor (major alteration from high status) and (v) bad (severe alteration from high status). For each stream type, 10 and usually 12 sites were sampled of which, in each set, at least three were pre-classified as of high status and the rest spread more or less evenly over the other four quality status classes.

Organism-specific metrics
Ten metrics were calculated for each organism group and stream type (Table 1). These metrics were the 10 'best' metrics for each organism groups according to Hering et al. (submitted). In brief, correlation was used to reduce the number of candidate metrics to a similar number for each organism group. For instance, if two metric pairs had a Spearman Rank Correlation coefficient of >0.8 or <−0.8, one of the two metrics was excluded. For fish assemblages, candidate metrics ($n = 120$) were calculated, covering species number, habitat preferences, feeding type composition, percentage of native species, and percentage of tolerant species. The 47 candidate metrics calculated for benthic macroinvertebrates covered measures of taxonomic richness (e.g., taxa numbers of certain organism groups), composition measures (e.g., the percentage of a taxa group), sensitivity/tolerance measures (e.g., Saprobic indices), and functional measures (e.g., feeding type composition). For macrophytes, 35 candidate metrics were calculated, covering richness measures (e.g., Shannon diversity), tolerance measures such as Mean Trophic Ranking, and functional measure (e.g., growing forms). For benthic diatoms, 14 metrics were calculated using the OMNIDIA software (Le Cointe et al., 1993) and 11 additional metrics were calculated with national databases. The maximum variation explained (r^2, coefficient of variation, Spearman rank correlation) for all correlation analyses (comparison of each metric with a number of stressors gradient for different stream type groups) was calculated for each metric. The 10 metrics included here were those with the highest maximum r^2 selected for each organism group. This procedure resulted in somewhat different metrics for mountain streams and lowland streams as selection was done separately for the mountain and lowland stream type groups.

Environmental stress and indicator response
Principal components analysis (PCA) was used to construct complex stress gradients by reducing the dimensionality of the physical-chemical, hydromorphological and land use/type characteristics for each of the sites. The environmental variables used are listed in Table 3.

To measure the response of the organism group/metric to stress and calculate the frequency of false negative errors, the PC stress gradients (the 1st and 2nd PC axes) were divided into two quality classes. The two quality classes (here referred to as 'best available' representing high quality and 'perturbed' representing poor quality) were created by removing sites between the 25th and 75th percentiles of the PC axis 1 and 2 distributions. Correlation of PC-axis scores with selected environmental variables (e.g., TP, woody debris) was used to determine which of the two classes (the lower 25th or upper 75th percentile) represented best available or perturbed quality classes (Fig. 2).

Metric response (discriminatory power) between the two quality classes was tested using Wilcoxon rank-sum tests. For those metrics that did not show a significant ($\alpha = 5\%$) response between best available and perturbed conditions power estimates were calculated. Statistical power is the probability of getting significance at or below a given p-value alpha for a given situation. Power is a function of the sample size, the effect size, the standard deviation of the error and the significance level. Power was calculated using the statistical program JMP (SAS, 1994).

For those metrics that showed a significant response, the frequency of type II or false negative

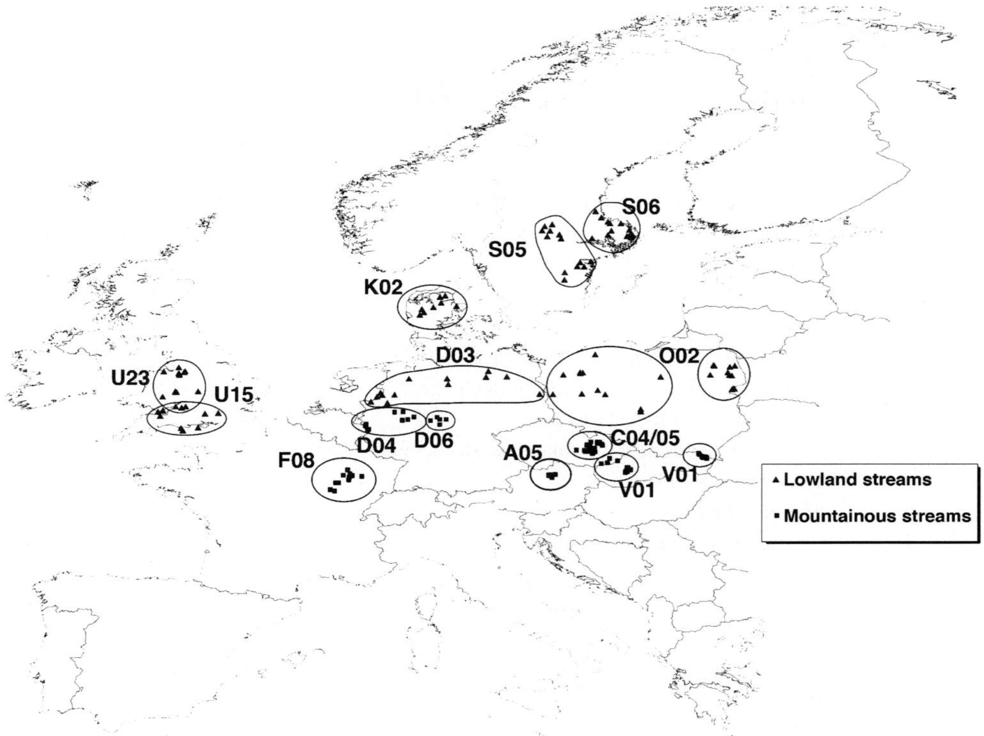

Figure 1. Location of the streams sampled as part of the EU-STAR project.

error was calculated. The 25th or 75th percentiles of the high quality sites was arbitrarily selected as the cutoff (critical threshold) for determining an ecological effect. Thus the type I error was fixed at 25%. If the predicted response of the metric to stress was negative, sites in the low quality (perturbed) class according to the environmental gradients that had metric values above the 25th percentile of the high quality (best available) sites were considered to indicate type II errors, if the predicted response of the metric to stress was positive (e.g., % gatherers increase with nutrient enrichment), then the 75th percentile of the high quality sites was used as the critical threshold, and sites in the perturbed class that were below this value were considered to be false negatives. Type II error rates for a metric were calculated as the proportion of sites in the perturbed class whose values failed to indicate change from 'best available'. Ideally, these tests should be hypothesis-based, i.e., the direction of change should be determined *a priori*. However, given the complex gradients and the uncertainty associated with predicting the expected response of the organism group/metric to the stress gradient we simply chose to use the observed response as the expected response and calculate the type II error frequencies accordingly.

All tests were performed using the statistical program JMP (version 3.1) (SAS 1994).

Results

Mountain streams ($n = 77$ streams) were situated at higher altitude (mean 337 m a.s.l.) and had smaller catchments (mean = 57 km^2) compared to lowland streams ($n = 85$ streams, mean = 57 m a.s.l. and 199 km^2, Table 2). Mountain streams were often situated in forested catchments (e.g., mean = 58% forest and 10% of the streams had >80% of their catchments classified as native deciduous forest). Lowland streams had more of their catchments classified as cropland (mean = 30%) or pasture (mean = 15%). The two stream types also differed regarding the predominant substratum type: For

Table 1. Biological metrics included into the analysis

Code	Description	M/L	Source
Diatom metrics			
IPS	IPS (polluo-sensibilité, Zelinka and Marvan Index modified by Cemagref)	M	Coste in Cemagref (1982)
She	Trophic conditions acc. Steinberg and Schiefele	L	Steinberg & Schiefele (1988)
EPI-D	EPI-D	M/L	Dell'Uomo (1996)
ROTT	ROTT calculated with Omnidia	M/L	
IDG	IDG	L	Rumeau & Coste (1988)
CEE	CEE	M	Descy & Coste (1990)
TDI DVWK	Trophic Diatom Index acc. Deutscher Verband für Wasserwirtschaft und Kulturbau	M/L	
TDI Rott	Trophic Diatom Index acc. Rott	M/L	Rott et al. (1999)
TDI lakes	Trophic Diatom Index for lakes acc. Hofmann	M	Hofmann (1999)
DI Swi	Diatom index Switzerland	L	
SI Rott	Saprobic Index Rott	M/L	Rott et al. (1997)
Halo	Halobienindex (targeting salinity)	M/L	Ziemann (1999)
PHYLIP	PHYLIP multimetric diatom index for German river type 4 (small siliceous mountain streams)	M/L	Schaumburg et al. (2004)
Macrophyte metrics			
n_sp_subm	Number of submerged species	M	
n_sp_float	Number of floating free species	M/L	
n_sp_amph	Number of amphibious species	M/L	
n_sp_terr	Number of terrestrial species	M	
cov_mo_li	Cover mosses and liverworts	M	
cov_subm	Cover submerged species	M	
cov_amph	Cover amphibious species	M/L	
MTR	Mean Trophic Rank	M/L	Holmes et al. (1999)
IBMR	Macrophyte Biological Index for Rivers (IBMR)	M/L	Haury et al. (2002)
sp_n	Number of of all occurring species	L	
ge_n	Number of of all occurring genera	L	
fa_n	Number of of all occurring families	L	
Species number (S*)	Shannon–Wiener diversity all species	L	Shannon & Weaver (1949)
Species number (S**)	Domination (all species)	L	McNaughton (1967)
Ellenberg_N*	Ellenberg_N, typical macrophytes only	M	Ellenberg et al. (1992)
Macroinvertebrate metrics			
ASPT	Average Score Per Taxon	M/L	Armitage et al. (1983)
DSFI	Danish Stream Fauna Index	L	Skriver et al. (2001)
epirhithral	Epirhithral preferring taxa [%]	L	Hering et al. (2004)
EPT-taxa [%]	Ephemeroptera, Plecoptera, Trichoptera [%]	M/L	
gatherers	Gatherers [%]	M/L	Hering et al. (2004)
GFID01	German Fauna Index D01	M/L	Lorenz et al. (2004)
GFID05	German Fauna Index D05	M/L	Lorenz et al. (2004)
MAS_IC	Mayfly Average Score Integrity Class	M	Buffagni (1997, 1999)
metarhithral	Metarhithral preferring taxa [%]	M	Hering et al. (2004)
pelal	Pelal preferring taxa [%]	L	Hering et al. (2004)
Plecoptera [%]	Plecoptera [%]	M/L	
SI(ZM)	Saprobic Index (Zelinka and Marvan)	M/L	Zelinka & Marvan (1961)

Continued on p. 121

Table 1. (Continued)

Code	Description	M/L	Source
xeno	Xenosaprobic taxa [%]	M	Zelinka & Marvan (1961)
Fish metrics			
n_sp_tol	Number of tolerant species	M	Noble & Cowx (2002)
n_sp_intol	Number of intolerant species	L	Noble & Cowx (2002)
Perc_sp_intol	Native intolerant species [% species]	L	Noble & Cowx (2002)
Perc_sp_tol	Native tolerant species [% species]	M/L	Noble & Cowx (2002)
n_ha_hab_wc	Density of species preferring the water column [n/ha]	M	Noble & Cowx (2002)
n_sp_hab_b	Number of native benthic species	L	Noble & Cowx (2002)
n_ha_hab_b	Density of native benthic species [n/ha]	L	Noble & Cowx (2002)
perc_nha_hab_b	Native benthic species [% individuals of density]	L	Noble & Cowx (2002)
n_sp_hab_rh	Number of rheophilic species	M	Noble & Cowx (2002)
n_ha_hab_rh	Density of rheophilic species [n/ha]	M	Noble & Cowx (2002)
n_sp_hab_eury	Number of eurytopic species	M	Noble & Cowx (2002)
n_ha_hab_eury	Density of eurytopic species [n/ha]	L	Noble & Cowx (2002)
perc_nha_re_lith	Lithophilic species [% individuals of density]	M	Noble & Cowx (2002)
n_ha_re_phyt	Density of phytophilic species [n/ha]	L	Noble & Cowx (2002)
n_sp_lon_ll	Number of long living species	M	Noble & Cowx (2002)
n_sp_fe_insev	Number of insectivorous species	L	Noble & Cowx (2002)
n_ha_fe_omni	Density of omnivorous species [n/ha]	M	Noble & Cowx (2002)
n_sp_mi_long	Number of long distance migrating species	L	Noble & Cowx (2002)
n_sp_mi_potad	Number of potamodromous species	M	Noble & Cowx (2002)

M = metric used in the mountain; L = lowland stream type groups.

mountain streams, both cobbles and coarse gravel were most common types of substratum (38 and 23%, respectively), whereas lowland streams had a high frequency of soft-bottom substratum (36% sand). Clear differences were also noted regarding nutrient concentrations. Total phosphorus concentrations were on average $>5\times$ higher in lowland (mean TP = 1091 mg/l) compared to mountain (mean TP = 193 mg/l) streams. Similarly, ammonium concentrations were also much higher in lowland (mean = 321 mg/l) than in mountain (mean = 0.166 mg/l), although NH_4 concentration varied markedly among the individual streams (e.g., 10th and 90th percentiles were 0 and 184 mg NH_4/l, respectively for lowland streams).

Environmental gradients

The first three axes of principal component analysis explained 38.4% of the variation in catchment land use/cover and physico-chemical variables in mountain streams, and 40.1% of the variation in lowland streams (Table 3). The primary environmental gradient for mountain streams (1st PC axis) explained 17.3% of the variation and was related to catchment land use (e.g., eigenvector loading for % cropland = 0.32) and nutrient concentration (e.g., eigenvector loadings for ortho- and total P were 0.25 and 0.24, respectively). Conversely, % forest in the catchment (−0.20) and riparian variables (e.g., woody riparian vegetation = −0.22) were negatively correlated with the 1st PC axis. The secondary environmental gradient (2nd PC axis) explained another 11.8% of the residual variance and was related to habitat variables. For example, the amount of woody riparian vegetation (eigenvector loading = 0.27) and number of debris dams (0.29) and number of logs (0.24) were positively correlated with this axis, whereas the amount submerged vegetation (−0.27) was negatively related to this axis. The 3rd PC axis explained 9.2% of the variance and was related to hydromorphological variables such as no bed or bank fixation (eigenvector loadings = 0.23 and

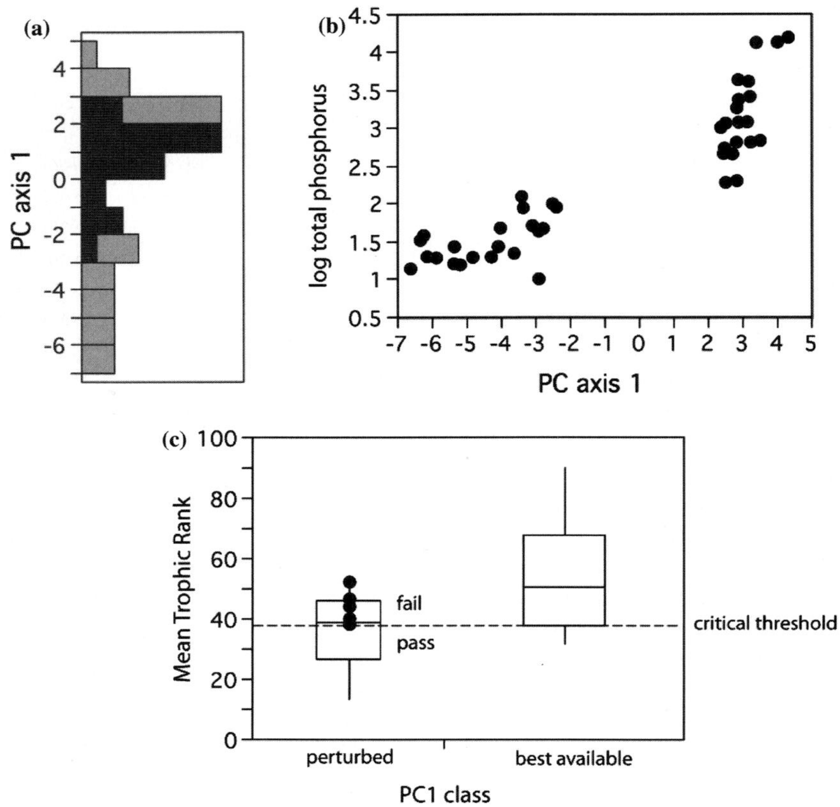

Figure 2. Steps used in assessing the discriminatory power of selected biological metrics: (a) construction of latent PC stress gradients and use of 'best available' (e.g., top 75th-percentile) and 'perturbed' (e.g., 25th-percentile) sites. Samples between these two percentiles were omitted from the analyses; (b) interpretation of gradients using PC loadings and correlation and (c) determination of false negative (type II error) error rates (calculated as the proportion of samples in the disturbed class whose metric values would not indicate fail).

0.38, respectively), and coarse gravel substratum (0.25), catchment land use such as % pasture (−0.20) and nitrogen concentration (e.g., ammonium concentration = −0.27).

For lowland streams the 1st PC axis explained 20.4% of the variance in land use and site-specific descriptors, and, similar to mountain streams, the 1st PC axis extracted a gradient representing agricultural land use (% cropland and % pasture had eigenvector loadings of 0.19 and 0.17, respectively) and nutrient concentration (ortho- and total P had eigenvector loadings of 0.28 and 0.26, respectively; Table 3). Percent forest in the catchment and hard-bottom substratum were negatively correlated with this axis. The 2nd PC axis explained another 10.3% of the variance and was related to hydromorphological variables such as the absence of bank or bed fixation, both of which were positively correlated (loadings of 0.26 and 0.30, respectively), and straightening (−0.25), which was negatively correlated with this secondary gradient. The 3rd PC axis explained 9.3% of the variance and was related to the amount of woody debris both beside (e.g., average width of woody riparian zone loading = 0.36) and within the stream (e.g., number of logs (0.27) and debris dam (0.22)).

Organism/metric response to stress

The response of the different organism groups/metrics to stress (PC gradients) varied among organism groups and with stress and stream type (Fig. 3a,b). For mountain streams the majority (36 of 39) of metrics showed a significant response to the 1st PC gradient (Table 4). One fish metric

Table 2. Selected physico-chemical and catchment characteristics of mountain and lowland streams

	Mountain streams (n=77)		Lowland streams (n=85)	
Altitude m a.s.l.	337 ± 84	(220–448)	57 ± 60	(4.6–144)
Catchment area (km^2)	57 ± 77	(13–136)	199 ± 199	(20–418)
Catchment classification (%)				
Urban (sum)	0.14 ± 0.24	(0–0.322)	5.9 ± 13	(0–10)
Forest (sum)	58 ± 23	(30–90)	36 ± 28	(10–70)
Native deciduous	32 ± 33	(0–80)	3.13 ± 5.82	(0–10)
Native coniferous	11 ± 17	(0–40)	3.9 ± 10	(0–16)
Cropland	24 ± 26	(0–60)	30 ± 28	(0–76)
Pasture	8.4 ± 16	(0–20)	15 ± 22	(0–50)
Substratum (%)				
Large cobbles, boulders (> 40 cm)	6.9 ± 18	(0–15)	5.8 ± 13	(0–22)
Coarse blocks, cobbles (> 20–40 cm)	12.7 ± 16	(0–35)	7.9 ± 13	(0–30)
Cobbles (> 6–20 cm)	38 ± 23	(0–70)	12 ± 20	(0–40)
Coarse gravel (> 2–6 cm)	23 ± 18	(0–46)	13 ± 21	(0–47)
Fine gravel (> 0.2–2 cm)	9.3 ± 12	(0–22)	13 ± 23	(0–47)
Sand (> 6 μm–2 mm)	7.5 ± 15	(0–25)	36 ± 37	(0–100)
Silt (< 6 m)	1.6 ± 5	(0–5)	9 ± 25	(0–42)
Physico-chemistry				
pH	7.9 ± 0.57	(7.02–8.6)	7.55 ± 0.40	(6.95–8.01)
Conductivity (μS/cm)	315 ± 261	(88.2–631)	390 ± 236	(49–676)
BOD5 (mg O_2/L)	2.25 ± 1.58	(1.2–3.72)	2.58 ± 1.50	(1.12–4.7)
Ammonium (mg NH_4/L)	0.166 ± 0.360	(0.01–0.38)	321 ± 1918	(0–184)
Nitrate (mg NO_3/L)	9.45 ± 9.77	(0.944–21.9)	13 ± 14	(0.121–34)
Total phosphorus (μg TP/L)	193 ± 270	(50–474)	1091 ± 2747	(19–2510)

Mean ± 1 sd and in parenthesis 10th and 90th percentile.

('number of rheophilic species') and two macrophyte metrics ('cover mosses and liverworts' and 'Ellenberg_N') did not show a significant response to this stress gradient. Coefficients of determination (R^2 values), measuring the proportion of total variability in metric values explained by differences between best available and perturbed sites, varied markedly among the organism groups and metrics. For example, the nine fish metrics had R^2 values <0.187 and the eight significant macrophytes metrics had R^2 values between 0.036 ('cover amphibious species') and 0.335 ('number of amphibious species'). By contrast, both macroinvertebrate and benthic diatom metrics had relatively high R^2 values. Coefficients of determination for macroinvertebrate metrics ranged from 0.149 for the 'German Fauna Index D01' to 0.652 for the 'Mayfly Average Score Integrity Class', and seven of the 10 metrics had R^2 values >0.50. Similarly, six of the 10 diatom metrics had R^2 values >0.50; values ranged from 0.301 for the 'Halobienindex' (a metric indicating salinization) to 0.669 for the 'Trophic Diatom Index'.

Only 14 of the 39 metrics showed a significant response with the 2nd PC axis; namely, six fish metrics, one macroinvertebrate and seven macrophyte metrics (Fig. 3b). The six significant fish metrics had R^2 values that ranged from 0.04 ('density of omnivorous species') to 0.533 ('number of potamodromous species'), the one macroinvertebrate metric ('% xenosaprobic taxa') had an R^2 of 0.328 and the seven macrophyte metrics had R^2 values between 0.052 for 'cover amphibious species' to 0.537 for 'number of submerged species'.

Different patterns were noted regarding organism group/metric response in lowland compared to mountain streams. Only 23 of the 39 metrics showed a significant response to the 1st PC gradient in lowland streams (Fig. 3c, Table 5). The

Table 3. Eigenvectors (loadings) of physico-chemical, substratum and catchment land use/cover

	Mountain streams			Lowland streams		
Eigenvalue	6.9	4.7	3.7	8.8	4.4	4.0
Percent	17.3	11.8	9.2	20.4	10.3	9.3
Cumulative percent	17.3	29.1	38.4	20.4	30.7	40.1
	Eigenvectors					
Total forest	**−0.20**	0.02	0.02	**−0.24**	−0.03	**0.25**
Total urban	−0.10	0.00	−0.13	**0.15**	−0.13	−0.07
Wetland (mire)				**−0.19**	−0.02	0.00
Open grass-/bushland	−0.11	0.06	−0.07	0.04	**0.24**	**−0.21**
Standing water				**−0.20**	−0.07	0.05
Cropland	**0.32**	0.01	0.14	**0.19**	−0.13	−0.08
Pasture	−0.11	−0.06	**−0.20**	**0.17**	**0.23**	−0.12
Clear-cutting				**−0.26**	−0.01	0.07
Shading at zenith (foliage cover)	**−0.19**	**0.20**	0.12	0.04	**0.18**	**0.35**
Average width of woody riparian vegetation [m]	**−0.21**	**0.17**	0.09	−0.05	0.06	**0.36**
No. of debris dams	0.02	**0.29**	**0.18**	0.11	**0.22**	**0.22**
No. of logs	−0.08	**0.24**	**0.20**	0.13	**0.16**	**0.27**
Shoreline covered with woody riparian Vegetation left	**−0.22**	**0.27**	**0.17**	0.07	**0.15**	**0.30**
No bank fixation	**−0.16**	0.03	**0.23**	−0.02	**0.26**	0.08
No bed fixation	−0.08	0.08	**0.38**	−0.05	**0.30**	0.04
Stagnation	0.11	**0.20**	−0.08	0.09	−0.13	−0.07
Straightening	**0.20**	0.10	**−0.27**	0.12	**−0.25**	**−0.16**
Hygropetric	0.08	−0.06	0.10	**−0.16**	−0.02	0.03
Large cobbles	0.01	−0.08	**−0.34**	**−0.26**	−0.03	0.01
Coarse blocks	**−0.17**	**−0.15**	−0.11	**−0.26**	−0.06	−0.06
Cobbles	**−0.25**	0.07	0.02	**−0.21**	0.03	−0.11
Coarse gravel	0.11	0.05	**0.25**	−0.03	0.13	−0.03
Fine gravel	**0.18**	**0.16**	0.12	0.09	**−0.16**	**0.23**
Sand	**0.15**	0.03	0.06	**0.20**	0.00	0.00
Silt	**0.19**	**0.18**	−0.05	0.01	0.09	−0.06
Submerged Macrophytes	0.09	**−0.27**	0.16	−0.05	**−0.16**	0.03
Emergent macrophytes	0.10	−0.08	0.09	0.07	**−0.19**	−0.08
Xylal	0.02	**0.24**	0.02	0.12	**0.22**	**0.22**
CPOM	−0.03	0.13	0.14	0.13	**−0.29**	**0.24**
FPOM	0.13	0.13	−0.01	0.14	**−0.27**	**0.22**
pH	**0.17**	**0.20**	0.11	**0.19**	0.07	0.03
Conductivity	**0.28**	−0.02	0.14	**0.29**	0.02	−0.07
BOD5	0.12	**0.23**	**−0.21**	−0.03	−0.05	0.07
Ammonium	−0.10	**0.31**	**−0.27**	0.12	**−0.31**	**0.17**
Nitrite	**−0.17**	**0.31**	**−0.26**			
Nitrate	**0.24**	0.10	−0.03	**0.18**	**0.22**	**−0.22**
Ortho-phosphate	**0.25**	0.11	−0.06	**0.28**	0.01	−0.02
Total phosphate	**0.24**	**0.21**	**−0.13**	**0.26**	−0.02	−0.02

Only variables with loadings >0.15 on at least one of the first three PC axes are shown and those with loading >0.15 are shown with bold text.

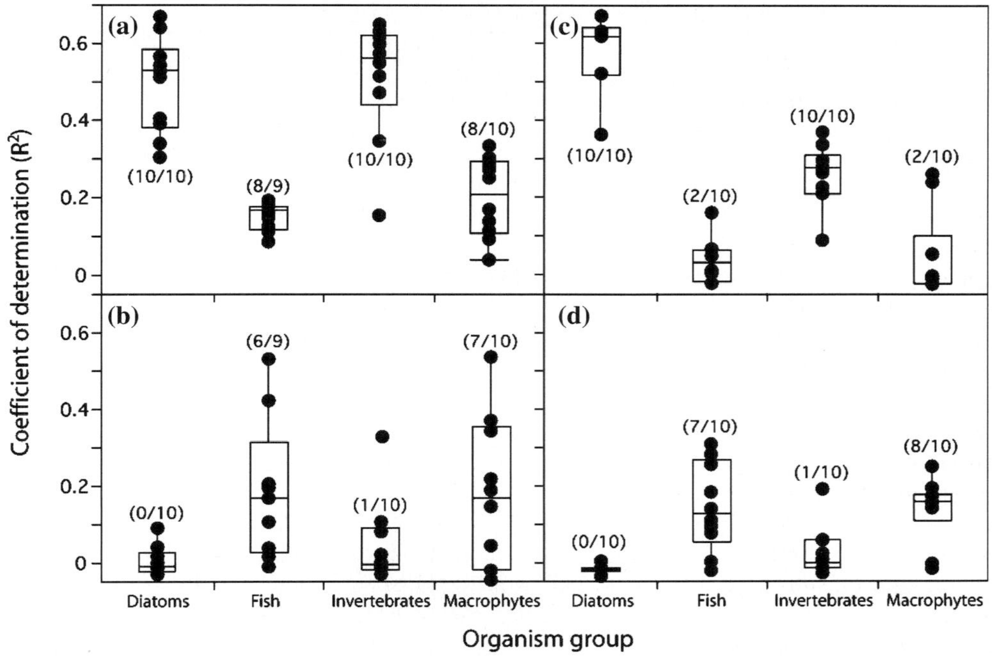

Figure 3. Coefficient of determination (R^2) for *t*-tests between best available and perturbed mountain (a & b) and lowland (c & d) streams vs PC1 (a & c) and PC2 (b & d). Numbers in parenthesis show the number of metrics tested (denominator) and the number that showed a significant response (numerator). Box plots show the median, 25th and 75th and 10th and 90th percentiles.

only fish metric that showed a significant response to this (nutrient) gradient ('% native intolerant species') exhibited a weak relationship ($R^2 = 0.063$). All 10 of the macroinvertebrate metrics showed significant responses to the 1st PC gradient, with R^2 values ranging from 0.085 for the 'German Fauna Index D05' to 0.369 for '% gatherers'. By contrast, only two of the 10 macrophyte metrics showed a significant response to this gradient; namely the 'Mean Trophic Rank' and 'Macrophyte Biological Index for Rivers' indices, with R^2 values of 0.246 and 0.26, respectively. Similar to the macroinvertebrate metrics, all 10 benthic diatom metrics showed robust relationships with the 1st PC axis. Coefficients of determination were also relatively high, ranging from 0.358 for 'Halobien index' to 0.677 for the 'EPI-D', and nine of the 10 metrics had R^2 values >0.50.

Only 16 of the 39 metrics showed a significant response to the 2nd PC (hydromorphological) stress gradient; seven fish, one macroinvertebrate and eight macrophyte metrics (Fig. 3d). Both fish and macrophyte metrics were more strongly related to the second than the first PC stress gradient. Generally, R^2 values were higher for the 2nd compared to the 1st PC gradient and seven of the fish metrics showed a significant response with the 2nd PC gradient compared to one with the 1st PC gradient. Likewise, eight of the macrophyte metrics showed a significant response to the 2nd stress gradient, while only two responded significantly to the 1st PC gradient.

Organism/metric response and uncertainty

The percentage frequency of type II errors differed markedly among organism groups/metrics and stream types (Fig. 4). For mountain streams, both fish and macrophyte metrics exhibited higher false negative error frequencies than either macroinverbrate or benthic diatom metrics (Fig. 4a, b, Table 6). Fish metric response to the 1st PC stress gradient was generally positive; all except one metric ('% native intolerant species') showed a significant positive response to stress (Fig. 4a). Consequently, the upper 75th percentile of the best available sites was used as the critical threshold value for seven of the metrics and sites in the perturbed class that had values less than this

Table 4. Response of organism group/metric to two PC stress gradients for mountain streams

	PC 1 mountain streams						PC 2 mountain streams					
	Best available	Perturbed	*p*-value	Change	Power	R^2	Best available	Perturbed	*p*-value	Change	Power	R^2
Fish metrics												
n_sp_tol	0.0526	1.789	**	+		0.184	1.947	0.0625	***	–		0.209
n_ha_hab_wc	1974	9854	**	+		0.168	10716	4031	ns		0.593	0.108
n_sp_hab_rh	3.21	4.68	ns		0.536	0.085	5.683	2.437	***	–		0.424
n_ha_hab_rh	2969	13114	*	+		0.121	14873	9907	ns		0.127	−0.009
n_sp_hab_eury	0.0526	1.6316	**	+		0.164	11.842	0.0625	***	–		0.2
perc_nha_Re_lith	99.9	82.7	***	–		0.187	88.9	95.56	ns		0.225	0.015
n_sp_lon_ll	0.1053	1.316	**	+		0.161	1.4737	0.1875	**	–		0.171
n_ha_fe_omni	5.79	3981	***	+		0.108	3599	635	**	–		0.04
n_sp_mi_potad	0.4211	1.158	**	+		0.147	469	162	***	–		0.533
Macroivertebrate metrics												
ASPT	7.17	5.52	***	–		0.618	6.33	6.22	ns		0.067	−0.023
EPT-Taxa [%]	42.3	16.44	***	–		0.516	29.3	23.5	ns		0.182	0.004
gatherers	22.84	61.79	***	+		0.599	51.6	41.8	ns		0.271	0.024
GFID01	0.6989	−0.0832	**	–		0.149	0.5419	0.2634	ns		0.145	−0.004
GFID05	0.9278	−0.0019	***	–		0.634	0.4094	0.289	ns		0.096	−0.016
MAS_IC	1	3.8	***	+		0.652	3.1579	2	ns		0.621	0.107
metarhithral	31.23	19.88	***	–		0.573	23.39	21.44	ns		0.165	0
Plecoptera [%]	8.039	0.3605	***	–		0.553	3.74	3.53	ns		0.052	−0.027
SI(ZM)	1.503	1.931	***	+		0.469	1.827	1.647	ns		0.539	0.086
xeno	7.043	2.532	***	–		0.346	2.091	7.148	***	+		0.328
Macrophyte metrics												
n_sp_subm	0.5	4.684	***	+		0.249	6.5	0.4117	***	–		0.537
n_sp_floating	0	0.5263	*	+		0.161	0.5555	0	**	–		0.219
n_sp_amphi	0.0833	3.158	***	+		0.335	3.555	0.2941	***	–		0.377
n_sp_terr	0.4167	5.263	***	+		0.298	6.5	0.882	***	–		0.345
cover_moss_liv	17.15	3.761	ns		0.490	0.091	11.96	0.3441	**	–		0.148
cover_sp_subm	0.5	34.8	***	+		0.138	36.5	0.488	***	–		0.19
cover_sp_amphi	0.0042	1.171	***	+		0.036	1.1806	0.0147	***	–		0.052
MTR	65.87	36.52	**	–		0.269	41.51	45.26	ns		0.096	−0.017
IBMR	13.3	9.446	*	–		0.289	10.122	10.59	ns		0.083	−0.021
Ellenberg_N*	5.57	6.767	ns		0.345	0.107	6.649	6.8	ns		0.068	−0.043
Benthic diatom metrics												
IPS	18.54	15.04	***	–		0.566	15.85	16.61	ns		0.162	0
EPI-D	12.51	9.855	***	–		0.534	11	10.81	ns		0.062	−0.025
ROTT	14.77	12.64	***	–		0.401	13.82	13.62	ns		0.071	−0.022
CEE	17.43	12.98	***	–		0.545	14.79	14.23	ns		0.088	−0.018
TDI DVWK	1.947	2.67	***	+		0.645	2.365	2.479	ns		0.142	−0.005
TDI Rott	1.638	2.993	***	+		0.669	2.463	2.708	ns		0.180	0.004
TDI lakes	3.737	4.612	***	+		0.339	4.141	4.495	ns		0.528	0.09

Continued on p. 127

Table 4. (Continued)

	PC 1 mountain streams						PC 2 mountain streams					
	Best available	Perturbed	*p*-value	Change	Power	R^2	Best available	Perturbed	*p*-value	Change	Power	R^2
SI Rott	1.785	2.163	***	+		0.522	1.929	2.055	ns		0.367	0.046
Halob	−0.0347	9.976	***	+		0.301	6.168	4.157	ns		0.105	−0.014
PHYLIP DI	1.2637	0.4155	***	−		0.39	0.8378	0.5721	ns		0.265	0.022

Mean values for best available (upper 75th percentile of PC gradient) and perturbed (lower 25th percentile of PC gradient) sites, significance using a non-paramentric Wilcoxon rank-sum test, if significant direction of change, if not significant, power and coefficient of determination (R^2). ns = Non significant; $*p < 0.05$; $**p < 0.01$; $***p < 0.001$.

critical threshold level were classified as having failed to detect change. Error frequencies for fish metrics ranged from 21.1% ('density of species preferring the water column') to 63.2% ('number of potamodromous species'). Similar to fish metrics, the majority of the macrophyte metrics (six of eight) showed a positive response to the 1st PC gradient. Error frequencies were generally lower than those noted for fish metrics, with the lowest error found for the 'number of submerged species' (10.5%) and 'Mean Trophic Rank' (21.1%). One metric in particular, the 'number of floating free species', had a very high error frequency (57.9%). All 10 of the macroinvertebrate metrics showed a significant response to the 1st PC stress gradient and error frequencies were generally low. Indeed, six of the metrics had an error frequency 5% or less (Table 6). One metric ('German Fauna Index D01') showed, however, a high error frequency (40%). Diatom metrics had the lowest frequency of false negative errors of the four organism groups tested here. All 10 metrics had error

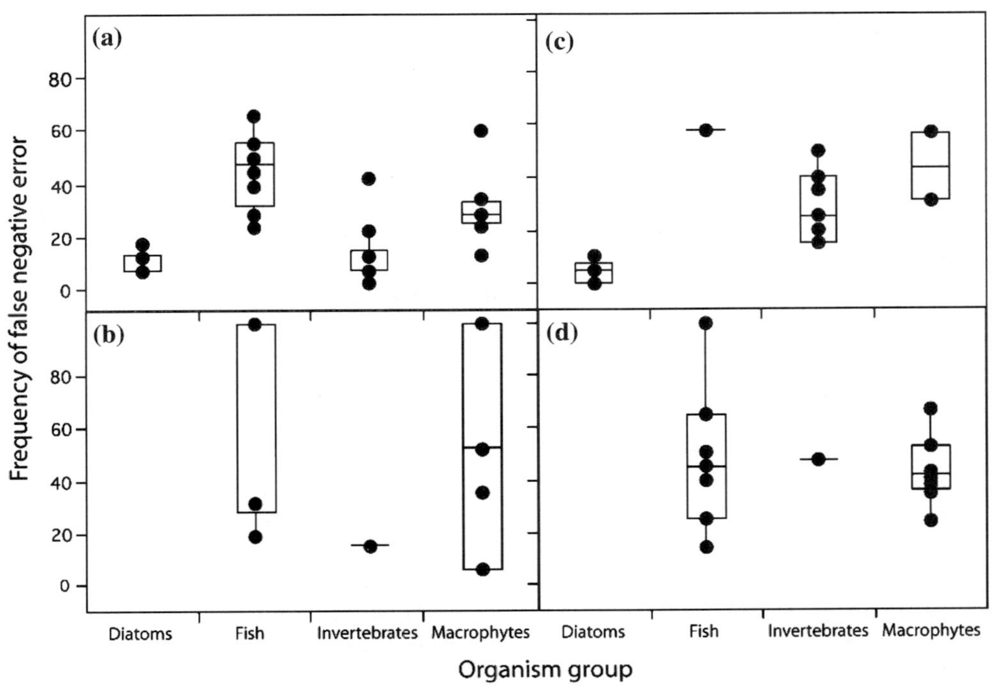

Figure 4. Frequency of false negative error for *t*-tests between best available and perturbed mountain (a & b) and lowland (c & d) streams vs. PC1 (a & c) and PC2 (b & d). Error frequencies were calculated as the percentage of sites in the disturbed class whose metric values did not indicate fail. Only metrics that showed a significant response (numerator values shown in Fig. 3) were used to calculate error frequencies. Box plots show the median, 25th and 75th and 10th and 90th percentiles.

Table 5. Response of organism group/metric to two PC stress gradients for lowland streams

	PC 1 lowland streams						PC 2 lowland streams					
	Best available	Perturbed	*p*-value	Change	Power	R^2	Best available	Perturbed	*p*-value	Change	Power	R^2
Fish metrics												
n_sp_intol	1.286	0.9474	ns		0.183	0.004	2.095	0.7	***	–		0.285
perc_sp_intol	32.95	18.16	*	–		0.063	41.238	20.2	**	–		0.109
n_sp_tol	1.238	1.947	ns		0.408	0.052	1.047	1.95	ns		0.513	0.074
n_sp_hab_b	1.667	2.474	ns		0.452	0.062	2.1905	2	ns		0.066	−0.022
n_ha_hab_b	677	1227	ns		0.395	0.049	1840	146	***	–		0.309
perc_nha_hab_b	40	44.8	ns		0.076	−0.02	48.2	28.15	*	–		0.099
n_ha_hab_eury	615	2239	ns		0.206	0.009	339	1727	ns		0.169	0.001
n_ha_re_phyt	37.5	43.4	ns		0.054	−0.025	0.9048	73.3	***	+		0.142
n_sp_fe_insev	1.095	0.9474	ns		0.078	−0.019	1.952	0.65	***	–		0.258
n_sp_mi_long	0.0952	0.4737	**			0.157	0.7142	0.15	**	–		0.182
Macroinvertebrate metrics												
ASPT	6.099	4.64	***	–		0.293	5.64	5.089	ns		0.259	0.019
DSFI	5.85	3.706	**	–		0.268	5.842	4	**	–		0.193
Epirhithral	11.14	6.428	***	–		0.282	8.593	7.16	ns		0.226	0.013
EPT-Taxa [%]	32.77	10.66	***	–		0.339	17.52	18.39	ns		0.054	−0.024
Gatherers	28.84	50.26	***	+		0.369	47.08	42.71	ns		0.113	−0.011
GFID01	0.0811	−0.7056	**	–		0.207	−0.1769	−0.03286	ns		0.088	−0.016
GFID05	0.3344	−0.20135	*	–		0.085	0.1806	0.02119	ns		0.105	−0.013
Pelal	7.476	27.609	***	+		0.297	14.38	24.41	ns		0.450	0.058
Plecoptera [%]	3.2567	0.4196	***	–		0.208	1.238	0.776	ns		0.087	−0.016
SI(ZM)	1.9737	2.3718	***	+		0.23	2.0706	2.293	ns		0.459	0.06
Macrophyte metrcs												
n_sp_floating	0.45	0.7	ns		0.149	−0.003	0.3333	1.0952	*	+		0.146
n_sp_amphi	2.65	2.15	ns		0.102	−0.014	2.238	4.523	*	+		0.169
cover_sp_amphi	4.465	1.585	ns		0.405	0.051	2.109	8.886	**	+		0.156
MTR	54.39	35.82	**	–		0.246	38.5133	36.97	ns		0.074	−0.022
IBMR	12.087	9.1539	***	–		0.26	10.0179	9.466	ns		0.152	−0.003
sp_n	9.05	8.7	ns		0.054	−0.025	8.381	14.6667	*	+		0.159
ge_n	8.5	8.2	ns		0.054	−0.026	7.952	13.333	*	+		0.159
fa_n	8.25	7.2	ns		0.107	−0.013	7.1905	11.52	*	+		0.152
Species number (S*)	7.4	7.105	ns		0.054	−0.026	6.1905	12.4	**	+		0.255
Species number (S**)	4.9	5	ns		0.051	−0.027	5.2	9.25	**	+		0.197
Benthic diatom metrics												
She	15.7	11.945	***	–		0.519	12.7143	12.47	ns		0.068	−0.021
EPI-D	12.67	9.225	***	–		0.677	10	10.357	ns		0.105	−0.013
ROTT	16.87	12.19	***	–		0.62	13.386	13.43	ns		0.051	−0.025
IDG	15.89	11.955	***	–		0.522	12.957	12.58	ns		0.101	−0.034
TDI DVWK	1.829	3.07	***	+		0.625	2.752	2.8095	ns		0.066	−0.021
TDI Rott	1.69	3.13	***	+		0.674	2.9142	2.7619	ns		0.188	0.005

Continued on p. 129

Table 5. (Continued)

	PC 1 lowland streams						PC 2 lowland streams					
	Best available	Perturbed	*p*-value	Change	Power	R^2	Best available	Perturbed	*p*-value	Change	Power	R^2
DI Swi	3.352	5.3	***	+		0.524	4.952	5.009	ns		0.057	−0.023
SI Rott	1.376	2.2	***	+		0.633	2.047	1.976	ns		0.112	−0.011
Halob	−6.062	12.24	***	+		0.358	9.076	8.961	ns		0.050	−0.025
PHYLIP	1.491	0.375	***	−		0.632	0.6666	0.7333	ns		0.073	−0.019

Mean values for best available (upper 75th percentile of PC gradient) and perturbed (lower 25th percentile of PC gradient) sites, significance using a non-paramentric Wilcoxcn rank-sum test, if significant direction of change, if not significant, power and coefficient of determination (R^2). ns = Non significant; $^*p<0.05$; $^{**}p<0.01$; $^{***}p<0.001$.

frequencies ≤15%, and six metrics had errors of only 5% (1 in 20).

Although several fish and macrophyte metrics showed a significant response to the 2nd PC stress gradient, error frequencies were high (Fig. 4b). For example, four of six fish metrics had 100% type II errors (Table 6) (This was because the lower 25 percentile values for high quality sites used to set the threshold for these metrics was zero, which made it impossible to obtain lower values and thus be classified as disturbed). Three of the seven macrophyte metrics had type II error frequencies of 100%, for similar reasons. Two macrophyte metrics the 'number of submerged species' and 'cover submerged species' had, however, low error frequencies <6%.

For lowland streams, only one of the fish metrics '% native intolerant species' and two macrophyte metrics, 'Mean Trophic Rank' and 'Macrophyte Biological Index for Rivers', showed a significant response to the 1st PC stress gradient (Fig. 4c, Table 7). Error frequencies for these three metrics were, however, high (all had type II errors >30% and two had errors >55%). In contrast to fish and macrophyte metrics, all macroinvertebrate and diatom metrics showed a significant response to the 1st PC gradient. Error frequencies for the diatom metrics were, however, much lower (<10%) than those noted for the macroinvertebrate metrics (range = 15–50%).

Similar to mountain streams, both fish and macrophyte metrics for lowland streams showed a significant response with the second PC stress gradient (Fig. 4d, Table 7). Error frequencies were high, in particular for the fish metrics, with values ranging from 15% for 'density of native benthic species' to 100% error for the 'number of long distance migrating species' where the lower inclusive limit for high quality sites was zero (i.e., at least 25% of high quality sites had no such species). Macrophyte metrics had error frequencies that ranged from 23.8% for the 'cover amphibious species' to 66.7% for the 'number of submerged species', and seven of the eight metrics had errors >35%.

Discussion

Organisms respond differently to stress depending on how they 'perceive' their environment (e.g., Southwood, 1977). Streams, in particular riffle habitats, are considered as harsh environments, where flow-related movements of the substratum strongly regulate the structure of biotic communities (e.g., Resh et al., 1988; Lake, 2000). However, whether stream assemblages are influenced more by abiotic or biotic factors (or both) remains unclear. A number of studies have shown the importance of abiotic environmental factors, such as water movement and substratum and riparian vegetation as strong predictors of assemblage composition (e.g., Ormerod et al., 1993; Allan, 1995; Richards et al., 1997; Johnson et al., 2004). By contrast, other studies have emphasised the importance of biotic interactions in structuring communities at both small (e.g., Englund, 1997) and large (e.g., Kohler & Wiley, 1997; Hildrew et al., 2004) spatial scales. It is likely that the scale of the study is important for the outcome (e.g., Wiley et al. 1997), and both abiotic and biotic variables may be important along varying and interweaving spatial and temporal templates.

Table 6. Type II error estimates for metrics that differed between best available and perturbed conditions for mountain streams

	PC 1 mountain streams					PC 2 mountain streams				
	Critical percentile of best available	Critical threshold value	No perturbed sites	Number of sites failed	% Type II error	Critical percentile of best available	Critical threshold value	No perturbed sites	Number of sites failed	% Type II error
Fish metrics										
n_sp_tol	75th	0	19	10	52.6	25th	0	16	16	100.0
n_ha_hab_wc	75th	2546	19	4	21.1					
n_sp_hab_rh						25th	4	16	3	18.8
n_ha_hab_rh	75th	3048	19	5	26.3					
n_sp_hab_eury	75th	0	19	10	52.6	25th	0	16	16	100.0
perc_nha_re_lith	25th	100	19	7	36.8					
n_sp_lon_ll	75th	0	19	9	47.4	25th	0	16	16	100.0
n_ha_fe_omni	75th	0	19	8	42.1	25th	0	16	16	100.0
n_sp_mi_potad	75th	1	19	12	63.2	25th	1	16	5	31.3
Macroivertebrate metrics										
ASPT	25th	7	20	1	5.0					
EPT-Taxa [%]	25th	35.45	20	2	10.0					
gatherers	75th	30.1	20	1	5.0					
GFID01	25th	0	20	8	40.0					
GFID05	25th	0.806	20	1	5.0					
MAS_IC	75th	1	20	1	5.0					
metarhithral	25th	28.2	20	2	10.0					
Plecoptera [%]	25th	3.817	20	0	0.0					
SI(ZM)	75th	1.598	20	1	5.0					
xeno	25th	4.13	20	4	20.0	75th	3.28	19	3	15.8

Macrophyte metrics										
n_sp_subm	75th	0.75	19	2	10.5	25th	2.75	17	1	5.9
n_sp_floating	75th	0	19	11	57.9	25th	0	17	17	100.0
n_sp_amphi	75th	0	19	5	26.3	25th	0	17	17	100.0
n_sp_terr	75th	1	19	5	26.3	25th	1	17	9	52.9
cover_moss_liv						25th	0.125	17	6	35.3
cover_sp_subm	75th	0.4125	19	6	31.6	25th	1.075	17	1	5.9
cover_sp_amphi	75th	0	19	5	26.3	25th	0	17	17	100.0
MTR	25th	49.2	19	4	21.1					
IBMR	25th	10.8	19	5	26.3					
Ellenberg_N*										
Benthic diatom metrics										
IPS	25th	18	20	1	5.0					
EPI-D	25th	12	20	3	15.0					
ROTT	25th	14	20	1	5.0					
CEE	25th	17.3	20	1	5.0					
TDI DVWK	75th	2.08	20	1	5.0					
TDI Rott	75th	1.95	20	1	5.0					
TDI lakes	75th	4.11	20	2	10.0					
SI Rott	75th	1.89	20	1	5.0					
Halob	75th	0	20	2	10.0					
PHYLIP	25th	0.86	20	3	15.0					

Critical threshold percentile of best available sites (25th percentile if change is negative and 75th percentile if change is positive), threshold value, number of sites in the perturbed class, number of sites failed and % type II error estimate.

Table 7. Type II error estimates for metrics that differed between best available and perturbed conditions for lowland streams

	PC 1 lowland streams					PC 2 lowland streams				
	Critical percentile of best available	Critical threshold value	No perturbed sites	Number of sites failed	% type II error	Critical percentile of best available	Critical threshold value	No perturbed sites	Number of sites failed	% type II error
Fish metrics										
n_sp_intol						25th	1	20	10	50.0
perc_sp_intol	25th	7	19	11	57.9	25th	25	20	5	25.0
n_sp_tol										
n_sp_hab_b										
n_ha_hab_b						25th	237	20	3	15.0
perc_nha_hab_b						25th	18	20	13	65.0
n_ha_hab_eury										
n_ha_re_phyt						75th	0	20	8	40.0
n_sp_fe_insev						25th	1	20	9	45.0
n_sp_mi_long						25th	0	20	20	100.0
Macroinvertebrate metrics										
ASPT	25th	5.76	20	3	15.0					
DSFI	25th	5	17	6	35.3	25th	5	21	10	47.6
epirhithral	25th	8.84	20	8	40.0					
EPT-Taxa [%]	25th	19.4	20	5	25.0					
gatherers	75th	34.7	20	3	15.0					
GFID01	25th	−0.783	20	8	40.0					
GFID05	25th	−0.027	20	10	50.0					
pelal	75th	12.1	20	3	15.0					
Plecoptera [%]	25th	0.432	20	4	20.0					
SI(ZM)	75th	2.05	20	5	25.0					

Macrophyte metrics										
n_sp_floating						75th	1	21	14	66.7
n_sp_amphi						75th	3	21	9	42.9
cover_sp_amphi						75th	1.83	21	5	23.8
MTR	25th	38	16	9	56.3					
IBMR	25th	10.6	16	5	31.3					
sp_n						75th	12	21	11	52.4
ge_n						75th	12	21	11	52.4
fa_n						75th	9.5	21	8	38.1
Species number (S*)						75th	8.5	20	7	35.0
Species number (S**)						75th	7.8	20	8	40.0
Benthic diatom metrics										
She	25th	14.2	20	1	5.0					
EPI-D	25th	12.5	20	1	5.0					
ROTT	25th	15.2	20	0	0.0					
IDG	25th	13.7	20	2	10.0					
TDI DVWK	75th	2.5	20	2	10.0					
TDI Rott	75th									
DI Swi	75th	2.45	20	1	5.0					
SI Rott	75th	4.15	20	0	0.0					
Halob	75th	0.5	20	1	5.0					
PHYLIP	25th	1.05	20	0	0.0					

Critical threshold percentile of best available sites (25th percentile if change is negative and 75th percentile if change is positive), threshold value, number of sites in the perturbed class, number of sites failed and % type II error estimate.

Given differences in life history strategies, we anticipated that the different organism groups studied here would respond differently to different environmental gradients. For example, due to their relatively high mobility we expected that fish metrics might be more correlated with large-scale, catchment- or reach-scale variables, whereas, conversely, due to their relative sedentary behavior we anticipated that macroinvertebrate metrics would be more related to habitat (substratum) conditions. Moreover, we expected that organisms that are more directly related (first order principle) to an environmental variable (e.g., benthic diatom reliance on nutrient concentrations for growth) would be more related to an important primary gradient (such as nutrient concentration) than organism-groups which rely on the gradient indirectly (e.g., invertebrates that rely on diatoms for food).

Our findings that benthic diatom metrics were strongly related to the 1st PC axis, interpreted as primarily representing nutrient levels for both mountain and lowland streams, supports the conjecture that this organism group (or the metrics tested here) is a robust indicator for monitoring changes in nutrient enrichment (e.g., Hering et al. submitted). Moreover, benthic diatom metrics often had the lowest frequency of false negative error (or uncertainty) associated with their response. In mountain streams the order of response (ranked by mean R^2 values, see Table 4) was macroinvertebrates (0.511) > benthic diatoms (0.491) > > macrophytes (0.197) > fish (0.147). For lowland streams, where differences in nutrient levels between best available and perturbed sites were even greater than those for mountain streams, the order of importance switched between macroinvertebrates and diatoms; namely, benthic diatoms (0.578) > macroinvertebrates (0.258) ≫ macrophytes (0.042) ≈ fish (0.033). These findings, in particular the response of diatom and macroinvertebrate metrics to the 1st PC gradient in both mountain and lowland streams, lends support to the use of these organism groups/metrics for monitoring the effects of catchment land use on stream integrity. Our findings that macroinvertebrates were placed either first (in mountain streams) or second (in lowland streams) in order of ranked importance indicates a close relation between benthic diatoms and macroinvertebrates. This is not surprising since diatoms are considered as a high quality food source, supplying a number of essential fatty acids. Moreover, these findings agree with studies by Hering et al. (submitted) who, using subsets of this same project data, showed that benthic diatoms and macroinvertebrate assemblages were reliable indicators of changes in stream nutrient status. The impact of nutrients on macroinvertebrates might, however, be indirect, since most likely macroinvertebrates most probably react to related changes in oxygen content.

The second most prevalent gradient (PC2) was interpreted as being either related to habitat (mountain streams) or hydromorphological (lowland streams) characteristics. For example, the number of debris dams averaged 1.6 at the best available mountain sites, compared to 0.10 at the perturbed sites. Habitat characteristics are known be important predictors of stream communities (e.g., Allan, 1995), hence we anticipated that loss (e.g., decrease in the number of debris dams) or alteration (e.g., siltation) of in-stream habitat quality/quantity should negatively affect most if not all of the four organism groups studied here. For example, high habitat heterogeneity often results in high diversity by increasing niche space. Earlier studies have shown both macrophytes and fish to be reliable indicators of alterations in flow and habitat quality (e.g., Tremp & Kohler, 1995; Gorman & Karr, 1978; Bain et al., 1988), hence we anticipated that fish and macrophytes would be more dependent on habitat variability than benthic diatoms and macroinvertebrates. Both fish and macrophyte metrics were better correlated to the 2nd PC gradient than either benthic diatoms or macroinvertebrates, thereby supporting this conjecture. For mountain streams the ranked order for the 2nd PC gradient was fish (mean $R^2 = 0.188$) ≈ macrophytes (0.179) ≫ macroinvertebrates (0.048) > benthic diatoms (0.008). Alterations in habitat and hydromorphology are difficult, if not impossible, to disentangle and both might be expected to result in similar behavioural responses of the four organism groups studied here. Consequently, we were not too surprised to find that the ranking of organism-group/metrics for lowland streams was similar to that noted for mountain streams; fish (0.144) ≈ macrophytes (0.137) ≫ macroinvertebrates (0.026) > 0.019). Hering et al. (submitted), who used the same dataset, found the same order of organism group's response to nutrient enrichment and related saprobity. However, when excluding heavily polluted sites and correlating the same metrics to hydromorphology gradients the response of invertebrate metrics was much stronger, even stronger than those of fish and macrophyte metrics, while diatom metrics did not respond to hydromorphology gradients. Thus, the results of this paper might be influenced by several heavily polluted sites, which weaken the response of metrics to less obvious gradients.

Although the organism groups/metrics seemed to respond in predictable way to the main environmental gradients studied here, the statistical power and error associated with determining differences between the two quality classes varied markedly with organism group and metric. In general, levels of uncertainty followed the reverse order of ranked importance to the gradients. For example, benthic diatom and macroinvertebrate metrics had on averaged lower frequencies of false negative errors than macrophyte or fish metrics. For mountain streams, fish and macrophyte metrics had approximately three (macrophyte) to four (fish) times higher error frequencies than benthic diatom or macroinvertebrate metrics (1st PC axis). This implies that choice of the 'best' organism group and/or metric can be crucial for detecting (or not detecting) human-induced change in stream integrity. Although few studies have compared the error associated with different methods used in bioassessment, these error frequencies are not too uncommon. For example, Johnson (1998) in a study of the response of macroinvertebrate metrics to acidification stress in lakes showed that statistical power and type II error varied with habitat type and metric. The most robust metrics were those that were stress-specific (using species tolerance/sensitivity to stress), followed by richness-based metrics. Metrics based on species abundance were found to have lower power to detect degradation. Similar findings were found by Sandin & Johnson (2000) for stream macroinvertebrate metrics. Common to both of these studies was the surprisingly low power (or high type II error) associated with measures of diversity. These findings imply that considerable ecological change may go undetected (i.e., false negative error), and that not only the organism group to be monitored but also the metric used should be given be given careful consideration at the outset of a study.

Ideally, the response variables used in ecological assessment to detect anthropogenic effects should be stressor specific and have low levels of uncertainty. Organism/metric response to stress varied among organism groups/metrics and with different types of stress. Our study also showed, however, that often two or more of the organism groups were significantly related to the same stressor, which implies a certain degree of redundancy among the organism groups/metrics tested here. The strength of the relationship (response) and the potential for false negative error varied considerably, indicating that consideration should be given to the type of impact that is expected to occur when selecting the 'best' indicator organism/metric. For example, if nutrient enrichment is the main stressor affecting stream integrity then diatom or macroinvertebrate metrics might be given first consideration. Conversely, if habitat/hydromorphological alteration is the main stressor then fish or macrophytes might be considered.

Acknowledgements

This paper is a result of the EU-funded project STAR (5th Framework Programme; contract number: EVK1-CT-2001-00089). Parts of the data analysis were supported by the EU-funded Integrated Project Euro-limpacs (6th Framework Programme; contract number: GOCE-CT-2003-505540). We are most grateful to all STAR partners having provided data for this analysis.

References

Allan, J. D., 1995. Stream Ecology. Structure and Function Of Running Waters. Chapman and Hall, London.

Armitage, P. D., D. Moss, J. F. Wright & M. T. Furse, 1983. The performance of a new biological water quality score system based on macroinvertebrates over a wide range of unpolluted running-water sites. Water Research 17: 333–347.

Bain, M. B., J. T. Finn & H. E. Booke, 1988. Streamflow regulation and fish community structure. Ecology 69: 382–392.

Battarbee, R. W., R. J. Flower, S. Juggins, S. T. Patrick & A. C. Stevenson, 1997. The relationship between diatoms and surface water quality in the Hoylandet area of Nord-Trondelag, Norway. Hydrobiologia 348: 69–80.

Buffagni, A., 1997. Mayfly community composition and the biological quality of streams. In Landolt, P. & M. Sartori (eds), Ephemeroptera & Plecoptera: Biology-Ecology-Systematics. MTL, Fribourg, 235–246.

Buffagni, A., 1999. Pregio naturalistico, qualità ecologica e integrità della comunità degli Efemerotteri. Un indice per la classificazione dei fiumi italiani. Acqua & Aria 8: 99–107.

Buffagni, A., S. Erba, M. Cazzola & J. L. Kemp, 2004. The AQEM multimetric system for the southern Italian Apennines: assessing the impact of water quality and habitat degradation on pool macroinvertebrates in Mediterranean rivers. Hydrobiologia 516: 313–329.

CEMAGREF, 1982. Etude des méthodes biologiques d'appréciation quantitative de la qualité des eaux. Rapport Q. E. Lyon A. F. Bassin Rhône-Méditérannée-Corse, 218 pp.

Coring, E., 1999. Situation and developments of algal (diatom)-based techniques for monitoring rivers in Germany. In Prygiel, J., B. A. Whitton & J. Bukowska (eds), Use of Algae for Monitoring Rivers III. Agence de l'Eau Artois-Picardie, Douai, 122–127.

Dell'Uomo, A., 1996. Assessment of water quality of an Apennine river as a pilot study for diatom-based monitoring of Italian watercourses. In Whtitton, B. A. & E. Rott (eds), Use of Algae for Monitoring Rivers II, Institut für Botanik, Universität Innsbruck, 65–72.

Descy, J. P. & M. Coste, 1990. Utilisation des diatomées benthiques pour l'évaluation de la qualité des eaux courantes. Rapport final, UNECED, Namur, Cemagref, Bordeaux.

Ellenberg, H., H. E. Weber, R. Düll, V. Wirth, W. Werner & D. Paulißen, 1992. Zeigerwerte von Pflanzen in Mitteleuropa. Scripta Geobotanica 18: 1–257.

Englund, G., 1997. Importance of spatial scale and prey movements in predator caging experiments. Ecology 78: 2316–2325.

European Commission, 2000. Directive 2000/60/EC of the European Parliament and of the Council – Establishing a framework for Community action in the field of water policy. Brussels, Belgium, 23 October 2000.

Fore, L.S., J. R. Karr & R. W. Wisseman, 1996. Assessing invertebrate responses to human activities: evaluating alternative approaches. Journal of the North American Benthological Society 15: 212–231.

Furse, M. T., A. Schmidt-Kloiber, J. Strackbein, J. Davy-Bowker, A. Lorenz, J. van der Molen & P. Scarlett, 2004. Results of the sampling programme. A report to the European Commission. Framework V Project STAR (EVK1-CT-2001_00089).

Gorman, O. T. & J. R. Karr, 1978. Habitat structure and stream fish communities. Ecology 59: 507–515.

Haury, J., M. C. Peltre, M. Tremolieres, J. Barbe, G. Thiebaut, I. Berne, H. Daniel, P. Chatenet, S. Muller, A. Dutartre, C. Laplace-Treyture, A. Cazaubon & E. Lambert-Servien, 2002. A method involving macrophytes to assess water trophy and organic pollution: the Macrophyte Biological Index for Rivers (IBMR) – application to different types of rivers and pollutions. In Dutartre, A. & M. H. Montel (eds), Proceedings 11th EWRS Internat. Symp. Aquatic Weeds. Moliets Et Maa, France, 247–250.

Hering, D., C. Meier, C. Rawer-Jost, C. K. Feld, R. Biss, A. Zenker, A. Sundermann, S. Lohse & J. Böhmer, 2004. Assessing streams in Germany with benthic invertebrates: selection of candidate metrics. Limnologica 34: 398–415.

Hering, D. & J. Strackbein, 2002. STAR stream types and sampling sites. A report to the European Commission. Framework V Project STAR (EVK1-CT-2001_00089).

Hildrew, A. G., G. Woodward, J. H. Winterbottom & S. Orton, 2004. Strong density dependence in a predatory insect: large-scale experiments in a stream. Journal of Animal Ecology 73: 447–458.

Hirst, H., I. Jüttern & S. J. Ormerod, 2002. Comparing the responses of diatoms and macroinvertebrates to metals in upland streams of Wales and Cornwall. Freshwater Biology 47: 1752–1765.

Hoffmann, G., 1999. Trophiebewertung von Seen anhand von Aufwuchsdiatomeen. Von Tümpling, W. & G. Friedrich (eds), Biologische Gewässeruntersuchung, 2. Gustav Fischer Verlag, Jena: 319–333.

Holmes, N. T. H., J. R. Newman, S. Chadd, K. J. Rouen, L. Saint & F. H. Dawson, 1999. Mean Trophic Rank: A users manual. R&D Technical Report No. E 38, Environment Agency, Bristol, UK.

Johnson, R. K., 1998. Spatio-temporal variability of temperate lake macroinvertebrate communities: detection of impact. Ecological Applications 8: 61–70.

Johnson, R. K., T. Wiederholm & D. M. Rosenberg, 1993. Freshwater biomonitoring using individual organisms, populations and species assemblages of benthic macroinvertebrates. In Rosenberg, D. M. & V. H. Resh (eds), Freshwater Biomonitoring and Benthic Macroinvertebrates. Chapman and Hall, New York, 40–158.

Johnson, R. K., W. Goedkoop & L. Sandin, 2004. Spatial scale and ecological relationships between the macroinvertebrate communities of stony habitats of streams and lakes. Freshwater Biology 49: 1179–1194.

Knoben, R. A. E., C. Roos & M. C. M. van Oirschot, 1995. Biological Assessment Methods for Watercourses. UN/ECE Task Force on Monitoring and Assessment, Lelystad.

Kohler, S. L. & M. J. Wiley, 1997. Pathogen outbreaks reveal large-scale effects of competition in stream communities. Ecology 73: 2164–2176.

Kolkwitz, R. & M. Marsson, 1902. Grundsätse für die biologische Beurteiling des Wassers nach seiner Flora und Fauna. Mitt. Prüfungsanst. Wasserversorg. Abwasserreinig 1: 33–72.

Lake, P. S., 2000. Disturbance, patchiness, and diversity in streams. Journal of North American Benthological Society 19: 573–592.

LeCointe, C., M. Coste & J. Prygiel, 1993. "OMNIDIA" software for taxonomy, calculation of diatom indices and inventories management. Hydrobiologia 269/270: 509–513.

Lorenz, A., D. Hering, C. K. Feld & P. Rolauffs, 2004. A new method for assessing the impact of morphological degradation on the benthic invertebrate fauna for streams in Germany. Hydrobiologia 516: 107–127.

McNaughton, S. J., 1967. Relationships among functional properties of Californian grasslands. Nature 216: 168–169.

Metcalfe, J. L., 1989. Biological water-quality assessment of running waters based on macroinvertebrate communities –

history and present status in Europe. Environmental Pollution 60: 101–139.

Noble, R. & I. Cowx, 2002. Development of a river-type classification system (D1); Compilation and harmonization of fish species classification (D2). Report of the EU funded project Development, Evaluation & Implementation of a Standardised Fish-based Assessment Method for the Ecological Status of European Rivers – A Contribution to the Water Framework Directive (FAME). Available from http://fame.-boku.ac.at/downloads/D1_2_typology_and%20species_classification.pdf.

Ormerod, S. J., S. D. Rundle, E. C. Lloyd & A. A. Douglas, 1993. The influence of riparian management on the habitat structure and macroinvertebrate communities of upland streams draining plantation forests. Journal of Applied Ecology 30: 13–24.

Paavola, R., T. Muotka, R. Virtanen, J. Heino & P. Kreivi, 2003. Are biological classifications of headwater streams concordant across multiple taxonomic groups? Freshwater Biology 48: 1912–1923.

Raven P. J., N. T. H. Holmes, F. H. Dawson, P. J. A. Fox, M. Everard, I. R. Fozzard & K. J. Rouen, 1998. River Habitat Quality – the physical character of rivers and streams in the UK and Isle of Man. River Habitat Survey Report Number 2, Bristol: Environment Agency, Stirling: Scottish Environment Protection Agency. Belfast: Environment and Heritage Service, 84.

Resh, V. H., A. V. Brown, A. P. Covich, M. E. Gurtz, H. W. Li, G. W. Minshall, S. R. Reice, A. L. Sheldon, J. B. Wallace & R. C. Wissmar, 1988. The role of disturbance in stream ecology. Journal of the North American Benthological Society 7: 433–455.

Reynoldson, T. B., R. H. Norris, V. H. Resh, K. E. Day & D. M. Rosenberg, 1997. The reference condition: a comparison of multimetric and multivariate approaches to assess water-quality impairment using benthic macroinvertebrates. Journal of the North American Benthological Society 16: 833–852.

Richards, C., R. J. Haro, L. B. Johnson & G. E. Host, 1997. Catchment and reach-scale properties as indicators of macroinvertebrate species traits. Freshwater Biology 37: 219–230.

Rott, E., G. Hofmann, K. Pall, P. Pfister & E. Pipp, 1997. Indikationslisten für Aufwuchsalgen. Teil 1: Saprobielle Indikation. Bundesministerium für Land- und Forstwirtschaft, Wien, 73 pp.

Rott, E., P. Pfister, H. van Dam, E. Pipp, K. Pall, N. Binder & K. Ortler, 1999. Indikationslisten für Aufwuchsalgen. Teil 2: Trophieindikation sowie geochemische Präferenz, taxonomische und toxikologische Anmerkungen. Bundesministerium für Land- und Forstwirtschaft, Wien, 248 pp.

Rumeau, A. & M. Coste, 1988. Initiation à la systématique des diatomées d'eau douce pour l'utilisation pratique d'un indice diatomique générique. Bulletin Francais de la Peche et de la Pisciculture 309: 1–69.

Sandin, L. & R. K. Johnson, 2000. Statistical power of selected indicator metrics using macroinvertebrates for assessing acidification and eutrophication of running waters. Hydrobiologia 422/423: 233–243.

Sandin, L., J. Dahl & R. K. Johnson, 2004. Assessing acid stress in Swedish boreal and alpine streams using benthic macroinvertebrates. Hydrobiologia 516: 129–148.

SAS, 1994. JMP – Statistics Made Visual, Version 3.1. SAS Institute Inc, Cary, North Carolina, USA.

Southwood, T. B. E., 1977. Habitat, the template for ecological strategies?. Journal of Animal Ecology 46: 336–365.

Schaumburg, J., C. Schranz, J. Foerster, A Gutowski, G. Hofmann, P. Meilinger, S. Schneider & U. Schmedtje, 2004. Ecological classification of macrophytes and phytobenthos for rivers in Germany according to the Water Framework Directive. Limnologica 34: 283–301.

Shannon, C. E. & W. Weaver, 1949. The Mathematical Theory of Communication. University of Illinois Press, Urbana.

Skriver, J., N. Friberg & J. Kirkegaard, 2001. Biological assessment of running waters in Denmark: Introduction of the Danish Stream Fauna Index (DSFI). Verhandlungen der internationalen Vereineinigung für theoretische und angewandte Limnologie 27: 1822–1830.

Steinberg, C. & S. Schiefele, 1988. Biological indication of trophy and pollution of running waters. Zeitschrift für Wasser- und Abwasser-Forschung 21: 227–234.

Tremp, H. & A. Kohler, 1995. The usefulness of macrophyte monitoring-systems, exemplified on eutrophication and acidification of running waters. Acta Botanica Gallica 142: 541–550.

Wiley, M. J., S. L. Kohler & P. W. Seelbach, 1997. Reconciling landscape and local views of aquatic communities: lessons from Michigan trout streams. Freshwater Biology 37: 133–148.

Zelinka, M. & P. Marvan, 1961. Zur Präzisierung der biologischen Klassifikation der Reinheit fließender Gewässer. Archiv für Hydrobiologie 57: 389–407.

Ziemann, H., 1999. Bestimmung des Halobienindex. Tümpling W. & G. Friedrich (eds), Biologische Gewässeruntersuchung. Methoden der Biologischen Gewässeruntersuchung, 2: 310–313.

Hydrobiologia (2006) 566:139–152
M.T. Furse, D. Hering, K. Brabec, A. Buffagni, L. Sandin & P.F.M. Verdonschot (eds), The Ecological Status of European Rivers: Evaluation and Intercalibration of Assessment Methods
DOI 10.1007/s10750-006-0100-9

Indicators of ecological change: comparison of the early response of four organism groups to stress gradients

Richard K. Johnson[1,*], Daniel Hering[2], Mike T. Furse[3] & Piet F.M. Verdonschot[4]
[1]*Department of Environmental Assessment, Swedish University of Agricultural Sciences, P.O. Box 7050 SE-750 07 Uppsala, Sweden*
[2]*Department of Hydrobiology, University of Duisburg-Essen, D-45117 Essen, Germany*
[3]*Centre for Ecology and Hydrology, Winfrith Technology Centre, Winfrith Newburgh, DT2 8ZD, Dorchester, Dorset, UK*
[4]*Alterra Green World Research, Freshwater Ecology, Droevendaalsesteeg 3a, NL-36700 AA Wageningen, The Netherlands*
*(*Author for correspondence: E-mail: richard.johnson@ma.slu.se)*
Received 25 July 2005; accepted 3 October 2005

Key words: early response, streams, bioassessment, monitoring

Abstract

A central goal in monitoring and assessment programs is to detect change early before costly or irreversible damage occurs. To design robust early-warning monitoring programs requires knowledge of indicator response to stress as well as the uncertainty associated with the indicator(s) selected. Using a dataset consisting of four organism groups (fish, macrophytes, benthic diatoms and macroinvertebrates) and catchment, riparian and in-stream physico-chemical variables from 77 mountain and 85 lowland streams we determined the relationships between indicator response and complex environmental gradients. The upper (> 75th percentile) and lower (< 25th percentiles) tails of principal component (PC) gradients were used to study the early response of the four organism groups to stress. An organism/metric was considered as an early warning indicator if the response to the short gradients was more robust (higher R^2 values, steeper slope and lower error) than the null model (organism response to the full PC gradient). For mountain streams, both fish and macrophyte CA scores were shown to exhibit an early warning response to the upper tail of the 1st PC gradient when compared to the null model. Five of the eight metrics showed better response to the upper tail of the 2nd PC gradient compared to the null model, while only one metric (macrophyte CA scores) showed improvement when compared to the lower tail of the 2nd PC gradient. For lowland streams all four organism-groups showed better response (CA scores) to the upper tail of the PC gradient when compared to the null model. Only one metric (fish CA scores) regressed against the lower tail of the 2nd PC gradient was found to be more robust than the PC2 null model. These findings indicate that the nonlinear relationships of organism/metric response to stress can be used to select potentially robust early warning indicators for monitoring and assessment.

Introduction

Humans have altered the landscape of Europe for centuries resulting in a substantial loss of habitats and biodiversity. Aquatic resources in general, and stream habitats in particular, are some of the most threatened on Earth. Recognizing that biodiversity as well as the functions and services provided by aquatic ecosystems have changed markedly over the years, the European Community recently agreed upon a number of measures to impede degradation and improve quality of inland and coastal waters (European Commission, 2000). One of the innovative aspects of the Water Framework

Directive is the use of multiple indicators (organism groups and metrics) in monitoring and assessment programs. Advocates of the approach argue that the use of multiple indicators increases the probability of detecting change if/when change occurs (a.k.a. multiple lines of evidence). This presumption is based on the premise that the indicators selected are not redundant, but supply complementary information. A second advantage of using multiple indicators to detect ecological impairment is that not only the trajectories, but also the rates of change may differ between indicators selected. Knowledge of how organisms respond to different types of stress can and should be used to design more robust and cost-effective monitoring programs.

Biological response variables are often selected over physical–chemical variables because they represent valued ecosystem attributes such as species richness or ecosystem productivity (e.g., Stevenson et al., 2004). The use of biological variables in European monitoring and assessment programmes has a long history (e.g., Metcalfe, 1989), stemming from the early 1900s when German aquatic ecologists began using the Saprobien Index to assess the effects of organic pollution on streams (Kolkwitz & Marsson, 1902). Although benthic macroinvertebrates are probably the single most common organism group presently used in bioassessment, other groups such as fish and periphyton are being used more frequently. In North America, for example, benthic diatoms, macroinvertebrates and fish are commonly used together to assess the ecological quality of streams (Barbour et al., 1998).

Ideally, the selection of what or which organism group(s) to use in bioassessment should not be arbitrary, but should be based on conceptual models and empirical (e.g., dose–response) relationships that characterize the response of the indicator to the stressor of interest as well as quantify the levels of uncertainty associated with the stressor–response relationship. Organism response to stress varies with a number of abiotic and biotic factors, such as an organism's life history stage. Because responses to environmental stress originate at the biochemical and physiological levels of the organism, changes at the suborganism-level may provide the earliest warning of possible adverse effects (Johnson et al., 1993). Unfortunately, knowledge of the normal background variability often limits the use of biochemical and physiological indicators in biomonitoring. The idea of using an indicator that provides an early indication of change has, however, many benefits, not the least being the socioeconomic aspects of failing to detect an ecological change early. For example, considerable costs may be incurred if human-induced damage is allowed to proceed undetected.

Organism response to human-induced stress is not always linear, and selection of indicators that respond rapidly at the outset of impairment is one way of determining and quantifying early-on the effects that humans may have on ecosystem integrity. Our working hypothesis is that organisms that show a nonlinear response to the stress will show more rapid response (higher slopes) with moderate than high levels of stress. At high levels of stress, dose–response relationships often show a leveling off (e.g. low variance around the regression line) resulting in typical funnel-shaped response curves. Here we use stress–response relations of four organism groups to determine if the organisms differ in their response to stress. In doing so, we hope to better our understanding of the use of early warning indicators for detecting ecological change if/when it occurs.

Methods

Study sites

Some 162 streams were sampled in 2003 and 2004 as part of the European funded STAR project (Hering & Strackbein, 2002; Furse et al., 2004). Two common stream types (mountain, $n = 77$ and lowland, $n = 85$) are used here to study the response of different organism groups to human-induced stress (Fig. 1). To ensure adequate sampling of stressor gradients, prior to sampling, all sites were pre-classified into five classes of ecological status using physico-chemical and in some instances biological information and/or expert opinion: (i) high (no or only minimal disturbance), (ii) good (slight deviation from high status), (iii) moderate (moderate deviation from high status and significantly more disturbed than good), (iv) poor (major alteration from high status) and

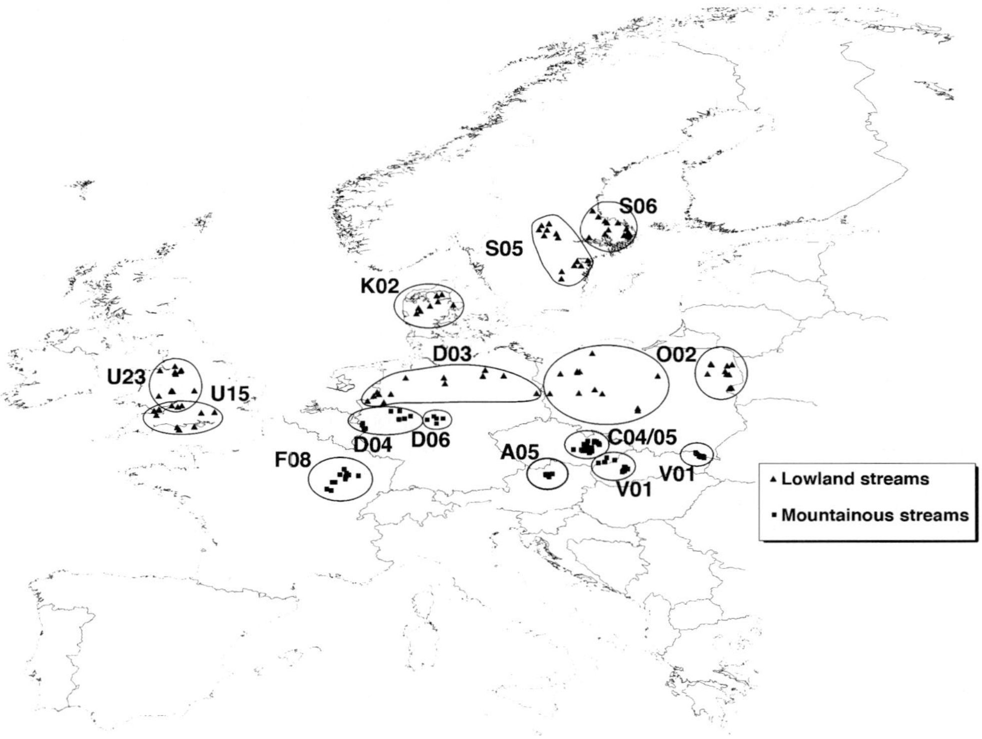

Figure 1. Location of streams and types sampled as part of the EU-STAR project.

(v) bad (severe alteration from high status). For each stream type, 10 and usually 12 sites were sampled of which, in each set, at least three were pre-classified as of high status and the rest spread more or less evenly over the other four quality status classes. In addition, a large number of physical, chemical and geographical variables were sampled or obtained (e.g., using GIS) for each of the streams sampled (Furse et al., 2004). The substratum of each of the sampling sites was classified (in percent) according to seven inorganic size classes (i.e., from silt/clay $< 6\ \mu m$ to large cobbles, boulders and blocks > 40 cm) and 10 organic classes/fractions such as the amount of algae (macro & micro), vegetation (aquatic submerged and emergent and living parts of terrestrial plants) and detritus (e.g., woody debris). A number of physical–chemical metrics representative of nutrient status (nitrogen and phosphorus fractions) and acidity (pH) status, as well as oxygen conditions (BOD5), were measured for each site. Catchment and riparian land use/type was classified according to 16 classes (e.g., forest type, cropland, pasture, urbanization). A number of measures of stream hydrology and morphology were recorded such as mean annual discharge valley and channel form (seven classes) and stream width and depth using the RHS survey technique (Raven et al., 1998). Also included in hydromorphological classification were measures of bank and bed fixation and the number of debris dams and woody debris in and along the stream channel.

Biological samples

Four organism groups were sampled at each stream site; namely, fish, macroinvertebrates, macrophytes and periphytic diatoms. A brief description of the sampling method used is given here, for more detailed information refer to the STAR website (www.eu-star.at).

Fish were normally sampled by electric fishing in accordance with the procedures set out in CEN prestandard PrEN 14011. Fishing was undertaken along two runs of a stop-netted area on a single occasion in late summer or early autumn. The recommended sampling length was 10× the stream width, with a minimum of a 100 m stream length sampled. The fish variables recorded were number of species, life history stage (young of the year per species), density (number of fish per m^2) and assessment of the degree of infestation of external parasites or other diseases.

Benthic macroinvertebrates were sampled in spring and either summer or autumn using a Surber sampler or by standardized kick-sampling with a handnet (area 625 cm^2, mesh size 500 μm). Generally the sampling section consisted of 20–50 m in length in small (1–100 km^2) and 50–100 m in length in medium (100–1000 km^2) sized streams. Each sampling site encompassed the whole width of the stream and was deemed to be representative of a minimum area surveyed (i.e., 500 m of length or 100× average width). Before sampling, the sampling site was first classified according to the coverage of all microhabitats with at least 5% cover. A multi-habitat sampling strategy was then adopted that reflected the proportion of different habitat types present at each stream site. Each complete sample consisted of 20 sample units of dimensions 25 cm × 25 cm. These sampling units were proportionally situated in all microhabitats with > 5% coverage. The 20 sample units resulted in ca. 1.25 m^2 of stream bottom being sampled. Each composite sample was preserved with formalin (4% final concentration) or 95% ethanol to a final concentration of ca. 70%. Macroinvertebrates were sorted (subsampling with the target of 700 individuals) and identified (usually to genus/species).

Macrophytes were sampled using a single survey in late summer or early autumn. Macrophytes included higher aquatic plants, vascular cryptograms, bryophytes as well as groups of algae. A 100 m stream length was surveyed in each stream by wading, walking along the bank or by boat according to the MTR method described by Holmes et al. (1999). All macrophyte species were recorded as well as the percent cover of the overall macrophyte growth. Submerged vegetation was observed using a glass-bottom bucket. If identification was uncertain, representative samples were collected and identified later.

Periphytic diatoms were sampled from hard (usually cobbles or pebbles) or soft (sand/silt) substratum or macrophytes. Wherever possible periphyton samples were collected within the same sampling area as benthic macroinvertebrates. In brief, a minimum of five cobbles were arbitrarily selected at each site (the combined exposed surface area comprised ca. 100 cm^2). The stones were individually placed in a plastic tray and 100–200 ml of distilled or filtered water added to the tray. The upper part of the stone substratum was washed using a toothbrush, and the dislodged material was decanted into a sample bottle and a composite sample was preserved (using formalin or Lugol's iodine solution) if the sample could not be processed within 24 h. Submerged macrophytes and parts of emergent ones were collected, placed in a wide-mouth 1-l container, ca. 100–200 ml of distilled/filtered water added and the container was shaken vigorously for about 60 s. A 250 ml aliquot of the sample was decanted to a sample bottle and preserved as above if not analyzed within 14 h. Mineral sediments were sampled using a glass tube submerged in the sediment and extracting sediment and interstitial water. Replicate samples were collected until volume of ca. 200 ml was obtained. Light microscopy was used to identify the living and dead diatom cells. The diatom species were counted (a minimum of 300 diatom valves) and identified to species at 400× and 1000× magnification.

Analyses

Environmental gradients and organism response

Principal components analysis (PCA) was used to construct complex stress gradients by reducing the dimensionality of the physical–chemical, hydromorphological and land use/type characteristics for each of the sites. Most environmental variables were log10 or arcsine square-root transformed before the analyses to approximate normal distributions. To test the early response of the different organism groups to stress, the upper and lower tails of the PC gradients were used as 'short' environmental gradients. The short environmental gradients were constructed by using the two tails of the first two PC axes for the mountain and lowland streams (Fig. 2). The 75th-percentile was arbitrarily selected as the cutoff for 'best available' sites and the 25th-percentile as the cutoff for 'perturbed' sites, resulting in four environmental

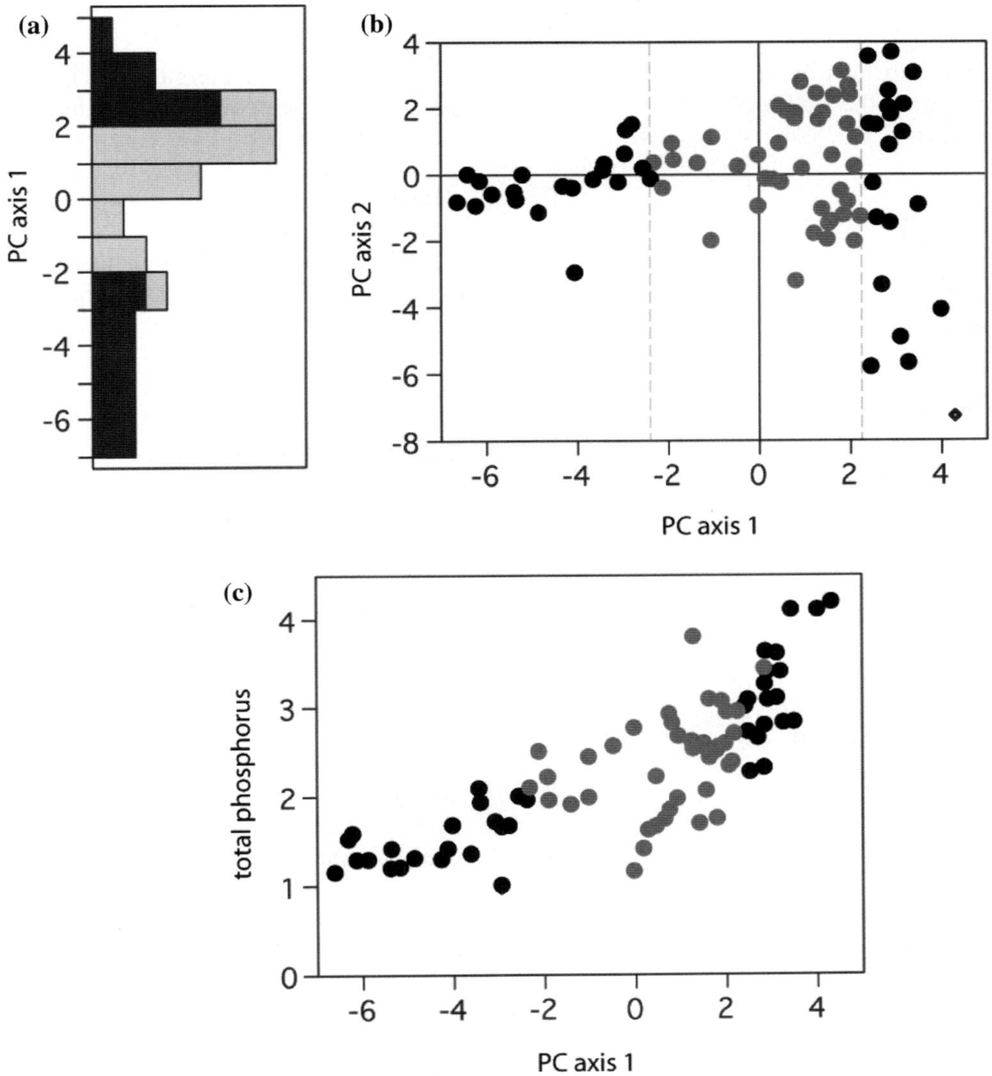

Figure 2. Example of the selection of upper and lower tail sites of the 1st and 2nd principal component gradients (PC axis 1 and PC axis 2, respectively). (a) Distribution of PC axis 1 scores, (b) PC axis 1 plotted against PC axis 2 showing the two tails of the distribution that were used in short-gradient regression analyses, (c) relationship between stream log total phosphate concentration and PC axis 1.

datasets or gradients. Sites with PC-axes scores between the 25th- and 75th-percentiles were omitted from the analyses.

Two biological metrics (correspondence scores and Hill's N2-diversity) were used to compare the response of the different organism groups to stress. To obtain correspondence scores, fish, macroinvertebrate, macrophyte and diatom abundances were ordinated separately for the two stream types using correspondence analysis (ter Braak, 1988, 1990). Correspondence analysis was run on square-root transformed species abundance, with the downweighting of rare taxa option invoked. The ordination scores on the first CA axis (CA1) and Hill's N2-diversity (Hill, 1973) were used as dependent variables and related to environmental stress gradients.

Linear regression was used to determine the relationship between the two metrics for the four organism groups and their response to the four short gradients (Fig. 3). For the null model we used the PC1 and PC2 gradients for all mountain

(n = 77) and lowland (n = 85) streams. Regression results of the response of the four organism groups to the two tails of the PC gradients were then compared to the null model. Coefficients of variation (adjusted R^2), slope, error (RMSEP) and

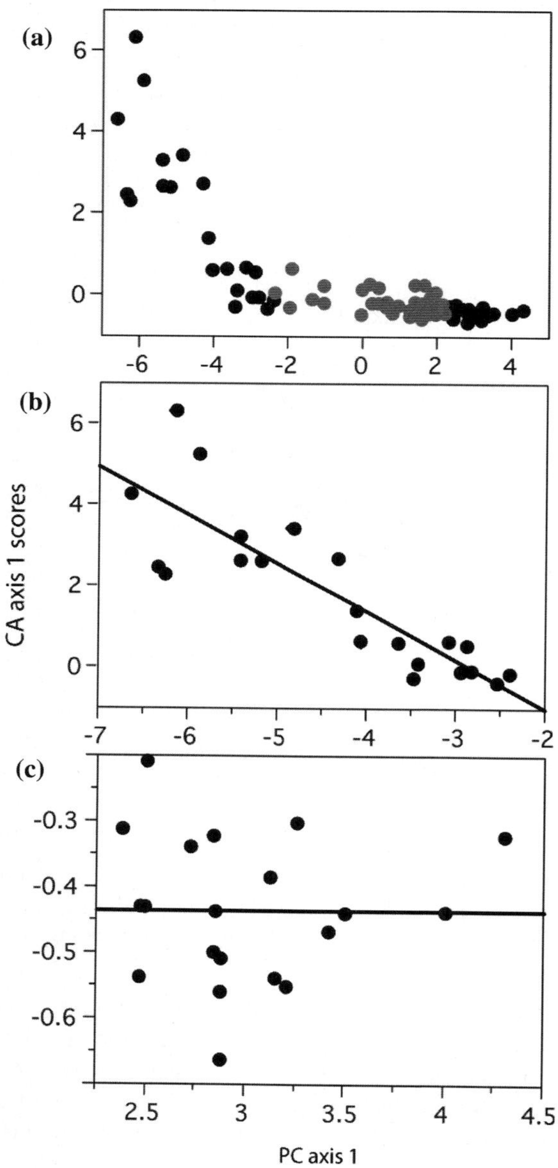

Figure 3. Example of benthic diatom response (CA axis 1 scores) to the 1st principal component gradient (PC axis 1, representing nutrient enrichment) for lowland streams. (a) Response using all stream sites, (b) response using upper tail (> 75th percentile of PC axis 1 gradient) sites, (c) response using lower tail (< 25th percentile of PC axis 1 gradient) sites.

p-values were used to compare the response of the four organism groups to the PC gradients. Coefficients of variation were used as a measure of the precision, slope provided an estimate of the magnitude of change and the root mean square error of the prediction was an estimate of the standard deviation of the random error associated with the response model.

All tests were performed using the statistical program JMP (version 3.1) (SAS, 1994).

Results

The two stream types studied here, small, shallow mountain streams and medium-sized, lowland streams, differed regarding a number of physico-chemical variables. Mountain streams were generally situated at higher altitude (mean 337 m a.s.l.) and had smaller catchments (mean = 57 km^2) compared to lowland streams (mean 57 m a.s.l. and 199 km^2) (Table 1). Moreover, mountain streams were often situated in forested catchments (e.g., mean = 58% forest), whilst lowland streams had more of their catchments classified as cropland (mean = 30%) or pasture (mean = 15%). The two stream types also differed regarding the predominant substratum type; cobbles and coarse gravel were most common types of substratum (38 and 23%, respectively) in mountain streams, whereas lowland streams had a high frequency of soft-bottom substratum (sand = 36%). Clear differences were also noted regarding nutrient concentrations. For example, total phosphorus concentrations were on average > 5× higher in lowland (mean TP = 1091 mg/l) compared to mountain (mean TP = 193 mg/l) streams.

Environmental gradients

The first two axes of principal component analysis explained ca. 30% of the variation in catchment land use/cover and physico-chemical variables in mountain and lowland streams (Table 2). The primary environmental gradient for mountain (explained 17.3% of the variance) and lowland (20.4%) streams was related to catchment land use and nutrient concentration. For example, total phosphate was positively correlated (eigenvector

Table 1. Selected physico-chemical and catchment characteristics of mountain and lowland streams

	Mountain (n = 77)	Lowland (n = 85)
Altitude (m a.s.l.)	337 ± 84	57 ± 60
Catchment area (km^2)	57 ± 77	199 ± 199
Catchment classification (%)		
Urban (sum)	0.14 ± 0.24	5.9 ± 13
Forest (sum)	58 ± 23	36 ± 28
Native deciduous	32 ± 33	3.13 ± 5.82
Native coniferous	11 ± 17	3.9 ± 10
Cropland	24 ± 26	30 ± 28
Pasture	8.4 ± 16	15 ± 22
Substratum (%)		
Large cobbles, boulders (> 40 cm)	6.9 ± 18	5.8 ± 13
Coarse blocks, cobbles (> 20–40 cm)	12.7 ± 16	7.9 ± 13
Cobbles (> 6–20 cm)	38 ± 23	12 ± 20
Coarse gravel (> 2–6 cm)	23 ± 18	13 ± 21
Fine gravel (> 0.2–2 cm)	9.3 ± 12	13 ± 23
Sand (> 6 μm–2 mm)	7.5 ± 15	36 ± 37
Silt (< 6 m)	1.6 ± 5	9 ± 25
Physico-chemistry		
pH	7.9 ± 0.57	7.55 ± 0.40
Conductivity (μS/cm)	315 ± 261	390 ± 236
BOD5 (mg O_2/l)	2.25 ± 1.58	2.58 ± 1.50
Ammonium (mg NH_4/l)	0.166 ± 0.360	321 ± 1918
Nitrate (mg NO_3/l)	9.45 ± 9.77	13 ± 14
Total phosphate (μg TP/l)	193 ± 270	1091 ± 2747

Mean ± 1 standard deviation.

loadings = 0.24 and 0.26 for mountain and lowland streams, respectively) and % total forest in the catchment (–0.20 and –0.24, respectively) was negatively correlated with the 1st PC axis. The 2nd PC axis explained another 11.8% (mountain) or 10.3% (lowland) of the variance, and was seemingly related to habitat quality (e.g., number of debris dams) and/or hydromorphological alteration.

Dividing the PC gradients into shorter environmental gradients resulted in clear differences in mean values and ranges of a number of environmental variables (Table 3). For the 1st PC gradient, the most marked among-group differences were related to differences in land use/cover and nutrients. Mountain streams in the best available PC1 group (upper tail) had on averaged 14% of their catchments classified as pasture (range = 0–80%) compared to 2.5% (range = 0–20%) for streams in the perturbed PC1 group (lower tail). Nutrient concentrations also varied between the two groups. For mountain streams total phosphate averaged 65 μg/l (range = 50–182 μg/l) for sites in the PC1 best available group compared to 431 μg/l (range = 30–1270 μg/l) for sites in the PC1 perturbed group. In contrast to mountain streams, lowland streams exhibited stronger gradients in percent catchment land use classified as cropland. Streams in the best available group had on average 7.6% (range 0–40%) of their catchments classified as cropland compared to 39% (range = 0–80%) for streams in the perturbed group. Total phosphate averaged 41 μg/l (range = 8.6–127 μg/l) for streams in the best available group compared to 3315 μg/l (range = 186–15430 μg/l) for streams in the perturbed group.

The 2nd PC gradient was interpreted as being either related to habitat quality or alterations in hydromorphology or both (concomitant changes in habitat/hydromorphology). For both mountain and lowland streams the percentage of substratum classified as coarse blocks and cobbles or coarse gravel changed markedly between the upper (best available) and lower (perturbed) tails of the PC gradient. For mountain streams, coarse blocks and cobbles substratum averaged 14% cover (range = 0–60%) for streams in the best available group compared to 2.4% (range = 0–10%) for sites in the perturbed group. For lowland streams, coarse gravel substratum in the best available group averaged 26% cover (range = 0–80%) compared to 5% (range = 0–50%) for sites in the perturbed group.

Organism/metric response to stress

Three of the four organism groups showed a significant response to the 1st PC (null model) gradient for mountain streams; the exception being diatoms, which did not show a significant response to this stressor gradient (Table 4). Coefficients of variation for the various null model regressions varied markedly among the organism groups. Macroinvertebrates showed the strongest response, with R^2 values for CA scores and N2-diversity of 0.422 and 0.259, respectively, followed by macrophytes (0.306 and 0.247) and fish (0.118 and 0.142). Comparison of organism-group response along the short gradients to the null model showed only two

Table 2. Eigenvectors (loadings) of physico-chemical, substratum and catchment land use/cover

	Mountain streams		Lowland streams	
	PC1	PC2	PC1	PC2
Eigenvalue	6.9	4.7	8.8	4.4
Percent	17.3	11.8	20.4	10.3
Cum percent	17.3	29.1	20.4	30.7
Eigenvectors				
Total forest	**−0.20**	0.02	**−0.24**	−0.03
Total urban	−0.10	0.00	**0.15**	−0.13
Wetland (mire)			−0.19	−0.02
Open grass/bushland	−0.11	0.06	0.04	**0.24**
Standing water			**−0.20**	−0.07
Cropland	**0.32**	0.01	**0.19**	−0.13
Pasture	−0.11	−0.06	**0.17**	**0.23**
Clear-cutting			**−0.26**	−0.01
Shading at zenith (foliage cover)	**−0.19**	**0.20**	0.04	0.18
Average width of woody riparian vegetation (m)	**−0.21**	**0.17**	−0.05	0.06
no. of debris dams	0.02	**0.29**	0.11	**0.22**
no. of logs	−0.08	**0.24**	0.13	**0.16**
Shoreline covered with woody riparian vegetation left	**−0.22**	**0.27**	0.07	**0.15**
No bank fixation	**−0.16**	0.03	−0.02	**0.26**
No bed fixation	−0.08	0.08	−0.05	**0.30**
Stagnation	0.11	**0.20**	0.09	−0.13
Straightening	**0.20**	0.10	0.12	**−0.25**
Hygropetric	0.08	−0.06	**−0.16**	−0.02
Large cobble	0.01	−0.08	**−0.26**	−0.03
Coarse blocks	**−0.17**	**−0.15**	**−0.26**	−0.06
Cobbles	**−0.25**	0.07	**−0.21**	0.03
Coarse gravel	0.11	0.05	−0.03	0.13
Fine gravel	0.18	**0.16**	0.09	**−0.16**
Sand	0.15	0.03	**0.20**	0.00
Silt	0.19	**0.18**	0.01	0.09
Submerged macrophytes	0.09	**−0.27**	−0.05	**−0.16**
Emergent macrophytes	0.10	−0.08	0.07	**−0.19**
Xylal	0.02	**0.24**	0.12	**0.22**
CPOM	−0.03	0.13	0.13	**−0.29**
FPOM	0.13	0.13	0.14	**−0.27**
pH	**0.17**	**0.20**	**0.19**	0.07
Conductivity	**0.28**	−0.02	**0.29**	0.02
BOD5	0.12	**0.23**	−0.03	−0.05
Ammonium	−0.10	**0.31**	0.12	**−0.31**
Nitrite	**−0.17**	**0.31**		
Nitrate	**0.24**	0.10	**0.18**	**0.22**
ortho-Phosphate	**0.25**	0.11	**0.28**	0.01
Total phosphate	**0.24**	**0.21**	**0.26**	−0.02

Only variables with loadings >0.15 on either PC1 or PC2 are shown, and loadings >0.15 are shown in bold.

Table 3. Selected physico-chemical variables and catchment characteristics of the PC gradient-ends for mountain and lowland streams

	Upper tail PC1	Lower tail PC1	Upper tail PC2	Lower tail PC2
Mountain streams				
n	19	20	19	19
Altitude (m a.s.l.)	388 (250–534)	309 (174–485)	295 (160–430)	346 (220–485)
Catchment area (km^2)	25 (10–95)	117 (16–450)	138 (23–450)	30 (9.3–63)
Native deciduous forest (%)	46 (0–100)	16 (0–50)	34 (0–80)	18 (0–80)
Native coniferous forest (%)	7.4 (0–70)	12 (0–40)	3.2 (0–40)	22 (0–60)
Cropland (%)	1.6 (0–10)	56 (10–90)	39 (0–90)	36 (0–80)
Pasture (%)	14 (0–80)	2.5 (0–20)	7.9 (0–50)	2.1 (0–20)
Coarse blocks (%)	28 (0–55)	7.5 (0–60)	14 (0–60)	2.4 (0–10)
Cobbles (%)	52 (25–95)	15 (0–45)	30 (0–95)	36 (5–60)
Coarse gravel (%)	10 (0–20)	29 (0–80)	25 (0–80)	34 (5–60)
Conductivity (μS/cm)	156 (69–272)	592 (118–1662)	504 (90–1662)	367 (134–710)
Nitrate (mg/l)	2.9 (0.63–11)	18 (4.4–45)	9.7 (0.74–23)	15 (1.5–38)
Total phosphate (μg/l)	65 (50–182)	431 (30–1270)	175 (20–910)	347 (91–1270)
Lowland streams				
n	21	20	21	21
Altitude (m a.s.l.)	88 (2–261)	50 (7.5–180)	47 (0–120)	77 (4–239)
Catchment area (km^2)	236 (45–1139)	147 (8.8–459)	103 (8.8–413)	301 (57–883)
Native deciduous forest (%)	0 (0–0)	6 (0–20)	9.5 (0–30)	1.0 (0–10)
Native coniferous forest (%)	0 (0–0)	0.5 (0–10)	4.8 (0–60)	6.7 (0–30)
Cropland (%)	7.6 (0–40)	39 (0–80)	22 (0–70)	41 (0–80)
Pasture (%)	0 (0–0)	32 (0–80)	38 (0–70)	5.2 (0–40)
Coarse blocks (%)	21 (0–40)	0.3 (0–5)	0 (0–0)	5 (0–50)
Cobbles (%)	23 (5–65)	0 (0–0)	8.1 (0–50)	6.2 (0–50)
Coarse gravel (%)	11 (0–30)	8.3 (0–80)	26 (0–80)	5 (0–50)
Conductivity (μS/cm)	143 (24–375)	652 (122–1022)	484 (205–879)	455 (26–1022)
Nitrate (mg/l)	2.1 (0.04–23)	19.6 (0.2–45)	22 (6.6–45)	6.6 (0.04–41)
Total phosphate (μg/l)	41 (8.6–127)	3315 (186–15430)	1784 (45–13201)	1841 (19–15430)

Upper tail (best available) = sites above the 75th-percentile; lower tail (perturbed) = sites below the 25th-percentile. Mean values and in parenthesis min and max values.

groups that responded significantly to the upper tail of the PC1 gradient (best available of the PC gradient), and none of the groups showed a significant response to the lower (perturbed) tail of the PC1 gradient. Both fish and macrophyte CA scores indicated an early response. Coefficients of variation for fish increased from 0.118 to 0.306, and the slope changed from −0.1376 to −1.188 when CA scores were regressed against the upper tail of the PC1 gradient. For macrophyte CA scores, the R^2 increased only marginally (from 0.306 to 0.359), but the slope changed from –0.3904 to –4.583.

All four organism-groups showed a significant response to the 2nd PC (null model) gradient. The strongest relationship was found for macroinvertebrate CA scores ($R^2 = 0.475$, $p < 0.0001$) and macrophyte ($R^2 = 0.435$, $p < 0.0001$) and fish ($R^2 = 0.311$, $p < 0.001$) diversity. Fish CA scores and macroinvertebrate and diatom diversity were also significantly related to the 2nd PC gradient, albeit weakly (R^2 value < 0.16). Comparison of organism-group response using the upper tail of the PC2 gradient with the null model showed that R^2 values and/or regression slopes of all organism groups (and four of the six regressions) increased, indicating a significant early warning response. Neither macrophyte nor diatom CA scores were significantly related to the 2nd PC (null model)

Table 4. Summary statistics for regression of organism group CA scores and N2-diversity and PC gradients for mountain streams

	Fish		Macrophytes		Macroinvertebrates		Diatoms	
	CA axis 1 scores	Diversity	CA axis 1 scores	Diversity	CA axis 1 scores	Diversity	CA axis 1 scores	Diversity
PC1 gradient (null model)								
n	**72**	**72**	**58**	**58**	**76**	**76**	76	76
R^2	**0.118**	**0.143**	**0.306**	**0.247**	**0.259**	**0.422**	−0.013	−0.012
RMSEP	**0.937**	**1.236**	**1.577**	**7.674**	**0.921**	**8.550**	0.931	5.326
Slope	**−0.138**	**0.201**	**-0.390**	**1.650**	**0.211**	**-2.802**	−0.009	0.077
p value	**0.0018**	**0.0006**	**0.0001**	**0.0001**	**0.0001**	**0.0001**	0.8333	0.7416
PC1 upper tail (values > 75th percentile)								
n	**19**	**19**	**12**	**12**	19	19	19	19
R^2	**0.306**	-0.059	**0.359**	0.031	0.107	-0.057	−0.056	−0.004
RMSEP	**0.730**	0.718	**2.425**	2.799	0.190	8.530	1.163	3.385
Slope	**−1.188**	−0.012	**−4.583**	2.495	0.184	-0.887	−0.145	−1.793
p value	**0.0082**	0.9529	**0.0234**	0.2358	0.0933	0.8507	0.8211	0.3439
PC1 lower tail (values < 25th percentile)								
n	19	19	19	19	20	20	20	20
R^2	0.099	0.124	0.146	-0.052	−0.024	0.048	-0.039	−0.049
RMSEP	0.952	1.887	0.237	10.482	1.530	9.433	0.565	6.881
Slope	−0.288	0.623	-0.079	0.562	0.188	-2.166	−0.049	0.380
p value	0.102	0.0766	0.0595	0.7497	0.4628	0.1781	0.6019	0.7399
PC2 gradient (null model)								
n	**72**	**72**	58	**58**	**76**	**76**	76	**76**
R^2	**0.142**	**0.311**	-0.015	**0.435**	**0.475**	**0.053**	-0.012	**0.163**
RMSEP	**0.923**	**1.108**	1.907	**6.644**	**0.775**	**10.942**	0.930	**4.845**
Slope	**0.186**	**-0.358**	−0.038	**−2.449**	**−0.341**	**1.324**	-0.015	**−1.015**
p value	**0.0006**	**0.0001**	0.7161	**0.0001**	**0.0001**	**0.0254**	0.7559	**0.0002**
PC2 upper tail (values > 75th percentile)								
n	**19**	**19**	**18**	18	**19**	**19**	**19**	**19**
R^2	**0.206**	0.057	**0.248**	**0.500**	**0.352**	0.146	**0.220**	0.029
RMSEP	**0.705**	1.587	**0.192**	**7.159**	**1.021**	10.211	**1.005**	6.931
Slope	**0.370**	-0.504	**0.112**	**-6.898**	**−0.738**	4.535	**0.545**	1.888
p value	**0.0291**	0.1671	**0.0205**	**0.0006**	**0.0044**	0.0596	**0.0247**	0.2326
PC2 lower tail (values < 25th percentile)								
n	16	16	**17**	17	19	19	19	19
R^2	−0.068	−0.032	**0.485**	-0.065	0.050	0.111	-0.040	−0.048
RMSEP	0.963	1.001	**0.259**	1.773	0.113	10.022	0.466	3.222
Slope	0.053	-0.199	**−0.224**	0.052	-0.033	−3.816	0.055	-0.285
p value	0.842	0.4765	**0.0011**	0.8935	0.1802	0.0898	0.5858	0.6818

Values shown in bold text are significant ($p < 0.05$).

gradient, whereas relatively strong relationships (R^2 values of 0.248 and 0.220, respectively) were noted when these metrics were regressed using the upper tail of the PC2 gradient (slopes increased from −0.0383 to 0.1119 for macrophyte CA scores and from −0.0154 to 0.5447 for diatom CA scores). Only one metric, macrophyte CA scores, showed a significant relationship using the lower tail of the PC2 gradient ($R^2 = 0.485$, $p = 0.0011$).

Five of the eight metrics showed a significant response to the 1st PC (null model) gradient for lowland streams (Table 5). The strongest relationship was found between diatom CA scores and the stress gradient ($R^2 = 0.606$, $p < 0.0001$), followed by macrophyte CA scores ($R^2 = 0.366$, $p < 0.0001$). Although the slopes of the other three regressions were significant, the relations were relatively weak (R^2 values < 0.181). Comparison of organism groups/metric response of the upper tail of the PC1 gradient (best available sites) and the null model showed that four of the eight relationships were significant. The response of two organism groups, in particular, improved suggesting that these organism groups/metrics might be considered as early warning indicators of stress: R^2 values for fish and macroinvertebrate CA scores increased from 0.049 to 0.327 and from 0.148 to 0.565 and the slopes changed from −0.089 to 0.522 and from 0.116 to 0.170, respectively. Diatom CA scores showed only a modest increased response (R^2 value increased from 0.606 to 0.724), and the R^2 value for macrophyte CA scores was actually lower (0.366–0.290) compared to the null model. However, the slopes of both relationships increased markedly from −0.355 to −1.188 for diatoms and from 0.1401 to −1.115 for macrophytes. None of the metrics showed significant relationships using the perturbed sites.

Three of the four organism groups showed a significant response to the 2nd PC (null model) gradient, however R^2 values were low (<0.177). Neither of the two diatom metrics showed a significant response to this stressor gradient. Comparison of organism/metric response using the short gradients with the null model revealed no significant relationships using the best available sites. The relationship between fish CA scores using the lower tail of the PC2 gradient (perturbed sites) was slightly better than the null model; R^2 values increased from 0.177 to 0.273 and slopes changed from −0.2272 to −0.3041.

Discussion

Assessing the ecological integrity of running water ecosystems, and being confident that if change occurs it will be detected, is a fundamental objective of most monitoring programmes as well as the underpinning aim of the recently adopted European Water Framework Directive (European Commission, 2000). The major stressors affecting the integrity of European surface waters are overexploitation, nutrient enrichment and organic pollution, acidification and alterations of hydrology and morphology (Stanner & Bordeau, 1995). Our results support this view; the two main stress gradients were interpreted as being related to land use and nutrient concentrations (the primary gradient) and alterations in habitat quality and hydromorphology (the secondary gradient). Streams, in particular, are affected (simultaneously) by a multitude of human-generated pressures. For example, agricultural land use can result in several different types of stress, which may singly or in concert affect the structure and function of stream assemblages. For instance, increased runoff due to agricultural activity can result in changes in hydrology, increased siltation and changes in habitat quality/quantity, while inputs of nutrients from agriculture can result in eutrophication effects. The single and combined effects of stress on the organism assemblages inhabiting the ecosystem will vary, depending on the response of the organism to the stress. Here we show that organism response to stress was in some cases asymmetrical (thereby supporting the conjecture that organism-responses are not redundant), and several organism groups/metrics responded differently to the environmental gradients tested here. This information is useful for designing more cost-effective monitoring programs, where the use of early warning indicators can potentially signal change before deterioration is allowed to proceed too far.

Since environmental stress gradients are often correlated (multiple stressors), principal components analysis was used to construct complex stressor gradients. Comparison of the response of the four organism groups to the tails of the PC stressor gradients was used to evaluate organism-specific response to stress. In particular, we were interested in the slope and error of the response within the top end (upper tail) of the environmental gradient, where ecological impairment may be considered as changing from high to lower quality along the gradient. Higher slope and lower error than the null model would imply that the organism can be considered as an early warning indicator for the stressors studied here.

The primary gradient for both mountain and lowland streams was interpreted as representing a

Table 5. Summary statistics for regression of organism group CA scores and N2-diversity and PC gradients for lowland streams

	Fish		Macrophytes		Macroinvertebrates		Diatoms	
	CA axis 1 scores	Diversity	CA axis 1 scores	Diversity	CA axis 1 scores	Diversity	CA axis 1 scores	Diversity
PC1 gradient (null model)								
n	**82**	82	**81**	81	**71**	71	**81**	**81**
R^2	**0.049**	0.030	**0.366**	−0.011	**0.148**	0.011	**0.606**	**0.181**
RMSEP	**1.047**	1.638	**1.462**	5.327	**0.828**	7.114	**0.854**	**7.849**
Slope	**−0.089**	0.114	**0.140**	0.083	**0.116**	−0.380	**−0.355**	**1.269**
p value	**0.026**	0.0662	**0.0001**	0.6924	**0.0006**	0.184	**0.0001**	**0.0001**
PC1 upper tail (values > 75th percentile)								
n	**21**	21	**19**	19	**21**	21	**21**	21
R^2	**0.327**	−0.001	**0.290**	0.109	**0.565**	−0.028	**0.724**	0.054
RMSEP	**0.998**	1.156	**2.281**	3.660	**0.205**	6.927	**1.016**	4.324
Slope	**0.522**	−0.184	**−1.115**	1.109	**0.170**	0.743	**−1.188**	1.011
p value	**0.004**	0.333	**0.0102**	0.0912	**0.0001**	0.5102	**0.0001**	0.1599
PC1 lower tail (values < 25th percentile)								
n	19	19	20	20	20	20	15	15
R^2	0.000	0.003	−0.040	−0.052	−0.056	0.104	−0.073	−0.011
RMSEP	0.921	1.711	1.192	6.325	0.114	7.471	1.332	9.251
Slope	0.414	−0.789	0.275	0.691	0.234	−9.552	0.001	3.685
p value	0.3306	0.3184	0.6131	0.8101	0.9844	0.1291	0.8276	0.3865
PC2 gradient (null model)								
n	**82**	82	**81**	**81**	**71**	71	81	81
R^2	**0.177**	−0.013	**0.053**	**0.111**	**0.089**	−0.013	−0.005	0.021
RMSEP	**0.974**	1.673	**1.788**	**4.995**	**0.856**	7.202	1.362	8.586
Slope	**−0.227**	0.002	**−0.219**	**−0.871**	**0.187**	0.159	−0.056	−0.737
p value	**0.0001**	0.985	**0.0222**	**0.0014**	**0.0067**	0.7794	0.4403	0.1059
PC2 upper tail (values > 75th percentile)								
n	21	21	21	21	21	21	21	21
R^2	0.031	0.017	0.112	0.050	0.061	0.008	−0.041	−0.026
RMSEP	0.257	1.428	1.226	3.093	1.053	7.852	0.228	7.495
Slope	−0.115	−0.577	−0.807	−1.544	0.556	2.962	−0.370	−1.829
p value	0.2157	0.2614	0.075	0.1691	0.1468	0.2935	0.6469	0.4931
PC2 lower tail (values < 25th percentile)								
n	**20**	20	21	21	19	19	21	21
R^2	**0.273**	−0.053	0.003	−0.053	−0.115	0.051	−0.007	−0.045
RMSEP	**0.840**	2.071	0.661	7.150	1.263	8.138	0.856	9.490
Slope	**−0.304**	−0.059	0.083	0.024	0.199	−3.642	0.097	−0.445
p value	**0.0106**	0.825	0.3148	0.9781	0.6869	0.2714	0.3653	0.7043

Values shown in bold text are significant ($p < 0.05$).

gradient in land use and in-stream nutrient concentrations. Benthic diatoms rely on nutrients (especially P) for growth. Therefore, we expected that diatoms would react strongly to changes in the upper tail of the PC gradient, where nutrients might be limiting (e.g., for lowland streams the upper tail represented a gradient from 8.6 to 127 μg TP/l). Likewise, as many benthic macroinvertebrates (e.g. grazers and scrapers) rely on diatoms for food we might expect a close, albeit weaker, relation between macroinvertebrate community composition and the upper tail of the

PC gradient. Our findings of the response of benthic diatoms and macroinvertebrates to the 1st PC gradient were, however, equivocal. Neither diatom CA scores nor diversity were significantly related to the 1st PC (the null model) gradient for mountain streams and the slopes of the two metrics were not significant when regressed against the short gradients (upper and lower tails) of the 1st PC axis. By contrast, for lowland streams both metrics were significantly related to the 1st PC (null model) gradient. The relation between diatom CA scores and the null model was highly significant (CA scores had an R^2 value of 0.606), and this relation improved when CA score were regressed against the upper tail of the PC gradient (R^2 = 0.724). The fit between macroinvertebrate CA scores also improved when regressed against the upper tail of the 1st PC gradient (R^2 = 0.148 for the null model compared to 0.565 for the upper tail of the gradient). These findings, in particular the first principle relation between benthic diatom response and the PC (nutrient) gradients, supports the conjecture that benthic diatoms, and to some extent even macroinvertebrates, may be considered as early warning organisms of nutrient enrichment. However, the finding that neither diatoms nor macroinvertebrates showed better improvement when regressed against the upper tail of the 1st PC gradient for mountain streams implies that caution should be exercised when extrapolating these finding to other stream types.

Both fish and macrophyte CA scores for mountain streams and fish CA scores for lowland streams showed improved response (higher R^2 values and steeper slopes) compared to the null model. Although macrophyte growth in streams might be expected to be related to increased nutrient concentrations, fish response would not unless there is a bottom-up effect where an increase in diatom biomass results in an increase in macroinvertebrate biomass and subsequently changes in the fish community. For mountain streams we find no support for this conjecture, since neither diatom nor macroinvertebrates were significantly related to the upper tail of the PC gradient. Other factors may, however, be affecting the responses noted here. For example, although we interpreted the primary PC gradients in both mountain and lowland streams to represent nutrient enrichment, other factors, like characteristics of the riparian foliage (PC1 mountain streams) covary with nutrients along these gradients.

Clear differences were noted not only among the four organism groups studied here, but also between the two metrics used to assess their response to stress. For null model predictions in lowland streams, for example, diversity did not show a significant response for three of the four organism groups (only diatom diversity responses were significant). Conversely, for null model predictions in mountain streams (PC2 gradient) diversity metrics responded more clearly than CA scores (all four for diversity compared to two of four for CA scores). This finding implies that consideration should be given not only to the organism group but also to the metric selected to monitor the effects of the stressor of interest. Recent studies comparing the multiple organism groups and metrics lend support to this finding (e.g., Hering et al., submitted; Johnson et al., 2006).

Evaluating organism–response relations along short environmental gradients revealed interesting findings. We anticipated that benthic diatoms would respond strongly when sites became more impaired (nutrient enriched), and that diatoms would be an appropriate 'first choice' indicator for monitoring early changes in nutrient levels. Data from lowland streams strongly support the use of diatoms (and also macroinvertebrates) for monitoring the effects of agricultural land use. However, for mountain streams we found no such support for this relation. Although nutrient concentrations were strongly correlated with the 1st PC gradients for both mountain and lowland streams, other factors may be confounding the nutrient–diatom response signal. Our finding that fish and macrophytes responded to the 1st PC gradient for mountain streams lends support to this conjecture. In summary, our results showed that rates of organism response to the environmental gradients studied here varied among the four groups, implying that certain organisms/metrics can be considered as early warning indicators of ecological change. Selection of organisms that respond more rapidly at the outset of impairment is one way of determining (quantifying) the potential harmful, human-induced effects on ecosystem integrity before degradation is allowed to proceed to the point where the damage is either too costly or impossible to restore. Another commonly used approach is to 'create' early-warning metrics (or pollution-specific metrics) by weighting taxa according to their tolerance or sensitivity to a known stressor (e.g. the Saprobien index). Clearly, both approaches should be used together in designing robust methods for detecting ecological change.

Acknowledgements

This paper is a result of the EU-funded project STAR (5th Framework Programme; contract number: EVK1-CT-2001-00089). Parts of the data analysis were supported by the EU-funded Integrated Project Euro-limpacs (6th Framework Programme; contract number: GOCE-CT-2003-505540). We are most grateful to all STAR partners having provided data for this analysis.

References

Barbour, M. T., J. Gerritsen, B. D. Snyder & J. B. Stribling, 1998. Rapid bioassessment protocols for use in streams and wadeable rivers: periphyton, benthic macroinvertebrates and fish (2nd edn.). EPA/841/B/98-010 U.S. Environmental Protection Agency Office of Water, Washington, DC.

European Commission, 2000. Directive 2000/60/EC of the European Parliament and of the Council – Establishing a framework for Community action in the field of water policy, Brussels, Belgium, 23 October 2000.

Furse, M. T., A. Schmidt-Kloiber, J. Strackbein, J. Davy-Bowker, A. Lorenz, J. van der Molen J. & P. Scarlett, 2004. Results of the sampling programme. A report to the European Commission. Framework V Project STAR (EVK1-CT-2001_00089).

Hill, M. O., 1973. Reciprocal averaging: an eigenvector method of ordination. Journal of Ecology 61: 237–249.

Hering, D., R. K. Johnson, S. Kramm, S. Schmutz, K. Szoszkiewicz & P. F. M. Verdonschot. Assessment of European rivers with diatoms, macrophytes, invertebrates and fish: a comparative metric-based analysis of organism response to stress, submitted manuscript.

Hering, D. & J. Strackbein, 2002. STAR stream types and sampling sites. A report to the European Commission. Framework V Project STAR (EVK1-CT-2001_00089).

Holmes, N. T. H., J. R. Newman, S. Chadd, K. J. Rouen, L. Saint & F. H. Dawson, 1999. Mean Trophic Rank: A users manual. R & D Technical Report No. E 38, Environment Agency, Bristol, UK.

Johnson, R. K., T. Wiederholm & D. M. Rosenberg, 1993. Freshwater biomonitoring using individual organisms, populations and species assemblages of benthic macroinvertebrates. In Rosenberg, D. M. & V. H. Resh (eds), Freshwater Biomonitoring and Benthic Macroinvertebrates. Chapman and Hall, New York, 40–158.

Johnson, R. K., D. Hering, M. T. Furse & R. T. Clarke, 2006. Detection of ecological change using multiple organism groups: metrics and uncertainty. Hydrobiologia 566: 115–137.

Kolkwitz, R. & M. Marsson, 1902. Grundsätze für die biologische Beurteilung des Wassers nach seiner Flora und Fauna. Mitteilungen Prüfungsanstalt Wasserversorgung und Abwasserreinigung 1: 33–72.

Metcalfe, J. L., 1989. Biological water-quality assessment of running waters based on macroinvertebrate communities – history and present status in Europe. Environmental Pollution 60: 101–139.

Raven, P. J., N. T. H. Holmes, F. H. Dawson, P. J. A. Fox, M. Everard, I. R. Fozzard & K. J. Rouen, 1998. River habitat quality – the physical character of rivers and streams in the UK and Isle of Man. River Habitat Survey Report Number 2, Environment Agency, Bristol, Scottish Environment Protection Agency, Stirling, Environment and Heritage Service, Belfast, 84 pp.

SAS., 1994. JMP – Statistics Made Visual, Version 3.1. SAS Institute Inc, Cary, NC, USA.

Stanner, D. & P. Bordeau, 1995. Europe's Environment: The Dobris Assessment. European Environment Agency, Luxembourg 712 pp.

Stevenson, R. J., R. C. Bailey, M. C. Harrass, C. P. Hawkins, J. Alba-Tercedor, C. Couch, S. Dyer, F. A. Fulk, J. M. Harrington, C. T. Hunsaker & R. K. Johnson, 2004. Designing data collection for ecological assessments. In Barbour, M. -T., S. B. Norton, H. R. Preston & K. W. Thornton (eds), Ecological Assessment of Aquatic Resources: Linking science to decision making. SETAC, Pensacola, FL, USA, 55–84.

ter Braak, C. F. J., 1988. CANOCO – a FORTRAN program for canonical community ordination by [partial] [detrended] [canonical] correspondence analysis, principal component analysis and redundancy analysis (version 3.15). Agricultural Mathematics Group, Wageningen, The Netherlands.

ter Braak, C. F. J., 1990. Update Notes: CANOCO Version 3.10. Agricultural Mathematics Group, Wageningen, The Netherlands.

Hydrobiologia (2006) 566:153–172
M.T. Furse, D. Hering, K. Brabec, A. Buffagni, L. Sandin & P.F.M. Verdonschot (eds), The Ecological Status of European Rivers: Evaluation and Intercalibration of Assessment Methods
DOI 10.1007/s10750-006-0099-y

Biological quality metrics: their variability and appropriate scale for assessing streams

Gunta Springe[1,*], Leonard Sandin[2], Agrita Briede[1] & Agnija Skuja[1]
[1]*Institute of Biology, University of Latvia, 3 Miera St., LV 2169 Salaspils, Latvia*
[2]*Department of Environmental Assessment, Swedish University of Agricultural Sciences, 7050, SE-750 07 Uppsala, Sweden*
*(*Author for correspondence: E-mail: gspringe@email.lubi.edu.lv)*

Key words: biological quality elements, Water Framework Directive, metric variability, spatial scale, medium-sized lowland streams, high quality sites

Abstract

The concept of spatial scale is at the research frontier in ecology, and although focus has been placed on trying to determine the role of spatial scale in structuring communities, there still is a further need to standardize which organism groups are to be used at which scale and under which circumstances in environmental assessment. This paper contributes to the understanding of the variability at different spatial scales (reach, stream, river basin) of metrics characterizing communities of different biological quality elements (macrophytes, fishes, macroinvertebrates and benthic diatoms) as defined by the Water Framework Directive. For this purpose, high-quality reaches from medium-sized lowland streams of Latvia, Ecoregion 15 (Baltic) were sampled using a nested hierarchical sampling design: (river basin → stream → reach). The variability of metrics within the different groups of biological quality elements confirmed that large-bodied organisms (macrophytes and fish) were less variable than small-bodied organisms (macroinvertebrates and benthic diatoms) at reach, stream and river basin scales. Single metrics of biological quality elements had the largest variation at the reach scale compared with stream and basin scales. There were no significant correlations between biodiversity indices of the different organism groups. The correlation between diversity indices (Shannon's and Simpson's) of the biological quality elements (macrophytes, fish, benthic macroinvertebrates and benthic diatoms) and a number of measured environmental variables varied among the different organism groups. Relationships between diversity indices and environmental factors were established for all groups of biological quality elements. Our results showed that metrics of macrophytes and fish could be used for assessing ecological quality at the river basin scale, whereas metrics of macroinvertebrates and benthic diatoms were most appropriate at a smaller scale.

Introduction

The EU Water Framework Directive (Directive 2000/60/EC – Establishing a Framework for Community Action in the Field of Water Policy) (Anonymous, 2000) defines a framework for assessing all kinds of waterbodies using biotic indicators (metrics) from different organism groups – macrobenthos, fish, macrophytes and benthic diatoms. The organism groups proposed by the Water Framework Directive presumably indicate environmental change at different spatial and temporal scales. It is generally assumed that the scale at which communities exhibit the greatest variation is the scale over which important physical/chemical gradients or biotic interactions control assemblage composition (Li et al., 2001). According to Thompson et al. (2001), the topic of spatial scale is one of the four paramount frontiers in ecology for "understanding how biological and physical processes interact over multiple spatial and temporal scales to shape the earths' biodiversity". Although focus has been placed on trying to determine if stream ecosystems are structured by abiotic (e.g., physico-chemical), biotic (e.g., predation) or by a combination of abiotic/biotic factors, contention still exists as to whether large-scale (regional or catchment) or small-scale (local or habitat) environmental factors have the main importance for structuring the communities (e.g., Lammert & Allan, 1999; Sandin & Johnson, 2000a, b).

In bioassessment, our ability to detect change is often confounded by natural spatial and temporal variability. In the selection of robust indicators of biodiversity and ecological status, effort should be placed on selecting indicators that exhibit low natural, but high human-induced variance (Johnson, 1995). Simply put, our ability to detect change if/when the change occurs is, for the most part, a function of indicator variance and observed change (Johnson, 1998; Sandin & Johnson, 2000c). Accordingly, robust biodiversity indicators or metrics must have a low spatial and temporal variability compared to the change in the index value caused by human perturbation (Johnson, 1998; Sandin, 2001).

There is a need to standardize which organism group or groups are to be used at which scale and under which circumstances. Therefore some of the aims of the project "Standardisation of River Classifications: Framework method for calibrating different biological survey results against ecological quality classifications to be developed for the Water Framework Directive" (www.eu-star.at) were to contribute to our understanding of the use of different taxonomic groups for the assessment of ecological status of streams and how they can be used in implementing the WFD. The aim of this paper was therefore to (i) compare the variability in commonly used metrics for the four organism groups at three sampled spatial scales (reach, stream, basin) and (ii) to relate these metrics to environmental variables.

This type of research is necessary to develop recommendations for integrated monitoring programmes and sampling networks that deliver cost-effective assessments at appropriate levels of spatial scale resolution, and with a low type II error (thus with a high statistical power). Our hypothesis is that large-bodied organisms (fish and macrophytes) are less variable at the smaller spatial scales as opposed to the small-bodied organisms (benthic diatoms and macroinvertebrates) and that locally measured environmental variables are more correlated with the small as opposed to large-bodied organisms.

Materials and methods

Scheme of sampling sites

According to the System A typology (WFD, Annex II; anonymous, 2000), high quality reaches from medium-sized (catchment area 100–1000 km^2), deeper lowland (<200 m) streams of Latvia, Ecoregion 15 (Baltic) were sampled in the spatial scale study.

A nested hierarchical sampling design (river basin → stream → reach, Fig. 1) was used to test at which ecological scale the different taxonomic groups (fish, macrophytes, benthic macroinvertebrates and benthic diatoms) were most variable. From three selected catchments, three streams in each with the best available high ecological status were sampled, and within each stream three reaches were sampled. In total 27 sites

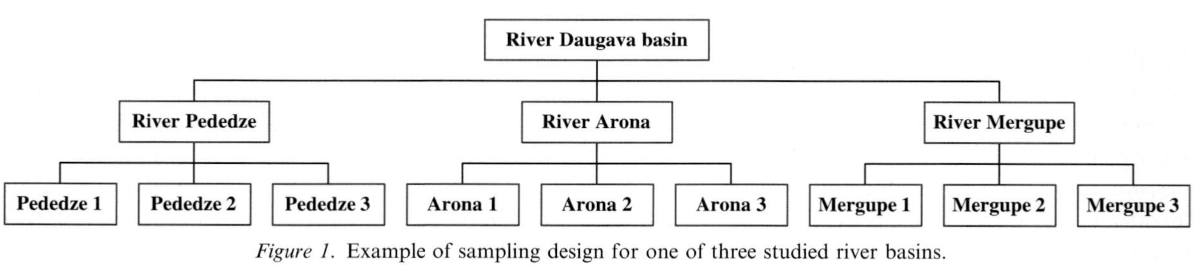

Figure 1. Example of sampling design for one of three studied river basins.

were sampled (Fig. 2 and Table 1). At the lowest sampling reach in each stream, two additional replicates were sampled for benthic macroinvertebrates.

Selection of sampling sites

Sampling sites were selected considering criteria for reference site selection according to STAR field protocols (Furse et al., 2006) and were based on existing information and expert judgement. Preliminary selection of high quality river sites was based on information from topographical maps (1:50, 000), e.g., existence of dams, accessibility, point source pollutants, and land use patterns. Reynoldson et al. (1997) stated that a reference condition is the condition that is representative of a group of minimally disturbed sites, i.e., reference site, described by selected physical, chemical and biological characteristics. In this study the problem was that the percentage of agricultural lands in several cases was larger than 20% and did not correspond with the criteria for reference sites, but at the same time the land-use is not intensive. In Latvia a very low level of fertilizers is typical in comparison with that in European countries. For example, in 2001, the use of N was 36 kg/ha, P was 9 kg/ha, and K was 14 kg/ha (Jansons et al., 2002). The presence of agricultural lands in river basin in most cases was not related to diffuse pollution from the catchments area. All preliminary selected sites were inspected in the field for local pollutant sources and hydromorpohological stress. All chosen reaches were wadeable.

Environmental data and River Habitat Survey (RHS)

About 40 parameters from the site protocol of the STAR project (Furse et al., 2006) describing morphology (character of mineral substrates and percent coverage of biotic microhabitats), hydrology (discharge, velocity, width, depth), chemistry (pH, conductivity, oxygen concentration, alkalinity, hardness, chloride, BOD_5, ammonium, nitrite, nitrate, phosphate, total phosphorus), catchment characteristics (catchment size, altitude, gradient slope, distance from source, percentage of forests and agricultural lands) as well as the Habitat Quality Assessment index (HQA), and Habitat Modification Score (HMS) from the River Habitat Survey (Raven et al., 1997, 1998) were used for correlation and multiple linear regression analyses. Samples for chemical analyses were processed in the laboratory according to 'Standard Methods' (Anonymous, 1992).

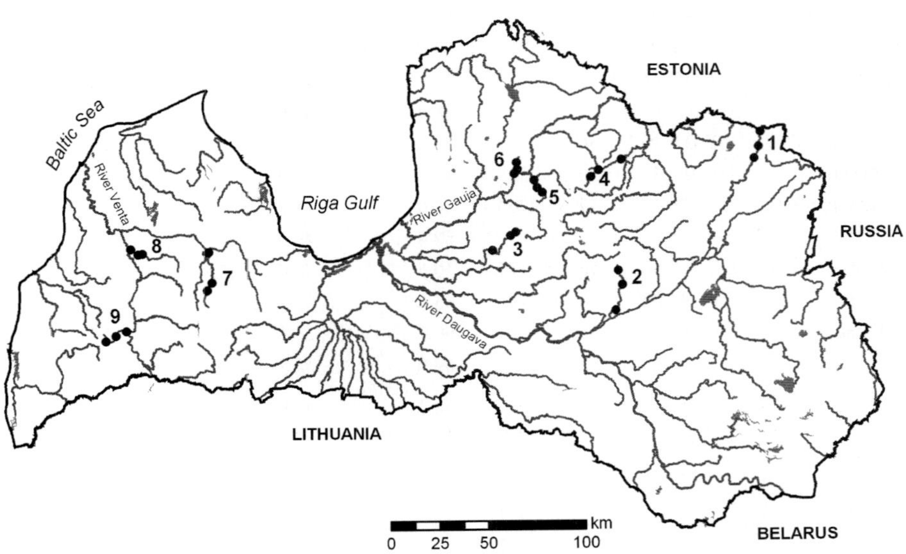

Figure 2. Location of sampling sites within the territory of Latvia.

Table 1. Coordinates of sampling sites and percentage coverage of land-use types

Stream basin	Sampling site	Coordinates		Land use pattern			
		Longitude	Latitude	Forests (%)	Agricult. land (%)	Bog area (%)	Others (%)
Pededze	Pededze 1	27° 20′ 58″	57° 30′ 51″	75.7	23.6	0.7	0.00
	Pededze 2	27° 19′ 43″	57° 26′ 35″	57.8	40.1	1.2	0.87
	Pededze 3	27° 17′ 06″	57° 23′ 29″	62.6	33.1	4.0	0.31
				65.4	**32.3**	**2.0**	**0.47**
Arona	Arona 1	26° 05′ 28″	56° 53′ 49″	51.6	46.9	0.7	0.81
	Arona 2	26° 07′ 41″	56° 49′ 44″	57.0	41.9	0.1	1.03
	Arona 3	26° 02′ 49″	56° 42′ 49″	55.3	41.7	0.5	2.47
				54.6	**43.5**	**0.4**	**1.4**
Mergupe	Mergupe 1	25° 14′ 36″	57° 05′ 27″	53.2	46.0	0.1	0.67
	Mergupe 2	25° 12′ 03″	57° 04′ 30″	65.0	30.3	4.4	0.50
	Mergupe 3	25° 02′ 39″	57° 00′ 20″	58.2	39.6	1.5	0.68
				58.8	**38.6**	**2.0**	**0.6**
Daugava basin				**59.6**	**38.1**	**1.5**	**0.87**
Rauza	Rauza 1	25° 52′ 56″	57° 19′ 58″	57.6	40.7	0.0	1.73
	Rauza 2	25° 57′ 06″	57° 21′ 59″	40.0	59.2	0.0	0.72
	Rauza 3	26° 08′ 53″	57° 24′ 46″	80.7	18.5	0.3	0.53
				59.4	**39.5**	**0.1**	**0.99**
Raunis	Raunis 1	25° 28′ 32″	57° 16′ 04″	56.5	42.0	0.2	1.20
	Raunis 2	25° 26′ 05″	57° 17′ 29″	31.7	64.0	0.0	4.64
	Raunis 3	25° 24′ 26″	57° 19′ 33″	47.2	52.4	0.0	0.41
				45.1	**52.8**	**0.1**	**2.1**
Strikupe	Strikupe 1	25° 15′ 23″	57° 24′ 50″	46.4	49.7	0.2	3.73
	Strikupe 2	25° 15′ 52″	57° 23′ 00″	80.5	19.5	0	0.00
	Strikupe 3	25° 14′ 31″	57° 21′ 46″	74.8	25.1	0	0.04
				67.3	**31.4**	**0.1**	**1.3**
Gauja basin				**57.3**	**41.2**	**0.1**	**1.44**
Amula	Amula 1	22° 38′ 26″	56° 49′ 19″	54.8	44.0	0.4	0.80
	Amula 2	22° 40′ 27″	56° 51′ 32″	40.9	54.6	1.9	2.59
	Amula 3	22° 38′ 44″	56° 59′ 58‴	51.0	48.1	0.7	0.19
				48.9	**48.9**	**1.0**	**1.2**
Riezupe	Riezupe 1	22° 05′ 19″	56° 59′ 15″	58.4	38.0	0.9	2.74
	Riezupe 2	22° 03′ 16″	56° 59′ 22″	42.7	55.0	2.20	0.02
	Riezupe 3	21° 59′ 17″	57° 00′ 26″	50.0	45.0	2.2	3.31
				50.4	**46.0**	**1.8**	**2.0**
Koja	Koja 1	21° 47′ 44″	56° 34′ 47″	50.5	48.0	1.2	0.37
	Koja 2	21° 50′ 33″	56° 34′ 44″	83.6	15.7	0	0.70
	Koja 3	21° 57′ 48″	56° 37′ 38″	60.7	37.5	1.2	0.63
				64.9	**33.7**	**0.8**	**0.7**
Venta basin				**54.7**	**42.9**	**1.2**	**1.38**

Metrics and indices of Biological Quality Elements (BQEs)

For the assessment of stream biological quality, benthic macroinvertebrates that are most frequently used in EU countries as well as benthic diatoms (diatoms), fish and macrophytes were sampled.

For all groups of BQEs, Shannon's (H′) and Simpson's (D) diversity indices (Krebs, 1999) were calculated to compare assemblage composition patterns.

For aquatic macrophytes, a number of different metrics were selected for indicating composition, tolerance and trophic status. Metrics of richness used were: number of species, number of genera, and number of families. Evenness and domination characterized macrophytes composition. Tolerance metrics were represented by hemeroby index, which is an integrative measure for anthropogenic impacts on ecosystems, which compares present vegetation with pristine reference vegetation (Jalas, 1955). Mean Trophic Rank (MTR) (Dawson et al., 1999), Ellenberg Nitrophyllous index (Ellenberg, 1979) and Macrophyte Biological Index for Rivers (IBMR) (Haury et al., 2002) were used as trophic metrics. Fish species were assigned to functional guilds, describing the requirements and behaviour of each species in terms of habitat, reproduction, feeding and tolerance. Fish metrics selected in refer to species composition (including overall composition and functional structure), density and population structure. They were calculated in the EC Framework V project (EVK1-CT-2001-00094) called "Development, evaluation and implementation of a standardized fish-based assessment method for the ecological status of European rivers – A contribution to the Water Framework Directive" (FAME) (http://fame.boku.ac.at).

The number of native species *N*, abundance by both density (*n*/ha) and biomass (kg/ha), functional guilds characterized by general tolerance (intolerant, tolerant), habitat preference (water column, benthic, rheophilic, limnophilic, eurytopic), reproduction mode (lithophilic, phytophilic), longevity (long lived, short lived), feeding type (piscivorous, insectivorous/invertivorous, omnivorous), migration preference (long distance, potamodromons), historical metrics and sentinel species were analysed. The European Fish Index (EFI) developed in the FAME project, and the Lithuanian fish index (Lith_FI) proposed for rivers of Ecoregion 15 (Kesminas & Virbickas, 2000) were also used.

Macroinvertebrate metrics were grouped into three subgroups: (i) eutrophication metrics – Saprobic Index (Zelinka & Marvan, 1961), Biological Monitoring Working Party (BMWP) (Friedrich et al., 1996); (ii) richness metrics – number of families, number of genera, and abundance (ind/m^2); and (iii) composition metrics – number of taxa, abundance of taxonomic groups, EPT-taxa, EPT/OL, EPT/Diptera, OD/Total-Taxa, EP-Taxa, EPTCOB, EPT-Taxa (%), EPT/OL (%), EP (%), EP number of individuals/to total number of individuals (%) and EPT (%) (abundance classes) (Lenat & Penrose, 1996); taxonomic group (%), where E is Ephemeroptera, P is Plecoptera, T is Trichoptera, OL is Oligochaeta, C is Coleoptera, O is Odonata, and B is Bivalvia.

Species number, density, and fourteen diatom indices based on relative abundance of epilithic diatom species were used: Specific Pollution Sensitivity Index (IPS) (CEMAGREF, 1982), Sládecek's pollution index (SLAD) (Sládeček, 1986), Descy's pollution index (DESCY) (Descy, 1979), Leclercq & Maquet's pollution index (L&M) (Leclercq & Maquet, 1987), Steinberg & Schiefele trophic index (SHE) (Steinberg & Schiefele, 1988), Watanabe et al pollution index (WAT), Generic Diatom Index (IDG), Indice Diatomique Artois Picardie (IDAP) (Lecointe et al., 1993), Trophic Diatom index (TDI), % pollution tolerant taxa (%PT) (Kelly & Whitton, 1995), Pollution index based on diatoms (EPI_D) (Dell'Uomo, 1996), Trophic index (ROTT) (Rott, 1999), Commission for Economical Community index (CEE) (Descy & Coste, 1991), Biological Diatom Index (IBD) (Prygiel & Coste, 1993).

Field sampling and samples processing in laboratory

Sampling and sample processing was conducted according to STAR protocols (http://www.eu-star.at). Sampling of benthic macroinvertebrates and diatoms, and measurements of physicochemical parameters were performed during spring to early summer in 2003. Macrophytes were sampled in July 2003, and fish July–September 2003.

In total, 44 macroinvertebrate samples (26 main samples (one sample was spoiled) and 18 replicates), 54 benthic diatoms samples (27 from stones and 27 from sand/silt), 27 macrophyte samples and 26 fish samples (one sampling site was too deep for sampling) were collected and analysed. At four sampling sites (reaches) less than 30 individuals of fish were collected, precluding further calculations.

Data handling and analytical approach

The values of macrophyte metrics were calculated using the STAR project software (http://www.eu-star.at). The FAME project programme was used to calculate fish metrics (http://fame.boku.ac.at). Lithuanian fish index (Lith_FI) for rivers adapted to the conditions of Ecoregion 15 (Lithuania) was calculated for Latvian fish communities using an algorithm of ecological status class assessment in accordance with WFD (Kesminas & Virbickas, 2000). To obtain consistent macroinvertebrate data and to ensure unambiguous data processing, exported data from AQEM Dip database were taxonomically adjusted according to the AQEM guidelines (Hering et al., 2004). Macroinvertebrate metrics were calculated using the AQEM assessment software Version 2.3 (http://www.aqem.de).

Diatom indices based on relative abundance of epilithic diatom species were calculated using OMNIDIA software version 3.2 (Lecointe et al., 1993). Mean values of diatoms from stones and sand/silt substrata were analysed.

Descriptive statistics and coefficients of variation (CV) were calculated for single metrics of all BQEs, both from reaches within streams and basins. As only 27 sites were sampled, similarity among replicates, stream reaches, streams and river basins were not analysed for single metrics, but for subgroups of metrics using the Sign test (Gibbons, 1985). Significant differences among samples were not stated if Exact Sig. (two-tailed) >0.05.

Relationships between biological metrics and environmental factors were determined by the Sign test, correlation coefficients (significant correlations $r = \pm 0.66$; $\alpha = 0.05$) and linear regression analysis after standardisation of environmental variables. Since the environmental variables conformed to a normal distribution, the relationships with metrics were assessed using linear regression and Pearson's correlation coefficients. Only environmental variables that had significant correlations with BQEs metrics (Shannon's and Simpson's indices) were used for the linear regression model. All of the statistical analyses were carried out using SPSS 12.0.1(SPSS, 2004) and PC-ORD (McCune & Mefford, 1999).

Results

Reach-scale variation of metrics within organism groups

In general, single metrics of BQEs had the largest variation at the reach scale compared with stream and basin scales (Tables 2–5), therefore, further analyses were focused on the reach-scale variations.

Macrophytes

Stream reaches were inhabited by 1–20 macrophyte species, from 1 to 17 genera and 1–16 families, which were distributed rather unevenly among the reaches. There was a strong correlation between number of species and number of genera ($r = 0.99$; $\alpha = 0.01$) as well as between number of species and number of families ($r = 0.98$; $\alpha = 0.001$).

Among the macrophyte composition and diversity metrics the largest CV was for Shannon's diversity index, followed by evenness, domination and species number. This group of metrics was more variable in comparison with trophic and tolerance metrics (hemeroby index), except Simpson's diversity index, which was least variable of all calculated macrophyte metrics. The most variable of all metrics was cover of macrophytes (Table 2).

Fish

The number of native fish species varied from 3 to 10 species at the reach scale (CV = 30.7). The CV of Shannon's diversity index at the reach scale was 37.1, while that of Simpson's diversity

Table 2. Coefficients of variation (CV) for macrophyte metrics

Spatial scale	Richness	Composition		Diversity		Abundance	Trophic			Tolerance
	N_species	Evenness	Domination	Shannon's	Simpson's	Σ cover	Ellenberg_N	MTR	IBMR	Hemeroby
Reach	49.8	77.9	59.0	89.53	3.78	134.9	11.1	17.8	13.6	5.7
Stream	33.4	61.4	48.7	71.92	2.47	101.6	7.4	9.9	10.1	3.8
Basin	15.0	23.2	33.0	25.31	1.35	45.1	5.5	6.6	7.1	2.9

Table 3. Coefficients of variation (CV) for fish metrics

Spatial scale	EFI	Richness metrics	Diversity metrics		Abundance metrics		Tolerance metrics	
		N_species	Shannon's diversity index	Simpson's diversity index	Density_species_all	Biomass sp_all	n_sp_Intol	n_sp_tol
Reaches	24.6	30.7	37.1	33.3	89.4	71.1	50.5	98.6
Streams	17.5	20.9	21.3	18.6	42.6	52.8	29.9	76.1
Basins	2.8	7.8	20.6	12.9	28.0	30.9	9.8	45.8

index was 33.3. Abundance and tolerance metrics were generally more variable than richness and diversity metrics (Table 3).

Rheophilic species richness was the least variable habitat preference metric (CV = 29.4) and limnophilic species richness was the most variable (CV = 198.2). With respect to reproduction, lithophilic species richness (CV = 34.1) was less variable than phytophilic species richness (CV = 220.8). Short-lived species (CV = 39.0) were less variable in terms of richness than long-lived species (CV = 126.1). The number of insectivorous/invertivorous species varied less (CV = 50.5) than omnivorous (CV = 110.9) and piscivorous (CV = 186.2) fishes. Richness of fish migration metrics was highly variable (CV = 148.5 for long distance fish, CV = 168.0 for potamodromons).

In general, fish guild density metrics CVs varied from 80.1 (number per ha of benthic habitat preferring species) to 484.7 (number per ha of limnophilic habitat preferring species). CVs of richness metrics varied from 29.4 (number of rheophilic habitat preferring species) to 220.8 (number of phytophilic mode reproduction species).

EFI was the least variable fish metric (Table 3). According to the EFI, sampling sites were classified from poor (one case, value 0.22) to good (highest value 0.65). The Lithuanian fish index ranked the sites as moderate to high ecological status.

Macroinvertebrates

Stream reaches were inhabited by 19–64 species, from 18–42 genera and 17–36 families. The organic

Table 4. Coefficients of variation (CV) for macroinvertebrate metrics

Macroinvertebrates												
Spatial scale	Eutrophication metrics		Richness metrics				Composition metrics				Diversity metrics	
	SI	BMWP	Families	Genera	Species number	Abundance	Evenness	EPT-taxa	EP-taxa	EPTCOB	Shannon's diversity index	Simpson's diversity index
Reaches	17.499	22.748	16.268	19.329	32.26	47.05	22.477	32.4	32.3	30.9	26.22	23.49
Streams	13.866	14.965	7.987	11.337	10.53	35.12	19.421	21.8	24.3	23.5	23.25	19.48
Basins	10.755	9.031	4.213	5.597	7.7	20.47	6.77	13.4	15.7	14.4	9.33	5.70

pollution metric BMWP was more variable than the saprobity index (Table 4). Among the richness metrics, the abundance of macroinvertebrates was most variable, followed by number of species, number of genera and number of families (Table 4). Shannon's diversity index ranged from 1.1 to 2.7 (CV = 26.2) and Simpson's diversity index from 0.4 to 0.9 (CV = 23.5). The EPT taxa metrics were more variable than the eutrophication and diversity metrics (Table 4). The mean abundance of taxonomic groups ranged from 0.1 ind/m^2 (or 0.002% of all individuals) for Nematomorpha to 2787.6 ind/m^2 (or 54.1% of all individuals) for Diptera. The values of CV exceeded 100% for most of the taxa, and the highest values were typical for taxa with low abundances or represented by few species. The more abundant macroinvertebrates were less variable (CV for Ephemeroptera was 93.1, CV for Trichoptera was 48.6 and CV for Diptera was 70.0). The mean number of taxa ranged from 0.01 for Nematomorpha to 7.3 for Trichoptera on the reach scale. The CV of this metric varied from 28.0 for Diptera to 509.9 for Turbellaria and Nematomorpha.

Among the macroinvertebrate groups the more variable metrics were those related to taxonomic composition such as number of EPT taxa, and especially the ratio of taxonomic groups and number of taxa, in comparison with eutrophication and diversity metrics (Table 4).

Benthic diatoms

Benthic diatoms had high species numbers (64–102 species per reach, CV = 11.4) and values of Shannon's (mean value 3.44, CV = 9.85) and Simpson's (mean value 0.91, CV = 6.22) diversity indices. The abundance of benthic diatoms varied more than the number of species. Among the fourteen trophic indices the least variable were L&M, ROTT and WAT, and the most variable was %PT followed by TDI (Table 5).

Comparison of metrics among different organism groups

The number of species, Shannon's diversity index and Simpson's diversity index were compared between BQEs groups. The largest number of species among all investigated BQEs was observed for benthic diatoms followed by macroinvertebrates. These groups were also less variable in species number at all scales than macrophytes and fishes, which were represented by only a few species (Tables 2–5). Benthic diatoms had the highest Shannon's diversity index values, followed by macroinvertebrates, fish and macrophytes. Macrophytes had the most variable Shannon diversity index at all scales followed by macroinvertebrates at the stream scale and fish at the reach and basin scales. The Shannon's diversity index values for benthic diatoms were least variable at all scales (Tables 2–5).

Usually macrophytes and benthic diatoms had the highest Simpson's diversity index values, followed by macroinvertebrates and fish. The most variable was Simpson's diversity index for fishes at the reach and the basin scales, and for macroinvertebrates at the stream scale. Macrophytes had the least variable Simpson's diversity index values. Variability of Simpson's diversity index values for BQEs were in all cases less than those for Shannon's diversity index (Tables 2–5). In general, no correlations were found in Shannon's and Simpson's index values among the groups of BQEs at the reach scale. The only positive tie at the reach scale was found between macrophyte and macroinvertebrate Simpson's diversity indices ($r = 0.41$, $p < 0.05$). At the stream scale negative correlation occurred between macroinvertebrates and benthic

Table 5. Coefficients of variation (CV) for phytobenthos metrics

Spatial scale	Number of species	Abundance	Diversity metrics		Trophic metrics													
			Shannon's	Simpson's	IDG	IPS	Descy	SLAD	L&M	SHE	WAT	TDI	%PT	EPI-D	ROTT	CEE	IBD	IDAP
Reach	11.4	14.0	9.85	6.22	7.3	7.2	5.9	6.9	5.7	6.0	5.9	15.1	48.8	6.0	5.7	7.0	10.9	8.0
Stream	8.8	13.4	7.19	4.05	5.8	5.9	4.5	5.8	4.9	4.7	4.2	9.3	41.2	5.0	5.0	6.0	8.0	6.5
Basin	4.3	6.8	3.15	1.88	3.6	4.6	4.5	4.7	3.3	1.9	3.1	8.6	28.0	3.6	3.3	4.5	6.9	2.5

diatoms Shannon's diversity indices ($r = -0.68$, $p < 0.05$). As only three river basins were studied, correlations between them were not calculated.

Comparison of BQEs metric groups among different spatial scales

Comparison of BQE metric indicative groups (not single metrics) was made at different spatial scales (within reach [only macroinvertebrates], reaches, streams and river basins) by the Sign test.

Macrophytes

The groups of macrophyte trophic metrics, composition metrics and combined trophic and composition metrics indicated that there were usually no statistically significant differences among reaches, streams and river basins. Differences between samples occurred for composition metrics in two cases: one at the reach scale and one at the stream scale.

Fish

No significant differences were found among reaches, streams and river basins when number of fish species, biomass and density metrics were compared. Differences between samples from two streams in one basin occurred for feeding metrics (piscivorous, insectivorous/invertivorous and omnivorous). The largest differences were found among fish habitat metric, in 6 of 27 cases at the reach scale and 2 of 9 cases at the stream scale. No differences were found for feeding or habitat metrics at the basin scale.

Macroinvertebrate

Significant differences between replicate samples for EPT taxa metrics occurred in 4 of 26 cases. In 3 cases out of 26, differences were found for number of taxa, and in one case for the percentage of a taxonomic group. The Sign test showed no significant differences for eutrophication metrics, diversity indices, diversity metrics, or abundance of taxonomic groups.

For stream reaches significant differences were found in 9 of 15 cases for EPT-taxa metrics. A difference between abundance of taxonomic groups and number of taxa was found only in one case. There were no significant differences between stream reaches according to the eutrophication metrics, diversity indices, diversity metrics, and taxonomic groups (%). Regarding EPT-taxa metrics at the stream scale, significant differences were found between 4 cases of 6 in two basins. In the third basin, 2 streams differed in the abundance of individuals.

At the river basin scale, one basin differed significantly from two others according to EPT-taxa. There were no significant differences between diversity indices, eutrophication metrics, diversity metrics, abundance of taxonomic groups, number of taxa and % taxonomic groups among stream basins.

Benthic diatoms

Benthic diatoms trophic metrics of reach scale differed in 8 out of 27 cases. No differences were found at the stream scale, but were again observed at the river basin scale (one basin [Venta basin] differed from the two others).

Comparison of BQEs metrics within and among river basins

The most and the least variable metrics for all BQEs as well as Shannon's and Simpson's diversity indices were compared among river basins.

Macrophytes

Regarding macrophyte metrics the most variable within catchments was Shannon's diversity index and the least variable was Simpson's diversity index. The values of Shannon's diversity index varied considerably among river basins, but differences of Simpson's diversity index were negligible (Table 6).

Fish

Variability of Shannon's and Simpson's diversity indices within river basins was relatively high, especially in one of the river basins (Table 6). Variability of the most and least variable fish metrics i.e. number of limnophilic species per ha (n_ha_Hab_li) and EFI were compared among river basins. Values of n_ha_Hab_li varied largely, but values of EFI were more similar (Table 7).

Macroinvertebrates

Variability of Shannon's and Simpson's diversity indices was not considerable within river basins (Table 6).

The number of families was the least variable macroinvertebrate metric among river basins, and the most variable metrics were %, abundance and number of taxa of some taxonomic groups (Table 7).

Phytobenthos

Values of diversity indices of phytobenthos within river basins were relatively less variables, especially those for Simpson's diversity index (Table 6).

The lowest CV of the phytobenthos metrics occurred for the ROTT, and this index varied slightly among basins. Highly variable %PT differed more among basins (Table 7).

Relationships of diversity indices with environmental variables

In general, the relationships among Shannon's and Simpson's diversity indices (calculated for all groups of biological elements) and environmental variables were inconsistent. We found relationships between macrophytes and nutrient enrichment, alkalinity and dissolved oxygen. Fish diversity was linked with stream morphometry (slope, altitude, maximum depth) and substrata. Relationships between macroinvertebrates and chemical parameters linked with river basin genesis, such as hardness, alkalinity, chloride, slope, stream velocity and oxygen. However, such relationships were typical only in one of three basins. Maximal depth, conductivity and ammonium influenced phytobenthos in one of the basins, but chemical parameters such as phosphorus, alkalinity, hardness and conductivity were more important in two other basins (Tables 8 and 9). Compared to other BQEs, relationships of benthic diatoms to environmental variables were the most pronounced, followed by macroinvertebrate, macrophyte and fish. In total, relationships of environmental factors with BQEs differed among the river basins (Tables 8 and 9).

Discussion

Variation in metrics and assessment systems at different spatial scales

Generally, we found that all metrics of macrophytes, fish, benthic macroinvertebrates and benthic diatoms showed the largest variations at the reach scale in comparison with stream and basin scales. This agrees to the theory that the relative variation of species richness may be expected to be greatest at small-to-intermediate spatial scales (Loreau et al., 2001). A hierarchical analysis approach to biodiversity studies in ecosystems enhances our understanding of ecological phenomena operating at different scales along multidimensional environmental gradients (Ward & Tockner, 2001) and physical and biological variables on a small spatial scale are influenced by

Table 6. Comparison of coefficient of variation (CV) for the biological quality elements (BQEs) Shannon's (H′) and Simpson's (D) diversity indices among river basins

River basin	Gauja		Daugava		Venta	
	H′	D	H′	D	H′	D
BQE						
Macrophyte	112.70	6.14	57.03	1.11	46.17	1.07
Fish	27.17	25.78	39.02	41.63	34.85	28.11
Phytobenthos	7.15	3.05	10.85	6.48	11.32	8.30
Macroinvertebrate	20.63	21.29	29.42	27.72	27.38	21.42

Table 7. Comparison of coefficient of variation (CV) for least and most variable indices of the biological quality elements (BQEs) among river basins

BQE	Fish		Phytobenthos		Macroinvertebrates	
	EFI	n_ha_Hab_li	ROTT	%PT	Number of families	TG (%), N_taxa, Abund_TG
River basins						
Gauja	26.8	218.1	4.2	34.7	11.8	135.1 Megaloptera (%)
Daugava	17.6	242.0	5.9	52.1	12.4	173.2 Nematoda (%), N_Nematoda_ taxa, Abund_ Nematoda
Venta	29.6	277.9	5.2	39.1	24.6	173.2 TG (%),N_taxa, Abund_TG , Turbellaria, Nematoda, Nematomorpha

EFI – European Fish Index; n_ha_Hab_li – Number of limnophilic species per ha, ROTT – Trophic index, % PT – % Pollution tolerant taxa, TG (%) – Taxonomic groups (%), N_taxa – Number of taxa of taxonomic groups, Abund_TG – Abundance of taxonomic groups

variables on larger spatial scales (Sandin & Johnson 2004). Spatial-scale has been investigated for macroinvertebates (Sandin & Johnson, 2000a, b; Sandin, 2001), fish (Legendre & Fortin, 1989; Tonn et al. 1990; McCormick & Hughes, 1998; Lammert & Allan, 1999; Van Sickle & Hughes, 2000), aquatic macrophytes (Gantes & Caro, 2001; Mackay et al., 2003, Vis et al., 2003) and diatom communities (Whittier et al., 1988, Pan et al., 1999; Soininen, 2004). We found that variability of metrics differed within different assemblages, but typically, the abundance of organisms varied more than species richness.

Macrophytes

In a study of three rivers in the UK, a decrease in MTR value was observed from upstream to downstream sites, reflecting changes in nutrient load, substrate and flow (Rivers Ouse, Ure and Wharfe Macrophyte Surveys, 2001). Such patterns were not found in our study. In most cases there was no statistically proven dissimilarity among reaches, streams or river basins when macrophyte groups were compared. At the same time an uneven distribution of single macrophyte metrics, including MTR, was observed instead. The most variable was cover of macrophytes followed by composition metrics. Trophic and especially tolerance metrics (hemeroby index) were less variable, likely because all sampling sites were high quality sites.

Fish

No significant differences among the reaches, streams and river basins were found when groups of fish composition and abundance were compared. Some differences among streams existed for feeding metrics. The largest differences were found for habitat metrics at the reach and stream scales. This is not surprising, since as fish usually are sensitive to habitat degradation (Gorman & Karr, 1978). Metrics within fish guilds varied differently. The migration metrics were comparatively more variable, followed by feeding, reproduction, habitat, longevity, abundance, tolerance and composition metrics. The most robust was EFI. Comparison of EFI and the local ecoregional Lithuanian fish index showed that the Lithuanian index is more suitable for the ecoregion. This corresponds to conclusion of Van Sickle & Hughes (2000) that fish communities were generally more similar within ecoregions than between ecoregions. At the same time they also reported that ecoregions accounted for very little of the classification difference. IBI scores differed significantly among the geological areas (Joy & Death, 2004).

Macroinvertebrates

Comparison of macroinvertebrate groups revealed that replicate samples differed for EPT-taxa metrics, number of taxa, and percentage of

Table 8. Significant Pearson's correlation coefficients (in bold) for environmental variables and BQEs indices in the river basins

Parameters	Daugava		Gauja		Venta	
	Shannon's index	Simpson's index	Shannon's index	Simpson's index	Shannon's index	Simpson's index
Fish						
Agricultural land	−0.54	−0.49	−0.33	−0.07	**−0.66**	−0.52
Max depth	0.20	0.15	0.42	0.34	**-0.66**	**−0.78**
Velocity	**0.69**	**0.71**	-0.05	−0.05	−0.08	−0.14
Discharge	0.46	0.44	0.36	0.17	0.65	**0.66**
Mesolithal	−0.48	−0.53	**−0.67**	−0.58	−0.60	−0.60
FPOM	−0.33	−0.34	0.61	**0.68**	-0.12	−0.16
Macrophyte						
HQA score	0.25	**-0.68**	−0.19	−0.2	0.08	-0.03
Slope	0.02	**-0.75**	−38	0.28	-0.15	0.31
Velocity	0.5	**-0.73**	0.11	0.1	0.23	0.2
Altitude	−0.08	−0.06	**−0.67**	0.45	0.15	-0.53
Psammal	0.38	-0.09	**0.68**	-0.37	−0.2	−0.41
CPOM	0.54	-0.11	−0.1	−0.03	**−0.74**	0.07
Oxygen	**0.66**	**-0.76**	−0.04	0.17	-0.05	0.63
Alkalinity	−0.33	**0.69**	0.11	-0.15	0	**-0.66**
Hardness	−0.3	**0.67**	-0.01	−0.04	−0.05	−0.55
Chloride	−0.53	**0.74**	-0.27	0.16	-0.26	0.02
Ammonium	0.03	0.01	**-0.79**	0.54	-0.09	0.38
Phosphate	0.11	-0.36	−0.15	0.04	0.35	**-0.8**
Tot- phosphorus	−0.09	0.25	0.27	-0.52	0.09	**-0.68**
Nitrite	**0.84**	-0.52	0.1	0.08	0.1	0.43
Macrozoobenthos						
HQA score	**−0.75**	**−0.78**	−0.17	−0.22	0.55	0.46
Catchment's size	0.28	0.27	0.34	0.13	**0.88**	0.59
Width	0.13	0.10	**0.78**	**0.78**	0.59	0.21
Source	0.05	0.05	0.27	0.10	**0.74**	0.39
Oxygen	**−0.89**	**−0.90**	−0.20	−0.24	0.48	0.33
Alkalinity	0.57	**0.68**	-0.35	−0.37	0.04	-0.03
Hardness	0.58	**0.66**	-0.29	−0.32	0.32	0.22
Chloride	**0.83**	**0.88**	-0.01	0.05	0.12	0.24
BOD_5	**−0.68**	**−0.68**	−0.26	−0.25	0.16	0.20
Nitrite	**−0.69**	**−0.76**	0.57	0.54	0.53	**0.70**
Phytobenthos						
HQA score	0.25	**-0.68**	−0.19	−0.2	0.08	-0.03
Slope	0.02	**-0.75**	−0.38	0.28	-0.15	0.31
Velocity	0.5	**-0.73**	0.11	0.1	0.23	0.2
Altitude	−0.08	−0.06	**−0.67**	0.45	0.15	-0.53
Psammal	0.38	-0.09	**0.68**	-0.37	−0.2	−0.41
CPOM	0.54	-0.11	−0.1	−0.03	**−0.74**	0.07
Oxygen	**0.66**	**-0.76**	−0.04	0.17	-0.05	0.63
Alkalinity	−0.33	**0.69**	0.11	-0.15	0	**-0.66**
Hardness	−0.3	**0.67**	-0.01	−0.04	−0.05	−0.55
Chloride	−0.53	**0.74**	-0.27	0.16	-0.26	0.02

Continued on p. 165

Table 8. (Continued)

Parameters	Daugava		Gauja		Venta	
	Shannon's index	Simpson's index	Shannon's index	Simpson's index	Shannon's index	Simpson's index
Ammonium	0.03	0.01	**-0.79**	0.54	-0.09	0.38
Phosphate	0.11	-0.36	−0.15	0.04	0.35	**-0.8**
Tot-phosphorus	−0.09	0.25	0.27	-0.52	0.09	**-0.68**
Nitrite	**0.84**	-0.52	0.1	0.08	0.1	0.43

taxonomic groups. EPT-taxa metrics also differed at the stream and basin scales. Single metrics connected with taxonomic composition like EPT taxa, and especially, taxonomic groups (%) and number of taxa, were most variable, in comparison with eutrophication and diversity metrics. The observed variability in the metrics contradicts the theory that each ecoregion has a predictable benthic macroinvertebrate assemblage (Barbour et al., 1999). However, others have found that local-scale variables such as in-stream substrata, riparian vegetation, and some chemical variables were most strongly associated with the among-site differences (Sandin & Johnson, 2004). The heterogeneity of microhabitats could be the reason for the variability in macroinvertebrate metrics, as the spatial pattern of habitats is a crucially important attribute of local conditions and significantly influences the trophic structure of communities (Bis et al., 2000).

Benthic diatoms

Benthic diatoms are traditionally considered as being regulated more by local than larger scale factors (Pan et al., 1999). However, it has recently been found that in fact there is no strict evidence confirming that unicellular diatoms have higher local species richness than metazoans (Hillebrand et al., 2001). Large-scale spatial factors, such as climate, geology and vegetation also influence diatom community structure (Leland, 1995). Currently it is believed that spatial variation in algal communities is the result of factors acting at multiple scales. Diatom communities in Finland exhibit a rather strong spatial component, and the proportion of variation explained by spatial factors was about 25% (Soininen, 2004). Some taxa exhibit regionally restricted distributions, thus contradicting the view that diatom communities have high dispersal abilities (Muotka et al., 2004). We found that among single metrics of combined stone and sand/silt substrata, the most variable metrics were % PT and TDI, as the least variable were ROTT, L&M and Descy. Our examination of differences among combined benthic diatoms groups by the Sign test showed that they differed at the reach, stream and basin scales. This confirms Soininen's (2003) statement that considerable heterogeneity occurs in benthic diatom communities at different scales.

Relationships between groups of biological quality elements

In this study benthic diatoms had the largest and less variable species richness among all investigated BQEs, followed by macroinvertebrates, macrophytes and lastly fish which were represented by few species. In boreal streams the spatial patterns exhibited by benthic diatoms were found to correspond quite closely with those of stream macroinvertebrates (Heino et al., 2002). However, Soininen & Könönen (2004) demonstrated clear separation of diatom community structure between sampling stations, but the corresponding macroinvertebrate communities were more similar to each other. Correlation between diatom and macroinvertebrate pollution indices in that study was rather low and insignificant. On the other hand, community similarity between the replicate samples was slightly lower among macroinvertebrates probably due to their larger local scale spatial variation, sampling of more habitats and lower density compared to diatoms (Soininen & Könönen, 2004). We compared Shannon's and

Table 9. Multiple linear regression model for BQEs diversity indices (dependent variables) and significance of environmental variables (predictors)

Daugava basin		Gauja basin		Venta basin	
Shannon's fish diversity index					
Env. Variable	Sign. of coeff	Env. Variable	Sign. of coeff	Env. variable	Sign. of coeff
Velocity	0.045	Mesolithal	0.032	microlithal	0.013
Psammal	0.140	Catchment size	0.051	Altitude	0.049
Conductivity	0.251	Depth	0.274	Chloride	0.194
Sign. of model: 0.054		Sign. of model: 0.053		Sign. of model: 0.018	
R^2: 0.76		R^2: 0.76		R^2: 0.90	
Simpson's fish diversity index					
Env. Variable	Sign. of coeff	Env. variable	Sign. of coeff	Env. variable	Sign. of coeff
Velocity	0.097	FPOM	0.066	Altitude	0.012
Conductivity	0.567	Chloride	0.179	microlithal	0.009
Psammal	0.171				
Sign. of model: 0.086		Sign. of model: 0.057		Sign. of model: 0.010	
R^2: 0.84		R^2: 0.61		R^2: 0.84	
Shannon's macrophyte diversity index					
Env. Variable	Sign. of coeff	Env. variable	Sign. of coeff	Env. variable	Sign. of coeff
Oxygen sat.	0.028	Ammonium	0.166	Catchment size	0.03
Nitrite	0.007	Psammal	0.912	Macrolithal	0.59
Sign. of model: 0.003		Sign. of mode: 0.058		Sign. of model: 0.004	
R^2: 0.90		R^2: 0.61		R^2: 0.95	
Simpson's macrophyte diversity index					
Env. variable	Sign. of coeff	Env. variable	Sign. of coeff	Env. variable	Sign. of coeff
Slope	0.003	Tot- P	0.047	Phosphate	0.017
oxygen	0.003	Psammal	0.072	Alkalinity	0.049
Sign. of model: 0.004		Sign. of model:0.073		Sign. of model: 0.009	
R^2: 0.95		R^2: 0.58		R^2: 0.85	
Shannon's macrozoobenthos diversity index					
Env. variable	Sign. of coeff	Env. Variable	Sign. of coeff	Env. variable	Sign. of coeff
Oxygen	0.004	Width	0.031	Catchment size	0.031
HQA score	0.037	Forest	0.520	Macrolithal	0.598
Sign. of model: 0.001		Sign. of model: 0.04		Sign. of model: 0.022	
R^2: 0.90		R^2: 0.64		R^2: 0.78	
Simpson's macrozoobenthos diversity index					
Env. variable	Sign. of coeff	Env. Variable	Sign. of coeff	Env. variable	Sign. of coeff
HQA score	0.010	Width	0.040	Nitrite	0.049
Oxygen	0.001	Macrolithal	0.670	Catchment size	0.098
Sign. of model: 0.00		Sign. of model: 0.05		Sign. of model: 0.04	
R^2: 0.94		R^2: 0.62		R^2: 0.72	

Continued on p. 167

Table 9. (Continued)

Daugava basin		Gauja basin		Venta basin	
Shannon's phytobenthos diversity index					
Env. variable	Sign. of coeff	Env. variable	Sign. of coeff	Env. variable	Sign. of coeff
mesolithal	0.02	Max depth	0.03	Max depth	0.006
Ammonium	0.02	Conductivity	0.03	Xylal	0.006
Sign. of model: 0.012		Sign. of model: 0.003		Sign. of model: 0.005	
R^2: 0.77		R^2: 0.85		R^2: 0.82	
Simpson's phytobenthos diversity index					
Env. variable	Sign. of coeff	Env. variable	Sign. of coeff	Env. Variable	Sign. of coeff
Mesolithal	0.038	Max depth	0.012	PH value	0.013
Hardness	0.238	Ammonium	0.050	Discharge	0.007
Sign. of model: 0.028		Sign. of model: 0.004		Sign. of model:0.008	
R^2: 0.81		R^2: 0.83		R^2: 0.8	

Simpson's diversity indices calculated for all studied organism groups, and found no correlations at the reach scale for macrophytes, fish, macroinvertebrates and benthic diatoms. The only positive correlation was found between macrophytes and macroinvertebrates Simpson's diversity indices ($r = 0.41$, $p < 0.05$) at the reach scale. Negative correlation occurred between benthic diatoms and macroinvertebrates Shannon's diversity indices ($r = -0.68$, $p < 0.05$) at the stream scale. It agrees with the conclusion of Soininen & Könönen (2004) that similarity in diversity among main biological groups is often rather low, especially at the smaller, within basin spatial scales and therefore it is advisable to base stream biomonitoring on macroinvertebrates and benthic diatoms. This is probably because different taxonomic groups showed different relationships to environmental gradients, leading to relatively low levels of concordance (Muotka et al., 2004). Regional EMAP-SW surveys also indicated the importance of assessing multiple biological assemblages because each assemblage was differentially sensitive to different stressors and at different spatial scales (Hughes et al., 2000).

Relations between diversity metrics and environmental variables

One of our objectives was to assess the importance of environmental variables to biological quality elements for different river basins. The strength of observed patterns depends on the extent to which various mechanisms act in concert; clear patterns arise when several processes act in one direction, and in general observed patterns can have multiple explanations (Gaston & Blackburn, 1999). Thus the central question is not which explanation is the correct one, but what are their relative roles (Soininen, 2004).

The role of current velocity and light regime (Westlake, 1975), pH (Tremp & Kohler, 1995) and especially eutrophication (Westlake, 1975; Holmes et al., 1999; Haury et al., 2002) have been found important for aquatic macrophytes. In Australia investigations of spatial variation in submersed macrophytes confirmed the influence of geomorphology at the catchment and at the reach scale (Mackay et al., 2003). Our findings confirmed the role of nutrient enrichment, alkalinity and dissolved oxygen in structuring the macrophytes diversity.

Many investigations of fish community in relation to environmental variables revealed a role of different factors contributing to fish distribution such as altitude, distance from the source, stream width, substrate (Jowett & Richardson, 2003), flow (Lammert & Allan, 1999), conductivity (Frenzel & Swanson, 1996; Meador & Goldstein, 2003) and especially land use pattern (Frenzel & Swanson, 1996; Lammert & Allan, 1999; Jowett & Richardson, 2003; Meador

& Goldstein, 2003; Dauwalter & Jackson, 2004). At the river scale, distance from the source, altitude and surface of catchment area (km^2) could explained 70% of the total variation in fish species richness (Mastrorillo et al., 1998). Fish are closely linked with habitat (Gorman & Karr, 1978), and in general the composition and relative abundance of fish species can be considered as the result of sequence of 'filters' ranging from continental to local spatial scale (Tonn et al., 1990). As in this study we compared only high quality sites, our findings linked fish diversity mainly with stream morphometry (slope, altitude, maximum depth) and substrata.

A lot of investigations have shown that benthic macroinvertebrate assemblages are structured by factors such as stream hydraulics, substrate, water chemistry, and riparian vegetation (Minshall, 1984; Statzner et al., 1988; Richards et al., 1993, 1997: Leland & Fend, 1998; Lammert & Allan, 1999, Bis et al., 2000; Sandin & Johnson, 2004; Soininen & Könönen, 2004). Basin geology also contributes significantly, but the explained variance associated with this factor is less than that related to land use (Leland, 1995, Leland & Porter, 2000). Local physical and chemical variables were found to explain the largest part of the among-site variability of macroinvertebrate community assemblages in Sweden (Sandin & Johnson, 2004). We also found a relationship between macroinvertebrates and chemical parameters linked with river basin genesis such as hardness, alkalinity, chloride, slope, stream velocity and oxygen. However, such relationships were typical only in one of three basins.

The development of benthic diatoms in streams is determined by complex local (e.g., discharge) and regional (catchments, ecoregions) factors (e.g., geology, topography, climate) that act as environmental filters (Poff, 1997). At the basin scale the ionic composition and major nitrogen and phosphorus concentration, salinity, substrate type and physiognomic form of dominant species were primary factors contributing to variation in benthic-algal assemblages. Diatoms are considered to be better indicators of changes in water chemistry than macroinvertebrates due to their shorter life cycles and larger sensitivity (Steinberg & Schiefele, 1988), but large-scale spatial factors, such as climate, geology and vegetation also influence diatom community structure (Leland, 1995). In the USA, up to one-third of the total explainable variation in diatom species data was attributed solely to geographical factors (latitude and altitude) not correlated with measured environmental characteristics (Potapova & Charles, 2002). Conductivity was found to be the strongest environmental gradient explaining diatom distribution patterns in Finland, with total P and latitude also being important (Soininen & Könönen, 2004). In southern Poland, all diatom indices calculated using OMNIDIA software (except for the Sládeček's index), correlated significantly with organic load (COD) and oxygen concentration. Some indices showed a significant negative correlation with NH_4-N and P-PO_4 (Kwandrans et al., 1998). Discharge (channel width, depth, current velocity) frequently plays an overriding role in the regulation of development of benthic organisms in general (Hart & Finelli, 1999). We found that maximal depth and ammonium influenced phytobenthos in one basin, but phosphorus, alkalinity, and hardness were more important in two others basins. In all basins relationships between conductivity and benthic diatoms occurred. Compared to other BQEs, relationships of benthic diatoms to environmental variables were more pronounced.

In general, the relationships between diversity indices of macrophytes, fish, macroinvertebrates and benthic diatoms, and environmental variables differed by basins. These relationships were stronger for small bodied organisms – benthic diatoms and macrophytes, followed by large bodied organisms – macrophytes and fish.

Conclusions

A comparison of single metrics of different biological quality elements demonstrated that in general they varied most at the reach scale in comparison with stream and basin scales.

Variability of metrics within groups of biological quality elements:

- Macrophytes composition metrics were the most variable, and trophic and especially tolerance metrics (hemeroby index) were least variable.

- Fish migration metrics were most variable, followed by feeding, reproduction, habitat, longevity, abundance, tolerance and composition metrics. The variability of fish metrics expressed by density was considerably larger than corresponding metrics characterized by species richness. The most robust indicator was EFI.
- The most variable macroinvertebrate metrics were those connected to taxonomic composition like EPT taxa, especially taxonomic groups (%) and number of taxa. Eutrophication and diversity metrics were less variable.
- The %PT and TDI were the most variable benthic diatoms metrics, while ROTT, L&M and Descy were the least variable.

In general, the least variable assemblages were the large bodied organisms, macrophytes and fish, followed by the small-bodied organisms – macroinvertebrates and benthic diatoms.

The disparity among reaches, streams and river basins was evaluated by Sign test using different indicative groups of the biological quality elements.

- In most cases there was no statistically proven dissimilarity among reaches, streams or river basins if macrophyte groups were compared.
- No significant differences among the reaches, streams and river basins were found if fish composition and abundance metrics were compared. Some differences among streams existed for feeding metrics. The largest differences were found for habitat metrics at the reach and stream scales. No differences were found for fish guilds at the basin scale.
- Comparison of macroinvertebrate metrics revealed that replicate samples differed for EPT-Taxa metrics, number of taxa, and percentage of taxonomic groups. EPT-Taxa metrics also differed at the stream and basin scales.
- Combined benthic diatoms metrics differed at the reach, stream and basin scales, especially those on soft substrate.

Shannon's diversity index varied considerably, while Simpson's diversity index differed minimally. In general, no correlations in community structure (Shannon's and Simpson's diversity indices) were found among groups of organisms.

Analysis of the relationships between environmental variables and diversity indices (Shannon's and Simpson's) of all BQEs did not reveal consistent relationships and differed by river basins. These relationships were stronger for small-bodied organisms – benthic diatoms and macrophytes, followed by large-bodied organisms – macrophytes and fish. We concluded that macrophytes and fish assemblages are most useful for assessing stream ecological status at the river basin scale, while macroinvertebrates and benthic diatoms assemblages are better at indicating smaller scale patterns.

Acknowledgements

The authors acknowledge the European Commission, 5th Framework Program, Energy, Environment and Sustainable Development, Key Action Water, Contract no. EVK1-CT–2001–00027 for financial support and Dr. M. T. Furse for perfect project leading. We thank all our colleagues for contribution to the spatial scale study, and Prof. B. Kawecka and Dr. J. Kwandrans for preparation and identification of benthic diatoms samples. We are grateful to two anonymous reviewers for their highly useful comments on the earlier version of the manuscript.

References

Anonymous, 1992. Standard Methods for Examination of Water and Wastewater. 18th edn. APHA, AWWA, WEF.

Anonymous., 2000. European Commission Directive 2000/60/EC of the European Parliament and of the Council of 23 October 2000 establishing a framework for community action in the field of water policy. Official Journal L 327, 22/12/2000 P : 0001–0073.

Barbour, M. T., J. Gerritsen, B. D. Snyder & J. B. Stribling, 1999. Rapid Bioassessment Protocols for use in Streams and Wadeable Rivers: Periphyton, Benthic Macroinvertebrates, and Fish. EPA 841-B-99-002. 2nd edn. US Environmental Protection Agency, Office of Water, Washington DC.

Bis, B., A. Zdanovic & M. Zalewski, 2000. Effects of catchment properties on hydrochemistry, habitat complexity and invertebrate structure in a lowland river. Hydrobiologia 422/423: 369.

CEMAGREF, 1982. Etude des méthodes biologiques d'appréciation quantitative de la qualité des eaux. Rapport Q. E. Lyon A. F. Bassin Rhōne-Méditérannée-Corse, 218.

Dawson, F. H., J. R. Newman, M. J. Gravelle, K. J. Rouen & P. Henville, 1999. Assessment of the Trophic Status of Rivers Using Macrophytes – Evaluation of the Mean

Trophic Rank. R&D Technical Report E39. Environment Agency, Bristol, 177 pp.

Dauwalter, D. C. & J. R. Jackson, 2004. A provisional fish index of biotic integrity for assessing Ouachita Mountains streams in Arkansas, U.S.A. Environmental Monitoring Assessment 91(1–3): 27–57.

Dell'Uomo, A., 1996. Assessment of water quality of an Apennine river as a pilot study. In Whitton, B. A. & T. Rott (eds), Use of Algae for Monitoring Rivers. II Institute für Botanik, Univerität Innsbruck, 65–73.

Descy, J. P., 1979. A new approach to water quality estimation using diatoms. Nova Hedwigia 64: 305–323.

Descy, J. P. & M. Coste, 1991. A test of methods for assessing water quality based on diatoms. Verhandlungen der Internationalischen Vereinigung für Theoretische und Angewandte Limnologie 24: 2112–2116.

Ellenberg, H., 1979. Die Zeigerwerte der Gewässpflanzen Mitteleuropas. Scripta Geobotanica 9: 1–122.

Frenzel, S. A. & R. B. Swanson, 1996. Relations of Fish Community Composition to Environmental Variables in Streams of Central Nebraska, USA. Environmental Management 20(5): 689–705.

Furse M., D. Hering, O. Moog, P. Verdonschot, R. K. Johnson, K. Brabec, K. Gritzalis, A. Buffagni, P. Pinto, N. Friberg, J. Murray-Bligh, J. Kokes, R. Alber, P. Usseglio-Polatera, P. Haase, R. Sweeting, B. Bis, K. Szoszkiewicz, H. Soszka, G. Springe, F. Sporka & I. Krno, 2006. The STAR project: context, objectives and approaches. Hydrobiologia 566: 3–29.

Friedrich, G., D. Chapman & A. Beim, 1996. The use of biological material. In Chapman, D. (ed.), Water Quality Assessments. A Guide to the Use of Biota, Sediments and Water in Environmental Monitoring. Published on behalf of UNESCO, WHO and UNEP by Chapman & Hall, London, 175–242.

Gantes, H. P. & A. S. Caro, 2001. Environmental heterogeneity and spatial distribution of macrophytes in plain streams. Aquatic Botany 70(3): 225–236.

Gaston, K. J. & T. M. Blackburn, 1999. A critique for macroecology. Oikos 84: 353–368.

Gibbons, J. D., 1985. Nonparametric Statistical Inference, 2nd edn. M. Dekker.

Gorman, O. T. & J. R. Karr, 1978. Habitat structure and stream fish communities. Ecology 59: 507–515.

Hart, D. D. & C. M. Finelli, 1999. Physical-biological coupling in streams: the pervasive effects of flow on benthic organisms. Annual Review of Ecology and Systematics 30: 363–395.

Haury, J., M. C. Peltre, M. Tremolieres, J. Barbe, G. Thiebaut, I. Bernez, H. Daniel, P. Chatenet, S. Muller, A. Dutartre, C. Laplace-Treyture, A. Cazaubon & E. Lambert-Servien, 2002. A method involving macrophytes to assess water trophy and organic pollution: the Macrophyte Biological Index for Rivers (IBMR) – application to different types of rivers and pollutions. Proc. 11th ewrs internat'l. symp. aquatic weeds, Moliets et Maa, France, (eds), A. Dutartre & M. -H. Montel, 247–250.

Heino, J., T. Muotka, R. Paavoal, H. Hämäläinen & E. Koskemmiemi, 2002. Correspondence between regional delineations and spatial patterns in macroinvertebrate assemblages of boreal headwater streams. Journal of the North American Benthological Society 21: 397–413.

Hering, D., O. Moog, L. Sandin & P. F. M. Verdonschot, 2004. Overview and application of the AQEM assessment system. Hydrobiologia 516: 1–21.

Hillebrand, H., R. Waterman, R. Karez & U. G. Berninger, 2001. Differences in species richness patterns between unicellular and multicellular organisms. Oekologia 126(1): 114–124.

Holmes, N. T. H., J. R. Newman, S. Chadd, K. J. Rouen, L. Saint & F. H. Dawson, 1999. Mean Trophic Rank: A Users Manual. R&D Technical Report No. E38. Environment Agency, Bristol, UK.

Hughes, R. M., S. G. Paulsen & J. L. Stoddard, 2000. EMAP-Surface Waters: a national, multiassemblage, probability survey of ecological integrity. Hydrobiologia 423: 429–443.

Jalas, J., 1955. Hemerobe und hemerochore Pflanzenarten – ein terminologischer Reformversuch. Acta Societatis pro Fauna et Flora Fennica 72(11): 1–15.

Jansons, V., N. Vagstad, R. Sudars, J. Deelstra, I. Dzalbe & D. Kirsteina, 2002. Nutrient losses from point and diffuse agricultural sources in Latvia. Landbauforschnung Volkenrode. 1(52): 9–17.

Johnson, R. K., 1995. The indicator concept in freshwater biomonitoring Thienemann lecture. In Cranston, P. S. (ed.), Chironomids – from Genes to Ecosystems, Proceedings of the 12th International Symposium on Chironomidae, Canberra, Australia. CSIRO, Melbourne, 11–27.

Johnson, R. K., 1998. Spatial-temporal variability of temperate lake macroinvertebrate communities: detection of impact. Ecological Applications 8: 61–70.

Jowett, I. G. & J. Richardson, 2003. Fish communities in New Zealand rivers and their relationship to environmental variables. New Zealand Journal of Marine and Freshwater Research 37: 347–366.

Joy, M. K. & R. G. Death, 2004. Application of the index of biotic integrity methodology to New Zealand freshwater fish communities. Environmental Management 34(3): 415–28.

Kelly, M. G. & B. A. Whitton, 1995. The Trophic Diatom Index: a new index for monitoring eutrophication in rivers. Journal of Applied Phycology 7: 433–444.

Kesminas, V. & T. Virbickas, 2000. Application of an adapted index of biotic integrity to rivers of Lithuania. Hydrobiologia 422/423: 257–270.

Krebs, C. J., 1999. Ecological Methodology (2nd edn). Addison Wesley Longman, Inc, Menlo Park, California, 620 pp.

Kwandrans, J., P. Eloranta, B. Kawecka & K. Wojtan, 1998. Use of benthic diatom communities to evaluate water quality in rivers of southern Poland. Journal of Applied Phycology 10(2): 193–201.

Lammert, M. & J. D. Allan, 1999. Assessing the biotic integrity of streams: effects of scale in measuring the influence of land use/cover and habitat structure on fish and macroinvertebrates. Environmental Management 23: 257–270.

Lecointe, C., M. Coste & J. Prygiel, 1993. 'Omnidia' Software for taxonomy, calculation of diatom indices and inventories management. Hydrobiologia 269/270: 509–513.

Leclercq, L. & B. Maquet, 1987. Deux nouveaux indices chimique et diatomique de qualité d'eau courante. Application au Samson et à ses affluents (bassin de la Meuse belge). Comparaison avec d'autres indices chimiques, biocènotiques et diatomiques. Institut Royal des Sciences Naturelles de Belgique, document de travail 28: 113.

Legendre, P. & M. J. Fortin, 1989. Spatial pattern and ecological analysis. Vegetation 80: 107–138.

Leland, H. V., 1995. Distribution of benthic diatoms in the Yakima River Basin, Washington, in relation to geology, land use, and other environmental factors. Canadian Journal of Fisheries and Aquatic Sciences 52: 1108–1129.

Leland, H. V. & S. V. Fend, 1998. Benthic invertebrate distributions in the San Joaquin River, California, in relation to physical and chemical factors. Canadian Journal of Fisheries and Aquatic Sciences 55: 1051–1067.

Leland, H. V. & S. D. Porter, 2000. Distribution of benthic algae in the upper Illinois River basin in relation to geology and land use. Freshwater Biology 44: 279–301.

Lenat, D. R. & D. L. Penrose, 1996. History of the EPT taxa richness metric. Bulletin North American Benthological Society 13: 305–307.

Li, J., A. Herlihy, W. Gerth, P. Kaufmann, S. Gregory, S. Urquhart & D. P. Larsen, 2001. Variability in stream macroinvertebrates at multiple spatial scales. Freshwater Biology 46: 87–97.

Loreau, M., S. Naeem, P. Inchusti, J. Bengtsson, J. P. Grime & A. Hector, 2001. Biodiversity and ecosystem functioning: current knowledge and future challenge. Science 294: 804–808.

McCune, B. & M. J. Mefford, 1999. PC-ORD. Multivariate Analysis of Ecological Data, version 4. MjM Software Design, Gleneden Beach, Oregon, USA.

Mackay, S. J., A. H. Arthington, M. J. Kennard & J. Pusey, 2003. Spatial variation in the distribution and abundance of submersed macrophytes in an Australian subtropical river. Aquatic Botany 77: 169–186.

Mastrorillo, S., F. Dauba, T. Oberdorff, J. F. Guégan & S. Lek, 1998. Predicting local fish species richness in the Garonne river basin. Comptes Rendus de l'Académie des Sciences de Paris 321: 423–428.

McCormik, F. H. & R. M. Hughes, 1998. Aquatic vertebrates. In Lazorchak, J. L., D. J. Klemm & D. V. Peck (eds), Environmental Monitoring and Assessment Program – Surface Waters: Field Operations and Methods for Measuring the Ecological Condition of Wadable Streams. EPA/620/R-94/004F US EPA, Washington DC: 161–182.

Meador, M. R. & R. M. Goldstein, 2003. Assessing water quality at large geographic scales: relations among land use, water physicochemistry, riparian condition, and fish community structure. Environmental Management 31(4): 504–517.

Muotka, T., J. Heino, R. Paavola & J. Soininen, 2004. Large scale biodiversity patterns of boreal stream communities. In Eloranta, P. (ed.), Inland and Coastal Waters of Finland 116–119.

Minshall, G. W., 1984. Aquatic insect-substratum relationships. In Resh, V. H. & D. M. Rosenberg (eds), The ecology of aquatic insects. Praeger Scientific, New York, USA: 358–400.

Pan, Y., R. J. Stevenson, B. H. Hill, P. R. Kaufmann & A. T. Herlihy, 1999. Spatial patterns and ecological determinants of benthic algal assemblages in Mid-Atlantic streams, USA. Journal of Applied Phycology 35: 460–468.

Poff, N. L., 1997. Landscape filters and species traits: towards mechanistic understand and prediction in stream ecology. Journal of North American Benthological Society 16: 391–409.

Prygiel, J. & M. Coste, 1993. The assessment of water quality in the Artois-Picardie water basin (France) by the use of diatom indices. Hydrobiologia 302: 179–188.

Potapova, M. & D. F. Charles, 2002. Benthic diatoms in USA Rivers: distributions along spatial and environmental gradients. Journal of Biogeography 29: 167–187.

Raven, P. J., P. Fox, M. Everald, N. T. H. Holmes & F. H. Dawson, 1997. River Habitat Survey: a new method for classifying rivers according to their habitat quality. In Boon, P. J. & D. L. Howell (eds), Freshwater Quality: Defining the Indefinable? The Stationery Office, Edinburgh, 215–234.

Raven, P. J., N. T. H. Holmes, F. H. Dawson & M. Everald, 1998. Quality assessment using River Habitat Survey data. Aquatic Conservation: Marine And Freshwater Ecosystems : 477–499.

Reynoldson, T. B., R. H. Norris, V. H. Resh, K. E. Day & D. M. Rosenberg, 1997. The reference condition: a comparison of multimetric and multivariate approaches to assess water-quality impairment using benthic macroinvertebrates. Journal of the North American Benthological Society 16: 833–852.

Richards, C., G. E. Host & J. W. Arthur, 1993. Identification of predominant environmental factors structuring stream macroinvertebrate communities within a large agricultural catchment. Freshwater Biology 29: 285–294.

Richards, C., R. J. Haro, L. B. Johnston & G. E. Host, 1997. Catchment and reach-scale properties as indicators of macroinvertebrate species traits. Freshwater Biology 37: 219–230.

Rivers Ouse, Ure and Wharfe Macrophyte Surveys, 2001. Report for Yorkshire Services Ltd by Bullen Consultants.

Rott, E. (ed.), 1999. Indikationslisten für Aufwuchsalgen in Österreichischen fliessgewässern. Teil 2: Trophienindikation sowie geochemische Präferenz, taxonomische und toxikologische Anmerkungen Bundesministerium für Land- und Forstwirschaft, Wasserwirtschaftskataster Wien.

Sandin,, L., 2001. Spatial and temporal variability of stream benthic macrroinvertebrates. Implications for environmental assessment. Doctoral thesis, Silvestria 172, Swedish University of Agricultural Sciences.

Sandin, L. & R. K. Johnson, 2000a. Spatial scale of benthic macroinvertebrate communities in Swedish streams: variation partitioning using partial Canonical Correspondence Analysis. Verhandlungen der Internationalischen Vereinigung für Theoretische und Angewandte Limnologie 27: 382–383.

Sandin, L. & R. K. Johnson, 2000b. Ecoregions and benthic macroinvertebrates in Swedish streams. Journal of North American Benthological Society 19: 462–474.

Sandin, L. & R. K. Johnson, 2000c. Statistical power of selected indicator metrics using macroinvertebrates for assessing acidification and eutrophication of running waters. Hydrobiologia 422/423: 233–243.

Sandin, L. & R. K. Johnson, 2004. Local, landscape and regional factors structuring benthic macroinvertebrate assemblages in Swedish streams. Landscape Ecology 19: 501–514.

Sládeček, V., 1986. Diatoms as indicators of organic pollution. Acta Hydrochimica et Hydrobiologica 14: 555–566.

Soininen, J., 2003. Heterogeneity of benthic diatom communities in different spatial scales and current velocities in a turbid river. Archiv für Hydrobiologie 156: 551–564.

Soininen, J., 2004. Benthic Diatom Community Structure in Boreal Streams. Distrubution Patterns Along Environmental And Spatial Gradients. Academic dissertation in limnology, Helsinki 46 pp.

Soininen, J. & K. Könönen, 2004. Comparative study of monitoring South-Finnish rivers and streams using macroinvertebrate and benthic diatom community structure. Aquatic Ecology 38(1): 63–75.

SPSS for Windows Rel. 12.0.1., 2004. Chicago: SPSS Inc.

Statzner, B., J. A. Gore & V. H. Resh, 1988. Hydraulic Stream Ecology: observed patterns and potential applications. Journal of the North American Benthological Society 7(4): 307–360.

Steinberg, C. & S. Schiefele, 1988. Biological indication of trophy and pollution of running waters. Zeitschrift für Wasser-Abwasser Forschung 21: 227–234.

Tonn, W. M., J. J. Magnuson, M. Rask & J. Toivonen, 1990. Intercontinental comparison of small-lake fish assemblages: the balance between local and regional processes. American Naturalist 136: 345–375.

Thompson, J. N., O. J. Reichman, P. J. Morin, G. A. Polis, M. E. Power, R. W. Sterner, C. A. Couch, L. Gough, R. Holt, D. U. Hooper, F. Keesing, C. R. Lovell, B. T. Milne, M. C. Molles, D. W. Roberts & S. Y. Strauss, 2001. Frontiers of Ecology. Bioscience 5: 15–24.

Tremp, H. & A. Kohler, 1995. The usefulness of macrophyte monitoring-systems, exemplified on eutrophication and acidification of running waters. Acta botanica Gallica 142: 541–550.

Van Sickle, J. & R. M. Hughes, 2000. Classification strengths of ecoregions, catchments and geographical clusters for aquatic vertebrates in Oregon. Journal of the North American Benthological Society 19(3): 370–384.

Vis, C., C. Hudon & R. Carignan, 2003. An evaluation of approaches used to determine the distribution and biomass of emergent and submerged aquatic macrophytes over larger spatial scale. Aquatic Botany 77: 187–201.

Ward, J. V. & K. Tockner, 2001. Biodiversity: towards a unifying theme for river ecology. Freshwater Biology 46: 807–819.

Westlake, D. F., 1975. Macrophytes. In Whitton, B. A. (ed.), River Ecology: Studies in Ecology Vol. 2. University of California Press, Berkeley, 106–128.

Whittier, T. R., R. M. Hughes & D. P. Larsen, 1988. Correspondence between ecoregions and spatial patterns in stream ecosystems in Oregon. Canadian Journal of Fisheries and Aquatic Sciences 45: 1264–1278.

Zelinka, M. & P. Marvan, 1961. Zur Präzisierung der biologischen Klassifikation der Reinheit fliessender Gewässer. Archiv für Hydrobioliogie 57: 389–407.

Macrophytes and Diatoms

Hydrobiologia (2006) 566:175–178

M.T. Furse, D. Hering, K. Brabec, A. Buffagni, L. Sandin & P.F.M. Verdonschot (eds), The Ecological Status of European Rivers: Evaluation and Intercalibration of Assessment Methods
DOI 10.1007/s10750-006-0097-0

Macrophytes and diatoms – major results and conclusions from the STAR project

Karel Brabec[1,]* & Krzysztof Szoszkiewicz[2]
[1]*Institute of Botany and Zoology, Faculty of Science, Masaryk University, Kotlarska 2, 611 37, Brno, Czech Republic*
[2]*Department of Ecology and Environmental Protection, Agricultural University of August Cieszkowski, ul. Piątkowska 94C, 61-691 Poznan, Poland*
*(*Author for correspondence: E-mail: brabec@sci.muni.cz)*

Key words: Water Framework Directive, aquatic plants, assessment

Abstract

The interactions between sensitivity and variability of macrophyte and diatom communities were evaluated as a research support of methodologies required by the Water Framework Directive. Slope and shading were identified as additional typological parameters improving links between unimpacted macrophyte communities and running water types. Two other studies demonstrated indication value of macrophytes for assessment of nutrient enrichment and hydromorphological degradation. The special exercises were realized within the STAR project to evaluate sources of variability/uncertainty in assessment methods based on macrophytes and diatoms. Sampling period and shading of the site were found as major factors affecting variability in macrophyte assessment results. Uncertainty of diatom assessment is predominantly associated with selection of site, substrate type and taxonomic identification. Further extension of indication systems and definition of macrophyte/diatom-specific typology of running waters are considered as the main aims of subsequent investigations.

Introduction

Aquatic macrophytes and algae are primarily sensitive to human impacts changing the light and nutrient regimes, physical characteristics of habitats and organic matter spiralling (Murphy, 1998). Agriculture, forest management, urban development, river regulation and other activities affect aquatic primary production and structure of autotrophic communities. Macrophyte communities can be viewed as indicators of flow conditions, but they can also act as factor affecting flow significantly or they are utilised as the substrate by algae and macroinvertebrates.

Scale of response varies from short-term conditions existing at microhabitat through response to characteristics of channel or riparian zone to large-scale changes in watershed. Phytobenthos is usually considered as early warning indicator while macrophytes due to their longer life cycles and tolerance to short-term changes of environmental conditions indicate more persistent impairment.

Aquatic plants are considered as valuable indicators of ecological status and their monitoring is required by Water Framework Directive (Council of the European Communities, 2000). General advantages of using macrophytes for assessment are that they are stationary, are visible by naked eye, relatively low manpower demands, quantity can be expressed as cover and in some cases remote sensing bring information about quantity of macrophytes. Advantages of diatoms are well-established relationships with certain stressors (particularly eutrophication), they are common and diverse organisms and fossil records provide information about past environments (Downes et al., 2002).

Preparation of monitoring methodologies for the biological quality elements (phytoplankton, phytobenthos, macrophytes, benthic invertebrates and fish) requires reviewing existing knowledge, development of new methods, calibration assessment systems and testing their uncertainty/robustness.

In this chapter we present the STAR project contributions oriented to macrophytes and diatoms in terms of relation to abiotic river typology, response to various impairments and sources of uncertainty.

Community pattern in relation to stream types

The paper prepared by Baattrup-Pedersen et al. (2006) develops the knowledge on macrophyte assemblage patterns in unimpacted streams and rivers. This issue plays a substantial role in ecological quality classification. In the presented paper 60 unimpacted and less impacted stream and river sites situated throughout Europe were analysed concerning community structure, i.e. composition, richness and diversity measures.

A considerable variation of macrophyte community structure was found. Moving from small-sized, shallow mountain streams to medium-sized, lowland streams there was a clear transition in species richness, diversity and community structure. Especially there was a shift from a predominance of species poor mosses and liverwort, which dominated communities in the small-sized, shallow mountain streams to more species rich communities dominated by vascular plants in the medium-sized, lowland streams. It was found that the macrophyte communities responded to most of the features underlying the typological framework defined in WFD.

Slope and shading affected basic macrophyte community characteristics (taxa richness, diversity) and indices based on sensitivity to stressors as well (MTR, IBMR). The typology used for definition of type-specific reference conditions should incorporate these environmental parameters.

Relation of macrophyte metrics to stressors

The study by Szoszkiewicz et al. (2006) used most of STAR macrophyte records representing 17 stream types from four geographical groups: mountain streams (72 sites, 7 stream types), lowland streams (101 sites, 7 stream types) and Southern European streams (31 sites, 3 stream types).

Macrophyte field survey methods were standardised for the STAR project integrating several national methodologies (Dawson, 2002) and were most closely related to the Mean Trophic Rank (MTR) methodology (Holmes et al., 1999).

The relationship between several macrophyte metrics and nutrient enrichment was highly significant although the correlation ratio was not satisfactory to implement any of the existing metrics as a uniform for all the river types on the pan-European scale. Analysis based on the STAR database enabled to propose redesigned Mean Trophic Rank (MTR) for lowland rivers and slightly different version of the same index for the mountain streams. It is supposed that enlarged core group of macrophyte species can form a part of an improved pan-European macrophyte-based bioassessment system, although regional modifications may be required to adequately describe the nutrient status of certain stream types.

The investigation of O'Hare et al. (2006) was focused on the impact of hydromorphological degradation on three macrophyte community types. The study utilised 107 European stream sites where River Habitat Survey was used as habitat survey technique. Macrophyte attribute groups and structural metrics such as species richness were successfully linked to hydromorphological variables indicative of impact. Most links were specific to each macrophyte community type, e.g., the attribute group liverworts, mosses and lichens decreased in abundance with increasing homogeneity of depth and decreasing substrate size at lowland sites but not at upland sites. The potential of a macrophyte tool indicative for hydromorphological impact was discussed. It is concluded one could be constructed by combining indicator species and metrics such as species richness and evenness.

Sources of variability in macrophyte assessment systems

Patterns of uncertainty in macrophytes survey were analysed by Staniszewski et al. (2006). This unique study estimated major sources of variance of macrophyte based methods and related them with the probability of wrong assignment to ecological quality classes.

Generally, the inter-surveyor differences had a limited effect, whereas temporal factor (years and seasons) and shading were more important. The analysis showed that some of macrophyte-based metrics (notably MTR and IBMR) are of sufficient precision in terms of sampling uncertainty so that they could be useful for estimating the ecological status of rivers in accordance with the aims of the Water Framework Directive.

Uncertainty in diatom assessment

The use of diatoms (phytobenthos) for monitoring of running waters can benefit from existing information on sensitivity of indicator taxa to various impairments, experience with monometric assessment systems (indices) and taxomomic expertise developed in the past decades. Sensitivity to eutrophication/organic pollution, acidity, salinity and current velocity are well established and they are incorporated into new methodological approaches, standards and practice of water management.

The STAR project contributed by:

a. updates of taxalist and its incorporation into a data input software
b. ring test and audit of analytical quality investigating the role of sampling, processing, identification and counting techniques in variation/uncertainty of diatom community characteristics
c. possibility to evaluate results of assessment based on diatoms together with data of another quality elements

Besse-Lototskaya et al. (2006) found that diatom samples taken from macrophyte substrates reveal the lowest inter-partner and inter-replicate variability. However, the recommendation to prefer this substrate is valid only for rivers where macrophytes occur widespread and it also depends on metrics used for assessment. The general conclusion of that study is that choice of sampling site, substrate type and taxonomic identification contribute the most to the uncertainty in assessment metrics. The study showed that standardisation of counting technique of diatom valves can significantly reduce uncertainty in metric results. Taxalists with sensitivity information should be considered as living documents periodically updated and extended by co-operating diatom taxonomists.

Conclusions

1. Macrophyte communities responded to most of the features underlying the typological framework defined in WFD.
2. The response of macrophytes to disturbance was confirmed and trophic (nutrient) status and hydromorphological degradation appeared as the most important impairment detected by of aquatic plant communities.
3. Macrophyte-based metrics, notably IBMR and MTR (especially in adjusted form) are of sufficient precision in terms of correlation to nutrient concentration and sampling uncertainty and could be useful for estimating the ecological status of rivers in accordance with the aims of the Water Framework Directive.
4. The robustness of diatom metrics in relation to the sources of uncertainty was evaluated.

Needs for further investigations:

a. to develop new metrics, to increase number of scoring taxa, and to extend sensitivity/trait information
b. to define a typology of running waters, which is better suitable for macrophyte and diatom metrics used in assessment systems

References

Baattrup-Pedersen, A., K. Szoszkiewicz, R. Nijboer, M. O'Hare & T. Ferreira, 2006. Macrophyte communities in unimpacted European streams: variability in assemblage patterns, abundance and diversity. Hydrobiologia 566: 179–196.

Besse-Lototskaya, A., P. F. M. Verdonschot & J. A. Sinkeldam, 2006. Uncertainty in diatom assessment: Sampling, identification and counting variation. Hydrobiologia 566: 247–260.

Council of the European Communities, 2000. Directive 2000/60/EC, Establishing a framework for community action in the field of water policy. European Commission PE-CONS 3639/1/100 Rev 1, Luxembourg.

Dawson, F. H., 2002. Guidance for the field assessment of macrophytes of rivers within the STAR Project. http://www.eu-star.at/frameset.htm.

Downes, B. J., L. A. Barmuta, P. G. Fairweather, D. P. Faith, M. J. Keough, P. S. Lake, B. D. Mapstone & G. P. Quinn, 2002. Monitoring ecological impacts: concepts and practice in flowing waters. Cambridge University Press, Cambridge, 434 pp.

Holmes, N. T. H., J. R. Newman, S. Chadd, K. J. Rouen, L. Saint & F. H. Dawson, 1999. Mean Trophic Rank: A Users Manual. R&D Technical Report E38, Environment Agency of England & Wales, Bristol, 134 pp.

Murphy, M. L., 1998. Primary production. In Naiman, R. J. & R. E. Bilby (eds), River ecology and management. Lessons from the Pacific Coastal Ecoregion. Springer Verlag, New York: 144–168.

O'Hare M. T., A. Baattrup-Pedersen, R. Nijboer, K. Szoszkiewicz & T. Ferreira, 2006. Macrophyte communities of European streams with altered physical habitat. Hydrobiologia 566: 197–210.

Staniszewski, R., K. Szoszkiewicz, J. Zbierska, J. Lesny, S. Jusik & R. T. Clarke, 2006. Assessment of sources of uncertainty in macrophyte surveys and the consequences for river classification. Hydrobiologia 566: 235–246.

Szoszkiewicz K., T. Ferreira, T. Korte, A. Baattrup-Pedersen, J. Davy-Bowker & M. O'Hare, 2006. European river plant communities: the importance of organic pollution and the usefulness of existing macrophyte metrics. Hydrobiologia 566: 211–234.

Hydrobiologia (2006) 566:179–196
M.T. Furse, D. Hering, K. Brabec, A. Buffagni, L. Sandin & P.F.M. Verdonschot (eds), The Ecological Status of European Rivers: Evaluation and Intercalibration of Assessment Methods
DOI 10.1007/s10750-006-0096-1

Macrophyte communities in unimpacted European streams: variability in assemblage patterns, abundance and diversity

Annette Baattrup-Pedersen[1,*], Krzysztof Szoszkiewicz[2], Rebi Nijboer[3], Mattie O'Hare[4] & Teresa Ferreira[5]
[1]*Department of Freshwater Ecology, National Environmental Research Institute, Vejlsøvej 25, P.O. Box 314, DK-8600 Silkeborg, Denmark*
[2]*Department of Ecology and Environmental Protection, Agricultural University of August Cieszkowski, ul. Piątkowska 94C, 61-691 Poznan, Poland*
[3]*Green World Research, Alterra, P.O. Box 47, 6700 AA Wageningen, The Netherlands*
[4]*Winfrith Technology Centre, Centre for Ecology and Hydrology, DT2, 8ZD Dorchester, Dorset, UK*
[5]*Forestry Department, Agronomy Institute, Technical University of Lisbon, Tapada da Ajuda, 1349-017 Lisboa, Portugal*
*(*Author for correspondence: E-mail: abp@dmu.dk)*

Key words: WFD, vegetation, stream, classification, reference

Abstract

Macrophytes are an important component of aquatic ecosystems and are used widely within the Water Framework Directive (WFD) to establish ecological quality. In the present paper we investigated macrophyte community structure, i.e., composition, richness and diversity measures in 60 unimpacted stream and river sites throughout Europe. The objectives were to describe assemblage patterns in different types of streams and to assess the variability in various structural and ecological metrics within these types to provide a basis for an evaluation of their suitability in ecological quality assessment. Macrophyte assemblage patterns varied considerably among the main stream types. Moving from small-sized, shallow mountain streams to medium-sized, lowland streams there was a clear transition in species richness, diversity and community structure. There was especially a shift from a predominance of species-poor mosses and communities dominated by liverwort in the small-sized, shallow mountain streams to more species-rich communities dominated by vascular plants in the medium-sized, lowland streams. The macrophyte communities responded to most of the features underlying the typological framework defined in WFD. The present interpretation of the WFD typology may not, however, be adequate for an evaluation of stream quality based on macrophytes. First and most important, by using this typology we may overlook an important community type, which is characteristic of small-sized, relatively steep-gradient streams that are an intermediate type between the small-sized, shallow mountain streams and the medium-sized, lowland streams. Second, the variability in most of the calculated metrics was slightly higher when using the pre-defined typology. The consistency of these results should be investigated by analysing a larger number of sites. Particularly the need of re-defining the typology to improve the ability to detect impacts on streams and rivers from macrophyte assemblage patterns should be investigated.

Introduction

Historically, vegetation has changed in streams and rivers in all Europe. Before the vast tree clearances in the Neolithic, the vegetation must have been much sparser than now because of more intensive shading from the riparian area. Even at that time, however, macrophyte communities may have been an important biological characteristic of many streams and rivers. Thus, recent studies (Svenning, 2002 and references herein) have documented that open vegetation was widespread in river floodplains throughout north-western Europe in past oceanic interglacials and the pre-agricultural Holocene, i.e., before the onset of strong human impact. Consequently, the conditions may have been suitable for macrophyte growth in many stream and river reaches, and variable physical conditions and good water quality may have supported rich vegetation.

While paleoecological evidence adds to our knowledge of past conditions in floodplains, we are entirely dependent on published records of stream vegetation to improve our understanding of assemblage patterns before the onset of strong human impact. The first published records of macrophyte surveys in Europe are from the late 1800s (Baggøe & Ravn, 1896; Raunkiær, 1895–1999; Mountford, 1994; work cited in Preston, 1995). These records give an indication of very rich and abundant vegetation. Many slow-growing species, such as broad-leaved Potamogeton species (*P. lucens* L., *P. natans* L., *P. polygonifolius* Pourret, *P. praelongus* Wulf., *P. alpinus* Balbis), for example, were very common at that time in many European lowland streams and rivers (Riis & Sand-Jensen, 2001 and references herein). During recent decades the vegetation has undergone pronounced changes. Physical degradation of the stream channel (channelisation, regulation for hydropower and navigation purposes, weed cutting and dredging) and eutrophication have resulted in a loss of many particularly slow-growing species, whereas many fast-growing species with a high dispersal capacity have increased in abundance (Carbiener et al., 1990; Mesters, 1995; Riis & Sand-Jensen, 2001).

Despite our knowledge of the adverse effects of various human activities on the vegetation in streams and rivers, no investigations have deliberately distinguished between unimpacted, slightly or highly impacted stream and river sites in previous macrophyte classifications (e.g., Butcher, 1933; Holmes et al., 1998; Riis et al., 2000). Therefore, the existing knowledge on macrophyte assemblage patterns in unimpacted European streams and rivers is limited, particularly regarding stream and river types that are situated in highly impacted areas (particularly lowland sites). In the present paper we will characterise the macrophyte communities in different types of unimpacted streams and rivers in Europe. We will use the stream typology defined in a previous EU project (AQEM) (http://www.aqem.de), which is based on ecoregion (according to Illies, 1978), size class (based on catchment area), geology of the catchment, and altitude class (Hering et al., 2004) and extended in the STAR project (Hering & Strackbein, 2001). This typology has proven useful for an assessment system based on macroinvertebrates (Verdonschot & Nijboer, 2004), but no attempts have been made to evaluate this typology for the macrophyte communities. The first objective of this study was to describe community assemblage patterns and their variation within and among these *a priori* defined stream types and to evaluate the typology by characterising assemblage patterns independently using ordination techniques. The second objective was to assess the natural variability in macrophyte-based metrics also to provide a basis for an evaluation of their suitability in ecological quality assessment.

Methods

Site selection

A total of 288 stream sites were selected in the STAR project. These sites were classified using the stream typology defined in a previous EU project (AQEM) (http://www.aqem.de), which is based on ecoregion (according to Illies, 1978), size class (based on catchment area), geology of the catchment, and altitude class (Hering et al., 2004), and extended in the STAR project (Hering & Strackbein, 2001). The sites covered an impact gradient from sites of high ecological quality (*sensu* WFD) to sites of poor or bad ecological quality (*sensu* WFD). Sites were chosen so that only one major impact was allocated to each site being either organic pollution, toxic pollution or habitat degradation. For the purpose of our study, we only included unimpacted stream sites (ecological quality class 5) in the analyses. A total of 64 sites were identified as being unimpacted and four of these sites were without growth of macrophytes.

The unimpacted sites in the STAR project were identified onsite by comparing site characteristics with a list of *a priori* exclusion criteria (Hering et al., 2003, Nijboer et al., 2004). In addition, pre-existing data on site conditions or GIS information were compared with the list of criteria for reference sites, when available. Table 1 gives an overview of the investigated unimpacted stream sites included in the analyses and their location in terms of ecoregion, country, latitude, longitude and altitude.

Macrophyte sampling

Macrophyte surveys were undertaken using the protocols associated with the Mean Trophic Rank (MTR) indexation method (Holmes et al., 1999). This method is the standard procedure used in the United Kingdom in association with the implementation of the European Union Urban Wastewater Directive and is compatible with methodologies used in several of the other Member States participating in STAR. The term

Table 1. An overview of the investigated unimpacted stream sites within each stream type and their location in terms of ecoregion, country, latitude and altitude

Stream type	Number of observations	Ecoregion	Country	Latitude (average)	Longitude (average)	Altitude (average)
Small-sized, shallow mountain streams	8	8, 9, 10	Austria, Czech Republic, Germany	49	13.9	399
Small-sized, lowland calcareous streams	3	18	United Kingdom	51	−1.6	46
Small-sized streams in the Central, sub-alpine mountains	3	9	Czech Republic	50	17.3	361
Small-sized, shallow headwater streams in Eastern France	3	8	France	48	5.4	344
Small-sized, calcareous mountain streams in Western, Central and Southern Greece	3	6	Greece	38	22.2	528
Small-sized Buntsandstein streams	2	9, 14	Germany	50	9.6	220
Small-sized, calcareous streams in the Central Apennines	3	3	Italy	43	11.4	393
Small-sized calcareous mountain streams in the Eastern Carpathians*	3*	10	Slovak Republic	49	22.3	413
Small-sized siliceous mountain streams in the Western Carpathians*	5*	10	Slovak Republic	49	18.6	408
Medium-sized, lowland streams	23	14, 15, 16, 18, 22	United Kingdom, Sweden, Poland, Denmark, Latvia	55	17.5	84
Medium-sized, lowland calcareous streams	3	18	United Kingdom	52	−2.7	119
Medium-sized streams in the lower mountainous areas of Southern Portugal	3	1	Portugal	39	−7.6	234
Medium-sized streams on calcareous soils	2	14	Sweden	60	17.8	9

*Two sampling sites were without growth of macrophytes in each of the two stream types.

macrophyte includes all higher plants that grow submerged or partly submerged, vascular cryptograms and bryophytes, together with groups of algae which can be seen to be composed predominantly of a single species. Therefore the term macrophyte also encompasses terrestrial species growing partly submerged in the stream channel. The sampling reach was 100 m in length. Macrophyte sampling was undertaken in late summer/early autumn 2002 or 2003. Macrophyte abundance was expressed in terms of the percentage of the survey length covered. A cover score was allocated to each macrophyte species present using the following scale 1: <0.1%, 2: 0.1–1%, 3: 1–2.5%, 4: 2.5–5%, 5: 5–10%, 6: 10–25%, 7: 25–50%, 8: 50–75%, 9: >75%. For all percentage cover estimates, the whole survey area surveyed equals 100%, i.e. the individual species percentage cover estimates are a percentage of the whole survey area and not of the overall percentage cover estimated. For wadeable surveys a glass-bottom bucket was used to aid observations. A grapnel was used to retrieve submerged macrophytes for identification from small areas of deep water. For non-wadeable areas a grapnel was used to retrieve macrophyte specimens from the banks. Particular care was taken to examine all small niches within the survey site to look for small patches of species. For a more detailed description, see Holmes et al. (1999) or the STAR website (http://www.eu-star.at) under the public-access section "Protocols". If identification to species could not be done due to absence of seasonal diagnostic features, e.g. *Ranunculus* and *Callitriche*, the record was only performed to the genus level (for species names and authors see Supplementary material).[1]

Site characteristics

The River Habitat Survey was also undertaken in late summer/early autumn 2002 or 2003 together with supporting chemical, physico-chemical and geographical elements. All relevant protocols, i.e. the AQEM and STAR site protocol, the river habitat survey (RHS) protocol and MTR protocol, are accessible at the STAR website (http://www.eu-star.at) under the public-access section "Protocols".

Data analysis

The pan-European taxonomic standardisation of the macrophyte data was used for all analyses performed (Furse et al., 2004). To analyse assemblage patterns in the *a priori* defined stream types a Detrended Correspondence Analysis (DCA) was performed (PC-ORD; McCune & Mefford, 1999) and DCA site scores were used to summarise the variability in assemblage patterns among the stream sites within each stream type. An Indicator Species Analysis (Dufrene & Legendre, 1997) was performed to identify indicator species (PC-ORD; McCune & Mefford, 1999). This analysis could only be performed for small-sized shallow mountain streams and medium-sized lowland streams, however, as the number of sampling sites was restricted to 2–3 in the other stream types. For each species encountered in the two stream types, an indicator value was calculated ranging from zero (no indication) to 100 (perfect indication). The indicator values were tested for statistical significance using a Monte Carlo permutation test. Only significant indicator species ($p < 0.05$) were used in data interpretation.

To further describe the variability within and among the stream type, mean values and ranges for a number of structural and ecological metrics were calculated. The structural metrics are mathematical expressions of community structure and the ecological metrics are based on the information of ecological tolerance of indicator species. In the present context the term macrophyte community is used broad and encompasses the complex of communities that may exist along the 100 m stream reaches studied.

The structural metrics used were species, genus and family richness, Shannon and Simpson diversity (Margalef, 1958) and domination and evenness. The index C that was used as a measure of domination was calculated as:

$$C = \sum_{i=1}^{S} \left(\frac{p_i}{\sum_{i=1}^{S} p_i^2} \right)^2$$

[1] Supplementary material is available for this article at http://www.dx.doi.org/10.1007/s10750-006-0096-1 and accessible for authorised users.

where s is the number of species and p_i the abundance (share of the cover) of species i.

The index $E_{1/D}$ that was used as a measure of evenness was calculated as:

$$E_{1/D} = \frac{\left(1 - \sum_{i=1}^{S} \frac{N_i}{N}\right)}{S}$$

where S is the number of species, N the total abundance, and N_i is the abundance of species i.

In supplement to the above described diversity measures, species-area curves for the main stream types (i.e., small-sized, shallow mountain streams and medium-sized, lowland streams) were generated from the sample plots, and the overall species richness using the jackknife method was estimated (PC-ORD; McCune & Mefford, 1999).

The ecological metrics calculated were Mean Trophic Rank (MTR; Holmes et al., 1999) and Macrophytical Biological Index for Rivers (IBMR; Haury et al., 2002). These metrics are based on information of tolerance of species to eutrophication. MTR scores lie in the range 10–100, where low values (< 25) indicate eutrophic conditions and values between 25 and 65 indicate either eutrophic conditions or that the site is at risk of becoming eutrophic (Holmes et al., 1999). IBMR was recently developed in France to assess water trophy and organic pollution in rivers. The IBMR scores vary between 0 (degraded) and 20 (high quality) (Haury et al., 2002). We did not statistically test for differences in DCA site scores or metric values among the *a priori* defined stream types because the number of sampling sites was low for most stream types invalidating the analysis.

Assemblage patterns were characterised independently from the *a priori* defined stream typology. A TWINSPAN classification of the 60 sampling sites was performed using default options in PC-ORD (McCune & Mefford, 1997). The significance of the classification was tested by comparing DCA coordinates among the major end-clusters (including more than six sites) using ANOSIM (Analysis of Similarities; Clarke & Green, 1988). We also calculated diversity and distributional metrics as well as ecological metrics (MTR and IBMR) for the major end-clusters.

An Indicator Species Analysis (Dufrene & Legendre, 1997) for the major TWINSPAN end-clusters was performed using cluster membership (cluster 1–8) as a grouping variable. The indicator values were tested for statistical significance as described above. The clusters were further characterised in terms of number of sampling sites present, their relation to the *a priori* defined types, species richness, dominant taxonomic groups, growth morphology and species abundance. The relative distribution of coverage classes was used as a measure of species abundance for the major end-clusters and these were tested statistically using a Kruskal–Wallis test. The distribution of species abundance was also evaluated using rank-abundance curves. The logarithm of the relative abundance of species was plotted as a function of the rank number (x) in each group. The rank number was scaled as x/S, where S is the number of species in the groups, so that the most abundant species had the lowest rank of $1/S$ close to zero, while the rarest species had the highest rank of 1 (Wilson, 1991).

The relationships between the major TWINSPAN end-clusters and stream site characteristics at various scales (ecoregion, catchment, riparian, habitat) were further analysed. In doing that an integrated measure of shade from riparian vegetation was calculated (weighted shade index, WSI). The WSI takes values in the interval [0; 200] and is defined as:

$$\mathrm{WSI} = \sum_{i=1}^{3} k_i \sum_{j=1}^{2} s_{ij},$$

where i is the degree of shading ($i=1$: no shading; $i=2$: 33%, $i=3$: greater than 33% shading) and j stands for left ($j=1$) or right bank ($j=2$). Finally, $k_1=0$, $k_2=25$ and $k_3=100$.

A variance analysis (ANOVA with Bonferroni correction) was performed to test for differences among the major end-clusters regarding macrophyte community characteristics and sampling site characteristics. Differences among categorical variables were tested using X^2. Relations between clusters and variables were analysed using Spearman rank correlation analysis. Some of the categorical variables, i.e. planform, flow category and water clarity, were treated as non-categorial variables in this analysis as the values assigned represented gradients. Thus increasing planform value implied increasing channel complexity

(1 = straight, 2 = sinuous, 3 = irregular meanders, 4 = regular meanders), increasing discharge values implied increasing discharge (1: $<0.31\ m^3\ s^{-1}$; 2: $>0.31–0.62\ m^3\ s^{-1}$; 3: $>0.62–1.25\ m^3\ s^{-1}$; 4: $>1.25–2.50\ m^3\ s^{-1}$; 5: $>2.5–5.0\ m^3\ s^{-1}$; 6: $>5.0–10.0\ m^3\ s^{-1}$; 7: $>10–20\ m^3\ s^{-1}$; 8: $>20–40\ m^3\ s^{-1}$; 9: $>40–80\ m^3\ s^{-1}$ and 10: $>80\ m^3\ s^{-1}$) and increasing clarity implied decreasing water clarity (1 = clear; 2 = cloudy; 3 = turbid). We chose to perform only correlation analysis to relate stream site characteristics to community variables as the environmental data for some of the variables were too incomplete to allow multivariate methods to be applied. Particularly the chemistry data were incomplete. The chemistry data from the sites in the Czech Republic were not included in the analysis as the detection limit was too high compared to detection limits in the other sites.

Results

Richness, diversity and metrics

Macrophytes were present in all *a priori* defined stream types, but the number of species, genera and families increased from small to middle-sized streams with the exception of small-sized, shallow headwater streams in Eastern France that were very species-rich (Table 2). Numbers of species, genera and families were all positively correlated ($r = 0.987$; $p < 0.05$). The jackknife estimates for overall species richness were 23 for small-sized, shallow mountain streams and 145 for medium-sized, lowland streams. The total number of species actually encountered was 14 in the small-sized, shallow mountain streams and 98 in medium-sized, lowland streams, which indicates that the number of investigated sampling sites of both stream types was too low to adequately estimate the average species richness.

Both the Shannon and Simpson diversity indices were also generally lower in the small-sized streams compared to the middle-sized streams (Table 2), again with the exception of small-sized, shallow headwater streams in Eastern France. The Shannon and Simpson diversity indices were positively correlated with the number of species, genera and families ($r = 0.973$; $p < 0.05$). The distributional indices, domination and evenness, also varied within and among the stream types. The domination index was negatively correlated with all indices ($r = -0.603$; $p < 0.05$), whereas the evenness index was unrelated to the other indices ($p > 0.05$).

The mean MTR was generally highest in the small-sized mountainous streams (58–80) except for the small-sized mountain streams in Western, Central and Southern Greece (45) (Table 2). The MTR was lower in the medium-sized mountainous streams (64) and lowest in the medium-sized lowland streams (37–46). In small-sized, shallow mountain streams the MTR varied between 50 and 100, and in medium-sized lowland streams the MTR varied between 28 and 79. The IBMR performed similarly to the MTR (Table 2). Both indices correlated negatively with the number of species, genera and families and with the diversity indices. The MTR correlated positively with the domination index. The IBMR and MTR indices were positively inter-correlated ($r = 0.586$; $p < 0.05$).

Macrophyte assemblage patterns

Some of the stream types showed a high degree of dispersion along the DCA ordination axes (small-sized, calcareous streams in the Central Apennines, medium-sized streams on calcareous soils), whereas other stream types were more uniformly distributed (small-sized streams in the Central, sub-alpine mountains, medium-sized lowland calcareous streams) (Figs 1a and b). The small-sized, shallow mountain streams were positioned in the middle of the ordination diagrams (Figs 2a and b). The medium-sized lowland streams were also positioned in the middle of the ordination diagrams, but to the left of small-sized, shallow mountain streams (Figs 1a and b). These sites were mainly distributed along DCA 2 (Fig. 2d). Only medium-sized, lowland stream sites from Sweden were clearly distinguishable from the other medium-sized, lowland stream sites (Germany, Latvia, Poland and Denmark) (Figs 1c and d). The medium-sized, lowland calcareous streams and the small-sized lowland calcareous streams were not clearly distinguishable from the other medium-sized, lowland stream sites (Figs 1a and b).

In total, 8 end-clusters were identified from the TWINSPAN classification (Fig. 2). Of these only

Table 2. Summary of the mean and ranges of various structural and ecological metrics in different *a priori* defined reference stream types

Stream type	Species number	Shannon diversity	Simpson diversity	Domination	Evenness	MTR	IBMR
Small-sized, shallow mountain streams	3 (1–6)	0.890 (0–1.792)	0.496 (0–0.833)	0.645 (0.166–1)	0.721 (0–1)	62 (50–100)	15 (13–18)
Small-sized, lowland calcareous streams	9 (1–14)	1.590 (0–2.480)	0.596 (0–0.907)	0.475 (0.157–1)	0.620 (0–0.940)	42 (38–46)	11 (11–11)
Small-sized streams in the Central, sub-alpine mountains	4.3 (4.0–5.0)	1.398 (1.311–1.550)	0.737 (0.716–0.776)	0.469 (0.383–0.632)	0.957 (0.946–0.963)	58 (50–62)	13 (10–15)
Small-sized, shallow headwater streams in Eastern France	33 (23–39)	3.279 (2.904–3.475)	0.951 (0.927–0.963)	0.195 (0.108–0.331)	0.938 (0.914–0.951)	42 (41–42)	11 (10–12)
Small-sized, calcareous mountain streams in Western, Central and Southern Greece	9 (4–14)	1.913 (1.332–2.375)	0.821 (0.720–0.891)	0.411 (0.272–0.633)	0.937 (0.925–0.961)	45 (43–47)	15 (14–16)
Small-sized Buntsandstein streams	7 (3–11)	1.673 (1.012–2.333)	0.756 (0.615–0.898)	0.366 (0.177–0.556)	0.947 (0.921–0.973)	60 (59–62)	10 (10–10)
Small-sized, calcareous streams in the Central Apennines	3 (2–5)	1.037 (0.693–1.519)	0.596 (0.500–0.757)	0.658 (0.500–0.862)	0.921 (0.819–1.000)	55 (45–70)	11 (9–13)
Small-sized calcareous mountain streams in the Eastern Carpathians	1						
Small-sized siliceous mountain streams in the Western Carpathians	1.3 (1–2)	0.230 (0–0.691)	0.166 (0–0.498)	0.844 (0.531–1.000)	0.332 (0–0.997)	80 (80–80)	15 (15–15)
Medium-sized, lowland streams	13 (2–32)	2.195 (0.693–3.330)	0.844 (0.500–0.959)	0.370 (0.102–0.831)	0.945 (0.860–1.000)	42 (28–79)	11 (9–14)
Medium-sized, lowland calcareous streams	12 (6–18)	2.289 (1.705–2.810)	0.877 (0.806–0.934)	0.252 (0.203–0.305)	0.956 (0.946–0.972)	37 (33–42)	10 (9–10)
Medium-sized streams in the lower mountainous areas of Southern Portugal	7 (5–8)	1.742 (1.359–2.007)	0.791 (0.680–0.858)	0.425 (0.206–0.841)	0.922 (0.845–0.965)	64 (60–70)	12 (12–12)
Medium-sized streams on calcareous soils	9 (2–16)	1.859 (1.040–2.678)	0.774 (0.625–0.924)	0.456 (0.279–0.633)	0.956 (0.946–0.966)	46 (45–47)	11 (11–11)

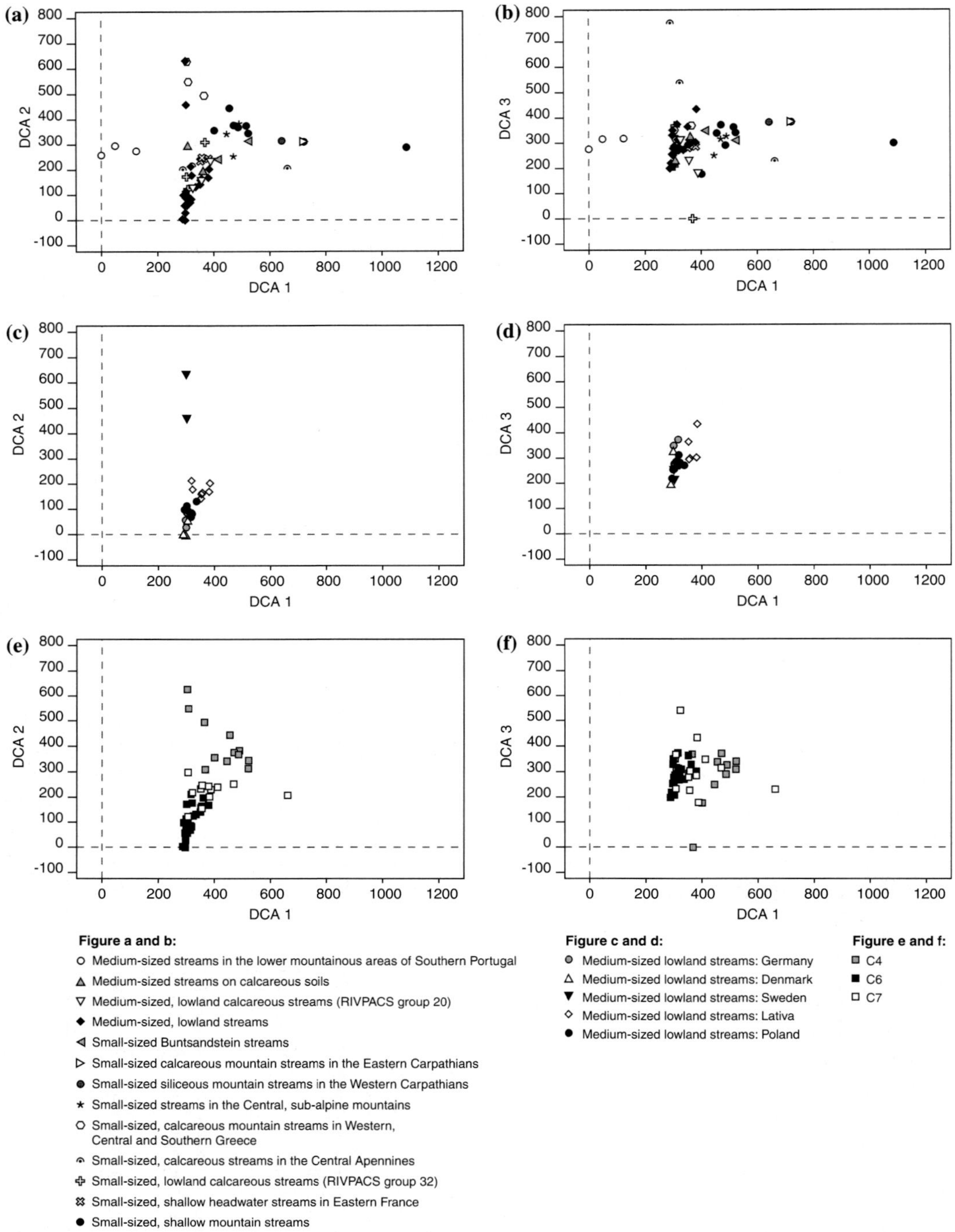

Figure 1. Detrended correspondence analysis (DCA) of 60 sample plots distributed in different streams situated all over Europe. The analysis was performed with downweighting of rare species. In total, 182 species were included in the analysis. (a) and (b) include all sampling sites and different symbols are used for different *a priori* defined stream types. (c) and (d) include medium-sized, lowland streams sites with different symbols for different countries. E and F include C4, C6 and C7 sites identified from the TWINSPAN classification of sampling sites.

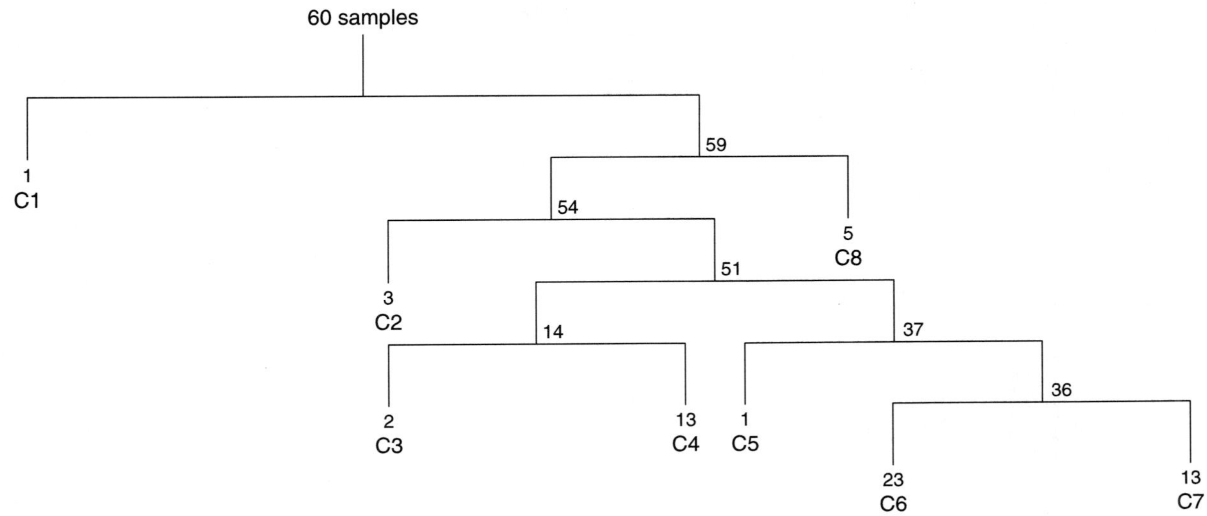

Figure 2. TWINSPAN tree with 8 end-clusters. The number of samples is shown above each node. End-clusters are named C1 to C8. Only C4, C6 and C7 ($n > 6$) were subjected to further analyses.

three end-clusters (C4, C6 and C7) were sufficiently large (>6 sites) for subsequent analysis. Species abundance and indicator values for the three end-clusters are given in Supplementary material. A large group of small-sized streams (C4) consisted of several of the *a priori* defined small-stream types (Table 3). This end-cluster was relatively species-poor with predominant growth of mosses (e.g., *Rhynchostegium riparioides* and *Cratoneuron filicinum*) (Table 3; Supplementary material). A large group of medium-sized lowland streams (C6) was species-rich and consisted of primarily vascular plants (e.g., *Sparganium emersum*, *Phalaris arundinacea*, *Berula erecta* and *Elodea canadensis*) (Table 3; Supplementary material). Finally, a mixed group of small- and medium-sized streams (C7) displayed intermediate characteristics. As opposed to C4 this end-cluster was very species-rich with growth of both mosses (e.g., *Fontinalis antipyretica*) and many amphibious (*Veronica anagallis-aquatica* and *Myosotis palustris*) and terrestrial dicots (Table 3; Supplementary material).

End-cluster C4, C6 and C7

The three end-clusters C4, C6 and C7 were found to differ significantly based on their DCA coordinates (ANOSIM, $p < 0.05$; Figs 1e and f). The number of species, genera and families encountered, and diversity indices also varied among the three clusters (Fig. 4; ANOVA $p < 0.05$). C4 had fewer species, genera and families compared to C6 and C7, and the Shannon and Simpson diversity were also lower. In contrast, the domination index was higher for this cluster (Fig. 4; ANOVA $p < 0.05$). This was mainly related to a high abundance of *Rhynchostegium riparioides* in many C4 sites. C6 and C7, on the other hand, were very similar regarding species, genera and family richness as well as diversity and domination indices (ANOVA $p > 0.05$), but C7 possessed a distinct community that shared characteristics with both small-sized, shallow mountain streams and middle-sized, lowland streams (Table 3; Supplementary material). The MTR also varied among C4, C6 and C7 (Fig. 4). C4 had the highest and most variable MTR scores and this cluster was clearly distinguishable from C6 and C7 (ANOVA $p < 0.05$). In contrast, C6 and C7 had more similar and less variable MTR scores. The IBMR scores gradually declined from C4 to C6 and C7 (ANOVA $p < 0.05$).

There was a high degree of overlap between the main stream types, i.e., small-sized shallow mountain streams and medium-sized lowland streams and end-cluster 4 and 6 (Figs 1e and f and Table 3). This was reflected in a high percentage of

Table 3. Characteristics of the TWINSPAN end-clusters and their relations to the *a priori* defined stream types

> Identified clusters	Typology	Number of species	Indicator species: small-sized, shallow mountain streams	Indicator species: medium-sized, lowland streams	Dominant group	Dominant growth morphology	DCA score axis 1	DCA score axis 2	DCA score axis 3
End-cluster 1	Small-sized shallow mountain streams (1)	5	0	0	Moss, algae	Submerged	1066 (860–1118)	280 (204–299)	281 (101–326)
End-cluster 2	Medium-sized streams in the lower mountainous areas of Southern Portugal (3)	14	0	0	Dicots, monocots	Submerged, terrestrial	82 (−47–291)	279 (96–540)	307 (196–356)
End-cluster 3	Medium-sized lowland streams (Sweden) (2)	12	2	1			299 (290–332)	603 (289–751)	237 (180–344)
End-cluster 4	Small-sized shallow mountain streams (6)	17	2		Moss, liverworts	Submerged	377 (289–315)	464 (263–613)	300 (0–427)
	Small-sized, lowland calcareous streams (1)								
	Small-sized streams in the Central, sub-alpine mountains (2)								
	Small-sized, calcareous mountain streams in Western, Central and Southern Greece (3)								
	Small-sized Buntsandstein streams (1)								
End-cluster 5	Small-sized, calcareous streams in the Central Apennines (1)	3	0	0	Algae	Submerged	290	205	777
End-cluster 6	Medium-sized, lowland streams (20)	69		11	Dicots	Amphibious, submerged	305 (260–381)	56 (−203–227)	272 (−79–539)
	Medium-sized, lowland calcareous streams) (2)								
	Medium-sized streams on calcareous soils (1)								
End-cluster 7	Small-sized, lowland calcareous streams (1)	63		2	Moss, dicots	Submerged, amphibious	351 (286–860)	236 (6–490)	295 (50–844)
	Medium-sized, lowland calcareous streams (2)								
	Small-sized streams in the Central, sub-alpine mountains (1)								
	Small-sized, shallow headwater streams in Eastern France (3)								
	Small-sized Buntsandstein streams (1)								
	Small-sized, calcareous streams in the Central Apennines (2)								
	Medium-sized streams on calcareous soils (1)								
	Medium-sized, lowland streams (1)								
End-cluster 8	Small-sized shallow mountain streams (1)	2	1	0	Liverworts, moss	Submerged	644 (566–722)	316 (312–320)	383 (382–384)
	Small-sized calcareous mountain streams in the Eastern Carpathians (2)								
	Small-sized siliceous mountain streams in the Western Carpathians (3)								

The number of sites of the *a priori* defined stream types within each end-cluster is given in brackets.

overlap between the sites of 75% and 87%, respectively, and in an overlap of indicator species identified for small-sized, shallow mountain streams and C4 (*Rhynchostegium riparioides*) and for medium-sized, lowland streams and C6 (e.g., *Sparganium emersum*, *Berula erecta*, *Elodea canadensis*) (Table 3). There was no indicator species overlap between C4 and medium-sized, lowland streams, or between C6 and small-sized, shallow mountain streams. End-cluster 7 was a very broad group and eight *a priori* defined stream types were represented in this cluster (Table 3). The identified indicator species for this end-cluster included both species identified as indicator species for small-sized, shallow mountain streams and for medium-sized, lowland streams (*Veronica anagallis-aquatica* and *Myosotis palustris*) (Table 3; Supplementary material). Two of the most abundant species (*Hygroamblystegium fluviatile* and *Fontinalis antipyretica*) were also found in both small-sized, shallow mountain streams and medium-sized, lowland streams.

The species occurring in the identified clusters were all very low in abundance (Fig. 3a) and the distribution of coverage classes for the three clusters did not vary significantly (Kruskal–Wallis, $p > 0.05$). Distribution of species abundance in C4, C6 and C7 was further evaluated using rank-abundance curves (Fig. 3b). *Rhynchostegium riparioides* was very abundant in C4 sites, counteracting the general abundance pattern.

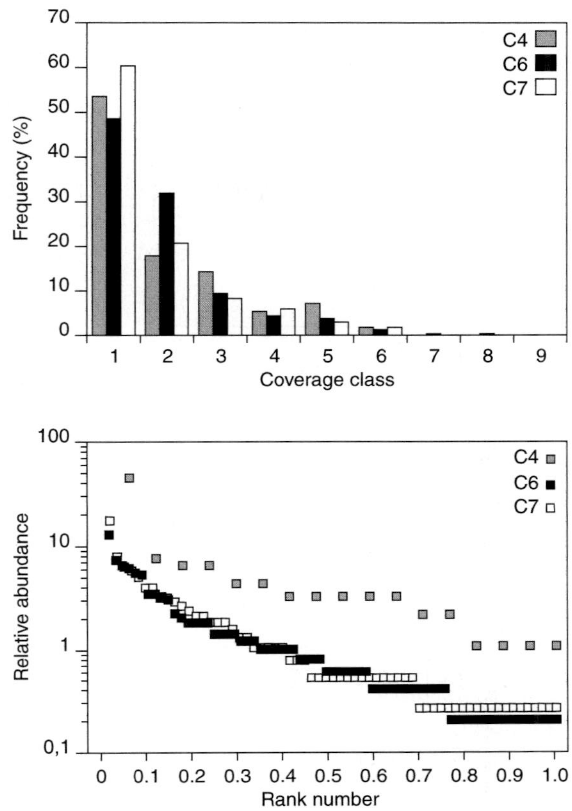

Figure 3. (a) Distribution of coverage classes expressed as the relative frequency of coverage class 1–9 divided by the total number of species allocated to a coverage class in end-cluster C4, C6 and C7 respectively. The coverage classes were 1: <0.1 %, 2: 0.1–1%, 3: 1–2.5%, 4: 2.5–5%, 5: 5–10%, 6: 10–25%, 7: 25–50%, 8: 50–75%, 9: >75%. (b) Rank-abundance curves expressing the relative distribution of species abundance according to their rank. The relative abundance was calculated as the sum of coverage classes allocated to a species divided by the total sum of coverage classes in end-cluster C4, C6 and C7, respectively.

End-cluster C4, C6 and C7 and stream site characteristics

Several stream site characteristics differed among end-cluster C4, C6 and C7 (Table 4; ANOVA, $p < 0.05$; χ^2, $p < 0.05$). In general, C4 and C6 were very distinct as to both ecoregion, catchment, riparian and habitat variables, whereas C7 shared characteristics with both C4 and C6 (Table 4). C4 sites, in particular, were positioned at higher altitudes and lower latitudes, they had steeper slopes (ANOVA $p < 0.05$) and were positioned closer to the source compared to C6 sites (ANOVA $p < 0.05$). C4 sites were also less wide and more shaded from riparian vegetation compared to C6 sites, and with predominantly alluvial deposits in the valley (Table 4). C7 sites were at higher altitudes than C6 sites, they were closer to the source, less wide, and moraine deposits were less widespread in the valleys compared to C6 sites (Table 4).

The variability in the calculated metrics was slightly higher in the *a priori* defined middle-sized, lowland streams compared to the corresponding end-cluster C6 (Table 5) and also in the small-sized, shallow mountain streams compared to the end-cluster C4 with the exception of IBMR and species number. The variability in MTR, in particular, was higher in the *a priori* defined stream types compared to the corresponding TWINSPAN end-clusters (Table 5).

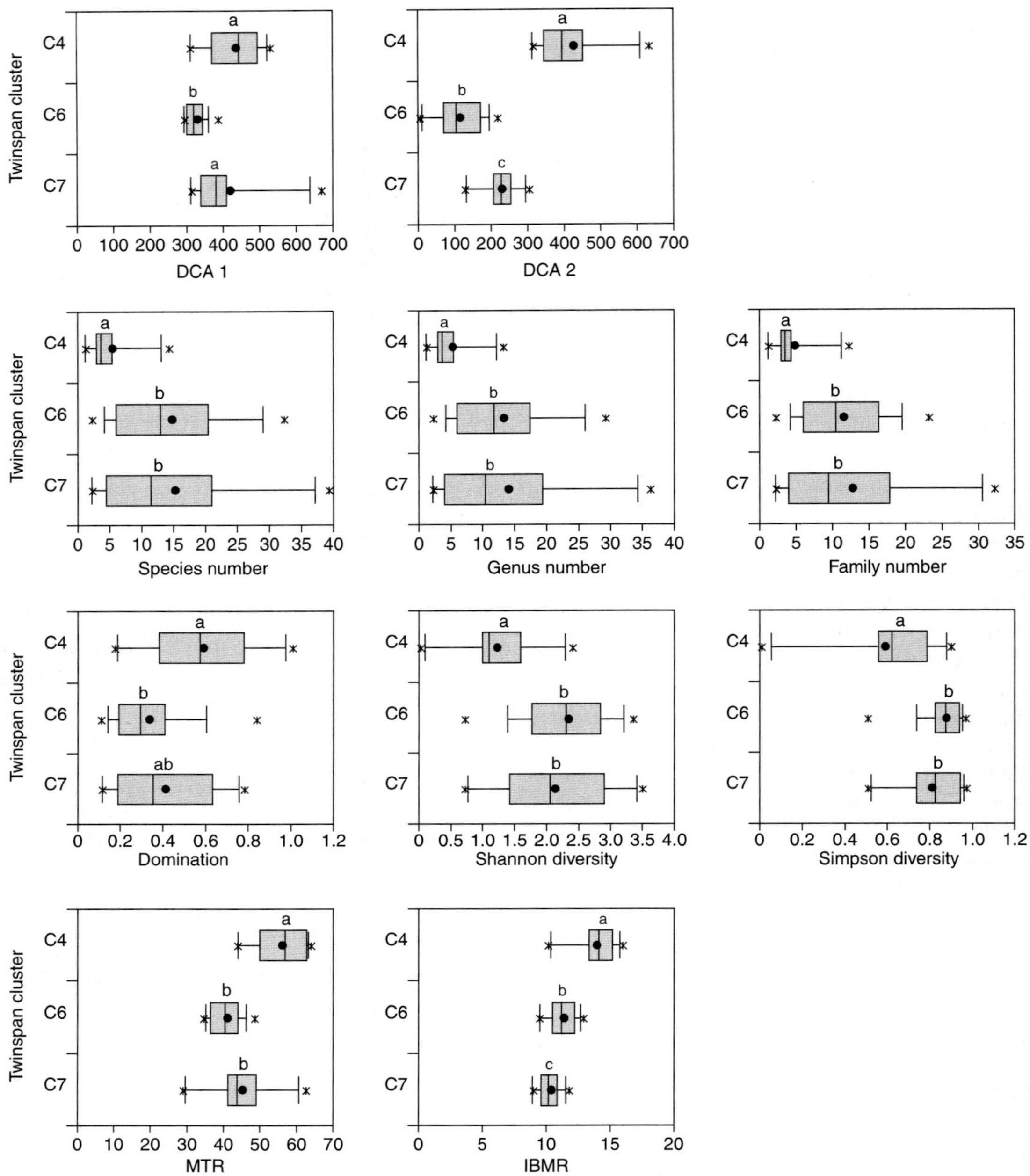

Figure 4. Box-whisker plots of macrophyte community characteristics in end-clusters C4, C6 and C7. Letters a, b and c signify differences between mean values (ANOVA with Bonferroni correction, $p < 0.05$). The box represents 10, 25%, median, 75 and 90% and the symbol the mean value. Error bars represent the 5% and 95% and the star (*) the minimum and maximum value.

Significant correlations were found between many macrophyte community characteristics (i.e., DCA scores, richness, diversity, MTR and IBMR) and stream site characteristics at various scales (Table 6). The separation on the DCA axis was mainly related to ecoregion, catchment and riparian variables (e.g., altitude, slope, discharge, and catchment area). Habitat variables were mainly of

Table 4. Ecoregion, catchment, riparian and habitat characteristics of the major TWINSPAN end-clusters 4 ($n=13$), 6 ($n=23$) and 7 ($n=12$)

	Cluster 4		Cluster 6		Cluster 7	
	Mean/Median	SD	Mean/Median	SD	Mean/Median	SD
Ecoregion variables						
Latitude	43.3[a]	14	54.7[b]	2.4	50.3[b]	5.1
Altitude (m)	379.9[a]	205.4	71.7[b]	44.9	221.6[c]	144
Catchment variables						
Slope (m km^{-1})	42.8[a]	54.6	4.2[b]	7.6	5.5[b]	6.3
Distance to source (km)	10.4[a]	4.4	31.2[b]	16.1	16.9[a]	9.3
Height of source (m)	769.8[a]	406.2	137.4[b]	77.9	421.5[c]	252.8
Catchment area (km^2)	26.7[a]	15.0	179.9[b]	119.0	89.8[a]	73.8
Riparian variables						
Drift geology*[1)]	1[a]		3 [b]		1[ab]	
Planform*[2)]	2 [a]		3[b]		3 [b]	
Habitat variables, physical						
Width (m)	4.41[a]	2.03	8.23[b]	3.27	5.86[ab]	2.04
Discharge*[3)]	1[a]		4[b]		3 [b]	
Clarity*[4)]	3 [a]		2 [b]		3 [ab]	
Shading – integrated measure [5)]	171.2[a]	70.6	96.7[b]	71.6	116.7[b]	82.8
Habitat variables, chemical						
Chloride (mg l^{-1})	6.5[a]	3.6	29.1[b]	12.2	15.7[a]	7.6
BOD5 (mg l^{-1})	2.1	1.1	1.5	1.0	1.6	0.8
Ortho-phosphate (mg l^{-1})	24.3	11.5	14.2	13.9	8.5	1.7
Total phosphate (mg l^{-1})	42.9[a]	16.2	28.8[ab]	8.8	18.0[b]	6.0

(1) Drift geology: 1: alluvial, 2: lacustrine, 3: moraine, 4: sandar, 5: marine, 6: organic.(2) Planform: 1 = straight; 2 = sinuous; 3 = meanders, irregular; 4 = meanders, regular.(3) Discharge: 1: $<0.31\ m^3\ s^{-1}$, 2: >0.31–$0.62\ m^3\ s^{-1}$, 3: >0.62–$1.25\ m^3\ s^{-1}$, 4: >1.25–$2.50\ m^3\ s^{-1}$, 5: >2.5–$5.0\ m^3\ s^{-1}$, 6: >5.0–$10.0\ m^3\ s^{-1}$, 7: >10–$20\ m^3\ s^{-1}$, 8: >20–$40\ m^3\ s^{-1}$, 9: >40–$80\ m^3\ s^{-1}$ and 10: $>80\ m^3\ s^{-1}$.(4) Water clarity: 1 = clear; 2 = cloudy; 3 = turbid.(5) Shade: see Data analysis section for formula.Only variables that differed among clusters are included (ANOVA $p<0.05$ or X^2 for categorical variables see*). a, b and c signifies differences between mean/median values ($p<0.05$).

significance for the separation of C6 and C7 sites on DCA axis 2 (Figs 1e and f, Table 6). Richness of species, genera and families were negatively related to slope and positively to discharge, increasing channel complexity and stream width. The Shannon diversity was also negatively related to slope and to shading, whereas it was positively related to stream width (Table 6). The MTR was also correlated to many variables at various scales (e.g., altitude, slope, distance to source, discharge). The IBMR was less correlated to the ecoregion variables than MTR but to several of the habitat variables (e.g., depth, substrate type). The correlation between IBMR and ortho-phosphate is based on only 13 measurements and the found relation should therefore be considered with caution.

Discussion

General patterns in community structure

We found that macrophytes were present in almost all the investigated stream and river types but also found that their abundance was limited, which probably relates to unfavourable habitat conditions (i.e., high disturbance levels in upland regions and shading from riparian vegetation in both upland and lowland regions). We also found that there was a high degree of variability in community structure among the main stream types investigated. Moving from the small streams in upland areas (small-sized, shallow mountain streams) to middle-sized lowland streams there was a clear

Table 5. Percentage coefficient of variation (CV) for various metrics describing structural and ecological characteristics of the macrophyte community for small-sized, shallow mountain streams and middle-sized lowland streams and for the corresponding TWINSPAN end-cluster 4 and 6

	CV%			
	Small-sized, shallow mountain streams ($n=8$)	TWINSPAN C4 ($n=13$)	Middle-sized, lowland streams ($n=23$)	TWINSPAN C6 ($n=23$)
MTR	27	23	23	10
IBMR	11	12	12	9
Species number	59	80	70	63
Shannon diversity	72	68	32	29
Simpson diversity	64	59	13	12
Domination	58	49	57	56
Evenness	62	57	4	3

transition from a predominance of species-poor mosses and liverwort-dominated communities to more species-rich communities dominated by vascular plants (Tables 2, 3 and Fig. 1). We could not clearly distinguish among the various different small-sized or medium-sized stream types on the basis of the macrophyte community structure (Fig. 1 and Table 2). This result may indicate that the number of sites investigated within the stream types was too low to give an adequate description of the macrophyte communities, or that the typology used is unsuited to describe macrophyte assemblage patterns (discussed later).

To characterise assemblage patterns independently from the typology a TWINSPAN classification was performed. Three distinct groups of plant species were identified (C4, C6 and C7, see Supplementary materials) of which two turned out to largely support two of the pre-defined main typologies. Thus there was a very good agreement between small-sized, shallow mountain streams and end-cluster C4 and between medium-sized, lowland streams and end-cluster C6 (75 and 87%, respectively). The last major end-cluster, C7, possessed a distinct community that shared characteristics with both the small-sized, shallow mountain streams and the middle-sized lowland streams. This community can be characterised as an intermediate community with growth of both many different species of non-vascular plants (e.g., *Fontinalis antipyretica*, *Amblystegium riparium*, *Fissidens carssipes*) and vascular plants (e.g., many amphibious and terrestrial species).

The result of the classification performed has many similarities with that previously performed in Great Britain (Holmes, 1998). Holmes identified four groups (A–D) based on the classification of more than 1500 British stream and river sites. The C6 end-cluster identified in this study is comparable to his A group, which was defined as eutrophic lowland streams comprising both low gradient lowland rivers, clay-dominated lowland rivers, chalk rivers and other base-rich rivers with stable flows and finally, impoverished lowland rivers. The C6 end-cluster was species-rich and contained many more truly aquatic species than the other clusters. Similarly to Holmes (1998), we also found that some of the calcareous stream sites were located within this cluster (i.e., medium-sized, lowland calcareous streams and medium-sized streams on calcareous soils). The C7 end-cluster was comparable to the B group identified by Holmes (1998). The B group comprises sandstone, mudstone and hard limestone rivers of England, Wales and Scotland. In our study the C7 end-cluster also comprised streams sites with varying geology (Table 3). In this cluster the submerged habitats were often dominated by non-vascular plants, whereas a wide array of vascular amphibious and terrestrial species grew emergent in the stream channel. The C4 end-cluster was not really comparable to the other groups defined by Holmes (1998), as non-vascular species were more predominant in this cluster compared to the two remaining groups identified in Great Britain. Holmes (1998) performed further sub-divisions of

Table 6. Significant Spearman rank correlation coefficients between various macrophyte community characteristics and stream site variables at different scales: ecoregion, catchment, riparian and habitat for TWINSPAN end-clusters 4, 6 and 7

Variables	Number of observations	DCA1	DCA2	Species number	Shannon	Domination	Evenness	MTR	IBMR
Ecoregion variables									
Latitude	44–48	−0.384**	−0.641***					−0.418**	
Altitude (m)	44–48	0.549***	0.674***					0.469**	
Catchment variables									
Slope (m km^{-1})	42–46	0.467**	0.609***	−0.443*	−0.467*	−0.492*	−0.325*	0.675***	0.490**
Distance to source (km)	44–48		−0.587***	0.390*	0.406*	−0.361*		0.573***	
Height of source (m)	44–48	0.590***	0.740***	−0.285*				0.433*	
Discharge$^{(1)}$	38–41	−0.450*	−0.529**	0.617***	0.635***			−0.526**	
Catchment area	29–33	−0.477*	−0.666***	0.452*	0.465*			0.656***	
Riparian variables									
Planform (1–9)$^{2)}$	44–48		−0.356*	0.396*	0.411*	−0.324*			
Shading – integrated measure$^{(3)}$	44–48	0.353*				0.398*			
Habitat variables, physical									
Clarity $^{(4)}$	43–47	0.308*				0.398*		0.315*	
Width (m)	44–48		−0.495**	0.458**	0.472*	0.404*		−0.431*	
Bed stability, unstable (%)	43–47		−0.347*						
Bed stability, stable (%)	43–47		0.338*						
Depth: 0.25–0.50 (%)	43–47		0.328*						
Depth > 1.0 (%)	43–47								0.318*
Substrate: Bedrock (%)	44–48							−0.417*	−0.400*
Substrate: Boulders/copples (%)	44–48	0.341*	0.350*						
Habitat variables, chemical									
Chloride (mg l^{-1})	21–24	−0.521**	−0.810***					−0.663**	
Nitrate (mg l^{-1})	10–11		−0.694*						
Ortho-phosphate (mg l^{-1})	12–13								0.681*

(1) Discharge: 1: <0.31 m^3 s^{-1}, 2: >0.31–0.62 m^3 s^{-1}, 3: >0.62–1.25 m^3 s^{-1}, 4: >1.25–2.50 m^3 s^{-1}, 5: >2.5–5.0 m^3 s^{-1}, 6: >5.0–10.0 m^3 s^{-1}, 7: >10–20 m^3 s^{-1}, 8: >20–40 m^3 s^{-1}, 9: > 40–80 m^3 s^{-1} and 10: >80 m^3 s^{-1}.(2) Planform: 1=straight; 2=sinuous; 3=meanders, irregular; 4=meanders, regular.(3) Shade: see Data analysis section.(4) Water clarity: 1 = clear; 2 = cloudy; 3 = turbid. *$p<0.05$. **$p<0.001$. ***$p<0.0001$.

the identified groups into 38 sub-types. Many of these sub-types are not or only poorly represented in the present investigation and therefore we will not compare the two classifications at a more detailed level.

As stated above, the general trend moving from C4 to C6 over C7 relates to a gradient from high altitude, high gradient, and small stream sites with steep slopes to lower altitude small and medium-sized stream sites (Tables 4 and 6). Thus there was a clear gradation of mean site altitude with C4 sites being 81% higher than C6 sites, and C7 sites being 42% higher than C6 sites. Altitude has also previously been identified as being of primary importance for the distribution of plant communities in European streams (Haslam, 1987). Particularly mosses and liverworts that were predominant in C4 sites (see Supplementary material) were abundant in these streams, probably reflecting their preference for stable substrates and low water depths (Haslam, 1987; Scarlett & O'Hare, 2006). This was also evident from the correlations found between DCA axes and both streambed stability and the predominance of coarse substrata, i.e. boulders/cobbles (Table 4). The C7 sites were intermediate altitude streams that were positioned closer to the source compared to C6 sites. The C7 sites can be considered intermediate between strictly upland stream sites and lowland sites (Haslam, 1987; Holmes, 1998). This group consisted of a mixture of mosses and vascular plants (see Supplementary material). We found that amphibious species were an important feature of this group compared to C4 and C6, which probably relates to the stream size. Thus amphibious species tended to be most abundant in shallow water and the relative abundance of this group of species will therefore decrease from upstream to downstream reaches as the water depth increases (Riis et al., 2001). Finally, the C6 sites were true lowland sites with moraines deposits in the valley in several of the investigated sites and soft streambeds. These sites were unique in being dominated by vascular aquatic plant species (see Supplementary material) of which four were identified as indicator species, namely *Sparganium emersum*, *Berula erecta*, *Elodea canadensis* and *Lemna minor*.

Several smaller-scale habitat variables also affected the segregation of the macrophyte communities (Table 6), but many of these co-correlated with the large-scale variables mentioned above. However, shading and water clarity were important small-scale variables (Table 6). Thus moving from C4 over C7 to C6 sites the shading got less intense, which is likely to relate to the wider stream reaches that diminish the degree of shading from the riparian vegetation.

Metrics and their variation within C4, C6 and C7

The here-performed analysis suggests that great care should be taken in comparing macrophyte-based metrics in an evaluation of ecological quality *sensu* WFD without a detailed knowledge of stream type characteristics. This concerns both metrics that are mathematical expressions of the community structure (based on taxa richness, evenness and abundance patterns) and metrics based on the information of ecological tolerance of indicator taxa. We found that both types of metrics exhibited an intrinsic variability among the community types identified. The C4 community was less diverse (both richness and diversity measures) than the C6 and C7 communities, and the domination of a single or a few species was more typical here than in the other stream types. Similarly, the MTR and IBMR indices, both developed as assessment methods of the trophic status of streams and rivers, exhibited a marked variability among the three community types. The C4 community exhibited higher MTR scores (43–63) than the C6 (34–48) and C7 community (28–62). This result indicates that a natural shift in macrophyte abundance patterns takes place moving from upland to lowland sites (discussed below). This result also emphasises the general recommendation that the use of MTR should be restricted to making comparisons between streams and rivers that are of the same physico-chemical type (Dawson et al., 1999). Otherwise the lower MTR scores in C6 and C7 sites observed can incorrectly be inferred as more enriched conditions compared to C4 sites.

We found that much of the above-mentioned variation in metrics can be ascribed to differences in the physical stream environment among the identified types moving from upland to lowland regions (Table 6). Thus a low species-richness and diversity seems to be an inherent feature of small stream reaches as both parameters were negatively correlated with slope and discharge. Similarly we found that the MTR and IBMR were positively correlated with these environmental parameters, which suggests that these indices will be higher in upland reaches compared to wider lowland reaches with lower flow velocities. This is not surprising as species that typically grow in upland reaches, i.e., mosses and liverworts are high-scoring MTR species (Dawson et al., 1999). Thus the median STR (Species Trophic Rank), which is a value assigned to a species on a scale from 1 to 10 designed to reflect the tolerance of that species to eutrophication, is 8, 10 and 5 for liverworts, mosses and vascular plants, respectively (Dawson et al., 1999).

Species richness was associated with higher discharges and wider reaches that are less shaded from riparian vegetation. The increase in species richness with increasing stream size confirms the general positive correlation between species richness and area (Rosenzweig, 1995). In addition, the increase in species richness probably relates to habitat characteristics (French & Chambers, 1996). Thus, middle-sized streams are likely to be physically more heterogeneous and experience lower levels of disturbance than small-sized streams, which may promote the co-existence of a wider array of species (Vannote et al., 1980). In addition, when moving from upstream to downstream reaches a continuously larger upstream area is likely to enhance the propagule supply with the current and thereby species recruitment locally, which may also increase species richness (Barrat-Segretain, 1996).

Conclusions

We found that macrophyte communities in unimpacted European streams responded to most of the characteristics underlying the typological framework defined in the EU Water Framework Directive (WFD: European Commission, 2000). The present interpretation of the WFD typology may not, however, be adequate for an evaluation of stream quality based on macrophytes. First and most important, by using this typology we may overlook an important community type (C7), which is characterised as small-sized relatively steep-gradient streams being an intermediate type between small-sized, shallow mountain streams and medium-sized, lowland streams. This stream type is species-rich and consists of a mixture of non-vascular and vascular plant species. Second, the natural variability in most structural and ecological metrics appeared to be higher when using the pre-defined typology compared to a typology based on macrophyte assemblage patterns particularly regarding MTR. The consistency of these results should be examined by analysing a larger number of sites. Particularly the need of re-defining the typology to improve the ability to detect impacts in streams and rivers from macrophyte assemblage patterns should be investigated.

Acknowledgements

We thank Mike Furse for a very good organisation of the STAR project and the large number of participants who have taken part in data collection, treatment and analysis. We thank the European Union for financial support (project no. EVK1-CT 2001-00089).

References

Baggøe, J. & F. K. Ravn, 1896. Excursioner til jyske søer og vandløb i sommeren 1895 (in Danish). Botanisk Tidskrift 20: 288–236.

Barrat-Segretain, M. H., 1996. Strategies of reproduction, dispersion and competition in river plants: a review. Vegetatio 123: 13–37.

Butcher, R. W., 1933. Studies on the ecology of rivers I. On the vegetation distribution of macrophyte vegetation in the rivers of Britain. Journal of Ecology 21: 58–91.

Carbiener, R., M. Tremolieres, J. L. Mercier & A. Ortscheit, 1990. Aquatic macrophyte communities as bioindicators of eutrophication in calcareous oligosaprobe stream waters Upper Rhine plain, Alsace. Vegetatio 86: 71–88.

Clarke, K. R. & R. H. Green, 1988. Statistical design and analysis for a 'biological effects' study. Marine Ecology-Progress Series 46: 213–226.

Dawson, F. H., J. R. Newman, M. J. Gravelle, K. J. Rouen & P. Henville, 1999. *Assessment of the Trophic Status of Rivers Using Macrophytes. Evaluation of the Mean Trophic Rank*. R&D Technical Report E39, Environment Agency.

Dufrene, M. & P. Legendre, 1997. Species assemblages and indicator species: the need for a flexible asymmetrical approach. Ecological Monographs 67: 345–366.

European Commission, 2000. Directive 2000/60/EC of the European Parliament and of the Council – Establishing a framework for Community action in the field of water policy. Brussels, Belgium, 23 October 2000.

French, T. D. & P. A. Chambers, 1996. Habitat partitioning in riverine macrophyte communities. Freshwater Biology 36: 509–520.

Furse, M., A. Schmidt-Kloiber, J. Strackbein, J. Davy-Bowker, A. Lorenz, J., van der Molen & P. Scarlett, 2004. Standardisation f river classifications: Framework method for calibrating different biological survey results against ecological quality classifications to be developed for the Water Framework Directive. 6th deliverable, due 31/07/04. Results of the sampling programme. Accessible at the STAR website (http://www.eu-star.at).

Haslam, S. M., 1987. River Plants of Western Europe. Cambridge University Press, Cambridge.

Haury, J., M. -C., Peltre, M., Tremolieres & J., Barbe, 2002. A method involving macrophytes to assess water trophy and organic pollution: the Macrophyte Biological Index for Rivers (IBMR) – application to different types of rivers and pollutions. Proceedings of 11th EWRS International Symposium on Aquatic Weeds: 247–250.

Hering, D., A. Buffagni, O. Moog, L. Sandin, M. Sommerhäuser, I. Stubauer, C. Feld, R. Johnson, P. Pinto, N. Skoulikidis, P. Verdonschot & S. Zahrádková, 2003. The development of a system to assess the ecological quality of streams based on macroinvertebrates – design of the sampling programme within the AQEM project. International Review of Hydrobiology 88: 345–361.

Hering, D., O. Moog, L. Sandin & P. F. M. Verdonschot, 2004. Overview and application of the AQEM assessment system. Hydrobiologia 516: 1–20.

Hering, D. & J. Strackbein, 2001. Standardisation of river classifications: Framework method for calibrating different biological survey results against ecological quality classifications to be developed for the Water Framework Directive. 1st deliverable, due 30/06/02. STAR stream types and sampling sites. Accessible at the STAR website (http://www.eu-star.at).

Holmes, N. T. H., P. J. Boon & T. A. Rowell, 1998. A revised classification system for British rivers based on their aquatic plant communities. Aquatic Conservation 8: 555–578.

Holmes, N. T. H., J. R. Newman, S. Chadd, K. J. Rouen, L. Saint & F. H. Dawson, 1999. Mean Trophic Rank: A Users Manual, R&D Technical Report E38, Environment Agency.

Illies, J., 1978. Limnofauna Europaea. Gustav Fisher Verlag, Stuttgart 532 pp.

Margalef, D. R., 1958. Informatione theory in ecology. General Systems 3: 36–71.

Mesters, C. M. L., 1995. Shifts in macrophyte species composition as a result of eutrophication and pollution in Dutch transboundary streams over the past decades. Journal of Aquatic Ecosystem Health 4: 295–305.

McCune, B. & M. J. Mefford, 1999. PC-ORD for Windows (4.01). Multivariate Analysis of Ecological Data. GMjM Software, leneden Beach, Oregon, USA.

Mountford, J. O., 1994. Floristic change in English grazing marshes: the impact of 150 years of drainage and land-use change. Watsonia 20: 3–24.

Nijboer, R. C., R. K. Johnson, P. F. M. Verdonschot, M. Sommerhauser & A. Buffagni, 2004. Establishing reference conditions for European streams. Hydrobiologia 516: 91–105.

Preston, C. D., 1995. Pondweeds of Great Britain and Ireland. Botanical Society of the British Isles, Handbook No. 8. London.

Raunkiær, C., 1895–1999. De danske blomsterplanters naturhistorie. Enkimbladede (in Danish). Gyldendal, Copenhagen.

Riis, T., K. Sand-Jensen & O. Vestergaard, 2000. Plant communities in lowland Danish streams: species composition and environmental factors. Aquatic Botany 66: 255–272.

Riis, T. & K. Sand-Jensen, 2001. Historical changes of species composition and richness accompanying disturbance and eutrophication of lowland streams over 100 years. Freshwater Biology 46: 269–280.

Riis, T., K. Sand-Jensen & S. E. Larsen, 2001. Plant distribution and abundance in relation to physical conditions and location within Danish stream systems. Hydrobiologia 448: 217–228.

Rosenzweig, M. L., 1995. Species Diversity in Space and Time. Cambridge University Press, Cambridge.

Scarlett, P., & O'Hare M., 2006. Community structure of instream bryophytes in English and Welsh rivers. Hydrobiologia 553: 143–152.

Svenning, J. -C., 2002. A review of natural vegetation openness in north-western Europe. Biological Conservation 104: 133–148.

Vannote, R. L., G. W. Minshall, K. W. Cummins, J. R. Sedell & C. E. Cushing, 1980. The river continuum concept. Canadian Journal of Fisheries and Aquatic Sciences 37: 130–137.

Verdonschot, F. M. & R. C. Nijboer, 2004. Testing the European stream typology of the Water Framework Directive for macroinvertebrates. Hydrobiologia 516: 35–54.

Wilson, J. B., 1991. Methods of fitting dominance/diversity curves. Journal of Vegetation Science 2: 35–46.

Hydrobiologia (2006) 566:197–210
M.T. Furse, D. Hering, K. Brabec, A. Buffagni, L. Sandin & P.F.M. Verdonschot (eds), The Ecological Status of European Rivers: Evaluation and Intercalibration of Assessment Methods
DOI 10.1007/s10750-006-0095-2

Macrophyte communities of European streams with altered physical habitat

Mattie T. O'Hare[1,*], Annette Baattrup-Pedersen[2], Rebi Nijboer[3], Krzysztof Szoszkiewicz[4] & Teresa Ferreira[5]
[1]*Centre for Ecology and Hydrology, Winfrith Technology Centre, Dorchester, DT2 8ZD Dorset, UK*
[2]*Department of Freshwater Biology, National Environmental Research Institute, Vejlsøvej 25, P. O. Box 314, DK-8600 Silkeborg, Denmark*
[3]*Alterra, Green World Research, P.O. Box 476700 AA, Wageningen, The Netherlands*
[4]*Department of Ecology and Environmental Protection, Agricultural University of August Cieszkowski, ul. Piątkowska 94C, 61-691 Poznan, Poland*
[5]*Agronomy Institute, Forestry Department, Technical University of Lisbon, Tapada da Ajuda, 1349-017 Lisboa, Portugal*
*(*Author for correspondence: moha@ceh.ac.uk)*

Key words: plant, vegetation, stream, damming, realignment

Abstract

The impact of altering hydro-morphology on three macrophyte community types was investigated at 107 European stream sites. Sites were surveyed using standard macrophyte and habitat survey techniques (Mean Trophic Rank Methodology and River Habitat Survey respectively). Principal Components Analysis shows the macrophyte community of upland streams live in a more structurally diverse physical habitat than lowland communities. Variables representing the homogeneity and diversity of the physical environment were used to successfully separate un-impacted from impacted sites, e.g. homogeneity of depth and substrate increased with decreasing quality class for lowland sites (ANOVA $p < 0.05$). Macrophyte attribute groups and structural metrics such as species richness were successfully linked to hydro-morphological variables indicative of impact. Most links were specific to each macrophyte community type, e.g., the attribute group liverworts, mosses and lichens decreased in abundance with increasing homogeneity of depth and decreasing substrate size at lowland sites but not at upland sites. *Elodea canadensis*, *Sparganium emersum* and *Potamogeton crispus* were indicative of impacted lowland sites. Many of the indicator species are also known to be tolerant to other forms of impact. The potential for a macrophyte tool indicative of hydro-morphological impact is discussed. It is concluded one could be constructed by combining indicator species and metrics such as species richness and evenness.

Introduction

Aquatic macrophytes are considered as sensitive to physical alteration in streams. Here, in response to management needs, that sensitivity is assessed on a pan-European basis for the first time. The European Union (EU) requires member states to categorise the quality of their rivers, primarily using aquatic organisms (European Commission, 2000). Macrophytes are included on the list of organisms, as are fish, invertebrates, phytobenthos and phytoplankton. Alterations to a river, including physical alteration, that degrades the biota and causes a site to be categorised as impacted must be mitigated against.

The underlying aim of the new legislation, the Water Framework Directive (WFD), is to manage aquatic systems by catchment using measures of ecosystem health to assess success (Pollard & Huxham, 1998). The inclusion of hydro-morphology in the assessment of ecological status is significant. In the past monitoring in running water has focused on chemical parameters and benthic invertebrates. The WFD now widens that focus and implicitly requires that habitats are linked to biota, including macrophytes to physical habitat quality (Logan & Furse, 2002). There is therefore a clear management need to appraise the sensitivity of European macrophytes to physical habitat alteration.

Man's alterations to rivers through impoundments, realignment of channels, and in-stream engineering works can alter depth, velocity, substrate type, flow types and flow variability (Petts, 1984a; Brookes, 1988). These variables define the physical niches in rivers. Macrophytes have known preferences for these variables (Haslam, 1978; Fox, 1992). Since historic times macrophytes have been grouped by depth preference as emergent, marginal and submerged (Sculthorpe, 1967). In recent times niche separation and range preferences for other physical variables have been demonstrated for many macrophytes (Westlake, 1975; Chambers et al., 1991; French & Chambers, 1996; Dawson et al., 1999a). It is therefore not surprising that studies of physically altered rivers show impacts to macrophyte community structure. Following impoundment and canalisation changes include loss of species, altered species dominance and relative abundance (Petts, 1984b; Baattrup-Pedersen & Riis, 1999).

The point has been strongly made that WFD monitoring programmes need to take into account natural variation if they are not to provide data which leads to misclassification of sites (Irvine, 2004). Irvine argues that only the most sensitive and reliable groups should be monitored. This study aims to address the basic questions as to how macrophyte assemblages vary naturally in relation to physical parameters and do they have potential as indicators of impact to the physical habitat.

We ask a series of inter-related questions. Are known stream macrophyte assemblages associated with different types of physical habitat? Within each macrophyte assemblage can sites of different quality be identified using physical habitat variables? Can sites be assigned to previously identified macrophyte assemblages using site characteristics unlikely to be affected by man? Are macrophyte metrics sensitive to physical habitat alteration and is it possible to identify indicator species?

This work, is part of a much wider study supporting the implementation of the WFD, the EU funded Standardisation of River Classification (STAR) Project which has the aim of developing standardised, statistically robust monitoring methods, for use across Europe Furse et al. (2006). To answer the questions outlined above we analysed survey data from 107 stream sites across Europe collected during the STAR project. The sites were known to represent a wide range of physical habitat quality from highly degraded to un-impacted.

All three major stream macrophyte assemblages were represented which are; (C4) mountain streams poor in species and dominated by mosses and liverworts, (C6) lowland streams dominated by *Phalaris arundinacea* and *Sparganium emersum* and (C7) an intermediate group rich in species with many amphibious species, terrestrial dicotyledons and mosses (Baattrup-Pedersen et al., 2006).

Methods

Study sites

The sites included in the present investigation covered an impact gradient from sites having high ecological quality to sites having poor and bad ecological quality (*sensu* WFD). The WFD has 5 classes bad, poor, moderate, good and high (reference), which we coded 1 to 5 respectively. Sites were chosen that were either un-impacted (ecological quality class 5) or the major impact was hydro-morphological degradation (ecological quality class 1–4). Hydro-morphologically degraded sites included realigned, impounded and over-deepened reaches. The allocation of sites to an ecological degradation class was performed *a priori* according to criteria described in Furse et al. (2006). A total of 107 sites were included in the analysis. They were located in Austria, Czech Republic, Germany, United Kingdom, Denmark, France, Greece, Italy, Latvia and Poland, see map in Furse et al. (2006).

A system, River Habitat Survey (RHS), for assessing the character and quality of rivers based on their physical structure has been developed in the UK (Raven et al., 1997). Any 500 m length of river surveyed using RHS methodology can be categorised and its habitat quality assessed by comparison with other sites of a similar physical character. RHS was used within the STAR project and surveying was undertaken in late summer/early autumn together with supporting chemical, physico-chemical and geographical elements. The RHS survey records information on macrophytes as the abundance of ten attribute groups. The groups are, Liverworts/mosses/lichens, emergent broad leaved herbs, emergent reeds/sedges/rushes/grasses/horsetails, floating-leaved (rooted), free-floating, amphibious, submerged broad-leaved, submerged linear-leaved, submerged fine-leaved and filamentous algae. The attribute groups were used in the analysis.

Species level macrophyte data was also recorded using a separate survey technique at the same site. The macrophyte survey method was a version of the Mean Trophic Rank Methodology developed for the STAR project (Dawson et al., 1999b). Species present in mainland Europe, but not in the UK where the survey was originally developed, were added to the form. The survey methods are available at the STAR website (www.eu-star.at) under the public-access section "Protocols". As well as recording species data the MTR survey records categories of bed stability, substrate type, river width, and water depth.

Data analysis

Allocation of stream sites to biological groupings

Discriminant analysis was used to allocate sites to the macrophyte groupings previously identified for un-impacted European streams (TWINSPAN predictor group C4 (mainly mountain sites), C6 (mainly lowland sites) and C7 (intermediate)); see Baattrup-Pedersen et al. (2006); Jongman et al. (1987). Four ecoregion/catchment scale variables, which showed significant differences between the TWINSPAN predictor group C4, C6 and C7 reference sites, were used. These were altitude, reach slope, distance to source and height of source. The ability of the variables to discriminate between the groups was tested individually and in combination. Height of source performed better than the other individual variables and better than the variables in combination. It correctly allocated 79% of reference sites to their correct group.

The new groups contain both un-impacted and impacted sites. To differentiate between them and the original groups which only contain un-impacted sites, the new groups were named, Discrim4 (near source), Discrim6 (far from source) and Discrim7 (intermediate).

Sites from a particular *a priori* defined stream type were usually all assigned to the same TWINSPAN group. Sites from some *a priori* stream types were split between TWINSPAN groups 4 and 7 (see Baattrup-Pedersen et al., 2006). They were all small mountain streams in the Czech Republic (Type 5), Germany (Type 4) and Italy (Type 6).

Hydro-morphological site characteristics

Three main types of hydro-morphological impacts, from the same sites, were distinguished in the RHS. These included channel realignment, over-deepening and impoundment. The level of impact varied among the affected reaches depending on the channel length being affected. In addition some sites were subjected to more than one type of impact. Table 1 gives an overview of the number of sites within each of the predicted groups that were impacted by the different impact types.

The three types of impacts may affect macrophyte communities differently. For example average flow velocity may decrease in over-deepened reaches which may stimulate growth of emergent species at the edges of the channel. In contrast the velocity may increase and be more homogeneous in straightened reaches, which may stimulate growth of submerged species that are highly resistant to flow e.g. species with linear growth morphologies such as *Ranunculus pencillatus pseudofluitans*. To get a thorough description of the major impact types we therefore needed to calculate several hydro-morphological variables based on the RHS & MTR data.

We used 4 different types of hydro-morphological variables with the aim of including measures of system complexity i.e. domination, diversity, score and homogeneity. These variables were calculated for each of the 5 main hydro-morphological descriptors, e.g. bed stability, water depth, substrate, RHS flow types and (wetted) width. These descriptors are recorded as categories. A value between 1 and 9 was allocated to the possible categories within each of the descriptors

Table 1. Overview of the number of sites impacted by the three major impact types in the RHS within each of the predicted groups

	Discrim4 (near source)	Discrim6 (far from source)	Discrim7 (Intermediate)
Impounded	1	0	0
Realigned	2	3	2
Over-deepened	5	6	3
Impounded and realigned	2	0	1
Impounded and over-deepened	2	1	0
Realigned and over-deepened	7	2	8
Realigned, over-deepened and impounded	1	2	1

(Table 2). Domination expresses the dominant category, diversity the number of categories represented, score the weighted average of the categories represented and homogeneity the distribution of the categories represented.

The score was calculated as:

$$\mu = \frac{\sum_{i=1}^{N} i \cdot n_i}{\sum_{i=}^{N} n_i},$$

where
N = the number of categories represented, and
n_i = the percentage of reach in category i

The homogeneity was calculated as:

$$I = N \cdot \mu^2 \sum_{i=1}^{N} \frac{(n_i - \hat{m}_i)^2}{m_i},$$

where

$$\hat{m}_i = \frac{\sum_{i=1}^{N} n_i}{N}$$

is the mean percentage for each category in the case of equal representation.

A high homogeneity therefore implies that the distribution of categories is very uniform whereas a low homogeneity implies that the distribution is heterogeneous.

A Principal Components Analysis (PCA) ordination on the calculated hydro-morphological variables was performed to describe habitat characteristics of the investigated stream sites Jongman et al. (1987). To test whether the calculated hydro-morphological variables were able to distinguish between the macrophyte assemblages predicted using discriminant analysis earlier, as Discrim4 (near source), Discrim6 (far from source) and Discrim7 (intermediate), an ANOVA was performed. Relationships between PCA axis 1 and 2 and the hydro-morphological variables were furthermore analysed by Spearman Rank correlation analyses. In the PCA diagram the environmental vectors are exaggerated by 5 to make their relative importance (their lengths) obvious.

Additionally, we tested the ability of the calculated hydro-morphological variables to detect and assess hydro-morphological degradation. Each site was *a priori* allocated to an ecological

Table 2. Categories of hydro-morphological descriptors derived from RHS and MTR. The allocated values are on an ordinal scale and are used in domination, diversity, score and homogeneity calculations

Hydormor-phological descriptor	Allocated value								
	1	2	3	4	5	6	7	8	9
MTR Bed Stability	Firm	Stable	Unstable	Soft					
MTR Depth	<0.25 m	0.25–0.5 m	0.5–1.0 m	>1.0 m					
MTR Substrate	Bedrock	Boulders/cobbles	Pebbles/gravel	Sand	Silt				
RHS Flow	Free-fall	Chute	Chaotic	Unbroken standing waves	Broken standing waves	Upwelling	Rippled	Smooth	No perceptible flow
MTR Width	<1 m	1–5 m	5–10 m	10–20 m	>20 m				

quality class (1–5) according to Furse et al. (2006). To test whether the hydro-morphological variables varied among the ecological quality classes a One Way ANOVA with Bonferroni corrections was performed separately on Discrim4, Discrim6 and Discrim7 sites. To attain comparable classes in terms of number of sites included we chose to divide the *a priori* site classification into three groups i.e. ecological quality class 1, 2 and 3 (EQ1–3), ecological quality class 4 (EQ4) and ecological quality class 5 (EQ5). Achieving sufficient sample size for analyses and homogeneity of response within groups were the main criteria for dividing the groups.

Macrophyte communities
Various macrophyte attribute groups were derived directly from the RHS and include amphibious species, emergent broad-leaved herbs, emergent reeds/sedges/rushes, filamentous algae, floating-leaved (rooted) species, free-floating species, liverworts/mosses/lichens, submerged broad-leaved, submerged fine leaved and submerged linear leaved species (Environment Agency, 2003). In addition the structural metrics, species richness, domination and evenness were used to describe community structure, see Baattrup-Pedersen et al. (2006) for definitions of structural metrics. These were all derived from the MTR indexation (Dawson et al., 1999b; Szoszkiewicz et al., 2006).

To analyse relationships between hydro-morphological degradation and macrophyte communities, Spearman Rank correlation analyses were performed between hydro-morphological variables that separated ecological quality classes and attribute groups and structural metrics. In addition, a Canonical Correspondence Analysis (CCA) using CANOCO 4.5 (ter Braak & Smilauer, 1998) was performed to detect individual species or groups of species indicative of hydro-morphological degradation. A down-weighting of rare species was chosen in the analysis. All calculated hydro-morphological variables were included in this analysis initially and the best predictors were selected by forward selection, which is a multivariate extension of the stepwise regression method.

All other statistical analyses were carried out using Minitab software, Minitab (2004).

Results

Hydro-morphological site characteristics

The first three components of the PCA applied to the 20 hydro-morphological variables explained 62% of the variance in the system (PCA 1 40%, PCA 2 13% and PCA 3 9%). Stream sites from discriminant group 4 (near source), 6 (far from source) and 7 (intermediate) were primarily separated along PCA axis 1 (Fig. 1). Discrim6 sites were significantly separated from Discrim4 and Discrim7 sites (ANOVA, $p < 0.05$). Discrim4 and Discrim7 on the other hand were not significantly separated (ANOVA, $p > 0.05$). The hydro-morphological characteristics separating the predicted groups were related to several of the calculated variables (Fig. 1; Table 3). We found that Discrim4 sites were much more hydro-morphologically diverse than Discrim6 sites in terms of substrate, flow and stability characteristics. In accordance the homogeneity in both water depth and substrate characteristics was higher in Discrim6 sites. Discrim6 sites were also deeper and had a predominance of finer substrates compared to Discrim4 sites (data not shown).

The overall environmental variability, across all the hydro-morphological variables, assessed from PCA 1 site scores was highest in un-impacted Discrim6 and Discrim7 sites (EQ 5) (Fig. 2; Table 4). PCA site scores could not be used to differentiate ecological quality in Discrim4 and Discrim7 (ANOVA, $p > 0.05$). In contrast, PCA site scores could be used to differentiate ecological quality in Discrim6 (Fig. 2). Thus, PCA 1 site scores were significantly higher in EQ1–3 compared to EQ 5 (ANOVA $p < 0.05$; Table 4). Further EQ4 PCA 1 site scores were intermediate between EQ1–3 and EQ5 site scores (Table 4).

Several of the calculated hydro-morphological variables could also be used separately to distinguish among ecological quality classes. In Discrim4 the substrate diversity decreased significantly with decreasing ecological quality (Fig. 3a; ANOVA $p < 0.05$). In Discrim6 the homogeneity in substrate and water depth characteristics increased with decreasing ecological quality class (ANOVA $p < 0.05$; Fig. 3b). In addition, the substrate was also finer and less diverse in sites with low ecological quality compared to sites

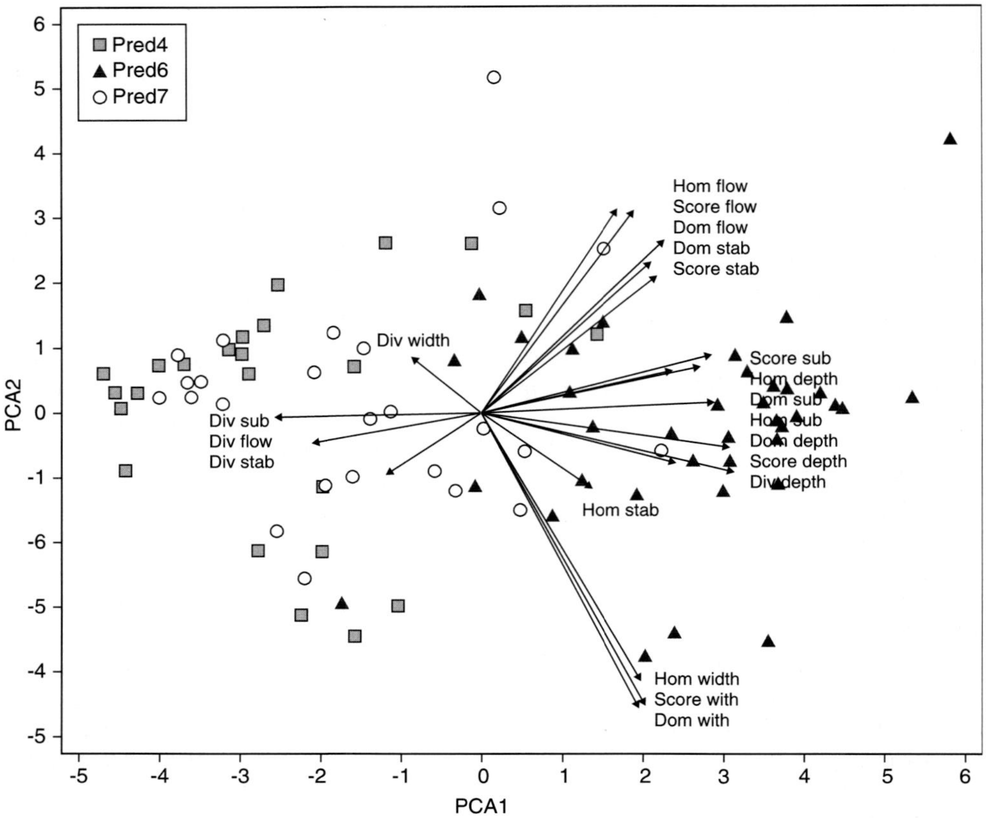

Figure 1. Principal component analysis (PCA) ordination of the calculated hydro-morphological variables from RHS/MTR in 107 stream sites distributed throughout Europe. Different symbols denote different discriminant groups 4, 6 and 7 respectively. These groups were identified from a discriminant analysis performed to predict biological communites (TWINSPAN groups – see data analysis section and Baattrup-Pederesen et al., 2006). The discriminant analysis was based on altitude, distance to source and height of source. Environmental vectors are exaggerated 5 times. Depth = water depth, Div = diversity, Dom = Domination, Flow = RHS flow category, Hom = homogeneity, Stab = bed stability, Score = weighted average of categories represented Sub = Substrate, and width = wetted width.

with high ecological quality (ANOVA $p < 0.05$; Fig. 3b). In Discrim7 depth characteristics changed in response to hydro-morphological impact. Both the score and the dominant depth decreased in sites with low ecological quality compared to sites with high ecological quality (Fig. 3c).

Linkages between macrophytes and hydro-morphological site characteristics

Spearman rank correlation analyses were used to identify linkages between macrophyte attribute groups/structural metrics and hydro-morphological site degradation. We only performed the analysis with hydro-morphological descriptors that separated ecological quality classes e.g. diversity in substrate types in Discrim4, homogeneity in depth and substrate characteristics and dominant depth and substrate type in Discrim6, and score and dominant water depth in Discrim7 (see Fig. 2). In Discrim4, we did not find any significant relations between diversity in substrate and the various attribute groups or structural metrics (see data analysis section). In Discrim6 the attribute group liverworts/mosses/lichens correlated negatively to both homogeneity in depth ($r = -0.452$), dominant depth ($r = -0.519$) and dominant substrate ($r = -0.465$) ($p < 0.05$; Table 5). Species richness decreased with increasing substrate homogeneity ($r = -0.324$) and the

Table 3. Significant correlation coefficients ($p < 0.05$) between PCA axis scores and hydro-morphological variables calculated from RHS

	PCA1	PCA2	PCA3
Bed Stability_Domination	0.623	0.213	−0.283
Bed Stability _Score	0.587	0.246	−0.257
Bed Stability _Diversity	−0.322		
Bed Stability_Homogeneity	0.361	−0.281	−0.409
Depth _Domination	**0.854**		
Depth_Score	**0.875**	−0.234	
Depth_Diversity	**0.688**		
Depth_Homogeneity	0.630		
Substrate_Domination	**0.770**		−0.267
Substrate_Score	**0.793**		−0.362
Substrate_Diversity	−0.709		
Substrate_ Homogeneity	**0.824**		
Flow_ Domination	0.548	0.293	0.297
Flow_ Score	0.540	0.346	0.358
Flow_ Diversity	−0.594		−0.305
Flow_ Homogeneity	0.643	0.326	0.372
Width_Domination	0.583	−0.732	
Width_ Score	0.603	−0.672	
Width_ Diversity	−0.212		−0.557
Width_Homogeneity	0.562	−0.602	0.350

See data analysis section for further explanation. Correlation coefficients above 0.650 (arbitrary threshold) are marked in bold.

evenness in species distribution increased with increasing homogeneity in water depth ($r = 0.397$; $p < 0.05$; 5). In Discrim7 several attribute groups correlated significantly with the depth score (submerged broad-leaved, liverworts/mosses/lichens, emergent reeds/sedges/rushes and amphibious species; Table 5). Similarly emergent reeds/sedges/rushes correlated significantly with dominant depth ($p < 0.05$; Table 5).

We performed a CCA to identify species or groups of species associated with hydro-morphological degradation. The eigenvalues of the CCA were 0.569, 0.263 and 0.209 for axis 1, 2 and 3 respectively (Fig. 4). Of the 20 hydro-morphological variables initially considered in the CCA (Table 3), 7 were retained in the analysis by forward selection. In Figure 4 significant variables are included as arrows that point in the direction of maximum change. We did not find any clear separation of ecological quality sites along the CCA axes in Discrim4 and Discrim7. In Discrim6, on the contrary, increasing water depth and substrate fineness were highly related to hydro-morphological degradation (Fig. 4a). Species associated with these variables were *Elodea canadensis*, *Sparganium emersum* and *Potamogeton crispus* (Fig. 4b). They are all submerged species.

Discussion

The physical habitat of the three-macrophyte assemblages, separated by discriminant analysis using the variable 'distance to source' examined is different. Discrim7 sites were intermediate in physical character to those in Discrim4 and Discrim6. A similar pattern exists for reference sites alone; C4 sites are small, shallow, upland streams, C6 sites medium sized lowland streams and C7 sites were intermediate in nature Baattrup-Pedersen et al. (2006).

Impacted sites are less physically diverse than non-impacted sites. This observation is consistent with physical characteristics associated with impounded waters, channel realignment and over-deepening (Environment Agency, 2003). The accumulation of fines at impacted Discrim6 sites also implies the system is characteristic of water

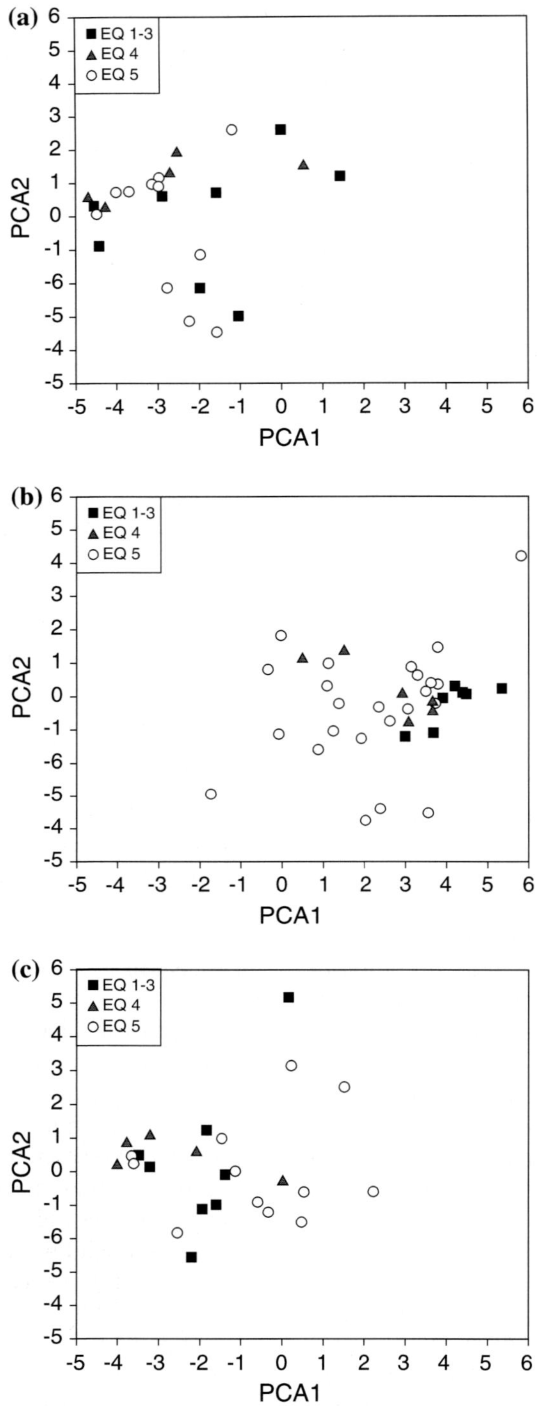

Figure 2. Site scores from PCA (see Fig. 1) superimposed by ecological quality (EQ) class in Discrim4 (plot a), Discrim6 (plot b) and Discrim7 (plot c) respectively. The ecological quality class was predicted in each site prior to the investigation from the degree of hydromorphological degradation (see method section). Ecological quality classes 1, 2 and 3 being moderate, poor and bad respectively, are grouped together.

Table 4. Mean and Standard Deviation of PCA axis scores in ecological quality class 1–3 (EQ 1–3), ecological quality class 4 (EQ 4) and ecological quality class 5 (EQ 5 = reference) in Predicted group 4, 6 and 7

	PCA1	PCA2	PCA3
Discrim4			
EQ 1–3	−1.898 ± 5.998	−0.077 ± 5.610	−0.610 ± 3.855
EQ 4	−2.734 ± 5.254	1.161 ± 1.666	0.848 ± 2.056
EQ5	−2.823 ± 3.288	−0.242 ± 6.085	−0.358 ± 4.835
Discrim6			
EQ 1–3	**4.158 ± 2.362a**	−0.227 ± 1.522	−0.122 ± 1.501
EQ 4	**2.567 ± 3.178ab**	0.243 ± 2.145	−0.867 ± 1.610
EQ5	**2.090 ± 7.561b**	−0.328 ± 7.975	0.084 ± 4.819
Discrim7			
EQ 1–3	−1.938 ± 3.646	0.288 ± 7.741	0.247 ± 4.894
EQ 4	−2.612 ± 4.035	0.532 ± 1.365	−0.189 ± 2.274
EQ5	−0.691 ± 5.895	0.070 ± 4.971	0.626 ± 4.894

For further explanation see data analysis section. Means in bold are significantly different (ANOVA with Bonferroni correction; $p < 0.05$).

where flow has been reduced because of downstream impoundment or over-deepening. Habitat diversity increases the number of niches available to aquatic organisms in freshwaters (French & Chambers, 1996; Vinson & Hawkins, 1998). The loss of habitat diversity is expected to lead to a loss in macrophyte diversity.

Metrics were successfully linked to hydromorphological factors associated with site degradation. The attribute group liverworts/mosses/lichens was correlated with water depth and substrate characteristics for Discrim6 and Discrim7 sites. This result accords with studies, over a wide geographic area, that show the diversity of bryophytes in rivers is associated with depth and substrate (Suren & Duncan, 1999; Scarlett & O'Hare, 2006).

The negative correlation of liverworts/mosses/lichens with homogeneity of water depth, deep water (the dominant depth) and fine particle sub-

Table 5. Significant Spearman rank correlation coefficients between hydro-morphological variables that separates ecological quality classes (EQ 1–3, 4 and 5) calculated from RHS and macrophyte attribute groups/structural metrics within Discriminant group 6 and 7. $N = 38–39$

RHS calculated variable	Attribute group/metric	Correlation coefficient	*p*
Discrim6			
Hom_depth	Liverworts/mosses/lichens	−0.452	0.0039
	Evenness	0.397	0.0135
Hom_sub	Species richness	−0.324	0.0439
Dom_sub	Liverworts/mosses/lichens	−0.465	0.0029
	Filamentous algae	−0.363	0.0233
Dom_depth	Liverworts/mosses/lichens	−0.519	0.0007
Discrim7			
Score_depth	Submerged broad-leaved	0.407	0.0353
	Liverworts/mosses/lichens	0.383	0.0486
	Emergent reeds/sedges/rushes	0.495	0.0086
	Amphibious species	0.400	0.0384
Dom_depth	Emergent reeds/sedges/rushes	0.358	0.0483

No significant correlation coefficients were found in Discrim4.

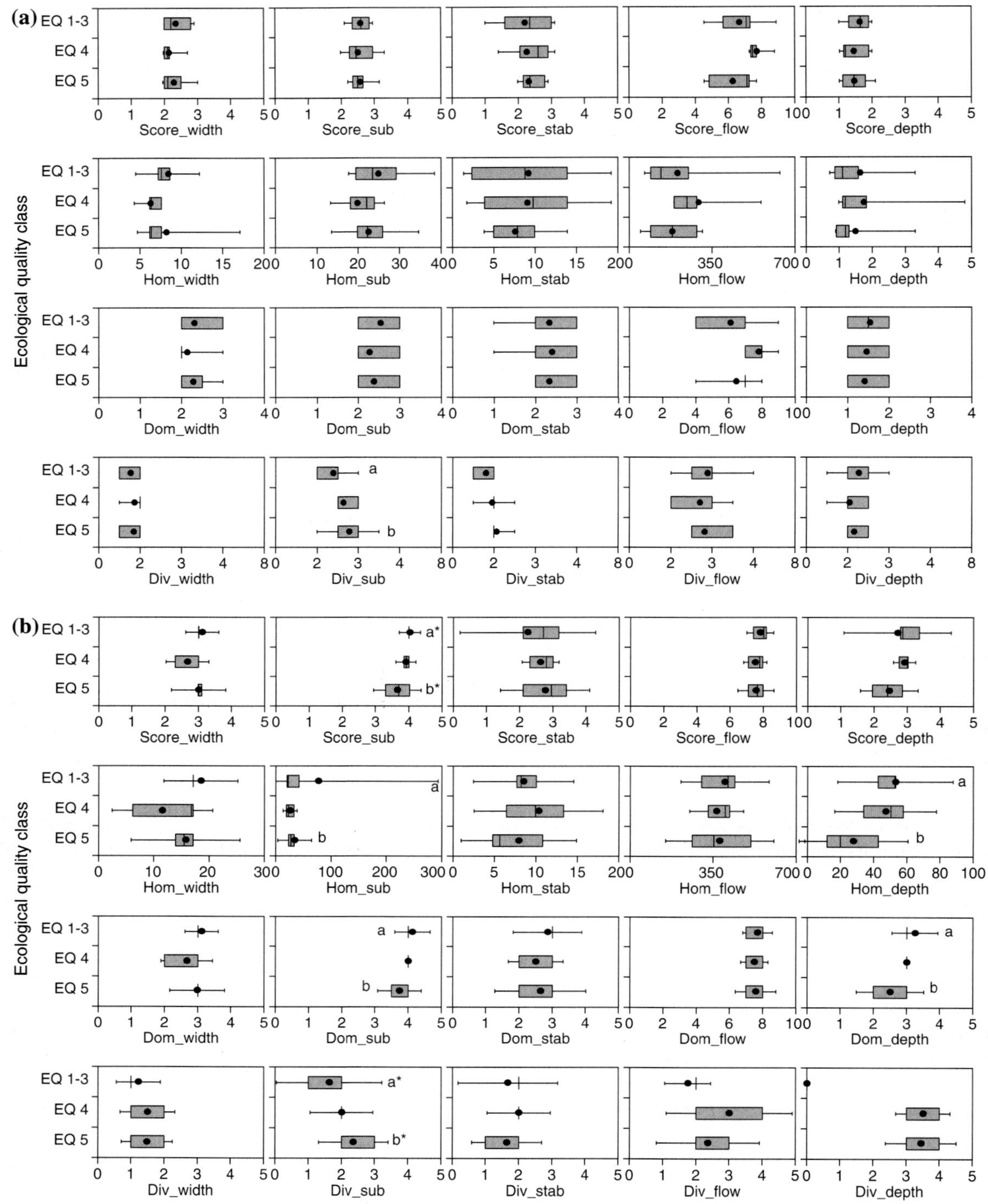

Figure 3. Box-whisker plots of hydro-morphological variables calculated from RHS/MTR for Discrim4 (plot a), Discrim6 (plot b) and Discrim7 (plot c) respectively. Letters signify differences between mean values (ANOVA with Bonferroni correction, $p < 0.05$). Letters with *indicate that $p < 0.10$. The box represents 10%, 25%, 75% and 90% and the symbol the mean value. Error bars represent the 5% and 95% percentiles. Abbreviations of environmental variables are given in Table 2. The homogeneity score was divided by 1000 in all cases to make the graphic presentation easy to read. The legend to Fig. 1 provides a key to the codes used in the diagram.

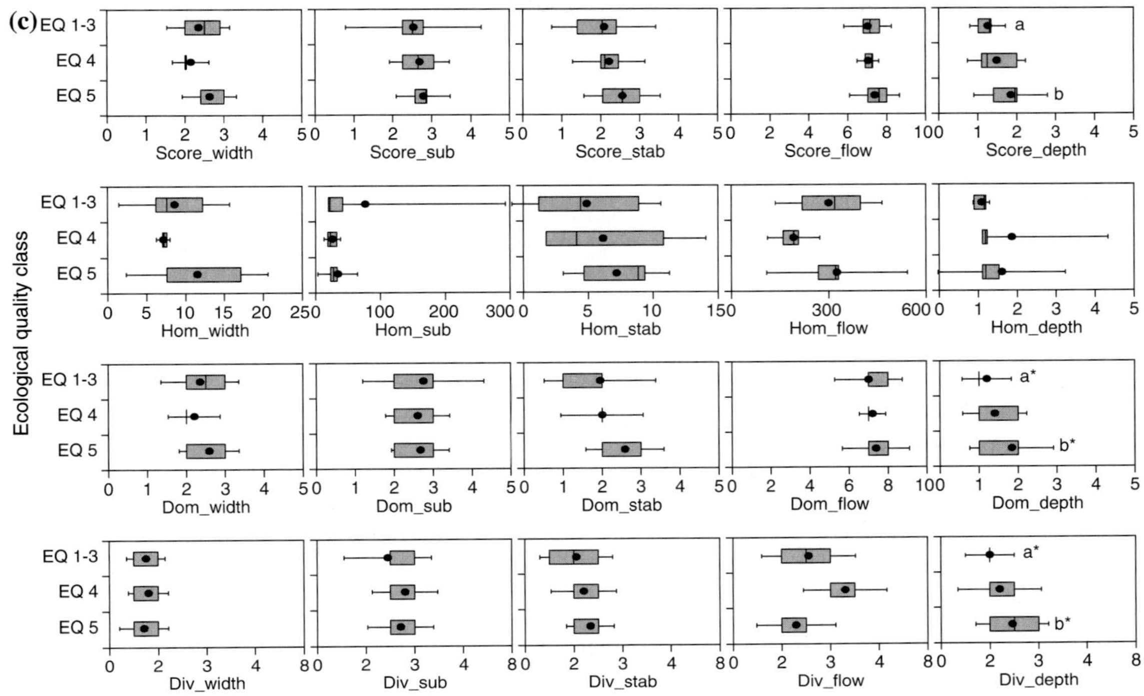

Figure 3. (Continued)

strate (the dominant substrate) implies increasing abundance of this group is indicative of sites which are probably not over-deepened. Equally, the responses of species richness and evenness to depth and substrate homogeneity also suggest they have the potential to be used as metrics.

The ordination analysis indicates that a number of species are tolerant to habitat degradation, e.g. *Sparganium emersum, Potomogeton crispus*, and *Elodea canadensis*. These species are also tolerant to other types of impacts, such as organic pollution and weed cutting (Battrup-Pedersen et al., 2003; Dawson et al., 1999b; Scneider & Melzer, 2003). Their value as indicators of degradation is therefore unspecific and should be augmented by combining them with other measures like evenness in species distribution.

Most of the species associated with physical variables that distinguish between ecological quality classes are present in both impacted and un-impacted stream sites (see Baattrup-Pedersen et al., 2006). However, these species may exhibit different abundances and spatial distributions in impacted as compared to un-impacted stream sites. To analyse this question properly would require the relationship to be tested on a larger dataset.

Other future work could potentially include using a revised sampling strategy. Previous investigations demonstrated that the spatial distribution of macrophytes in lowland stream reaches changes in response to physical degradation or impact (Baattrup-Pedersen et al., 2002; Wright et al., 2003). The STAR sampling methodology (MTR & RHS) does not allow distribution changes within reaches to be examined. This issue should be investigated in more detail, e.g. by applying the 'rectangle method' described by (Wright et al., 1981).

In conclusion, the presence of some species like *Fontinalis antipyretica* and metrics such as the presence of liverworts/mosses/lichens may indicate that a site is un-impacted by hydro-morphological degradation. Equally other taxa and metrics have been shown to be tolerant to degradation. Therefore, there is the basis for the evolution of a combined expression using both tolerant and sensitive species which distinguish degraded from un-impacted sites.

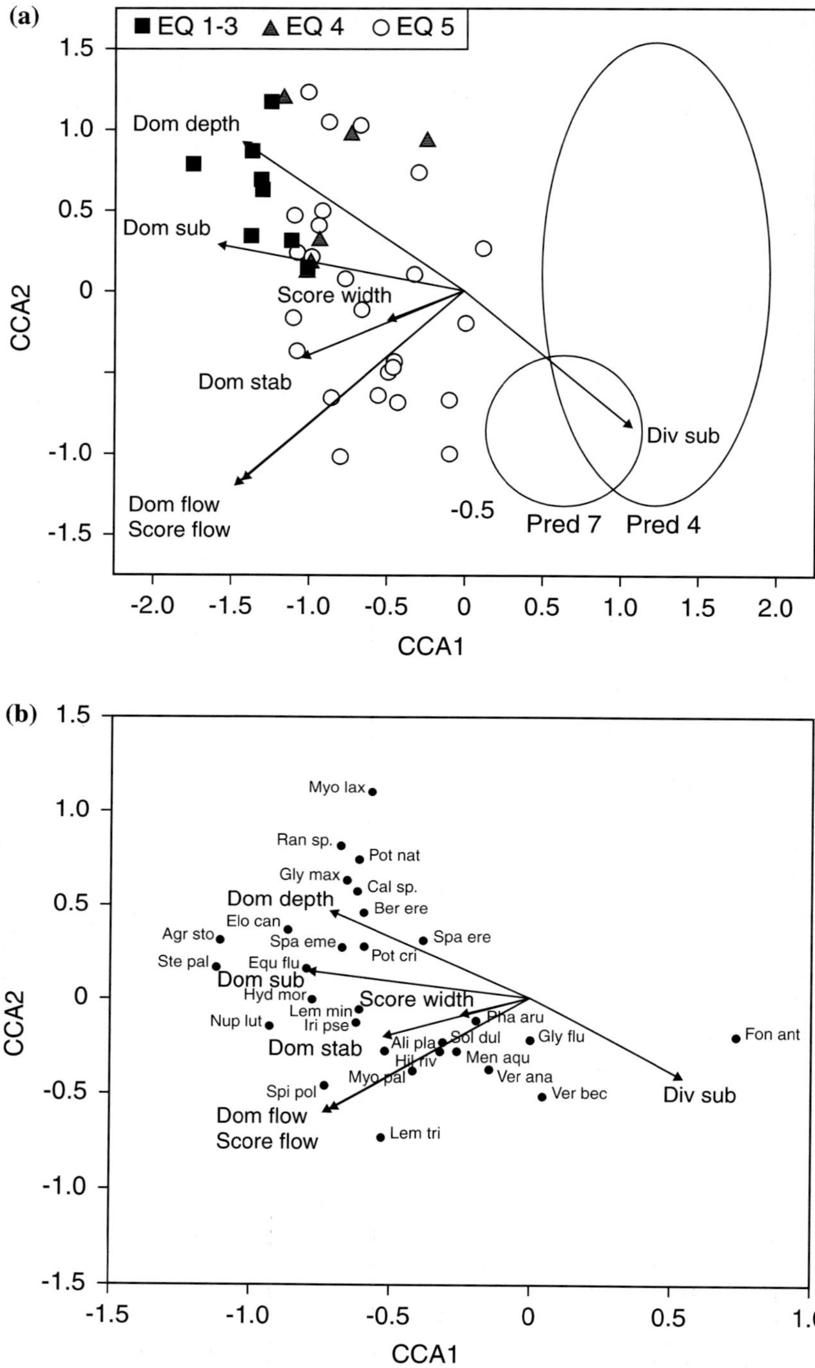

Figure 4. Canonical Correspondence Analysis (CCA) ordination of 107 stream sites distributed throughout Europe. (a) Sample scores of Discrim6 sites are shown with symbols, whereas mean values and one standard deviation of sample plot scores for Discrim4 and Discrim7 sites are shown as spheres. Different ecological quality classes are superimposed on the figure for Discrim6. (b) Species scores of species present in at least 4 Discrim6 stream sites. Only significant vectors are included on the figure forward selected by CANOCO version 4.5. The legend to Fig. 1 provides a key to the codes used in the diagram.

Acknowledgements

We thank Søren E. Larsen for his support in describing the homogeneity of the hydromorpholgical stream environment. We thank the European Union for financial support (project no. EVK1-CT 2001-00089 and for ABP also project no. SSPI-CT-2003-502158).

References

Baattrup-Pedersen, A., K. Szoszkiewicz, R. Nijboer, M. O'Hare & T. Ferreira, 2006. Macrophyte communities in unimpacted European streams: variability in assemblage patterns, abundance and diversity. Hydrobiologia 566: 179–196.

Baattrup-Pedersen, A., S. E. Larsen & T. Riis, 2003. Composition and richness of macrophyte communities in small Danish streams – influence of environmental factors and weed cutting. Hydrobiologia 495: 171–179.

Baattrup-Pedersen, A., S. E. Larsen & T. Riis, 2002. Long-term effects of stream management on plant communities in two Danish lowland streams. Hydrobiologia 481: 33–45.

Baattrup-Pedersen, A. & T. Riis, 1999. Macrophyte diversity and composition in relation to substratum characteristics in regulated and unregulated Danish streams. Freshwater Biology 42: 375–385.

Brookes, A., 1988. Channelized rivers: perspectives for environmental management. John Wiley, Chichester.

Chambers, P. A., E. E. Prepas, H. R. Hamilton & M. L. Bothwell, 1991. Current Velocity and Its Effect on Aquatic Macrophytes in Flowing Waters. Ecological Applications 1: 249–257.

Dawson, F. H., P. J. Raven & M. J. Gravelle, 1999a. Distribution of the morphological groups of aquatic plants for rivers in the U.K. Hydrobiologia 415: 123–130.

Dawson, F. H., J. R. Newman, M. J. Gravelle, K. J. Rouen & P. Henville, 1999b. Assessment of the Trophic Status of Rivers using Macrophytes: Evaluation of the Mean Trophic Rank. R&D Technical Report E39, Environment Agency of England & Wales, Bristol.

Environment Agency, 2003. River Habitat Survey in Britain and Ireland. Field Survey Guidance Manual. Environment Agency.

European Commission, 2000. Directive of the European Parliament and of the Council 2000/60/EC establishing a framework for Community action in the field of water policy. European Commission PE-CONS 3639/1/00 REV 1, Luxembourg.

Fox, A. M., 1992. Macrophytes. In Calow, P. & G. E. Petts (eds), The Rivers Handbook Hydrological and Ecological Principles. Blackwell Scientific Publications, Oxford: 216–233.

French, T. D. & P. A. Chambers, 1996. Habitat partitioning in riverine macrophyte communities. Freshwater Biology 36: 509–520.

Furse, M., D. Hering, O. Moog, P. Verdonschot, R. K. Johnson, K. Brabec, K. Gritzalis, A. Buffagni, P. Pinto, N. Friberg, J. Murray-Bligh, J. Kokes, R. Alber, P. Usseglio-Polatera, P. Haase, R. Sweeting, B. Bis, K. Szoszkiewicz, H. Soszka, G. Springe, F. Sporka & I. Krno, 2006. The STAR project: context, objectives and approaches. Hydrobiologia 566: 3–29.

Haslam, S. M., 1978. River Plants: The macrophytic vegetation of watercourses. Cambridge University Press, Cambridge.

Irvine, K., 2004. Classifying ecological status under the European Water Framework Directive: the need for monitoring to account for natural variability. Aquatic conservation: Marine and Freshwater Ecosystems 14: 107–112.

Jongman, R. H. G., C. J. F. Ter Braak & O. F. R. Van Tongeren, 1987. Data Analysis in Community and Landscape Ecology. Cambridge University Press, Cambridge.

Logan, P. & M. Furse, 2002. Preparing for the European Water Framework Directive - making the links between habitat and aquatic biota. Aquatic Conservation: Marine and Freshwater Ecosystems 12: 425–437.

Minitab, 2004. Minitab 14.12.0 Release 14 Statistical software. Minitab Inc.

Petts, G. E., 1984a. The quality of reservoir releases. In Petts, G.E. (ed), Impounded Rivers: Perspectives for Ecological Management. John Wiley & Sons, Chichester: 54–83.

Petts, G. E., 1984b. Vegetation reaction and structure. In Petts, G.E. (ed), Impounded Rivers: Perspectives for Ecological Management. John Wiley & Sons, Chichester: 150–173.

Pollard, P. & M. Huxham, 1998. The European Water Framework Directive: a new era in the management of aquatic ecosystem health?. Aquatic Conservation: Marine and Freshwater Research 8: 773–792.

Raven, P. J., P. Fox, M. Everard, H. T. H. Holmes & F. H. Dawson, 1997. River Habitat Survey: a new system to classify rivers according to their habitat quality. In Boon, P. J. (ed), Freshwater Quality: Defining the Indefinable. Scottish Natural Heritage, HMSO, London: 215–234.

Scarlett P. & M. T. O'Hare, 2006. Community structure of instream bryophytes in English and Welsh rivers. Hydrobiologia 553: 143–152.

Schneider, S. & A. Melzer, 2003. The Trophic Index of Macrophytes (TIM) – A new tool for indicating the trophic status of running waters. International Revue Hydrobiologie 88: 49–67.

Sculthorpe, C. D., 1967. The salient features of aquatic vascular plants. In Sculthorpe, C. D. (ed), The Biology of Aquatic Vascular Plants. Spottiswode, Ballantyne, London.

Suren, A. M. & M. J. Duncan, 1999. Rolling stones and mosses: effect of substrate stability on bryophyte communities in streams. Journal of the North American Benthological Society 18: 457–467.

Szoszkiewicz, K., T. Ferreira, T. Korte, A. Baattrup-Pedersen, J. Davy-Bowker & M. O'Hare, 2006. European river plant communities: the importance of organic pollution and the usefulness of existing macrophyte metrics. Hydrobiologia 566: 211–234.

ter Braak, C. J. F. & P. Smilauer, 1998. Canoco for Windows. Microcomputer Power, Ithaca.

Vinson, M. R. & C. P. Hawkins, 1998. Biodiversity of stream insects: Variation at local, basin, and regional scales. Annual Review of Entomology 43: 271–293.

Westlake, D. F., 1975. Macrophytes. In Whitton, B. A. (ed), River Ecology. Blackwell Scientific Publications, Oxford.

Wright, J. F., R. T. Clarke, R. J. M. Gunn, J. M. Winder, N. T. Kneebone & J. Davy-Bowker 2003. Response of the flora and macroinvertebrate fauna of a chalk stream site to changes in management. Freshwater Biology 48: 894–911.

Wright, J. F., P. D. Hiley, S. F. Ham & A. D. Berrie, 1981. Comparision of three mapping procedures developed for river macrophytes. Freshwater Biology 11: 369–379.

Hydrobiologia (2006) 566:211–234

M.T. Furse, D. Hering, K. Brabec, A. Buffagni, L. Sandin & P.F.M. Verdonschot (eds), The Ecological Status of European Rivers: Evaluation and Intercalibration of Assessment Methods
DOI 10.1007/s10750-006-0094-3

European river plant communities: the importance of organic pollution and the usefulness of existing macrophyte metrics

Krzysztof Szoszkiewicz[1,*], Teresa Ferreira[2], Thomas Korte[3], Annette Baattrup-Pedersen[4], John Davy-Bowker[5] & Mattie O'Hare[5]
[1]*Department of Ecology and Environmental Protection, Agricultural University of August Cieszkowski, ul. Piątkowska 94C, 61-691 Poznan, Poland*
[2]*Forestry Department, Agronomy Institute, Technical University of Lisbon, Tapada da Ajuda, 1349-017 Lisboa, Portugal*
[3]*Department of Hydrobiology, University of Duisburg-Essen, Campus Essen, Universitätstr. 2, 45117 Essen, Germany*
[4]*Department of Freshwater Biology, National Environmental Research Institute, Vejlsøvej 25, P.O. Box, 314, DK-8600 Silkeborg, Denmark*
[5]*Centre for Ecology and Hydrology, Winfrith Technology Centre, DT2 8ZD Dorchester, Dorset, UK*
*(*Author for correspondence: E-mail: kszoszk@au.poznan.pl)*

Key words: Water Framework Directive, macrophytes, river, trophy, MTR, IBMR, biological indicators, STAR

Abstract

The macrophyte surveys undertaken as part of the EU-funded STAR project are a unique resource allowing aquatic plant communities to be studied at a Pan-European scale (211 stream sites with macrophytes in 14 countries). Using this dataset, we examined the influence of organic pollution in relation to other environmental correlates of river plant community variation across Europe. We examined the relationships between several existing macrophyte metrics and nutrient enrichment, and we also explored the possibility of developing a pan-European macrophyte-based assessment system. We showed that trophic (nutrient) status is an important driver of aquatic plant communities in European rivers. We found that while most existing macrophyte metrics are useful, none can be applied at a pan-European scale in their current form. Our attempt to redesign the Mean Trophic Rank (MTR) index by the addition of further species, and the re-scoring of existing species, resulted in a considerable improvement in the relationship between MTR scores and nutrient variables. We conclude that an enlarged core group of macrophyte species can form part of an improved pan-European macrophyte-based bioassessment system, although regional modifications may be required to adequately describe the nutrient status of certain stream types.

Introduction

The development of macrophytes is dependent upon a variety of abiotic and biotic factors the most important of which are nutrient concentrations, flow velocity, hydrological conditions, substrate, pH, carbonate hardness, shading and anthropogenic impacts. In running waters, vegetation is strongly influenced by current velocity (Westlake, 1975; Dawson, 1988; Fennesy et al., 1994). Hydrological conditions and substrate can be major determinand for many taxa (Westlake, 1975; Haslam, 1978, 1987; Baattrup-Pedersen & Riis, 1999). Hydraulic patterns affect macrophyte growth and morphological phenotypes directly, and also indirectly (e.g., at low flow rates) through sedimentation of particulate material on the riverbed and over the photosynthetic apparatus (Remy, 1993). Different light regimes, due to water depth and artificial or natural turbidity, also strongly influence the appearance of macrophytes (Westlake, 1975; Dawson & Kern-Hansen, 1979; Remy, 1993). The pH of water, as well as hardness and alkalinity, affect the distribution of aquatic species, especially bryophytes (Butcher, 1933; Ellenberg et al., 1992; Tremp & Kohler, 1995).

The effect of anthropogenic nutrient enrichment (and other pollutants) on river plants has been demonstrated in many studies (Westlake, 1975; Kohler & Schiele, 1985; Allan, 1995; Robach et al., 1996; Demars & Harper, 1998; Thiebaut & Muller, 1998; Holmes et al., 1999; Schneider et al., 2000). Other than abiotic drivers, macrophyte assemblages are also shaped by inter- and intra-specific competition, 'diaspore' reservoir, herbivory (van de Weyer, 1997; Pott & Remy, 2000; Sipos et al., 2000) and natural dispersal within the water body, leading to small-scale patchiness at plant stand scale on the streambed and temporal fluctuation.

There are several advantages in using macrophytes for biological monitoring. Macrophytes are non-mobile and therefore present responses to local environmental changes. They can also integrate environmental changes over periods of a few years, and the cumulative effects of successive disturbances (Tremp & Kohler, 1995). The perennial species are particularly good indicators of persistent and long-term habitat change. Macrophytes are also relatively widespread and considered to be easy to identify.

A number of factors however make the use of macrophytes problematic in running water' ecological assessment. Natural differences in macrophyte communities from different river systems make comparisons between different types of river difficult (Werle, 1982; Wiegleb, 1988). The distribution of macrophytes is also dynamic so that macrophyte stands can enlarge or contract in what appear to be constant conditions (Veit et al., 2003). Many species also grow outside their optima and established macrophytes can tolerate various degrees of stress and persist outside their ecological optima as long as interspecific competition remains weak (Haslam, 1978). The complex small-scale physicochemical heterogeneity of the river habitats can also allow macrophytes with differing ecological requirements to form patchy distributions within the reach. Macrophytes can also be slow to re-establish after an improvement in general environmental conditions (this is dependent upon the upstream diaspore reservoir). Similarly, if environmental conditions deteriorate (e.g., eutrophication) it can also take time for nutrient-tolerant species to appear (Kohler, 1982; Pott & Remy, 2000).

While in the past the use of macrophytes for river bioassessment has been scarce, recently several systems have been developed and some of these have become integrated into national monitoring programmes, e.g., in Denmark (Svendsen & Rebsdorf, 1994), Britain (Holmes et al., 1999), Germany (Schneider et al., 2000) and France (Haury et al., 1996, 2002). These macrophyte based methods mainly focused on detecting eutrophication (Haury et al., 1996, 2002; Holmes et al., 1999; Schneider et al., 2000) and acidification (Tremp & Kohler, 1995), although macrophyte methods also exist to assess river degradation in a more holistic or integrative way (Ferreira et al., 2002; Passauer et al., 2002; Van De Weyer, 2003; Schaumburg et al., 2004) as required by the Water Framework Directive (European Commission, 2000).

The macrophyte surveys undertaken as part of the STAR project present a unique resource enabling us to challenge existing knowledge on the ecology of river plants at a pan-European scale. The standardised botanical surveys were coordinated to provide temporal synchrony across a wide geographical area including many stream types and supplemented by an extensive amount of environmental information. This dataset allowed us the undertake analyses over a greater spatial extent than has been hitherto possible. Our analysis had three main aims. Firstly, we attempted to examine correlations between macrophyte communities and trophic (nutrient) gradients across Europe. Secondly we examined the relationships between several existing macrophyte metrics and chemical measures of nutrient status. Thirdly, we explored the possibility of developing a pan-European Mean Trophic Rank (MTR) index by the addition of further scoring species and by the rescoring of existing species.

Methods

Stream types and sites

The study was based on 211 stream sites where both macrophyte surveys were undertaken and environmental data was recorded (Supplementary materials).[1] These were grouped into 17 stream types representing four geographical groups: mountain streams (72 sites, seven stream types), lowland streams (101 sites, seven stream types) and southern European streams (31 sites, three stream types). While the full STAR database consisted of further survey sites, these were dropped from the analysis due to the absence of a macrophyte surveys or incomplete environmental data.

Macrophyte and environmental database

Macrophyte surveys were undertaken in summer 2003 and 2004. Field survey methods were standardised for the STAR project integrating several national methodologies (Dawson, 2002) and were closely related to the Mean Trophic Rank (MTR) methodology (Dawson et al., 1999; Holmes et al., 1999). In addition to all submerged, free-floating, amphibious and emergent monocotyledonous and dicotyledonous plant species, the macrophyte assessment also included filamentous algae, liverworts, mosses and pteridophytes. The assessment included macrophytes attached or rooted in parts of the river bank that are likely to be submerged for more than 85% of the year. The presence of each species within the standard MTR survey river length (100 m) was recorded together with their percentage cover using the MTR nine-point Species Cover Value (SCV) scale (Holmes et al., 1999).

The STAR environmental dataset is very extensive having been collated from River Habitat Survey (RHS), the STAR site protocol, the MTR habitat description and hydrochemical analyses performed at every STAR site (Supplementary materials).

[1] Electronic supplementary material is available for this article at <http://dx.doi.org/10.1007/s10750-006-0094-3> and accessible for authorised users.

Relation of community structure to trophic and other characteristics

Canonical ordination techniques (Canoco 4.5) were used to investigate the role of chemical variables related to trophic (nutrient) gradients in comparison with other chemical and physical variables (e.g., hydromorphological) as correlates of macrophyte community composition. Mountain stream datasets (30 sites belonging to stream types C04 (6 sites), D04 (6), D06 (6) and F08 (12)) and lowland stream datasets (81 sites belonging to stream types D03 (7 sites), K02 (11), O02 (23), S05 (9), S06 (11), U15 (8), U23 (12) (Table 1), were analysed separately. Only taxa that occurred in more than 5% of surveys, and only those sites where macrophyte cover of a species exceeded 1% were included. Two missing environmental values (forest and crop in stream type U23) were replaced by mean values. We reduced the high number of available variables by removing those which were correlated. This was done by grouping variables of the same type (e.g., physicochemical variables or catchment area variables) and then testing the degree to which these were correlated within the group. Highly correlated variables (Spearman rank correlation, $p<0.05$, $R>0.49$) were removed, leaving 37 lowland and 28 mountain stream variables. The variables retained for use in our analyses are shown in Tables 2 and 3.

Detrended correspondence analysis (DCA) was used to examine the rate of turnover of macrophytes taxa across the sites on the first axis of variation. For both stream types this was >4 and CCA was therefore considered appropriate for use with these datasets (ter Braak & Prentice, 1988). DCA was also used to interpret the results of CCA. Prior to both types of ordination analysis, the data were transformed. Species cover values were log-transformed ($x' = \log(x+1)$), while the environmental variables were transformed as shown in Supplementary materials. The environmental variables were also automatically centred and standardized by the Canoco analysis program (mean = 0, variance = 1). Monte Carlo permutation tests (number permutations 499) and forward selection were used within CCA to detect significant ($p=0.05$ probability threshold level) and independent environmental variables.

Table 1. STAR macrophyte database

Stream types	STAR sites	Sites with macrophytes present: Total	Sites with macrophytes present: Reference sites	Dominant perturbation of disturbed sites: Organic	Dominant perturbation of disturbed sites: Hydro	Dominant perturbation of disturbed sites: General	Dominant perturbation of disturbed sites: Metals	Countries	Altitude
Lowlands									
K02 Medium-sized lowland streams	14	12	5		7			Denmark	<200
S05 Medium-sized lowland streams	16	14	2	7			5	Sweden	<200
O02 Medium-sized lowland streams	25	25	7	18				Poland	<200
S06 Medium sized lowland streams on calcareous soils	11	13	3	10				Sweden	<200
D03 Medium-sized lowland streams	10	12	4		8			Germany	<200
U15 Small-sized lowland calcareous streams (RIVPACS group 32)	13	13	3	9	1			UK	<200
U23 Medium-sized lowland calcareous streams (RIVPACS group 20)	12	12	3	9				UK	<200
Mountains									
C04 Small sized shallow mountain streams	14	13	3	10				Czech R.	200–800
C05 Small-sized streams in the Central sub-alpine Mountains	10	10	3		7			Czech R.	200–800
F08 Small sized shallow headwater streams in Eastern France	12	12	3	9				France	200–800
D04 Small-sized, shallow mountain streams	10	11	2		9			Germany	200–800
D06 Small-sized Buntsandstein-streams	6	6	2			4		Germany	200–800
A05 Small-sized, shallow mountain streams	11	8	3		5			Austria	200–800
V01 Small-sizes siliceous mountains streams in the West Carpathians	25	9	0	9				Slovakia	200–800
South European									
H04 Small-sized calcareous mountain streams in Western, Central and Southern Greece	10	10	3	7				Greece	200–800
P04 Medium-sized streams in lower mountainous areas of Southern Portugal	10	10	3	7				Portugal	<200
I06 Small-sized calcareous streams in the Central Apennines	11	11	3			8		Italy	200–800
Alpine									
A06 Small-sized crystalline streams of the ridges of the Central Alps	7	7	3		4			Austria	>800

Table 2. Observed environmental variables and conditional effects obtained from the summary for lowland rivers

Variable	LambdaA	*p*	Variable	LambdaA	*p*
Dis Sour	0.26	0.002	no_bedf	0.06	0.366
Nitrate	0.24	0.002	SI	0.05	0.348
Ortho_P	0.23	0.002	GP	0.07	0.080
Slope	0.15	0.004	UP	0.06	0.348
NP	0.13	0.002	catchm	0.04	0.522
Meanwidt	0.12	0.006	BFaceSC	0.05	0.450
SA	0.12	0.002	LUsIG	0.05	0.514
pH	0.09	0.006	LUsTH	0.05	0.612
Crop	0.09	0.016	Sha	0.04	0.630
Forest	0.10	0.012	SM	0.05	0.592
LusSU	0.09	0.034	HQA scor	0.04	0.662
Oxygen	0.08	0.044	n_cl_WD	0.06	0.420
Stable	0.07	0.096	LUsSC	0.03	0.790
Unstable	0.07	0.128	HMS scor	0.04	0.742
CL	0.07	0.142	Firm	0.04	0.904
LUsWL/MH	0.07	0.184	sum_cD	0.03	0.962
n_cl_flo	0.06	0.142	RS	0.06	0.250
LusRP	0.06	0.260	UW	0.02	0.978
n_cl-sub	0.06	0.264			

Lambda A, additional variance explained; *p*, significance of variable. See Supplementary materials for full names of abbreviated variables.

Trophy indication by macrophytes

Relationships between macrophyte metrics and chemical parameters were analysed using Spearman rank correlation (StatSoft, Inc. 2004) (Tables 4–6). This analysis used a subset of 147 sites (80 lowland sites, 44 mountain sites and 23 southern European sites) that formed an organic

Table 3. Observed environmental variables and conditional effects obtained from the summary for mountain rivers

Variable	LambdaA	*p*	Variable	LambdaA	*p*
Slope	0.55	0.002	n_cl_WD	0.11	0.418
ortho_P	0.29	0.010	BW	0.13	0.184
pH	0.22	0.034	BOD5	0.13	0.158
LusBL	0.20	0.044	n_cl_flo	0.10	0.406
RI	0.20	0.064	GP	0.12	0.190
HQA scor	0.15	0.176	BFaceSC	0.10	0.362
no_bankf	0.14	0.174	Sha	0.11	0.234
BO/CO	0.12	0.320	SM	0.09	0.384
BModRS	0.13	0.254	Nitrate	0.10	0.254
LusSC	0.14	0.206	Urban	0.08	0.460
LusTH	0.11	0.370	SA	0.06	0.522
HMS scor	0.13	0.240	Forest	0.13	0.248
n_cl-sub	0.11	0.320	11 altit	0.07	0.440
ChFevMB	0.09	0.394	BE	0.06	0.556

Lambda A, additional variance explained; *p*, significance of variable. See Supplementary materials for full names of abbreviated variables.

pollution gradient. Twenty-four macrophyte metrics were used: MTR – Mean Trophic Rank (Holmes et al., 1999), IBMR – Macrophyte Biological Index for Rivers (Haury et al., 2002), nitrogen affiliation based on the plant scores delivered by Ellenberg (Ellenberg et al., 1992) and TIM – Trophic Index of Macrophytes (Schneider et al., 2000), Hemoroby index (Jalas, 1955), total plant cover, coverage and number of species representing different ecological groups (bryophytes, submerged plants, anchored with floating leaves or heterophyllus, floating free, amphibious and ecotonal plants) and several diversity indices; Shannon diversity (Shannon & Weaver, 1949), Simpson diversity (Simpson, 1949), domination (McNaughton, 1967), evenness (Pielou, 1966), and number of species, genera and families.

The performance of the MTR index was investigated in more detail. The number of indicator species and proportion of indicator species to the total number of species was examined using both the 129 organically polluted sites and a subset of un-impacted sites. Separate analyses were also performed on subsets of sites representing different geographical regions, river types and levels of organic pollution.

The presence of indicator species from a group of common macrophyte metrics (MTR, IBMR, and nitrogen affiliation based on both Ellenberg and TIM) was examined across several geographical regions. Species scores assigned by these metrics were standardised to a 10-class scale (1 as the worst and 10 as the highest quality) and the rescaled species scores were then compared to reveal disparities between the metrics as standard deviation between four metrics (SD1) and separately standard deviation between two major trophy metrics MTR and IBMR (SD2).

Non-scoring recorded MTR species were tested as potential new scoring species. These analyses were undertaken separately for lowland and mountain rivers datasets (the sample of south European was too small for analysis). CCA was used to examine relationships between these species and relevant nutrient parameters. New STR scores were proposed and the influence of these additional taxa on the correlation between total MTR score and nutrient parameters was tested. Macrophytes found to improve the relationship between MTR score and nutrient parameters are highlighted.

Results

Eutrophication factors in comparison to other environmental factors

Lowland streams

Within the lowland stream type group there is no dominant environmental variable explaining species variability in the first canonical axis of CCA, but rather several independent parameters, as shown by the relatively high eigenvalues of the successive axes (2nd axis: 0.359, 3rd axis: 0.297, 4th axis: 0.254) and species-environmental correlations (2nd axis: 0.922, 3rd axis: 0.923, 4th axis: 0.876). The environmental variables distance from source, nitrate, ortho-phosphate, slope, proportion of flow type 'no perceptible flow' (NP), mean width, proportion of riverbed substrata type sand (SA), pH, proportion of crop in catchment, proportion of forest within catchment, land use suburban/urban development (LusSU) within 50 m of bank top and oxygen were significantly correlated with species composition in the surveyed river reaches (Table 2).

The biplot of the species and the forward selected 12 environmental variables is shown in Figure 1 for lowland streams. The first axis explains 7.3% of the species variation. Nitrate ($R=-0.57$), forest ($R=0.54$) and ortho-phosphate ($R=-0.53$) are significantly correlated with the negative part of this axis. The second axis (explaining 5% of the species variation) is most strongly correlated with distance from source. The remaining eight variables each have correlation coefficients of less than $R<0.5$ with the first two CCA axes. The sum of canonical eigenvalues of all 12 environmental variables is 1.708. Compared to the sum of canonical eigenvalues for all 37 environmental variables (3.119) it is apparent that the 12 environmental variables chosen by forward selection are the strongest correlates of community variation (together describing 55% of the explained variation in the macrophyte communities) while the 25 non-selected variables were weaker

Table 4. Significant Spearman rank correlation coefficients between environmental characteristics and selected metrics (see methods) using lowland sites impacted by organic pollution

	Cond	BOD5	Amm	Nitrite	Nitrate	Ortho_P	Total_P
Cover %							
Bryophytes			−0.39 ***				
Submerged plants		0.31 *					
Floating leaves/heterophyllus							
Floating free			0.42 ***	−0.34 *	−0.36 **		
Amphibious				−0.53 ***	−0.29 *		
Ecotonal plants			0.32 **		−0.25 *		
Total cover				−0.46 **	−0.36 **		
Number of species							
Bryophytes			−0.34 **				
Submerged plants					−0.24 *		
Floating leaves/heterophyllus							
Floating free			0.50 ***	−0.29 *	−0.36 **		0.24*
Amphibious			0.34 **	−0.58 ***	−0.38 ***		
Ecotonal			0.46 ***		−0.34 **		
Diversity metrics							
Species number			0.41 ***	−0.40 **	−0.38 ***		
Genus number			0.40 ***	−0.39 **	−0.38 **		
Family number			0.36 **	−0.37 *	−0.39 ***		
Shannon diversity			0.28 *	−0.39 **	−0.38 ***		
Simpson diversity				0.40 **	0.35 **		
Domination							
Evenness				−0.36 *	−0.37 **		
Other metrics							
Hemoroby index	0.31 **	−0.36 *	0.24 *		0.36 **	0.31 *	0.32 **
IBMR	−0.54 ***	−0.33 *	−0.55 ***	−0.34 *		−0.68 ***	−0.70 ***

Continued on p. 218

Table 4. (Continued)

	Cond	BOD5	Amm	Nitrite	Nitrate	Ortho_P	Total_P
Ellenberg index	0.43 ***		0.36 **	0.31 *	0.36 **	0.43 ***	0.45 ***
MTR	−0.49 ***		−0.45 ***	−0.39 *	−0.24 *	−0.63 ***	−0.66 ***
MTR (ind. sp. > 2)[2]	−0.44 ***		−0.43 ***	−0.36 *		−0.61 ***	−0.64 ***
MTR (ind. sp. > 3)[3]	−0.35 **		−0.36 **			−0.48 ***	−0.52 ***
Reconstructed MTR							
MTR	−0.62 ***	−0.46 **	−0.65 ***	−0.46 **	−0.30 *	−0.82 ***	−0.82 ***
MTR (ind. sp. > 2)[1]	−0.58 ***	−0.46 **	−0.64 ***	−0.45 **	−0.27 *	−0.81 ***	−0.81 ***
MTR (ind. sp. > 3)[2]	−0.56 ***	−0.42	−0.65 ***	−0.38 *		−0.81 ***	−0.81 ***

Probability levels used: $*p < 0.05$; $**p < 0.01$; $***p < 0.001$.
[1](ind. sp. > 2) - analysis based on sites with the number of indicator species larger than 2.
[2](ind. sp. > 3) - analysis based on sites with the number of indicator species larger than 3.

Table 5. Significant Spearman rank correlation coefficients between environmental characteristics and selected metrics (see methods) using mountain sites

	Cond	BOD5	Amm	Nitrite	Nitrate	Ortho_P	Total_P
Cover %							
Bryophytes	−0.50 ***	−0.39 *		−0.49		−0.62 ***	−0.61 ***
Submerged plants	0.71 ***	−0.33 *	−0.56 ***	−0.49 ***			
Floating leaves/heterophyllus	0.65 ***	−0.31 *	−0.76 ***	−0.76 *			
Floating free	0.59 ***		−0.54 ***	−0.55 **		0.42 **	
Amphibious	0.69 ***		−0.63 ***	−0.63 **			
Ecotonal plants	0.58 ***		−0.50 ***	−0.47 **	0.32 *		
Total cover		−0.38 *	−0.59 ***	−0.77 **			
Number of species							
Bryophytes		−0.44 **	−0.32 *	−0.31 **		−0.43 **	−0.37 *
Submerged plants	0.69 ***	−0.35 *	−0.60 ***	−0.54 ***			
Floating leaves/heterophyllus	0.64 ***		−0.76 ***	−0.77 ***			
Floating free	0.58 ***		−0.54 ***	−0.57 ***		0.39 *	
Amphibious	0.69 ***		−0.63 ***	−0.63 ***			
Ecotonal	0.62 ***		−0.52 ***	−0.50 ***			
Diversity metrics							
Species number	0.57 ***	−0.31 *	−0.50 ***	−0.46 ***	0.36 *		
Genus number	0.57 ***	−0.31 *	−0.50 ***	−0.46 ***	0.36 *		
Family number	0.57 ***	−0.31 *	−0.50 ***	−0.46 ***	0.36 *		
Shannon diversity		−0.40 **	−0.63 ***	−0.80 ***			
Simpson diversity		0.31 *	0.53 ***	0.73 *			
Domination	−0.33 *				−0.50 ***		
Evenness			−0.40 **	−0.54 *			
Other metrics							
Hemoroby index						−0.44 **	
IBMR	−0.82 ***					−0.52 **	−0.40 *
Ellenberg index							
MTR	−0.84 ***					−0.47 **	
MTR (ind. sp. > 2)[1]	−0.95 ***					−0.78 ***	−0.68 **
MTR (ind. sp. > 3)[2]	−0.92 ***	−0.71 *	−0.77 **	−0.77 **		−0.82 **	−0.77 **
Reconstructed MTR							
MTR	−0.87 ***					−0.56 ***	−0.41 *
MTR (ind. sp. > 2)[1]	−0.84 ***					−0.70 ***	−0.59 **
MTR (ind. sp. > 3)[2]	−0.92 ***	−0.74 **	−0.79 ***	−0.84 ***	0.69 **	−0.84 ***	−0.79 **

Probability levels used: *$p < 0.05$; **$p < 0.01$; ***$p < 0.001$.

[1](ind. sp. > 2) - analysis based on sites with the number of indicator species larger than 2.

[2](ind. sp. > 3) - analysis based on sites with the number of indicator species larger than 3.

Table 6. Significant Spearman rank correlation coefficients between environmental characteristics and selected metrics (see methods) using South European sites (H04 stream type – see Table 1) impacted by organic pollution

	Cond	BOD5	Amm	Nitrite	Nitrate	Ortho_P	Total_P
Cover %							
Bryophytes			0.68 **		−0.59 **		
Submerged plants							
Floating leaves/heterophyllus			−0.62 **	−0.65 **	0.57 **		
Floating free					0.45 *		
Amphibious			−0.50 *				
Ecotonal plants							
Total cover							
Number of species							
Bryophytes			0.62 **		−0.59 **		
Submerged plants					0.45 *		
Floating leaves/heterophyllus			−0.63 **	−0.63 **	0.59 **		
Floating free					0.44 *		
Amphibious			−0.44*				
Ecotonal							
Diversity metrics							
Species number							
Genus number							
Family number							
Shannon diversity							
Simpson diversity							
Domination							
Evenness							
Other metrics							
Hemoroby index			0.53 *	0.52 *	−0.58 **		
IBMR				−0.71 ***			
Ellenberg index							
MTR			−0.61 **	−0.85 ***	0.47 *		
MTR (ind. sp. > 2)[2]			−0.69 **	−0.87 ***	0.55 *		

Probability levels used: $*p<0.05$; $**p<0.01$; $***p<0.001$.
[1](ind. sp. > 2) - analysis based on sites with the number of indicator species larger than 2.
[2](ind. sp. > 3) - analysis based on sites with the number of indicator species larger than 3.

correlates (together describing only 45% of the explained variation) (Tables 2 and 3).

Mountain streams

Similarly, there is no dominant environmental variable within the observed mountain streams explaining the species variability in the first canonical axis of CCA, but rather several independent parameters, as shown by the relatively high eigenvalues of the successive axes (2nd axis: 0.63, 3rd axis: 0.58, 4th axis: 0.40) and species–environmental correlations (2nd axis: 0.97, 3rd axis: 1.00, 4th axis: 1.00). Four environmental variables were chosen by forward selection: slope, *ortho*-phosphate, pH and land use – broadleaf/mixed woodland (LUsBL) within 50 m of bank top (Table 3 and Fig. 2). Slope and *ortho*-phosphate were highly correlated with the first CCA axis ($R=0.83$ and −0.58, respectively). pH and LusBL had correlation coefficients of $R<0.5$ with the first two axes of the CCA. The sum of canonical eigenvalues for the four

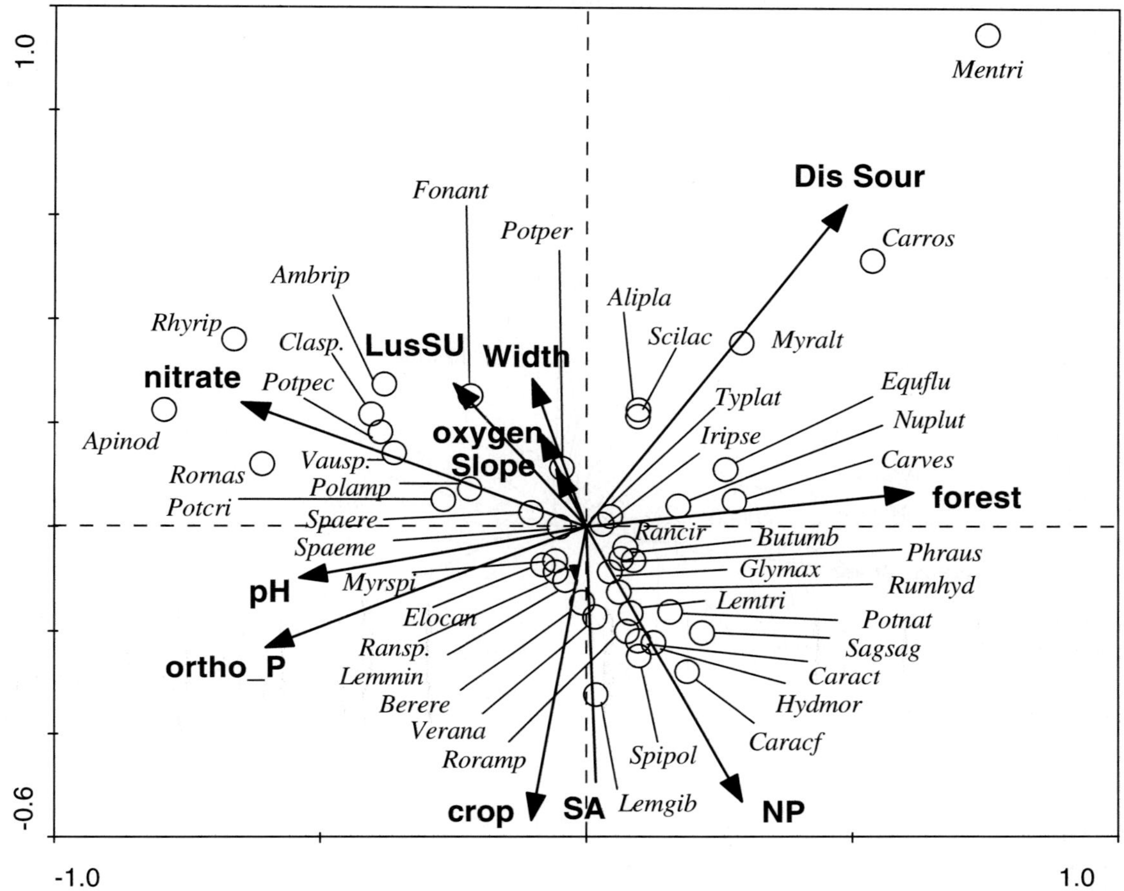

Figure 1. CCA-ordination diagram of macrophyte species and environmental variables in lowland streams. See Table 8 for full names of abbreviated species and Supplementary materials for full names of abbreviated variables.

significant variables was 1.26. Compared to the sum of the canonical eigenvalues with all 28 environmental variables (4.07) four independent and significant variables alone are not sufficient to predict the main variation in species composition.

Macrophyte indices and trophic status

Correlations between existing macrophyte metrics and chemical variables indicating eutrophication varied but were generally weak (Tables 4–6). *Lowland river analysis* (Table 4) showed that metrics expected to be related to trophic status (IBMR and MTR) were highly significant ($p < 0.001$) although the Spearman rank correlation coefficient (r) was quite low. The strongest correlations were between both forms of phosphorous and IBMR ($r = -0.70$ for total P and $r = -0.68$ for *ortho*-phosphates) and between both forms of phosphorous and MTR (−0.66 and −0.63, respectively). Development of different ecomorphological groups was related to the level of nitrogen and to an even greater extent nitrite and ammonia.

In the *mountain stream analysis* (Table 5), macrophyte metrics had relatively weak correlations with phosphorous (IBMR and orthophosphate being the most highly correlated). The correlation between MTR score improved when more indicative species were present. For example, correlations between soluble phosphorous and MTR were higher ($r = -0.82$) when only those sites with more than three indicator species were considered in comparison to when only those sites

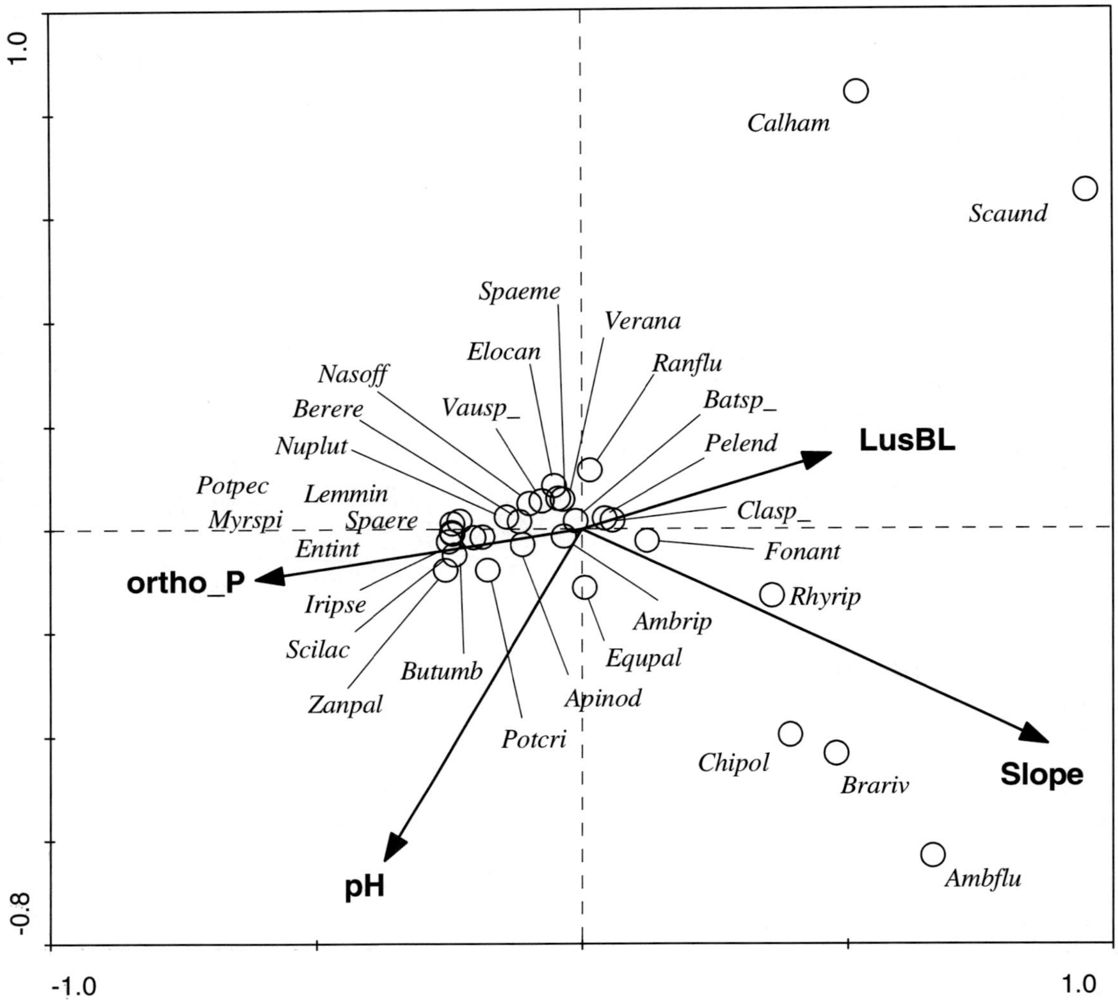

Figure 2. CCA-ordination diagram of macrophyte species and environmental variables in mountain streams. See Table 8 for full names of abbreviated species and Supplementary materials for full names of abbreviated variables.

with more than two indicator species were present ($r = -0.78$). A relatively strong correlation existed between conductivity and both trophic indices (MTR and IBMR) in the mountain streams. The analysis also shows that the species diversity of mountain streams is markedly limited by excess nitrite and ammonia, which both reduce the number of species and thereby allow a few tolerant species to dominate (Table 5).

In the analysis of southern European streams, correlations between macrophyte metrics and water quality parameters indicative of eutrophication were very weak. However, it should be noted that the number of southern European sites was comparatively low ($n \leq 22$) because not every site had an adequate environmental dataset and some sites were excluded due to a limited number of macrophytes in the channel itself where the MTR survey took place and consequent low number of metric indicator species (Table 6).

Plant species indicative of trophic status

Within the STAR dataset there is a total of 204 plant taxa that are regarded as nutrient indicators by the trophic metrics considered (Table 8). Most of them (132 taxa) have Ellenberg nitrogen indicator values. Similarly, a large number of taxa used in the IBMR and MTR metrics were recorded (104 and 93, respectively). The TIM indicators were less common (37 species). The largest number of indicator plants was found in lowland rivers (196) and the smallest in the South European streams (75).

Analysis of the indicator values of plant species (rescaled into 10-classes) showed that only 5 (out of the 104 included in at least two systems) had exactly the same score (Table 8). Differences were quite minor for 58 species (SD < 1), while for 46 species the differences in indicator values were larger (SD > 1) resulting in a greater than two class disparity in the 10-class scales for different metrics.

The MTR indicator plants were most frequent in lowland rivers (where 81 taxa were recorded) while the lowest number of indicator species occurred in the southern European dataset (22) (Table 7). The number of MTR indicator plants as a proportion of the total species richness (Fig. 3) was highest in mountain streams (followed by lowland and then southern European rivers). Examination of the growth forms of indicator plants showed that emergent and ecotonal species

Table 7. A summary of aquatic plants in the STAR macrophyte database

Number of species	Lowland	Mountain	South-European
General (taxa richness)	196	132	75
MTR indicators - total	81	51	22
STR = 1	5	2	1
STR = 2	8	7	3
STR = 3	17	6	3
STR = 4	12	9	3
STR = 5	16	9	2
STR = 6	9	3	0
STR = 7	6	2	1
STR = 8	4	6	3
STR = 9	3	4	3
STR = 10	2	4	3
IBMR indicators	85	63	24
not utilised by MTR	22	22	6
TIM indicators	36	25	8
not utilised by MTR	5	4	3
Ellenberg index indicators	133	62	41
not utilised by MTR	75	31	29
Morphological groups			
Filamentous algae	15	13	3
Bryophytes	27	36	11
Submerged	33	20	7
Anchored but with floating leaves or heterophyllus	14	7	3
Floating free	4	2	2
Amphibious	32	15	13
Ecotonal	79	45	38
Taxonomic groups			
Algae	15	13	3
Bryophytes	27	36	11
Polypodiophyta	2	4	3
Monocotyledons	67	26	27
Dicotyledons	85	53	31

Table 8. Presence/absence of species indicators in geographical regions and their species scores in different metrics. Species scores are standardised to a 10-point scale

Species	Abbrev.	Region			Trophy metrics			Variability		
		Low-land	Mountain	South Europe	MTR	IBMR	TIM	Ellenberg	SD1	SD2
Acorus calamus L.	*Acocal*	*			2.0	3.5	2.0	3.3	0.81	1.06
Agrostis canina L.	*Agrcan*	*						8.9		
Agrostis stolonifera L.	*Agrsto*	*	*			5.0				
Alisma lanceolatum With.	*Alilan*	*			3.0	4.5		5.6	1.31	1.06
Alisma plantago-aquatica L.	*Alipla*	*		*	3.0	4.0		2.2	0.90	0.71
Alopecurus geniculatus Sobol.	*Alogen*	*						3.3		
Alopecurus pratensis L.	*Alopra*	*						3.3		
Amblystegium riparium (Hedw.) B.S.G.	*Ambrip*	*	*		1.0	2.5			1.06	1.06
Angelica sylvestris L.	*Angsyl*	*						6.7		
Apium nodiflorum (L.) Lag.	*Apinod*	*	*	*	4.0	5.0		4.4	0.50	0.71
Baldellia ranunculoides L. Parl.	*Balran*			*				8.9		
Batrachospermum sp. Roth	*Batsp_*	*	*		6.0	8.0			1.41	1.41
Berula erecta (Huds.) Coville	*Berere*	*	*		5.0	7.0	3.4	4.4	1.52	1.41
Bidens cernua L.	*Bidcer*	*						1.1		
Bidens frondosa L.	*Bidfro*		*					2.2		
Bidens tripartita L.	*Bidtri*	*		*				2.2		
Brachythecium rivulare Schimp.	*Brariv*		*		8.0	7.5			0.35	0.35
Brachythecium rutabalum Bruch et Schimp.	*Brarut*	*			3.0					
Butomus umbellatus L.	*Butumb*	*	*		5.0	4.5	2.6	3.3	1.10	0.35
Calla palustris L.	*Calpas*	*						6.7		
Callitriche brutia Pet.	*Calbru*			*				5.6		
Callitriche cophocarpa Sendtn.	*Calcop*	*					3.8	5.6	1.27	
Callitriche hamulata Kutz ex W.D.J. Koch	*Calham*	*	*		9.0	6.0	5.5	6.7	1.55	2.12
Callitriche obtusangula Le Gall	*Calobt*	*	*		5.0	4.0	3.8	3.3	0.71	0.71
Callitriche platycarpa Kütz.	*Calpla*	*	*			5.0		3.3	1.20	
Callitriche stagnalis Scop.	*Calsta*	*		*		6.0		6.7	0.49	
Caltha palustris L.	*Calpal*	*	*					4.4		
Calystegia sepium (L.) R. Br.	*Calsep*	*	*					1.1		
Campanula ranunculoides L.	*Camran*		*					6.7		
Cardamine amara L.	*Carama*	*	*					6.7		
Carex acuta L.	*Caract*	*			5.0			6.7	1.20	
Carex acutiformis Ehrh.	*Caracf*	*			3.0			5.6	1.84	
Carex elata All.	*Carela*	*		*				5.6		
Carex flava L. s.str.	*Carfla*	*						8.9		
Carex hirta L.	*Carhit*	*	*					5.6		
Carex paniculata L.	*Carpai*	*		*				5.6		
Carex pendula Huds.	*Carpen*			*				4.4		
Carex pseudocyperus L.	*Carpse*	*						5.6		
Carex riparia Curtis	*Carrip*	*	*		4.0			6.7	1.91	
Carex rostrata Stokes	*Carros*	*			7.0	7.5		7.8	0.40	0.35
Carex vesicaria L.	*Carves*	*			6.0	6.0		5.6	0.23	0.00
Ceratophyllum demersum L.	*Cerdem*	*		*	2.0	2.5	2.1	2.2	0.22	0.35

Continued on p. 225

Table 8. (Continued)

Species	Abbrev.	Region			Trophy metrics			Variability		
		Low-land	Mountain	South Europe	MTR	IBMR	TIM	Ellenberg	SD1	SD2
Ceratophyllum submersum L.	*Cersub*	*				1.0		3.3	1.63	
Chara sp. L. ex Vaillant	*Chasp_*	*	*			6.0				
Chiloscyphus polyanthos (L.) Corda.	*Chipol*	*	*		8.0	7.5			0.35	0.35
Cicuta virosa L.	*Cicvir*	*						5.6		
Cinclidotus danubicus Sciffn. & Baumg.	*Cindan*		*			6.5				
Cinclidotus fontinaloides (Hedw.) P. Beauv.	*Cinfon*		*		5.0	6.0			0.71	0.71
Cinclidotus riparius Web. & Arnot	*Cinrip*		*			6.5				
Cladophora sp. Kütz.	*Clasp_*	*	*	*	1.0	3.0			1.41	1.41
Cratoneuron commutatum (Hedw.) G. Roth.	*Cracom*			*		7.5				
Cratoneuron filicinum (Hedw.) Spruce	*Crafil*		*	*		9.0				
Cyperus fuscus L.	*Cypfus*			*				6.7		
Cyperus longus L.	*Cyplon*			*				5.6		
Draparnaldia sp. Bory de St Vincent	*Drasp_*		*			9.0				
Echinochloa crus-galli (L.) P. Beauv.	*Echcrg*			*				2.2		
Eleocharis acicularis (L) Roem et Schult	*Eleaci*	*						8.9		
Eleocharis palustris (L.) Roem et Schult	*Elepal*	*	*		6.0	6.0			0.00	0.00
Elodea canadensis Michx.	*Elocan*	*	*		5.0	5.0	3.6	3.3	0.90	0.00
Enteromorpha sp. Link	*Entsp_*	*	*		1.0					
Epilobium hirsutum L.	*Epihir*	*	*					2.2		
Epilobium palustre L.	*Epiple*	*						8.9		
Epilobium parviflorum Schreb.	*Epipar*			*				4.4		
Epilobium roseum Schreb.	*Epiros*		*					2.2		
Equisetum arvense L.	*Equarv*		*	*				7.8		
Equisetum fluviatile L.	*Equflu*	*	*		5.0	6.0		7.8	1.42	0.71
Equisetum palustre L.	*Equpal*	*	*		5.0	5.0		5.6	0.35	0.00
Equisetum telmateia Ehrh.	*Equtel*			*				5.6		
Eupatorium cannabinum L.	*Eupcan*	*	*					2.2		
Filipendula ulmaria (L.) Maxim.	*Filulm*	*	*					5.6		
Fissidens crassipes Wils ex B.S.G.	*Fiscra*	*	*	*		6.0				
Fontinalis antipyretica Hedw.	*Fonant*	*	*	*	5.0	5.0			0.00	0.00
Fontinalis dalecarlica B.S.G.	*Fondal*	*								
Fontinalis squamosa Hedw.	*Fonsqu*	*			8.0	8.0			0.00	0.00
Galium palustre L.	*Galpal*	*	*					6.7		
Glechoma hederacea L.	*Glehed*		*					3.3		
Glyceria fluitans (L.) R. Br.	*Glyflu*	*	*			7.0		3.3	2.62	
Glyceria maxima (Hartm.) Holmb	*Glymax*	*			3.0		2.5	1.1	0.98	
Groenlandia densa (L.) Fourr.	*Groden*		*		5.0	5.5	5.7	5.6	0.31	0.35
Hildenbrandia rivularis (Liebm.) J.Agardh	*Hilriv*	*			6.0	7.5			1.06	1.06
Hippuris vulgaris L.	*Hipvul*	*			4.0	6.0	6.3		1.25	1.41
Hottonia palustris L.	*Hotpal*	*				6.0		6.7	0.49	
Humulus lupulus L.	*Humlup*	*						2.2		
Hydrocharis morsus-ranae L.	*Hydmor*	*			6.0	5.5		4.4	0.82	0.35

Continued on p. 226

Table 8. (Continued)

Species	Abbrev.	Region			Trophy metrics			Variability		
		Low-land	Mountain	South Europe	MTR	IBMR	TIM	Ellenberg	SD1	SD2
Hydrodictyon reticulatum (L.) Lagerh.	*Hydret*	*			3.0	3.0			0.00	0.00
Hygroamblystegium fluviatile (Hedw.)Loeske	*Hygflu*	*	*		5.0					
Hygrohypnum luridum (Hedw.) Jenn.	*Hyglur*		*		8.0	9.5			1.06	1.06
Hygrohypnum ochraceum (Wils.) Loeske	*Hygoch*		*		7.0	9.5			1.77	1.77
Hyocomium armoricum (Brid.) Wijk et Marg.	*Hyoarm*		*		2.0	10.0			5.66	5.66
Impatiens glandulifera Royle	*Impgla*	*						3.3		
Iris pseudacorus L.	*Iripse*	*	*		5.0	5.0		3.3	0.98	0.00
Juncus articulatus L.	*Junart*	*	*	*				8.9		
Juncus bufonius L.	*Junbuf*	*						6.7		
Juncus bulbosus L.	*Junbul*	*			10.0	8.0		8.9	1.00	1.41
Juncus conglomeratus L.	*Juncon*	*						7.8		
Juncus effusus L.	*Juneff*	*						6.7		
Juncus filiformis L.	*Junfil*	*						7.8		
Juncus inflexus L.	*Juninf*	*		*				6.7		
Lemanea fluviatilis (L.) C.Agardh	*Lemflu*		*	*	7.0	7.5			0.35	0.35
Lemna gibba L.	*Lemgib*	*		*	2.0	2.5		2.2	0.25	0.35
Lemna minor L.	*Lemmin*	*	*	*	4.0	5.0		4.4	0.50	0.71
Lemna minuta Kunth	*Lemmiu*	*			3.0					
Lemna trisulca L.	*Lemtri*	*	*		4.0	6.0		5.6	1.06	1.41
Littorella uniflora (L.) Ascherson	*Lituni*			*	8.0	7.5		8.9	0.71	0.35
Ludwigia palustris (L.) Elliott	*Ludpal*			*				6.7		
Lycopus europaeus L.	*Lyceur*	*	*	*		5.5		3.3	1.56	
Lysimachia thyrsiflora L.	*Lysthy*	*						6.7		
Mentha aquatica L.	*Menaqu*	*	*	*		6.0	5.0	5.6	0.50	
Mentha longifolia (L.) Huds. em. Harley	*Menlon*		*					3.3		
Mentha pulegium L.	*Menpul*			*				3.3		
Menyanthes trifoliata L.	*Mentrf*	*			9.0	8.0		7.8	0.64	0.71
Monostroma sp. Thuret	*Monsp_*		*			6.5				
Myosotis palustris (l.) L. em Rchb.	*Myopal*	*	*			6.0		5.6	0.28	
Myosoton aquaticum (L.) Moench	*Myoaqu*	*						2.2		
Myrica gale L.	*Myrgal*	*						7.8		
Myriophyllum alterniflorum DC.	*Myralt*	*			8.0	6.5	6.1	7.8	0.94	1.06
Myriophyllum spicatum L.	*Myrspi*	*	*	*	3.0	4.0	2.9	3.3	0.50	0.71
Myriophyllum verticillatum L.	*Myrvet*	*				6.0		2.2	2.69	
Nuphar lutea (L.) Sibth. & Sm.	*Nuplut*	*	*		3.0	4.5	2.1	4.4	1.16	1.06
Nymphaea alba L.	*Nymalb*	*			6.0	6.0		5.6	0.23	0.00
Octodiceras fontanum (La Pyl.) Lindb.	*Octfon*		*			3.5				
Oedogonium sp. Link	*Oedsp_*	*	*			3.0				
Oenanthe aquatica (L.) Poiret	*Oenaqu*	*				5.5		4.4	0.78	
Oenanthe crocata L.	*Oencro*	*		*	7.0	6.0			0.71	0.71

Continued on p. 227

Table 8. (Continued)

Species	Abbrev.	Region			Trophy metrics			Variability		
		Low-land	Mountain	South Europe	MTR	IBMR	TIM	Ellenberg	SD1	SD2
Pellia endiviifolia (Dicks) Dumort	*Pelend*	*	*		6.0					
Persicaria amphibia (L.) Gray	*Peramp*	*	*		4.0			6.7	1.91	
Persicaria hydropiper L. Delarbe	*Perhyd*	*	*	*				2.2		
Persicaria lapathifolia L. Gray	*Perlap*			*				2.2		
Petasites hybridus (L.) Gaertn., Mey. & Scherb.	*Pethyb*	*	*	*				2.2		
Peucedanum palustre (L.) Moench	*Peupal*	*						6.7		
Polygonum mite Schrank	*Polmit*			*				3.3		
Phalaris arundinacea L.	*Phaaru*	*	*			5.0		3.3	1.20	
Phragmites australis (Cav.) Trin. ex Steud	*Phraus*	*	*	*	4.0	4.5		3.3	0.60	0.35
Poa annua L.	*Poaann*	*						2.2		
Poa palustris L.	*Poapal*	*						3.3		
Poa trivialis L.	*Poatri*		*					3.3		
Polygonum persicaria L.	*Polper*	*						3.3		
Potamogeton alpinus Balbis	*Potalp*	*			7.0	6.5	6.1	4.4	1.13	0.35
Potamogeton berchtoldii Fieber	*Potber*	*	*		4.0	4.5	4.0	5.6	0.75	0.35
Potamogeton compressus L.	*Potcom*	*				3.0		6.7	2.62	
Potamogeton crispus L.	*Potcri*	*	*		3.0	3.5	2.8	5.6	1.28	0.35
Potamogeton filiformis Pers.	*Potfil*	*					5.8	7.8	1.41	
Potamogeton gramineus L.	*Potgra*	*			7.0	6.5		5.6	0.71	0.35
Potamogeton lucens L.	*Potluc*	*		*	3.0	3.5	3.4	3.3	0.22	0.35
Potamogeton natans L.	*Potnat*	*			5.0	6.0	5.0	5.6	0.49	0.71
Potamogeton nodosus Poir.	*Potnod*	*	*			2.0	2.3	5.6	2.00	
Potamogeton obtusifolius Mert. & Koch	*Potobt*	*			5.0	5.0		4.4	0.35	0.00
Potamogeton pectinatus L.	*Potpec*	*	*		1.0	1.0	2.8	2.2	0.90	0.00
Potamogeton perfoliatus L.	*Potper*	*	*		4.0	4.5	4.1	4.4	0.24	0.35
Potamogeton polygonifolius Pourret	*Potpol*	*			10.0	8.5	7.2	8.9	1.16	1.06
Potamogeton pusillus L.	*Potpus*	*	*		4.0		4.0		0.00	
Potamogeton trichoides Cham. & Schltdl	*Pottri*	*			2.0			6.7	3.32	
Potentilla palustris (L.) Scop.	*Potpal*	*				8.0		8.9	0.64	
Ranunculus aquatilis L.	*Ranaqu*	*	*		8.0	5.5	4.5	4.4	1.68	1.77
Ranunculus baudotii Godron	*Ranbau*	*		*				3.3		
Ranunculus circinatus Sibth	*Rancir*	*	*		4.0	5.0	4.4	2.2	1.21	0.71
Ranunculus flammula L.	*Ranfla*	*			7.0	8.0		8.9	0.95	0.71
Ranunculus fluitans Lamk	*Ranflu*	*	*		7.0	5.0	2.5	2.2	2.26	1.41
Ranunculus lingua L.	*Ranlin*	*						3.3		
Ranunculus peltatus ssp. *peltatus*	*Ranpet*			*	4.0			4.4	0.28	
Ranunculus pen. ssp. *pseudofluitans* Webster	*Ranpse*	*	*		5.0					
Ranunculus penicillatus Dum	*Ranpen*	*	*		5.0	6.0			0.71	0.71
Ranunculus repens L.	*Ranrep*	*	*	*				3.3		
Ranunculus sceleratus L.	*Ransce*	*			2.0			1.1	0.64	
Ranunculus sp. L.	*Ransp_*	*			6.0					

Continued on p. 228

Table 8. (Continued)

Species	Abbrev.	Region			Trophy metrics			Variability		
		Low-land	Mountain	South Europe	MTR	IBMR	TIM	Ellenberg	SD1	SD2
Ranunculus trichophyllus Chaix	*Rantri*	*	*		6.0		3.3	3.3	1.56	
Rhynchostegium riparioides (Hedw.) Cardo	*Rhyrip*	*	*		5.0					
Riccardia sinuata (Dicks.) Trev.	*Ricsin*		*			7.5				
Rorippa amphibia (L.) Besser	*Roramp*	*	*		3.0	4.5		2.2	1.17	1.06
Rorippa nasturtium-aquaticum L.	*Rornas*	*	*	*	5.0	5.5	4.4		0.55	0.35
Rumex aquaticus L.	*Rumaqu*	*						2.2		
Rumex crispus L.	*Rumcri*	*		*				4.4		
Rumex hydrolapathum Huds.	*Rumhyd*	*			3.0			3.3	0.21	
Sagittaria sagittifolia L.	*Sagsag*	*			3.0	3.0	2.6	4.4	0.79	0.00
Scapania undulata (L.) Dum	*Scaund*	*	*		9.0	8.5			0.35	0.35
Schoenoplectus lacustris (L.) Palla	*Schlac*	*	*	*	3.0	4.0		4.4	0.72	0.71
Schoenus nigricans L.	*Schnig*			*				8.9		
Scirpus sylvaticus L.	*Scisyl*	*				5.0		6.7	1.20	
Scrophularia auriculata L.	*Scraur*	*						3.3		
Scrophularia umbrosa Dum.	*Scrumb*	*	*					3.3		
Scutellaria galericulata L.	*Scugac*	*						4.4		
Sium latifolium L.	*Siulat*	*						3.3		
Solanum dulcamara L.	*Soldul*	*	*					2.2		
Sparganium emersum Rehmann	*Spaeme*	*	*		3.0	3.5	3.1	3.3	0.22	0.35
Sparganium erectum L.	*Spaere*	*	*		3.0	5.0	2.5	3.3	1.08	1.41
Spirodela polyrhiza (L.) Schleid	*Spipol*	*	*		2.0	3.0		4.4	1.21	0.71
Spirogyra sp. Link	*Spisp_*	*	*			5.0				
Stachys palustris L.	*Stapal*	*	*					4.4		
Stachys sylvatica L.	*Stasyl*	*						3.3		
Stigeoclonium sp. Kütz.	*Stisp_*		*			6.5				
Stratiotes aloides L.	*Stralo*	*						4.4		
Stellaria alsine Grimm.	*Steals*		*					6.7		
Symphytum officinale L.	*Symoff*	*						2.2		
Thamnobryum alopecurum (Hedw.) Gang.	*Thaalo*		*		7.0	7.5			0.35	0.35
Typha angustifolia L.	*Typang*			*	5.0	3.0		3.3	1.08	1.41
Typha latifolia L.	*Typlat*	*			2.0	4.0		2.2	1.10	1.41
Urtica dioica L.	*Urtdio*	*						1.1		
Utricularia intermedia Hayne	*Utrint*	*						10.0		
Utricularia vulgaris L.	*Utrvul*	*						6.7		
Vaucheria sp. De Candolle	*Vausp_*	*	*		1.0	2.0			0.71	0.71
Veronica anagallis-aquatica L.	*Verana*	*	*	*	4.0		3.6	4.4	0.40	
Veronica anagalloides Guss	*Verang*	*				5.5		5.6	0.07	
Veronica beccabunga L.	*Verbec*	*	*	*		5.0	4.0	4.4	0.50	
Zannichellia palustris L.	*Zanpal*	*	*		2.0	2.5	2.7	2.2	0.31	0.35

Variability between macrophyte metrics for particular species presented as standard deviation (SD1 – four metrics considered, SD2 MTR and IBMR considered only).

(1 = low quality, 10 = high quality).Variability between macrophyte metrics for particular species presented as standard deviation (SDI – four metrics considered, SD2 MTR and IBMR considered only)

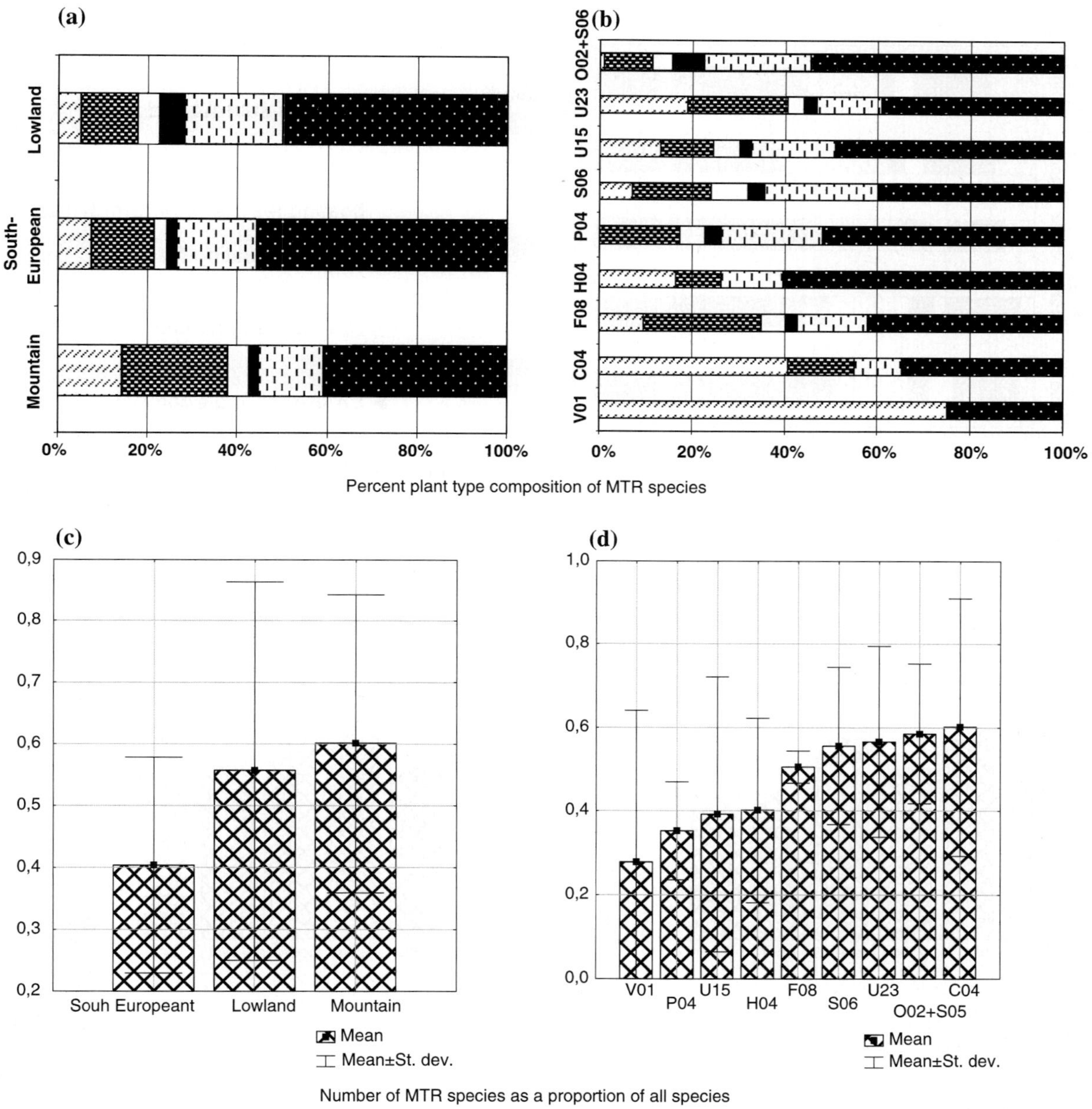

Figure 3. Based on organically polluted sites, percent plant type composition of MTR species in (a) geographical regions, and (b) stream types, and number of MTR species as a proportion of all species in (c) geographical regions and (d) stream types. See Figure 4 for key to plant types, and Table 1 for stream types.

were dominant in southern Europe (56%) while mountain streams had relatively large proportions of indicator bryophytes (15%) and submerged species (23.5%).

The presence of MTR indicator species varied considerably between different types of river (Fig. 3). Differences in the relative richness of indicator species were observed within both the lowland and mountain analyses. For instance, the mountain river type '*Small-sized siliceous mountains streams in the West Carpathians*' (V01) was poor in indicator species whereas the high share of indicator species was observed in a mountain stream river type ('*Small sized shallow mountain*

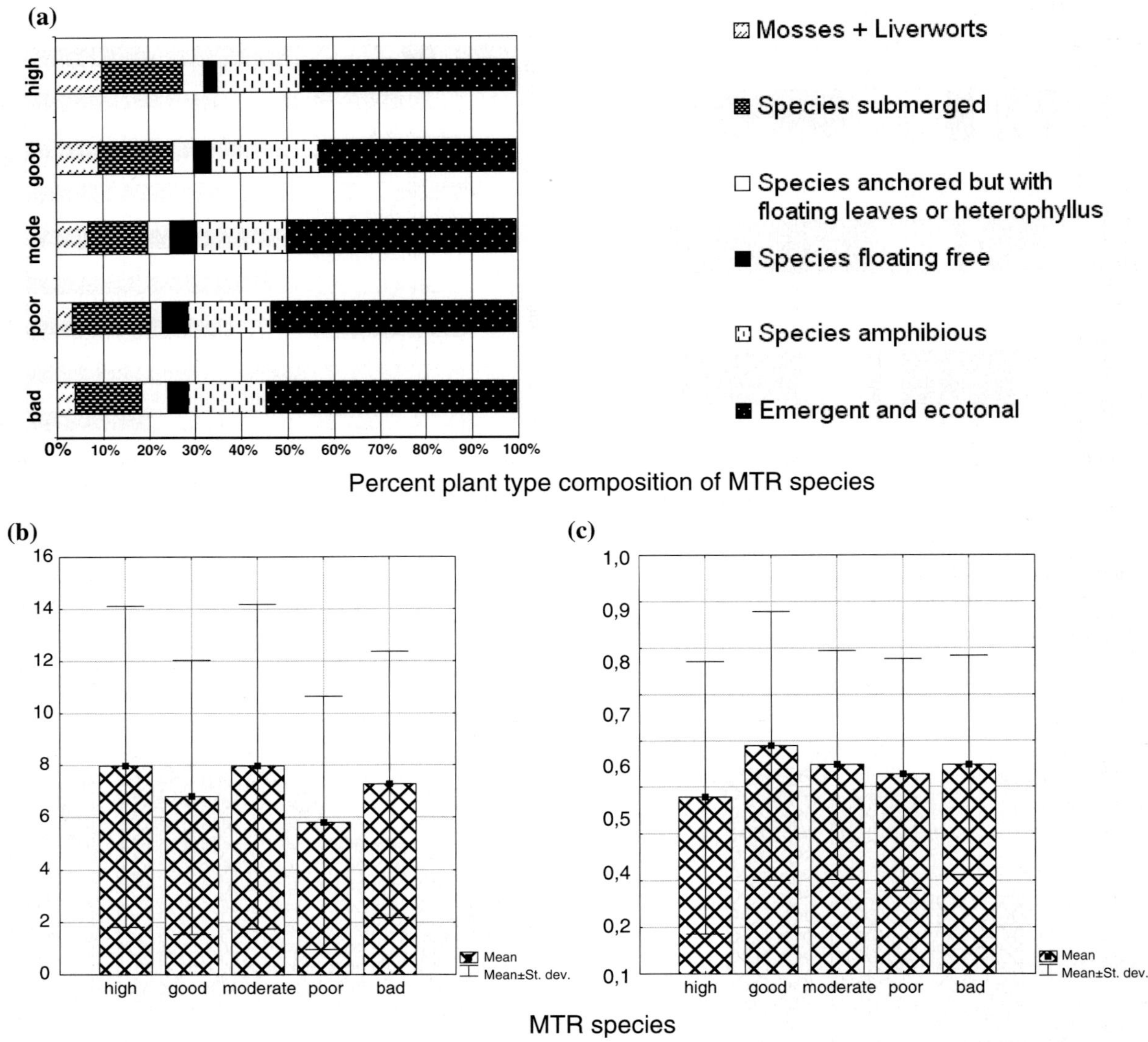

Figure 4. Based on five separate classes or organic pollution, (a) percent plant type composition of MTR species in geographical regions, (b) number of MTR species, and (c) number of MTR species as a proportion of all species.

streams', C04). Similarly, the lowland stream type '*Small sized lowland calcareous streams*' (U15) was poor in indicator species, whereas a high proportion of indicators were found in other lowland stream types (medium-sized rivers, types O02, S05 and S06).

The growth forms of indicator plants were more effective as a means of differentiating between particular river types than between rivers from different geographical regions. The extreme domination of bryophytes in two mountain stream types (C04 and V01) and the domination by emergent and ecotonal species in southern Europe (60.5% in case of H02) are interesting.

Variation in the occurrence of indicator plants across different levels of organic pollution was relatively limited although some types varied in consistent way (Fig. 4). For example, the proportion of bryophytes was highest in unpolluted rivers (10%) and lowest in the most degraded streams (4%), and conversely, the proportion of emergent and ecotonal flora varied from 54.5% (polluted) to

47% (unpolluted). Overall, the proportion of indicator plants to the total number of species was lowest in unpolluted conditions.

MTR enlargement and refinement

Within the lowland river dataset, of the 201 taxa recorded, 163 are used for biomonitoring (Tables 7 and 8). Among these, 82 are used in MTR. Typical macrophytes not included in MTR scoring were then tested for possible inclusion by examining their Spearman rank correlation with the trophic indicators. Taxa that further improved the positive correlation between MTR score and trophic related chemical parameters (Table 4) were selected as potential new MTR species. For *lowland rivers* the following plants are proposed: *Callitriche cophocarpa* (STR = 5), *Ceratophylum submersum* (3), *Cicuta virosa* (5), *Eleocharis acicularis* (8), *Fontinalis dalecarlica* (10), *Glyceria fluitans* (5), *Hottonia palustris* (7), *Myosotis palustris* (4), *Leptodictyum riparum* (10), *Lycopus europaeus* (4), *Lysimachia thyrsiflora* (8), *Mentha aquatica* (5), *Oedogonium* sp. (1), *Oenanthe aquatica* (6), *Persicaria hydropiper* (3), *Phalaris arundinacea* (2), *Potamogeton compressus* (4), *Potamogeton nodosus* (4), *Ranunculus lingua* (8), *Ricciocarpus natans* (7), *Sium latifolium* (7), *Scirpus sylvaticus* (4), *Utricularia intermedia* (10), *Utricularia vulgaris* (5), and *Veronica beccabunga* (4). In case of four species the STR scores were proposed to be changed: *Fontinalis antipyretica* (STR from 5 to 6), *Lemna minor* (4 to 3), *Alisma lanceolatum* (3 to 4) and *Sparganium emersum* (3 to 4).

In the mountain river dataset the same process was undertaken to select new potential MTR scoring species and STR scores (Tables 7 and 8). Of the 132 recorded taxa, 99 are used in biomonitoring and 51 are used in MTR. Of the remainder not used by MTR, eight species had good relationships with trophic status related water quality parameters and improved the positive relationship of the overall MTR index with trophic status related water quality parameters (Table 5): *Fissidens crassipes* (6), *Lemanea* sp. (8), *Mentha aquatica* (6), *Myosotis palustris* (4), *Oedogonium* sp. (2), *Persicaria hydropiper* (3), *Phalaris arundinacea* (2) and *Potamogeton nodosus* (4). The existing STR scores of four species were adjusted in the same way as in the lowland river dataset. Generally, the proposed indicative values of new species as well as modifications of existing scores were identical for mountain and lowland rivers as they correspondingly improved the positive correlation between MTR score and trophic related chemical parameters.

Discussion

Ortho-phosphate was clearly identified as an important driver of species selection aquatic plant communities in European rivers, both in lowland and mountain stream types. Our analyses included sites with various degrees of degradation (organic and other) while a similar analysis based solely upon unpolluted reference condition rivers reached somewhat different conclusions (Baattrup-Pedersen et al., 2006), in which the macrophyte communities were strongly related to ecoregion, catchment and habitat variables rather than to eutrophication factors. Our analysis also showed that while environmental variables could explain a part of the variability of macrophyte communities across Europe, still a considerable proportion remains unexplained. Nonetheless, the subsets of significant environmental variables we have isolated in both the lowland and mountain stream types are a sufficient set of predictors.

Our analysis showed that none of the existing macrophyte systems are entirely appropriate for river assessment at a pan-European scale. Although the plant metrics presented significant relationships with hydrochemical parameters indicating trophic status, the correlation coefficients were too low to make these systems reliable for biomonitoring applications across this broad spatial scale. However, our analyses have shown that the addition of 25 and eight new species to the MTR index (for lowland rivers and mountain rivers, respectively) and indicative value changes for four species, appreciably improved the ability of our expanded MTR indices to detect phosphorous gradients. It seems likely that further development of macrophyte indices could result in useful improvements in precision since the macrophyte metrics identified as the most effective in this study only incorporate part of the recorded list of macrophyte species.

Furthermore, the Europe-wide application of existing macrophyte metrics (MTR, IBMR and TIM) is confounded by the large differences that exist between the indicative values ascribed to the same macrophyte species between the indices (Table 8). These problems represent two potential avenues for future index development (incorporation of more species and refinement of indicator values). This is perhaps exemplified best by the MTR system. MTR was developed for use in the United Kingdom, and as such does not incorporate a broad spectrum of European plants that do not occur in the UK. The indicator values are also derived from analyses based on UK datasets. Enlargement of the MTR system to include a higher proportion of European plant species and refinement of indicator scores could make the MTR system more effective. This has been demonstrated both in the analyses here and in other studies in Poland (Szoszkiewicz et al., 2002).

The size of the macrophyte species pool suggests that there is considerable scope for the development of macrophyte-based bioassessment methods based on the species composition. Within the STAR survey dataset, 204 taxa were recorded that are also utilized by indicative systems. These represent the great majority of taxa that can be considered in river monitoring. Most of these are incorporated in the Ellenberg method (132 taxa) and many are also used by IBMR (104), and by MTR (93). Only 37 TIM index indicators were recorded. However, a high proportion of species are common to these indices, and our example of expanding the MTR to include additional species shows how utilization of a greater proportion of the macrophyte flora can improve the usefulness of overall index scores.

In this study the newly added species respond similarly with respect to nutrient levels in both lowland and mountain stream analyses. We have also shown that the plant species already incorporated in existing metrics have comparable indicator scores between the different metrics (although small disparities exist indicating that further refinement may be possible). We therefore conclude that a quite extensive core group of plant indicators can be used across European streams, and that regional versions of indices may only be needed for specific various types of rivers. Since analysis showed that the growth forms of indicator plants were more effective as a means of differentiating between particular river types than between rivers from different geographical regions the local versions should be limited to certain specific stream types.

The particular approach must be implemented for the southern European rivers, where the usefulness of strictly aquatic species as a means of identifying eutrophication was compromised in the result of the low number of scoring plant species combined with their low abundances. The solution could either be the inclusion of additional ecotonal species that are not recorded in current macrophyte methodologies and inhabit outside the channel (Ferreira et al., 2002) and/or also the adoption of macrophyte's bioassessment systems based on communities' functional characteristics rather than plant composition (Ferreira et al., in press).

The IBMR and MTR systems were identified as the most useful metrics while the index estimated based on nitrate Ellenberg values for plants was less effective. However, Ellenberg numbers were designed for terrestrial plants that absorb nitrate from the soil (Ellenberg et al., 1992) so their failure to out perform the specifically aquatic IBMR and MTR systems is perhaps not unexpected. Another trophic assessment system – TIM, was also shown to have disadvantages arising from a short indicative species list. TIM (and Ellenberg) only included seed-producing plants whereas IBMR and MTR also incorporate pteridophytes, bryophytes and algae. These non-flowering plants actually represent a significant proportion of the plants recorded in rivers (42 of the 195 species in lowland rivers – 22%, and 17 of the 75 species in southern European rivers – 23%). In mountain streams the proportion of non-flowering plants is even higher (53 of the 132 species – 40%).

We have shown that trophic status is an important driver of aquatic plant communities in European rivers. We found that none of the existing macrophyte systems are entirely appropriate for the assessment of trophic status at a pan-European scale. However, the existing metrics do indicate the usefulness of macrophytes in bioassessment systems. The redesigned Mean Trophic Rank system, incorporating additional species and re-evaluated species scores, was highly correlated with hydro-chemical variables associated with trophic status, and as such is better suited for use in a biomonitoring role. We conclude that a extensive core group of plant species indicators could be developed and used across European streams, and that regional versions of indices may only be needed for certain stream types.

Acknowledgements

We would like to thank all members of the STAR consortium involved in macrophyte field surveys, data entry and validation and all surveyors who have collected the physicochemical variables that supported our analyses.

References

Allan, J. D., 1995. Stream Ecology. Structure and Function of Running Waters. Chapman and Hall, London, 388 pp.

Baattrup-Pedersen, A. & T. Riis, 1999. Macrophyte diversity and composition in relation to substratum characteristics in regulated and unregulated Danish streams. Freshwater Biology 42: 375–385.

Baattrup-Pedersen, A., K. Szoszkiewicz, R. Nijboer, M. O'Hare & T. Ferreira, 2006. Macrophyte communities in unimpacted European streams: variability in assemblage patterns, abundance and diversity. Hydrobiologia 566: 179–196.

Butcher, R. W., 1933. Studies on ecology of rivers. I. On the distribution of macrophytic vegetation in the rivers of Britain. Journal of Ecology 21: 58–91.

Dawson, F. H. & U. Kern-Hansen, 1979. The effect of natural and Artificial Shade on Macrophytes of Lowland Streams and the Use of Shade as Management Technique. Internationale Revue der gesamten Hydrobiologie 64: 437–455.

Dawson, F. H., 1988. Water flow and the vegetation of running waters. In Symoens, J. J. (ed.), Vegetation of Inland Waters. Kluwer Academic Publishers, Dordrecht, 283–309.

Dawson, F. H., 2002. Guidance for the field assessment of macrophytes of rivers within the STAR Project. http://www.eu-star.at/frameset.htm.

Dawson, F. H., J. R. Newman, M. J. Gravelle, K. J. Rouen & P. Henville, 1999. Assessment of the Trophic Status of Rivers using Macrophytes: Evaluation of the Mean Trophic Rank. R&D Technical Report E39. Environment Agency of England & Wales, Bristol, 108 pp.

Demars, B. O. L. & D. H. Harper, 1998. The aquatic macrophytes of an English lowland river system: assessing response to nutrient enrichment. Hydrobiologia 384: 75–88.

Ellenberg, H., H. E. Weber, R. Dull, V. Wirth, W. Werner & D. Baulissen, 1992. Zeigerwerte von Pflanzen in Mitteleuropa. Scripta Geobotanica 18: 1–257.

European Commission, 2000. Directive 2000/60/EC of the European Parliament and of the Council – Establishing a framework for Community action in the field of water policy. Brussels, Belgium, 23 October 2000.

Fennessy, M. S., J. K. Cronk & W. J. Mitsch, 1994. Macrophytes productivity and community development in created freshwater wetlands under experimental hydrological conditions. Ecological Engineering 3: 469–484.

Ferreira, M. T., A. Albuquerque, F. C. Aguiar & N. Sidorkewicz, 2002. Assessing reference sites and ecological quality of river plant assemblages from an Iberian basin using a multivariate approach. Archive für Hydrobiologie 155: 121–145.

Ferreira, M. T., P. Rodríguez-González, F. C. Aguiar & A. Albuquerque, 2005. Assessing biotic integrity in Iberian rivers: Development of a multimetric plant indice. Ecological Indicators 5: 137–149.

Haslam, S., 1978. River Plants: The Macrophytic Vegetation of Watercourses. Cambridge University Press, Cambridge, 396 pp.

Haslam, S., 1987. River Plants of Western Europe. Cambridge University Press, Cambridge, 512 pp.

Haury, J., M. C. Peltre, S. Muller, M. Trémolières, J. Barbe, A. Dutartre & M. Guerlesquin, 1996. Des indices macrophytiques pour estimer la qualité des cours d'eau français: premières propositions. Ecologie 27: 233–244.

Haury, J., M. C. Peltre, M. Tremolieres, J. Barbe, G. Thiebaut, I. Berne, H. Daniel, P. Chatenet, S. Muller, A. Dutartre, C. Laplace-Treyture, A. Cazaubon & E. Lambert-Servien, 2002. A method involving macrophytes to assess water trophy and organic pollution: the Macrophyte Biological Index for Rivers (IBMR) – application to different types of rivers and pollutions. Proc. 11th EWRS International Symposium on Aquatic Weeds, Moliets Et Maa, France, eds. A. Dutartre & M.-H. Montel, 247–250.

Holmes, N. T. H., J. R. Newman, S. Chadd, K. J. Rouen, L. Saint & F. H. Dawson, 1999. Mean Trophic Rank: A Users Manual. R&D Technical Report E38. Environment Agency of England & Wales, Bristol, 134 pp.

Jalas, J., 1955. Hemorobe und hemerokore Pflanzenarten. Ein terminologischer Reformversuch. Acta Societas pro Fauna et Flora. Fennica 72: 1–15.

Kohler, A., 1982. Wasserpanzen als Belastungsindikatoren. Descheniana-Beihefte 26: 31–42.

Kohler, A. & S. Schiele, 1985. Veränderungen von Flora und Vegetation in den kalkreichen Fließgewässern der Friedberger Au (bei Augsburg) von 1972 bis 1982 unter veränderten Belastungsbedingungen. Archive für Hydrobiologie 103: 137–199.

McNaughton, S. J., 1967. Relationships among functional properties of Californian Grasslands. Nature 216: 168–169.

Passauer, B., P. Meilinger, A. Melzer & S. Schneider, 2002. Does the structural quality of running waters affect the occurrence of macrophytes? Acta Hydrochimica et Hydrobiologica 30: 197–206.

Pielou, E. C., 1966. The measurement of diversity in different types of biological collections. Journal of Theoretical Biology 13: 131–144.

Pott, R. & D. Remy, 2000. Gewässer des Binnenlandes. Eugen Ulmer Verlag, Stuttgart, 255 pp.

Remy, D., 1993. Pflanzensoziologische und standortkundliche Untersuchungen an Fließgewässern Nordwestdeutschlands.

Abhandlungen aus dem Westfälischen Museum für Naturkunde 55: 117.

Robach, F., G. Thiébaut, M. Trémolières & S. Muller, 1996. A reference system for continental running waters: plant communities as bioindicators of increasing eutrophication in alkaline and acidic waters in north-east France. Hydrobiologia 340: 67–76.

Schaumburg, J., C. Schranz, J. Foerster, A. Gutowski, G. Hofmann, P. Meilinger, S. Schneider & U. Schmedtje, 2004. Ecological classification of macrophytes and phytobenthos for rivers in Germany according to the Water Framework Directive. Limnologica 34: 283–301.

Schneider, S., T. Krumpholz & A. Melzer, 2000. Trophäeindikation in Fliessgewässern mit Hilfe des TIM (Trophäe-Index Macrophyten) – Erprobung eines neu entwickelten Index im Inniger Bach. Acta Hydrochimica et Hydrobiologica 28: 241–249.

Shannon, C. E. & W. Weaver, 1949. The Mathematical Theory of Communication. University of Illinois Press, Urbana 117 pp.

Simpson, E. H., 1949. Measurement of diversity. Nature 163: 688.

Sipos, V., A. Kohler & S. Björg, 2000. Makrophyten-Vegetation und Standorte im eutrophen Björka-Fluss (Südschweden). Botanische Jahrbücher für Systematik 122: 93–152.

StatSoft, Inc., 2004. STATISTICA (data analysis software system), version 6. www.statsoft.com.

Svendsen, L. & A. Rebsdorf, 1994. Kvalitetssikring af Overvågningsdata. Retningslinier for Kvalitetssikring af Ferskvandskemiske Data i Vandmiljøplanenes Overvågningsprogram. (Quality Assurance of Monitoring Data). Teknisk Anvisning fra DMU, 7: 87 pp.

Szoszkiewicz, K., K. Karolewicz, A. Lawniczak & F. H. Dawson, 2002. An assessment of the MTR aquatic plant bioindication system for determining the trophic status of Polish rivers. Polish Journal of Environmental Studies 11: 421–427.

ter Braak, C. J. F. & I. C. Prentice, 1988. A theory of gradient analysis. Advances in Ecological Research. 18: 271–317.

Thiebaut, G. & S. Muller, 1998. Aquatic macrophyte communities as water quality indicators: example of the river Moder (North-East France) Annales de limnologie. International Journal of Limnology 34: 141–153.

Tremp, H. & A. Kohler, 1995. The usefulness of macrophyte monitoring-systems, exemplified on eutrophication and acidification of running waters. Acta Botanica Gallica 142: 541–550.

Van de Weyer, K., 1997. Untersuchungen zur Biologie und Ökologie von Potamogeton polygonifolius im Niederrheinischen Tiefland. Dissertationes Botanicae 278: 178 pp.

Van De Weyer, K., 2003. Kartieranleitung zur Erfassung und Bewertung der aquatischen Makrophyten der Fließgewässer in NRW gemäß den Vorgaben der EU-Wasser-Rahmenrichtlinie. Landesumweltamt Nordrhein-Westfalen (LUA). Merkblätter 39: 60 pp.

Veit, U., K. Penksza & A. Kohler, 2003. Beurteilung von Fließgewässern am Beispiel einer Langzeituntersuchung der Makrophyten-Vegetation in der Friedberger Au (bei Augsburg). Deutsche Gesellschaft für Limnologie Tagungsbericht 2002 (Braunschweig), 263–268.

Werle, W., 1982. Eignung von submersen Makrophyten als Bioindikatoren in Fließgewässern. Mitteilungen der Pollichia. 70: 125–168.

Westlake, D. F., 1975. Macrophytes. In Whitton, B.A. (ed.), River Ecology. University of California Press, Berkeley, California: 106–128.

Wiegleb, G., 1988. Analysis of flora and vegetation in rivers: concepts and applications. In Symoens, J. (ed.), Vegetation of Inland Waters. Handbook of Vegetation Science 15. Kluwer Academic Publishers, Dordrecht, 311–340.

Hydrobiologia (2006) 566:235–246
M.T. Furse, D. Hering, K. Brabec, A. Buffagni, L. Sandin & P.F.M. Verdonschot (eds), The Ecological Status of European Rivers: Evaluation and Intercalibration of Assessment Methods
DOI 10.1007/s10750-006-0093-4

Assessment of sources of uncertainty in macrophyte surveys and the consequences for river classification

Ryszard Staniszewski[1,*], Krzysztof Szoszkiewicz[1], Janina Zbierska[1], Jacek Lesny[2], Szymon Jusik[1] & Ralph T. Clarke[3]
[1]*Department of Ecology and Environment Protection, August Cieszkowski Agricultural University, Piatkowska 94C, 61-691 Poznan, Poland*
[2]*Department of Agrometeorology, August Cieszkowski Agricultural University, Piatkowska 94B, 61-691 Poznan, Poland*
[3]*Centre for Ecology and Hydrology, Winfrith Technology Centre, DT2 8ZD Dorchester, Dorset, England*
*(*Author for correspondence: Tel: 48-61-8466510; Fax: 48-61-8466510; E-mail: erstan@au.poznan.pl)*

Key words: macrophytes, Mean Trophic Rank, aquatic vegetation, error assessment, biodiversity, river classification

Abstract

The application of macrophytes in freshwater monitoring is still relatively limited and studies on their intercalibration and sources of variation are required. Therefore, the aim of the study was to compare selected indices and metrics based on macrophytes and to quantify their variability. During the STAR project, several aspects influencing uncertainty in estimation of the ecological quality of river were assessed. Results showed that several metrics based on the indicative value of plant species can be used in evaluation of the ecological status of rivers. Among estimated sources of variance in metric values the inter-surveyor differences had the lowest effect and slightly stronger were the influences of temporal variation (years and seasons) and shading. The impact of habitat modification was the most important factor. Analysis showed that some of macrophyte-based metrics (notably MTR and IBMR) are of sufficient precision in terms of sampling uncertainty, that they could be useful for estimating the ecological status of rivers in accordance with the aims of the Water Framework Directive.

Introduction

The purpose of Water Framework Directive – WFD was to established an European framework for the protection of surface waters, transitional waters, coastal waters and groundwater (Directive 2000/60/EC). It formalises the need to obtain more standardised, comparable and widespread data about aquatic ecosystems in Europe, together with estimation of the uncertainty in assessing the ecological status of water bodies. Macrophytes are one of the major groups of organisms upon which the WFD prescribes that such assessments should be made. The application of macrophytes in monitoring of freshwaters is still relatively limited and studies on their intercalibration and sources of variation are required. Therefore, the aim of the study was to compare selected indices and metrics based on macrophytes and to quantify their variability.

Studies on macrophyte variability based on replicate sampling experiments were carried out in Polish lowland rivers in the years 2003 and 2004, in the period when river vegetation was well developed (5th June–30th September). The sites selected for replicate sampling were rich in macrophyte species and encompassed a wide environmental gradient, that improved the quality of the study. Each site was pre-classified within the STAR project into one of the five WFD ecological status classes. Diversity (Shannon's index) of aquatic plant species showed unimodal distribution along the ecological status gradient, peaking at intermediate site qualities and with detectable differences between status classes.

Methods based on the indicative values of plant species along environmental or stressor gradients are often used in studies of terrestial and aquatic ecosystems (Ellenberg et al., 1992; Haury et al., 1996, 1998; Allan, 1995; Dawson et al., 1998; Dawson & Szoszkiewicz, 1999; Schneider et al., 2000; Staniszewski, 2001; Szoszkiewicz et al., 2002 and others).

Variation of estimation of the ecological quality of water ecosystem depends on many factors like field sampling, personal judgement, selection of reference sites and the choice and estimation of metrics representing the biological Reference Conditions for the site (Clarke, 2000). Several aspects influencing uncertainty in estimation of the ecological quality of river were detected and assessed during the STAR project. The study was designed to assess the variation due to inter-surveyor differences, temporal variation, and the influence of stressors such as shading and hydromorphological degradation.

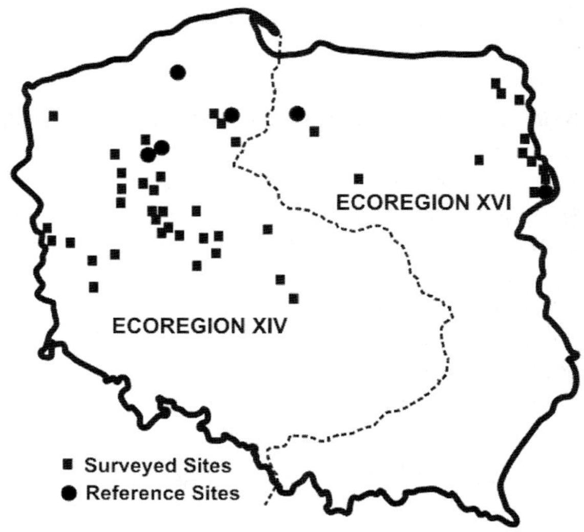

Figure 1. Distribution of surveyed river sites.

Methods

Site selection and sampling

The main criteria used in site selection were to cover a wide range of the eutrophication gradient and wide geographic distribution (Fig. 1,Electronic supplementary material is available for this article at <http://dx.doi.org/ 10.1007/s10750-006-0093-4> and accessible for authorised users.). An additional factor influencing site selection was the overall abundance of aquatic plants. Important part of studies was selection of reference sites according to the needs of WFD and characterised with undisturbed by human activities river valley and high hydrochemical quality (Table 1). All selected sites were from comparable conditions in the sense of being from the WFD System A stream type. These streams were characterised with an upstream watershed area between 100 and 1000^2 km and a site altitude of up to 200 m a.s.l.

The water samples were collected three times: summer 2003, autumn 2003 and summer 2004. Several water quality parameters were examined as like total phosphorus, soluble reactive phosphates, nitrates, conductivity and others. Samples were stored in ice boxes and analyses were made within 24 h. Water samples for soluble reactive phosphates and nitrates were filtered using 0.45 μm pore size.

Table 1. Differentiation of trophic parameters (TP – total phosphorus, SRP – soluble reactive phosphates, nitrates and conductivity) between reference sites and others

	TP mg P dm^{-3}		SRP mg PO_4 dm^{-3}		Nitrates mg N–NO_3 dm^{-3}		Conductivity mS cm^{-1}	
	Reference sites	Other sites	Reference sites	Other sites	Reference sites	Other sites	Reference sites	Other sites
Maximum	0.54	15.53	0.35	15.50	0.40	12.75	0.42	2.99
Mean	0.33	1.82	0.19	1.34	0.15	1.06	0.35	0.56
Median	0.31	0.62	0.20	0.37	0.13	0.21	0.32	0.46
Minimum	0.01	0.14	0.01	0.05	0.01	0.03	0.24	0.27

Selected indices and metrics as indicators of site quality

Studies on macrophytes were undertaken in Polish lowland rivers in 2003 and 2004, in the period when river vegetation was well developed (5 June–30 September). Field surveys were carried out according to STAR guidance for field assessment of macrophytes (Dawson, 2002) being closely related to the Mean Trophic Rank (MTR) methodology (Newman et al., 1997; Dawson et al., 1999; Holmes et al., 1999).

The macrophyte assessment was based on the presence of algae, mosses, horsetails, liverworths, monocotyledonous and dicotyledonous plant species which have value as biological indicators of water quality. All submerged, free floating, amphibious and emerged plants were considered. The assessment also included the macrophytes attached or rooted on parts of the river bank substrate where they were likely to be submerged for more than 85% of the year. The presence of each species on the standard MTR survey river length of 100 m was recorded together with their percentages of area covered using the standard MTR nine point scale (Holmes et al., 1999). All of the sites were surveyed by wading along the river channel except for the Sokolda River, which was too deep to wade and for which a grapnel was used to collect plant species. All taxa were identified individually by each surveyor and additionally, samples of algae were checked by algologists from University of Lodz to confirm results.

The STAR procedures for macrophyte field surveys were designed primarily to obtain MTR score but created database enabled the estimation of other macrophyte based metrics which are widely applied in the vegetation sciences. In this experiment based on the STAR protocol, in addition to MTR scores, the following four other metrics were also calculated:

1. Macrophyte Biological Index for Rivers – IBMR (Haury et al., 2002),
2. Index based on Ellenberg nitrogen values for plant species (Ellenberg et al., 1992),
3. Number of species,
4. and Shannon diversity index (Shannon & Weaver, 1949).

Sensitivity of indicators to various sources of error

The replicate survey of plant species was carried out by the group of six trained surveyors. The aim was to provide unaltered unbiased conditions for subsequent surveys by avoiding plant removal (especially for scarce species) whilst still providing accurate identification of difficult taxa. In all surveyed river sites the physical characteristics of river channel composition were recorded as percentage covers according to the STAR methodology (Dawson, 2002).

During the field surveys four experiments were conducted to assess different sources of variation (Supplementary material):

- inter-surveyor variation,
- temporal variation,
- influence of shading,
- impact of hydromorphological degradation.

Inter-surveyor variation was estimated during summer 2003 by comparing the macrophyte scores achieved by three independent and fully trained surveyors for the same 26 river sites.

Temporal variation was assessed by surveying 26 sites in summer and autumn 2003 and again in summer 2004. The variation between years was assessed by comparing the field surveys in June/July 2003 with those in summer 2004 on the same 26 river sites. Impact of seasonal variation was estimated by comparing the vegetation and derived metric values recorded in early summer (June/July) 2003 with the plant cover and values in early autumn (September) 2003. For any particular site, the same person carried out all three surveys, so that differences were focused on temporal sources of variation; and to avoid effects of spatial differences between surveys, the starting point was coordinated with GPS, maps and detailed drawn plans.

Influence of shading and hydromorphological degradation on river ecosystem was estimated in the separate experiments in 2004 (5 June–10 August). The sensitivity of the MTR method to shading (caused mainly by trees growing along rivers) was estimated by surveying matched pairs of sites on the same rivers in summer 2004. On each of 23 river stretches, macrophytes were surveyed in two sites within several hundred meters of each other, one unshaded and one shaded. The two matched sites had very similar environmental conditions in terms of water depth and width, current velocity, hydromorphological conditions and substrate. Absence of pollution discharge between pairs was checked.

Sixteen matched pairs of sites (modified and unmodified) on different rivers or river sections were selected to test the impact of physical modifications of river channel (e.g., bridges, reinforcements, regulations) on the MTR score. The two sites within each pair were selected to be within 1 km of each other but representing different classes of hydromorphological degradation.

The series of experiments enabled the assessment of natural background variation focusing mainly on temporal sources of variation (differences between years and seasons of the year) and influence of physical parameters like hydromorphological degradation and shading.

To estimate the influence of individual factors (surveyors, seasons, years, shading, modifications) on trophic indices and biological diversity, the Wilcoxon Matched-Pairs Signed-Ranks Test was used on the appropriate set of paired sites (Siegel & Castellan, 1988). It is a nonparametric test to estimate whether the parts of a pair differ in size and does not require population with normal distribution.

The software programme STARBUGS – STAR Bioassessment Uncertainty Guidance Software (Clarke, 2004) was used to test the effect of using particular status class boundaries on the status obtained for sites (initially without any assessment of uncertainty). The ecological status class assessment for individual metrics is evaluated as normalised Ecological Quality Ratios (EQRs) involving the ratio of the observed metric values (O) to the Reference Condition values (E_1) of the metric (Formula 1).

Formula 1. Ecological Quality Ratio.

$$\mathrm{EQR} = \frac{O - E_0}{E_1 - E_0}$$

where: O – observed value, E_1 – value of metric for which EQR = 1 (Reference Condition value), E_0 – value of metric for which EQR = 0 (Extreme bad status value).

By setting the E_0 values to zero, and the E_1 values to the RIVPACS-type model expected value under Reference Conditions, the EQR values become RIVPACS-type O/E ratios of the observed (O) to expected (E) values of metrics or biotic indices (Formula 2) (Clarke et al., 1996; Wright et al., 2000; Clarke et al., 2002; Clarke, 2004).

The probability of misbanding a site was tested for the whole possible range of EQR values using STARBUGS software. Class limits and level of uncertainty based on undertaken field studies were used in the setup of the programme. Class limits for the Ecological Quality Ratios (EQRs) were proposed for three macrophyte metrics (MTR, IBMR and Ellenberg index) as specified in Table 5. The criteria of setting the particular class limits was the evenness of probability of misbanding a site of each EQR class. The probability of misbanding a site of each true EQR was simulated using observed value (O) designed to provide EQR values covering the full range from extreme bad (EQR = 0) to reference condition (EQR = 1). The uncertainty standard deviations in the observed metric values were based on the estimated standard deviation due to the effects of one or more of inter-surveyor and temporal variation, shading and morphological modifications. The standard error (SE) of the mean of a metric's values for the reference sites was used as the uncertainty standard deviation in the estimated expected value (E) (i.e., reference condition) for the metric to simulate errors (R) in the expected value (Formula 2).

Formula 2. Simulated O/E ratio of the observed (O) to expected (E) values.

$$O/E = \frac{O + S}{E + R}$$

where: O/E – estimated ratio of the observed (O) to expected (E) values, O – observed value of metric, S – random value due to sampling or other sources of variation, E – expected (reference condition) value of metric, R – random error for reference condition value.

The estimated probabilities of misbanding a site based on the STARBUGS simulations for the whole possible range of true EQR are presented graphically (Fig. 4). Each figure shows the effect of considering a different source of detected uncertainty (inter-surveyor and temporal variation, shading and morphological transformations).

Results

Comparison of selected indices and metrics as indicators of site quality

In total, 227 plant species were recorded during the macrophyte surveys and monocotylodynes and dicotylodynes were the dominant groups. Among Pteridiophytes three species were found, and seven Bryophytes. Seventy of the identified taxa are indicators listed in the MTR protocol and thus have been assigned values of STR (Species Trophic Rank) covering the full range of the water trophic gradient – from oligotrophic (STR = 9 and 10) to eutrophic species (STR = 1 and 2). The most common species were *Elodea canadensis* Michaux., *Lemna minor* Linné, *Glyceria maxima* (Hartman) Holmberg, *Phalaris arundinacea* Linné, *Sparganium emersum* Rehmann and *Nuphar lutea* (Linné) Smith.

The surveyed rivers were rich in plant species (from 17 to 85 taxa per site). There were statistically significant differences between status classes in the number of plant species present (ANOVA $F_{4,198} = 8.61$, $p < 0.00001$, $n = 203$) and Shannon's diversity index (ANOVA $F_{4,198} = 7.12$, $p < 0.00002$, $n = 203$) (StatSoft Inc., 2004). The largest numbers of species were found on sites representing moderate status of water quality whereas more oligotrophic conditions as well as eutrophic were poorer in species (Fig. 2).

The study sites encompass the wide range of trophic conditions present in Polish Lowlands and the scale of the gradient is evident according to biological trophic indices (Table 1, Fig. 3).

Results of field experiments for estimation of the influence of temporal and inter-surveyor variation, shading conditions and morphological transformations are summarised in Table 2. For each macrophyte metric and factor, the standard deviation (SD) of the values due to the certain factor were calculated for each site and then averaged across sites to derive the values labelled 'SD mean' in Table 2. For example, the inter-surveyor SD mean is based on calculating the SD of the three surveyors values for each site in turn and then averaging across the 26 sites. In addition, the average coefficient of variation (CV mean) due to each factor was also calculated for each metric by dividing the SD by the average value for each site in turn and then averaging across sites.

For the Ellenberg index, IBMR and number of species the highest standard deviation was observed for pairs of sites differing in levels of anthropogenic modification. Variation in the Shannon diversity index was strongly influenced by shading and the largest component of variation in MTR values was found between surveys carried out in different years. In general, for all indices (except Shannon index) differences between surveyors caused the lowest variability (i.e., lowest

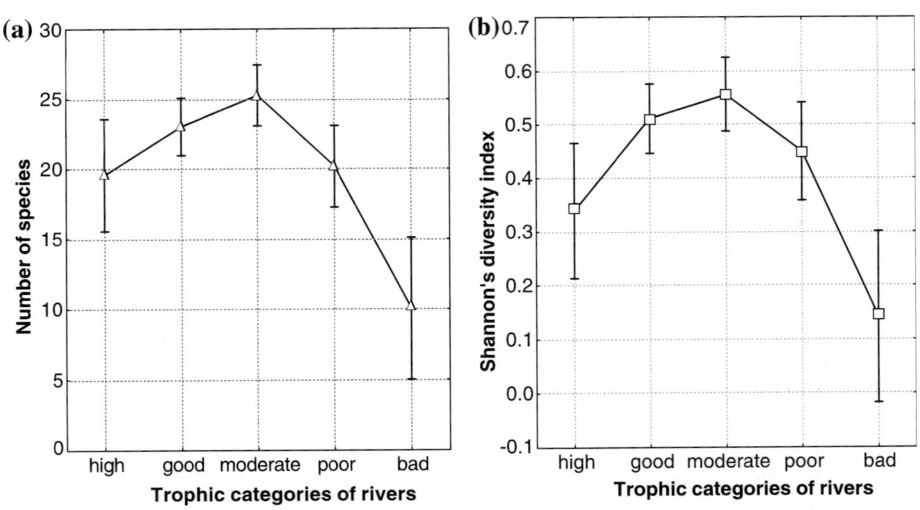

Figure 2. Mean (and its 0.95 confidence interval) of Shannon's index (squares) and the number of species (triangles) in relation to the ecological status pre-classification of sites. (triangle symbol) Number of species (square symbol) Shannon's diversity index.

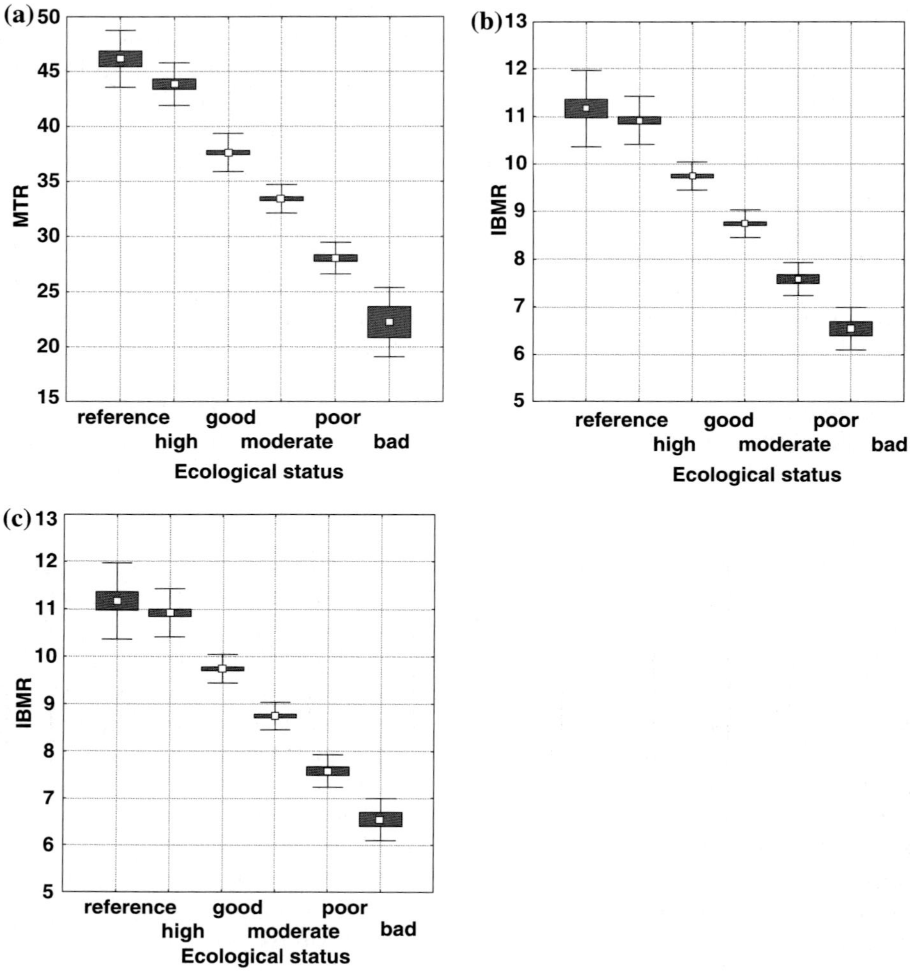

Figure 3. Distribution of the trophic metrics (MTR, Ellenberg index, IBMR) for different ecological status of rivers. (□) mean, (■) mean ± SE, (⌶) mean ± SD.

SD) in metric values amongst all of the factors assessed in this study.

Assessment of the sensitivity of indicators to various sources of error

To make comparisons between metrics of the relative size of different sources of variation (inter-surveyor, temporal, shade and habitat modification) the variance (the SD squared) of each metric's values caused by each factor was divided by the total variance amongst all values of that metric (Table 3). This expresses how the total variability of metric values across all sites of different types and conditions can attributed to each source of variation. If a very high proportion of total variability is due to inter-surveyor differences, then many apparent differences between sites could depend on the fact that different people did the surveys. It should be noted that these relative variance estimates are not strict analysis of variance (ANOVA) because they are based on simply SD derived from different subsets of the dataset.

The overall detected level of variance of selected factors was generally the lowest for MTR and IBMR, which suggests that these metrics are less susceptible to the effects of these factors. Although the Ellenberg index was relatively resistant to inter-surveyor variability and was relatively

Table 2. Basic statistics for selected metrics

Metrics	Surveyors	Years	Season	Shading	Modific.
MTR					
MTR SD mean	1.62	2.47	2.18	1.78	1.95
MTR CV mean	4.75	6.89	6.07	5.20	5.48
Ellenberg index					
Ellenberg index SD mean	0.11	0.13	0.18	0.19	0.24
Ellenberg index CV mean	1.78	1.94	2.82	2.97	3.57
IBMR					
IBMR SD mean	0.34	0.36	0.34	0.45	0.52
IBMR CV mean	3.83	3.83	3.81	5.08	5.34
Number of species					
Number of species SD mean	2.67	5.09	4.27	4.77	5.57
Number of species CV mean	14.37	24.48	20.92	25.68	26.29
Shannon's index					
Shannon's index SD mean	0.13	0.09	0.10	0.25	0.17
Shannon's index CV mean	29.23	23.96	23.25	53.44	39.84

repeatable between years, it was strongly disturbed by habitat modification. Changes in vegetation between seasons also influence its score very much. The two diversity metrics were strongly influenced by shading, especially Shannon's index, which was also very sensitive to habitat modifications. Shannon's diversity was, on average, much lower in shaded sites (mean = 0.29) than in unshaded sites (mean = 0.63).

Analysis showed that metrics react differently to particular factors. The Wilcoxon Matched-Pairs Signed-Ranks Test shows statistical significance of different metrics depending on selected factors, as like shading (in case of IBMR, number of species and Shannon index), season (Ellenberg index) and year of study (IBMR) (Table 4). There were no significant differences in MTR for any factor.

The estimated effect of sources of variation in metric values on assessments of site condition was assessed using the STARBUGS programme (Clarke, 2004) to estimate the uncertainty in classifying sites into ecological status classes and probabilities of mis-classification. The probability of mis-classifying a site is based on the simulation ideas of Clarke (2000), as used in the UK RIVPACS software system for site assessment (Wright et al., 2000) and developed into the STARBUGS software. Site assessments were based on just three of the metrics (MTR, IBMR and Ellenberg) because the two other metrics (number of species and Shannon index) did not seem applicable in the detection of eutrophication. Class limits for the Ecological Quality Ratios (EQRs) for trophic metrics were estimated as follows: 0.9 (high/good), 0.7 (good/moderate), 0.5 (moderate/poor) and 0.3 (poor/bad). Such a distribution of EQR limits gives the lowest probability of misbanding a site in each class.

Table 3. Estimated percentages of total variance for selected indices due to different factors (total variance excludes 10% of outlier values)

Source of variability	MTR	IBMR	Ellenberg Index	Shannon's index	Number of species
Surveyors	8	5	12	16	6
Years	16	8	15	11	22
Season	19	7	33	10	23
Modifications	8	12	54	36	15
Shading	12	15	23	78	33

Table 4. Results of Wilkoxon's test ($^{*}p < 0.05$; $^{**}p < 0.01$; $^{***}p < 0.001$)

Parameter	Factors combination	*p* Level
MTR	Summer 2003 vs. Summer 2004	
	Summer 2003 vs. Autumn 2003	
	Modified vs. Unmodified	
	Shaded vs. Unshaded	
Ellenberg's index	Summer 2003 vs. Summer 2004	
	Summer 2003 vs. Autumn 2003	*
	Modified vs. Unmodified	
	Shaded vs. Unshaded	
IBMR	Summer 2003 vs. Summer 2004	**
	Summer 2003 vs. Autumn 2003	
	Modified vs. Unmodified	
	Shaded vs. Unshaded	*
Number of species	Summer 2003 vs. Summer 2004	
	Summer 2003 vs. Autumn 2003	
	Modified vs. Unmodified	
	Shaded vs. Unshaded	**
Shannon's index	Summer 2003 vs. Summer 2004	
	Summer 2003 vs. Autumn 2003	
	Modified vs. Unmodified	
	Shaded vs. Unshaded	***

The whole range of settings implemented in the analysis (STARBUGS software) is given in Table 5.

The effect of inter-surveyor variation was tested in the first stage of studies (Fig. 4). The probability of misbanding a site because of inter-surveyor variability in macrophyte assessment is much higher if Ellenberg index is used than using either the MTR or IBMR methods (which gives very similar results).

Uncertainty in estimation of status class due to the effects of temporal variability was the highest for the Ellenberg index, but lower for the IBMR and MTR. Uncertainty in status class assignment due to the effect shading and modification was the highest for the Ellenberg index and the lowest for the MTR in both cases (Fig. 4).

Discussion

The sites selected for replicate sampling in Polish Lowlands were rich in macrophyte species and encompassed a wide environmental gradient. Having a large number of recorded species as well as large numbers of taxa for many sites improved the quality of the study. Species diversity positively influences river assessment preciseness (Environment Agency, 1996; Dawson et al., 1999; Holmes et al., 1999). The scale of the gradient is evident according to biological trophic indexes and such a conditions can be regarded as favourable for high quality studies.

Diversity of aquatic plant species was differentiated between sites representing different ecological status and showed unimodal distribution along the status gradient and peaking at intermediate site qualities. Similar relationship can be observed when several factors together are disturbing the ecosystem (Boyce, 1984; Jusik & Zgola, 2004).

The detected level of variance caused by the studied factors was low in the case of MTR and IBMR and it can be assumed that they are predictable trophic metrics of relatively high sampling precision and robustness. For the MTR the impact of every factor was not significant with the mean CV always less than 10%. The Ellenberg index which was originally calibrated on nitrogen level detection (Ellenberg et al., 1992) was resistant for

Table 5. Status class boundaries and uncertainty of sampling variation for selected metrics analysed using STARBUGS software

Metric name	Status class boundaries (for O or O/E)				Sampling variation (S) of different sources of uncertainty in observed (O) value					Refe-rence condition value (E)	Random error (R) for reference condition (E)
	Proposed class limit (EQR)				Inter-surveyor variation	Years variation	Season variation	Influence of channel shading	Influence of channel modifi-cations		
	high	good	mode-rate	poor	SD	SD	SD	SD	SD		
MTR	0.9	0.7	0.5	0.3	1.62	2.47	2.18	1.78	1.95	42.61	0.79
Ellenberg index	0.9	0.7	0.5	0.3	0.11	0.13	0.18	0.19	0.24	2.67	0.09
IBMR	0.9	0.7	0.5	0.3	0.34	0.36	0.34	0.45	0.52	10.77	0.17

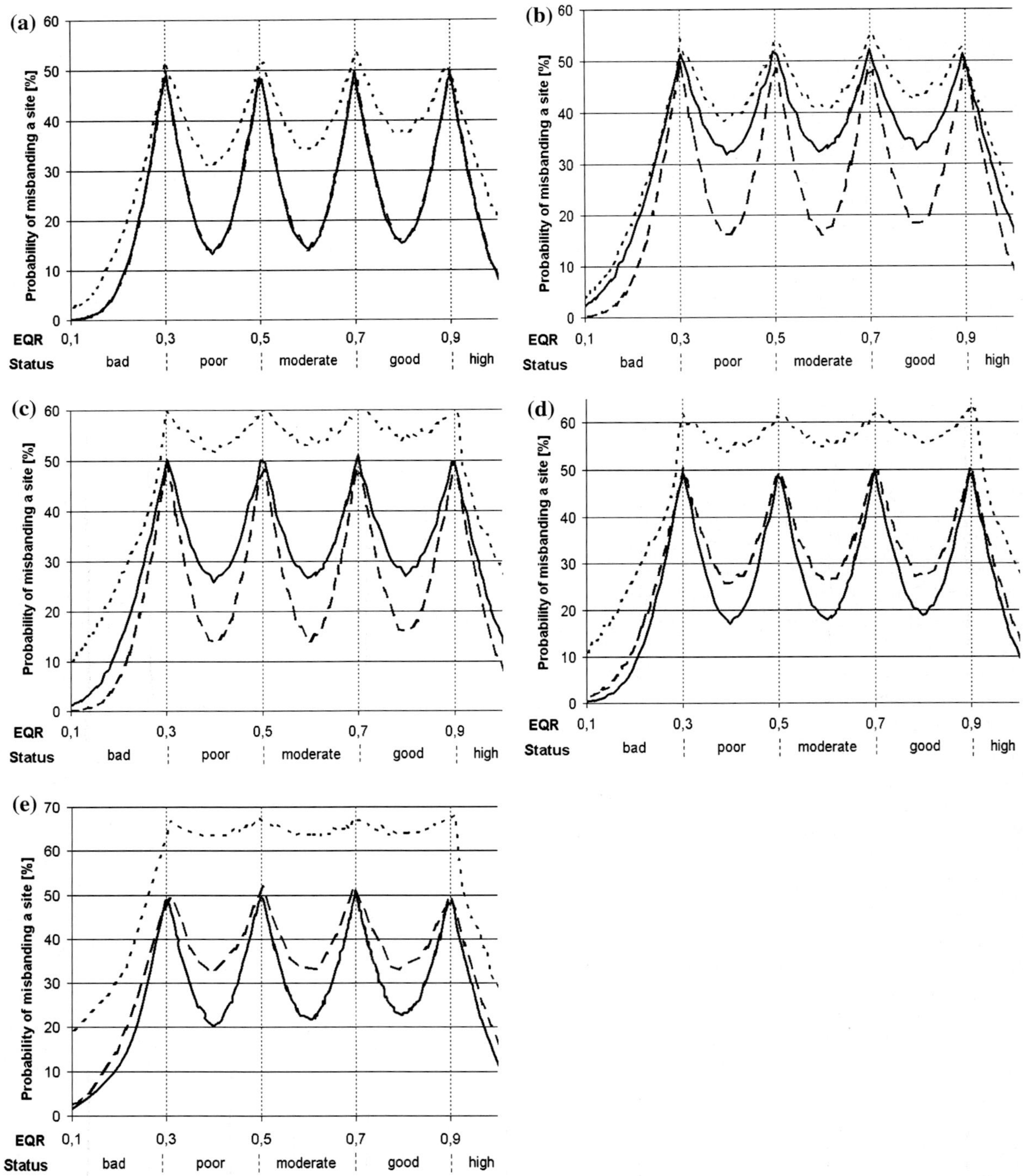

Figure 4. Plots of the probability of misbanding of site in relation to the true value of the EQR for macrophyte metrics: a – estimation of sampling variation due to the inter-surveyor effects, b – estimation of sampling variation due to temporal variation (different years), c – estimation of sampling variation due to the temporal variation (different seasons), d – estimation of sampling variation due to the extent of shading, e – estimation of sampling variation due to the effect of modifications, MTR – solid line, IBMR – dashed line, Ellenberg – dotted line.

inter-surveyor variability and was repeatable between years but it was strongly affected by habitat modification. Changes in vegetation between seasons also influenced its score very much. The diversity metrics are strongly influenced by shading, especially Shannon's index, which was also very sensitive to habitat modifications. The Ellenberg nitrogen index as well as the tested diversity metrics seem to be highly susceptible to various sources of variation and thus may not be sufficiently precise and reliable as tools in the classification of river site ecological status and their application requires further development.

The uncertainty associated with individual sources of variation (influence of temporal and inter-surveyor variation, shading conditions and morphological transformations) were assessed separately. The combined effect of all existing error sources on uncertainty in site classification and rates of mis-classification is higher than any single source but the mis-classification rates due to individual sources cannot simply be added together. Quantifying their collective effect on overall uncertainty was not feasible within this study although the variance components due to the various sources of variation in metric values should in theory be additive (Dawson et al., 1999).

The detected range of the MTR and IBMR values on Polish lowland was unfortunately low and it can influence reliability of classification (Clarke at al., 1996). In the case of Mean Trophic Rank the potential scale is between 10 and 100 but on the Polish lowland rivers (Core stream type 2) the smallest score was 18 and the largest 52. Other surveys on lowland rivers in Europe have recorded higher scores (Holmes et al., 1999). Further studies are required in wider range of river types with established class boundaries.

Analysis showed that sampling uncertainty was low enough to utilise macrophyte based metrics for estimating ecological status of rivers according to WFD (Directive, 2000). The detected level of variance can be utilised to estimate potential effects of sampling variation on uncertainty of site status class assessment and probability of mis-grading a site (for example using the STARBUGS uncertainty software). It must be underlined that assessment of macrophyte methods raised in this paper concerns only the aspect of uncertainty and the development of preciseness of these methods should also focus on calibration against particular stressors.

Conclusions

- Our results assessing the sources and extent of variation showed that several metrics based on the indicative value of plant species surveyed using the MTR method are sufficiently robust to potential sources of inherent variation that they can be used for reasonable precision in the assessment of the ecological status of river sites.
- Variances due to the various sources of uncertainty are relatively low compared with the total variance for MTR and IBMR especially. The estimated minimum probability of mis-grading a site in a middle ecological status class due to individual sources of variation varied from 15% (MTR and IBMR) to 50% in case of Ellenberg index.
- Among the estimated sources of variance the smallest role was indicated by the inter-surveyor factor. Slightly stronger was the influence of temporal variation (years and seasons) and shading. The impact of habitat modification was the most important.
- Obtained results based on the data from Polish rivers shows the major sources of variation and assess their impact on the final values of commonly used macrophyte metrics. It facilitates evaluation of the limitation of confidence in assessments based on aquatic plants, providing hints applicable in designing future field surveys.
- Analysis showed that some of macrophyte-based metrics (notably MTR and IBMR) are of sufficient precision in terms of sampling uncertainty, that they could be useful for estimating the ecological status of rivers in accordance with the aims of the WFD.

Acknowledgements

We want to thank to Jerzy Kupiec, Dominik Mendyk and Tomasz Zgola for their help during field studies and to Klaudia Borowiak and Justyna Urbaniak for laboratory analysis. This work was supported by EC project STAR EVK1-CT-2001-00089.

References

Allan, J. D., 1995. Stream Ecology. Structure and Function of Running Waters. Chapman and Hall, London, UK.

Boyce, M. S., 1984. Restitution of r – and K – strategy selection as a model of density – dependent natural selection. Annual Review of Ecology and Systematics 15: 427–447.

Clarke, R. T., M. T. Furse, J. F. Wright & D. Moss, 1996. Derivation of a biological quality index for river sites: comparison of the observed with the expected fauna. Journal of Applied Statistics 23: 311–332.

Clarke, R. T., 2000. Uncertainty in estimates of river quality based on RIVPACS. In Wright, J. F., D. W. Sutcliffe & M. T. Furse (eds), Assessing the Biological Quality of Freshwaters: RIVPACS and Similar Techniques. Freshwater Biological Association, Ambleside, 39–54.

Clarke, R. T., M. T. Furse, R. J. M. Gunn, J. M. Winder & J. F. Wright, 2002. Sampling variation in macroinvertebrate data and implications for river quality indices. Freshwater Biology 47: 1735–1751.

Clarke, R. T., 2004. STARBUGS 1.1 (STAR Bioassessment Uncertainty Guidance Software). Error/Uncertainty module software, Centre for Hydrology and Ecology in Dorset (CEH).

Dawson, F. H., P. J. Raven & N. T. H. Holmes, 1998. Distribution of aquatic plants by morphological group for rivers in the UK. 10th EWRS Symposium on Aquatic Weeds 1998, Lisbon, 1–5.

Dawson, F. H. & K. Szoszkiewicz, 1999. Relationships of some ecological factors with the associations of vegetation in British rivers. Hydrobiologia 415: 117–122.

Dawson, F. H., J. R. Newman, M. J. Gravelle, K. J. Rouen & P. Henville, 1999. Assessment of the Trophic Status of Rivers using Macrophytes: Evaluation of the Mean Trophic Rank. R&D Technical Report E39. Environment Agency of England & Wales, Bristol, UK.

Dawson, F. H., 2002. Guidance for the field assessment of macrophytes of rivers within the STAR Project. http://www.eu-star.at/frameset.htm.

Directive, 2000/60/EC. Water Framework Directive of the European Parliament and of the Council of 23 October 2000.

Ellenberg, H., H. E. Weber, R. Dull, V. Wirth, W. Werner & D. Baulissen, 1992. Zeigerwerte von Pflanzen in Mitteleuropa. Scripta Geobotanica, Vol. 18.

Environment Agency, 1996. Methodology for the assessment of freshwater riverine macrophytes for the purposes of the Urban Waste Water Treatment Directive. Enviroment Agency, May 1996 Version 2.

Haury, J., M. -C. Peltre, S. Muller, M. Trémolières, J. Barbe, A. Dutartre & M. Guerlesquin, 1996. Des indices macrophytiques pour estimer la qualité des cours d'eau français: premières propositions. Ecologie 27: 233–244.

Haury, J., M. Jaffre, A. Dutartre, M. C. Peltre, J. Barbe, M. Tremolieres, M. Guerlesquin & S. Muller, 1998. Application of the standardized protocol "Milieu Et Vegetaux aquatiques fixes" to 12 French rivers: preliminary floristic typology. International Journal of Limnology 34: 129–139.

Haury, J., M. -C. Peltre, M. Tremolieres, J. Barbe, G. Thiebaut, I. Berne, H. Daniel, P. Chatenet, S. Muller, A. Dutartre, C. Laplace-Treyture, A. Cazaubon & E. Lambert-Servien, 2002. A method involving macrophytes to assess water trophy and organic pollution: the Macrophyte Biological Index for Rivers (IBMR) – application to different types of rivers and pollutions. In Dutartre A. & M.-H. Montel (eds), Proc. 11th EWRS International Symposium on Aquatic Weeds, Moliets Et Maa, France, 247–250.

Holmes, N. T. H., J. R. Newman, S. Chadd, K. J. Rouen, L. Saint & F. H. Dawson, 1999. Mean Trophic Rank: A users manual. R&D Technical Report E38. Environment Agency of England & Wales, Bristol, UK.

Jusik, S. & T. Zgola, 2004. Influence of morphological transformation of littoral zone on species diversity of macrophytes. Zeszyty Naukowe AR w Krakowie, Inzynieria Srodowiska 412: 311–320.

Newman, J. R., F. H. Dawson, N. T. H. Holmes, S. Chadd, K. J. Rouen & L. Sharp, 1997. Mean Trophic Rank: A User's Manual. R&D Technical Report E38. Environment Agency, Bristol, 1–130.

Schneider, S., T. Krumpholz & A. Melzer, 2000. Trophäeindikation in Fliessgewässern mit Hilfe des TIM (Trophäe-Index Macrophyten) – Erprobung eines neu entwickelten Index im Inniger Bach. Acta hydrochimica et hydrobiologica 28: 241–249.

Shannon, C. E. & W. Weaver, 1949. The Mathematical Theory of Communication. University of Illinois Press, Urbana.

Siegel, S. & N. J. Castellan, 1988. Nonparametric Statistics for the Behavioral Sciences (2nd edn.). McGraw-Hill, New York.

Staniszewski, R., 2001. Estimation of river trophy in the Kujawskie Lakeland using Mean Trophic Rank and Chemical Index of Trophy. Roczniki AR Poznan 334, Botanika 4: 165–173.

StatSoft, Inc., 2004. STATISTICA (data analysis software system), version 6. www.statsoft.com.

Szoszkiewicz, K., K. Karolewicz, A. Lawniczak & F. H. Dawson, 2002. An assessment of the MTR aquatic plant bioindication system for determining the trophic status of Polish rivers. Polish Journal of Environmental Studies 11: 421–427.

Wright, J. F., D. W. Sutcliffe & M. T. Furse (eds), 2000. Assessing the Biological Quality of Fresh Waters: RIVPACS and Other Techniques. Freshwater Biological Association, Ambleside.

Hydrobiologia (2006) 566:247–260
M.T. Furse, D. Hering, K. Brabec, A. Buffagni, L. Sandin & P.F.M. Verdonschot (eds), The Ecological Status of European Rivers: Evaluation and Intercalibration of Assessment Methods
DOI 10.1007/s10750-006-0092-5

Uncertainty in diatom assessment: Sampling, identification and counting variation

Anna Besse-Lototskaya*, Piet F.M. Verdonschot & Jos A. Sinkeldam
Freshwater Biology, Alterra, Wageningen University and Research, P.O. Box 47, 6700 AA Wageningen, The Netherlands
*(*Author for correspondence: E-mail: anna.besse@wur.nl)*

Key words: diatoms, multimetric index, standardisation, ecological assessment

Abstract

Despite the widespread application of periphytic diatoms to water quality assessment at a regional level, there is no standard European sampling protocol or associated assessment metrics. Furthermore, relatively little is known about the uncertainty in the results of such assessments. One of the objectives of the European project for the Standardisation of River Classifications (STAR) is to improve and standardise diatom assessment methods. An extensive diatom ring test, together with an audit of the project results, provided a better understanding and quantification of the uncertainty in quality assessment of running waters using diatoms. The variation in multimetric analysis shows that the choice of site and substrate for sampling, the inter-operator differences in diatom taxonomy and the counting techniques are the primary sources of uncertainty. To some extent, this variation also reveals the robustness of specific metrics in relation to the sources of uncertainty. Of the three most common substrate types tested (stone, macrophyte and sediment), macrophytes emerge as the most preferred substrate for diatom sampling when performing multimetric water quality assessment.

Introduction

Periphytic diatoms have been included in the assessment of river water quality since the early 1900s, are known to be reliable indicators of water conditions, and can be used successfully for assessing water quality in running waters (e.g., Coring, 1999; Ector & Rimet, 2005). It has been shown that diatoms react to changes in the intensity of eutrophication, acidity, saprobity, nitrogen, salinity and current velocity (e.g., Denys, 1991a, b; Battarbee et al., 1997; van Dam, 1997; Kelly, 1998; Coring, 1999). Many studies of running water diatom assemblages by different research schools present and compare the results from different regions, water types and microhabitats. However, there relatively little is still known about the comparability of these results and the degree of uncertainty in diatom assessment caused by differences in methodology for sampling, identification and counting techniques. Several studies show that, for example, the choice of substrate for sampling can play an important role in the assessment of the diatom community (Stevenson & Hashim, 1989; Snoeijs, 1991; Rolland et al., 1997; Rothfritz et al., 1997; Kelly et al., 1998). However, substrate type differs from stream type to stream type. For example, sampling stones in lowland streams is difficult or impossible because they are simply not present. The use of macrophytes as substrate induces problems caused by differences in the composition and abundance of diatom species colonising different parts of macrophytes, such as leaves, stem or root (Cazaubon, 1996). In contrast, Gomez & Licursi (2001) conclude that soft sediment provides the most appropriate diatom community for monitoring in lotic systems. Furthermore, factors such as the choice of sampling site, and the methods for preparing and processing the sample and identifying the taxa can be crucial to the assessment results.

This paper attempts to identify and quantify the sources of uncertainty in the assessment of diatoms from running waters by comparing the results of an extensive diatom ring test that included simultaneous sampling from multiple sites and substrates, and replicate sampling and slide preparation, performed by different operators. The results of the ring test are also used to evaluate the reliability and precision of metrics derived from the diatom community. The objective of this paper was to test the diatom assessment methodology that was selected by the STAR project as a standard for studying European stream and river systems (Furse et al., this volume), and to audit the quality of diatom results from the project.

Materials

Diatom ring test

In order to identify and quantify the error in diatom assessment introduced by sampling, identification and counting, a ring test was performed during a training course on the Plaine River, France. A total of 116 samples were taken and analysed by 10 of the 14 partners that participated in the STAR diatom studies according to a standard diatom field and laboratory protocol (Furse et al., this volume). The samples were taken from 2 sites (PL0 and PL5), and from 3 substrate types (stones (H), macrophytes (M) and sediments (S)). The most common substrate type at both sampling sites was stone. Generally, each partner collected and analysed 3 samples from two out of three habitat types at each site (Table 1). Additionally, another test was performed where one participant (Alterra) prepared and analysed replicate slides from two of the ring test samples (4 in one sample, and 5 in another sample).

Audit of STAR diatom results

The main diatom assessment for the STAR project was performed by 14 partners according to the standardised diatom field and laboratory protocol (Furse et al., this volume). The samples were collected from various stream types, processed and analysed (i.e., identified and counted) by the partners in their respective laboratories.

The quality of the partners' analyses was audited by Alterra. Thirty eight percent of the STAR diatom samples were randomly selected from all the samples taken by each project partner. A total of 107 slides were analysed a second time by the auditor, where a new count and identification was undertaken according to the protocol.

Table 1. The distribution of the diatom samples per site (PL0 and PL5), substrate (H = stone, M = macrophyte, S = sediment) and partner

Partner	Code	Institute	PL0H	PL0M	PL0S	PL5H	PL5M	PL5S	Total
1	UK	Centre for Ecology and Hydrology, UK	3		3	3			9
2	D	University of Essen & Research & Institute Senkenberg, Germany		3	2		3	3	11
3	A	University of Agricultural Sciences, Vienna, Austria	3		3	3		3	12
5	S	Swedish University of Agricultural Sciences, Sweden	3	3		3	3		12
6	C	Masaryk University, Brno, Czech Republic	3		3	3		3	12
8	I1	Istituto di Recerca sulle Acque (IRSA-CNR), Italy	3		3	3		3	12
9	P	University of Evora, Portugal	3	3		3	3		12
10	DK	National Environmental Research Institute, Denmark		3	3		3	3	12
13	I2	Province of Bolzano (LABBIO), Italy	3	3		3	3		12
14	F	University of Metz, France		3	3		3	3	12
		Total	21	18	20	21	18	18	116

Methods

Taxonomic adjustment

In order to compare the diatom results between the partners, the nomenclatural differences between the partners first had to be resolved. Taxa were identified to the lowest achievable taxonomic level (species and/or variety/forma). By exchanging the results amongst the STAR partners through a round of comments, the level of identification was raised and the results improved. The taxonomic nomenclature used for all the results was then adjusted to the standardised STAR diatom taxa list that was agreed by all partners and experts.

Diatom metrics

The comparison of the diatom results was based on a total of 17 metrics. The OMNIDIA program (Lecointe et al., 2003) was used to compute 14 different diatom metrics that are regularly used to assess several aspects of water quality, mainly in running waters (Table 2). Other parameters such as number of taxa, Shannon diversity (Zar, 1996) and evenness (Zar, 1996) were also used in the comparison. The values of the metrics IPS, SLAD, DSECY, L&M, SHE, WAT, TDI, EPI-D, ROTT, IDG, CEE, IBD and IDAP were transformed by the OMNIDIA program to a scale from 0 to 20; the scale of number of taxa and Shannon diversity is infinite; the evenness and %PT values range between 0–1 and 0–100, respectively.

Data analyses

The values of diatom metrics for all samples were compared using the average value and the standard deviation per respective group of samples.

The diatom ring test results were split into the following sets, for which the average value and the standard deviation were calculated:

- the whole database (1 set of 116 samples)
- per site (2 sets of 57 and 59 samples, respectively)
- per substrate (3 sets of 42, 36, and 38 samples, respectively)
- per partner (10 sets of 9–12 samples each)
- per replicate sample (39 sets of 2–3 samples each)

The average value and the standard deviation for the replicate slide dataset were plotted for each sample (2 sets of 4 and 5 samples).

The relative diatom counts were ordinated by redundancy analysis (RDA) with the program

Table 2. Metrics used for the comparison of the diatom results

Abbreviation	Full name	Reference
No. taxa	Number of taxa	
Diversity	Shannon diversity	(MVSP, 2001)
Evenness	Evenness	(MVSP, 2001)
IPS	Specific Pollution Sensitivity Metric	(Coste, 1987)
SLAD	Sládeček's pollution metric	(Sládeček, 1986)
DESCY	Descy's pollution metric	(Descy, 1979)
L&M	Leclercq & Maquet's pollution metric	(Leclercq & Maquet, 1987)
SHE	Steinberg & Schiefele trophic metric	(Steinberg & Schiefele, 1988)
WAT	Watanabe et al. pollution metric	(Lecointe et al., 2003)
TDI	Trophic Diatom metric	(Kelly & Whitton, 1995)
%PT	% pollution tolerant taxa	(Kelly & Whitton, 1995)
EPI-D	Pollution metric based on diatoms	(Dell'Uomo, 1996)
ROTT	Trophic metric	(Rott et al., 1999)
IDG	Generic Diatom Metric	(Lecointe et al., 2003)
CEE	Commission for Economical Community metric	(Descy & Coste, 1991)
IBD	Biological Diatom Metric	(Prygiel & Coste, 1999)
IDAP	Indice Diatomique Artois Picardie	(Lecointe et al., 2003)

CANOCO 4.5 (Ter Braak & Ŝmilauer, 2002). The data analysis is fully described by Verdonschot & Ter Braak (1994). RDA assumes a linear model for the relationship between the response of each taxon and the ordination axes and is used if the gradient length in the data is short (< 4 units of standard deviation [SD]; Ter Braak, 1988). In our case the gradient length was smaller then 3 SD (axis 1: 2.1 and axis 2: 3.0) which implies that the data are quite homogeneous. RDA is the constraint form of PCA of taxon data, in which the components (axes) are constrained by linear combinations of environmental variables. The ordination results are presented as correlation biplots of sites and environmental variables (Verdonschot & Ter Braak, 1994). The eigenvalue of an ordination axis in RDA is the proportion of the total variance explained by that axis and indicates its relative importance. An unrestricted permutation test is used to test the validity of the total ordination. This technique is fully explained by Ter Braak & Ŝmilauer (2002) and Verdonschot & Ter Braak (1994). For this ordination the full diatom dataset was used, and the parameters partner, site, substrate and replicate were defined as nominal and included as environmental parameters.

For the audit database, the results from the partners were compared with the audit results. The average value and standard deviation were calculated for each partner versus audit sample (in total 107 sets each of 2 samples).

Furthermore, the average value of the SD was calculated for each of the above-mentioned datasets, in order to compare the methodological errors.

Diatom metrics from the ring test were also compared between partners and between replicates using an analysis of variance (ANOVA). The variance components were estimated by averages of restricted maximum likelihood (Patterson & Thompson, 1971). The hypothesis that there were no differences in the variance of metric values between replicate samples regardless of substrate type or sample site was tested with a chi-squared test. This test employs deviances differences as produced by restricted maximum likelihood. Analyses were performed with GenStat 8.11 (VSN International Ltd, 2002). If the probability (p) was less than 0.05, it was assumed that the hypothesis was not true, indicating that there was a relationship between the results and the sampling substrate or/and sampling site.

The correlation between each partner's and the auditor's metric was calculated, assuming that the correlation between the two is linear. The coefficient of determination (R^2) for each of the metrics is based on a dataset of 117 partner–audit samples. The R^2 above 0.5 is considered to indicate a relationship.

Results and discussion

The RDA ordination (total ordination is significant ($p < 0.002$)) of all samples resulted in a biplot (Fig. 1) that shows a clear separation of samples taken at site PL0 versus samples of site PL5. The two groups of samples are fully separated while at the same time each of the three replicate samples is plotted close to each other, while each of the groups of three replicate samples is clearly separated from the other groups of replicates.

Next, two ordination were run to establish the importance of either partner, or substrate, or

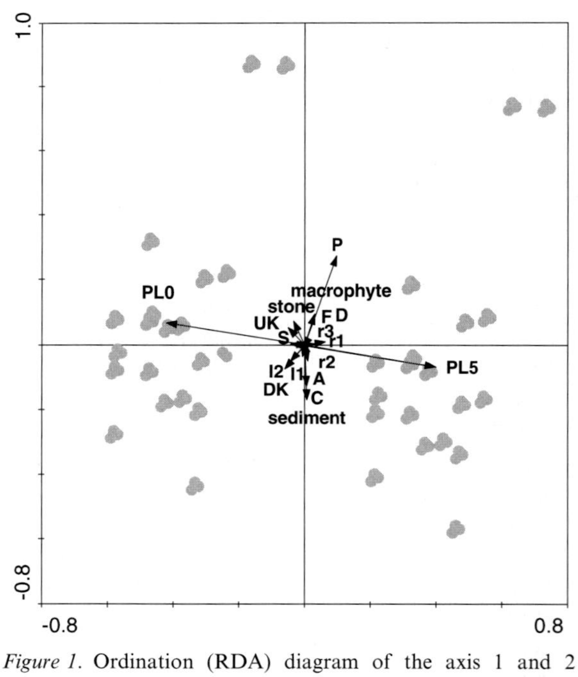

Figure 1. Ordination (RDA) diagram of the axis 1 and 2 (eigenvalues 0.18 and 0.11, respectively) showing the variation in the distribution of samples (grey dots) among environmental variables (arrows) in the two ring test sites (PL0 and PL5). Partner codes are given in Table 1. Replicates are coded r1, r2 and r3.

replicate as explanatory variable. The ordination of all samples of site PL0 (total ordination is significant ($p < 0.002$)) shows a clear separation based on the substrate parameters (Fig. 2). The grouping of replicates remains. The ordination of all samples taken at site PL5 (total ordination is significant ($p < 0.002$)) shows a different pattern (Fig. 3). Here the differences between partners, especially Czech Republic, Portugal, Denmark, Italy (second partner) and Sweden are most explaining. Again, the grouping of replicates remains. The differences found between sites PL0 and PL5 are due to the differences in homogeneity, in the sense of variation in environmental conditions between the habitats present, between the two sites. At the more homogeneous site PL0 the samples collected by the different partners were more alike. At this site the differences between substrates prevailed. The more heterogeneous site PL5, where local variation between habitats was much larger, resulted in an in-between partner variation. The differences between partners, who collected their replicate samples from individual spots within the stream site, can possibly be due to the instream

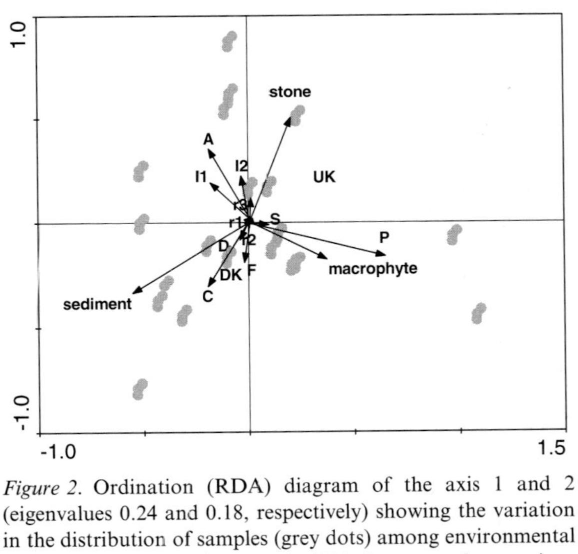

Figure 2. Ordination (RDA) diagram of the axis 1 and 2 (eigenvalues 0.24 and 0.18, respectively) showing the variation in the distribution of samples (grey dots) among environmental variables (arrows) at ring test site PL0. Partner codes are given in Table 1. Replicates are coded r1, r2 and r3.

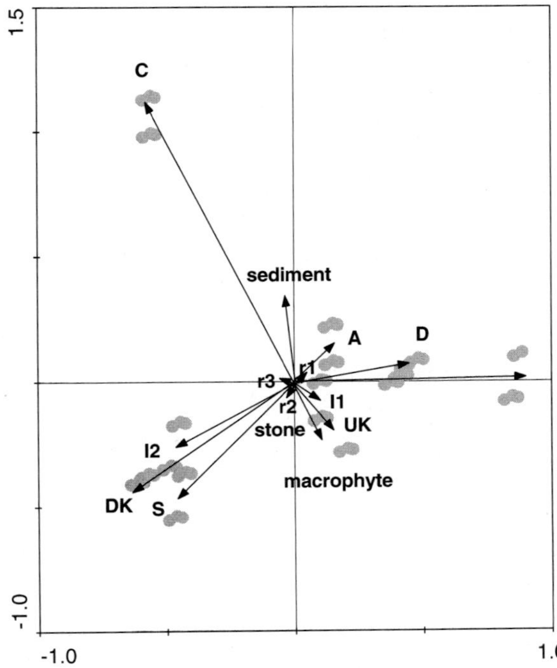

Figure 3. Ordination (RDA) diagram of the axis 1 and 2 (eigenvalues 0.18 and 0.16, respectively) showing the variation in the distribution of samples (grey dots) among environmental variables (arrows) at ring test site PL5. Partner codes are given in Table 1. Replicates are coded r1, r2 and r3.

variation. The consequence is that to sample diatoms in a representative way one should collect subsamples spread over various spots of, for the eye, more or less the same substrates within a larger transect of the stream site.

The metric results of the ring test are presented in Figure 4. Ideally, all 116 samples of the ring test should show the same value for each metric if the sampling site or substrate type fully represents the composition of the diatom community in the river and when all partners use exactly the same sampling, sample processing, identification and counting techniques. However, our results show differences between samples taken from different sites and substrates, by different partners and between replicate samples. It is thus important to find out:

Figure 4. The diatom ring test results, per metric. The horizontal axe indicates the sample number and is sorted by substrate type. The data in each substrate zone are sorted by site and further by partner. Each sample includes multiple values from replicate samples. "p" is the metric of the chi-squared test examining the hypothesis of variance in values between replicates regardless the sampled substrate. "sd" is the average standard deviation of metric values depending on the substrate type. Dashed line indicates the average metric value, per substrate type.

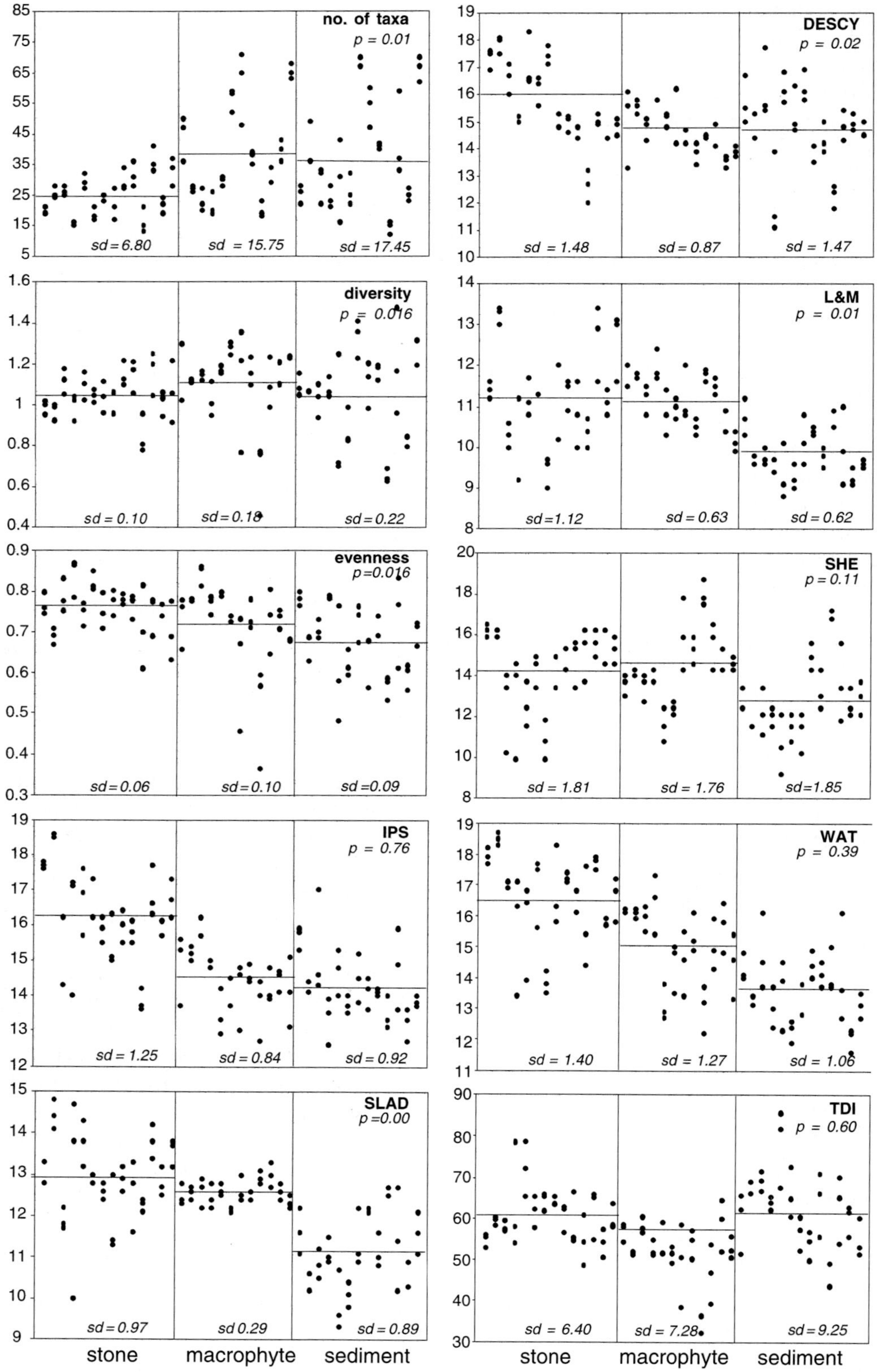

no. of taxa
p = 0.01
sd = 6.80
sd = 15.75
sd = 17.45
DESCY
p = 0.02
sd = 1.48
sd = 0.87
sd = 1.47
diversity
p = 0.016
sd = 0.10
sd = 0.18
sd = 0.22
L&M
p = 0.01
sd = 1.12
sd = 0.63
sd = 0.62
evenness
p = 0.016
sd = 0.06
sd = 0.10
sd = 0.09
SHE
p = 0.11
sd = 1.81
sd = 1.76
sd = 1.85
IPS
p = 0.76
sd = 1.25
sd = 0.84
sd = 0.92
WAT
p = 0.39
sd = 1.40
sd = 1.27
sd = 1.06
SLAD
p = 0.00
sd = 0.97
sd 0.29
sd = 0.89
TDI
p = 0.60
sd = 6.40
sd = 7.28
sd = 9.25
stone
macrophyte
sediment
stone
macrophyte
sediment

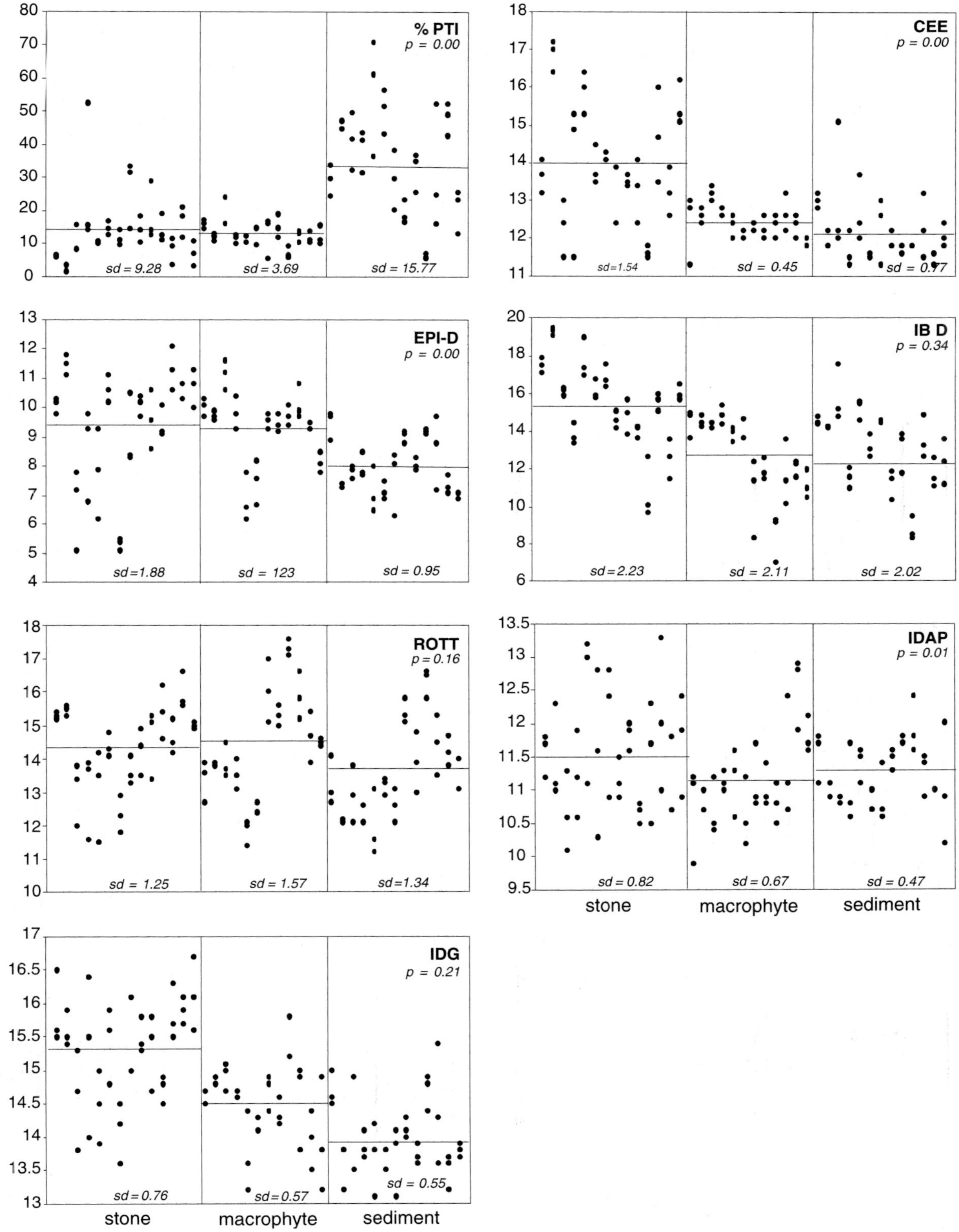

Figure 4. (Continued)

- what the sources of differences/uncertainty are;
- rank the sources of differences/uncertainty in order of importance and quantify the uncertainty;
- apply knowledge from the above to improve diatom assessments.

Choice of substrate for sampling

One of the most interesting and important questions that was posed for our diatom ring test was the choice of substrate for sampling and its influence on the diatom assessment results. Up to now, there is no standard method for sampling periphytic diatoms in running waters, and previous diatom studies have used various substrate types for sampling (among others Descy & Coste, 1991; Gomez, 1998; Rott et al., 1998). The results of studies testing the relationship between the diatom community and the substrate type were often conflicting. While several (Rothfritz et al., 1997; Winter & Duthie, 2000) showed that there is no consistent difference in the results of water quality monitoring using diatom community structure from different substrates, Snoeijs (1991) found that different types of substrates were colonised differently. Studies by Stevenson and Hashim (1989), Rolland et al. (1997) and Rothfritz et al. (1997) revealed significant differences in diatom diversity between different substrate types. Recent studies suggest hard natural or artificial substrates are the most suitable and reliable for ecological studies of periphytic diatoms (e.g., Kelly et al., 1998). Our diatom ring test included sampling from stone, macrophyte and sediment substrates and these are the most common substrates in running waters. Two sites were sampled by 10 partners in 2–3 replicates. When sorted according to substrate type, the variation and values of the resulting metric values between replicate samples and partners were compared by substrate type. This was done in order to relate the choice of substrate to the reliability of the results (Fig. 4).

Variation

In order to establish a possible relationship between the type of the substrate sampled and the results of the ring test, we compared the variation in each metric (Fig. 4, Table 3). The chi-squared test revealed that 8 metrics varied with habitat, namely number of taxa, SLAD, DESCY, L&M, % PT, EPI-D, CEE and IDAP ($p < 0.05$). The variation between replicate samples and between the partners, expressed as SDs, is also considerably lower for one (or several) substrate types than the other(s) (Table 3). The number of taxa varied least in the samples taken from the stone substrate; SLAD, DESCY, % PT and CEE, varied least in the macrophyte substrate; EPI-D and IDAP varied least in the sediment substrate; and L&M varied least in macrophyte and sediment samples. There was no relationship between the type of substrate and the variation of the nine other metrics (Table 3).

The relationship established between type of substrate and the variability in the results of 8 of the 17 water quality assessment metrics confirms the importance of the substrate choice when sampling diatoms. When using one of the diatom metrics, one should consider sampling the substrate providing the least variability. The macrophyte substrate is thus preferred for sampling diatoms when water quality is assessed using SLAD, DESCY, % PT and CEE. Sediment offers the most reliable substrate when the assessment is based on EPI-D and IDAP. Either macrophyte, or sediment substrate is preferred to stone when using L&M. The relatively high inter-partner variability in number of taxa for all substrates (especially for the macrophytes and the sediments) can be explained by the diatom counting technique. When we decided the sampling protocol for STAR (Furse et al., this volume), we agreed that 300 valves of diatoms (where possible) should be identified and counted from each sample. However, in a number of cases the protocol was not followed strictly; more valves were counted and/or the slide was surveyed for additional (rare) taxa after the count was completed. These technical discrepancies probably led to significant differences in the number of taxa found between the partners. The stone substrate, however, shows here the least variability of all substrates, and is thus the most representative substrate for sampling when using the number of taxa.

In the case of diversity, evenness, IPS, SHE, WAT, TDI, IDG and IBD, the choice of substrate for sampling does not play a significant role.

The macrophyte substrate gives the least variability (Table 3) and therefore appears to be the

Table 3. Variation (inter-partner and replicate) by substrate type, results of the chi-squared test of variance in replicate samples, and preferred substrate type, for the 17 diatom water quality assessment metrics

Metric	Standard deviation (sd)			p	Preferred substrate
	Stone	Macrophyte	Sediment		
Number of taxa	6.80	15.75	17.45	0.01	Stone
Diversity	1.05	1.11	1.05	0.16	None
Evenness	0.06	0.10	0.09	0.16	None
IPS	1.25	0.84	0.92	0.76	None
SLAD	0.97	0.29	0.89	0.00	Macrophytes
DESCY	1.48	0.87	1.47	0.02	Macrophytes
L&M	1.12	0.63	0.62	0.01	Macrophytes/sediment
SHE	1.81	1.76	1.85	0.11	None
WAT	1.40	1.27	1.06	0.39	None
TDI	6.40	7.28	9.25	0.60	None
%PT	9.28	3.69	15.77	0.00	Macrophytes
EPI-D	1.88	1.23	0.95	0.00	Sediment
ROTT	1.25	1.57	1.34	0.16	None
IDG	0.76	0.57	0.55	0.21	None
CEE	1.54	0.45	0.77	0.00	Macrophytes
IBD	2.23	2.11	2.02	0.34	None
IDAP	0.82	0.67	0.47	0.01	Sediment

most appropriate for diatom sampling in 5 out of 8 cases. In the remaining 3 cases the results do not differ between any of the substrate types. Thus, we conclude that if one uses a multiple metric approach, the macrophyte substrate is preferred more than the stone or the sediment as the substrate to be sampled.

Average value

The differences in the metric average values of each metric in each substrate type can also indicate the reliability of the substrate type and should be taken into account when choosing a substrate for sampling. To compare the discrepancies between the diatom metrics directly, the metrics should be given a common scale, as each metric is based on a different range of scores. Scaling is difficult because some metrics can score infinitely, so we only perform a qualitative comparison.

The average values of 11 of the 17 metrics vary significantly between substrates (number of taxa, IPS, SLAD, L&M, WAT, TDI, %PT, EPI-D, IDG, CEE and IBD; Fig. 4). This sequence of metrics is not related to the degree inter-substrate variation.

In the case of the number of taxa, the mean is considerably lower for the stone substrate compared to the macrophyte or the sediment substrate. The samples collected from stone never contained more than 40 taxa, whereas the samples from macrophytes and sediment included up to 68–70 taxa. The difference is due to the rare taxa collected by a number of partners on the macrophyte and the sediment substrates. In general, the partners collected less taxa on the stone substrate compared to the other substrates, although one would assume that the stone substrate supports a diatom community of multiple seasons, and is therefore richer in diversity. Our data oppose this hypothesis and suggest that if one strives to collect a high variety in diatom taxa, one should not focus on stone. However, stone is still the most reliable substrate (least variance) when comparing the number of taxa between the partners, independent of the counting technique used. All other diatom water quality assessment metrics are based on the relative abundance of different taxa and should therefore either be less or unaffected affected by the presence or absence of rare taxa in the samples.

The average values of all the other metrics (diversity, evenness, DESCY, SHE, ROTT, IDAP) (Fig. 4) did not vary significantly between substrates.

For the majority of the metrics where average values varied significantly between substrates, the average values from samples collected from macrophytes seldom exceed or are less than those of the other substrates. This is interpreted as a favourable feature of sampling from macrophytes.

Summarising the results of our ring test, the samples from the macrophyte substrate generally reveal the lowest inter-partner and inter-replicate variability, and show average values in comparison to the samples from the stone and sediment, and therefore should be used as preferred substrate for diatom sampling when performing multimetric water quality assessments.

Sources of uncertainty in diatom assessment

In order to assess and reduce the impact of error and uncertainty on diatom metrics, it is crucial to understand the sources of uncertainty, to quantify the error for each step of diatom sampling, and to identify measures that can decrease this uncertainty.

Uncertainty is introduced during all steps of sampling and can be quantified (Table 4). Our diatom ring test, together with the audit of analytical quality, provided a dataset that we used to quantify the sources of uncertainty and to set them in the order of significance. The variations (standard deviation and average standard deviation) in the diatom metric results in total, by site, substrate type, partner, replicate sample and by replicate slide are listed in Table 5.

The variation between samples from different sites (ring test) represents the uncertainty caused by the choice of sampling site; the variation between samples from different substrates (ring test) shows the contribution of the substrate choice during sampling to the uncertainty; the variation between replicate samples (ring test) adds to the uncertainty during sample collection. The variation between replicate slides in the ring test (the comparison of diatom composition from different slides taken from the same sample and identified and counted by the same operator) is similar to the variation between the partner and the auditor (diatom compositions taken from the same slide and identified and counted by different operators) because the random fields chosen for diatom identification and counting are different between both. The variation between partners (ring test) represents the uncertainty over all steps of sampling.

The total variation, expressed in the average SD between all samples of the ring test, comprises the sum of errors introduced at all steps of sampling and is highest in 12 out of 17 diatom metrics. The variation between the samples from different sites is generally the second highest (8 out of 17 cases) and is always lower than or equal to the total variation. The variation due to partner and substrate is generally ranked third or fourth. Partner is ranked third 9 times and fourth 5 times and substrate is ranked third 4 times and fourth 8 times. For the remaining audit is ranked fifth 7 times, replicate samples sixth 10 times and replicate slides seventh 16 times. Thus the order

Table 4. Linking diatom sampling steps to measures of uncertainty

<table>
<tr><th>sampling steps</th><th colspan="3">quantification of uncertainty</th></tr>
<tr><td>choice of sampling site</td><td colspan="2">variation between samples from different sites</td><td rowspan="4">variation between partners</td></tr>
<tr><td>choice of substrate for sampling</td><td colspan="2">variation between samples from different substrates</td></tr>
<tr><td>sample collection</td><td colspan="2">variation in replicate samples</td></tr>
<tr><td>slide preparation</td><td>variation in replicate slides</td><td>Variation partner-auditor</td></tr>
</table>

All sampling steps include uncertainty due to taxonomic identification and counting.

Table 5. Average standard deviation values (sd) for the ring test and the project audit results (after correction)

Average sd	Total	Site	Partner	Substrate	Audit	Repl. samples	Repl. slides
Number of taxa	15.11 (1)	14.81 (2)	6.15 (5)	13.33 (3)	11.60 (4)	4.05 (6)	1.97 (7)
Diversity	0.17 (2)	0.17 (3)	0.13 (5)	0.17 (4)	0.34 (1)	0.08 (7)	0.10 (6)
Evenness	0.09 (1)	0.08 (2)	0.07 (4)	0.08 (3)	0.05 (5)	0.04 (6)	0.03 (7)
IPS	1.39 (1)	1.34 (2)	1.12 (3)	1.00 (4)	0.64 (5)	0.57 (6)	0.32 (7)
SLAD	1.11 (1)	1.08 (2)	0.95 (3)	0.72 (4)	0.39 (6)	0.41 (5)	0.19 (7)
DESCY	1.38 (1)	1.10 (4)	1.13 (3)	1.27 (2)	0.50 (5)	0.49 (6)	0.34 (7)
L&M	1.04 (2)	1.04 (1)	0.94 (3)	0.79 (4)	0.41 (5)	0.39 (6)	0.16 (7)
SHE	1.94 (1)	1.63 (4)	1.84 (3)	1.81 (2)	0.52 (6)	0.83 (5)	0.36 (7)
WAT	1.76 (2)	1.76 (1)	1.54 (3)	1.24 (4)	0.79 (5)	0.69 (6)	0.45 (7)
TDI	8.78 (1)	8.37 (2)	7.24 (4)	7.64 (3)	5.91 (5)	4.35 (6)	3.50 (7)
% PT	14.57 (1)	14.15 (2)	12.36 (3)	9.58 (4)	2.14 (6)	5.12 (5)	1.97 (7)
EPI-D	1.55 (1)	1.48 (2)	1.29 (4)	1.35 (3)	0.48 (6)	0.55 (5)	0.30 (7)
ROTT	1.41 (1)	1.10 (4)	1.31 (3)	1.38 (2)	0.38 (6)	0.55 (5)	0.11 (7)
IDG	0.85 (1)	0.84 (2)	0.71 (3)	0.63 (4)	0.45 (5)	0.41 (6)	0.31 (7)
CEE	1.34 (2)	1.30 (3)	1.02 (4)	0.92 (5)	2.35 (1)	0.49 (6)	0.31 (7)
IBD	2.45 (1)	2.01 (3)	1.98 (4)	2.12 (2)	0.55 (6)	0.71 (5)	0.47 (7)
IDAP	0.69 (2)	0.68 (3)	0.59 (5)	0.65 (4)	1.21 (1)	0.49 (6)	0.38 (7)

of magnitude of variation is (from highest to smallest): total variation > sampling site variation > partner variation > substrate type variation > audit variation > sample collection variation (replicate sample) > slide preparation variation (replicate slide).

The variation between partners is larger than between substrates as it also includes replicate sample and slide variation. The audit variation is greater than the replicate sample and slide variation. Audit variation = slide variation + variation in analytical quality between partners. Some partners' analysts may consistently over or under count. Besides, as the audit comprised a variation in sites and water qualities, the identification uncertainty increased in comparison to the assessment of homogenous two ring test sites, despite the fact that identification uncertainty was reduced as much as possible by performing taxonomic adjustments to the standardised list of taxa.

In our experiment, we tested all steps of the sample processing procedure for diatom assessment. One must realise that, in the approach chosen, site variation includes partner, substrate, replicate sample and replicate slide variation, and substrate variation includes replicate sample and replicate slide variation. The variation is cumulative, in the order indicated. This is large in sampling site, partner variation, and substrate type, as all three approach the value for the total variation. The variation is small for slide and somewhat larger but still small for sample variation. The audit variation exceeds sample variation slightly. In general, uncertainty increases when inter-partner variation is introduced. This variation is related to the differences in environmental circumstances at the sampled sites, and also includes differences, though small, in sample treatment and preparation of slides between laboratories. All variation also includes uncertainty due to variation in identification (number of taxa and number of rare species detected with or without an extra survey for rare taxa) despite the adjustment to standard taxonomic list, and due to the number of valves counted.

Audit of the project results

The audit was performed to test whether partners kept to the protocol and whether the identification was correct. Ideally, project and audit results should be highly correlated because the auditor uses the same slide as the project partner did, and thus, only replicate slide variation is still present. The correlation between the partner and audit diatom metric results is generally high ($R^2 > 0.5$) (Table 6). However, the audit results show that the

Table 6. Correlation of diatom metrics between the STAR results and the audit

Metric	R^2 coefficient
Number of taxa	0.575
Diversity	0.557
Evenness	0.512
IPS	0.699
SLAD	0.751
DESCY	0.570
L&M	0.696
SHE	0.732
WAT	0.623
TDI	0.528
%PT	0.896
EPI-D	0.620
ROTT	0.748
IDG	0.818
CEE	0.288
IBD	0.771
IDAP	0.004

average variation between the audit and the project partner is always higher than the replicate slide variation but is generally lower than the total variation of the diatom ring test (Table 5). Unfortunately, during the STAR project the diatom protocol was not always strictly followed. Sometimes, different number of valves was counted and different substrates were sampled. Together with the different stream types surveyed these differences caused the observed variation.

The coefficient of determination R^2 is indicative for the susceptibility of a diatom metric to the determination differences between operators. Thus, the metric % PT appears to be the most robust ($R^2=0.90$), whereas metrics CEE ($R^2=0.29$) and IDAP ($R^2=0.00$) are probably sensitive to the identification and counting techniques (Table 6).

Conclusions and suggestions

This study based on an extensive diatom ring test and an audit of analytical quality, showed that sampling protocol plays a crucial role in the assessment of water quality using diatoms. The choice of sampling site and substrate type, and the taxonomic identification contribute the most to the uncertainty in the resulting water quality metrics. There is much controversy in the literature about the relationship between the substrate type and the composition of diatom assemblages, and its influence on water quality assessments. Of the three most common substrate types (stone, macrophyte and sediment), macrophyte generally gave the most consistent results, and thus should be used as preferred substrate for diatom sampling (if used with care) when performing multimetric water quality assessment. However, some diatom metrics perform better with samples from other substrates. In order to standardise the substrate choice in the sampling protocol, a further evaluation is needed in which the multimetric results are tested in relation to the substrate type only, while excluding all other variables such as site and partner.

Besides the choice of the substrate type, taxonomic identification proves to be another important contributor to the uncertainty in diatom assessment. Some partners are more experienced and more careful than others, and some have more skill. Some metrics are affected more by the degree of skill of the analysts, whilst others are more robust and do not require such highly skilled staff. This skill includes the ability to identify diatoms correctly and to differentiate between them, and to could accurately evaluate the sample without under- or over-estimation. These skills are often undervalued and the time that they take to develop is often not realised. If staff are not given sufficient time, the results will suffer. Good operational (c.f. research) laboratory management is a balance between analytical quality and number of samples analysed.

Furthermore, the ordination showed that if one would like to get a full and representative picture of the diatoms present at a stream stretch, one should collect subsamples. In order to overcome the within stream variation, the subsamples should be spread over various spots of, for the eye, more less the same substrates within a larger stretch of the stream site.

The taxonomic adjustment to a standardised list, incorporated in the protocol (Furse et al., this volume) is an excellent tool to reduce the uncertainty. However, the audit of the project results demonstrated that continuous active co-operation between diatom taxonomists is necessary in order to further improve the standard taxonomic list and thus, reduce the uncertainty in the assessment results. Our experiment also proved that the standardisation of the counting technique of diatom valves could significantly reduce the uncertainty in metric results. It is essential to follow strictly the protocol of counting 300 valves and afterwards searching the slide for rare taxa in order to assess the total diversity.

Diatom metrics that are currently used for assessing water quality usually have regional character, mainly because of the limited geographical variation in the data used. Comparing results between the different ecoregions is therefore problematic, but can be improved by using a multimetric analysis. Our study shows that some metrics are more sensitive to application by different operators and at different sampling sites in different geographical regions and substrate types than the others. The computation and evaluation existing data from different European stream sites in multimetric context could possibly lead to the establishment of one or several metrics that could be used to assess water quality conditions at a European scale.

Acknowledgements

This paper is a result of the EU-funded project STAR (6th Framework Programme; contract number: EVK1-CT-2001–00089). We thank J. van der Molen for the initiation of the experiment and co-ordinating the collecting of the data. We thank all our STAR partners for their data collection, identification and input. We are grateful to our colleagues at Wageningen University and Research M. van den Hoorn for technical assistance in processing the data, and to P. Goedhart for the statistical analyses. We also thank John Murray-Bligh and an anonymous reviewer for their useful comments.

References

Battarbee, R. W., R. J. Flower, S. Juggins, S. T. Patrick & A. C. Stevenson, 1997. The relationship between diatoms and surface water quality in the Hoylandet area of Nord-Trondelag, Norway. Hydrobiologia 348: 69–80.

Cazoubon, A., 1996. Algal epiphytes, a methodological problem in river monitoring. In Whitton, B. A. & E. Rott (eds), Use of Algae for Monitoring Rivers II. Innsbruck, Institut für Botanik, Universität Innsbruck, 47–50.

Coring, E., 1999. Situation and developments of algal (diatom)-based techniques for monitoring rivers in Germany. In Prygiel, J., B.A. Whitton & J. Bukowska (eds), Use of Algae for Monitoring Rivers III. Agence de l'Eau Artois-Picardie, Douai, 122–127.

Coste, M., 1987. Etude des méthods biologique quantitatives d'appréciation de la qualité des eaux. Rapport Division Qualité des Eaux Lyon. Agence de l'Eau Rhône, 28.

Dell'Uomo, A., 1996. Assessment of water quality of an Appanine river as a pilot study for diatom-based monitoring of Italian water courses. In Whitton, B. A. & E. Rott (eds), Use of Algae for Monitoring Rivers II. Innsbruck, Institut für Botanik, Universität Innsbruck, 64–72.

Denys, L., 1991a. A check-list of the diatoms in the holocene deposits of the Western Belgian coastal plain with a survey of their apparent ecological requirements. I. Introduction, ecological code and complete list. Ministère des Affaires Economiques – Service Géologique de Belgique.

Denys, L., 1991b. A check-list of the diatoms in the holocene deposits of the Western Belgian coastal plain with a survey of their apparent ecological requirements. II. Centrales. Ministère des Affaires Economiques – Service Géologique de Belgique.

Descy, J.-P., 1979. A new approach to water quality estimation using diatoms. Nova Hedwigia 64: 305–323.

Descy, J.-P. & M. Coste, 1991. A test of methods for assessing water quality based on diatoms. Verhandlung Internationale Vereingung de Limnologie 24: 2112–2116.

Ector, L. & F. Rimet, 2005. Using bioindicators to assess rivers in Europe: an overview. In Lek, S., M. Scardi, P. F. M. Verdonschot, J. -P. Descy & Y. -S. Park (eds), Modelling Community Structure in Freshwater Ecosystems. Springer: 7–19.

Furse, M., D. Hering, O. Moog, P. Verdonschot, R. K. Johnson, K. Barbec, K. Gritzalis, A. Buffagni, P. Pinto, N. Friberg, J. Murray-Bligh, J. Kokes, R. Alber, P. Usseglio-Polatera, P. Haase, R. Sweeting, B. Bis, K. Szoszkiewicz, H. Soszka, G. Springe, F. Sporka & I. Krno, 2006. The STAR project: context, objectives and approaches. Hydrobiologia 566: 3–29.

Gomez, N., 1998. Use of epipelic diatoms for evaluation of water quality in the Matanza-Riachuelo (Argentina), a pampean plain river. Water Research 32: 2029–2034.

Gomez, N. & M. Licursi, 2001. The Pampean Diatom Index (IDP) for assessment of rivers and streams in Argentina. Aquatic Ecology 35: 173–181.

Kelly, M. G., 1998. Use of the trophic diatom index to monitor eutrophication in rivers. Water Research 32: 236–242.

Kelly, M. G., A. Cazaubon, E. Coring, A. Dell'Uomo, L. Ector, B. Goldsmith, H. Guasch, J. Hürlimann, A. Jarlman, B. Kawecka, J. Kwandrans, R. Laugaste, E.-A. Lindstrøm, M. Leitao, P. Zarvan, J. Padisák, E. Pipp, J. Prygiel, E. Rott, S. Sabater, H. van Dam & J. Vizinet, 1998. Recommendations for the routine sampling of diatoms for water quality assessments in Europe. Journal of Applied Phycology 10: 215–224.

Kelly, M. G. & B. A. Whitton, 1995. The Trophic Diatom Index: a new index for monitoring eutrophication in rivers. Journal of Applied Phycology 7: 433–444.

Leclercq, L. & B. Maquet, 1987. Deux noveaux metrics diatomique et the qualite chimique des eaux courantes. Comparison avec differents metrics existants. Cahiers de Biologie Marine 28: 303–310.

Lecointe, C., M. Coste & J. Prygiel, 2003. Omnidia 3.2. Diatom Index Software Including Diatom Database with Taxonomic Names, References and Codes of 11645 Diatom Taxa.

MVSP: Multi-Variate Statistical Package, 1986–2001. Kovach Computing Services, Pentraeth, Wales, UK.

Patterson, H. D. & R. Thompson, 1971. Recovery of interblock information when block sizes are unequal. Biometrika 58: 545–554.

Prygiel, J. & M. Coste, 1999. Progress in the use of diatoms for monitoring rivers in France. In Prygiel, J., B. A. Whitton & J. Bukowska (eds), Use of Algae for Monitoring Rivers III. Douai, Agence de l'Eau Artois-Picardie: 165–179.

Rolland, T., S. Fayolle, A. Cazaubon & S. Pagnetti, 1997. Methodical approach to distribution of epilithic and drifting algae communities in a French subalpine river: inferences on water quality assessment. Aquatic Sciences 59: 57–73.

Rott, E., H. C. Duthie & E. Pipp, 1998. Monitoring organic pollution and eutrophication in the Grand River, Ontario, by means of diatoms. Canadian Journal of Fisheries and Aquatic Sciences 55: 1443–1453.

Rott, E., P. Pfister, H. Van Dam, E. Pipp, K. Pall, N. Binder & K. Ortler, 1999. Indikationslisten für aufwuchsalgen. Wien. Bundesministerium für Land - und Forstwirtschaft 248.

Sládeček, V., 1986. Diatoms as indicators of organic pollution. Acta Hydrochimica et Hydrobiologica 14(5): 555–566.

Snoeijs, P. J. M., 1991. Monitoring pollution effects by diatom community composition: a comparison of sampling methods. Archiv für Hydrobiologie 121: 497–510.

Steinberg, C. & S. Schiefele, 1988. Biological indication of trophy and pollution of running waters. Zeitschrift für Wasser- und Abwasser-Forschung 21: 227–234.

Stevenson, R. J. & S. Hashim, 1989. Variation in diatom community structure among habitats in sandy streams. Journal of Phycology 25: 678–686.

Ter Braak, C. J. F., 1988. CANOCO – A FORTRAN Program for canonical community ordination by [partial] [detrended] [canonical] correspondence analysis, principal component analysis and redundancy analysis (version 2.1). Report LWA-88-02. Agricultural Mathematics Group, Wageningen.

Ter Braak, C. J. F. & P. Ŝmilauer, 2002. CANOCO Reference Manual and Users Guide to Canoco for Windows. Software for Canonical Community Ordination (version 4.5). Centre for Biometry, Wageningen, the Netherlands.

Van Dam, H., 1997. Partial recovery of moorland pools from acidification: indications by chemistry and diatoms. Netherlands Journal of Aquatic Ecology 30: 203–218.

Verdonschot, P. F. M. & C. J. F. Ter Braak, 1994. An experimental manipulation of oligochaete communities in mesocosms treated with chlorpyrifos or nutrient additions: multivariate analyses with Monte Carlo permutation tests. Hydrobiologia 278: 251–266.

Winter, J. G. & H. C. Duthie, 2000. Stream epilithic, epipelic and epiphytic diatoms: habitat fidelity and use in biomonitoring. Aquatic Ecology 34: 345–353.

Hydromorphology

Hydrobiologia (2006) 566:263–265

M.T. Furse, D. Hering, K. Brabec, A. Buffagni, L. Sandin & P.F.M. Verdonschot (eds), The Ecological Status of European Rivers: Evaluation and Intercalibration of Assessment Methods
DOI 10.1007/s10750-006-0091-6

Hydromorphology – major results and conclusions from the STAR project

John Davy-Bowker* & Mike T. Furse
Centre for Ecology & Hydrology, Winfrith Technology Centre, Dorchester, Dorset DT2 8ZD, United Kingdom
*(*Author for correspondence: E-mail: jobo@ceh.ac.uk)*

Key words: Water Framework Directive, hydromorphology, River Habitat Survey, data quality, macroinvertebrates.

Abstract

The major results and conclusions of the two papers in the hydromorphology section of the Hydrobiologia special issue on the EU STAR project are summarised. Several key findings have emerged from this research. Firstly, the hydromorphological characteristics of rivers between different geographical regions of Europe were found to vary considerably with rivers in each region possessing distinctive hydromorphological characteristics. Secondly, the hydromorphological attributes that most strongly influence two existing hydromorphological indices (the Habitat Quality Assessment and the Habitat Modification Score) were identified and attention was drawn to the accurate definition and recording of these attributes in field surveys and training courses. Thirdly, links between hydromorphological characteristics and macroinvertebrate quality indices were investigated. Two types of bank modification (resectioning and reinforcement) were significantly correlated with two biotic indices (EPT taxa and MTS), while channel modifications were negatively correlated with ASPT. While biotic indices were often strongly correlated with Habitat Quality Assessment they were less strongly related to Habitat Modification Score suggesting that physical habitat diversity may be more important in determining macroinvertebrate community structure than morphological alteration. The papers in this section provide important underpinning research for the implementation of the European Water Framework Directive. In both papers suggestions are made for further research on the hydromorphology of European rivers.

Introduction

The European Union (EU) Water Framework Directive (Council of the European Communities, 2000), hereafter referred to as the WFD, requires Member States of the EU to assess, monitor and, where necessary, improve the ecological quality of its surface waters. The WFD sets out definitions (called normative definitions) of various surface water quality classes. These describe the biological and physico-chemical standards expected of 'high', 'good' and 'moderate' quality streams and rivers for four biological quality elements (phytoplankton, macrophytes and phytobenthos, benthic invertebrates and fish) and a variety of hydromorphological and physiochemical quality elements. The hydromorphological quality of a river in 'high' status class is defined as follows:

> Channel patterns, width and depth variations, flow velocities, substrate conditions and both the structure and condition of the riparian zones correspond totally or nearly totally to undisturbed conditions.

The WFD also states that hydromorphological assessment of streams and rivers should form part of the operational monitoring programmes of EU Member States (at 6 year intervals) and that hydromorphological assessment methodologies should adhere to the relevant CEN/ISO standard or equivalent national or international protocol.

Hydromorphological assessment therefore forms an integral part of the WFD monitoring and quality assessment process.

Synopsis of papers

The two papers presented in this section provide important underpinning research for the implementation of the WFD by addressing three key questions:

1. How does the hydromorphology of streams and rivers vary across Europe?
2. Which hydromorphological attributes most strongly influence current hydromorphological indices?
3. To what extent are hydromorphological characteristics linked to macroinvertebrate communities?

Both papers make use of the extensive and rigorously standardised STAR dataset covering 13 EU Member States. This new and unique data set (comprising phytobenthos, macrophytes, macroinvertebrates, fish and hydromorphology) for the first time allows analyses of this type to be undertaken at a European scale.

In the first paper, Szoszkiewicz et al. (2006) used 216 STAR sampling sites to examine differences in the hydromorphological characteristics of rivers across Europe. Using River Habitat Survey (RHS) attributes and indices (Environment Agency, 1997), they detailed the essential hydromorphological differences between rivers in four major European regions (lowlands, mountains, the Alps and southern Europe). Their analysis showed that the major direction of environmental variability is along a lowland to mountain/alpine gradient, while a secondary gradient distinguishes southern European rivers from the other regions. Rivers in the four major European regions can be distinguished by quite distinct hydromorphological characteristics. Alpine and mountain rivers were typified by high energy flow types, considerable amounts of bank reinforcement and distinctive land cover categories including 'rock, scree and sand dunes'. Lowland rivers typically had smooth flows and soft banks composed of peat or earth together with natural berms and bank poaching, while the surrounding land cover was often 'wetland' or 'irrigated'. Southern European rivers could be distinguished from both the preceding types in having water levels indicative of low flow conditions, attributes often indicating marked changes in water levels (two-stage channel, composite bank profile, cobble and discrete gravel deposits) and surrounding land cover categories including 'orchards' and 'moorland/heath'. This analysis is unique in its spatial scale and clearly demonstrates that the hydromorphological characteristics of rivers differ considerably across Europe. These results suggest that streams with differing hydromorphological characterises should have different targets for hydromorphological quality and hence the restoration of hydromorphologically degraded rivers should take these natural sources of variation into account.

Szoszkiewicz et al. also highlighted the RHS attributes that most influence the Habitat Quality Assessment (HQA) and Habitat Modification Score (HMS) indices. They recommend that special care and attention should be devoted to the accurate definition and recording of attributes in field surveys and that particular attention should be given to the highest scoring and most influential attributes in RHS field survey training courses. The authors recommend that quality assurance routines need to be included in RHS database software, flagging unusual or extreme values for those attributes that affect the RHS indices.

The second paper (Erba et al., 2006) made use of a subset of 79 hydromorphologically impacted STAR sampling sites, in seven stream types, in order to examine two key issues. Firstly, the component parts of the RHS system were examined to investigate which hydromorphological characteristics were most strongly related to the overall assessment of hydromorphological quality. Bank resectioning and reinforcement were both found to be common sources of alteration resulting in reduction of the hydromorphological quality of the rivers studied.

Secondly, links between hydromorphological characteristics and macroinvertebrate quality were assessed by identifying the hydromorphological features that were most strongly correlated with macroinvertebrate communities. Bank resectioning and reinforcement were found to be significantly correlated with two biotic indices (EPT taxa and MTS). This finding is in agreement with several other studies (see Erba et al., 2006) where bank structure has been found to influence in-stream benthic macroinvertebrate communities. The extent of channel modification was found to be negatively correlated with the ASPT index and the ICMi index was also found to be a good general indicator of morphological alteration.

Erba et al. also investigated relationships between hydromorphological and biotic indices. The biotic indices were found to be strongly correlated with HQA score but less strongly related to HMS. HQA gives high index values to physically diverse sites, suggesting that physical habitat diversity is more influential in determining macroinvertebrate community structure than the extent of hydromorphological modification.

This paper makes an important contribution to our knowledge of the principle sources of hydromorphological impairment across several European countries. The paper also increases our understanding of the importance of hydromorphological alterations for stream macroinvertebrate communities. Both these issues are of growing importance as national agencies seek to understand why streams and rivers may fail to meet their targets as the full implementation of the WFD grows closer. By examining the ecological significance of the hydromorphological features recorded in the RHS system this work also contributes to the refinement of international standards for hydromorphological assessment systems.

Conclusions

The authors of both papers in this section suggest areas where further work on the hydromorphology of European rivers is required to:

1. incorporate quality assurance routines into hydromorphological databases and software to flag unusual or extreme values for those attributes that affect hydromorphological indices.
2. achieve a greater understanding of the complex hydromorphology of southern European streams, both through the extension of existing survey techniques such as the southern European RHS (Buffagni & Kemp, 2002) and through new hydromorphological indices such as the Lentic-lotic River Descriptor – LRD (Buffagni et al., 2004).
3. differentiate between the impact of organic and hydromorphological stresses where these occur at the same sites.
4. relate autecological data for macroinvertebrate species to particular hydromorphological features.

References

Buffagni, A., S. Erba, D. Armanini, D. De Martini & S. Somaré, 2004. Aspetti idromorfologici e carattere lentico-lotico dei fiumi mediterranei: River Habitat Survey e descrittore LRD. In 'Classificazione ecologica e carattere lentico-lotico in fiumi mediterranei'. Quaderni Istituto di Ricerca sulle Acque, Roma 122: 41–63.

Buffagni, A. & J. L. Kemp, 2002. Looking beyond the shores of the United Kingdom: addenda for the application of River Habitat Survey in Southern European rivers. Journal of Limnology 61: 199–214.

Council of the European Communities, 2000. Directive 2000/60/EC, Establishing a framework for community action in the field of water policy. European Commission PE-CONS 3639/1/100 Rev 1, Luxembourg.

Environment Agency, 1997. River Habitat Survey – Field Guidance Manual. Environment Agency, Bristol.

Erba, S., A. Buffagni, N. Holmes, M. O'Hare, P. Scarlett & A. Stenico, 2006. Preliminary testing of River Habitat Survey features for the aims of the WFD hydromorphological assessment: an overview from the STAR Project. Hydrobiologia 566: 281–296.

Szoszkiewicz, K., A. Buffagni, J. Davy-Bowker, J. Lesny, B. H. Chojnicki, J. Zbierska, R. Staniszewski & T. Zgola, 2006. Occurrence and variability of River Habitat Survey features across Europe and the consequences for data collection and evaluation. Hydrobiologia 566: 267–280.

Hydrobiologia (2006) 566:267–280
M.T. Furse, D. Hering, K. Brabec, A. Buffagni, L. Sandin & P.F.M. Verdonschot (eds), The Ecological Status of European Rivers: Evaluation and Intercalibration of Assessment Methods
DOI 10.1007/s10750-006-0090-7

Occurrence and variability of River Habitat Survey features across Europe and the consequences for data collection and evaluation

Krzysztof Szoszkiewicz[1,*], Andrea Buffagni[2], John Davy-Bowker[3], Jacek Lesny[4], Bogdan H. Chojnicki[4], Janina Zbierska[1], Ryszard Staniszewski[1] & Tomasz Zgola[1]
[1]*Department of Ecology and Environmental Protection, August Cieszkowski Agricultural University, ul. Piatkowska 94C, 61-691 Poznan, Poland*
[2]*CNR-IRSA Water Research Institute, Via della Mornera 25, I-20047 Brugherio (Milan), Italy*
[3]*Centre for Ecology and Hydrology, Winfrith Technology Centre, DT2 8ZD Dorchester, Dorset, United Kingdom*
[4]*Department of Agrometeorology, August Cieszkowski Agricultural University, ul. Piatkowska 94B, 61-691 Poznan, Poland*
(**Author for correspondence: E-mail: kszoszk@au.poznan.pl*)

Key words: River Habitat Survey, hydromorphology, variability, data quality, river assessment

Abstract

River Habitat Survey (RHS) data collected for the EU-funded STAR project was used to identify hydromorphological characteristic features of rivers in four European regions namely: lowlands; mountain; the Alps; and the Mediterranean. Using RHS attributes, Habitat Quality Assessment (HQA) – a measure of natural habitat diversity, and Habitat Modification Score (HMS) – a measure of anthropogenic modification, we identified considerable differences in frequency, diversity and evenness of features between the regions. A relatively small subset of features clearly distinguish the hydromorphological characters of lowland, Alpine and southern European rivers. It was more difficult to distinguish mountain rivers from Alpine rivers. The highest statistical differences are observed between Lowland and Mountain region. Within the four regions studied the RHS attributes that most strongly influence the HQA and HMS indices were identified. We conclude that specific effort should be made to ensure these are recorded properly as part of the quality control of RHS data.

Introduction

Implementation of European Water Framework Directive (Directive 2000/60/EC, 2000), hereafter referred to as the WFD, requires estimation of the uncertainty in biological classifications (Clarke & Hering, 2006). Correspondingly, the uncertainty associated with physicochemical and hydromorphological datasets, which provide supporting information for the interpretation of biological results, should to be quantified and minimised. Some European countries already have relatively well developed national programs for hydromorphological river assessment that are suitable for WFD monitoring, although survey methods vary between countries (Raven et al., 2002). For the STAR Project, which sought to provide a standard Europe-wide dataset against which national methods and results could be compared and calibrated a common method was needed (Buffagni & Erba, 2002). As the most advanced method, River Habitat Survey (RHS) was chosen (Environment Agency, 1997; Raven et al., 1997, 1998a).

Several European countries have developed systems to assess river hydromorphology, notably: LAWA-vor-Ort in Germany (Schneider et al., 2003), SEQ-MP in France (Agence de l'Eau Rhin-Meuse, 1996) and RHS in the United Kingdom (Environment Agency, 1997). The RHS methodology enables qualitative and quantitative data on river hydromorphology, geomorphology, habitat and land use to be collected at several different scales in a structured and repeatable fashion. The data collected are well suited to statistical analyses and the calculation of indices useful for assessing river ecosystem quality and degradation. Among these indices, Habitat Quality Assessment (HQA) and Habitat Modification Score (HMS) quantify physical habitat quality and richness, and the degree of morphological degradation, respectively. Approximately 20,000 RHS surveys have been carried out in the UK (Environment Agency, 1997, 2003; Raven et al., 1998a). RHS is also used in other European countries (Buffagni & Kemp, 2002; Zbierska et al., 2002).

By summarizing complex information, river hydromorphogy indices therefore assist in site evaluation (Raven et al., 1998b). The HQA and HMS indices have also recently been supplemented by the Lentic–lotic River Descriptor index (LRD), an index designed to differentiate sites according to their lentic–lotic (slow flowing–fast flowing) character (Buffagni et al., 2004a). The LRD index is designed for use in rivers with high discharge variability (e.g., in the Mediterranean and Alpine regions). The accuracy and therefore use of these indices, derived from RHS data, is critically dependent upon the reliable recording of field survey data (Raven et al., 2002).

The hydromorphological nature of rivers across the European continent varies considerably between different types of rivers. Until now, the lack of a single standard Europe-wide hydromorphological dataset has prevented an analysis of this variability. Furthermore, the extent to which this variability influences the sensitivity of hydromorphological metrics is also unknown. This paper therefore has two main aims:

1) To analyse the distribution of river habitat attributes, their diversity and their evenness, in four major European river types namely the lowlands; mountains; the Alpine region and the Mediterranean region.
2) Within each of these four European river types, identify the RHS variables that influence the HQA and HMS indices most strongly.

Material and methods

River Habitat Survey

A STAR project protocol was used to standardise the collection of hydromorphological data across Europe (Buffagni & Erba, 2002). This was based on the RHS methodology (Environment Agency, 1997; Raven et al., 1998a). Each survey stretch was 500 m in length. Bank and channel features were recorded at 10 *spot-checks*, spaced every 50 m. At each *spot-check* physical features (e.g., flow type, substrate type, channel/bank modifications), land use and channel vegetation types were recorded. An extensive additional list of characteristics is recorded along the whole 500 m sampling unit, which are called "sweep-up information" in RHS methodology. The main categories recorded in sweep-up are: land use, bank profile, extent of trees, extent of channel features (e.g., pools, riffles, bars), channel dimensions, features of special interest and evidence of recent management. All data were stored in a Microsoft® Access database. For the purposes of our analysis we define 'category' as a group of attributes. For example, the category channel vegetation type contains the attributes Amphibious, Emergent broad lived, Emergent reeds etc. A full list of categories and their associated attributes is given in supplementary material[1].

Overall 216 sites in four geographical regions were investigated. Out of these, 100 of them were in the Lowland region, 67 in Mountains, 31 in the Mediterranean and 18 in the Alps. The STAR RHS protocol therefore facilitated the standardized collection and storage of a wide range of parameters useful for the characterization of a river. RHS also supports the classification of a survey site in terms of its morphological degradation and its habitat diversity and quality by calculation of the Habitat Quality Assessment

[1] Electronic supplementary material is available for this article at <http://www.dx.doi.org/10.1007/s10750-006-0090-7> and accessible for authorized users.

(HQA) and Habitat Modification Score (HMS) indices (Raven et al., 1998a). The HQA index incorporates a variety of measures of habitat quality such as the number of different flow types, channel substrates and deposition features. HQA is expressed as the sum of scores given to each single feature so that sites with numerous and diverse natural features score highly. Comparison of HQA scores is therefore meaningful only if made between sites of similar river type. Observed HQA values typically vary between 10 and 80 points, where 10 points indicate that a river has very few attributes characteristic of natural rivers and 80 points indicate that a river has many of the attributes indicative of a high degree of naturalness. HMS quantifies the extend and impact of anthropogenic modifications such as bank reinforcement, channel re-sectioning, culverts and the number of weirs. Modifications are scored according to their extent and weighted due to their impact. Observed HMS values typically vary between 0 and 100 points, where 0 points indicate that a river has none of the attributes characteristic of modification and 100 points indicate that a river has many attributes characteristic of modification.

Frequency of features between geographical regions

Variation in the occurrence of hydromorphological features found in rivers was based on four geographical regions across Europe, namely the lowlands, mountains, the Mediterranean and Alpine rivers (Herring & Strackbein, 2002) (Table 1). For example, for 'flow type', the frequency of eleven attributes within each region (free fall, rippled, smooth, etc.) was estimated.

Table 1. Stream types in different geographical regions

Stream type no.	Stream type (full name)	Country	Ecoregion	Altitude (masl)	No of RHS sites
Lowlands					
D03	Medium-sized lowland streams	Germany	14	<200	12
K02	Medium-sized lowland streams	Denmark	14	<200	11
O02	Medium-sized lowland streams	Poland	14, 16	<200	25
S05	Medium-sized lowland streams	Sweden	14	<200	16
S06	Medium sized lowland streams on calcareous soils	Sweden	14	<200	11
U15	Small sized lowland calcareous streams(RIVPACS group 32)	United Kingdom	18	<200	13
U23	Medium-sized lowland calcareous streams (RIVPACS group 20)	United Kingdom	18	<200	12
Mountains					
A05	Small-sized, shallow mountain streams	Austria	9	200–800	13
C04	Small sized shallow mountain streams	Czech Republic	9	200–800	14
C05	Small-sized streams in the Central sub-alpine Mountains	Czech Republic	9	200–800	10
D04	Small-sized, shallow mountain streams	Germany	9	200–800	12
D06	Small-sized Buntsandstein-streams	Germany	9	200–800	6
F08	Small sized shallow headwater streams in Eastern France	France	8	200–800	12
South-Europe					
H04	Small-sized calcareous mountain streams in Western, Central and Southern Greece	Greece	6	200–800	10
I06	Small-sized calcareous streams in the Central Apennines	Italy	3	200–800	11
P04	Medium-sized streams in lower mountainous areas of Southern Portugal	Portugal	1	<200	10
Alps					
A06	Small-sized crystalline streams of the ridges of the central Alps,	Austria	4	200–800	8
I05	Small-sized, streams in the southern calcareous Alps.	Italy	4	>800	10

Influence of attribute frequency on RHS indices

Attributes that contribute to the RHS indices were isolated and the influence of these attributes on the indices was then examined separately within the lowland, mountain, Mediterranean and Alpine geographical regions. This was done in two ways. Firstly a *share ratio* was calculated; this represents a ratio of the number of individual attribute records divided by the total number of attribute records. The share ratio therefore establishes the proportion of HQA and HMS index scores delivered by individual attributes. Secondly, for every feature recorded, an *impact ratio* was calculated by dividing the HQA and HMS scores by the number of records per feature. The impact ratio therefore shows the influence of an individual attribute on index score each time that attribute appears. For example, an attribute might have a high share ratio (i.e., may commonly contribute to index scores) but may have a low impact ratio because it makes little numerical difference to index scores. Conversely, an attribute might have a low share ratio (i.e., may only rarely contribute to index scores) but a high impact ratio by making a substantial numerical difference to index scores. These two ratios therefore highlight components of the RHS data that most frequently contribute to and most strongly affect index scores.

Correlation between attributes and geographical regions

Using the CANOCO software (ter Braak & Smilauer, 1998), Canonical Correspondence Analysis (CCA), a constrained unimodal ordination method (ter Braak, 1995), was used to examine variation in habitat variables between geographical regions. Using a matrix of geographical data (the four geographical regions) and a matrix of habitat attributes that contribute to the HQA and HMS indices, separate CCAs were performed for channel, bank and land-use attributes.

Within survey attribute diversity

Finally, we examined the variation of RHS attributes within each 500 m survey site. This was done separately for the four geographical regions. We calculated the *number of attributes* within the category (all attributes, which were recorded in each category), the *Shannon-Wiener index* (reflecting diversity of attributes in the RHS sample site) and the *Evenness index* (reflecting proportional share of analysed factors).

Results

Attributes influencing the HQA score in different regions in Europe

Attributes affecting HQA scores in the four geographical regions are presented in Table 2. Across all geographical regions, no individual attribute contributed more than 14% to the HQA score. In all four regions, a fairly consistent category of approximately seven attributes each contributed ~10% to the total score, and together contributed ~75% to the total. A common feature of the four regions was the role of bank vegetation (e.g., *Vegetation structure* at the *Bank face/Bank top* and *Bank trees*). These always delivered about 10% of total HQA score. Similarly important in all four regions were *Channel Substrate at spot-checks* and *Flow types at spot-checks*, although the individual scores were more variable (10.3–13.7%). In lowland rivers *in-stream plant vegetation* was very important, contributing on average 13.7% to the total HQA score (and delivering 963 points from only 759 records). *In-stream plant vegetation* was less important in the other geographical regions. The most important attribute of HQA score in mountain rivers was *bank features recorded in spot-checks* contributing 13.2% of the total score. This attribute was also important in Mediterranean rivers (10.7%). Many additional differences exist between the European regions in terms of the relative contributions of the less frequently recorded attributes.

Significant differences in impact ratio were identified between HQA attributes; this ratio was varied between 0.1 and 2.7. Impact ratios were especially low for *channel features only found in sweep-up, flow types only found in sweep-up* and *bank features only found in sweep-up*. Errors in the recording of these attributes would therefore not have a strong influence on total HQA scores. On the other hand, errors in the recording of high impact ratio attributes will strongly affect HQA scores. The high impact ratio attributes identified in Table 2 are: *flow types at spot-checks, channel substrate type at spot-checks, types of channel vegetation* and *features of special interest* for lowland and Alpine rivers only.

Attributes influencing HMS in different regions in Europe

Table 3 presents the influence of RHS attributes on HMS scores in the four geographical regions. The structure of the HMS index is very different from the HQA index, the main difference being the very uneven influence of a single attribute, *modifications at spot-checks*, which typically delivered the majority of points (between 61 and 78%) in an HMS score. Another difference is that *modifications at spot-checks* aggregates two categories of features recorded separately on the RHS survey sheets (*channel modifications* and *bank modifications*). The importance of *modifications at spot-checks* is essential in all of the European geographical areas but specially

Table 2. River Habitat Survey attributes influencing Habitat Quality Assessment (HQA) in different regions in Europe (tints: ■ – highest values, ■ – medium values, □ – lowest values individually for "Attribute impact on HQA score" category and "Share of attribute in HQA score [%]" category)

Attributes influencing HQA	Attribute impact on HQA score				Share of attribute in HQA score [%]			
	Lowlands	Mountains	South-European	Alpine	Lowlands	Mountains	South-European	Alpine
Flow type(s) (spot-checks)	2.7	2.2	1.7	2.1	10.3	12.8	12.1	13.7
Flow types only found in sweep up	0.3	0.6	0.8	0.6	3.2	4.6	4.0	4.1
Channel substrate (spot-checks)	2.3	2.0	1.8	1.8	10.9	11.0	11.7	11.8
Channel feature(s) (spot-checks)	0.5	0.9	1.1	1.4	1.3	2.8	3.8	6.0
Channel features only found in sweep-up	0.4	0.2	0.3	0.1	1.5	1.2	1.0	0.5
Marginal & Bank features (spot-checks)	0.9	1.3	1.5	1.0	6.1	13.2	10.7	7.4
Bank features only found in sweep up	0.4	0.2	0.3	0.3	1.7	0.8	1.9	0.8
Vegetation structure (Bank-face)	1.3	1.1	1.2	1.2	10.1	10.6	9.6	11.5
Vegetation structure (Bank-top)	1.4	1.1	1.5	1.1	10.5	9.8	10.4	10.0
Point bars	0.5	0.5	0.6	0.8	0.7	0.7	1.4	0.6
Channel vegetation types	1.4	1.4	1.4	1.9	13.7	3.4	4.7	3.7
Land-use within 50 m of banktop (Sweep-up)	0.5	0.5	0.5	0.5	6.6	6.2	7.2	6.3
Extent of trees (Sweep-up)	0.8	0.8	0.8	0.7	9.9	10.1	8.8	11.3
Associated features	1.1	1.1	1.0	0.9	8.7	8.7	6.9	7.4
Features of special interest	1.6	1.1	0.8	1.7	4.9	4.1	5.8	5.2

Table 3. River Habitat Survey attributes influencing Habitat Modification Score (HMS) in different regions in Europe (tints: ■ – highest values, ■ – medium values, □ – lowest values individually for "Attribute impact on HQA score" category and "Share of attribute in HQA score [%]" category)

Attributes influencing HMS	Attribute impact on HMS score				Share of attribute in HMS score [%]			
	Lowlands	Mountains	South-European	Alpine	Lowlands	Mountains	South-European	Alpine
Modifications at spot checks	1.1	1.1	**1.0**	1.4	72.15	62.78	**60.62**	78.53
Modifications only found in sweep-up	0.4	0.6	1.5	**0.2**	10.31	17.70	35.84	**5.51**
Artificial features	2.0	1.2	**0.3**	1.1	17.55	19.52	**3.54**	15.96

important for Alpine rivers where it delivers on average 78.5% of the total score (share ratio) and provides 1.4 points per recorded feature (impact ratio). Errors in the recording of *modifications at spot-checks* will therefore strongly affect HMS scores.

Correlations between attributes and geographical regions

Relationships between the three categories of river habitat attributes (channel, bank and land use attributes) and the four geographical areas (lowlands, mountains, the Mediterranean and Alpine rivers) are presented in Figures 1–3. Two main directions of differences are apparent. Firstly, the major direction of environmental variability is along lowland to mountain/Alpine gradient (correlated with the 1st CCA axis). A clear second gradient also distinguishes the southern European sites (correlated with the 2nd CCA axis).

Figure 1 presents the correlation between channel attributes (spot-checks and sweep-up) and geographical regions. Each of the geographical regions is highly correlated with specific channel attributes. Alpine and mountain regions are characterised by numerous high energy flow types including *chute flow, broken standing waves, chaotic flow* and *exposed boulders*. The lowland region is characterised just by two attributes, *smooth flow* and *no modifications*. The southern European region is characterised by a rather small number of low flow related attributes: *dry* (flow type), *not perceptible flow*, *mature island* and *marginal dead water*.

Relationships between bank attributes and geographical regions are presented in Figure 2. Each of the geographical regions is highly correlated with a specific bank category. For Alpine and mountain regions the most specific attributes are: *rip-rap, concrete, reinforced bank, bare bank-face* and *bank-top, embanked/reinforced bank* and *side-bar*. The lowland region is characterised by a rather small number of bank attributes: *peat, earth, natural berm* and *poached*. In the southern European region there are nine characteristic attributes, including *artificial, two-stage channel, composite bank profile, cobble* and *discrete gravel deposits*.

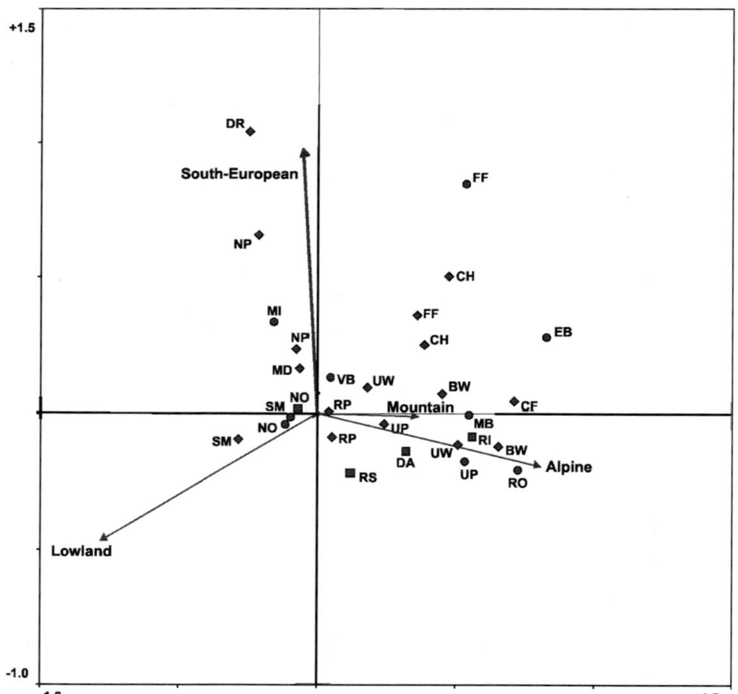

Figure 1. CCA ordination of channel categories and geographical regions. Diamonds – flow types; circles – channel features; squares – channel modifications (see supplementary material available for a key to symbols).

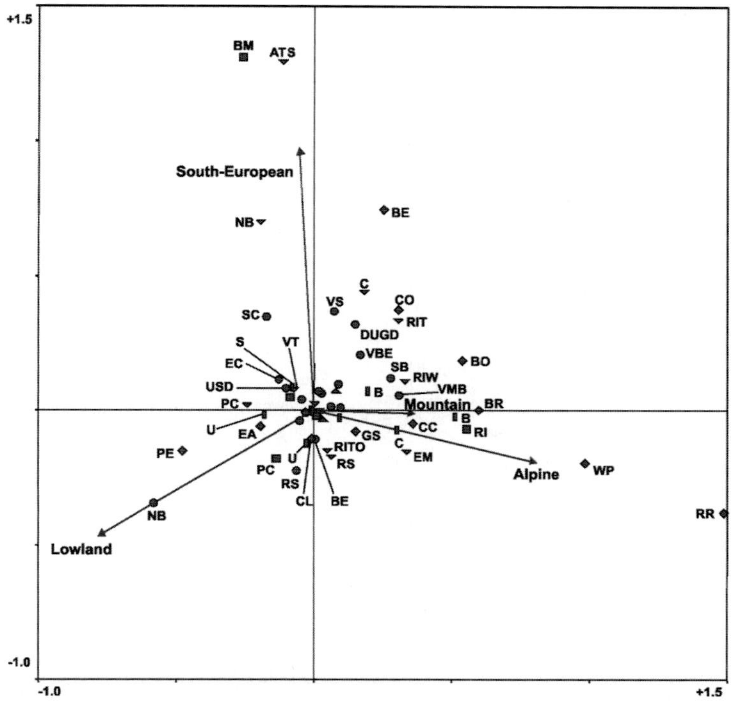

Figure 2. CCA ordination of bank categories and geographical regions. Diamonds – bank material; squares – bank modifications; circles – bank features; triangles – trees and associated features. Acronyms of the following attributes were excluded due to their location in the center of the plot (indicating most irrelevant relationship with the revealed gradients): UTR, SC, G, VP, EBR, V/U, BO, EC, SH, C, TR, BO, CSD, S (spot-checks), NO, DUSD, FT, S (sweep-up), NOcf, NO, PB (see supplementary material available for a key to symbols).

Figure 3 presents the results of the CCA analysis using land-use categories. Again, each of geographical regions is highly correlated with specific *land-use* types. In both the Alpine and mountain regions the most characteristic landscape were *suburban/urban development*, *rock, scree (& sand dunes)* and *parkland & gardens*. The lowland region was strongly represented by *wetland* and *irrigated land* land-use. For the southern European region, *orchard* and *moorland/heath* land-use categories were the most strongly correlated.

Within each of the four geographical areas, the attributes were also classified as *absent, rare* (<5% frequency), *abundant* (>5% frequency) or *most abundant* (see supplementary material). Flow-type *Dry* (DR) was rare in all four geographic areas, although it was clearly important in differentiating southern European streams from streams of other areas (Fig. 1). Sampling sites selected for the STAR project were also chosen for their suitability for fish, invertebrate, macrophyte and diatoms i.e., sites where all quality elements in the Water Framework Directive could be sampled. This probably biased the site selection process in favour of sites that do not dry up, so '*dry*' flow-type is probably underestimated by the STAR dataset. Indeed it has been shown that in the summer months, the '*dry*' attribute is quite common (Buffagni et al., 2004a, b).

Diversity of RHS attributes across geographical regions

The average number of different types of RHS attribute within each River Habitat Survey is shown in Table 4. The largest number of different attributes occurred in the Alpine sites, followed by the mountain and southern European sites. In all three of these geographical regions, the largest contribution to this diversity is from *channel* and *bank features*. Wilcoxon test for dependent

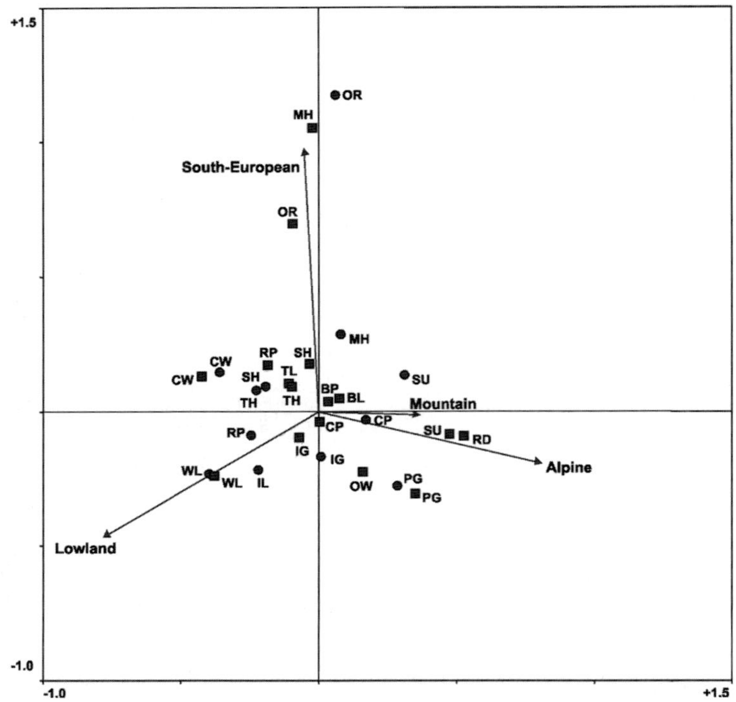

Figure 3. CCA ordination of land use categories and geographical regions. Circles – land use within 5 m of bank top (spot checks); squares – land use within 50 m of bank top (sweep up) (see supplementary material available for a key to symbols).

samples showed significant differences in attribute abundance in all analysed comparisons (Table 5).

Table 6 presents the Shannon diversity index for attribute types within each River Habitat Survey. High values indicate a high diversity of attributes in the RHS sample site. Attributes were most diverse in the Alpine region, followed by the mountain, and southern European regions; attribute diversity was considerably lower in the lowland region. Wilcoxon tests of the Shannon Index for all combinations of regions identified significant differences in attribute diversity between the lowland regions in mountain region (Table 7).

Evenness index values for attribute types in each River Habitat Survey are presented in Table 8. This index reflects the proportional share of analysed factors and a high value indicates proportional uniformity of attributes in the RHS section. As with diversity, attribute evenness was highest in the Alpine region, followed by the mountain, and southern European regions whilst attribute evenness was considerably lower in the lowland region. As with abundance, Wilcoxon tests of the Evenness index identified significant differences in attribute evenness between all combinations of regions (Table 9).

The gradient directions of the four geographical areas together with three main categories of attribute characteristics that best correlate with these geographical areas are summarised in Figure 4. The categories and attributes in Figure 4 capture the essential hydromorphological characteristics of rivers in the four geographical areas studied. A summary of the similarity between the four European regions (based on Wilcoxon tests of diversity indices) is presented in Figure 5. The mountain and Alpine regions and the mountain and southern European regions have the greatest similarity (thick arrows) while the lowland region is by far the most distinct in comparison with the other regions (dotted arrows).

Discussion

Our analysis has quantified the importance of each of the RHS attributes that contribute to HQA and HMS index scores in European rivers and has

Table 4. Attribute type richness within River Habitat Survey sections (tints: ■ – highest values, ■ – medium values, □ – lowest values)

Category	Region			
	Lowlands	Mountains	South-Europe	Alpine
Flow type	2.01	3.27	3.10	3.54
Channel substrate	2.48	3.05	3.08	3.54
Channel modifications	1.21	1.38	1.17	1.68
Channel features	1.27	1.46	1.75	1.65
Bank material	1.53	2.43	2.87	3.23
Riffles, pools, point bars	1.76	2.29	3.26	1.89
Extent of channel and bank features	9.89	8.88	6.67	10.46
Bank modifications	1.43	1.66	1.35	1.89
Marginal and bank features	1.87	2.70	2.87	2.02
Banktop structure	2.13	2.73	2.39	2.61
Bankface structure	2.13	2.74	2.56	2.61
Bank profiles (sweep-up)	3.52	4.07	4.10	3.86
Trees and other associated features	4.28	4.53	4.60	4.43
Land-use 50 m (sweep-up)	3.66	3.66	5.00	3.61
Land-use (5 m)	1.46	1.53	1.46	1.46
Mean value	**2.71**	**3.09**	**3.08**	**3.23**
Number of categories with highest value of parameter	**0**	**3**	**6**	**6**
Number of categories with lowest value of parameter	**11**	**0**	**4**	**2**

identified which attributes most influence HMS and HQA variability. To improve the quality of both RHS datasets in general and the reliability of the HQA and HMS index values in particular, we recommend that special care and attention should be devoted to the accurate recording of all scoring attributes in field surveys and that particular attention should be drawn to the significance of high-scoring attributes on RHS field survey training courses. Other studies of RHS variability (e.g., Szoszkiewicz et al., 2005) have demonstrated

Table 5. Attribute type richness within River Habitat Survey sections (Wilcoxon test)

Regions combination	*p*
Lowlands vs. Mountains	**0.01**
Lowlands vs. South Europe	**0.00**
Lowlands vs. Alpine	**0.00**
Mountains vs. South Europe	**0.00**
Mountains vs. Alpine	**0.00**
South Europe vs. Alpine	**0.00**

Table 6. The Shannon diversity of attribute types within River Habitat Survey sections (tints: ■ – highest values, ■ – medium values, □ – lowest values)

Category	Region			
	Lowlands	Mountains	South-Europe	Alpine
Flow type	0.21	0.40	0.39	0.43
Channel substrate	0.28	0.38	0.38	0.43
Channel modifications	0.04	0.07	0.04	0.12
Channel features	0.06	0.11	0.19	0.18
Bank modifications	0.09	0.13	0.07	0.19
Banktop structure	0.24	0.34	0.28	0.32
Bank material	0.10	0.26	0.34	0.39
Bankface structure	0.24	0.34	0.31	0.32
Land-use (5 m)	0.17	0.17	0.17	0.16
Mean value	**0.16**	**0.25**	**0.24**	**0.28**
Number of categories with lowest value of parameter	**1**	**3**	**2**	**5**
Number of categories with highest value of parameter	**7**	**0**	**2**	**0**

Table 7. Attribute type diversity within River Habitat Survey sections (Wilcoxon test)

Regions combination	*P*
Lowlands vs. Mountains	**0.01**
Lowlands vs. South Europe	0.14
Lowlands vs. Alpine	0.18
Mountains vs. South Europe	0.96
Mountains vs. Alpine	0.92
South Europe vs. Alpine	0.87

that errors in the recording of these attributes by experienced surveyors in particular can be frequent and the addition of quality assurance routines to RHS database software, perhaps flagging unusual or extreme values for those attributes that affect HQA and HMS indices, would prove very useful.

We found considerable differences in the average values of RHS attributes between the four major European geographical regions. We also found considerable differences in the diversity and evenness of these attributes between the regions. In many respects the geographical regions were therefore quite distinct from each other.

Table 8. The Evenness index for attribute types within River Habitat Survey sections (tints: ■ – highest values, ▒ – medium values, □ – lowest values)

Category	Region			
	Lowlands	Mountains	South-Europe	Alpine
Flow type	0.20	0.39	0.37	0.41
Channel substrate	0.26	0.35	0.36	0.40
Channel modifications	0.05	0.09	0.05	0.14
Channel features	0.06	0.12	0.20	0.19
Bank modifications	0.11	0.15	0.09	0.22
Banktop structure	0.34	0.49	0.39	0.46
Bank material	0.08	0.21	0.27	0.32
Bankface structure	0.34	0.49	0.44	0.46
Land-use (5 m)	0.13	0.14	0.14	0.13
Mean value	**0.18**	**0.27**	**0.26**	**0.30**
Number of categories with lowest value of parameter	**0**	**3**	**2**	**5**
Number of categories with highest value of parameter	**8**	**0**	**1**	**1**

Lowland region

Of the four geographical regions studied, the habitat of lowland rivers was the most distinct. Table 10 presents the most characteristic and distinctive attributes. Flow types were typically dominated by *smooth flow*, while *gravel* and *pebble* were the commonest substrate types recorded. The relatively small contribution of *silt* and *sand* distinguishes the lowland rivers in our study from other studies describing lowland river attributes in the United Kingdom (Raven et al., 1998a) and in Poland (Zbierska et al., 2002). *Natural berms* also appeared to be a particularly distinctive lowland channel feature in our study, contrasting with their rarity along lowland rivers in the United Kingdom (Raven et al., 1998a). The land-use category *wetland*, bank categories *natural berms* and *peat*, and channel category *smooth flow* were most strongly correlated with the direction of variability characteristic of lowland rivers. The variability of features within each survey site was also lowest in the lowland rivers, with only the largely anthropogenic bank modification category having any appreciable variability.

Alpine region

As one would expect, the Alpine and mountain regions were very similar to each other in terms of their RHS attributes. Both regions had highly correlated CCA vectors (Figs. 1–3) and the dissimilarity analyses (Wilcoxon tests) confirmed a high level of similarity between these rivers (Tables 6 and 8; Fig. 4). The most characteristic attributes (the most common and distinguishing) are presented in Table 10. Aquatic vegetation was very scarce in the Alpine streams, the commonest

Table 9. The attribute Evenness of River Habitat Survey sections (Wilcoxon test)

Regions combination	*p*
Lowlands vs. Mountains	**0.00**
Lowlands vs. South Europe	**0.00**
Lowlands vs. Alpine	**0.00**
Mountains vs. South Europe	**0.00**
Mountains vs. Alpine	**0.00**
South Europe vs. Alpine	**0.00**

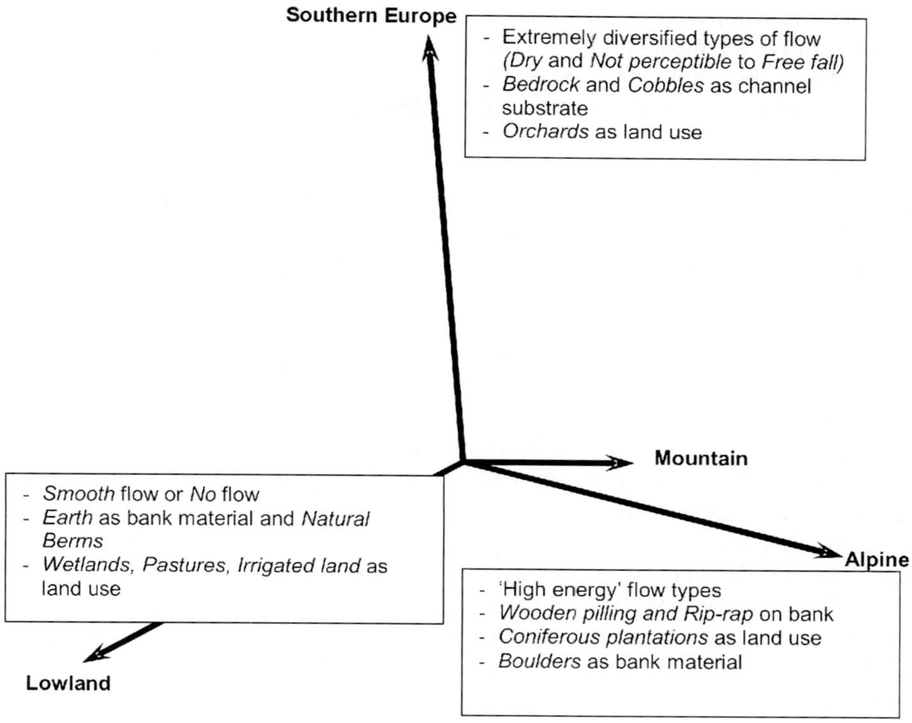

Figure 4. RHS features characteristic of European geographical regions.

category being *Liverworts* and *Mosses*. This differs from the mountain region where vegetation was generally absent. *Rip-rap* and *Wood piling* (bank modification categories) and *exposed boulders* and *chute flow*, were most strongly correlated with the CCA Alpine river vector. The Alpine region generally had the highest attribute variability, although *channel features* and *bank-top/bank-face structure* were somewhat less variable, and *land-use (within 5 m of banktop)* varied little within each survey site.

Mountain region

Mountain and Alpine rivers were very similar in terms of their RHS attributes. Both geographical regions had similar channel substrate types (mainly *boulders*) and similar flow types (mainly

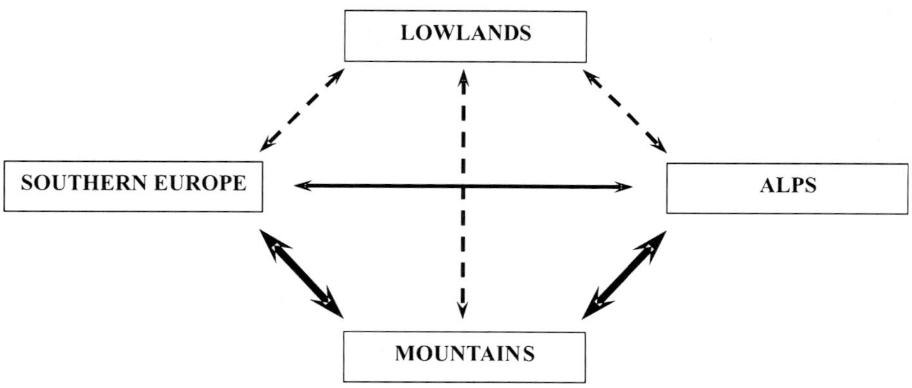

Figure 5. Similarity between major European regions based on attribute diversity within individual River Habitat Surveys (thick arrows – most similar; dotted arrows – least similar).

rippled). While our CCA indicated that the substrate type *boulders* was the most characteristic attribute of this region, the bank material *earth* was also unusually abundant in the mountain region. Interestingly, the flow type *free fall* was quite rare. This was in contrast to rivers in the United Kingdom where waterfalls were one of the most frequent flow types in upland and mountainous regions (Raven et al., 1998a). The commonest bank modification was the resectioned category. Again this contrasts with the analyses by Raven et al. (1998a) where upland and mountain streams in the United Kingdom that had modified banks were typically reinforced, and resectioned banks were only the second most abundant bank modification. As in the Alpine region, vegetation was very scarce in the mountain streams – this was also the case in the UK (Raven et al., 1998a). *Suburban/urban development* and *tall herbs* were the most common land use category. In contrast, land use around mountain streams in the United Kingdom was typically *Improved/semi-improved grassland* and *Broadleaf woodland* (Raven et al., 1998a) this disparity could in part be due to different interpretations of the land use categories by surveyors.

Mediterranean region

The southern European rivers and mountain rivers were generally quite similar to each other in terms of their RHS attributes. *Boulders* dominated the channel substrate and their most frequently recorded flow type was *rippled*. In Mediterranean rivers, *rippled* flow can be the predominant flow type (Buffagni & Kemp, 2002). Buffagni & Kemp (2002) reasoned that the number of sites at which *rippled* flow is likely to be missed by the application of the UK version of RHS (where only one flow type is recorded per transect) is low in Italian rivers. The findings of the present paper, based on rivers across Europe, support that conclusion. Surveying a group of sites in three different seasons, Buffagni & Kemp (2002), found that no perceptible flow and chute flow often occurred as secondary flow types, and consequently that they had a comparatively high probability of being missed by applying the UK version of RHS which only records the predominant flow at spot checks. The results of the present research, where these two flow types were abundant (see supplementary material), rather than contrasting with the interpretation of Buffagni & Kemp (2002), underlines

Table 10. The most characteristic attributes in four analysed geographical regions

Attribute's category	The most characteristic attributes			
	Lowlands	Mountains	Mediterranean	Alpine
Flow type	SM	RP, CH	RP, NP, CH	RP, CH
Channel substrate	GP(G), GP(P)	BO	BO	BO
Channel modifications	–	RS	RI	RS, DA
Channel features	–	MB	–	–
Bank material	–	RR, WP	RR, WP	RR, WP
Riffles. pools, point bars	–	–	–	–
Extent of channel and bank features	–	–	–	–
Bank modifications	–	–	–	–
Marginal and bank features	SB	USB, NB	USB, NB, VMB	BO, UMB
Banktop structure	–	–	S	–
Bankface structure	–	C	S	–
Bank profiles (sweep-up)	–	VU	VU	VU
Trees and other associated features	–	–	–	–
Land-use 50 m (sweep-up)	–	RO	BL, TH, RO	–
Land-use (5 m)	WL	SU, TH	–	SU
Vegetation	ER	–	–	LIV. MO
Channel material (Sweep-Up)	–	–	–	–

the need for a multi-season application of RHS in southern European rivers (e.g., Buffagni et al., 2004b). The STAR River Habitat Surveys were carried out in summer when the likelihood of finding no perceptible flow, chute flow and dry channels is greater than at other times of year. However, these surveys were carried out concurrently with the collection of biological samples (macroinvertebrates, macrophytes, diatoms and fish). These samples could not be collected from dry rivers. The need to take several types of biological samples, in addition to RHS therefore meant that RHS surveys were not carried out at times when the flow type *Dry* would have been commonest. Interestingly, even through typically categorized as rare (see supplementary material), *Dry* was found to be important factor in differentiating southern European rivers from streams of other regions (see Fig. 1). The southern European rivers were also quite distinct from rivers in the other regions in several important respects. The absence of vegetation was a very characteristic feature of the southern European studied streams as was the prevalence of the *no perceptible flow* attribute (Fig. 3).

Conclusions

Our analyses have identified the differences in hydromorphological character between rivers in four separate European geographical regions. The presence of different attributes as well as the diversity of attributes within each RHS survey has also been highlighted. We also found a subset of the RHS attributes, which most strongly affect the HQA and HMS indices. These attributes were identified and their numerical impact on HQA and HMS variability was quantified. We recommend that special care should be devoted to the accurate recording of these attributes in field surveys and that attention should be drawn to the significance of these attributes on RHS field survey training courses.

The differences observed between lowland and mountain/Alpine rivers are not surprising given their distinctive hydromorphological character. More surprising however was the distinct nature of the southern European rivers. Although being located at different latitude, in a sense the southern European rivers also belong to either lowland or mountain river types. The distinctness of the southern European rivers therefore confirms the value of using a standard methodology such as RHS for detecting differences in river morphology, but also highlights the need for ensuring that the attributes recorded adequately represent the various types of streams. As a first step in developing a more appropriate RHS methodology for southern European rivers, *addenda* to the standard RHS protocol have been proposed (Buffagni & Kemp, 2002). This has formed the basis for a southern European version of RHS system (Buffagni et al., 2005). In addition to the new southern European RHS form, the RHS indices (HQA and HMS) have also been tested in southern Europe. This has led to their general acceptance for WFD monitoring purposes (Balestrini et al., 2004). Nevertheless, the HMS scoring system still needs further testing and validation to be applied with confidence to the morphological degradation gradients found in southern Europe (Balestrini et al., 2004). New indices are also being proposed to better characterize Mediterranean rivers. One such index is the LRD (Lentic–lotic River Descriptor), an index designed to describe a river reach in terms of its local hydrological conditions (Buffagni et al., 2004a). The LRD index is also likely to be useful in river habitat assessment in southern European rivers. Training courses for the new southern European version of the RHS have already started so that field surveyors can gain a proper understanding of the unique hydromorphological features of Mediterranean streams.

Acknowledgements

We thank all STAR partners who contributed RHS data used in the present paper. The STAR Project (Standardisation of River Classifications: Framework method for calibrating different biological survey results against ecological quality classifications to be developed for the Water Framework Directive) was co-funded by the European Commission, 5th Framework Program, Energy, Environment and Sustainable Development, Key Action Water, E.U. Contract number: EVK1-CT 2001-00089).

References

Agence de l'Eau Rhin-Meuse, 1996. Outil d'évaluation de la qualité du milieu physique – synthèse. Metz.

Balestrini, R., M. Cazzola & A. Buffagni, 2004. Riparian ecotones and hydromorphological features of selected Italian rivers: a comparative application of environmental indices. In Hering, D., P.F.M. Verdonschot, O. Moog & L. Sandin (eds), Integrated Assessment of Running Waters in Europe. Kluwer Academic Publishers. Printed in the Netherlands. Hydrobiologia 516: 365–379.

Buffagni, A. & S. Erba, 2002. Guidance for the assessment of Hydromorphological features of rivers within the STAR Project. June 2002, 20 + 18 pp (Available at STAR web site, www.eu-star.at).

Buffagni, A. & J. L. Kemp, 2002. Looking beyond the shores of the United Kingdom: addenda for the application of River Habitat Survey in South-European rivers. Journal of Limnology 61: 199–214.

Buffagni, A., S. Erba, D. Armanini, D. De Martini & S. Somaré, 2004a. Aspetti idromorfologici e carattere Lentico-lotico dei fiumi mediterranei: River Habitat Survey e descrittore LRD. In: 'Classificazione ecologica e carattere lentico-lotico in fiumi mediterranei'. Quad. Ist. Ricerca Acque, Roma 122: 41–63.

Buffagni, A., S. Erba & R. Pagnotta, 2004b. Carattere Lentico-lotico dei fiumi mediterranei e classificazione biologica di qualità. In: 'Classificazione ecologica e carattere lentico-lotico in fiumi mediterranei'. Quad. Ist. Ricerca Acque, Roma 122: 157–178.

Buffagni, A., S. Erba & M. Ciampittiello, 2005. Il rilevamento idromorfologico e degli habitat fluviali nel contesto della Direttiva Europea sulle Acque (WFD): principi e schede di applicazione del metodo CARAVAGGIO. Notiziario dei Metodi Analitici Ist. Ric. Acque, Dicembre 2005 (2): 27–42.

Clarke, R. T. & D. Hering, 2006. Errors and uncertainty in bioassessment methods – major results and conclusions from the STAR project and their application using STARBUGS. Hydrobiologia 566: 433–439.

Directive 2000/60/EC. Water Framework Directive of the European Parliament and of the Council of 23 October 2000.

Environment Agency, 1997. River Habitat Survey – Field Guidance Manual, Bristol.

Environment Agency, 2003. River Habitat Survey in Britain and Ireland. Field Survey Guidance Manual. Environmental Agency, Bristol.

Hering, D. & J. Strackbein, 2002. STAR stream types and sampling sites. http://www.eu-star.at/pdf/FirstDeliverable.pdf.

Raven, P. J., P. J. A. Fox, M. Everard, N. T. H. Holmes & F. D. Dawson, 1997. River Habitat Survey: a new system for classifying rivers according to their habitat quality. In Boon, P. J. & D. L. Howell (eds), Freshwater Quality: Defining the Indefinable? The Stationery Office, Edinburgh, 215–234.

Raven, P. J., N. T. H. Holmes, F. D. Dawson, P. J. A. Fox, M. Everard, I. R. Fozzard & K. J. Rouen, 1998a. River Habitat Quality: the physical character of rivers and streams in the UK and Isle of Man. Environment Agency, Bristol.

Raven, P. J., N. T. H. Holmes, F. H. Dawson & M. Everard, 1998b. Quality assessment using River Habitat Survey data. Aquatic Conservation: Marine and Freshwater Ecosystems 8: 477–499.

Raven, P. J., N. T. H. Holmes, P. Charrier, F. H. Dawson, M. Naura & P. J. Boon, 2002. Towards a harmonised approach for hydromorphological assessment of rivers in Europe: a qualitative comparison of three survey methods. Aquatic Conservation: Marine and Freshwater Ecosystems 12: 405–424.

Schneider, P. J., M. Neitzel, M. Schaffrath & H. Schlumprecht, 2003. Physico-chemical assessment of the reference status in German surface waters: A contribution to the establishment of the EC Water Framework Directive 2000/60/EG in Germany. Acta Hydrochimica et Hydrobiologica 31: 49–63.

Szoszkiewicz K., Zbierska J., Staniszewski R., Jusik S., Zgoła T., Kupiec J., 2005. Errors and variation associated with field protocols for the collection and application of macrophyte and hydro-morphological data. STAR Deliverable N4.

ter Braak, C. J. F., 1995. Ordination. In Jongman, R. H. G., C. J. F. ter Braak & O. F. R. van Tongeren (eds), Data Analysis in Community and Landscape Ecology. Cambridge University Press, Cambridge, 109–115.

ter Braak, C. J. F. & P. Smilauer, 1998. CANOCO reference manual and user's guide to Canoco for Windows: software for canonical community ordination (version 4). Microcomputer Power, Ithaca, New York, USA.

Zbierska, J., S. Murat-Bazejewska, K. Szoszkiewicz & A. Lawniczak, 2002. Bilans biogenow w agroekosystemach Wielkopolski w aspekcie ochrony jakosci wod na przykladzie zlewni Samicy Steszewskiej. Wyd. AR Poznan.

Hydrobiologia (2006) 566:281–296
M.T. Furse, D. Hering, K. Brabec, A. Buffagni, L. Sandin & P.F.M. Verdonschot (eds), The Ecological Status of European Rivers: Evaluation and Intercalibration of Assessment Methods
DOI 10.1007/s10750-006-0089-0

Preliminary testing of River Habitat Survey features for the aims of the WFD hydro-morphological assessment: an overview from the STAR Project

Stefania Erba[1,]*, Andrea Buffagni[1], Nigel Holmes[2], Mattie O'Hare[3], Peter Scarlett[3]
Alberta Stenico[4]
[1]*CNR-IRSA, Water Research Institute, Via della Mornera, 25, I-20047 Brugherio (MI), Italy*
[2]*The Almonds, WARBOYS, PE28 2RW Huntingdon, Cambs, UK*
[3]*CEH, Centre for Ecology and Hydrology, Winfrith Technology Centre, DT2 8ZD Dorchester, Dorset, United Kingdom*
[4]*LABBIO, Provincia Autonoma di Bolzano, Agenzia Provinciale per la Protezione dell'Ambiente, Laboratorio biologico, Via Sottomonte, 2, I-39055 Laives, Italy*
(**Author for correspondence: E-mail: erba@irsa.cnr.it*)

Key words: river, macroinvertebrate, CEN, metrics, ICM

Abstract

The UK River Habitat Survey (RHS) method for the assessment of hydro-morphological features was applied within the EU STAR project simultaneously with the collection of biological data. A subset of data from 79 sites affected by hydro-morphological alteration and belonging to 7 different stream types was analysed. The different features recorded within RHS were evaluated separately considering the characteristics associated with banks, channel and riparian zone. Different scores were assigned to selected features representing hydro-morphological alteration and naturalness of habitat. The ability of the different compartments to represent the quality gradient of sites was investigated. In addition, the link between macroinvertebrate community and hydro-morphological data was investigated, directly relating indices and metrics calculated from taxa list collected in a site to scores assigned to the RHS features. The sections highly affected by morphological alteration were channel geometry and bank profile. Metrics showing the best correlation with the features selected were EPT taxa, ASPT and ICMi (Inter-calibration Common Metric index). Among the indices studied, the HQA score (Habitat Quality Assessment) apparently played the most important role in structuring biological communities and the lentic-lotic character of rivers was also important.

Introduction

The EU Water Framework Directive (Directive 2000/60/EC – Establishing a Framework for Community Action in the Field of Water Policy) defines the key principles for the protection of all water bodies recognizing the central role of biological communities. The WFD also encourage the countries of the EU to carry out hydro-morphological assessment activities to better understand biological data. The challenge proposed from the WFD rests on the integration of the biological and habitat components (Logan & Furse, 2002), into a holistic view of the ecological status of rivers. In this context, the STAR project (Furse et al., 2006) represents a rare attempt to simultaneously consider different Quality Elements (invertebrates, macrophytes, diatoms, fish, hydro-morphology and chemistry) over a wide geographical range covering different stream types distributed throughout Europe.

Europe has a long tradition of biological assessments, especially with regard to river invertebrates (Woodiwiss, 1964; Armitage et al., 1983; Metcalfe-Smith, 1994; Verdonschot, 2000; Hering et al., 2004). However, systems for the assessment of hydro-morphological characteristics and quality are far less developed. At present, different methods and indices are used in different countries. Austria uses the Austrian Habitat Survey (Werth, 1987; Muhar et al., 1996, 1998), Denmark uses the Danish Stream Habitat Index (Pedersen & Baattrup-Pedersen, 2003), France is testing the SEQ Physique (Agences de l'Eau & Ministère de l'Environment, 1998), Germany the Eco-morphological Survey for Large Rivers (Fleischhacker & Kern, 2002), and the UK uses the River Habitat Survey (RHS) (Raven et al., 1998). The methods use a number of parameters (channel, bank, floodplain, flow-related) and a scoring system to evaluate the hydro-morphological status of streams. In this context, the activity of the European Committee for Standardization (CEN) is aiming to define a common European framework for the assessment and interpretation of the hydro-morphological aspects of rivers. The STAR project has a formal link to CEN and one aim of the project is to provide relevant CEN working groups with draft methods. The CEN standard would provide a guidance on which hydro-morphological features have to be considered when studying and characterising river reaches. This would improve the comparability of hydro-morphological survey methods currently available, data processing, interpretation and presentation of results. Four areas are identified as a focus for the survey: (1) river channel, (2) banks, (3) riparian zones and (4) floodplains (Buffagni & Erba, 2002).

For the aims of the STAR project the UK RHS method was selected as a common method for the evaluation of hydro-morphological characteristics. RHS is a method designed in the UK to characterise and assess, in broad terms, the physical structure of freshwater streams and rivers (Raven et al., 1997, 1998). Some partners adopted a specially extended version of RHS, which provided a better description of the hydro-morphological complexity of South European rivers (Buffagni & Kemp, 2002). This version encompasses all the UK features and adds some new ones. Nevertheless, the present paper deals with the results obtained by applying the UK version of the method, provided by all partners.

The hydro-morphological assessment within STAR has the main objective of providing information that can be used to quantify the morphological degradation of STAR sites (Buffagni & Erba, 2002):

Additionally, a wider objective when applying RHS is to aim at providing:

- data to be quantitatively used to extend the information gained from the biological elements of different river reaches exhibiting similar habitat features (e.g., to predict taxa occurrence)
- information to investigate the functional relationships between taxa presence/abundance and single hydro-morphological features (e.g., to investigate the organism–response relationships).

The main aim of this paper is to quantify, according to the guidelines provided by CEN working groups, different aspects of hydro-morphological degradation through the application of the RHS in a selection of European sites. The data analysed here provide an overview, which covers a wide geographical gradient, of which river portions are mostly affected by morphological alteration and of the overall amount of modification that can be found at a site. Although RHS has been extensively applied not only in the UK, but also in some other European countries, there are few studies focusing on the relationship between the results of its application and biological communities. In general terms, the link between biota and hydro-morphology is not widely investigated, especially if aquatic invertebrates are concerned. Thus, a preliminary testing of the link between selected invertebrates metrics and the degree of modification of different portions of the river is presented here. This paper represents a starting point for further analysis aimed at linking biological data to morphological features and to evaluate which data are the most important in influencing community structure.

Methods and data analysis

Among the set of rivers investigated with the STAR project, RHS data from rivers mainly affected by hydro-morphological alteration were considered for a total of 79 sites belonging to 7 different stream types. The subset of data used is

Table 1. River types analysed and number of considered sites

Country	River type	No. of sites	No. of refs.	LRD (min–max)	LRD (average–std.dev.)	HMS (min–max)	HQA (min–max)
Austria	Small-sized, shallow mountain streams	15	3	(−31)–15	(−12.7)–12.7	1–62	18–54
Austria	Small-sized, crystalline streams of the ridges of the Central Alps	14	4	(−33)–9	(−17.9)–11.0	0–62	14–58
Czech Rep.	Small-sized streams in the Central sub-alpine Mountains	10	3	(−33.5)–1.5	(−19.4)–10.5	3–60	13–53
Germany	Small-sized, shallow mountain streams	8	2	(−23)–5.5	(−11.8)–9.3	0–45	27–59
Germany	Medium-sized lowland streams	10	2	(−4.5)–29	15.5–9.4	0–96	23–58
Italy	Small-sized streams in the southern calcareous Alps	10	3	(−51.5)–(−5)	(−25.5)–20.5	0–70	20–65
Denmark	Medium-sized lowland streams	12	5	(−13.5)–34	7.4–12.7	0–32	21–65
	Total	79	22				

from Austria, Czech Republic, Denmark, Germany and Italy (Table 1).

The RHS protocol requires the survey to be carried out over a 500-m long river stretch, with observations made at 10 equally spaced spot-checks along the channel (EA, 1997). At each spot-check, predominant substrate and flow-type, plus physical features of channel and banks, vegetation structure of banks and adjacent land, land-use and channel vegetation types are recorded. Other information on valley form, land-use in the river corridor, etc, is also collected. To improve links between RHS application and biological data, the hydro-morphological survey includes the river area selected for the invertebrate collection located close to the downstream end of the RHS survey stretch (Buffagni & Erba, 2002; Furse et al., 2006).

The different features recorded by RHS were evaluated separately considering banks, channel and land use according to the preliminary groups of quality attributes proposed within the CEN TC 230/WG 2/Task Group 5 (see Table 2). For this purpose, a standard of general character, *Water Quality: Guidance Standard for Assessing the Hydromorphological Features of Rivers* (EN14614:2004) has recently been proposed and a new one, including a broadly applicable scoring system, is under discussion. In the draft, a list of definitions is given for a number of river features relevant to the survey, some of which are of major importance when comparing results obtained from the application of different assessment methods. The standard will focus on the morphological features of rivers and on river continuity, without considering hydrological aspects. Different scores were assigned to some selected features associated with the different portions (i.e. channel, bank, substrate, etc) of the river. The features represent morphological alteration and/or naturalness of habitat structure. The scoring system used here is a slightly modified version of the one proposed to test CEN attributes (CEN TC 230/WG 2/TG 5: N45) (see supplementary material[1]). To assign scores to the different portions sometimes the same characteristics are considered, but different scores are given. For example, characteristics used to assess longitudinal continuity are similar to the ones considered for determining the effect of artificial structures on flow. Based on the RHS protocol, morphological impact (Habitat Modification Score) and habitat quality (Habitat Quality Assessment Score) were estimated for each site. To calculate the HMS and HQA indices, scores are assigned according to Raven et al. (1998). For the calculation of HMS different scores are given to each modification (e.g. bank re-sectioning, number of weirs, etc.), accordingly to the importance of the impact type and to the extent of its presence. HMS is thus the sum of all the individual scores. Increasing values of HMS and of all the CEN attributes (excluding land use and vegetation structure) indicate an increase in hydro-morpho-

[1] Electronic supplementary material is available for this article at http://dx.doi.org/10.1007/s10750-006-0089-0 and accessible for authorised users.

Table 2. Attributes considered for score assignation according to the CEN draft quality standard

CEN TG5 features	Attributes used	Acronym
Channel geometry	Presence of re-sectioning, reinforcement, culverting, dam, ford, artificial two stage channels	CHANN
Substrates	Presence of artificial substrate	SUBST
Flow	Presence of weirs, sluices, culverts, bridges, ford, deflectors	ART_STR
Longitudinal continuity as affected by artificial structures	Presence of weirs, sluices, culverts	Long_CON
Bank structure and modifications	Presence of re-sectioning, reinforcement, poaching, embankment	BANK Mod
Vegetation type/structure on banks and adjacent land	Land use within 5 m from bank top	Land5
Adjacent land-use and associated features	Land use within 50 m from bank top	Land50
Degree of (a) lateral connectivity of river and floodplain; (b) lateral movement of river channel	Presence of berm and embankment	Lat_CON

logical alteration. In contrast, the HQA index assesses the ecological quality of the site through the habitat richness evaluated on the basis of the extent and variety of natural features recorded (e.g. number of different flow types, different substrates and naturalness of land use). It is expressed as the sum of the scores given to each single feature. High HQA scores and high scores for vegetation structure and land use (indicating a high proportion of natural land use) represent the high naturalness of river sites, basically because it is supposed that natural sites are characterised by high diversification in habitat structure (Raven et al., 1998).

The Lentic–lotic River Descriptor (LRD: Buffagni et al., 2004a) was also calculated on the basis of RHS data. This descriptor is mainly derived by giving different scores to flow types, substrate types, depositional features, etc., in relation to their ability to indicate lotic or lentic characteristics (Buffagni et al., 2004a). LRD negative values represent rivers with a predominantly lotic character, positive values are reached when lentic habitats dominate.

The HMS, HQA and LRD scores were analysed in relation to the quality gradient assigned to the investigated sites, and Pearson correlation coefficients (Legendre & Legendre, 1998) were calculated with respect to some biological metrics. Sites were pre-classified, generally on the basis of historical data (e.g. biotic data, pressures data) and combined with expert judgement (Nijboer et al., 2004). This pre-classification represents a gradient of hydro-morphological alteration. The main parameters indicating water pollution present low/intermediate concentration (see Table 6). For the calculation of invertebrate metrics the software ASTERICS (AQEM/STAR Ecological RIver Classification System) was used. The identification level used was the best available within each country. All the metrics were normalized according to the maximum value reached in the reference sites of each country in order to compare the results between countries (Buffagni et al., 2005). The metrics selected for the analysis are among those most commonly included in the general assessment systems recently developed within the AQEM project (e.g., Buffagni et al., 2004b; Lorenz et al., 2004; Ofenböck et al., 2004; Pinto et al., 2004) also including a simple multimetric index specifically developed for European Inter-calibration purposes (ICMi: Buffagni et al., 2005). The biological metrics used in all the different stream types are reported in Table 3.

Some further metrics were tested for Austria, Germany and Czech Republic, for which identification level for most of the taxa was done at species level. These metrics are two German indices, specifically developed to assess morphological degradation in German river types i.e., German Fauna Indices (DIND in the tables, Lorenz et al., 2004) and metrics based on taxa habitat preference (preference for pelal, lithal and phytal: Lorenz et al., 2004).

Table 3. Considered biological metrics

Metric name	Taxa considered in the metric	Literature
ASPT	Whole community (Family level)	Armitage et al. (1983)
BMWP	Whole community (Family level)	Armitage et al. (1983)
MTS (Mayfly Total Score)	All Ephemeroptera (Genus, OU or species level)	Buffagni (1997)
Abundance	Whole community (Family/Genus/species level)	Hering et al. (2004)
Total number of taxa	Whole community (Family/Genus/species level)	Hering et al. (2004)
Number of EPT taxa	Number of Ephemroptera Plecoptera and Trichoptera taxa (Family/Genus/species level)	Hering et al. (2004)
Number of families	Whole community (Family level)	Buffagni et al. (2005)
Shannon diversity index	Whole community (Family/Genus/species level)	Krebs (1989)
Margalef Diversity Index	Whole community (Family/Genus/species level)	Margalef (1984)
Simpson Diversity Index	Whole community (Family/Genus/species level)	Krebs (1989)
Eveness Index	Whole community (Family/Genus/species level)	Krebs (1989)
Sel EPTD	Log(sum of Heptageniidae, Ephemeridae, Leptophlebiidae, Brachycentridae, Goeridae, Polycentropodidae, Limnephilidae, Odontoceridae, Dolichopodidae, Stratyomidae, Dixidae, Empididae, Athericidae & Nemouridae + 1)	Buffagni et al. (2005)
1-GOLD	1 – relative abundance of Gastropoda, Oligochaeta and Diptera (Family level)	Buffagni et al. (2005) and Pinto et al. (2004)
ICMi	Whole community (Family level)	Buffagni et al. (2005)

Results

Hydro-morphological features and ecological pre-classification

The sites analysed cover a wide range of hydromorphological alteration and show HMS values ranging from 0 (pristine sites) to 96 (highly modified sites). Minimum, maximum, average and standard deviation values derived from each attribute, including HMS, HQA and LRD, are presented in Table 4. Sites are grouped according to the ecological pre-classification provided by each country. This pre-classification is well reflected in the average scores of the different attributes the majority of which tend to increase with the worsening of ecological status. LRD values are not related to ecological quality of sites *per se*, being representative of slow flowing–fast flowing character of rivers. Nevertheless, at sites where the flow regime is seriously altered and data from nearly natural condition are available, it can be used to assess hydrological alteration (Buffagni et al., 2004a). When river flow is not artificially decreased or increased, LRD values are expected to vary slightly between sites from the same stream type. Thus, although the observed LRD value at a site might be affected by natural variability, e.g., among seasons, when considering LRD within a single river type variability is generally much lower than among different stream types.

Channel geometry was assessed using RHS data indicative of changes to natural channel geometry – re-sectioning, reinforcements, culverts and two stage channels. Not all the potential range is covered by the analysed data, 30 being the maximum reached out of a theoretical value of 50. The whole gradient was covered by considering channel substrate, assessed by allocating points for the presence and extent of artificial substrates, and bank modification. This modification was assessed by awarding points for re-sectioned or reinforced banks, culverts and embankments. High values indicating stream morphology alteration can also be reached in good status classes. Looking at average values, single attributes may fail in discriminating between adjacent classes, e.g., moderate vs. good or moderate vs. poor. The presence of artificial structures affecting flow character and the alteration of stream connectivity (both longitudinal and lateral) seem not to be

Table 4. Minimum, maximum, average scores and standard deviation for the different attributes at STAR sites affected by hydromorphological alteration (classified according to pre-classification of sites)

		No. of sites	CHANN	SUBST	BANK Mod	ART_STR	Lat_CON	Long_CON	Land5	Land50	HMS	HQA	LRD
Max. reachable			50	50	100	var	100	var	80	48	var	var	var
High	Min	22	0	0	0	0	0	0	4	2	0	40	−51.5
	Max		2	0	20	6	15	0	80	34	6	65	34.0
	Average		0.2	0.0	5.5	0.3	0.7	0.0	60.4	22.0	1.8	54.4	−16.0
	Std. Dev.		0.5	0.0	6.7	1.3	3.2	0.0	22.1	7.6	2.4	6.4	19.3
Good	Min	24	0	0	0	0	0	0	1	4	0	33	−37.5
	Max		30	50	100	65	15	65	68	40	62	59	21.0
	Average		4.7	2.9	34.9	6.1	1.7	4.1	31.8	20.2	20.7	45.3	−9.6
	Std. Dev.		7.9	10.2	33.1	15.1	4.1	13.4	18.4	7.9	17.7	8.5	16.7
Moderate	Min	18	0	0	0	0	0	0	0	3	0	22	−33.3
	Max		30	15	100	56	5	55	76	27	60	56	23.0
	Average		6.6	1.7	55.6	6.3	1.0	4.3	24.3	15.5	31.8	35.5	−6.8
	Std. Dev.		8.0	4.2	33.7	13.1	2.0	12.8	22.0	7.4	19.3	11.0	15.2
Poor	Min	8	0	0	0	0	0	0	0	0	4	13	−28.0
	Max		30	50	100	10	5	10	45	27	62	37	29.0
	Average		15.6	13.1	73.0	2.0	1.3	1.9	15.8	9.6	43.4	25.9	0.5
	Std. Dev.		11.8	22.8	33.4	3.7	2.3	3.7	16.3	9.2	18.5	9.1	20.6
Bad	Min	7	0	0	15	0	0	0	0	0	6	17	−19.5
	Max		30	50	100	16	0	5	72	27	96	52	18.0
	Average		8.4	16.4	75.6	4.7	0.0	1.4	20.0	10.0	49.9	28.3	−6.0
	Std. Dev.		10.2	19.1	31.6	5.5	0.0	2.4	24.2	9.2	29.0	14.4	14.2

related to a site's quality gradient, presenting varying values that do not depend on quality classes.

Some of the considered attributes show a high variability within the same ecological status class (i.e., high values of standard deviation). Usually the variation in high status sites is lower than the variation within other classes. The attributes that show a comparatively low variability are: channel geometry, lateral connectivity with floodplain and associated land-use and feature within 50 m. In these cases the standard deviation values are very similar among different classes. For artificial structures affecting flow character and longitudinal connectivity the standard deviation is high in intermediate status classes (i.e. good and moderate).

At first the variability of the analysed attributes had to be assessed, i.e. if the whole degradation gradient is covered within the studied datasets. Some attributes show a very low variability and are rarely found. The potential range for each of the attributes was defined and divided into five classes. A gradient where four classes were present was considered as adequately covered. This criterion was assessed jointly with the fact that more than half of all the sites have to achieve a score higher than 0. Only if these two criteria are met, is the attribute considered valid for future conclusions. Artificial substrates are not often present in the studied river types, with usually less than 25% of the sites including this alteration. In most cases, it relates to less than the 15% of the considered stretch (500 m). The percentage of sites including modification in lateral and longitudinal connectivity is also less than 25%. The percentage of sites affected by flow alterations due to the presence of artificial structures is higher (37%). In occurrence terms longitudinal continuity and flow alterations arising from the presence of artificial structures have a high impact only in the Italian Alps. In all the other studied stream types the presence of dams, weirs and sluices was not extensive. As far as channel modification is concerned, 53% of sites include this type of alteration but none of the considered sites reached the maximum potential score (see Table 4) and in most cases (4 out 7 types) the gradient covered is short (from 0 to less than 20). The main morphological impact affecting rivers is bank modification: this modification occurs at 88% of sites, with more than 40% sites affected by this kind of modification for more than half of the considered stretch.

With relation to attributes regarding river naturalness, land use within 5 m of the bank top covers the whole gradient, while land use within 50 m does not. The HQA index decreases the average values as ecological status worsens, presenting similar values for poor and bad status sites. High maximum values can also be reached in the bad status class.

As expected high status sites are characterised by the widest gradient covered by LRD, with values ranging from −51 (typical of very fast flowing water), to 34 (values characterizing sites with a high presence of lentic habitats). This is related to the fact that different river types are considered jointly. The shortest gradient is covered by bad status sites that present LRD values closer to 0 (from −19.5 to 18), indicating an equivalent presence of lentic and lotic habitats, whose character does not vary greatly among sites.

In general, the variability observed for single river types has the same trend as the variability found for all types together (see Table 4). This can be noticed looking at minimum and maximum values of HMS, HQA and LRD indices within each stream type (Table 1). For this reason, only summary results considering all the river types jointly are presented. Moreover, homogeneous trends for all the hydro-morphological variables were observed for different types.

Correlation between hydro-morphological, chemical and invertebrate attributes

In Table 5, Pearson correlation values among the different hydro-morphological attributes are reported. Bank modification shows the highest correlation both with parameters indicating alteration in stream morphology, as expected, as well as with parameters related to land use.

Generally only a few RHS attributes were related to variables representing water quality, indicating the relative independence of morphology from water quality in the selected datasets (Table 6). The LRD index is the parameter most closely correlated to the variables indicating water quality.

In Table 7, the minimum and maximum correlation values for all the biological metrics considered with respect to the hydro-morphological attributes are reported. Pearson correlation coefficients were calculated for all the considered attributes, even those scarcely present in the river types which are analysed here. Correlation coefficients were calculated by separately considering different stream types. The number of times (n in the table) R-values are ≥ 0.45 is reported. The number in brackets, beside the attribute names,

Table 5. Pearson auto-correlation values among the different considered attributes of RHS

	CHANN	SUBST	ART_STR	BANK_mod	Lat_CON	Long_CON	Land5	Land50	HMS	HQA
SUBST	**0.68**									
ART_STR	−0.03	0.02								
BANK_mod	0.65	0.42	0.08							
Lat_CON	0.06	−0.08	−0.05	0.06						
Long_CON	−0.04	−0.03	**0.90**	0.01	−0.04					
Land5	−0.33	−0.20	−0.10	−0.65	−0.10	−0.05				
Land50	−0.29	−0.25	0.12	−0.49	0.07	0.14	**0.66**			
HMS	0.70	0.53	0.21	**0.91**	0.06	0.11	−0.63	−0.47		
HQA	−0.62	−0.47	0.06	**−0.73**	−0.07	0.08	0.59	0.51	**−0.73**	
LRD	−0.05	−0.09	−0.01	−0.09	0.01	0.01	−0.05	−0.07	−0.03	**−0.26**

The highest values are indicated in bold; significant values ($p < 0.001$) are underlined.

Table 6. Pearson correlation coefficient among RHS attributes and some selected physico-chemical parameters

	pH	Conductivity	O_2 concentration	Chloride	BOD_5	NH_4	NO_3	PO_4
Average	7.6	306	9.46	2.48	1.44	0.06	0.9	78
Std. dev.	0.4	194	1.3	1.55	0.84	0.1	0.8	139
CHANN	0.10	0.17	−0.18	−0.14	−0.03	0.23	0.49	0.12
SUBST	0.07	0.32	−0.28	−0.02	0.02	0.06	**0.64**	0.02
ART_STR	0.34	−0.05	−0.18	−0.26	−0.11	−0.16	−0.19	−0.12
BANK_mod	0.27	0.04	0.03	−0.09	−0.12	0.23	0.16	0.12
Lat_CON	−0.15	−0.03	−0.11	−0.14	−0.04	−0.01	−0.12	0.08
Long_CON	0.25	−0.09	−0.13	−0.21	−0.06	−0.14	−0.14	−0.12
Land5	−0.18	0.11	−0.15	−0.23	−0.04	−0.29	−0.10	−0.23
Land50	0.10	0.09	−0.12	−0.20	−0.07	−0.25	−0.12	−0.07
HMS	0.28	0.13	−0.18	−0.08	−0.07	0.20	0.16	0.11
HQA	−0.11	−0.26	0.10	−0.07	−0.19	−0.35	−0.41	−0.33
LRD	−0.33	0.41	0.01	0.47	**0.64**	0.38	0.29	0.49

The highest values are indicated in bold; significant values ($p < 0.001$) are underlined.

indicates the number of river types (i.e., considered datasets) achieving a score for that feature. There are in fact some types for which given features were never recorded. The percentage of situations for which each biological metric achieves a value higher than 0.44 was also calculated, even considering not significant values. High correlation values, (even if not significant due to the low number of sites) might indicate which metrics can potentially be more useful in indicating a certain kind of hydro-morphological alteration.

Artificial structures affecting flow character and lateral and longitudinal connectivity present low correlation values with all the biological metrics. The lowest maximum correlation values were reached. For these attributes the number of observations for which *R*-values are ≥0.45 is also low. The metrics presenting the highest observed correlation value are the Number of families ($R = 0.95$) and the number of EPT taxa ($R = 0.94$) in relation to the presence of artificial substrates and the HQA index respectively. In general terms, ASPT, EPT taxa and ICMi are the metrics which present the best response with respect to the considered attributes, having a high percentage of cases for which *R*-values are ≥0.45.

In general, the number of families reaches higher maximum correlation values with all the attributes than the number of taxa where the identification is carried out at a more detailed level.

Pearson correlation coefficients were also calculated jointly considering the different stream types. In Table 8 the results for all the attributes with respect to the different biological metrics are summarised. Lower correlation coefficients are observed when combining the different stream types. Metrics presenting a high number of significant correlations and high coefficient values are BMWP, MTS and almost all the metrics included in the ICMi (ASPT, Number of EPT taxa, 1-GOLD and selected EPTD) as well as the ICMi itself. The highest values are observed with respect to HQA. In some cases high values are also reached with respect to bank modification and channel geometry modification. As examples, the Figures 1 and 2 represent maximum, median, minimum and interquartile range respectively for ICMi and EPT taxa in the different quality classes, according to pre-classification. Both graphs show that the classes are well separated, even when the variability is quite high and the median values have a linear trend.

For the metrics specifically tested in Austria, Germany and Czech Republic, the ones based on habitat preferences reached the highest correlation values with respect to HQA, LRD and the presence of artificial structures. Nevertheless, the best performing metric in relation to HQA remains the ICMi. The two German indices (DIND3, DIND4), specifically developed for the assessment of morphological degradation in the two German

Table 7. Minimum and maximum Pearson correlation values across stream types for all the biological metrics considered with respect to the hydromorphological attributes

	Abund.		#No. of taxa		BMWP		ASPT		MTS		Simpson		Shannon	
	n	Max. corr.	*n*	Max. corr.	*n*	Max. corr.	*n*	Max. corr.	*n*	Max. corr.	*n*	Max. corr.	*n*	Max. corr.
HMS (7)	2	0.51	2	−0.63	3	−0.72	**5**	−0.86	2	−0.72	1	−0.69	3	−0.60
CHANN (7)	3	0.61	2	−0.85	**4**	**−0.89**	**5**	−0.83	**4**	−0.79	1	−0.46	1	−0.66
SUBST (5)	1	0.68	1	−0.82	2	−0.85	3	−0.91	2	−0.64	2	−0.58	2	−0.71
ART_STR (5)	1	−0.57	0	0.22	0	0.31	1	−0.46	0	−0.36	2	0.51	1	0.52
BANK Mod	2	0.56	1	−0.67	4	−0.70	**5**	−0.84	**5**	−0.71	2	−0.72	2	−0.66
Lat_CON (7)	0	0.44	0	−0.29	1	0.51	0	−0.40	2	0.69	2	−0.60	1	−0.51
Long_CON (5)	1	−0.52	1	0.47	0	0.24	0	0.27	0	−0.40	1	−0.47	0	0.38
Land5 (7)	2	−0.77	1	0.63	3	0.76	**4**	0.88	2	0.78	1	0.67	2	0.57
Land50 (7)	3	−0.80	2	0.49	**4**	0.65	**4**	0.90	3	0.73	2	**0.72**	3	0.62
HQA (7)	1	−0.56	3	0.78	**4**	**0.89**	5	0.87	**5**	**0.85**	2	0.63	3	0.66
LRD (7)	2	0.56	2	−0.59	3	−0.66	1	0.76	1	−0.65	1	0.47	2	−0.55
% cases where $R \geq 0.45$	26		21		40		47		37		24		29	

	MARG.		EVENESS		EPT taxa		No. FAM		Sel EPTD		1-GOLD		ICMi	
HMS (7)	3	−0.58	2	−0.65	**5**	−0.80	2	−0.72	2	−0.76	**4**	−0.73	**5**	−0.80
CHANN (7)	2	−0.80	1	−0.47	**4**	−0.88	2	−0.83	2	−0.64	2	0.58	**5**	−0.77
SUBST (5)	1	−0.91	3	−0.56	2	−0.79	1	**−0.95**	1	−0.6	0	−0.37	1	−0.59
ART_STR (5)	0	0.28	1	0.61	1	−0.47	0	−0.26	1	−0.62	1	0.57	3	−0.66
BANK Mod	2	−0.56	2	−0.68	**6**	−0.78	2	−0.72	3	−0.57	2	−0.78	**4**	−0.78
Lat_CON (7)	0	−0.24	2	−0.55	1	0.48	0	0.44	0	0.39	0	0.37	1	−0.53
Long_CON (5)	1	0.49	0	0.31	0	−0.40	0	0.34	0	−0.38	2	−0.53	1	−0.77
Land5 (7)	1	0.64	1	0.63	3	0.84	2	0.55	2	0.72	3	0.73	2	0.85
Land50 (7)	3	0.55	2	0.65	**5**	0.86	2	0.67	1	0.47	**4**	0.75	**4**	0.70
HQA (7)	3	0.76	2	0.56	**4**	**0.94**	**4**	0.88	3	0.51	3	**0.86**	**5**	**0.86**
LRD (7)	2	0.56	2	0.86	3	−0.74	3	−0.76	1	−0.81	1	−0.69	2	−0.77
% cases where $R \geq 0.45$	28		28		49		26		23		31		47	

n represents the number of times in which *R*-values are ≥0.45 (even not significant values were considered). The highest values are indicated in bold; significant values ($p < 0.001$) are underlined.

river types tested here, have the highest correlation values with HMS.

Discussion

The results presented here show that bank re-sectioning and reinforcement are the most important alteration affecting the rivers of the study. In some cases, alterations to channel geometry can also appreciably affect river morphology. These considerations agree with the findings of Szoszkiewicz et al. (2006), according to whom the most typical modification found within EU rivers is bank re-sectioning. The percentage of bank reinforcement was also found to be one of the most important factors in describing morphological degradation in German rivers (Feld, 2004). Considering all the different stream types jointly (Table 8), bank modification was significantly correlated with EPT taxa, as well as MTS, an index dedicated to the assessment of the ecological integrity of the mayfly community (Buffagni, 1997), with a relatively high correlation coefficient value. The good relationship between this feature and these metrics is evident in all the different stream types that were considered: the *R*-values for all the stream types (except Germany medium-sized lowland streams for EPT and Italian Alps and

Table 8. Pearson correlation coefficients considering the different stream types jointly

	Abund.	No. of taxa	BMWP	ASPT	MTS	Simpson	Shannon	Margalef	Eveness	EPT taxa
CHANN	0.22	−0.11	−0.28	−0.35	−0.39	−0.10	−0.15	−0.10	−0.13	−0.41
SUBST	0.17	−0.26	−0.34	−0.29	−0.35	−0.10	−0.12	−0.25	−0.05	−0.35
ART_STR	0.11	0.12	−0.04	0.10	−0.04	0.09	0.10	0.09	0.09	0.05
BANK Mod	0.28	−0.10	−0.29	−0.29	−0.38	−0.13	−0.12	−0.06	−0.11	−0.44
Lat_CON	0.05	0.00	0.12	0.10	0.05	−0.03	−0.05	−0.03	−0.07	0.06
Long_CON	0.09	0.10	−0.03	0.07	−0.02	0.07	0.07	0.07	0.06	0.04
Land5	−0.26	0.01	0.20	0.19	0.14	−0.02	−0.03	−0.04	−0.05	0.26
Land50	−0.11	0.11	0.20	0.28	0.08	0.09	0.07	0.05	0.04	0.23
HMS	0.26	−0.08	−0.30	−0.29	−0.34	−0.06	−0.06	−0.04	−0.04	−0.39
HQA	−0.25	0.31	0.54	0.56	0.55	0.18	0.24	0.32	0.18	0.65
LRD	−0.14	−0.29	−0.32	−0.36	−0.09	−0.08	−0.15	−0.31	−0.09	−0.25

	No. of families	Sel EPTD	1-GOLD	ICMi	Pelal%*	Lithal%*	Phytal%*	DIND3*	DIND4*	Total significant values
CHANN	−0.16	−0.21	−0.18	−0.32	0.21	−0.04	0.15	−0.26	−0.24	4
SUBST	−0.29	−0.17	−0.28	−0.31	0.12	0.07	0.14	−0.36	−0.24	8
ART_STR	0.06	0.16	0.12	0.12	−0.09	−0.43	−0.32	−0.36	−0.40	4
BANK Mod	−0.20	−0.31	−0.30	−0.38	0.14	−0.20	0.07	−0.36	−0.51	10
Lat_CON	0.03	0.16	0.08	0.13	0.16	0.04	0.15	0.10	0.05	0
Long_CON	0.07	0.16	0.10	0.11	−0.14	−0.37	−0.33	−0.24	−0.29	1
Land5	0.17	0.34	0.34	0.33	0.02	0.19	−0.15	0.27	0.48	6
Land50	0.18	0.40	0.06	0.36	0.15	0.05	−0.30	0.29	0.39	5
HMS	−0.20	−0.29	−0.29	−0.37	0.21	−0.22	0.07	−0.40	−0.48	10
HQA	0.45	0.32	0.26	0.54	−0.40	0.25	−0.33	0.45	0.46	15
LRD	−0.34	−0.24	0.03	−0.33	0.17	−0.40	0.30	−0.26	−0.09	9

Significant values ($p < 0.001$) are underlined.*Indicates metrics only tested for Austria, Czech Republic and Germany.

Denmark for MTS) greater than 0.5. This supports results by Buffagni (1999), suggesting that MTS and EPT taxa can be used as metrics to indicate alteration in river morphology. These results confirm that the structure of banks can influence the invertebrate community, as was also demonstrated in the work of Naura & Robinson (1998), in which the ability of RHS features in predicting crayfish occurrence was investigated. Armitage et al. (2001) found that the bank side is a dynamic environment where communities change in relation to the growth of bank side vegetation and its concomitant effect on flow, with bank structure having a direct effect on invertebrates abundance and number of taxa.

ASPT is negatively related to channel geometry modification (i.e. reinforced or re-sectioned channel, culverting, etc.), even though this metric was designed to indicate organic pollution. River channelisation might have the effect of diminishing habitat diversification (Raven et al., 2000), thus influencing the self-depurating capacity of rivers, which in turn influences macroinvertebrate community structure.

Some of the attributes considered here do not change in correlation with changes in biological communities. In these cases the analysis of correlation with the HMS can furnish a more representative picture of the overall factors acting on the sites. It should also be noted that HMS needs to be adapted to different stream types, as it is a unique codification of scores and is not designed for representing the different river types that can be found throughout Europe (Balestrini et al., 2004; Szoszkiewicz et al., 2006).

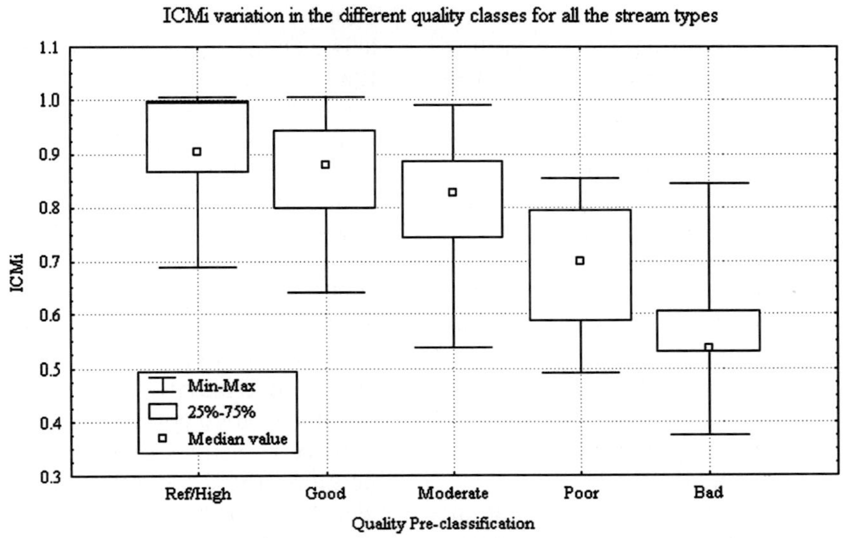

Figure 1. ICM index variation in relation to pre-classification in all the considered datasets.

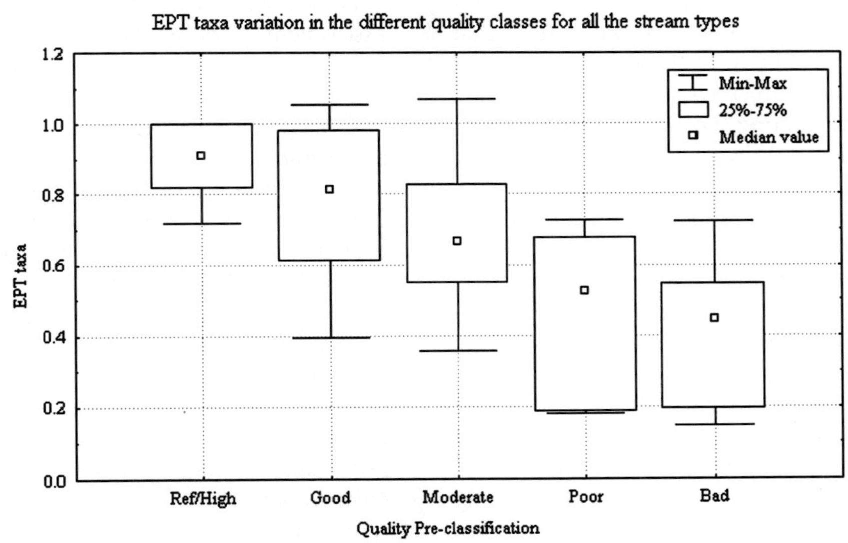

Figure 2. EPT taxa variation in relation to pre-classification in all the considered datasets.

It has been demonstrated that riparian habitats play a central role in determining river functioning (Pinay et al., 1990; Tabacchi et al., 1998). Results presented here show that vegetation structure within 5 and 50 m of the bank top seem to influence the metrics indicating the abundance of selected taxa. In-stream and riparian vegetation were also found to be the most important factors explaining intra-site variability in the species composition of stream invertebrates (Sandin & Johnson, 2004). On a larger spatial scale the results presented here indicate a low variability for land use in the floodplain (within 50 m of the banktop). A different codification for land use categories might help in underlining the effect of this variable on biological communities. Land use features often indirectly indicate alteration in stream morphology (Feld, 2004). From this work it seems possible to find specific metrics related to land use features: i.e., the abundance of selected taxa on the basis of all the types together (Table 8), 1-GOLD and taxa abundance considering river types separately (Table 7).

All the diversity indices expressed here do not seem to be adequate in indicating hydro-morphological quality, their correlation values being low and insignificant, when considering the different types together. On the other hand there is some work which demonstrates that for specific stream types diversity indices can well represent hydro-morphological impact (Lorenz et al., 2004).

The results show that not all the RHS variables considered, for the studied stream types, adequately represent the variability in the status of the different portions of the river (e.g., channel, banks, riparian habitats) as proposed for the CEN standard (EN14614:2004). This has been determined on the basis of the fact that some features are not extensive at the studied sites.

Consequently RHS is probably not ideal in determining lateral connectivity (a CEN requirement) as this may be affected by structures that are not present within the RHS site, or are not recorded. Nevertheless it should be stated that lateral connectivity is in general an important feature in relation to river functioning, being important in determining the occurrence of a wide variety of lentic and lotic habitats originating through fluvial dynamics (Ward & Wiens, 2001). The effect of constraining flow by embankments is evident in the changing of habitat patches related to floodplain, diminishing habitat heterogeneity and biodiversity (Ward & Wiens, 2001; Brunke, 2002). Appropriate measurements for this feature have therefore to be found.

Looking at the occurrence of characteristics influencing longitudinal connectivity and at the presence of artificial substrates and structures affecting flow character, it can be generally concluded that they were not important *per se* in the selection of sites analysed here. In addition, their correlation with invertebrate metrics is low. Thus it does not seem appropriate to derive any conclusions on their ability to represent different hydro-morphological conditions nor on the relationship with invertebrates metrics considered here. Studies investigating the effects of hydro-morphological alteration on biological communities have recently received wider consideration, nevertheless they are mostly focused on fish communities rather than on macroinvertebrates. Some studies demonstrate the effect of impoundments on macroinvertebrates (Jansen et al., 2000; Ogbeibu & Oribhabor, 2002; Ofenböck et al., 2004). Dams and weirs have the effect of altering flow conditions causing modification in trophic group ratios and functional groups. The metrics selected and investigated here were general metrics that did not result in adequately detecting this kind of alteration. More specific metrics, requiring a more detailed identification level and good ecological information, (that not always are available) (Buffagni et al., 2001), are probably required for the detection of this specific effect (i.e., impoundment). For example, when considering metrics related to invertebrates habitat preference (Table 8), a relatively good correlation can be found with the presence of artificial structures in the river. Furthermore, the detection of this kind of impact is related to the spatial scale analysed and to the position of the sampling area with respect to this alteration. Generally, the importance of spatial scale has been widely emphasized in relation to the assessment of habitat quality and biotic integrity (Allan & Johnson, 1997; Allan et al., 1997; Maddock, 1999).

This work also demonstrates the good general performance of the ICMi in indicating morphological alteration, in conjunction with the fact that the number of families shows a better response than the number of taxa (i.e. considered at the lowest level reached within each dataset) for the considered attributes.

The highest correlation values are reached from biological metrics with respect to HQA indicating that they are most influenced by habitat diversification and that morphological alteration is not as important as habitat quality, expressed here in terms of diversification. It has been demonstrated that habitat heterogeneity has a central role in structuring communities (Beisel et al., 2000; Griffith et al., 2003). It has also been demonstrated that HQA is related to the general degradation of sites (Balestrini et al., 2004), which partly explains the good correlation with biological metrics. Although HQA is better correlated than other RHS parameters, it is slightly related to variables indicating water quality. This confirms the ability of this index to quantify the overall quality of river sites i.e., including water quality. In addition, HQA has a strong correlation with HMS demonstrating its general capacity in indicating morphological degradation. The way in which information is coded in HQA is more relative to invertebrates than the one offered by HMS. This correlation is also high within the single datasets ($R > 0.80$ and always significant) and always higher with respect to the correlation with LRD ($R < 0.60$, not always significant). A part of the variability of HQA is linked to natural variability that can be better expressed by LRD, but in the datasets considered the hydro-morphological alteration seems to be more important in determining HQA variation. Another factor that has been demonstrated to have a strong influence on biology is the lentic–lotic character of rivers. In the present paper, the number of families and ASPT were found to be related to the LRD descriptor. From the results, it can also be argued that the metrics involved in some of the national assessment systems (e.g., IBE for Italy, IBGN for France and ASPT for UK) are potentially influenced by habitat diversification and lentic–lotic character of rivers. The number of families and the number of sensitive taxa (EPT) correlate well with HQA and LRD. The general good correlation among LRD and variables indicating water quality might be linked to self-depuration capacity of river stretches, which is likely to improve with increasing turbulence.

In summary, the best performing metrics were ASPT, EPT taxa and ICMi. In contrast, the worst performing metrics were diversity indices. In parallel, the most important factors influencing or correlating with benthic communities were bank modification and the hydro-morphological indices HQA, HMS and LRD.

Conclusion

The significant role of hydraulic factors influencing biological communities has been widely investigated. Less well known are the effects of morphological alteration on invertebrates even though the WFD recognises the important role of hydro-morphology in structuring biological communities. It is expected that factors such as re-sectioning and alterations to bank structure may cause a reduction in the biological diversity of aquatic ecosystems. Even if habitat assessment systems are being developed, the largest amount of work still to be done relates to the demonstration of the ecological relevance of the assessed features, possibly including the analysis of the different BQEs as requested by the WFD. As a first step it is important to compare the hydro-morphological assessments currently in use in different European countries (e.g., Raven et al., 2002). We are now in a stage of harmonisation of the information collected by different methods. One method is to try and codify different characteristics in the same way, linking them to a specific section of the river. River portions considered for CEN attributes are among those indicated in the REFCOND guidance (EC, 2003). In particular, features related to river hydrology were not considered, even if this aspect is particularly important especially in South European rivers. This paper has demonstrated that not all river sections carry the same weight in determining the hydro-morphological impact on the types studied. The sites discussed here are a representative selection chosen to give a comprehensive picture of the typical situations found in Europe, when considering sites affected by morphological alteration. For the types selected here it seems that some CEN features may play a marginal role in rivers. The general ability of RHS in characterising morphological attributes was demonstrated, even if refinements are needed to find a codification able to translate the recorded features into a meaningful score. As with HMS, the scores assigned to the different RHS features representing CEN attributes should be considered differently in relation to the various types, weighting scores on the basis of the importance of a specific feature within a specific river type. Features that are not well represented here may have a major impact on other stream types. The analysed spatial scale may also have an important role in determining the possibility of assessing a certain feature. In this context, the paired application of methods considering different scales can improve the overall characterization of the site. In addition, the results of other BQEs, being potentially very different, should be analysed.

Some questions remain open: how can organic and hydro-morphological impact be separated when they coexist? For this, assessment systems need to indicate local stream habitat features in order to identify which impact is acting. The data discussed here represent a first attempt at analysing the relationship between hydro-morphological features, especially linked to different river sections, and invertebrate communities. Due to the relatively low number of sites included in the analyses, all the results have to be considered preliminary. Within the data discussed here it can be seen that indices specifically developed for the detection of hydro-morphological impact (i.e., German indices) can have a good performance. Their ability is particularly useful when considering the regional scale and the type specific conditions under which these indices were developed. In general terms, the metrics that can be specifically linked to habitat alteration (e.g. invertebrates habitat preferences) seem not to have an appreciably better performance than more general indices. The scarce benefit in using species level metrics might be linked to the lack of autoecological information, which can still be important even in well-known areas such as Central Europe. More work should be addressed to the identification of selected taxa whose occurrence can be related to the presence and extent of specific features related to morphological alteration (e.g., presence of bars and extent of erosion). For these aspects taxonomic knowledge and auto-ecological information should be improved jointly with studies relating to the influence of habitat parameters on the distribution of invertebrates. A confirmation of the role of flow conditions in structuring biological communities was also found in relation to the correlation values found between LRD and some biological metrics. In particular, with the development of the southern European version of RHS (Buffagni & Kemp, 2002; Buffagni et al., in preparation) it will be easier to directly relate hydrological aspects to erosional/depositional features and channel stability.

Acknowledgements

We thank all STAR partners who contributed to the collection of data used in this paper. We are also indebted with CEN TC230/WG2/TG5 and namely to Phil Boon. The STAR Project (Standardisation of River Classifications: Framework method for calibrating different biological survey results against ecological quality classifications to be developed for the Water Framework Directive) was co-funded by the European Commission, 5th Framework Program, Energy, Environment and Sustainable Development, Key Action Water, E.U. Contract number: EVK1-CT 2001-00089).

References

Agences de l'Eau & Ministère de l'Environnement, 1998. SEQ-Physique: A System for the Evaluation of the Physical Quality of Watercourses, 15 pp.

Allan, J. D. & L. B. Johnson, 1997. Catchment-scale analysis of aquatic ecosystems. Freshwater Biology 37: 107–111.

Allan, J. D., D. A. L. Erickson & J. Fay, 1997. The influence of catchment land use on stream integrity across multiple spatial scales. Freshwater Biology 37: 149–161.

Armitage, P. D., D. Moss, J. F. Wright & M. T. Furse, 1983. The performance of a new biological water quality scores system based on macroinvertebrates over a wide range of unpolluted running-water sites. Water Research 17: 333–347.

Armitage, P. D., K. Lattmann, N. Kneebone & I. Harris, 2001. Bank profile and structure as determinants of macroinvertebrate assemblages – seasonal changes and management. Regulated Rivers: Research & Management 17: 543–556.

Balestrini, R., M. Cazzola & A. Buffagni, 2004. Riparian ecotones and hydromorphological features of selected Italian rivers: a comparative application of environmental indices. In Hering, D., P. F. M. Verdonschot, O. Moog & L. Sandin (eds), Integrated Assessment of Running Waters in Europe. Kluwer Academic Publishers, printed in the Netherlands. Hydrobiologia 516: 365–379.

Beisel, J., P. Usseglio-Polatera & J. Moreteau, 2000. The spatial heterogeneity of a river bottom: a key factor determining macroinvertebrate communities. Hydrobiologia 422/423: 163–171.

Brunke, M., 2002. Floodplains of a regulated southern alpine river (Brenno, Switzerland): ecological assessment and conservation options. Aquatic Conservation: Marine and Freshwater Ecosystems 12: 583–599.

Buffagni, A., 1997. Mayfly community composition and the biological quality of streams. In Landolt, P. & M. Sartori (eds), Ephemeroptera & Plecoptera: Biology-Ecology-Systematics. MTL, Fribourg, 235–246.

Buffagni, A., 1999. Pregio naturalistico, qualità ecologica e integrità della comunità degli Efemerotteri (Insecta Ephemeroptera): un indice per la classificazione dei fiumi italiani. Acqua&Aria 8: 99–107.

Buffagni, A., J. L. Kemp, S. Erba, C. Belfiore, D. Hering & O. Moog, 2001. A Europe wide system for assessing the quality of rivers using macroinvertebrates: the AQEM project and its importance for southern Europe (with special emphasis on Italy). Journal of Limnology 60(suppl.1): 39–48.

Buffagni, A. & S. Erba, 2002. Guidance for the assessment of hydromorphological features of rivers within the STAR Project. June 2002, 20 + 18 pp (Available at STAR web site, www.eu-star.at).

Buffagni, A. & J. L. Kemp, 2002. Looking beyond the shores of the United Kingdom: addenda for the application of River Habitat Survey in South-European rivers. Journal of Limnology 61: 199–214.

Buffagni, A., S. Erba, D. Armanini, D. De Martini & S. Somaré, 2004a. Aspetti idromorfologici e carattere Lentico-lotico dei fiumi mediterranei: River Habitat Survey e descrittore LRD. In 'Classificazione ecologica e carattere lentico-lotico in fiumi mediterranei'. Quad. Ist. Ricerca Acque, Roma 122: 41–63.

Buffagni, A., S. Erba, M. Cazzola & J. L. Kemp, 2004b. The AQEM multimetric system for the southern Italian Apennines: assessing the impact of water quality and habitat degradation on pool macroinvertebrates in Mediterranean rivers. In Hering, D., P. F. M. Verdonschot, O. Moog & L. Sandin (eds), Integrated Assessment of Running Waters in Europe. Kluwer Academic Publishers, printed in the Netherlands. Hydrobiologia 516: 313–329.

Buffagni, A., S. Erba, S. Birk, M. Cazzola, C. Feld, T. Ofenböck, J. Murray-Bligh, M. T. Furse, R. Clarke, D. Hering, H. Soszka & W. van de Bund, 2005. Towards European Inter-calibration for the Water Framework Directive: Procedures and examples for different river types from the E.C. project STAR. 11th STAR Deliverable. STAR Contract No: EVK1-CT 2001-00089. Rome (Italy), Quad. Ist. Ric. Acque 123, Rome (Italy), IRSA, 468 pp.

EN14614:2004. Water Quality: Guidance Standard for Assessing the Hydromorphological Features of Rivers. CEN TC 230/WG 2/TG 5: N47.

Environmental Agency, 1997. River Habitat Survey – Field Guidance Manual. Bristol.

European Commission, 2000. Directive 2000/60/EC of the European Parliament and of the Council of 23 October 2000 establishing a framework for Community action in the field of water policy. Official Journal of the European Communities L 327, 22.12.2000, 1–72.

European Commission, 2003. Common Implementation Strategy for the Water Framework Directive (2000/60/EC). Guidance Document No 10. Rivers and Lakes – Typology, Reference Conditions and Classification Systems Produced by Working Group 2.3 – REFCOND, 94 pp.

Feld, C. K., 2004. Identification and measure of hydromorphological degradation in Central European lowland streams. In Hering, D., P. F. M. Verdonschot, O. Moog & L. Sandin (eds), Integrated Assessment of Running Waters in Europe. Kluwer Academic Publishers, printed in the Netherlands. Hydrobiologia 516: 69–90.

Fleischhacker, T. & K. Kern, 2002. Ecomorphological Survey of Large Rivers. German Federal Institute of Hydrology, Postfach 200 253, D-56002 Koblenz, 41 pp.

Furse, M., D. Hering, O. Moog, P. Verdonschot, R. K. Johnson, K. Brabec, K. Gritzalis, A. Buffagni, P. Pinto, N. Friberg, J. Murray-Bligh, J. Kokes, R. Alber, P. Usseglio-Polatera, P. Haase, R. Sweeting, B. Bis, K. Szoszkiewicz, H. Soszka, G. Springe, F. Sporka & I. Krno, 2006. The STAR project: context, objectives and approaches. Hydrobiologia 566: 3–29.

Griffith, M. B., P. Husby, R. P. Hall, P. R. Kaufmann & B. Hill, 2003. Analysis of macroinvertebrate assemblages in relation to environmental gradients among lotic habitats of California's central valley. Environmental Monitoring and Assessment 82: 281–309.

Hering, D., O. Moog, L. Sandin & P. F. M. Verdonschot, 2004. Overview and application of the AQEM assessment system. In Hering, D., P. F. M. Verdonschot, O. Moog & L. Sandin (eds), Integrated Assessment of Running Waters in Europe. Kluwer Academic Publishers, printed in the Netherlands. Hydrobiologia 516: 1–20.

Jansen, W., J. Böhmer, B. Kappus, T. Beiter, B. Breitinger & C. Hock, 2000. Benthic invertebrate and fish communities as indicators of morphological integrity in the Enz River (south-west Germany). Hydrobiologia 422/423: 331–342.

Krebs, C. J., 1989. Ecological Methodology. University of British Columbia, Harper Collins Publishers, 357–367.

Legendre, P. & L. Legendre, 1998. Numerical Ecology. Developments in Environmental Modelling 20. Elsevier, Amsterdam, 853 pp.

Logan, P. & M. Furse, 2002. Preparing for the European Water Framework Directive – making the links between habitat and aquatic biota. Aquatic Conservation: Marine and Freshwater Ecosystems 12: 425–437.

Lorenz, A., D. Hering, C. K. Feld & P. Rolauffs, 2004. A new method for assessing the impact of hydromorphological degradation on the macroinvertebrate fauna of five German stream types In Hering, D., P. F. M. Verdonschot, O. Moog & L. Sandin (eds), Integrated Assessment of Running Waters in Europe. Kluwer Academic Publishers, printed in the Netherlands. Hydrobiologia 516: 107–127.

Maddock, I., 1999. The importance of physical habitat assessment for evaluating river health. Freshwater Biology 41: 373–391.

Margalef, R., 1984. Ecosystems: Diversity and Connectivity as measurable components of their complication. In AIDA, et al. (ed.), The Science and Praxis of Complexity. Tokyo: United Nations University, 1984, 228–244.

Metcalfe-Smith, J.L., 1994. Biological water-quality assessment of rivers: use of macroinvertebrate communities. In Calow, P. & G. E. Petts (eds), The River Handbook. Blackwell Scientific Publication, Oxford Vol 2: 144–170.

Muhar, S., M. Kainz, M. Kaufmann & M. Schwarz, 1996. Ausweisungflusstypspezifisch erhaltener Fliessgewässerabschnitte in Österreich (In German). Österreichische Bundesgewässer, BMLF, Wasserwirtschaftskataster, Wien, 176 pp.

Muhar, S., M. Kainz & M. Schwarz, 1998. Ausweisungflusstypspezifisch erhaltener Fliessgewässerabsc- hnitte in Österreich – Fliessgewässer mit einem Einzugsgebiet > 500 km^2 ohne Bundesflüsse (In German). BMLF, BMUJF, Wasserwirtschaftskataster, Wien: 177 pp.

Naura, M. & M. Robinson, 1998. Principles of using River Habitat Survey to predict the distribution of aquatic species: an example applied to the native white-clawed crayfish *Austropotamobius pallipes*. Aquatic Conservation: Marine and Freshwater Ecosystems 8: 515–527.

Nijboer, R. C., R. K. Johnson, P. F. M. Verdonschot, M. Sommerhäuser & A. Buffagni, 2004. Establishing reference conditions for European streams. In Hering, D., P. F. M. Verdonschot, O. Moog & L. Sandin (eds), Integrated Assessment of Running Waters in Europe. Kluwer Academic Publishers, printed in the Netherlands. Hydrobiologia 516: 91–105.

Ofenböck, T., O. Moog, J. Gerritsen & M. Barbour, 2004. A stressor specific multimetric approach for monitoring running waters in Austria using benthic macro-invertebrates. In Hering, D., P. F. M. Verdonschot, O. Moog & L. Sandin (eds), Integrated Assessment of Running Waters in Europe. Kluwer Academic Publishers, printed in the Netherlands. Hydrobiologia 516: 251–268.

Ogbeibu, A. E. & B. J. Oribhabor, 2002. Ecological impact of river impoundment using benthic macro-invertebrates as indicators. Water Research 36: 2427–2436.

Pedersen, M. L. & A. Baattrup-Pedersen, 2003. Økologisk overvågning i vandløb og på vandløbsnære arealer under NOVANA 2004–2009 (In Danish). Teknisk Anvisning fra DMU nr. 21. National Environmental Research Institute, 128 pp.

Pinay, G., H. Decamps, E. Chauvet & E. Fustec, 1990. Functions of ecotones in fluvial systems. In Naiman, R. J. & H. Decamps (eds), The Ecology and Management of Aquatic-Terrestrial Ecotones. The Parthenon Publishing Group, Paris, 141–164.

Pinto, P., J. Rosado, M. Morais & I. Antunes, 2004. Assessment methodology for southern siliceous basins in Portugal. In Hering, D., P. F. M. Verdonschot, O. Moog & L. Sandin (eds), Integrated Assessment of Running Waters in Europe. Kluwer Academic Publishers, printed in the Netherlands. Hydrobiologia, 516: 191–214.

Raven, P. J., P. J. A. Fox, M. Everard, N. T. H. Holmes & F. D. Dawson, 1997. River Habitat Survey: a new system for classifying rivers according to their habitat quality. In Boon, P. J. & D. L. Howell (eds), Freshwater Quality: Defining the Indefinable? The Stationery Office, Edinburgh, 215–234 .

Raven, P. J., T. H. Holmes, F. H. Dawson, P. J. A. Fox, M. Everard, I. R. Fozzard & K. J. Rouen, 1998. River Habitat Survey, the physical character of rivers and streams in the UK and Isle of man. River Habitat Survey No.2, May 1998. The Environment Agency, Bristol, 86 pp.

Raven, P. J., N. T. H. Holmes, M. Naura & F. H. Dawson, 2000. Using river habitat survey for environmental assessment and catchment plan in the UK. Hydrobiologia 422/423: 359–367.

Raven, P. J., N. T. H. Holmes, P. Charrier, F. H. Dawson, R. Naura & P. J. Boon, 2002. Towards a harmonized approach for hydromorphological assessment of rivers in Europe: a qualitative comparison of three survey methods. Aquatic Conservation: Marine and Freshwater Ecosystems 12: 405–424.

Sandin, L. & R. K. Johnson, 2004. Local, landscape and regional factors structuring benthic macroinvertebrate assemblages in Swedish streams. Landscape Ecology 19: 501–514.

Szoszkiewicz K., A. Buffagni, J. Davy-Bowker, J. Lesny, B. H. Chojnicki, J. Zbierska, R. Staniszewski & T. Zgola, 2006. Occurrence and variability of River Habitat Survey features across Europe and consequences on data quality evaluation. Hydrobiologia 566: 267–280.

Tabacchi, E., D. L. Correll, R. Hauer, G. Pinay, A. -M. Planty-Tabacchi & R. C. Wissmar, 1998. Development, maintenance, and role of riparian vegetation in the river landscape. Freshwater Biology 40: 497–516.

Verdonschot, P. F. M., 2000. Integrated ecological assessment methods as a basis for sustainable catchment management. Hydrobiologia 442/443: 389–412.

Ward, J. V. & J. A. Wiens, 2001. Ecotones of riverine ecosystems: role and typology, spatio-temporal dynamics, and river regulation. Ecohydrology Hydrobiology 1: 25–36.

Werth, W., 1987. Okomorphologische Gewasserbewertung (Ecomorphological survey of streams, in German). Osterreichische Wasserwirtschaft 39(5/6): 122–128.

Woodiwiss, F. S., 1964. The biological system of stream classification used by the Trent River Board. Chemistry and Industry 14: 443–447.

Tools for Assessing European Streams with Macroinvertebrates

Hydrobiologia (2006) 566:299–309
M.T. Furse, D. Hering, K. Brabec, A. Buffagni, L. Sandin & P.F.M. Verdonschot (eds), The Ecological Status of European Rivers: Evaluation and Intercalibration of Assessment Methods
DOI 10.1007/s10750-006-0088-1

Tools for assessing European streams with macroinvertebrates: major results and conclusions from the STAR project

Piet F.M. Verdonschot[1,*] & Otto Moog[2]
[1]*Alterra–Green World Research, P.O. Box 47, 6700 AA Wageningen, The Netherlands*
[2]*Department Water, Atmosphere, Environment, Institute of Hydrobiology and Aquatic Ecosystem Management, BOKU – University of Natural Resources & Applied Life Sciences, Max Emanuel Straße 17, A-1180 Vienna, Austria*
*(*Author for correspondence: E-mail: piet.verdonschot@wur.nl)*

Key words: autecology, database, metric, index, multimetrics, assessment

Abstract

This short paper summarises the information developed in the EU funded research project STAR on autecology databases, metrics, multimetrics and community approaches. For Europe the WFD implementation gave an important stimulus for the development of ecology based assessment techniques. Along with the development of metrics and multimetrics indices taxalists and autecological information were strongly improved. Recommendations are given to further develop ecological assessment in European streams and rivers.

Introduction

The systematic use of biological responses to evaluate changes in the environment with the intent to use this information in a water quality control program is defined biological assessment (Matthews et al., 1982). The biological response is measured by using biological indicators. In the first decades of the 20th century biological assessment mostly used simple straightforward techniques (Hynes, 1960; Hellawell, 1978) related to organic waste pollution. Within these methodologies the saprobic approaches and indices (as described by e.g., Kolkwitz & Marsson, 1902, 1908, 1909; Forbes & Richardson, 1913; Ellis, 1937; Liebmann, 1962; Sládecek, 1973) have been widely used in many central and eastern European countries. During the last decades of the 20th century these biological assessments more and more appeared to be low sensitive, superficial, robust to natural variation, and unpredictable as the impacts of organic pollution decrease due to enhanced wastewater treatment techniques and facilities. The traditional quality assessment approaches failed after the impairment of rivers got mixed with other environmental disturbances. Furthermore, using biotic indices (e.g., Chandler, 1970; Woodiwiss, 1980; Armitage et al., 1983) the condition of sites near the ends of the measuring scale were easy to judge but the middle part of the scale, thus the moderately degraded sites appeared not (Tolkamp, 1985). Traditional biological assessment methods no longer provided a sufficient tool for integrated water management due to their restricted approach to one or a few aspects of the aquatic ecosystem.

In ecological assessment the corresponding environment is added to the biological one (Odum, 1971) to reflect together the ecosystem as a whole. To assess a running water system one should use a high variety of parameters reflecting the structure and functioning of the ecosystem (Cairns, 1975; Frey, 1975; Karr et al., 1986) and also reflecting different types of disturbance (e.g., Nelson, 1990; Richter et al., 1996; Roth et al., 1996). On the other hand Fore et al. (1996) stated that it would be wrong to think the more parameters added, the higher probability of an accurate diagnoses.

Still, the presence, numbers, and condition of specific species of aquatic macroinvertebrates, fish, algae and macrophytes can provide accurate information about the quality of a specific stream or river. Ecologists try to understand this information and use it to support management. This summary shows in short the development of community approaches and multimetric indices, as both of these tools ecologists use to reach their objectives.

Single metrics and indices

The more conventional approach in using individual species composition and/or abundances related measures was to select a biological parameter that referred to a one factor range of change in environmental conditions, mostly related to a very strong and obvious stressor, and to evaluate that parameter (e.g., species distributions, abundance trends), such as the saprobic indices mentioned in the introduction. Such single biological parameter was interpreted with a summary statement about the water quality, by using an index or metric score. This approach is limited in that the key parameter emphasised may not reflect the overall ecological status.

Nevertheless the recent bio-monitoring is still based on biological indicators. A biological indicator reflects the biological response to chemical, physical or biological properties of a water body or to the overall ecological condition (Karr & Dudley, 1981; Rosenberg & Resh, 1993; Simon & Davies, 1995; Verdonschot, 2000). A biological indicator can be used to characterise the current status, can identify major ecosystem stress and can track or predict significant change of the status of a water body. In general, an indicator has a diagnostic feature. Indices or metrics make use of indicators and combine them into a numerical value or score.

Following Karr & Chu (1999) metrics are defined as "measurable parts or processes of a biological system empirically shown to change in value along a gradient of human influence". An index or metric is useful when it is:

(1) relevant to the ecosystem under study and to the specified objectives;
(2) sensitive to stressors;
(3) able to provide a response that can be discriminated from natural variation;
(4) environmentally benign to measure in the aquatic environment;
(5) cost-effective to sample.

A number of indices or metrics have been developed and subsequently tested in field surveys of different aquatic organism groups; from the early saprobic systems (e.g., Kolkwitz & Marsson, 1902, 1908, 1909; Pantle & Buck, 1955; Zelinka & Marvan, 1961; Liebmann, 1962; Sládecek, 1973), diversity indices (e.g., Shannon & Weaver, 1949; Washington, 1984; Boyle et al., 1990) to biotic indices (Woodiwiss, 1964; Tuffery &Verneaux, 1968; BMWP, 1979; De Pauw & Vanhoren, 1983; Andersen et al., 1994 and others). Biotic indices and biotic scores use both a saprobic rank and a diversity measure and thus combine a richness measure and a (mostly organic) pollution tolerance measure (Metcalfe, 1989; De Pauw et al., 1992). The biotic index therefore could be classified as a bimetric.

Multimetric indices

The next logical step in metric development was to combine a number of different metrics, each of which provides information on an ecosystem feature and when integrated, performs as an overall indicator of ecological conditions of a water body. The value of a multimetric index is that such an approach integrates information from different ecosystem components and evaluates, with reference to biogeography, a number of single ecologically based indices (Karr et al., 1986; Plafkin et al., 1989; Barbour et al., 1995). Such multimetric assessments provide detection capability over a broader range and nature of stressors and give a more complete picture of ecological conditions than single bio- or ecological indicators.

The US EPA defined a multimetric index as an index that combines indicators, or metrics, into a single index value. Each metric is tested and calibrated to a scale and transformed into a unitless score prior to being aggregated into a multimetric index. Both the index and the metrics are useful in assessing and diagnosing ecological condition.

A large number of metrics has been developed (e.g., Fausch et al., 1990; Karr, 1991; Karr & Kerans, 1992) The Index of Biotic Integrity (IBI) was probably the first and most original multimetric index (Karr, 1981), and was based on fish. It originally included 12 metrics that reflected fish species richness and composition, number and abundance of indicator species, trophic organisation and function, reproductive behaviour, fish abundance, and condition of individual fish. These metrics reflect the ecosystem characteristics of food source, water quality, habitat structure, flow regime and biotic interactions. Later on, other multimetrics were developed that included the benthic macroinvertebrate assemblage (e.g., Invertebrate Community Index (ICI); Ohio EPA, 1987; Plafkin et al., 1989; Kerans & Karr, 1994), or the macrophytes (Nelson, 1990). Barbour et al. (1992) presented the conceptual base for the multimetrics approach in which the benthic community health is composed of community structure, community balance and functional feeding groups, and in combination with habitat quality, an integrated assessment is obtained (Verdonschot, 2000).

Consequently, all multimetrics were and are based on ecological attributes of biological communities. Eight major groups of metrics can be distinguished (adapted after Resh & Jackson, 1993; Thorne & Williams, 1997):

- richness measures (e.g., number of taxa, number of EPT taxa, number of Chironomidae taxa); often these metrics are considered to be sensitive to organic pollution;
- enumerations or composition measures (e.g., number of individuals, % of the total EPT taxa (sensitive) and chironomids (tolerant), % dominant taxon, number of. intolerant taxa, % Oligochaeta, sediment tolerant taxa); often these metrics are considered to increase in dominance of one or more taxa due to pollution or disturbance;
- diversity measures (e.g., Shannon–Wiener Index, sequential comparison index); often these metrics are considered to decrease with increasing disturbance;
- similarity/loss measures (e.g., number of taxa in common, community loss index, Bray–Curtis index); these metrics use comparisons between sites (reference vs. disturbed sites);
- tolerance/intolerance measures or biotic indices (e.g., saprobic index and Hilsenhoff's family biotic index, BMWP score, ASPT score); the last two metrics rely on the assignment of (in-) tolerance values to taxa and include richness;
- functional and trophic measures (e.g., % of functional feeding groups, % habitat or current preferences, % locomotion types, longitudinal zonation index); these metrics use the alteration in food types, habitats and environmental conditions under different types of disturbance;
- (life) strategy metrics; the metrics use the biological life strategy features like, length of the lifecycle, number of eggs or diapause;
- condition metrics; these metrics use features of the condition or health of a specimen (e.g., percent of individuals that are diseased, deformed, or fish that have eroded fins, lesions, or tumours).

The common approach is to define a number of metrics that individually provide information on each ecosystem characteristic and when integrated, function as an overall indicator of biological condition.

The scores of the individual metrics are aggregated to calculate the multimetric index (e.g., Karr, 1981; Barbour et al., 1996). The multimetrics establish relative values for each single metric based on comparison of values for the best available habitat (with minimal human disturbance) to those areas which are strongly disturbed (see Verdonschot, 2000).

Autecology databases

The important base for many metrics or indices is the taxonomical status of each collected organism. The AQEM/STAR macroinvertebrate taxalist includes the updated taxonomical information of aquatic orders, families and species, as well as the species occurrences in 14 European countries (Schmidt-Kloiber et al., 2006).

Autecology databases for aquatic species cover ecological attributes of various ecological preferences (such as tolerances and preferences for current, acidity, organic load, substrate, trophic state and toxic substances) and of strategy or trait features like length of lifecycle, number of eggs, short-winged. These databases are crucial to support the use of metrics. Such lists were compiled already when the first saprobic indices were drafted (e.g., Kolkwitz & Marsson, 1908, 1909; Sládecek, 1973). In the last decade several databases were published, such as Verdonschot (1990), van der Hoek & Verdonschot (1994), Moog (1995), Schmedtje & Colling (1996), Šporka (2003). These list became more operational by including them in software packages, like ECOPROF (Moog et al., 2001) and the AQEM assessment program (AQEM consortium, 2002). Schmidt-Kloiber et al. (2006) list the available autecological information in the AQEM/STAR database. Most common in this database are the ecological attributes of oxygen demand (saprobic indices), stream zonation, current and substrate preferences, feeding and locomotion types. During the EU funded project, Euro-limpacs (www.eurolimpacs.ucl.ac.uk, Contract No: GOCE-CT-2003-505540) this database serves as a basic data source and will be extended to include ecological parameters, which are assumed to be sensitive to direct or indirect impacts of climate change. As a final outcome of this project all autecological parameters will be made available to the scientific public via a website for manifold multiple uses, e.g., the development of future assessment systems.

Community assessment

With the upcoming use of multivariate analysis techniques in ecology, aquatic ecologists started to explore relationships between whole taxa lists and accompanying environmental parameters (Verdonschot, 2000). Wright et al. (1984) used multivariate analysis techniques to classify unpolluted running water sites and to predict community types from environmental data (the River Invertebrate Prediction and Classification System (RIVPACS)). RIVPACS offered a prediction of the macroinvertebrate composition to be expected at a given site from a small number of environmental parameters recorded. By comparing the fauna observed (at species or at family level) with the expected or "target" fauna predicted, a measure of site quality was obtained (Wright et al., 1989, 2000).

The Australian River Assessment Scheme (AUSRIVAS) is based on the RIVPACS model. The differences are that the major habitats are sampled and modelled separately and that different models are used for different bioregions over Australia (Simpson & Norris, 2000).

Verdonschot (1990) conducted a large extensive data collection and multivariate analysis of macroinvertebrates in surface waters in the Netherlands. Verdonschot described macrofaunal site groups (cenotypes), which are recognised on the basis of environmental variables and the abundance of taxa. His cenotypes were described as overlapping entities with limited internal variation, no clear boundaries were provided only a recognisable centroid. The cenotypes are mutually related in terms of key factors, which represent major ecological processes. The cenotypes and their mutual relationships form a web that offers an ecological basis for the daily practice of water and nature management (Verdonschot, 1991). The web allows the development of water quality objectives, provides a tool to monitor and assess, indicates targets and guides the management and restoration of water bodies (Verdonschot & Nijboer, 2000).

Other multivariate approaches using different techniques are described by amongst others Johnson, 1998; Hawkins et al., 2000; Reynoldson et al., 2000. The PERLA system (Kokeš et al., 2006) involves a network of reference sites, a database of reference sites involving both respective biotic and abiotic data, and a prediction model. It is an expansion of the RIVPACS model. As most multivariate based approaches it assesses the overall condition of the ecosystem and as such is not stressor specific.

AQEM

The AQEM project (The Development and Testing of an Integrated Assessment System for the Ecological Quality of Streams and Rivers throughout Europe using Benthic Macroinvertebrates. Contract no.: EVK1-CT-1999-00027; www.aqem.de) was carried out between 2000 and 2002. The development of the AQEM ecological quality assessment system was based on newly collected data that covered both the benthic macroinvertebrate fauna and general stream characteristics. The data were collected by 8 countries (Austria, Czech Republic, Germany, Greece, Italy, The Netherlands, Portugal and Sweden). Generally, to develop the assessment system the following steps were taken (Hering et al., 2004):

- deriving a stream-type specific classification, which reflects the degradation of a site, based either on abiotic data recorded in a harmonised "site protocol" or on the biotic composition;
- testing of various attributes of the assemblage (i.e., metrics) with the goal to identify those most effective in measuring the degradation of the stream;
    - the starting point is the taxa list obtained from the sampling site, which is to be assessed;
    - based on this taxa list a number of metrics is calculated;
    - generally, the metric's results are individually converted into scores by comparing their values with the values of the same metrics in stream-type specific reference conditions;
- selecting those metrics that most strongly correlate with the site's state of degradation measured by chemical or hydromorphological parameters;
- aggregating these core metrics into a multimetric index;
    - the scores or results of the metrics are combined in a simple multimetric index (usually the average of all scores);
- calibrating the stream-type specific assessment systems with independent data;
- defining quality classes of "high", "good", "moderate", "poor" and "bad" ecological status for the selected stream types.

The consortium all together tested, independently for each of the 29 stream types, the correlation of a large number of metrics against the extent of degradation of a site as determined by assessment of the site protocol data (Ofenböck et al., 2004). Metrics that clearly respond to specific pollutants or stressors were considered most useful as a diagnostic tool (Karr & Chu, 1999). Furthermore, several criteria were followed:

- the metrics used should in combination cover diverse aspects of structure, composition, quality and function of the aquatic ecosystem;
- the metrics should deliver information on different components of the community;
- the metrics should be consistent with the country's traditions in stream monitoring.

The selection process resulted in up to 18 suitable core metrics for the individual AQEM stream types (Hering et al., 2004). Most interesting in the multimetric indices development within AQEM is the final use of the criteria. In all cases it was clear that those metrics selected to construct a multimetric index showed a significant correlation with the respective stressor gradient.

Further developments in STAR

The STAR project (Standardisation of river classifications: Framework method for calibrating different biological survey results against ecological quality classifications to be developed for the Water Framework Directive. Contract no.: EVK1-CT-1999-00027, www.eu-star.at) was carried out between 2003 and 2005. The STAR project used the AQEM multimetrics development. AQEM was restricted to macroinvertebrates, STAR included several new countries than AQEM (Denmark, France, Great-Britain, Latvia, Poland, Slovakia) as well as three additional organism groups (macrophytes, diatoms, fishes). The metric types used belong to the categories: composition/abundance metrics, richness/diversity measures, sensitivity/tolerance metrics and functional metrics (Hering et al., 2006). A general procedure to select the most suited metrics is fully described by Hering et al. (2006). Metrics are calculated using existing autecology data on species. Environmental gradients are extracted through ordination analyses procedures on selected parameters. Correlation is used to select metrics whereby a number of criteria (e.g., being robust, reflecting quantitatively an impact gradient, founded on ecological principles, representing different components of the ecosystem) are added to select the final metrics. Those selected metrics are used to construct the multimetric. For diatoms no multimetrics were developed. For fish the FAME approach has been adopted (www.fame.boku.ac.at). The new countries up to this point did not develop national multimetrics.

Germany and Austria have applied the experiences from STAR and AQEM and developed and integrated multimetric indices into their national assessment systems to achieve the EU Water Framework Directive demands for an integrated biological assessment for macroinvertebrates (Hering et al., 2004; Ofenböck et al., 2004).

In both countries modular and stream type-specific systems were generated, which are capable to distinguish the impact of different stressors. Due to the modular structure, the assessment systems integrate the impact of different stressors on the benthic invertebrates community and consist of three basic modules, developed to consider the main stressor types. The three modules are

- "organic pollution"
- "acidification"
- "general degradation"

For the module "Organic Pollution" the traditional Saprobic Index was adapted to a five class system and evaluated in relation to a stream type-specific reference value in both countries (the revised German Saprobic Index (DIN 38 410) (Rolauffs et al., 2003); Austrian "Guidelines Saprobiology" ÖNORM M 6232, 1997; Moog et al., 1999; Stubauer & Moog, 2002). For acidification an acidification index (Braukmann & Biss, 2004) is designated for bioregions at risk of acidification, while multimetric indices are used for the evaluation of "general degradation". Metrics for the indices were selected to address all major aspects (metric groups) of the biota, which are required in the WFD. In the Austrian classification scheme for the stressors "general degradation" furthermore distinguishes between two different indices for every stream type to address two diametrically opposed effects of stressors on running waters:

- "potamalisation" (e.g., caused by impoundment or siltation) and
- "rhithralisation/loss of diversity" (e.g., caused by river straightening (loss of habitats) or toxic contamination).

Metrics used for the multimetric indices are standardised in relation to metric values under stream type-specific reference conditions. Indices are calculated by averaging the standardised metrics. The class boundaries finally were defined to result in classes of equal width. The final Ecological Quality Class in both countries is determined by the worst case applying all relevant modules.

Discussion and conclusions

Databases

There is a common agreement that the performance of any biological assessment approach increases with the quality rating of its ecological background. Consequentially there was a remarkable increase of taxalists that related ecological information to indicator taxa in the last 10–15 years. These taxalists include functional ecosystem characteristics, species traits and others more in ecological assessment. Although the newly developed methodologies are quite promising we are still away from having assessment tools that can be applied robustly and area-wide. The results of the STAR project contributed valuably to ecological status assessment of rivers but also clearly indicated gaps in the knowledge on aquatic ecology that need to be closed. A strong cooperation of basic limnology with applied aquatic ecologists who translate the scientific knowledge into easy understandable and applicable tools is still needed for achieving the target goal of a good ecological status of rivers and streams.

Multimetric indices and community approaches

Metrics that relate to specific stressors or characteristics of the ecosystem functioning provide, individually, a strong diagnostic tool. The effects of various stressors on the behaviour of specific metrics strongly depend on the knowledge of the distribution and ecological requirements of the respective species. Hering et al. (2006) clearly postulate that a harmonisation in developing a multimetric assessment system in Europe is an inevitable must. The authors suggest a normative methodology for the development and application of multimetric indices which is composed of the following steps: (1) Selection of the most suitable form of a multimetric index; (2) metric selection, broken down into metric calculation, exclusion of numerically unsuitable metrics, definition of a stressor gradient, correlation of stressor gradients and metrics, selection of candidate metrics, selection of core metrics, distribution of metrics within the metric types, definition of upper and lower anchors and scaling; (3) generation of a Multimetric Index (general or stressor-specific approach); (4) Setting class boundaries; (5) interpretation of results.

On the other hand community approaches provide an integrated approach of the ecosystem but do not specifically point to a stressor or specific environmental condition.

In conclusion, using both a community approach together with a number of diagnostic metrics would provide a very strong tool for WFD proof water management.

Multiple stress and metric selection

In all cases it was clear that those metrics selected to construct a multimetric index showed a significant correlation with the respective stressor gradient. But selecting those metrics with a high correlation implies not including those metrics that provide information on weak stressors or healthy ecosystem components. For example, when a river is organically polluted a high correlation can be found between the organic load stressor with a saprobity metric. But at the same gradient a hydromorphological change can occur, and this is very often the case. Either metric selection is done along mono-stressor gradients which is not explicitly explained in most of the literature or the effects of those less dominant stressors are ignored following the selection procedure as described in the AQEM manual (AQEM consortium, 2002). This mono-stressor based selection procedure does not correspond to one of the most important criteria a multimetric is based on, namely telling the user about a number of features of the respective ecosystem. In such case, the multimetrics construction procedure should much more accurate test each metric within each of the groups of metric within each individual the ecosystem feature.

Organism groups

All multimetric approaches developed in the AQEM and STAR projects only used macroinvertebrates although hydromorphology indices were developed and diatom, fish (FAME) and macrophytes indices were used. For these organism groups there were no multimetric indices developed due to lack of sufficient data, noncoherence with the WFD stream typology, detection of a scale of response problem and elaboration time. In future these problems should be overcome. Furthermore, an authentic multimetrics approach should include metrics of different organism groups.

Ecosystem components

The theory states that metrics should cover all ecosystem components. Looking at these

Table 1. The overall list of AQEM metric categories (Hering et al., 2004)

Metric category	Examples
Richness measures	Total number of taxa, number of EPT taxa
Composition measures	% Dominant taxon, % Oligochaeta
Diversity measures	Shannon–Wiener diversity index
Similarity/loss measures	Species deficit, missing taxa
Tolerance/intolerance measures	Saprobic index, BMWP, ASPT
Functional and trophic measures (Feeding measures)	% Filterers, index of trophic completeness, RETI
Habitat/mode of existence measures	% of clinger, number of (semi)sessil taxa
Current preference measures	% Limnophil, % rheophil
Zonation measures	Zonation Index, % littoral
Generation turnover measures	% Bivoltin, % univoltin
Individual condition measures	Contaminant levels, % diseased individuals

Table 2. Number of metrics per ecosystem feature per stream according to the AQEM results (Hering et al., 2004)

Stream type	Richness measures	Composition measures	Diversity measures	Similarity/loss measures	Tolerance/intolerance measures	Functional and trophic measures (Feeding measures)	Habitat/mode of existence measures	Current preference measures	Zonation measures	Generation turnover measures	Individual condition measures	Total number of metrics	Number of metric categories
A01	2	3	1		2	1			1			10	6
A02	3	2	1		2							8	4
A03	2	3	1		2	1			1			10	6
A04	2	3			1	1	1		1			9	6
C01					2	1						3	2
C02	2				1							3	2
C03	1				1							2	2
D01		2			1	1		1	2			7	5
D02		1			1							2	2
D03		2			1	1		1	2			7	5
D04		1	1		2		2		2			8	5
D05	1	1	1		1	1	1					6	6
H01	1				1			2				4	3
H02	2		3		3	1	1	1	1			12	7
H03					1	2	1		1			5	4
I02	3	9			2	1						15	4
I03	2	6			1	1	2					12	5
I04	2	7	1		2							12	4
N01	1	4			1	1	1	1	1			10	7
N02	2	5			1	1			1			10	5
P01		1										1	1
P02		1										1	1
P03		1										1	1
S01	1				1		1					3	3
S02	1				1	1						3	3
S03	1				1							2	2
S04	1				1							2	2
S05			1		2							3	2
Total	30	52	10	0	35	15	10	6	13	0	0	171	8

components preferably metrics should cover system conditions (e.g., temperature regime), hydrology (e.g., current velocity conditions), physical structures (e.g., bank profile), water chemistry (e.g., nutrients), energy sources (e.g., production), biotic interactions (e.g., competition). Although the metric categories used (see Tables 1 and 2) list a number of ecological attributes, most relate to a restricted number of ecosystem components. Table 2 clearly shows that richness, composition and tolerance/intolerance measures dominate most multimetrics developed. Within these metrics the focus mainly goes for organic load and current conditions. Other attributes have not been used in composing a multimetric index due to sampling fuzziness (e.g., abundance measures) or lack of species based ecological knowledge (functional measures). A further development of more ecosystem and organism functioning related attributes and metrics is needed to fulfil the multimetric promise.

In the process of metric selection the criterion of correlation is discussable. If certain ecosystem components still are functioning in a more optimal way it does not mean that the related metric should be excluded from the assessment or multimetric. Because such metric also tells about the ecosystem condition. On the other hand one has to avoid an overemphasis of a single metric's type. Based on the STAR experience it is advised to embrace also a selection procedure that does include non-responding but informative metrics.

The STAR project paid attention to the need of the Water Framework Directive for monitoring the ecological status of rivers and streams in Europe. Based on the finding of the AQEM project a remarkable increase in the knowledge on bio-monitoring methodologies is achieved. With respect to multimetric approaches the output of AQEM and STAR has been successfully incorporated in the development of national bio-monitoring networks. But, as usual in scientific activities, each fissure that could be closed opened much more gaps that need to be filled. We therefore strongly encourage the European administration to "make hay while the sun shines" by utilising the scientific manpower of the AQEM and STAR consortium in follow-up research programme on many of the practical issues associated with the implementation of the Water Framework Directive.

References

Andersen, M. M., F. F. Riget & H. Sparholt, 1994. A modification of the Trent index for use in Denmark. Water Research 18: 145–151.

AQEM consortium, 2002. Manual for the application of the AQEM method. A comprehensive method to assess European streams using benthic macroinvertebrates, developed for the purpose of the Water Framework Directive. Version 1.0, February 2002.

Armitage, P. D., D. Moss, J. F. Wright & M. T. Furse, 1983. The performance of a new biological water quality score system based on macroinvertebrates over a wide range of unpolluted running water sites. Water Research 17(3): 333–347.

Barbour, M. T., J. Gerritsen, G. E. Griffith, R. Frydenborg, E. McCarron, J. S. White & M. L. Bastian, 1996. A framework for biological criteria for Florida streams using benthic macroinvertebrates. Journal of the North American Benthological Society 15: 185–211.

Barbour, M. T., J. L. Plafkin, B. P. Bradley, C. G. Graves & R. W. Wisseman, 1992. Evaluation of EPA's rapid bioassessment benthic metrics: metric redundancy and variability among reference stream sites. Environmental Toxicology and Chemistry 11: 437–449.

Barbour, M. T., J. B. Stribling & J. R. Karr, 1995. Multimetric approach for establishing biocriteria and measuring biological condition, chap. 6. In Davis, W. S. & T. P. Simon (eds), Biological Assessment and Criteria – Tools for Water Resources Planning and Decision Making. Lewis Publishers, Boca Raton, Florida, 6377.

BMWP, 1979. Biological Monitoring Working Party. The 1978 National Testing Exercise. Technical Memorandum 19. Water Data Unit, Reading, UK.

Boyle, T. P., G. M. Smillie, J. C. Anderson & D. R. Beeson, 1990. A sensitivity analysis of nine diversity and seven similarity indices. Journal of Water Pollution Control Federation 62: 749–762.

Braukmann, U. & R. Biss, 2004. Conceptual study – An improved method to assess acidification in German streams using benthic macroinvertebrates. Limnologica 34: 433–450.

Cairns, J. Jr., 1975. Quantification of biological integrity. In Kusler, J. A., M. L. Quammen & G. Brooks (eds), Mitigation of Impacts and Losses. Proceedings of the National Wetland Symposium, Berne, 276–282.

Chandler, R. J., 1970. A biological approach to water quality management. Journal of Water Pollution Control Federation 69: 415–422.

De Pauw, N. & G. Vanhoren, 1983. Method for biological quality assessment of watercourses in Belgium. Hydrobiologia 100: 153–168.

De Pauw, N., P. F. Ghetti & D. P. Manzini, 1992. Biological assessment methods for running waters. In Newman et al. (eds), River Water Quality: Ecological Assessment and Control. C.C.E., Bruxelles, pp. 217–248.

Ellis, M. M., 1937. Detection and measurement of stream pollution. Bulletin of the United States Bureau of Fisheries 48: 365–437.

Fausch, K. D., J. Lyons, J. R. Karr & P. L. Angermeier, 1990. Fish communities as indicators of environmental degradation. American Society Symposium 8: 123–144.

Forbes, S. A. & R. E. Richardson, 1913. Studies on the biology of the upper Illinois river. Bulletin of the Illinois State Laboratory of Natural History 9: 481–574.

Fore, L. S., J. R. Karr & R. W. Wisseman, 1996. Assessing invertebrate responses to human activities: evaluating alternative approaches. Journal of the North American Benthological Society 15(2): 212–231.

Frey, D., 1975. Biological integrity of water: an historical perspective. In Ballantine, R. K. & L. G. Guarraia (eds), The Integrity of Water. EPA, Washington, 127–139.

Hawkins, C. P., R. H. Norris, J. N. Hogue & J. Feminella, 2000. Development and evaluation of predictive models for measuring the biological integrity of streams. Ecological Applications 10: 1456–1477.

Hellawell, J. M., 1978. Biological Surveillance of Rivers. A biological monitoring handbook. NERC, Stevenage 333 pp.

Hering, D., C. K. Feld, O. Moog & T. Ofenböck, 2006. Cook book for the development of a Multimetric Index for biological condition of aquatic ecosystems: experiences from the European AQEM and STAR projects and related initiatives. Hydrobiologia 566: 311–324.

Hering, D., O. Moog, L Sandin & P. F. M. Verdonschot, 2004. Overview and application of the AQEM assessment system. Hydrobiologia 516: 1–20.

Hynes, H. B. N., 1960. Biology of Polluted Waters. Liverpool Univ. Press, Liverpool, 202 pp.

Johnson, R. K., 1998. Spatio-temporal variability of temperate lake macroinvertebrate communities: detection of impact. Ecological Applications 8: 61–70.

Karr, J. R. & B. L. Kerans, 1992, Components of biological integrity – Their definition and use in development of an invertebrate IBI, in Midwest Pollution Control Biologists Meeting, Chicago, Ill., 1991, Proceedings: U.S. Environmental Protection Agency, Region V, EPA-905/R-92-003: 1–16.

Karr, J. R. & D. R. Dudley, 1981. Ecological perspective on water quality goals. Environmental Management 5: 55–68.

Karr, J. R., 1981. Assessment of biotic integrity using fish communities. Fisheries 6: 21–27.

Karr, J. R., 1991. Biological integrity. A long-neglected aspect of water resource management. Ecological Applications 1: 66–84.

Karr, J. R. & E. W. Chu, 1999. Restoring Life in Running Waters: Better Biological Monitoring. Island Press, Washington, DC.

Karr, J. R., K. D. Fausch, P. L. Angermeier, P. R. Yant & I. J. Schlosser, 1986. Assessing Biological Integrity in Running Waters – A method and its Rationale. Illinois Natural History Survey Special Publication 5, 28 pp.

Kerans, B. L. & J. R. Karr, 1994. A benthic index of biotic integrity (B-IBI) for rivers of the Tennessee Valley. Ecological applications 4: 768–785.

Kokeš, J., S. Zahrádková, D. Němejcovál, J. Hodovský, J. Jarkovský & T. Soldán, 2006. The PERLA System in the Czech Republic: a multivariate approach for assessing the ecological status of running waters. Hydrobiologia 566: 343–354.

Kolkwitz, R. & M. Marsson, 1902. Grundsätze für die biologische Beurteilung des Wassers nach seiner Flora und Fauna. Mitt. Aus d. Kgl. Prüfungsanstalt für Wasserversorgung und Abwässerbeseitigung 1: 33–72.

Kolkwitz, R. & M. Marsson, 1908. Ökologie der planzlichen Saprobien. Berichten der deutshen botanische Gesellschaft 26: 505–519.

Kolkwitz, R. & M. Marsson, 1909. Ökologie der tierischen Saprobien. Internationale Revue der Hydrobiologie 2: 126–519.

Liebmann, H., 1962. Handbuch der Frischwasser und Abwasserbiologie. Band I, R. Oldenburg, Munich, pp. 588.

Matthews, R. A., A. L. Buikema, J. Cairns & J. H. Rodgers, 1982. Biological monitoring: Part IIa: receiving system functional methods, relationships and indices. Water Research 16: 129–139.

Metcalfe, J. L., 1989. Biological water quality assessment of running water based on macro-invertebrate communities: history and present status in Europe. Environmental Pollution 60: 101–139.

Moog, O. (ed.), 1995. Fauna Aquatica Austriaca – a comprehensive species inventory of Austrian aquatic organisms with ecological data. First edition, Wasserwirtschaftskataster, Bundesministerium für Land- und Forstwirtschaft, Wien.

Moog, O., A. Schmidt-Kloiber, R. Vogl & V. Koller-Kreimel, 2001. ECOPROF-Software. Wasserwirtschaftskataster, Bun desministeriums für Land- & Forstwirtschaft, Umwelt & Wasserwirtschaft, Wien.

Moog, O., A. Chovanec, J. Hinteregger & A. Römer, 1999. Richtlinie für die saprobiologische Gewässergütebeurteilung von Fließgewässern.- Wasserwirtschaftskataster. Bundesministerium für Land- und Forstwirtschaft, Wien, 144 pp.

Nelson, W. G., 1990. Prospects for development of an index of biotic integrity for evaluating habitat degradation in coastal systems. Chemistry and Ecology 4: 197–210.

Odum, E. P., 1971. Fundamentals of Ecology. Saunders Company, Philladelphia, 574 pp.

Ofenböck, T., O. Moog, J. Gerritsen & M. T. Barbour, 2004. The development of a macro-invertebrate based multimetric index for monitoring the ecological status of running waters in Austria. Hydrobiologia 516: 251–268.

Ohio EPA, 1987/1989. Biological criteria for the protection of aquatic life. Vol. I, II, III. Ohio Environmental Protection Agency, Columbus, OH.

ÖNORM M6232, 1997. Richtlinie für die ökologische Untersuchung und Bewertung von Fließgewässern.- Österreichisches Normungsinstitut Wien, 38 pp. Austrian Standards M 6232 (1997): Guidelines for the ecological study and assessment of rivers.

Pantle, E. & H. Buck, 1955. Die biologische Überwachung de Gewässer und die Darstellung der Ergebnisse. Gas und Wasserfach 96(18): 1–604.

Plafkin, J. L., M. T. Barbour, K. D. Porter, S. K. Gros & R. M. Hughes, 1989. Rapid bioassessment protocols for use in streams and rivers: benthic macroinvertebrates and fish. EPA 444/4-89-001. U.S. Environmental Protection Agency. Washington.

Resh, V. H. & J. K. Jackson, 1993. Rapid assessment approaches to biomonitoring using benthic macroinvertebrates. In Rosenberg, D. M. & V. H. Resh (eds), Freshwater Biomonitoring and Benthic Macroinvertebrates. Chapman and Hall, New York, 195–233.

Reynoldson, T. B., K. E. Day & T. Pascoe, 2000. The development of the BEAST: a predictive approach for assessing sediment quality in the North American Great Lakes. In Wright, J. F., D. W. Sutcliffe & M. T. Furse (eds), Assessing the Biological Quality of Fresh Waters: RIVPACS and Other Techniques. Freshwater Biological Association, Ambleside, Cumbria, UK. The RIVPACS International Workshop, 16–18 September 1997, Oxford, UK, 165–194.

Richter, B. D., J. V. Baumgartner, J. Powell & D. P. Braun, 1996. A method for assessing hydrologic alternation within ecosystems. Conservation Biology 10(4): 1163–1174.

Rolauffs, P., D. Hering, M. Sommerhäuser, S. Jähnig & S. Rödiger, 2003. Entwicklung eines leitbildorientierten Saprobienindexes für die biologische Fließgewässerbewertung. Umweltbundesamt Texte 11/03. Forschungsbericht 200 24 227, 137 pp.

Roth, N. E., J. D. Allan & D. L. Erickson, 1996. Landscape influences on stream biotic integrity assessed and multiple spatial scales. Landscape Ecology 11(3): 141–156.

Schmedtje, U. & M. Colling, 1996. Ökologische Typisierung der aquatischen Makrofauna. Informationsberichte des Bayerischen Landesamtes für Wasserwirtschaft 4/96.

Schmidt-Kloiber, A., W. Graf, A. Lorenz & O. Moog, 2006. The AQEM/STAR taxalist – a pan-European macroinvertebrate ecological database and taxa inventory. Hydrobiologie 566: 325–342.

Shannon, C. E. & W. Weaver, 1949. The Mathematical Theory of Communication. University of Illinois Press, Urbana.

Simpson, J. C. & R. H. Norris, 2000. Biological assessment of river quality: development of AUSRIVAS models and outputs. In Wright, J. F., D. W. Sutcliffe M. & T. Furse (eds), Assessing the Biological Quality of Fresh Waters: RIVPACS and Other Techniques. Freshwater Biological Association, Ambleside, Cumbria, UK. The RIVPACS International Workshop, 16–18 September 1997, Oxford, UK, pp. 125–142.

Sládecek, V., 1973. System of water quality from the biological point of view. Ergebnisse der Limnologie 7: 1–128.

Šporka, F., 2003. Vodné bezatavovce (makroenvertebrata) Slovenska, súpis druhov a autekologické charakteristiky. Slovenský hydrometeorologický ústav, Bratislava, 590 pp.

Stubauer, I. & O. Moog, 2002. Verfahren zur Anpassung des Saprobiensystems an die Vorgaben der EU- Wasserrahmenrichtlinie in Österreich. Deutsche Gesellschaft für Limnologie – Tagungsbericht der Jahrestagung 2001 (Kiel).

Thorne, R. S. T. J. & W. P. Williams, 1997. The response of benthic macroinvertebrates to pollution in developing countries: a multimetric system of bioassessment. Freshwater Biology 37: 671–686.

Tolkamp, H. H., 1985. Using several indices for biological assessment of water quality in running water. Verhandlungen der Internationale Vereinigung der Limnologie 22: 2281–2286.

Tuffery, G. & J. Verneaux, 1968. Methode de determination de la qualite biologique des eaux courantes. CERAFER, Paris, 21 pp.

Van der Hoek, W. F. & P. F. M. Verdonschot, 1994. Functionele karakterisering van aquatische ecotooptypen. IBN rapport (in Dutch) 072: 1–81.

Verdonschot, P. F. M., 1990. Ecological Characterization of Surface Waters in the Province of Overijssel (The Netherlands). Thesis, Wageningen, 1–255 pp.

Verdonschot, P. F. M., 1991. The web-approach: a tool in water management. In Ecological Water Managemenet in Practice (Proceedings of the technical meeting held in Ede, The Netherlands, 3 October 1990). Proceedings and Information of the CHO-TNO 45: 59–76.

Verdonschot, P. F. M., 2000. Integrated ecological assessment methods as a basis for sustainable catchment management. In Jungwirth, M., S. Muhar & S. Schmutz (eds), Assessing the Ecological Integrity of Running Waters. Proc. Int. Conf., Vienna, Austria. Developments in Hydrobiology 149. Hydrobiologia 422/423: 389–412.

Verdonschot, P. F. M. & R. C. Nijboer, 2000. Typology of macrofaunal assemblages applied to water and nature management: a Dutch approach. In Wright J. F., D. W. Sutcliffe & M. T. Furse (eds), Assessing the Biological Quality of Fresh Waters: RIVPACS and Other Techniques. Freshwater Biological Association, Ambleside, Cumbria, UK. The RIVPACS International Workshop, 16–18 September 1997, Oxford, UK. Chapter 17: 241–262.

Washington, H. G., 1984. Diversity, biotic and similarity indices. A review with special relevance to aquatic ecosystems. Water Research 18(6): 653–694.

Woodiwiss, F. S., 1964. The biological system of stream classification used by the Trent River Board. Chemistry and Industry 11: 443–447.

Woodiwiss, F. S., 1980. Biological Monitoring of Surface Water Quality. Summary Report, Commission of the European Communities. Severn Trent Water Authority, UK, 45 pp.

Wright, J. F., P. D. Armitage & M. T. Furse, 1989. Prediction of invertebrate communities using stream measurements. Regulated Rivers; Research & Management 4: 147–155.

Wright, J. F., D. Moss, P. D. Armitage & M. T. Furse, 1984. A preliminary classification of running-water sites in Great Britain based on macroinvertebrate species and the prediction of community type using environmental data. Freshwater Biology 14: 221–256.

Wright, J. F., D. W. Sutcliffe & M. T. Furse (eds), 2000. Assessing the biological quality of fresh waters: RIVPACS and other techniques. Freshwater Biological Association, Ambleside, Cumbria, UK. The RIVPACS International Workshop, 16–18 September 1997, Oxford, UK.

Zelinka, M. & P. Marvan, 1961. Zur Präzisierung der biologischen Klassification er Reinheit fliessender Gewässer. Archiv für Hydrobiologie 57: 389–407.

Hydrobiologia (2006) 566:311–324
M.T. Furse, D. Hering, K. Brabec, A. Buffagni, L. Sandin & P.F.M. Verdonschot (eds), The Ecological Status of European Rivers: Evaluation and Intercalibration of Assessment Methods
DOI 10.1007/s10750-006-0087-2

Cook book for the development of a Multimetric Index for biological condition of aquatic ecosystems: experiences from the European AQEM and STAR projects and related initiatives

Daniel Hering[1,*], Christian K. Feld[1], Otto Moog[2] & Thomas Ofenböck[2]
[1] *Department of Hydrobiology, University of Duisburg Essen, D-45117, Essen, Germany*
[2]*Department of Water, Atmosphere & Environment, BOKU – University of Natural Resources and Applied Life Sciences, Vienna, Max Emanuel-Strasse 17, 1180, Vienna, Austria*
*(*Author for correspondence: E-mail daniel.hering@uni-essen.de)*

Key words: biological quality elements, macroinvertebrates, bioassessment, ecological status, stressor-specific Multimetric Index, Water Framework Directive

Abstract

The requirements of the European Water Framework Directive (WFD), aimed at an integrative assessment methodology for evaluating the ecological status of water bodies are frequently being achieved through multimetric techniques, i.e. by combining several indices, which address different stressors or different components of the biocoenosis. This document suggests a normative methodology for the development and application of Multimetric Indices as a tool with which to evaluate the ecological status of running waters. The methodology has been derived from and tested on a European scale within the framework of the AQEM and STAR research projects, and projects on the implementation of the WFD in Austria and Germany. We suggest a procedure for the development of Multimetric Indices, which is composed of the following steps: (1) selection of the most suitable form of a Multimetric Index; (2) metric selection, broken down into metric calculation, exclusion of numerically unsuitable metrics, definition of a stressor gradient, correlation of stressor gradients and metrics, selection of candidate metrics, selection of core metrics, distribution of metrics within the metric types, definition of upper and lower anchors and scaling; (3) generation of a Multimetric Index (general or stressor-specific approach); (4) setting class boundaries; (5) interpretation of results. Each step is described by examples.

Introduction

The "ecological status" of rivers, which is mainly based on their biotic components, is an important parameter for European water management (European Water Framework Directive 2000/60/EC; WFD). To assess the ecological status of a water body the taxonomic composition, abundance, ratio of disturbance sensitive taxa to insensitive taxa, and the diversity of biological indicators, have to be considered and compared to respective target values under reference conditions. This ensures the adaptation of the assessment models to a stream typology based on typological descriptors, such as ecoregions, bioregions, catchment size, and altitude. Thus, the WFD forces a re-orientation of existing monitoring procedures towards an integrative type- and reference-specific approach (Heiskanen et al., 2004).

Going far beyond the traditional procedure of documenting biological water quality with respect to organic pollution, the assessment of the ecological status of water bodies under the WFD has to document the relationships between aquatic biota and manifold environmental pressures, particularly the hydrological, morphological, and physical–chemical components. The experiences of the EU-funded AQEM and STAR projects show that the multimetric approach is a valuable procedure for bridging the gap between the current methodologies and future need for evaluating the ecological status of water bodies (Hering et al., 2004a; Furse et al., 2006). Similar experiences have been made particularly in the United States, where Multimetric Indices are frequently used in routine water management (Davis & Simon, 1995; Hughes et al., 1998; Barbour et al., 1999; Karr & Chu, 1999).

Multimetric Indices are now a commonly used tool in regionalised assessment systems for describing the quality of fresh- and brackish water ecosystems (rivers, lakes, transitional waters, wetlands; Hughes & Oberdorff, 1999). The multimetric approach attempts to provide an integrated analysis of the biological community of a site by deriving a variety of biological measures and knowledge of a site's fauna (Karr & Chu, 1999). Within a multimetric index, each single component metric is predictably and reasonably related to specific impacts caused by environmental alterations. For example, while the proportion of different feeding types is suited to assess the trophic integrity of an ecosystem, saprobic or acid indices provide a measure with which to directly assess the impact of certain pollutants and acidification, respectively. Thus, the Multimetric Index considers multiple impacts and combines individual metrics (e.g. saprobic indices, diversity indices, feeding type composition, current preferences, etc.) into a unitless measure, which can be used to assess a site's overall condition. By combining different categories of metrics (e.g. taxa richness, diversity measures, proportion of sensitive and tolerant species, trophic structure) reflecting different environmental conditions and aspects of the community the multimetric assessment is regarded as a more reliable tool than assessment methods based on single metrics (Barbour et al., 1995, 1999; Klemm et al., 2003). The multimetric approach was first developed by Karr (1981) using fish as indicators to describe stream quality. Since the development of Karr's Index of Biotic Integrity (IBI) numerous multimetric indices have been developed (Plafkin et al., 1989; Resh et al., 2000; Hering et al., 2004b; Ofenböck et al., 2004). In principle, Multimetric Indices can be applied to different types of ecosystem (rivers, lakes, transitional waters, wetlands, forests) and to different Biological Quality Elements (fish, benthic invertebrates, macrophytes, phytoplankton, phytobenthos, or other biota) and provide a flexible tool with regard to the set of components.

Within the AQEM and STAR projects Multimetric Indices have been developed for various river types throughout Europe. The procedure has been intensively discussed with both the project consortia and the water authorities, particularly in Germany and Austria, where Multimetric Indices are currently applied in water management. This document is based on the experiences gained during this implementation process.

The experiences of the AQEM and STAR projects clearly show that to enhance comparability between assessment systems the procedure of developing and applying a Multimetric Index needs to be standardised. Aim of this paper is to condense the experiences in developing Multimetric Indices gained in the AQEM and STAR projects into a more generally applicable approach, which may also be useful for ecosystem types others than rivers. In a continent like Europe, where both river biota and political and economic conditions are heterogeneous, a single approach for river assessment, which is likely to be used by all water managers, is unrealistic. At least there is the need to distinguish between simple, unspecific methods, which are useful as a first attempt in areas with little experiences in assessment, and more complex, stressor-specific approaches. Thus, we suggest alternatives differing in their degree of precision, but which always use the same basic steps.

The principles of developing a Multimetric Index

A "metric" is defined as a measurable part or process of a biological system empirically shown to change in value along a gradient of human influence (Karr & Chu, 1999). It reflects specific and predictable responses of the community to human activities, either to a single impact factor or to the cumulative effects of multiple human impairments within a watershed. Metrics are addressing comparable ecological aspects of a community, regardless of the stressor they are responding to. The following metric types can be distinguished:

- *Composition/abundance metrics.* All metrics giving the relative proportion of a taxon or taxonomic group with respect to its total number or abundance, respectively.
- *Richness/diversity metrics.* All metrics giving the number of species, genera, or higher taxa within a certain taxonomical entity, including the total number of taxa, all diversity indices.
- *Sensitivity/tolerance metrics.* All metrics related to taxa known to respond sensitively or tolerantly to a stressor or a single aspect of the stressor, respectively, either using presence/absence or abundance information.
- *Functional metrics.* All metrics addressing the ecological function of taxa (other than their sensitivity to stress), such as feeding types, habitat and current preferences, ecosystem type preferences, life cycle parameters, biometric parameters. They can be based on taxa abundance.

The procedure of data analysis during the development of a Multimetric Index typically involves the following steps:

- Selection of the most suitable form of a Multimetric Index
- Metric selection
- Metric calculation
  - Exclusion of numerically unsuitable metrics
  - Definition of a stressor gradient
  - Correlation of stressor gradients and metrics
  - Selection of candidate metrics
  - Selection of core metrics
  - Distribution of metrics within the metric types
  - Definition of upper and lower anchors and scaling
- Generation of a Multimetric Index
  - Development of a Multimetric Index (general approach)
  - Development of a Multimetric Index (stressor-specific approach)
- Setting class boundaries
- Interpretation of results

Aspects of methods needed to gain comparable taxa lists, i.e. sampling, sorting, and proper determination of the sampled individuals (Schmidt-Kloiber & Nijboer, 2004) is not considered further here, since this paper aims at describing the procedure that starts with metric calculation.

Selection of the most suitable form of a Multimetric Index

Depending on purpose, ecosystem type, organism group and available data Multimetric Indices may be designed differently. In many cases a reliable assessment reflecting the integrity of an ecosystem is sufficient, in other cases more specific data on which stressor causes deterioration of the biota is required. Thus, we distinguish two main forms of Multimetric Indices: (1) the general approach and (2) the stressor-specific approach. Stressor-specific Multimetric Indices can only be derived if the development data set includes environmental data reflecting different specific stress types, if different environmental gradients are present in the development data set and if the autecology of the targeted organism group is well known.

Metric selection

Metric calculation

Due to the long-term tradition of macro-invertebrate research, which has led to extensive ecological knowledge of this group of aquatic organisms, numerous metrics and indices have been developed that can simply be derived from taxa lists (e.g. Moog, 1995; Merritt & Cummins, 1996; Schmedtje & Colling, 1996; Tachet et al., 2002). Several software packages (e.g. ECOPROF; Moog et al., 2001) aid the quick derivation of metrics from those taxa lists, among which the AQEM River Assessment Program (Hering et al., 2004a) provides a tool that is capable of calculating more than 200 macroinvertebrate metrics. Other tools are available for fish (EFI Software: Fame Consortium, 2004), macrophytes and phyto-benthos (Schaumburg et al., 2004).

Exclusion of numerically unsuitable metrics

In order to reduce the long lists of metrics that are quickly and easily processed by software packages, filter procedures have to be applied. These procedures include the identification and exclusion of numerically unsuitable measures, for example, measures with a narrow range of values or with many outliers and extreme values, which can be simply revealed by box-whisker plots (Figs. 1 and 2).

Definition of stressor gradients

It is mandatory that the data set used for development includes data on a gradient of sites, ideally including unimpacted (reference) sites and heavily degraded (poor) sites.

An environmental stressor gradient is ideally represented by a set of sites of one freshwater ecosystem type covering the whole range (high, good, moderate, poor, and bad sites) of the environmental stressor that is to be targeted by the Multimetric System. The gradient may be a continuous measure or may be classified into five classes or even into the two classes "unstressed" and "stressed", only. Stressor gradients provide an invaluable tool by which to minimize the subjective "expert judgement" in pre-classification of sites and the subsequent selection of candidate metrics, which is based on the pre-classification.

Analysis of the gradient may be restricted to a single stressor or may include the impact of multiple stressors. For description of the impact of a single stressor, physical, chemical, or hydromorphological data on the individual sites can be used. We propose to use:

- data on BOD_5 or oxygen content to describe the impact of organic pollution;
- data on BOD_5, $N–NO_2$, chloride, *Escherichia coli*, eventually combined, for a Multimetric Index addressing water pollution in general terms;

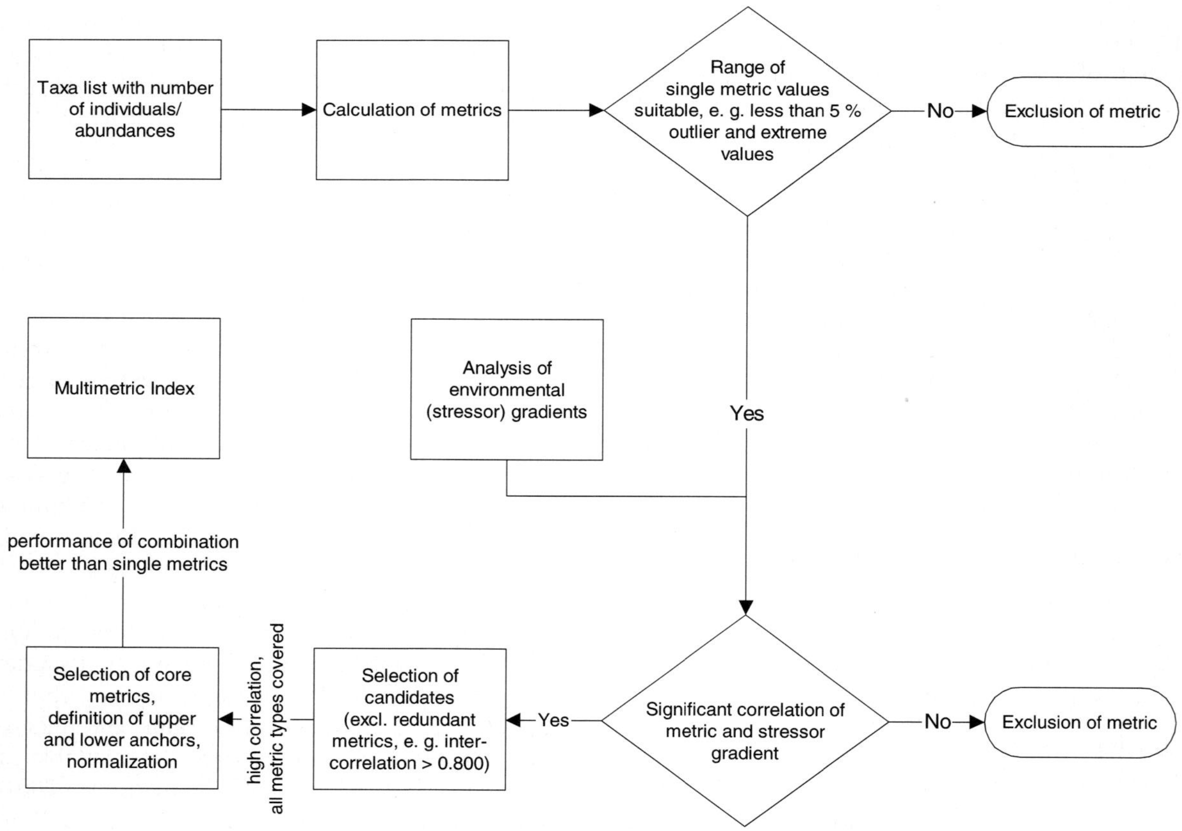

Figure 1. Schematic overview of the steps required to develop a Multimetric Index based on taxa lists.

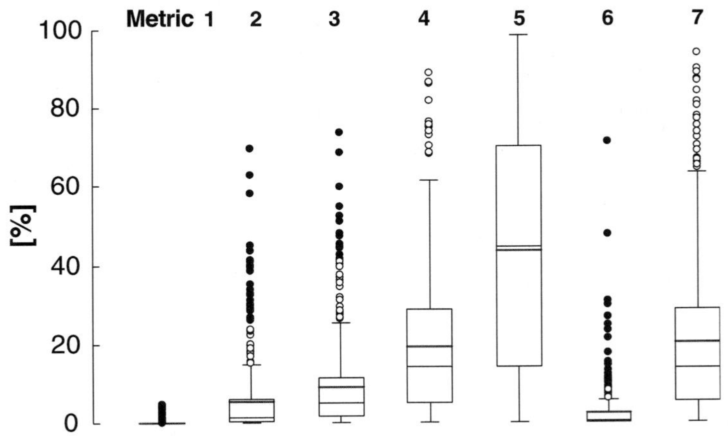

Figure 2. Example for numerically unsuitable (Metrics 1–3, 6) and suitable metrics (Metrics 4–5, 7). Circles indicate outliers (○) and extremes (●).

- data describing the trophic status of sites such as concentrations of phosphorus and nitrogen compounds;
- data that characterise the morphological situation of a site such as the German Structure-Index (Feld, 2004; Lorenz et al., 2004);
- data that characterize the hydrological and hydraulic situation of a site with information on alterations to the discharge regime, damming, residual flows, etc.;
- data on catchment land use for describing general stress gradients (Böhmer et al., 2004a);
- several of the above mentioned data to describe more general types of stress.

A statistical analysis such as PCA (Principal Component Analysis) can be used to reduce the number of variables by (i) calculating hypothetical main gradients of the environmental dataset and (ii) identifying redundant (co-correlating) variables. The direct analysis of metrics and abiotic environmental data is possible with Redundancy Analysis (RDA). The advantage of direct ordination procedures is their aim to fit the main abiotic and biotic gradients. Thus, if the existence of a strong stressor gradient is obvious, this method can be used simultaneously to identify the faunal response (Feld, 2005).

Johnson et al. (2006), and Hering et al. (submitted) defined stress gradients for a subset of the STAR sites by means of a PCA, with data indicating different sources of impairment. The selection of parameters (Table 1) was dependent on (1) their availability and completeness in the dataset (2), their relevance to the targeted stream type and (3) their relevance for the targeted stressor.

Correlation of stressor gradients and metrics

Correlating the results of a metric to the stressor gradient is a central part of the procedure, which can be processed either by looking for significant differences (*t*-test, *U*-test) or by running rank correlation analysis (e.g. Spearman, Kendall). It is also possible to use Pearson' product moment correlation in cases of large data sets, but, this coefficient is prone to partial correlation. Thus, a simple scatter plot may be used to aid the judgement on the strength and quality of metric-stressor correlations.

Selection of candidate metrics

An ideal metric should be responsive to stressors, have a low natural variability, provide a response that can be distinguished from natural variation, and be interpretable. A candidate metric's results must show a significant correlation to the stressor gradient. This correlation can be positive or negative, either across the whole stressor gradient or measured for a part thereof (e.g. only moderate to high quality sites). Metrics fulfilling this criterion are, in principal, suited to assessing the degradation of the freshwater ecosystem type and can be selected as candidate metrics. There are numerous examples

Table 1. Environmental parameters used for calculating stressor gradients

Parameter	Transformation	Lowlands				Mountains			
Pollution/eutrophication		G	P	H	M	G	P	H	M
pH		x	x			x	x		
Conductivity	log 10	x	x			x	x		
BOD5	log 10					x	x		
Oxygen [mg/l]	log 10		x				x		
Ammonium [mg/l]	log 10	x	x			x	x		
Nitrite [mg/l]	log 10					x	x		
Nitrate [mg/l]	log 10	x	x			x	x		
Ortho-phosphate [μg/l]	log 10	x	x			x	x		
Total phosphate [μg/l]	log 10		x				x		
Source pollution (yes/no)	log 10						x		
Non-source pollution (yes/no)	log 10						x		
Eutrophication (yes/no)	log 10						x		
Land use									
Forest catchment [%]	arcsin sq. root	x				x			
Urban sites catchment [%]	arcsin sq. root	x				x			
Natural grassland catchment [%]	arcsin sq. root	x				x			
Cropland catchment [%]	arcsin sq. root	x				x			
Pasture catchment [%]	arcsin sq. root	x				x			
Hydromorphology									
Shading at zenith (foliage cover)	arcsin sq. root	x		x		x		x	
Width woody rip. vegetation [m]	arcsin sq. root	x		x		x		x	
Number of debris dams		x		x		x		x	
Number of logs		x		x		x		x	
Shoreline covered with woody riparian vegetation [%]	arcsin sq. root	x		x		x		x	
No. bank fixation [%]	arcsin sq. root	x		x		x		x	
No. bed fixation [%]	arcsin sq. root	x		x		x		x	
Stagnation (yes/no)		x		x		x		x	
Straightening (yes/no)		x		x		x		x	
Microhabitats									
Hygropetric sites [%]	arcsin sq. root				x				x
Megalithal > 40 cm [%]	arcsin sq. root				x				x
Macrolithal > 20–40 cm [%]	arcsin sq. root				x				x
Mesolithal > 6–20 cm [%]	arcsin sq. root				x				x
Microlithal > 2–6 cm [%]	arcsin sq. root				x				x
Akal > 0.2–2 cm [%]	arcsin sq. root				x				x
Psammal/psammopelal [%]	arcsin sq. root				x				x
Argyllal < 6 μm [%]	arcsin sq. root				x				x
Macro-algae [%]	arcsin sq. root				x				x
Micro-algae [%]	arcsin sq. root				x				x
Submerged macrophytes [%]	arcsin sq. root				x				x
Emergent macrophytes [%]	arcsin sq. root				x				x
Living parts of ter. plants [%]	arcsin sq. root				x				x
Xylal [%]	arcsin sq. root				x				x

Continued on page 317

Table 1. (Continued)

Parameter	Transformation	Lowlands				Mountains			
Pollution/eutrophication		G	P	H	M	G	P	H	M
CPOM [%]	arcsin sq. root				x				x
FPOM [%]	arcsin sq. root				x				x
Debris [%]	arcsin sq. root				x				x

G = general degradation gradient; p = pollution/eutrophication gradient; H = hydromorphology gradient; M = microhabitat gradient (from Hering et al., submitted, altered).

in the literature for the process of selecting candidate metrics (Vlek et al., 2004; Böhmer et al., 2004b; Ofenböck et al., 2004; Buffagni et al., 2004). Three examples on how metrics relate to different stress gradients are given in Figure 3.

Numerous papers describe the possible approaches to metric selection (e.g. Holland, 1990; Barbour et al., 1992; Karr & Kerans,1992; Barbour et al., 1999; Karr & Chu, 1999; Buffagni et al., 2004; Hering et al., 2004b; Ofenböck et al., 2004; Vlek et al., 2004). Based on existing knowledge and literature information, the candidate metrics are selected on the basis of knowledge of the aquatic biota within a geographical entity, e.g. the metric "number of *Corbicula* individuals" would make no sense if this taxon does not occur in the targeted ecoregion. As another example, the inclusion of the metric "morphological deformation of chironomids" will be useless if the administrative framework would not allow financing of the necessary investigations. On the other hand, candidate metrics must fit the sampling method applied. If Chironomid pupae or Annelids are collected with 1000 μm-mesh samplers, metrics derived from those taxa are not likely to be reliable.

After having selected the candidate metrics they need to be evaluated for efficacy and validity. This means that inappropriate metrics have to be eliminated from the process. Metrics have to be considered as inappropriate if they (1) are less than robust and have a high temporal and/or spatial variability that does not allow discrimination between anthropogenic influences and natural variability, (2) do not reflect human impairment and have little relationship to the impacts, (3) are not well founded on ecological principles and understanding; for example, the correlation of land use and the feeding type miner.

Only those metrics that show a quantitative impact-response change across a stressor gradient that is reliable, interpretable and not diffused or obscured by natural variation, must be selected. Moreover, different types of metric should be considered (composition/abundance metrics, richness/diversity metrics; sensitivity/tolerance metrics; functional metrics).

Hering et al. (2004a), who aimed at designing a Multimetric Index for indicating "general degradation", restricted a more extensive list of metrics to those indices which are not explicitly designed to detect organic pollution. A further selection criterion was the taxonomic resolution needed for the metric (order/family vs. genus/species level), which should be achieved by, and comparable among, the majority of taxa lists (e.g. Eurolimpacs, 2004; Schmidt-Kloiber et al., 2006) used for the development process. These criteria resulted in restricting a list of almost 300 to 79 metrics.

Selection of core metrics

Candidate metrics, which can be identified as robust and most informative are scrutinised further in the process of selecting core metrics. To be selected as a core metric two major aspects have to be considered: (1) the metrics should cover the different metric types (Table 2) and (3 2) redundant metrics need to be excluded. Metrics that show strong inter-correlations (Spearman's $r > 0.8$) with one another are defined as redundant. The identification of redundant metrics is aided by triangular cross-correlation matrices and, in case of redundancy, the correlation of each of the pair of metrics with the other metrics is compared in order to finally omit the one that showed the higher overall mean correlation (see examples given in Table 4). For the selection of appropriate

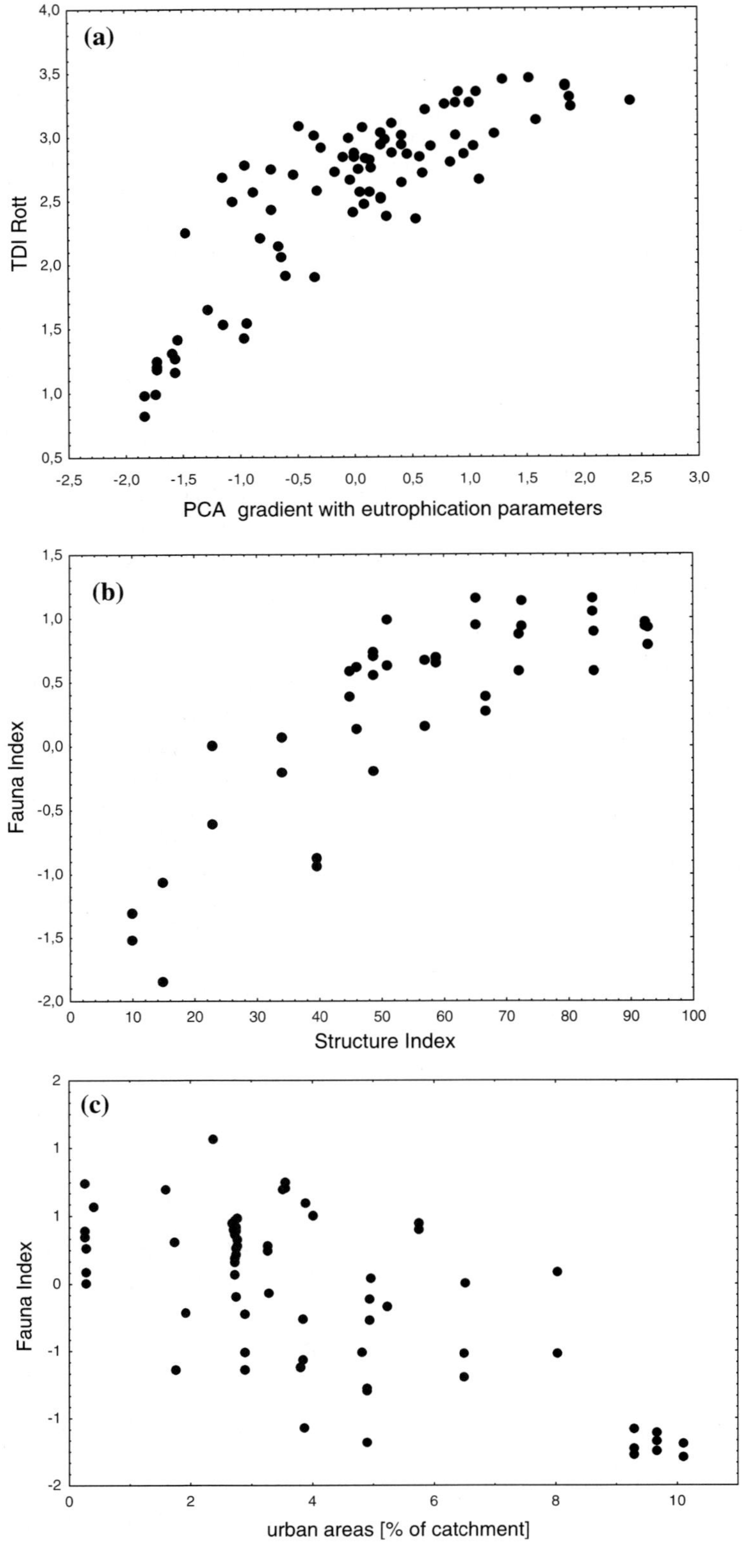
(a)
TDI Rott
4,0
3,5
3,0
2,5
2,0
1,5
1,0
0,5
-2,5 -2,0 -1,5 -1,0 -0,5 0,0 0,5 1,0 1,5 2,0 2,5 3,0
PCA gradient with eutrophication parameters
(b)
Fauna Index
1,5
1,0
0,5
0,0
-0,5
-1,0
-1,5
-2,0
0 10 20 30 40 50 60 70 80 90 100
Structure Index
(c)
Fauna Index
2
1
1
0
-1
-1
-2
0 2 4 6 8 10
urban areas [% of catchment]

Figure 3. (a) Correlation of a periphyton metric (Trophic Diatom Index according to Rott et al., 1999) to a eutrophication gradient. Samples from the STAR lowland rivers (data from Hering et al. in press). (b) Correlation of a benthic invertebrate metric (German Fauna Index D05 according to Lorenz et al., 2004) to hydromorphological quality measured with a structure index (data from Lorenz et al., 2004). (c) Correlation of a benthic invertebrate metric (German Fauna Index D03 according to Lorenz et al., 2004) to catchment land use (share of urban areas) in medium-sized lowland rivers in Germany.

Table 2. Examples for metrics used to assess individual Biological Quality Elements, assigned to metric types

	Composition/abundance metrics	Richness/diversity metrics	Sensitivity/tolerance metrics	Functional metrics
Fish	Population age structure	Diversity (Shannon-Weaver, Margalef)	Individuals of tolerant species	Number of rheophile species
	Population size	Number of river type specific species		Number of lithophile species
Benthic invertebrates	[%] EPT	Diversity (Shannon-Weaver Margalef)	Saprobic indices	[%] sand-preferring taxa
	[%] Trichoptera	Number of Trichoptera species	Acid Index (Henrikson & Medin, 1986)	[%] shredders, RETI (Schweder, 1992)
			German Fauna Index (Lorenz et al., 2004)	[%] rheophile species
Macrophytes	[%] *Potamogeton pectinatus*	Diversity (Shannon-Weaver, Margalef)	Mean Trophic Ranking (Holmes et al., 1999)	Ellenberg et al. (1992) numbers (humidity, light, salinity)
		Number of taxa		
Phytobenthos	[%] Pennales (volume) (Mischke & Behrendt, 2005)	Diversity (Shannon-Weaver, Margalef)	Trophic Diatom Index (Kelly & Whitton, 1995)	[%] planctonic taxa
		Number diatom of taxa	Trophic Index Austria (Rott et al., 1999)	
Phytoplankton	[%] Pennales (volume) (Mischke & Behrendt, 2005)	Diversity (Shannon-Weaver, Margalef,)	Rare taxa and indicative taxa (Coesel, 2001)	[%] planctonic taxa
		Number of Desmid taxa	"Index-20" (Mischke & Behrendt, 2005)	

core metrics, statistical analysis aimed at identifying those variables, which show the strongest relationship to certain environmental stressors, are recommended.

Distribution of metrics within the metric types

Well-constructed Multimetric Indices contain a suggested number of metrics from each type (Table 2) and therefore reflect multiple dimensions of biological systems (Karr & Chu, 1999). About three metrics per metric type is considered ideal. A higher (e.g. to more exhaustively describe the community attributes) or lower (e.g. if fewer suitable metrics can be identified) number of metrics can be included into a Multimetric Index. If there is at least one candidate metric of a particular metric type, then at least one of this metric type must be selected as a core metric to ensure that each metric type is represented in the Multimetric Index. This procedure makes Multimetric Indices more comparable and ensures that different aspects of the community are regarded. The possible combinations of metrics resulting from the selection of candidate metrics must be correlated to the stressor gradient used to select the candidate metrics. For this purpose, all metric results are first scaled by transformation into a score ranging from 0 to 1 (100%). This enables the calculation of means for all candidate metrics.

Those metrics whose combination results in the strongest significant correlation to the stressor gradient should be selected as core metrics.

Table 3. Example for the definition of upper anchors and lower anchors of candidate metrics in the stream type "medium-sized lowland rivers" in Germany (data from Hering et al., 2004a)

Metric	Shannon-diversity	[%] litoral preferring taxa	[%] rheophile taxa	[%] shredderes	German Fauna Index D03	[%] EPT	# Plecoptera taxa	# Trichoptera Taxa
Upper (95%) percentile	3.50	29.07	61.28	36.51	0.89	66.79	2.00	13.00
Lower (5%) percentile	1.54	2.52	6.04	2.88	−1.43	6.87	0.00	0.00
Correlation with land use index								
Correlation coefficient	−0.35	0.41	−0.54	0.39	−0.44	−0.65	−0.37	−0.48
Suggested upper anchor	3.39	−3.60	77.99	−0.40	1.26	82.12	1.69	16.01
Suggested lower anchor	2.29	24.27	11.71	17.93	−0.90	13.36	−0.40	1.71
Correlation with Structure Index								
Correlation coefficient	−0.18	0.55	−0.43	−0.44	−0.75	−0.14	−0.42	−0.53
Suggested upper anchor	2.71	8.06	46.61	20.31	0.59	38.30	0.46	7.35
Suggested lower anchor	2.50	18.95	32.05	7.59	−0.50	31.22	−0.10	3.95
Chosen upper anchor	3.50	3.00	70.00	35.00	1.50	70.00	3.00	15.00
Chosen lower anchor	1.00	25.00	10.00	3.00	−1.50	5.00	0.00	0.00

Three different methods for defining anchors have been applied: (1) 95% and 5% percentile of all data; (2) Spearman Rank Correlation with a land use index; (3) Spearman Rank Correlation with a structure index.

Table 4. Example for a correlation matrix of candidate metrics (invertebrate metrics, stream type "medium-sized lowland rivers" in Germany) (data from Hering et al., 2004a)

	Shannon diversity	[%] litoral preferring taxa	[%] rheophile taxa	[%] shredderes	German Fauna Index D03	[%] EPT	# Plecoptera taxa	# Trichoptera Taxa
Shannon diversity	1.0000							
[%] lioral preferring taxa	−0.1906	1.0000						
[%] rheophile taxa	0.1420	**−0.8505**	1.0000					
[%] shredderes	−0.3349	0.1195	−0.2393	1.0000				
German Fauna Index D03	0.2390	**−0.8020**	0.7911	−0.1247	1.0000			
[%] EPT	0.2874	−0.6495	0.7040	−0.2887	0.6855	1.0000		
# Plecoptera taxa	0.1467	−0.4450	0.4650	−0.0038	0.5168	0.5894	1.0000	
# Trichoptera taxa	0.6000	−0.5165	0.4733	−0.1627	0.6212	0.6881	0.5324	1.0000

Correlation coefficients of individual metrics are given. Bold: Correlation coefficient > 0.8 (one of these metrics needs to be excluded).

Definition of upper and lower anchors and scaling

The upper and lower anchors mark the indicative range of a metric, i.e. the values that are empirically set and defined as "1" (upper anchor) and "0" (lower anchor), respectively, to normalize a metric's result. The upper anchor corresponds to the upper limit of the metric's value under reference conditions. If data on reference sites are available, the upper anchor should be set as a percentile of all the metric values of the reference sites (e.g. 95%, 75% or median, depending on the quality of the reference sites). If few data (e.g. up to 5–10 samples) are available for reference sites, and the site classification is to some extent uncertain, the highest observed value can also be considered (excluding abundance metrics). If there are no data on reference sites but data on sites representing different degrees of stress are available, the upper anchor can be obtained by extrapolation.

The lower anchor corresponds to the lower limit of the metric's value under the worst attainable conditions. If data on sites of bad ecological quality are available, the lower anchor should be set as a percentile (e.g. 5 or 10%) of all metric values of the bad ecological quality sites, or at the lowest value obtained or obtainable. If there are no data on bad ecological quality sites but data on sites representing different degrees of stress are available, the Lower Anchor can be obtained by extrapolation. An example from the German invertebrate assessment system is given in Table 4.

The results of the various core metrics that have been selected for contributing to a Multimetric Index may vary between different ranges of values: while the metric "number of Plecoptera species" can have a value between 0 and *n*, the German Saprobic Index can range from 1.0 to 4.0 and the metric "[%] shredders" from 0 to 100. To combine these individual measures into an integrated Multimetric Index; it is essential to normalize the core metrics via transformation to unitless scores. In practice, each metric result must be translated into a value between 0 and 1 (Ecological Quality Ratio), using the following formula:

$$\text{Value} = \frac{\text{Metric\_result} - \text{Lower\_Anchor}}{\text{Upper\_Anchor} - \text{Lower\_Anchor}}$$

for metrics decreasing with increasing impairment, and

$$\text{Value} = 1 + \frac{\text{Metric\_result} - \text{Lower\_Anchor}}{\text{Upper\_Anchor} - \text{Lower\_Anchor}}$$

for metrics increasing with increasing impairment. Values > 1 are set to 1.

The resulting metric value for a given site is finally expressed as an ecological quality ratio (EQR). The EQR represents the relationship between the values of the biological parameters observed for a given body of surface water and the values for these parameters under the reference conditions applicable to that water body. The ratio is expressed as a numerical value between zero and one: high ecological status is represented by values close to one and bad ecological status by values close to zero.

Generation of a Multimetric Index

The aggregation of metric scores into an index simplifies decision making so that a single value can be used to determine the quality class of a river site. The action, which is potentially needed to improve the ecosystem (e.g. restoration, mitigation, pollution enforcement) is not inherently determined by the index value, but may be deduced from the single component metrics, in addition to the raw data, and consideration of other ecological information (Barbour et al., 1999).

We propose two ways of generating a Multimetric Index: a "general approach" and a "stressor-specific approach". In the "general approach", various metrics are calculated and the results are individually compared to the respective metric values under reference conditions. From this comparison, a score is derived for each metric. These scores are finally combined into a Multimetric Index. The "stressor-specific" approach sorts out the metrics forehand according to their ability to detect the effects of a certain stressor on the targeted biota. Thus, the scores of the metrics addressing a single stressor are first combined into a value reflecting the intensity of this stressor; the assessment results for all stressors are finally combined into the Multimetric Index.

Development of a Multimetric Index (general approach)

The aggregation of metrics into a Multimetric Index should ensure that each metric type is represented by a similar number of metrics (e.g. Karr & Chu, 1999). Nevertheless, the final selection of metrics for a Multimetric Index should produce the strongest multimetric view of biological condition. Therefore, we do not recommend a fixed number of metric types or measures per metric type.

The procedure described by Böhmer et al. (2004a) is based on the assumption that if the same number of metrics has been selected for each metric type, the Multimetric Index can be calculated as the mean of the 0–1 digit scores of all core metrics. This will attribute the same weight to each metric and metric type. If the number of core metrics belonging to different metric types is different, weighting factors can be used so that e.g.

each group of metrics (i.e. clustered within a type) has the same influence on the final Multimetric Index. If, within a metric type, the various core metrics are based on information of different confidence levels (e.g. one is based on the whole Invertebrate community, while the others on single insect orders) weighting factors can be applied to the metrics so that the more inclusive metrics contribute to a greater extent to the final score.

Development of a Multimetric Index (stressor-specific approach)

For the "stressor-specific approach" almost exactly the same steps as for the "general approach" are required. However, all the above described steps (from the generation of environmental gradients to the scaling of metrics) should be done separately for different environmental gradients, representing different stressors. This procedure results in a separate list of core metrics for each stressor, e.g. organic pollution, acidification or hydromorphological degradation.

The scores of those core metrics, which have been selected using the gradient of a single stressor, must the be combined into a Multimetric Index by calculating the mean of their 0–1 scores. This step results in a quality class for each stressor, e.g. "organic pollution" and "acidification".

If the same degree of confidence is expected for the different stressor-specific indices, the resulting stressor-specific quality classes are converted into the ecological quality class using the worst result of all stressor-specific quality classes. Otherwise, priority can be given to the most robust metric, the results of the other metric(s) being used to confirm the classification obtained. Weighting factors can be considered as explained above.

Setting Class boundaries

The final Multimetric Index provides a score that represents the overall relationship between the combined values of the biological parameters observed for a given site and the expected value under reference conditions. This score is – as for single metrics – expressed as a numerical value between zero and one. This range can be subdivided into any number of categories corresponding to various levels of impairment. Because the metrics are scaled to reference conditions and expectations for the stream classes, any decision on subdivision should reflect the distribution of the scores for the reference sites. We propose quality classes with equal ranges to provide five ordinal rating categories for assessment of impairment in accordance with the demands of the WFD, using the following scheme for setting class boundaries:

reference ≥ 0.8
good $\geq 0.6 < 0.8$
moderate $\geq 0.4 < 0.6$
poor $\geq 0.2 < 0.4$
bad < 0.2

The more metrics are included into the Multimetric Index, the more the index values under reference conditions will diverge from 1, because even under the most pristine conditions, not all metrics will reach maximum levels in a single site. Alternative: Therefore, it is recommended not to use the best available values as reference values, but e.g. to use the 25% percentile of index values from reference sites as the class boundary for reference conditions.

Interpretation of results

Multimetric Indices can be easily interpreted, which is regarded as a main advantage of this type of bioassessment. However, since European water managers have only little experience with Multimetric Indices, an aid for the interpretation of results is highly recommended, particularly if the "general approach" is applied, which does not inherently distinguish between stress types. An interpretation aid should include the values to be expected under reference conditions, the stress type the metric is most strongly reacting to and the restoration measures needed to improve the metric.

Conclusions

Multimetric Indices provide a valuable tool for assessing various types of freshwater ecosystems, since they integrate different stressors and different components of the community. Thus, they can be adapted to the specific conditions of a river type or lake type in an optimal way, by considering the most relevant stressors, and specific characters of the biocoenosis. However, to gain a certain level of comparability, the development of Multimetric Indices should be carried out in an analogous way. By considering the steps described in this paper, comparability of Multimetric Indices can be ensured, without loosing the degree of freedom which is necessary to cope with the natural variability of river and lake types and their communities. Thus, the procedure described here may be helpful for consideration as an international standard.

Acknowledgments

AQEM and STAR were funded by the European Commission, 5th Framework Program, Energy, Environment and Sustainable Development, Key Action Water, Contract no. EVK1-CT 1999-00027 and EVK1-CT-2001-00089. The authors acknowledge the support of all their colleagues who have contributed to the work described and want to express their special thanks to Mike Furse the terrific co-ordinator of STAR. We would also like to acknowledge Mary Burgis for linguistic help and two anonymous reviewers for valuable comments, which significantly improved the paper.

References

Barbour, M. T., J. L. Plafkin, B. P. Bradley, C. G. Graves & R. W. Wisseman, 1992. Evaluation of EPA's rapid bioassessment benthic metrics: Metric redundancy and variability among reference stream sites. Environmental Toxicology and Chemistry 11: 437–449.

Barbour, M. T., J. B. Stribling & J. R. Karr, 1995. Multimetric approach for establishing biocriteria. Biological assessment and criteria. In Davies, W. S. & T. P. Simon (eds), Tools for water resource planning and decision making. CRC Press, Boca Raton: 63–77.

Barbour, M. T., J. Gerritsen, B. D. Snyder & J. B. Stribling, 1999. Rapid Bioassessment Protocols for Use in Streams and Wadeable Rivers: Periphyton, Benthic Macroinvertebrates and Fish (2nd edn.nd ed.). U.S. EPA. Office of Water, Washington, DC, EPA/841-B-98-010.

Böhmer, J., C. Rawer-Jost & A. Zenker, 2004a. Multimetric assessment of data provided by water managers from Germany: assessment of several different types of stressors with macrozoobenthos communities. Hydrobiologia 516: 215–228.

Böhmer, J., C. Rawer-Jost, A. Zenker, C. Meier, C. K. Feld, R. Biss & D. Hering, 2004b. Assessing streams in Germany with benthic invertebrates: Development of a multimetric invertebrate based assessment system. Limnologica 34(4): 416–432.

Buffagni, A., S. Erba, M. Cazzola & J. L. Kemp, 2004. The AQEM multimetric system for the southern Italian Apennines: assessing the impact of water quality and habitat degradation on pool macroinvertebrates in Mediterranean rivers. Hydrobiologia 516: 315–331.

Coesel, P. F. M. 2001. A method for quantifying conservation value in lentic freshwater habitats using desmids as indicator organisms. Biodiversity and Conservation 10(2): 177–187.

Davis, W. S. & T. P. Simon, 1995. Biological Assessment and Criteria. Tools for Water Resource Planning and Decision Making. Lewis Publishers, Boca Raton, London, Tokyo.

Ellenberg, H., H. E. Weber, R. Düll, V. Wirth, W. Werner & D. Paulißen, 1992. Zeigerwerte von Pflanzen in Mitteleuropa, Scripta Geobotanica 18.

Eurolimpacs, 2004. Integrated project to evaluate the impacts of global change on European freshwater ecosystems. Macro-invertebrate Taxa and autecology database. www.freshwaterecology.info.

Fame Consortium, 2004. Manual for the application of the European Fish Index – EFI. A fish-based method to assess the ecological status of European rivers in support of the Water Framework Directive. Version 1.1, January 2005.

Feld, C. K., 2004. Identification and measure of hydromorphological degradation in Central European lowland streams. Hydrobiologia 516: 69–90.

Feld, C. K., 2005. Assessing hydromorphological degradation of sand-bottom lowland rivers in Central Europe using benthic macroinvertebrates. Dissertation, University of Duisburg-Essen, Essen, 127 pp + App.

Furse, M., D. Hering, O. Moog, P. Verdonschot, R. K. Johnson, K. Brabec, K. Gritzalis, A. Buffagni, P. Pinto, N. Friberg, J. Murray-Bligh, J. Kokes, R. Alber, P. Usseglio-Polatera, P. Haase, R. Sweeting, B. Bis, K. Szoszkiewicz, H. Soszka, G. Springe, F. Sporka & I. Krno, 2006. The STAR project: context, objectives and approaches. Hydrobiologia 566: 3–29.

Henrikson, L. & M. Medin, 1986. Biologisk bedömning av försurningspåverkan på Lelångens tillflöden och grundområden 1986. Aquaekologerna, Rapport till länsstyrelsen i Älvsborgs län.

Hering, D., J. Böhmer, C. Meier, R. Biss, C. Rawer-Jost, C. K. Feld & A. Zenker, 2004a. Development of a multimetric invertebrate based assessment system for German rivers. Limnologica 34(4): 398–415.

Hering, D., O. Moog, L. Sandin & P. F. M. Verdonschot, 2004b. Overview and application of the AQEM assessment system. Hydrobiologia 516: 1–20.

Hering, D., R. K. Johnson, S. Kramm, S. Schmutz, K. Szoszkiewicz & P. F. M. Verdonschot, submitted. Assessment of European rivers with diatoms, macrophytes, invertebrates and fish: A comparative metric-based analysis of organism response to stress.

Heiskanen, A. S., W. van de Bund, A. C. Cardoso & P. Noges, 2004. Towards good ecological status of surface waters in

Europe – interpretation and harmonisation of the concept. Water Science and Technology 49: 169–177.

Holland, A. F. (ed), 1990. Near Coastal Program Plan for 1990–Estuaries. U.S. Environmental Protection Agency Environmental Research Laboratory, Office of Research and Development, Washington, DC 259 pp, EPA-600/9-90-033.

Holmes, N. T. H., J. R. Newman, S. Chadd, K. J. Rouen, L. Saint & F. H. Dawson, 1999. Mean Trophic Rank: A Users Manual. R&D Technical Report No. E38. Environment Agency, Bristol, UK.

Hughes, R. M., P. R. Kaufmann, A. T. Herlihy, T. M. Kincaid, L. Reynolds & D. P. Larsen, 1998. A process for developing and evaluating indices of fish assemblage integrity. Canadian Journal of Fisheries and Aquatic Sciences 55: 1618–1631.

Hughes, R. M. & T. Oberdorff, 1999. Applications of IBI concepts and metrics to waters outside the United States. In T. P. Simon (ed.), Assessing the Sustainability and Biological Integrity of Water Resource Quality Using Fish Communities. CRC Press, Boca Raton, Florida: 79–96.

Johnson, R. K., D. Hering, M. T. Furse & R. T. Clarke, 2006. Detection of ecological change using multiple organism groups: metrics and uncertainty. Hydrobiologia 566: 115–137.

Karr, J. R., 1981. Assessment of biotic integrity using fish communities. Fisheries 6: 21–27.

Karr, J. R. & E. W. Chu, 1999. Restoring Life in Running Waters: Better Biological Monitoring. Island Press, Washington, DC 200 pp.

Karr, J. R. & B. L. Kerans, 1992. Components of biological integrity – Their definition and use in development of an invertebrate IBI, in Midwest Pollution Control Biologists Meeting, Chicago, Ill., 1991, Proceedings: U.S. Environmental Protection Agency, Region V, EPA-905/R-92-003, pp. 1–16.

Kelly, M. G. & B. A. Whitton, 1995. The Trophic Diatom Index: a new index for monitoring eutrophication in rivers. Journal of Applied Phycology 7: 433–444.

Klemm, D. J., K. A. Blocksom, F. A. Fulk, A. T. Herlihy, R. M. Hughes, P. R. Kaufmann, D. V. Peck, J. L. Stoddard, W. T. Thoeny, M. B. Griffith & W. S. Davis, 2003. Development and evaluation of a macroinvertebrate biotic integrity index (MBII) for regionally assessing Mid-Atlantic Highlands streams. Environmental Management 31: 656–669.

Lorenz, A., D. Hering, C. Feld & P. Rolauffs, 2004. A new method for assessing the impact of hydromorphological degradation on the macroinvertebrate fauna of five German stream types. Hydrobiologia 516: 107–127.

Merritt, R. W. & K. W. Cummins, 1996. An Introduction to the Aquatic Insects of North America. Dubuque, Kendall/Hunt 862 pp.

Mischke, U. & H. Behrendt, 2005. Vorschlag zur Bewertung ausgewählter Fließgewässertypen anhand des Phytoplanktons. In C. K. Feld, S. Rödiger, M. Sommerhäuser & G. Friedrich (eds), Typologie, Bewertung, Management von Oberflächengewässern. E. Schweizerbart'sche Verlagsbuchhandlung, Stuttgart: 46–62.

Moog, O. (ed), 1995. Fauna Aquatica Austriaca. Bundesministerium für Land- und Forstwirtschaft. Wasserwirtschaftskataster Wien.

Moog, O., A. Schmidt-Kloiber, R. Vogl & V. Koller-Kreimel, 2001. ECOPROF-Software. Wasserwirtschaftskataster, Bundesministeriums für Land- & Forstwirtschaft. Umwelt & Wasserwirtschaft, Wien.

Ofenböck, T., O. Moog, J. Gerritsen & M. T. Barbour, 2004. A stressor specific multimetric approach for monitoring running waters in Austria using benthic macro-invertebrates. Hydrobiologia 516: 253–270.

Plafkin, J. L., M. T. Barbour, K. D. Porter, S. K. Gross & R. M. Hughes, 1989. Rapid Bioassessment Protocols for use in Streams and Rivers: Benthic Macroinvertebrates and Fish. U.S. Environmental Protection Agency, Office of Water Regulations and Standards, Washington, DCEPA 440-4-89-001.

Resh, V. B., D. M. Rosenberg & T. B. Reynoldson, 2000. Selection of benthic macroinvertebrate metrics for monitoring water quality of the Fraser River, British Columbia: implications for both multimetric approaches and multivariate models. In J. F. Wright, D. W. Sutcliffe & M. T. Furse(eds), RIVPACS and Similar Techniques for Assessing the Biological Quality of Freshwaters. Freshwater Biological Association and Environment Agency, UK, Ableside, Cumbria, UK.

Rott, E., P. Pfister, H. Dam, E. Pipp, K. Pall, N. Binder & K. Ortler, 1999. Indikationslisten für Aufwuchsalgen. Bundesministerium für Land – und Forstwirtschaft, Wien 248 pp.

Schaumburg, J., C. Schranz, J. Foerster, A. Gutowski, G. Hofmann, P. Meilinger, S. Schneider & U. Schmedtje, 2004. Ecological classification of macrophytes and phytobenthos for rivers in Germany according to the Water Framework Directive. Limnologica 34: 283–301.

Schmedtje, U. & M. Colling, 1996. Ökologische Typisierung der aquatischen Makrofauna. Informationsberichte des Bayerischen Landesamtes für Wasserwirtschaft 4/96.

Schmidt-Kloiber, A. & R. C. Nijboer, 2004. The effect of taxonomic resolution on the assessment of ecological water quality classes. Hydrobiologia 516: 269–283.

Schmidt-Kloiber, A., W. Graf, A. Lorenz & O. Moog, 2006. The AQEM/STAR taxalist – a pan-European macro-invertebrate ecological database and taxa inventory. Hydrobiologia 566: 325–342.

Schweder, H., 1992. Neue Indices für die Bewertung des ökologischen Zustandes von Fließgewässern, abgeleitet aus der Makroinvertebraten-Ernährungstypologie. Limnologie Aktuell 3: 353–377.

Tachet, H., P. Richoux, M. Bournaud & P. Usseglio-Polatera, 2002. Invertébrés d'eau douce: Systematique, Biologie, Ecologie. CNRS Editiona, p. 587.

Vlek, H. E., P. F. M. Verdonschot & R. C. Nijboer, 2004. Towards a multimetric index for the assessment of Dutch streams using benthic macroinvertebrates. Hydrobiologia 516: 175–191.

Hydrobiologia (2006) 566:325–342
M.T. Furse, D. Hering, K. Brabec, A. Buffagni, L. Sandin & P.F.M. Verdonschot (eds), The Ecological Status of European Rivers: Evaluation and Intercalibration of Assessment Methods
DOI 10.1007/s10750-006-0086-3

The AQEM/STAR taxalist – a pan-European macro-invertebrate ecological database and taxa inventory

Astrid Schmidt-Kloiber[1,*], Wolfram Graf[1], Armin Lorenz[2] & Otto Moog[1]
[1]*BOKU – University of Natural Resources & Applied Life Sciences, Vienna, Department of Water, Atmosphere, Environment, Institute of Hydrobiology and Aquatic Ecosystem Management, Max Emanuel Straße 17, A-1180 Vienna, Austria*
[2]*Department of Hydrobiology, University of Duisburg-Essen, Universitätsstr. 5, D-45117 Essen, Germany*
*(*Author for correspondence: Tel.: +43-1-47654/5225; Fax: +43-1-47654/5217; E-mail: astrid.schmidt-kloiber@boku. ac.at)*

Key words: aquatic macro-invertebrates, European taxa inventory, autecological database, ecological classifications, biological assessment systems

Abstract

The European list of aquatic macro-invertebrate taxa, and its associated ecological database, originated within the context of the AQEM project and have been extended during the STAR project. The AQEM/STAR taxalist is a product of co-operation between applied freshwater ecologists and scientists from different zoological fields, applied partners and the administration. The basic idea is that a sound understanding of benthic invertebrate ecology is a prerequisite for the implementation of a biological approach to aquatic ecosystem management in Europe. The database has been generated under the management of BOKU (University of Natural Resources and Applied Life Sciences, Vienna) and UDE (University of Duisburg-Essen) and provides an important means of standardisation and unification of ecological classifications in Europe. This paper outlines the aims for setting up the AQEM/STAR macro-invertebrate taxalist and autecological database and provides a current summary of the numbers of aquatic orders, families, species, and species occurrences in 14 European countries. The number of available and applicable assignments of taxa to each ecological parameter is summarised and examples are given for different parameters and taxonomic groups. Gaps in the autecological information are identified and discussed. Besides its ecological relevance, the operational character of this database is underlined by the fact that it provides the associated taxon codes for each of five different European assessment systems for nearly 10,000 European macro-invertebrate taxa.

Introduction

The temporal and spatial distributions of freshwater organisms are tightly connected to aspects of zoogeography plus their physiological and behavioural responses to varying levels of environmental factors. The most frequently studied key factors, such as water temperature, flow velocity, oxygen balance, food composition and availability, and quality of habitat, are regarded as the main predictors of the community composition and distribution of benthic invertebrates. The comparatively good knowledge of their environmental needs, and of species' responses to various environmental factors, has led to these organisms being widely used as (bio)indicators in water management and in applied ecology (see Davis & Simon, 1995; Rosenberg & Resh, 1995).

Numerous commonly used biological assessment systems for rivers and streams across the USA and Europe are based on so-called "metrics" or – synonymously used – "measures" or "biological attributes". Following Karr & Chu (1999) metrics are defined as "measurable parts or processes of a biological system empirically shown to change in value along a gradient of human influence". The metrics of assessment systems use either, (1) taxonomic richness and composition (number of species/taxa, diversity indices, number of individuals, % Trichoptera, etc.), or (2) biological information on ecological functions or requirements (e.g., habits and species traits of the aquatic fauna, such as feeding types, stream zonation preferences, habitat preferences, tolerance/intolerance measures such as, e.g., saprobic indices, individual health and others) (Statzner et al., 1994; Barbour et al., 1999; Karr & Chu, 1999; Hering et al., 2004). The first type of metric depends only on species/taxa lists, whereas the second needs a profound knowledge of species' ecological demands. In order to use this ecological knowledge in a comprehensible system of bioindicators it needs to be "translated" into numerical values.

The requirements of the European Water Framework Directive (EC, 2000/60; WFD) for an integrated assessment methodology with which to evaluate the ecological status of water bodies is a big challenge for the applied limnological sciences. The "ecological status" of rivers, which is mainly based on their biotic components, is an important parameter for European water management. To assess the ecological status of a water body selected attributes of the biological indicators have to be considered, and compared to relevant target values under reference conditions. As a consequence, new assessment systems and evaluation techniques have had to be developed throughout Europe during the last few years. Among other approaches, the applicability of multi-metric techniques, i.e., combinations of several measures and indices addressing different stressors or different components of the biocoenosis, has been tested (Brabec et al., 2004; Buffagni et al., 2004; Lorenz et al., 2004; Ofenböck et al., 2004; Pinto et al., 2004; Sandin et al., 2004; Vlek et al., 2004). An important scientific input into this recently adopted approach has been the creation of taxa inventories with associated autecological databases.

Currently, the collections of data on European taxa used for this purpose are species/taxa checklists for single countries (e.g., Austria, Bavaria, Slovakia) or large-scale regions (e.g., Limnofauna Europaea, Illies, 1978). The most comprehensive inventory is the Fauna Europaea (Fauna Europaea Web Service, 2004), which was developed at the same time as the AQEM/STAR taxalist. It contains country-related occurrences of species from most freshwater groups at species level, checklists of species within genera and higher taxonomic units, plus comments on nomenclature and phylogenetics. Nevertheless, ecological data are not included because the Fauna Europaea is primarily focused on taxonomy and faunistics.

Compared to the compilation of national or international checklists (species inventories) the gathering of autecological information on macroinvertebrate species, and its transformation into numerical values, is a sophisticated, responsible and thus time consuming and costly task. Therefore only a few databases dealing with this kind of information have so far been established (Merrit & Cummins, 1984; Moog, 1995, 2002; Schmedtje & Colling, 1996; Usseglio-Polatera et al., 2000; Sporka, 2003).

Within the European Union the task of intercalibration seeks to harmonise the results of the different national assessment systems throughout the European countries. The necessity of evaluating streams and water courses in a wider perspective than national guidelines leads to the need for a standardised pan-European macro-invertebrate species list and for widely harmonised autecological data as a basis for ecological quality assessment. This paper presents a new collection of European data that fulfils these criteria and will be accessible not only to the scientific public but also to stakeholders and national monitoring institutions in a public WWW service under www.freshwaterecology.info.

Methods

History of the AQEM/STAR macro-invertebrate taxalist

The AQEM/STAR macro-invertebrate taxalist is a "living document" that was first set up for the purposes of the EU funded AQEM project (AQEM consortium, 2002; www.aqem.de). The aim of this project was the development and testing of an integrated system for assessment of the ecological quality of streams and rivers throughout Europe using benthic macro-invertebrates (Hering et al., 2004). The eight project member countries (Austria, Czech Republic, Germany, Greece, Italy, The Netherlands, Portugal and Sweden) developed multi-metric assessment systems for different stream types, which can be applied via a computer program (ASTERICS, to be downloaded at www.aqem.de). Because the assessment systems require ecological knowledge of the taxa it was essential to collect information on both occurrence and distribution of taxa within the partner countries, and ecological information on these taxa, to create a consistent and reliable database. To achieve this goal the AQEM/STAR macro-invertebrate taxalist builds on the scientific expertise of many scientists, universities, organisations and societies.

The steps taken towards the development of the AQEM/STAR database were:

- Election of persons responsible for the national checklists: from each partner country at least one person was selected to be responsible for collecting the national records of the targeted invertebrate groups. The minimum requirement for the quality of the national checklist was that it provides sufficient information to start developing the assessment system with which to evaluate the ecological status of a water body. In those cases of incomplete faunistic knowledge the national checklists were integrated into the database as "working taxalists". For information on national experts please consult www.freshwaterecology.info.
- Founding of a board of experts for the individual taxonomic groups: acknowledged and approved experts contributed to the checking of these national taxa inventories (Table 1). Besides the input of their own knowledge the

Table 1. Taxonomic experts who have contributed to the synthesis of the AQEM/STAR taxalist

Taxonomic group	Expert	Organisation
Turbellaria	Piet Verdonschot	Alterra Green World Research, Wageningen
Mollusca	Michal Horsak, Lubos Beran	Masaryk University, Brno
Oligochaeta	Piet Verdonschot	Alterra Green World Research, Wageningen
Polychaeta	Piet Verdonschot	Alterra Green World Research, Wageningen
Hirudinea, Branchiobdellida	Hasko Nesemann	BOKU Vienna
Hydrachnidia	Tj.H. van den Hoek	Alterra Green World Research, Wageningen
Crustacea	Tj.H. van den Hoek	Alterra Green World Reserach, Wageningen
Ephemeroptera	Tomas Soldan	Entomological Institute, AS, Ceske Budejovice
Odonata	Jiri Zeleny	Entomological Institute, AS, Ceske Budejovice
Plecoptera	Wolfram Graf	BOKU Vienna
Heteroptera	Tj.H. van den Hoek	Alterra Green World Research, Wageningen
Megaloptera	Tomas Soldan	Entomological Institute, AS, Ceske Budejovice
Planipennia	Jiri Zeleny	Entomological Institute, AS, Ceske Budejovice
Coleoptera	Wolfram Sondermann	private consultant
Trichoptera	Wolfram Graf	BOKU Vienna
Chironomidae	Karel Brabec	Masaryk University, Brno
Ceratopogonidae	Jan Knoz	Masaryk University, Brno
Simuliidae	Gunther Seitz, Ellen Kiel	Regierung Niederbayern, Landshut Hochschule Vechta
Pediciidae	Herbert Reusch	private consultant
Limoniidae	Herbert Reusch	private consultant
Tipulidae	Herbert Reusch	private consultant
Brachycera	Rudolf Rozkosny	Masaryk University, Brno

experts kept contact with other experts throughout Europe to collect data files on their parts of the targeted taxonomic groups and performed quality control with respect to species validity, species nomenclature and synonymy. Basically, the taxonomy follows present day international taxonomic standards and the experts consequently used comprehensive taxonomic sources.

- Compiling the database: the data were compiled into an MS Access database, using the proven structure of the Austrian software ECOPROF that has been developed for data storage and evaluation (Moog et al., 2001a; www.ecoprof.at).
- Compiling the autecological information: as a basic data source, existing ecological classifications were critically checked and adopted. They were (in order of prioritisation) the Fauna Aquatica Austriaca (Moog, 1995; 2002), the Bavarian List (Schmedtje & Colling, 1996) and other national lists (for example Verdonschot, 1990; Van den Hoek & Verdonschot, 1994). When possible, selected species were assigned to experts and project partners for amendment of their designations.
- Coding the new autecological information: depending on the parameter, data were given a numerical code using either a 10 points or single category assignment system (see section 'Systems for assigning the ecological information').

In the succeeding, EU funded, STAR project (www.eu-star.at) which aimed to standardise river classifications, the AQEM/STAR macro-invertebrate taxa and autecology database was extended with national checklists from Denmark, France, Great Britain, Latvia, Poland and Slovakia.

The main goal of the AQEM/STAR taxalist was, and still is, to provide a tool for the ecological assessment of water bodies. For this reason the list was designed as a "living document" that should be available to the scientific public at a comparatively early stage of its development. This means that the species inventories and/or the ecological rankings are in different stages of completeness for most of the targeted countries and taxonomic groups. Consequently the AQEM/STAR taxa inventory does not necessarily represent the state of the art of a country's recorded species: these lists must be understood as an operational tool for running bio-monitoring projects under the auspices of the new Directive. Nevertheless, besides its operational character, the final product should represent a numerically transformed, state of the art database of European zoogeographic and ecological knowledge on benthic invertebrates.

Database structure

The database is set up in MS Access. It is a relational database consisting of three main modules:

- Taxonomic tables: holding species, subfamilies, families, higher taxonomic groups and current synonyms. All the systematic units are number-coded. Beside the ID_Aqem (number code) the species/taxa are also linked with the Austrian ID_Ecoprof, an eight-letter shortcode, the German DV-number, the British Maitland code, the Dutch TCM code and the Czech Perla code.
- National checklist tables: containing the occurrence of species in different countries.
- Ecological information tables: holding the ecological attributes, as numerical classification values, of different taxonomic levels (species, genus, subfamily, family).

A mySQL database with a PHP-interface for presenting the data on the web is currently under construction (www.freshwaterecology.info).

Systems for assigning the ecological information

Since most of the scientific information on the environmental needs of the biota is recorded in narrative form, two different methods were used to transform the ecological knowledge into numerical values that can be processed for ecological quality assessment.

Ten point system

The ecological designations of the taxa used in the database are based on the known, or estimated, average distributions, occurrences or behaviours of the organisms within the environmental gradient under consideration. The 10-point system goes back to Zelinka & Marvan (1961) who introduced the saprobic valences approach into the calculation of a Saprobic Index. These authors used 10 points as a substitute for 100% occurrence of each taxon. Up to 10 points were allocated to the saprobic state of a water body according to the tolerance of a species for each of the five saprobic quality classes (xeno-, oligo-, beta-meso-, alpha-meso- and polysaprobic water quality). This 10-point system was extended by Moog (1995) to other ecological classifications, such as stream zonation preferences and feeding types. If, for example, 70% of a species' records were observed in spring brooks and 30% in the upper trout region, 7 out of 10 points will be allocated to spring brook preference (hypocrenal rivers) and 3 points to upper trout region reference to describe the expected occurrence of this species within the longitudinal zonation of a stream. The parameters for which the 10 point ranking system is used in the AQEM/STAR database are summarised in Table 2.

Single category assignment system

The single category assignment system is used if a taxon can be allocated to only one ecological parameter, criterion or zone. If a criterion applies to the species, "1" is assigned, if not "0" is used. The parameters for which the single category assignment system is used are summarised in Table 2.

Ecological parameters without an indicated assignment system in the column "Syst." in Table 2 represent different kinds of indices. For details see the relevant references.

Categories for assigning the ecological information

Different numbers of categories are used for the designation of ecological information (Table 2). The six main parameters, their categories, and their definitions are presented in Tables 3–8.

Results

Number of families, genera and species/taxa recorded

The AQEM/STAR taxa database currently (status 01/04/05) holds a total of 6971 European benthic invertebrate species, categorised into 1317 genera, 279 families and 28 higher taxonomic groups (mostly orders). Including working taxa like species-groups the list contains 9612 taxa. Table 9 shows the occurrence of species among the higher taxonomic units and countries. The taxa inventories (number of families, genera and species) per country are presented in Figure 1.

The numbers of species indicated in Table 9 need not reflect the current state of the zoogeographic art since the conceptual design of the AQEM/STAR list focuses on its operative character and the use of its ecological classifications in assessment systems. Nevertheless, Table 9 and Figure 1 clearly indicate that several taxonomic groups have been well investigated in most parts of Europe: for example for Coleoptera, Heteroptera, Odonata or Trichoptera the AQEM/STAR taxa-list does seem to cover the current state of the art of a countries' species spectrum. With respect to other groups there are clear gaps concerning the number of recorded species in individual countries, particularly in Southern Europe. For example, Bivalvia are poorly documented for Portugal, so they were only considered at higher taxonomic levels and are therefore not included in Table 9. But there are also "neglected" taxonomic groups in well investigated, Central European countries. These include species-rich taxonomic units such as Hydrachnidia, that are poorly documented in most European countries, but also groups such as aquatic Lepidoptera, Porifera or Polychaeta with a naturally low diversity. Basically, small numbers of species in Table 9 may indicate both missing data and/or deficient knowledge of species distribution (for example Crustacea, Diptera or Oligochaeta in Southern European countries).

Autecological information

The autecological information compiled into the AQEM/STAR taxa database serves as an essential resource for assessment systems. Currently 26 ecological parameters and indices, with varying numbers of classified taxa, are integrated into the database. The available autecological information is summarised in Table 2.

The six most commonly investigated ecological parameters included in the database are oxygen demand (saprobic indices), stream zonation, current and substrate preferences, as well as feeding and locomotion types. Table 10 shows the number and percentage of designated species and taxa respectively for which these parameters are available.

Table 2. Ecological parameters integrated into the current AQEM/STAR database including lowest taxonomic level of assignment (Level), assignment system (Syst.; 10p: 10 points system, sc: single category assignment system, –: different kind of indices), number of categories (Cat.), number of classified taxa (No. taxa) and references

Ecological Parameter & Indices	Level	Syst.	Cat.	No. taxa	References
Austrian saprobic valences, index & indicator weight	Species	10p	5	1244	Moog (1995, 2002)
Czech saprobic valences, index & indicator weight	Species	10p	5	923	CSN 75 7716 (1998)
Dutch saprobic valences	Species	10p	5	1329	Verdonschot (1990) Van der Hoek & Verdonschot (1994)
German saprobic index & indicator weight (1992 & 2003)	Species	–	–	146, 619	DEV (1992, 2003) Rolauffs et al. (2003)
Slovak saprobic valences, index & indicator weight	Species	10p	5	982	Sporka (2003)
Feeding types	Species	10p	10	2924	Moog (1995, 2002) Schmedtje & Colling (1996) AQEM consortium (2002) Wolf (2004)
Stream zonation preferences	Species	10p	10	1955	Moog (1995, 2002) Schmedtje & Colling (1996) AQEM consortium (2002) Wolf (2004)
Current preferences	Species	s.c.	7	1540	Schmedtje & Colling (1996) AQEM consortium (2002) Wolf (2004)
Substrate preferences	Species	10p	8	1452	Schmedtje & Colling (1996) AQEM consortium (2002) Wolf (2004) Bochert (2003)
Locomotion types	Species	10p	6	1006	Schmedtje & Colling (1996) AQEM consortium (2002) Wolf (2004)
German PTI (Potamon Typie Index)	Species	s.c.	5	342	Schöll et al. (2005)
German RTI (Rhithron Typie Index)	Species	s.c.	6	882	Biss et al. (2002)
Rheoindex	Species	s.c.	3	350	Banning (1998)
r/k-strategy	Species	s.c.	2	87	Schöll et al. (2005)
Acid Index Braukmann (2000 & 2003)	Species	s.c.	4, 5	89, 264	Braukmann (2000) Braukmann & Biss (2004)
Swedish Acid Index	Species	s.c.	3	56	Henrikson & Medin (1986)
MAS (Mayfly Average Score) small & large streams	Species/genus	–	–	299	Buffagni (1997) Buffagni (1999)
German Fauna Index stream type D01, D02, D03, D04, D05	Species	s.c.	5	335	Lorenz et al. (2004)
"Sensitive taxa" of Austrian rivers & streams	Species	s.c.	2	393	Moog et al. (2003)
DSFI (Danish Stream Fauna Index)	Genus/family	–	–	630	Skriver et al. (2001)
IBE (Indice Biotico Esteso)	Genus/family	–	–	1124	Ghetti (1997)
BBI (Belgian Biotic Index)	Genus/family	–	–	1415	De Pauw & Vanhooren (1983) De Pauw et al. (1992)
Portuguese Index	Subfamily	–	–	110	Pinto et al. (2004)
BMWP (Biological Monitoring Working Party)	Family	–	–	277	Armitage et al. (1983)
BMWP – Spanish Version	Family	–	–	299	Alba-Tercedor & Sanchez-Ortega (1988)
LIFE	Family	s.c.	6	337	Extence et al. (1999)

Table 3. Saprobic classes and definitions of the amount of decomposable, organic material at a recording site, according to Moog (1995)

Saprobic preference	Explanation
Xenosaprobic zone	Clean water (no organic pollution)
Oligosaprobic zone	Little organic pollution
Beta-mesosaprobic zone	Moderately polluted
Alpha-mesosaprobic zone	Heavily polluted
Polysaprobic zone	Extremely polluted

Table 4. Feeding types of invertebrates and their definitions according to Moog (1995)

Feeding type	Sources of food
Grazer and scrapers	Endo & epilithic algal tissues, biofilm, partially POM, partially tissues of living plants
Miners	Leaves of aquatic plants, algae & cells of aquatic plants
Xylophagous taxa	Woody debris
Shredders	Fallen leaves, plant tissue, CPOM
Gatherers/collectors	Sedimented FPOM
Active filter feeders	Food in water current is actively filtered: suspended FPOM, CPOM, micro prey is whirled
Passive filter feeders	Food brought by flowing water current: suspended FPOM, CPOM, prey
Predators	Prey
Parasites	Host
Other feeding types	Cannot be classified into this scheme or omnivorous

Table 5. Stream zonation preferences of invertebrates and their definitions according to Moog (1995)

Stream zonation	Region
Eucrenal	Spring region
Hypocrenal	Spring-brook
Epirhithral	Upper-trout region
Metarhithral	Lower-trout region
Hyporhithral	Grayling region
Epipotamal	Barbel region
Metapotamal	Bream region
Hypopotamal	Brackish water region
Littoral	Lake and stream shorelines, ponds, etc.
Profundal	Bottom of stratified lakes

Table 6. Current preferences of invertebrates and their definitions according to Schmedtje & Colling (1996)

Current preference	Explanation
Limnobiont	Occurring only in standing waters
Limnophil	Preferably occurring in standing waters; avoids current; rarely found in slowly flowing streams
Limno- to rheophil	Preferably occurring in standing waters but regularly occurring in slowly flowing streams
Rheo- to limnophil	Usually found in streams; prefers slowly flowing streams and lentic zones; also found in standing waters
Rheophil	Occurring in streams; prefers zones with moderate to high current
Rheobiont	Occurring in streams; bound to zones with high current
Indifferent	No preference for a certain current velocity

With respect to the ranking of functional feeding guilds 28.4% (i.e., 1980 species) of a total of 6971 species in the database are classified, followed by 26.3% (1833 species) designations for the stream zonation preferences, 21.1% (1569 species) for saprobic values, 17.5% (1217 species) for current preferences, 16.4% (1146 species) for substrate preferences and 10.3% (720 species) for locomotion types. This means – regarding those six main parameters – that numerically transformed autecological information is available for only 10 to 28% of all species and 10 to 30% of all taxa in the database. These values may vary considerably among the different countries and taxonomic groups. Compared to the averages of other countries, the proportion of ranked taxa is generally highest for the Austrian and German taxa inventories, reaching e.g., about 73% of Austrian taxa designated by feeding type (Lorenz & Schmidt-Kloiber, 2005).

The species of the orders Ephemeroptera, Plecoptera, Trichoptera and Coleoptera which have

Table 7. Habitat/substrate preferences of invertebrates and their definitions according to Schmedtje & Colling (1996)

Microhabitat preference	Explanation
Pelal	Mud; grain size <0.063 mm
Argyllal	Silt, loam, clay; grain size <0.063 mm
Psammal	Sand; grain size 0.063–2 mm
Akal	Fine to medium-sized gravel; grain size 0.2–2 cm
Lithal	Coarse gravel, stones, boulders; grain size >2 cm
Phytal	Algae, mosses and macrophytes including living parts of terrestrial plants
Particulate organic matter	Woody debris, CPOM, FPOM
Other habitats	Other habitats (e.g., host of a parasite)

Table 8. Locomotion types among invertebrates and definitions according to Schmedtje & Colling (1996)

Locomotion type	Explanation
Swimming/skating	Species, which float in lakes or drift in rivers passively
Swimming/diving	Species, which swim or dive actively
Burrowing/boring	Species, which burrow in soft substrates or bore in hard substrates
Sprawling/walking	Species, which sprawl or walk actively with legs, pseudopods or on a mucus
(Semi)Sessil	Species, which are tightened to hard substrates, plants or other animals
Other locomotion type	Other locomotion type like flying or jumping (mainly outside the water)

been classified indicate a relatively good state of knowledge of the ecological requirements of these groups. Mayflies show 55% feeding type designations and 50% stream zonation preferences. Most stoneflies are assigned saprobic values (40%) and, again, feeding types (38%). Thirty eight percent of the caddisflies are also classified according to feeding type and stream zonation preferences. For beetles, about 30% have been assigned for these two parameters (Table 11). The lowest percentages of classified species are stonefly substrate preferences (9%) and locomotion types (3%) as well as Coleoptera locomotion types (4%). In general, more ecological knowledge can be transformed into numerical classifications for Ephemeroptera and Trichoptera than for the other two taxonomic groups considered here.

Discussion

Taxa richness of families, genera and species in different countries

Table 9 and Figure 1 give the numbers of families, genera and species recorded in 14 European countries: Latvia, Sweden, Denmark, France, Great Britain, The Netherlands, Poland, Austria, Czech Republic, Germany, Slovakia, Greece, Italy, and Portugal. The results of the AQEM/STAR records have not as yet been compared with the Fauna Europaea list (www.faunaeur.org), because both projects ran simultaneously. Because the focus of the AQEM/STAR project was concentrated on completing the ecological information for the database, the species inventories may differ from those of the taxonomically and faunistically oriented Fauna Europaea. Table 9 gives clear evidence of existing gaps in the AQEM/STAR database. In general, "low" numbers of species in the table may be caused by:

- The checklist submitted by a countries' responsible person does not reflect faunistic or taxonomic knowledge of this country ("working list").
- There is not enough information about the occurrence of species in the country; the biodiversity of aquatic invertebrates is poorly known.
- The biodiversity is generally low due to climatic factors and/or zoogeographical reasons.

The number of species recorded per country ranges from 632 (Portugal) to 4136 (Germany), (Table 9). Although this broad span seems, remarkably, to be caused by nationally different

Table 9. Numbers of aquatic invertebrate species among the different higher taxonomic groups and countries according to national checklists or "working lists" (marked by a asterisk); EU: all AQEM/STAR countries, LV: Latvia, SE: Sweden, DK: Denmark, FR: France, GB: Great Britain, NL: The Netherlands, PL: Poland, AT: Austria, CZ: Czech Republic, DE: Germany, SK: Slovakia, GR: Greece, IT: Italy, PT: Portugal

	EU	LV	SE	DK	FR	GB	NL*	PL	AT	CZ	DE	SK	GR*	IT*	PT*
Area (in 1000 m^2)	10500	65	450	43	550	243	42	313	84	79	357	49	132	301	92
Araneae	1	1	1	1	1	1	1	1	1	1	1			1	
Bivalvia	58	29	31	22	32	29	24	32	36	31	47	28	15	18	
Branchiobdellida	7	2	1	2		1		4	6	1	4	1		4	
Bryozoa	14	6		7	11	11		11	10	10	13	9		9	
Cestoda	1				1			1					1		
Coelenterata	13	4		6	8	8	1	9	1	6	11				
Coleoptera	1112	202	354	294	578	410	201	420	358	287	480	349	269	342	225
Crustacea	125	15	23	18	51	39	18	55	44	11	108	25	9	6	3
Diptera	3236	194	857	1091	1511	1406	274	1571	1270	1365	2233	568	103	151	37
Ephemeroptera	241	49	60	43	147	51	35	114	116	97	141	123	58	113	42
Gastropoda	174	47	48	38	53	59	47	63	98	51	97	52	51	46	14
Heteroptera	123	23	64	60	79	56	65	66	61	62	73	37	60	89	52
Hirudinea	63	15	15	17	16	16	25	42	34	19	45	22	14	19	10
Hydrachnidia	253	116		159	202	175	252	205	35		163				
Hymenoptera	39	1				39			1		2				
Kamptozoa	1								1		1	1			
Lepidoptera	11	4		5	6	6	5	11	7	4	7				
Megaloptera	6	5	5	3	3	3	2	3	3	3	5	4	1		2
Nematomorpha	1	1		1	1			1	1	1	1			1	
Nemertini	1				1						1				
Odonata	125	54	59	52	97	52	53	73	77	69	84	71	84	87	54
Oligochaeta	232	71	144	98	111	102	56	158	111	127	126	102	17	15	18
Planipennia	6	1	5	5	5	3	1	4	4	4	5	1	1	2	
Plecoptera	313	10	37	25	143	34	10	113	123	101	126	101	92	144	36
Polychaeta	12	1		1	2	1		1	3	2	11	2			
Porifera	9	5		5	5	5		6	2	6	8	5	2		
Trichoptera	734	191	221	168	479	196	112	264	308	252	321	176	152	391	138
Turbellaria	60	7	12	11	32	34	11	14	27	36	22	13	15	3	1
Sum	6971	1054	1937	2132	3575	2737	1193	3242	2738	2546	4136	1690	944	1441	632

states of the taxonomic art, this variation cannot be explained only by the effects of taxonomic resolution. Basically, there is a tendency for the species richness of a country to be positively correlated to its area (Fig. 2) which confirms the fundamental ecological principle of the species-area-relationship. However, this general observation is masked by two other factors that affect a countries' biotic inventory. Firstly, the number of species accumulates as the topographic heterogeneity of a country increases. In Austria, for example, a total of 2738 species are recorded although the area of this country covers only 84,000 km^2, but it includes portions of six out of 27 European ecoregions according to Annex 11 of the WFD (Illies, 1978) and also has a wide altitudinal range. Secondly, species richness decreases from South to North, which may be explained by the history of glaciation (Rosenzweig, 1995). This postulate is confirmed by the comparatively small number of 1937 species recorded for Sweden (450,000 km^2). This fact may not be clearly apparent from Table 9, because most of the Southern European taxalists are "working lists" and do not reflect the complete species richness in these countries (see below).

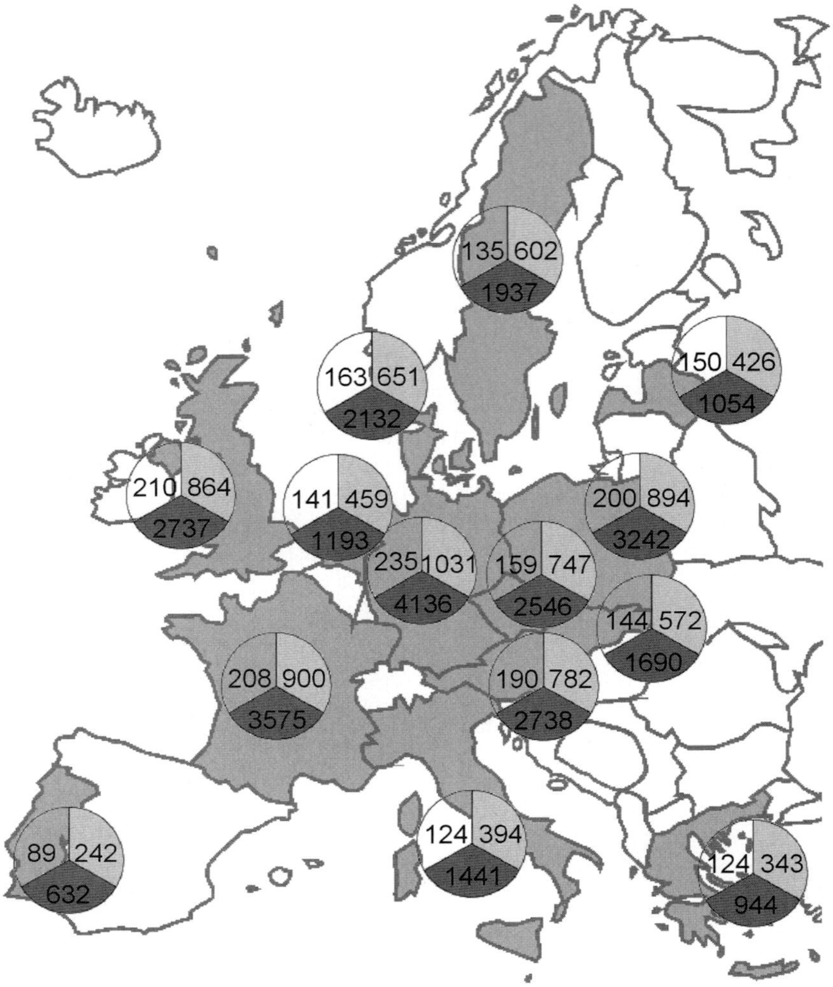

Figure 1. Numbers of families (left third of the circles), genera (right third of the circles) and species (lower third of the circles) of aquatic invertebrates within the different AQEM/STAR partner countries.

Table 10. Numbers and percentages of ecologically classified taxa/species of aquatic invertebrates for the six main ecological parameters in the AQEM/STAR database

	Species (6971)		Taxa (9612)	
	No.	%	No.	%
Saprobic classifications	1569	21.1	1855	19.3
Feeding types	1980	28.4	2924	30.4
Stream zonation preferences	1833	26.3	1955	20.3
Current preferences	1217	17.5	1540	16.0
Substrate preferences	1146	16.4	1452	15.1
Locomotion types	720	10.3	1006	10.5

Some of the national lists were primarily compiled for the development and use of new assessment systems containing the most frequently occurring taxa, regardless of whether or not these taxa reflect the complete national species inventory. These lists are still under construction and will have to be refined in future. For example, the Greek taxalist within the AQEM/STAR database included 152 species of Trichoptera. Recent studies provide evidence that this number is too low and Malicky (2005) expects there to be more than 300 species. Another question is, how to put this updated knowledge into use since most Southern

Table 11. Numbers and percentages of ecologically classified aquatic species within the orders Ephemeroptera, Plecoptera, Trichoptera and Coleoptera, for the six main ecological parameters in the AQEM/STAR database

	Ephemeroptera (241)		Plecoptera (313)		Trichoptera (734)		Coleoptera (1112)	
	No.	%	No.	%	No.	%	No.	%
Saprobic classifications	121	50.2	124	39.6	242	33.0	190	17.1
Feeding types	133	55.2	118	37.7	282	38.4	331	29.8
Stream zonation preferences	119	49.4	111	35.5	277	37.7	332	30.0
Current preferences	78	32.4	50	16.0	224	30.5	280	25.2
Substrate preferences	66	27.4	28	9.0	195	26.6	215	19.3
Locomotion types	57	23.7	9	2.9	88	12.0	47	4.2

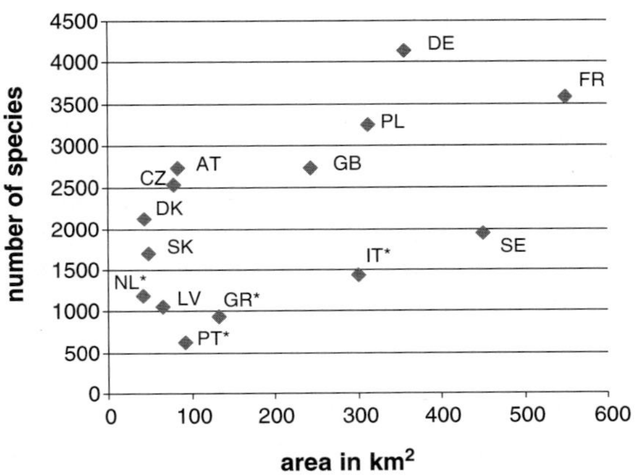

Figure 2. The relationship of numbers of species of aquatic invertebrates identified in individual AQEM/STAR partner countries to their areas.

European species have so far only been described as adult stages. The juvenile instars are still unknown and thus not available for routine monitoring and assessment. Gaps of that kind reflect the limits of the current state of nationally used assessment approaches. To overcome this lack of knowledge more emphasis needs to be given to larval taxonomic studies so that, by this means, "working taxalists" can be upgraded to national checklists. The Fauna Europaea may be a valuable tool in combination with intensified research in these fields. As species ranges are in the process of expansion as well as regression we regard the AQEM/STAR database as a "living document" that should stimulate national experts to check their results and contribute to the knowledge of local fauna.

Examples of sound metrics and identifying gaps within the ecological assignments

As a consequence of severe epidemics (e.g., cholera) in the 19th century, assessments of river quality historically focussed solely on impacts due to organic pollution. Yet, after the water quality of most European rivers has been restored, there is clear evidence that habitat impairment – mainly due to flood protection and hydropower generation – is a primary cause of degraded aquatic landscapes. Muhar et al. (2000) concluded that only 6% of the Austrian rivers with a catchment area >500 km^2 are of pristine character. Thus river restoration has evolved to include a crucial element of water management, and habitat quality has become an essential component of biological surveys. Therefore, in some countries special attention was given to the development of metrics that show the relationship between habitat quality and biological conditions (e.g., Feld, 2004). Nevertheless, we still see a gap between the need to improve the scientific quality of ecological surveys and the actual degree of attention that is given to biological sciences. Although the scientific literature in general is growing intensively the ecological potential of particular species across varying environmental factors is comparatively poorly known. On the one hand the current study of natural science has changed its focus towards other fields (e.g., genetics, biotechnology). On the other hand it is hardly possible to measure the complexity of species' responses in the field, especially considering the complex spatial and temporal distributions of all relevant factors. It has therefore become common practice to use surrogate parameters in ecological assessment, or to focus environmental evaluations on only a few, well studied and easily observable factors and measures, in order to transfer existing knowledge into applied practice most efficiently. Among these factors are the six most intensively investigated ecological attributes of aquatic species included in the database: oxygen requirements, stream zonation patterns, current and substrate preferences, as well as feeding and locomotion types. For these parameters our ability to classify species ranges from 10 to 28% of species. The situation is not even that good for other ecological parameters. The question whether the ecological designation of species can be extended to genus or even higher taxonomic levels is a controversially discussed topic (summarised for example in Schmidt-Kloiber & Nijboer, 2004). This approach is, nevertheless, common practice, in order to increase the number of ecological assignments for the assessment system. Also, within the AQEM/STAR taxa database, higher taxonomic units are classified ecologically, but only if the underlying references allow (e.g., if all species of a genus are classified identically then the genus gets the same designation as the included species). In this way the number of taxa classified by feeding type increases from 1980 to 2924 (28.4 to 30.4%) or from 720 to 1006 for locomotion types. Similar increases, but not that high, are true for all the other main parameters (Table 10).

The current activities in progress for the implementation of the WFD emphasise the general necessity of having ecological classifications available. During the AQEM project, for a total of 28 stream-types, multi-metric indices were developed for assessing different types of human impact. More than half (15) of these indices use a kind of saprobic assignment as metric, about half (13) apply the feeding behaviour, 10 use stream zonation preferences, 5 use current and substrate preferences (AQEM consortium, 2002). Generally, the analyses of functional feeding types are the most investigated ecological measures. Practical knowledge concerning trophic-relationships, food chains, food quotient and essential nutrients is widely available. There are not only a lot of individual publications, but also several substantial catalogues available, that are based on anatomical structures and behaviours concerned with food acquisition (Merritt & Cummins, 1984; Moog, 1995, 2002; Schmedtje & Colling, 1996). Discussing the distribution of functional feeding guilds within an assemblage permits a relatively dynamic view of the nutrient status of a particular river site. Changes in the composition of the feeding guild structure of a site, compared to the reference condition, may indicate a disturbance. Clear trends between the composition of feeding types in the community and an investigated stressor are graphed in Figure 3a; the left corner of the figure indicates the best ecological conditions (reference); the ecological quality of the river sites under

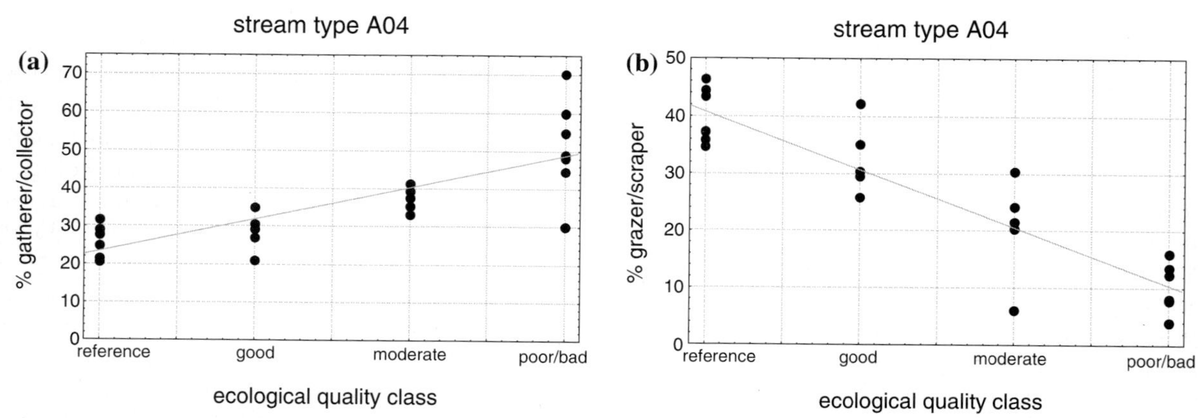

Figure 3. Different feeding types (% individuals) of aquatic invertebrates and their responses to environmental stress expressed in ecological quality classes; (a) feeding type "gatherer/collector", (b) feeding type "grazer/scraper".

investigation decreases to the right ending with bad sites. The classification of these sites follows the application of a multi-metric procedure as described in Hering et al. (2004), the calculation of the "% gatherer/collector" value follows the AQEM manual (AQEM consortium, 2002). Statistical analyses regarding the discrimination efficiency between the individual ecological quality classes are to be found in Ofenböck et al. (2004). The increase of the feeding type "gatherer/collector" is proportional to a decrease in river quality. Examples are given from the Austrian stream type "Mid-sized streams in the Bohemian Massif, ecoregion Central Highlands" (coded as A04) during investigations of river impoundment, due to hydropower generation, as a stressor. Values decrease from about 70% "gatherers/collectors" in bad ecological classes to 20% in reference sites. In the same way, sound responses to environmental exposure can be shown for the feeding type "grazer/scraper" in Figure 3b. The proportion of this functional guild increases with decrease of the stressor from about 5% in rivers with bad status to 45% in reference sites.

Analogous relationships exist between environmental stress and other parameters that describe ecological features or the needs of organisms, such as locomotion types, current or stream zonation preferences, and the predicted responses of the benthic assemblages are well known (Barbour et al., 1999). To illustrate the ability of functional measures to indicate the relationships between different morphological stressors and the benthic conditions, examples are given in Figure 4. Figure 4a shows the use of a selected locomotion type for visualising the effects

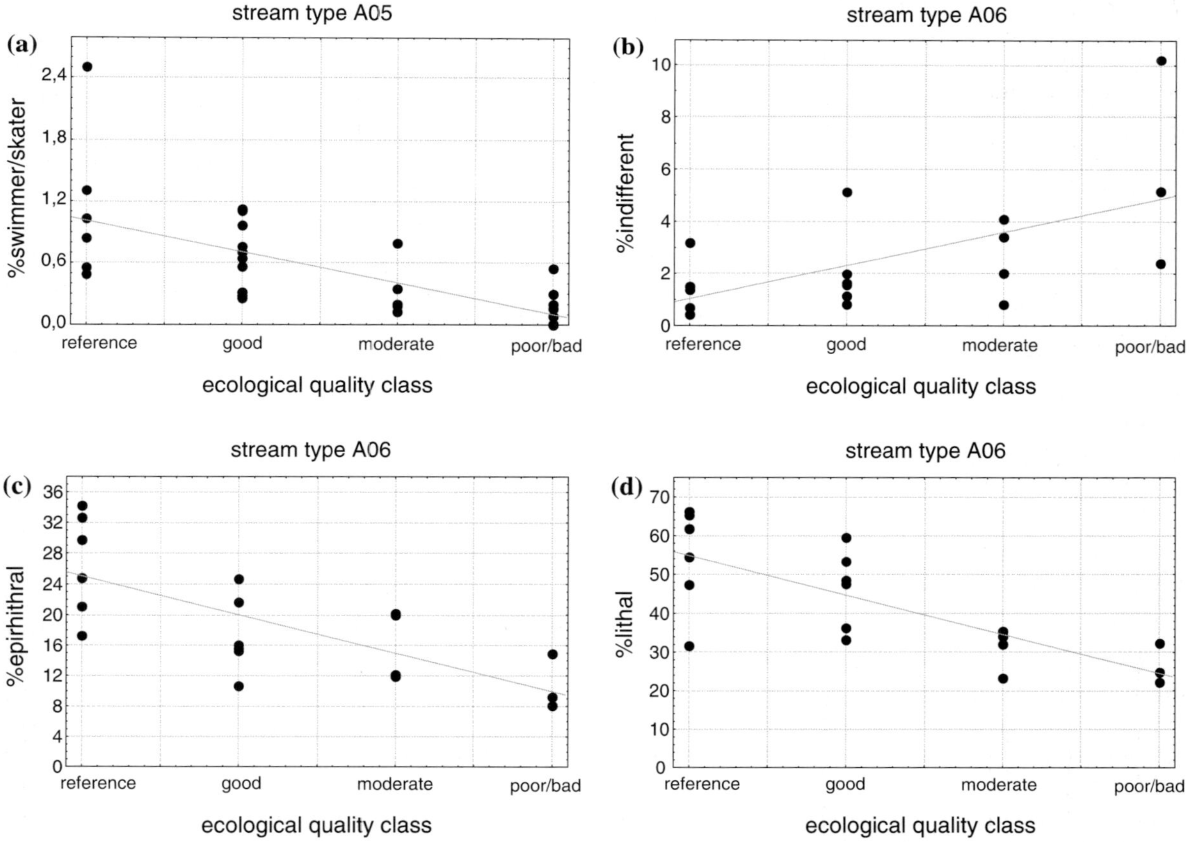

Figure 4. Different types of ecology-based metrics (% individuals) and their response to environmental stress expressed in ecological quality classes; (a) locomotion type "swimmer/skater", (b) current preference "indifferent", (c) stream zonation preference "epirhithral", (d) substrate preference "lithal".

of damming rivers (impoundment to facilitate hydropower generation) in the Austrian stream type A05 "Small-sized streams in the Bohemian Massif". Compared to reference conditions the proportion of the "swimmer/skaters" decreases with increasing degree of impairment (stress). The stressor classes have been defined according to conditions of changed current flow and composition of bed sediments, as described by Moog & Stubauer (2003). Other decreasing metrics – relating to river morphology degradation as a stressor in the Austrian stream type A06 "Small-sized crystalline streams of the ridges of the Central Alps" – are stream zonation (preference for epirhithral zones; Fig. 4c) and substrate preferences (share of lithal preferences; Fig. 4d). With respect to current preferences, taxa designated as "indifferent" increase with increasing stress (Fig. 4b).

These responses of the metrics to different kinds of stressors have generally been well investigated and documented, which makes them an indispensable part of assessment systems (Barbour et al., 1999; AQEM consortium, 2002). Basically, the power of a (multi-metric) assessment methodology to detect environmental influences increases the more ecologically classified species/taxa are included. Therefore it is seen as an important future task to extend the current knowledge of more ecological parameters. Gaps within the specification of species' ecological requirements are most often due to:

- existing but unavailable information (grey literature)
- lack of adequate biological/ecological studies
- missing translations of complex ecological responses into abstract scores which can be used for assessment systems
- deficient information due to taxonomic problems

These gaps could be filled, and the ecological information enlarged, through analysis of large, existing datasets such as the AQEM/STAR dataset. Verdonschot (2006) for instance, showed that for Oligochaeta it is possible to extend the knowledge of many ecological parameters within the autecological database by performing statistical analysis of the AQEM/STAR dataset. Another example was demonstrated by Moog & Schmidt-Kloiber (1999) with a statistical analysis of observed saprobic preferences, followed by a refinement of saprobic indices for selected groups of the Austrian stream fauna.

Conclusion and outlook

The composition of a stream community is the result of interaction between environmental and biological factors. The routine use of benthic invertebrates as sentinel organisms to monitor ongoing environmental impairment requires considerable understanding of the factors involved (Johnson et al., 1995). The presence of a taxon indicates that the habitat is suitable for that taxon and, because some of their environmental requirements are known for many species, their presence indicates something about the nature of the environment in which they are found. The community's response to the combined effects of single factors can be approximated in the form of biological measures (e.g., indices) that evaluate specific features such as, for example, pollution (saprobic) conditions and ecosystem functions such as e.g., the longitudinal distribution patterns of species and functional-feeding guilds. The saprobic indices indicate the reaction of the biota to the oxygen conditions in the system (Sladecek, 1973). The functional-feeding-guild classification allows assessment of the nutrient availability and the dominant bio-processing functions of the community (Cummins & Klug, 1979; Schweder, 1992). An analysis of the longitudinal zonation patterns of a community, based on the concept of uni-directional spatial succession within river systems, provides the opportunity to discuss the effect of serial discontinuities (Ward & Stanford, 1995) such as altered thermal conditions (temperature regimes) and current velocity (Moog, 1995). Thus, some of the key factors are adequately covered within the AQEM/STAR database. Other evaluation approaches based on e.g. habitat/substrate or current preferences (Schmedtje, 1995) provided promising results but still need to be developed, because only a few species can be ranked according to their preferences (17% and 16% for current and substrate preferences respectively). As one of the purposes of the WFD is to establish a framework for the protection of terrestrial ecosystems depending directly on aquatic ecosystems, the multidimensional functionality of aquatic ecosystems, which is determined in an essential way by surrounding wetlands, aquifers and connections to groundwater aspects of lateral and vertical connectivity, needs to be included in future bio-monitoring tasks (Ward, 1989; Ward et al., 1998; Jungwirth et al., 2002, 2003).

The AQEM/STAR database is a product of applied freshwater ecology in co-operation with scientists from different zoological fields, applied partners and the administration. The basic idea is that a sound understanding of benthic invertebrate ecology is a prerequisite for the implementation of a biological approach to European aquatic ecosystem management. Even though there are still gaps within the AQEM/STAR database, it is the first comprehensive taxalist integrating taxonomic knowledge and species distribution ranges with autecological information. It is a first big step towards improvement of the usability of macroinvertebrates in freshwater assessment on a pan-European scale.

The integrated national checklists (even if some of them are still "working lists") may serve as a first base on the way to realising an update of Illies (1978) widely available Limnofauna Europaea. In addition to listing the geographic distribution of 14,457 European aquatic species Illies outlined some ecological information on these taxa, such as their preference for types of water bodies. The current need for an updating of the Limnofauna Europaea is clearly reflected in the fact that these 27 zoogeographic regions were adopted as "European ecoregions" by the European Water Framework Directive (Annex 11) in the year 2000. Originating from the US nearly 20 years ago (Omernik, 1987; Hughes & Larsen, 1988), ecoregions are used as a spatial framework for environmental resource management on a worldwide scale. The main reasons for the use of ecoregions are because they (1) provide an ecological framework for organising environmental data, (2) are independent of political boundaries and (3) provide a logical approach to monitoring and assessment. Ecoregions exhibit similar features and environmental characteristics such as e.g., geology, climate, topography, soil and vegetation. The big advantage of using ecoregions as a geographical/typological framework for assessment models can be explained by the fact that within ecoregions the environmental conditions and the biota are relatively homogenous. Ecoregions have a high internal similarity of abiotic and biotic components compared to the conditions in adjacent ecoregions (Hawkins & Norris, 2000; Moog et al., 2001b). This relative homogeneity reduces the natural variation and makes it easier to distinguish between signal and noise when developing ecological assessment systems. Besides their scientific advantages, the use of ecoregions provides the opportunity for states or agencies to share resources. For this reason the inclusion of ecoregional aspects (integration of ecoregions and where necessary sub-units) will be a future focus of the AQEM/STAR database.

Concluding, from the analysis of gaps in the system the following future aims of the database development are defined as:

- Completion of national benthic invertebrate taxa inventories (checklists) for all European countries and adjustment with the findings of the Fauna Europaea group. Checklists from other European countries will be included (e.g., Norway, Finland, Spain).
- Amendment of the checklists on a zoogeographic scale according to Illies' ecoregions (1978).
- Filling the gaps in our knowledge of the ecological parameters treated so far.
- Inclusion of more ecological parameters such as temperature preference, resistance to droughts, hydrological preference, reproductive cycles and life cycle duration, altitude preference, and others. The focus will be on wetland–ground–water-interactions with the river corridor as an entity of aquatic systems in the sense of lateral and vertical connectivity.

Major steps towards the realisation of these aims will be achieved within the scope of another EU funded project, Euro-limpacs (www.euro-limpacs.ucl.ac.uk, Contract No: GOCE-CT-2003-505540). During the five-year project 37 European partner institutions will investigate and evaluate the impacts of climate change on European freshwater ecosystems. The Euro-limpacs consortium has agreed to adopt the AQEM/STAR database as a basic data source that will be extended to include ecological parameters which are assumed to be sensitive to direct or indirect impacts of climate change. As a final outcome of this project all parameters will be made available to the scientific public for multiple uses, e.g., the development of future assessment systems.

Within recent years the design of systems for the assessment of the ecological status of freshwater ecosystems has enormously increased to meet the requirements of the WFD. From traditional saprobic water quality monitoring, to the evaluation of various stressors and their integrated impact on benthic invertebrate assemblages, these assessment methodologies have become more and more complex and sophisticated. On the other hand, the performance of autecological studies seems to be decreasing due to the fashion for "up-to-date" sciences as mentioned above. The gap between our basic knowledge of indicators and the number of different so-called indicator-based assessment systems is, in fact, becoming greater, which seems to be contradictory. Fundamental and applied sciences need to develop synchronously. It is important to fill as many as possible of the taxonomic and autecological gaps identified in this paper. The more ecologically classified species are included in an assessment methodology, the more likely the model will become both quantitatively powerful and increasingly sensitive to the full range of possible environmental influences. Effective assessment programmes to evaluate the ecological status of freshwater systems can contribute to the overall health of the aquatic environment.

Acknowledgements

We greatly appreciate the work of all the taxonomic experts who volunteered for the taxonomic validity checking (Table 1). We also want to thank Robert Vogl for computer technical support and realisation, as well as Thomas Ofenböck for providing some graphs and analysis. Thanks to Daniel Hering and Mike Furse for the co-ordination of the EU funded AQEM (Contract No: EVK1-CT-1999-00027) and STAR (Contract No: EVK1-CT 2001-00089) projects as well as to all partners in these projects for their contribution to the database. This paper represents a deliverable of WP7 within the EU funded Euro-limpacs project (Contract No: GOCE-CT-2003-505540). Thanks to Rick Battarbee for co-ordinating this project. We want to express our gratitude to Mary Burgis for linguistic assistance.

References

Alba-Tercedor, J. & A. Sanchez-Ortega, 1988. Un metodo rapido y simple para evaluar la calidad biologica de las aguas corrientes basado en el de Hellawell (1978). Limnetica 4: 51–56.

AQEM Consortium, 2002. Manual for the application of the AQEM system. A comprehensive method to assess European streams using benthic macroinvertebrates, developed for the purpose of the Water Framework Directive. Version 1.0 (www.aqem.de), February (2002).

Armitage, P. D., D. Moss, J. F. Wright & M. T. Furse, 1983. The performance of a new biological water quality score system based on macroinvertebrates over a wide range of unpolluted running-water sites. Water Research 17: 333–347.

Barbour, M. T., J. Gerritsen, B. D. Snyder & J. B. Stribling, 1999. Rapid bioassessment protocols for use in streams and wadeable rivers: Periphyton, Benthic Macroinvertebrates and Fish. (2nd edn.) EPA/841-B-98-010. U.S. EPA. Office of Water, Washington, DC.

Banning, M., 1998. Auswirkungen des Aufstaus größerer Flüsse auf das Makrozoobenthos dargestellt am Beispiel der Donau. Essener ökologische Schriften 9. Westarp-Wissenschaften.

Biss, R., P. Kübler, I. Pinter & U. Braukmann, 2002. Leitbildbezogenes biologisches Bewertungsverfahren für Fließgewässer (aquatischer Bereich) in der Bundesrepublik Deutschland – Ein erster Beitrag zur integrierten ökologischen Fließgewässerbewertung – UBA-Texte 62/02 als CD-Rom, Hrsg. Umweltbundesamt Berlin.

Bochert, R., 2003. Datenbank zur Autökologie der Arten des deutschen Küstenraumes. Im Auftrag der Bundesanstalt für Gewässerkunde, Koblenz.

Brabec, K., S. Zahrádková, D. Nmejcová, P. Paril, J. Kokeš & J. Jarkovský, 2004. Assessment of organic pollution effect considering differences between lotic and lentic stream habitats. Hydrobiologia 516: 331–346.

Braukmann, U., 2000. Hydrochemische und biologische Merkmale regionaler Bachtypen in Baden-Württemberg. Landesanstalt für Umweltschutz Baden-Württemberg, Oberirdische Gewässer. Gewässerökologie 56: 1–501.

Braukmann, U. & R. Biss, 2004. Conceptual study – An improved method to assess acidification in German streams by using benthic macroinvertebrates. Limnologica 34: 433–450.

Buffagni, A., 1997. Mayfly community composition and the biological quality of streams. In Landolt, P. & M. Sartori (eds), Ephemeroptera & Plecoptera: Biology – Ecology – Systematics. MTL, Fribourg, 235–246.

Buffagni, A., 1999. Pregio naturalistico, qualità ecologica e integrità della comunità degli Efemerotteri. Un indice per la classificazione dei fiumi italiani. Acqua & Aria 8: 99–107.

Buffagni, A., S. Erba, M. Cazzola & J. L. Kemp, 2004. The AQEM multimetric system for the southern Italian Apennines: assessing the impact of water quality and habitat degradation onpool macroinvertebrates in Mediterranean rivers. Hydrobiologia 516: 313–329.

Cummins, K. W. & M. J. Klug, 1979. Feeding ecology of stream invertebrates. Annual Review of Ecology and Systematics 10: 147–172.

CSN 75 (7716), 1998. Water quality, biological analysis, determination of saprobic index. Czech Technical State Standard. Czech Standards Institute, Prague, 174 pp.

Davis, W. S. & T. P. Simon (eds), 1995. Biological Assessment and Criteria Tools for Water Resource Planning and Decision Making. Lewis Press, Boca Raton, Florida.

De Pauw, N. & G. Vanhooren, 1983. Method of biological quality assessment of watercourses in Belgium. Hydrobiologia 100: 153–168.

De Pauw, N., P. F. Ghetti, D. P. Manzini & D. R. Spaggiari, 1992. Biological assessment methods for running water. In Newman, P. J., M. A. Piavaux & R. A. Sweeting (eds), River Water Quality. Ecological Assessment and Control. Commission of the European Communities, EUR 14606 En-Fr: 217–248.

DEV (Deutsches Institut für Normung e.V.), 1992. Biologisch-ökologische Gewässergüteuntersuchung: Bestimmung des Saprobienindex (M2). Deutsche Einheitsverfahren zur Wasser-, Abwasser- und Schlammuntersuchung. VCH Verlagsgesellschaft mbH, Weinheim, 1–13.

DEV (Deutsches Institut für Normung e.V.), 2003. Biologisch-ökologische Gewässergüteuntersuchung: Bestimmung des Saprobienindex (revidierte Fassung). Deutsche Einheitsverfahren zur Wasser-, Abwasser- und Schlammuntersuchung. Biologisch-ökologische Gewässeruntersuchung (Gruppe M). Berlin.

European Commission, 2000. Directive 2000/60/EC. Establishing a framework for community action in the field of water policy. European Commission PE-CONS 3639/1/100 Rev 1, Luxemburg.

Extence, C. A., D. M. Balbi & R. P. Chadd, 1999. River flow indexing using British benthic macroinvertebrate: A framework for setting hydroecological objectives. Regulated Rivers: Research & Management 15: 543–574.

Fauna Europaea Web Service, 2004. Fauna Europaea version 1.1, Available online at http://www.faunaeur.org.

Feld, C. K., 2004. Identification and measure of hydromorphological degradation in Central European lowland streams. Hydrobiologia 516: 69–90.

Ghetti, P. F., (1997). Manuale di applicazione Indice Biotico Esteso (I.B.E.). I macroinvertebrati nel controllo della qualità degli ambienti di acque correnti. Provincia Autonoma di Trento, Agenzia provinciale per la protezione dell'ambiente. 1–22.

Hawkins, C. P. & R. H. Norris (eds), (2000). Landscape Classifications: Aquatic Biota and Bioassessments. Journal of the North American Benthological Society 19.

Henrikso, L. & M. Medin, 1986. Biologisk bedömning av försurningspåverkan på Lelångens tillflöden och grundområden 1986. Aquaekologerna, Rapport till länsstyrelsen i Älvsborgs län.

Hering, D., O. Moog, L. Sandin & P. F. M. Verdonschot, 2004. Overview and application of the AQEM assessment system. Hydrobiologia 175: 1–20.

Hughes, R. M. & D. P. Larsen, 1988. Ecoregions: An approach to surface water protection. Journal of the Water Pollution Control Federation 60: 486–493.

Ilies, J. (ed.), 1978. Limnofauna Europaea. Gustav Fischer Verlag, Stuttgart.

Johnson, R. K., T. Wiederholm & D. M. Rosenberg, 1995. Freshwater biomonitoring using individual organism, populations, and species assemblages of benthic macroinvertebrates. In Rosenberg, D. M. & V. H. Resh (eds), Freshwater Biomonitoring and Benthic Macroinvertebrates. Chapman & Hall, New York, London, 40–158.

Jungwirth, M., S. Muhar & S. Schmutz, 2002. Re-establishing and assessing ecological integrity in riverine landscapes. Freshwater Biology 47: 867–887.

Jungwirth, M., G. Haidvogl, O. Moog, S. Muhar & S. Schmutz, 2003. Angewandte Fischökologie an Fließgewässern. Facultas Verlag, UTB 2113, Wien.

Karr, J. R. & E. W. Chu, 1999. Restoring Life in Running Waters: Better Biological Monitoring. Island Press, Washington DC, 1–200.

Lorenz, A., D. Hering, C. K. Feld & P. Rolauffs, 2004. A new method for assessing the impact of hydromorphological degradation on the macroinvertebrate fauna of five German stream types. Hydrobiologia 516: 107–127.

Lorenz, A. & A. Schmidt-Kloiber, 2005. Die AQEM Autökologiedatenbank – "a living document". Deutsche Gesellschaft für Limnologie (DGL) – Tagungsbericht 2004 (Potsdam), Berlin 2005: 141–154.

Malicky, H., 2005. Ein kommentiertes Verzeichnis der Köcherfliegen (Trichoptera) Europas und des Mittelmeergebietes. Linzer biologische Beiträge 37/1: 533–596.

Merritt, R. W. & K. W. Cummins, 1984. An Introduction to the Aquatic Insects of North America. Kendall/Hunt Publishing Company, Dubuque.

Moog, O. (ed.), 1995. Fauna Aquatica Austriaca – A Comprehensive Species Inventory of Austrian Aquatic Organisms with Ecological Notes. Federal Ministry for Agriculture and Forestry, Wasserwirtschaftskataster Vienna: loose-leaf binder.

Moog, O. & A. Schmidt-Kloiber, 1999. Mathematisch unterstützte Verfahren zur Nachvollziehbarkeit der saprobiellen Einstufung von Makrozoobenthos-Arten. Deutsche Gesellschaft für Limnologie (DGL) – Tagungsbericht 1998 (Klagenfurt), Band II. Tutzing 1999: 541–545.

Moog, O., A. Schmidt-Kloiber, R. Vogl & Koller-Kreimel, 2001a. ECOPROF-Software. Wasserwirtschaftskataster, Bundesministeriums für Land- & Forstwirtschaft. Umwelt & Wasserwirtschaft, Wien.

Moog, O., A. Schmidt-Kloiber, T. Ofenböck & J. Gerritsen, 2001b. Aquatische Ökoregionen und Bioregionen Österreichs – eine Gliederung nach geoökologischen Milieufaktoren und Makrozoobenthos-Zönosen. Bundesministerium für Land und Forstwirtschaft, Umwelt und Wasserwirtschaft. Wasserwirtschaftskataster, Vienna.

Moog, O. (ed.), 2002. Fauna Aquatica Austriaca – A Comprehensive Species Inventory of Austrian Aquatic Organisms with Ecological Notes 2nd edn. Federal Ministry for Agriculture and Forestry, Wasserwirtschaftskataster Vienna: loose-leaf binder.

Moog, O., W. Graf, B. F. U. Janecek & T. Ofenböck, 2003. Inventory of "sensitive taxa" of Austrian rivers and streams – A valuable measure among the multimetric approaches and a tool for developing a rapid field screening method to assess the ecological status of rivers and streams in Austria. In Moog, O. (ed.), Fauna Aquatica Austriaca – Ergänzungen 2003.

Moog, O. & I. Stubauer, 2003. Activity 1.1.2 – Adapting and implementing common approaches and methodologies for stress and impact analysis with particular attention to hydromorphological conditions – Final Report, UNDP/GEF DANUBE REGIONAL PROJECT. Strengthening the implementation capacities for nutrient reduction and transboundary cooperation in the Danube River Basin; to be downloaded from http://www.icpdr.org/undp-drp.

Muhar, S., M. Schwarz, S. Schmutz & M. Jungwirth, 2000. Identification of rivers with high and good habitat quality: Methodological approach and applications in Austria. In Jungwirth, M., S. Muhar & S. Schmutz (eds), Assessing the Ecological Integrity of Running Waters. Kluwer Academic Publishers, Dortrecht Boston London, 343–358.

Ofenböck, T., O. Moog, J. Gerritsen & M. T. Barbour, 2004. A stressor specific multimetric approach for monitoring running waters in Austria using benthic macro-invertebrates. Hydrobiologia 516: 251–268.

Omernik, J. M., 1987. Ecoregions of the conterminous United States. Annals of the Association of American Geographers 77: 118–125.

Pinto, P., J. Rosado, M. Morais & I. Autunes, 2004. Assessment methodology for southern silicious basins in Portugal. Hydrobiologia 516: 191–214.

Rolauffs, P., D. Hering, M. Sommerhäuser, S. Rödiger & S. Jähnig, 2003. Entwicklung eines leitbildorientierten Saprobienindexes für die biologische Fließgewässerbewertung. Umweltbundesamt Texte 11/03.

Rosenberg, D. M. & V. H. Resh (eds), 1995. Freshwater Biomonitoring and Benthic Macroinvertebrates. Chapman & Hall, New York, London.

Rosenzweig, M. L., 1995. Species Diversity in Space and Time. Cambridge University press, 1–436.

Sandin, L., J. Dahl & R. K. Johnson, 2004. Assessing acid stress in Swedish boreal and alpine streams using benthic macroinvertebrates. Hydrobiologia 516: 129–148.

Schmedtje, U., 1995. Ökologische Grundlagen für die Beurteilung von Ausleitungsstrecken. Schriftenreihe des Bayrischen Landesamtes für Wasserwirtschaft, Heft 25.

Schmedtje, U. & M. Colling, 1996. Ökologische Typisierung der aquatischen Makrofauna. Informationsberichte des Bayerischen Landesamtes für Wasserwirtschaft 4/96.

Schmidt-Kloiber, A. & R. Nijboer, 2004. The effect of taxonomic resolution on the assessment of ecological water quality classes. Hydrobiologia 516: 269–283.

Schöll, F., A. Haybach & B. König, 2005. Das erweiterte Potamontypieverfahren zur ökologischen Bewertung von Bundeswasserstraßen (Fließgewässertypen 10 und 20: kies- und sandgeprägte Ströme, Qualitätskomponente Makrozoobenthos) nach Maßgabe der EU-Wasserrahmenrichtlinie. Hydrologie und Wasserbewirtschaftung 49(5): 234–247.

Schweder, H., 1992. Neue Indizes für die Bewertung des ökologischen Zustandes von Fließgewässern, abgeleitet aus der Makroinvertebraten-Ernährungstypologie. Limnologie aktuell Band 3: 353–377.

Skriver, J., N. Friberg & J. Kirkegaard, 2001. Biological assessment of running waters in Denmark: introduction of the Danish Stream Fauna Index (DSFI). Verhandlungen der Internationale Vereinigung für Theoretische und Angewandte Limnologie 27: 1822–1830.

Sládecek, V., 1973. System of water quality from the biological point of view. Archiv für. Hydrobiologie. Ergebnisse der Limnologie 7: 1–218.

Šporka, F. (ed.), 2003. Vodné bezstavovce (makroevertebráta) Slovenska, súpis druhova autekologické charakteristiky. Slovenský hydrometeorologický ústav, Bratislava.

Statzner, B., V. H. Resh & L. Roux, 1994. The synthesis of long-term ecological research in the context of concurrently developed ecological theory: design of a research strategy for the Upper Rhône River and its floodplain. Freshwater Biology 31: 253–263.

Usseglio-Polatera, P., M. Bournaud, P. Richoux & H. Tachet, 2000. Biological and ecological traits of benthic freshwater macroinvertebrates: relationship and definition of groups with similar traits. Freshwater Biology 43: 175–205.

Van der Hoek, W. F. & P. F. M. Verdonschot, 1994. Functionele karakterisering van aquatische ecotooptypen. IBN rapp 072: 1–81.

Verdonschot, P. F. M., 1990. Ecological characterization of surface waters in the province of Overijssel (The Netherlands). Thesis, Wageningen, 1–255.

Verdonschot, P. F. M., 2006. Beyond masses and blooms: the indicative value of oligochaetes. Hydrobiologia 564: 127–142.

Vlek, H. E., P. F. M. Verdonschot & R. C. Nijboer, 2004. Towards a multimetric index for the assessment of Dutch streams using benthic macroinvertebrates. Hydrobiologia 516: 173–189.

Ward, J. V., 1989. The four-dimensional nature of lotic ecosystems. Journal of North American Benthological Society 8: 2–8.

Ward, J. V. & J. A. Stanford, 1995. The serial discontinuity concept. Extending the model to floodplain rivers. Regulated Rivers: Research and Management 10: 159–169.

Ward, J. V., G. Bretschko, M. Brunke, D. Danielopol, J. Gibert, T. Gonser & A. G. Hildrew, 1998. The boundaries of river systems: the metazoan perspective. Freshwater Biology 40: 531–569.

Wolf, B., 2004. Datenbank zu den metrics des Makroozoobenthos der Marschengewässer Niedersachsens. Im Auftrag der Hochschule Vechta.

Zelinka, M. & P. Marvan, 1961. Zur Präzisierung der biologischen Klassifikation der Reinheit fließender Gewässer. Archiv für Hydrobiologie 57: 389–407.

Hydrobiologia (2006) 566:343–354
M.T. Furse, D. Hering, K. Brabec, A. Buffagni, L. Sandin & P.F.M. Verdonschot (eds), The Ecological Status of European Rivers: Evaluation and Intercalibration of Assessment Methods
DOI 10.1007/s10750-006-0085-4

The PERLA system in the Czech Republic: a multivariate approach for assessing the ecological status of running waters

Jiří Kokeš[1,*], Světlana Zahrádková[2], Denisa Němejcová[1], Jan Hodovský[3], Jiří Jarkovský[4] & Tomáš Soldán[5]

[1]*Brno Branch Department, Water Research Institute T.G.M. Prague, Dřevařská 12, 657 57 Brno, Czech Republic*
[2]*Dept. of Zoology and Ecology, Masaryk University Brno, Kotlářská 2, 611 37 Brno, Czech Republic*
[3]*Ministry of Environment of the Czech Republic, Vršovická 65, 100 10 Prague, Czech Republic*
[4]*Centre of Biostatistics and Analyses, Faculty of Medicine and Faculty of Science, Masaryk University, Kamenice 126/3, 625 00 Brno, Czech Republic*
[5]*Institute of Entomology, Academy of Sciences of the Czech Republic, Branišovská 31, 370 05 České Budějovice, Czech Republic*
*(*Author for correspondence: E-mail: jiri.kokes@wri.cz)*

Key words: benthic macroinvertebrates, ecological quality, stream assessment, typology, classification

Abstract

The assessment of running water quality has a long tradition in the Czech Republic, but in the past it focused on the evaluation of organic pollution using the saprobic system. Considering the modern trends of stream ecological status evaluation in water management a new assessment system named PERLA was developed. The system is a complex of biological methods of ecological status assessment of running waters and connected activities in the Czech Republic. It involves 300 reference sites with respective biotic and abiotic data and a prediction model using a newly developed software HOBENT. The model generally follows the published mathematical principles of RIVPACS and represents the site specific and stressor non-specific approaches. The HOBENT software allows the prediction of the target assemblage of benthic macroinvertebrates for any site based on a set of environmental variables (latitude, longitude, distance from source, altitude, slope, catchment area, and stream order) which characterise the site. The predicted assemblage can be compared with the fauna observed at the same site. The comparison makes it possible to evaluate the extent of disturbance, expressed by index B. The model allows to evaluate spring, summer, and autumn seasonal data of the majority of wadable streams in the Czech Republic. The practical application of the PERLA system has started in 2001.

Introduction

Assessment of the water quality of running waters based on biota has a long tradition in the Czech Republic. In relation to strong organic pollution, in Central Europe regarded as a cardinal problem of water management for the past century, a wide range of saprobiological methods have been applied (Bernardová et al., 1996). Nevertheless, wide ranging social and economic changes since 1989 have lead to a decrease of organic pollution in the Czech Republic (WRI, 1993, 2002). Owing to this fact and in accordance with European Union policies concerning the assessment of the ecological status of aquatic ecosystems (European Commission, 2000), there is a necessity for new methods for evaluating the impact of issues such as changed river morphology and unnatural discharge regimes. The British RIVPACS (Armitage et al., 1983; Wright, 1995; Wright et al., 1989,

1993) has been adopted as the most suitable approach, being based on a comparison with a reference status of benthic macroinvertebrate assemblages. However, the application of the system requires the compilation of a reference data set for the given geographical region. This condition has been fulfilled through PERLA, a newly constituted system for evaluating running water quality. It is named after the stonefly genus *Perla*, which occurs predominantly in clear running waters. The PERLA system is a complex of biological methods of ecological status assessment of running waters and connected activities in the Czech Republic, taking into account the official activities of the Czech Republic (Kokeš, 2002).

Material and methods

Study area

The Czech Republic is an inland state that is situated in the middle of a temperate climate zone of the Northern hemisphere of the central part of Europe. The total area is 78,864 km^2 with a population density of about 131 inhabitants/km^2. The climate of the Czech Republic is characterised by the mutual penetration and mixing of oceanic and continental influences. The oceanic influence is most evident in the western part of the country; increasing continental climate effects are more pronounced in the eastern areas. Elevations range from 116 to 1,602 m a.s.l. with the average altitude of 430 m a.s.l.. The lowlands (up to 200 m a.s.l.) are situated along the lower parts of large rivers and as well as mountain areas (altitudes above 800 m) cover only a small part of the country (Fig. 1a). From a geomorphological point of view (Demek, 1987), the mountains of the Hercynian orographic system form a ring along the state border in the western (Bohemian) part; the Outer Carpathian Ridge follows along the eastern border of the state. The Pannonian lowlands and the Polonium in Moravia and Silesia represent a band of lowland areas dividing the Hercynian and Carpathian mountain systems. Geological differences between the regions are expressed by a higher proportion of flysch and molasse and a lower share of acid silicate rocks in the Carpathian catchments. Consequently, water alkalinity and

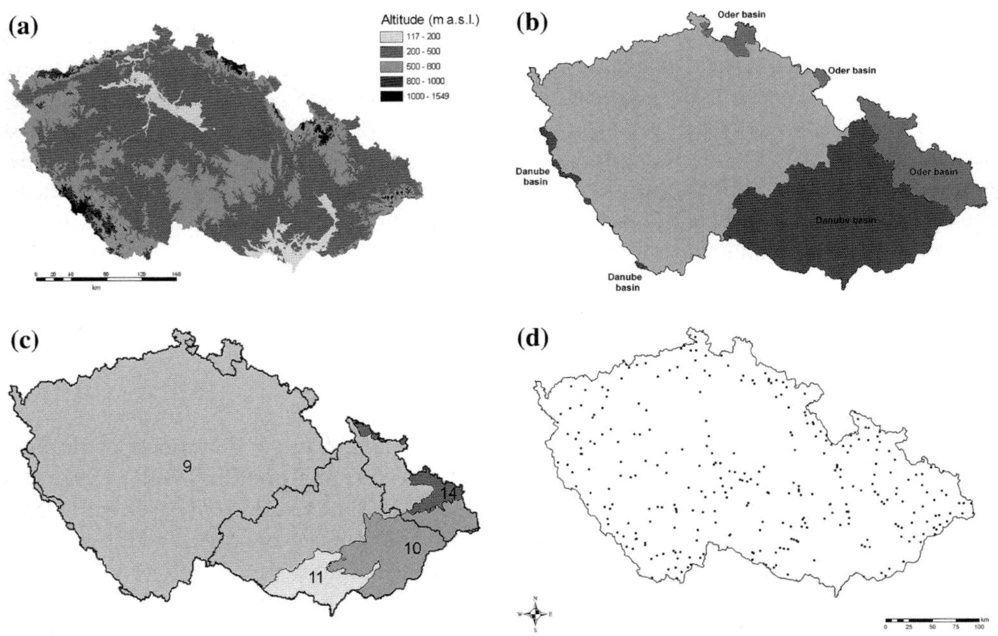

Figure 1. Map of the Czech Republic: (a) distribution of altitude categories, (b) main river basins, (c) ecoregions – detailed delineation after Culek (1996). 9 – Central Highlands, 10 – The Carpathians, 11 – Hungarian lowlands and 14 – Central plains, (d) distribution of reference sites.

total hardness are higher in the flysch and molasse region.

The Czech Republic is sometimes called the "Roof of Europe" because only atmospheric precipitation water supply harbours the three main river basins and/or sea drainage areas. These are the Labe (Elbe) River Basin (North Sea) - 51,399 km^2, the Odra (Oder) River Basin (Baltic Sea) – 4,721 km^2, and the Dunaj (Danube) River Basin (Black Sea) – 22,744 km^2 (Fig. 1b). Thanks to its geographic position, the Czech Republic is characterised by a vast majority of very small, small, and medium-sized permanent running waters (catchment areas <1,000 km^2 covering 94% of the territory). Very small streams with catchment areas <10 km^2 covering 20% of the territory play an important role in forming the conditions of densely inhabited landscape and intensive land use.

The study area is a part of four European ecoregions based on Illies (1978): No. 9 (Central Highlands), No. 10 (the Carpathians), No. 11 (Hungarian lowlands), and No. 14 (Central plains). A detailed delimitation of ecoregion borders was done by Culek (1996). The respective catchment areas of the individual ecoregions of the Czech Republic are the Elbe catchment belonging to European ecoregion No. 9, the Danube catchment belonging partly to No. 9, 10 and 11, and the Oder catchment belonging to No. 9, 10 and 14 (Fig. 1c).

Site selection and reference conditions

The network of potential reference sites was suggested on the basis of data published earlier (e.g., Landa & Soldán, 1989; Soldán et al., 1998), on the database of long-term saprobiological monitoring results, and on expert advice. More than 400 sampling sites were taken into account; this number was reduced to about 350 after detailed screening in the field. Laboratory analyses (both biological and chemical data) showed only 300 sampling sites that meet the requirements of European standard EN ISO 8689-1:2000, which states "A reference site is a site where only natural stresses are present and man-made stresses are considered to be insignificant. The community present at a reference site is a natural community when it is influenced only by natural stress (e.g. flood) and man-made stress is not significant." The following criteria were taken into consideration in order to meet the requirements of Czech National Standards:

- The degree of urbanisation, agriculture, and silviculture in a catchment must be as low as possible.
- A reference site floodplain should preferably not be cultivated. If possible, it should be covered with natural climax vegetation and unmanaged forest.
- Coarse woody debris must not be removed.
- Stream bottoms and stream banks must not be fixed (old river bank fixation by a belt of trees is acceptable).
- Natural riparian vegetation and floodplain conditions must still exist, making lateral connectivity between the stream and its floodplain possible.
- No alterations of the natural hydrographic and discharge regime.
- No hydrological alterations such as water diversion, abstraction, or pulse releases.
- No (or only minor) upstream impoundments, reservoirs, weirs, or reservoirs retaining sediments may be present (a dam 20 km upstream is acceptable for some stretches of mid-sized or large streams).
- Physical and chemical conditions close to natural background levels describing the baseload of a specific catchment area.
- No point sources of pollution or nutrients.
- No signs of acidification.
- No liming activities.
- No impairment due to physical conditions, especially the thermal conditions, which must be close to natural.
- Physical and chemical conditions are checked by physico-chemical and chemical analyses of water and sediment.
- There must not be any significant impairment of the allochthonous biota by introduced Crustacea or Mollusca.
- The value of the Czech saprobic index must not be higher than 2.2 (beta-mesosaprobity).

Naturally, it was not possible to determine real reference sites for all stream types present in the Czech Republic, since the landscape has generally been exploited for centuries. In such cases, the optimum sites within the corresponding stream type were considered as the reference sites.

Large lowland water flows seem to be extremely difficult to treat because of pronounced morphological changes and/or advanced eutrophication. Consequently, we are not able to identify any suitable reference sites for the largest rivers (e.g. lowland stretches of the Morava, Elbe, and Vltava Rivers).

Field and laboratory methods

Field sampling was done from 1997 to 2000. The PERLA sites were sampled 3 times a year in the spring (March–May), summer (July–August) and autumn periods (September–November) to meet the requirements of all the seasons.

A stream stretch typical for the watercourse in question was selected. In narrow streams, the length of this stretch was equal to 14 times the average stream width (a width of less than 5 m). In wider streams, the length of the characteristic stretch was 100 m. Because it was impossible to sample the characteristic stretch completely, a representative sampling section inside the characteristic one was chosen. Sampling points inside the sampling section were then sampled. The sampling section was sampled for benthic macroinvertebrates using a multi habitat sampling method. Semi-quantitative 3-minute kick samples gathered with a hand net (25×35 cm aperture and 500 μm mesh size) were taken. All habitats (riffle, pool, macrophytes, woody debris, etc.) were sampled in proportion to their area within the sampling section. Samples were pre-selected in the field (to preserve fragile organisms) and transferred to the laboratory where final sorting was done. Samples were preserved in 4% formaldehyde or 70% ethanol solution (Mollusca, Oligochaeta, Simuliidae). With some quantitatively extremely rich samples, their half or quarter was processed and the final number of individuals was estimated by simple multiplication. Taxonomic identification was done to the lowest level, preferably to a species level. However, in some cases (e.g., Oligochaeta, Hydracarina, and some Diptera), only genus or higher taxa could be identified. The following set of environmental variables was recorded at each site of the characteristic stretch: mean substratum – phi (Furse et al., 1986), mean current velocity, mean width and mean depth, ratio of riffles and pools, slope, shading, riparian vegetation, and surrounding terrestrial biotopes.

Other variables were obtained from respective hydrological maps and GIS layers (latitude, longitude, altitude, distance from the source, catchment area, stream order based on Strahler (Strahler, 1952), affiliation to catchment, ecoregion, geomorphologic unit, etc.). There were three series of physico-chemical and chemical analyses done of the water for a large range of parameters (pH, conductivity, alkalinity, total hardness, Ca^{2+}, Mg^{2+}, SO_4^{2-}, N–NO^{3-}, N–NH^{4-}, P_{tot}, DO, BOD, COD, TOC, etc.). One series of chemical analyses of the sediment was done for specific pollutants (PCB, PAU) and heavy metals (Pb, Cd, As, Hg) (Kokeš, 2002). All chemical analyses were done using international standards (ISO) or the Czech national standards according to the rules of quality assurance and quality control (QA/QC).

Data processing

Prior to any treatment, a taxonomical adjustment was made according to the abundance and frequencies of each taxonomical level (AQEM consortium, 2002). A taxonomic adjustment was done to prevent data inconsistency (Nijboer & Verdonschot, 2000). This means that there should be no taxa overlap, as taxonomic overlap results in the multiplication of the same information in one sample.

The adjusted taxonomic data was classified into groups by TWINSPAN (Hill, 1979). Five pseudospecies cut levels were defined (0, 3, 30, 120 and 300); minimum group size for division was 7 and the maximum level of division was 8. All other settings remained as default.

For the evaluation of the importance of environmental variables for benthic invertebrate communities, the forward selection analysis in Canonical Correspondence Analysis (CCA) was performed in CANOCO for Windows (ter Braak, 1986). Data were transformed ln $(x+1)$, 9999 permutations vas used.

Results

Reference sites

The 300 sites are more or less evenly distributed within the area (see Table 1 and Fig. 1d). For the basic characteristics of the 300 selected sites see Table 2. In compliance with the abiotic conditions

Table 1. Distribution of reference sites within WFD System A stream types

Site altitude (m a.s.l.)	Upstream catchment size (km²)	Ecoregion	Central highlands		The Carpathians		Hungarian lowlands		Central plains	
		Geology	Siliceous	Calcareous	Siliceous	Calcareous	Siliceous	Calcareous	Siliceous	Calcareous
≤ 200	≤ 10		1							
	> 10–100		1							
	> 100–1,000		1							
	> 1,000–10,000		1				2			
201–800	≤ 10		80	9	2	18		1	11	2
	> 10–100		61	4	1	7	3	1	5	
	> 100–1,000		62	1	3		1		2	
	> 1,000–10,000		13	1	2				1	
> 800	≤ 10		1							
	> 10–100		1							
	> 100–1,000		1							
	> 1,000–10,000									

of the Czech Republic, three of the most frequent abiotic stream types amongst the reference sites were very small streams (26.7%), small streams (20.3%), and medium streams (20.7%), all belonging to the altitude category of 200–800 m a.s.l. and with siliceous geology.

Altogether more than 1,500,000 individuals have been collected. After taxonomic adjustment, they belong to 419 taxa in the spring data set, 372 in the summer set, and 335 in the autumn set. The Chironomidae family was not included in the summer and autumn evaluations. The whole dataset based on 300 sites was used for the spring season evaluation. Some sites of very small streams sampled in spring dried up in summer (9 sites) and autumn (3 sites).

Classification

TWINSPAN classification resulted in 20 end groups in spring (Fig. 2), 18 end groups in summer (Fig. 3) and 20 groups in autumn (Fig. 4).

The selection of environmental variables

The environmental variables suitable as predictors for RIVPACS type model were identified by the forward selection of environmental variables in CCA. The spring season dataset was used for this analysis. The following 23 environmental variables were included in the analysis: distance from source, order of stream according to Strahler, BOD, mean width, mean depth, catchment area, slope, COD_{Cr}, N–NO_2, P_{tot}, mean annual air temperature, TOC, N–NO_3, altitude, conductivity, total hardness, SO_4^{2-}, mean substratum roughness , latitude, DO, longitude, pH and N–NH_4.

When an automatic forward selection was done (only 7 the best fitted), the following variables were chosen: stream order, mean substratum roughness, distance from source, latitude, BOD, conductivity, and N–NO_3 (Table 3).

Regarding the fact that the chemical and physico-chemical analyses represent only one analysis per sample and parameters like substratum roughness and mean depth and width are not suitable variables for prediction due to the man-made changes at evaluated localities, the manual forward selection was done with the aim to (i) avoid these problematic variables and (ii) to prefer a more practical one between the strongly correlated variables (e.g. altitude and mean annual air temperature). The final set of variables is as follows: distance from source, stream order, altitude, longitude, latitude, slope, and catchment area. For variance explained by the variable selected and p-value, see Table 4.

Discriminant analysis

Discriminant analysis (Klecka, 1980; Deichsel & Trampisch, 1985) is an important mathematical

Table 2. Characterisation of the reference sites data set – selected environmental variables

Variable	Minimum	Maximum	Median
Stream order (Strahler)	1.00	7.00	3.00
Distance from source [km]	0.50	220.40	7.65
Altitude [m a.s.l.]	125.00	888.00	417.00
Upstream catchment area [km^2]	0.53	7,522.35	16.11
Slope [m km^{-1}]	0.10	85.00	14.02

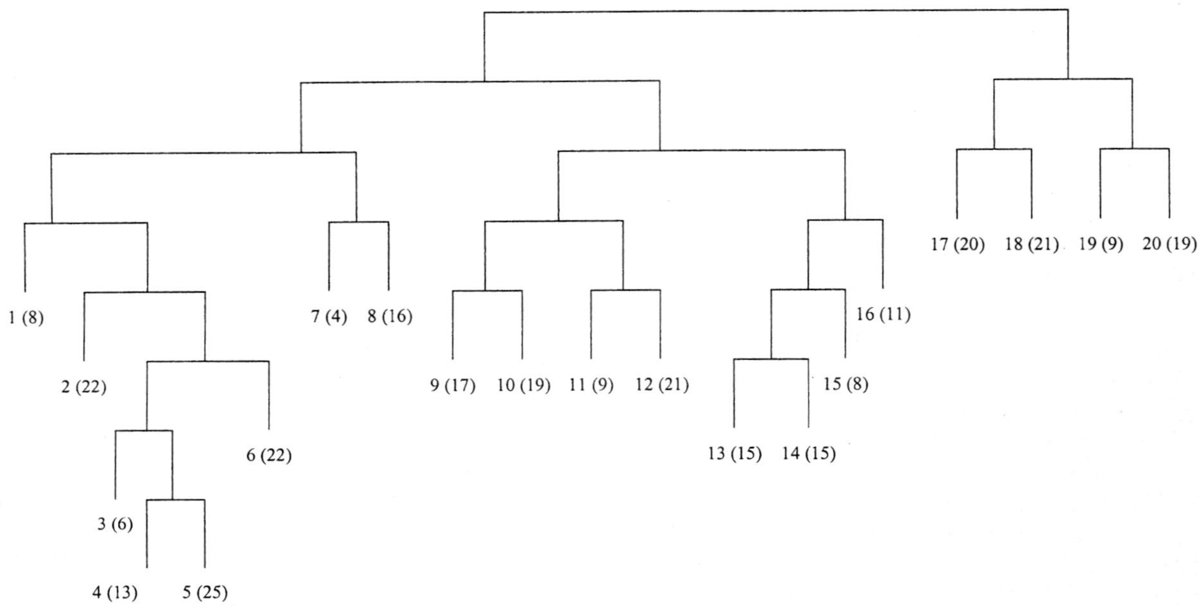

Figure 2. Dendrogram of TWINSPAN classification result – spring, 20 groups.

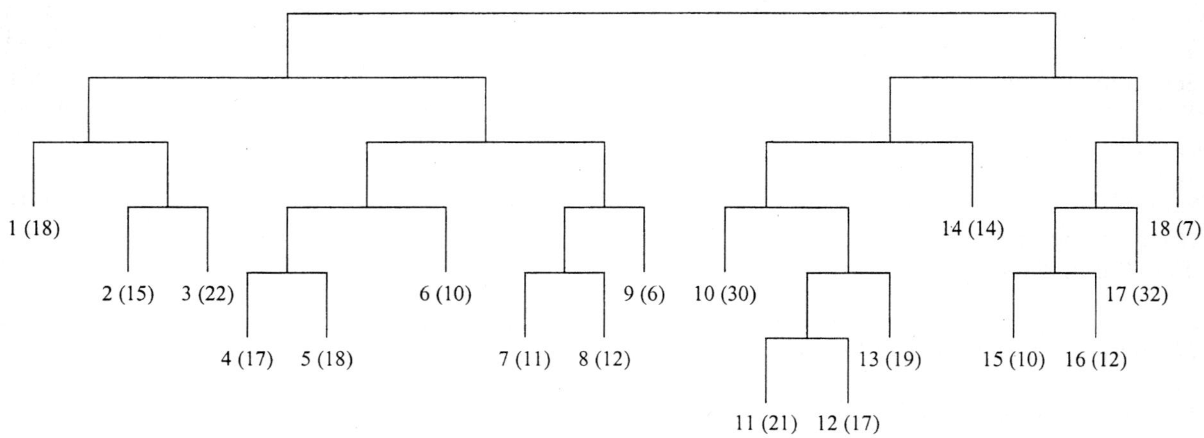

Figure 3. Dendrogram of TWINSPAN classification result – summer, 18 groups.

background of the RIVPACS type prediction model. The SPSS package was used for the computation of discriminant equations and another quantities which are used by software Hobent for the categorization of an observed site into groups of the reference database.

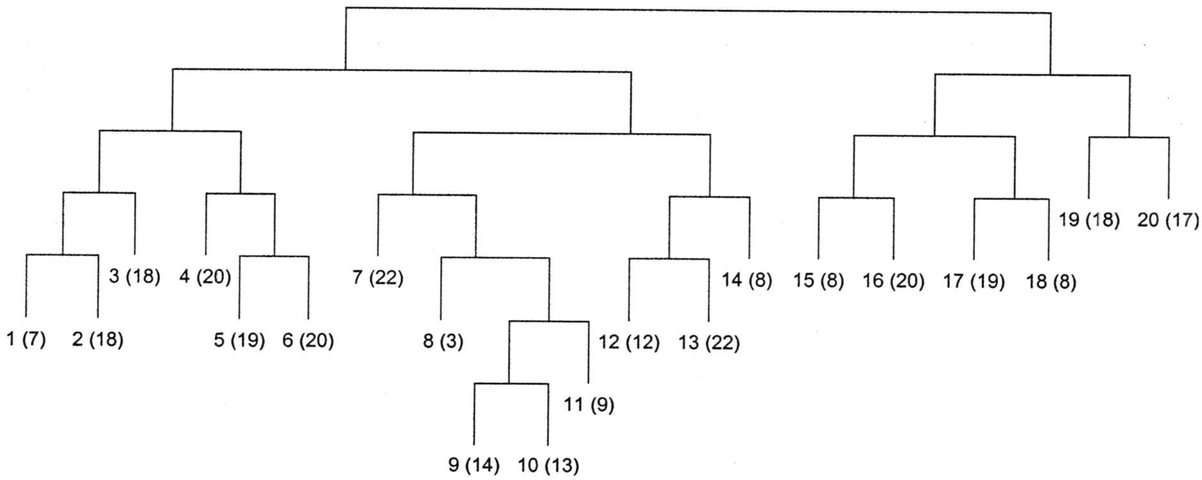

Figure 4. Dendrogram of TWINSPAN classification result – autumn, 20 groups.

Software HOBENT

The prediction is computed by HOBENT software, which was developed at the Water Research Institute (WRI) by Jiri Kokes. The mathematical part of HOBENT uses the same approach as that used in the RIVPACS system (Wright, 1995; Clarke et al., 1996). The software allows the prediction of the target assemblage of benthic macroinvertebrates for any site based on a set of environmental variables (latitude, longitude, distance from source, altitude, slope, catchment area, and stream order) which characterise the site. Then the predicted assemblage is compared with the fauna observed at the same site. The comparison makes it possible to evaluate the extent of disturbance, expressed by index B.

For the computation of the probabilities that the checked site belongs to groups of the reference database, Hobent uses the same formulas as SPSS package (Anonymous, 1997). On the base of discriminant equations and another quantities, which are a part of the reference database, and environmental variables values, Hobent computes discriminant scores and then Mahalanobis distances and uses them for the computation of the probabilities. The sizes of groups (expressed as the prior probabilities, which are also a part of the reference database) are also included in the computation.

Next, for every species of the reference database, the probability of capture at the observed site is computed according to the formula (Clarke et al., 1996):

$$C_s = \sum_{g=1}^{G} F_{sg} * P_g$$

$$C_s = \sum_{g=1}^{G} F_{sg} \times P_g$$

s = species, C_s = species probability captured at the observed site, g = group, G = number of groups, F_{sg} = frequency of occurrence of species s in group g, P_g = probability which the observed site belongs to group g with. All species are then ordered according their C_s and the number of species expected at the observed site is computed as:

$$N_E = \sum_{s=(C_s \geq C_s L)}^{S} C_s$$

$$N_E = \sum_{s=(C_s \geq C_s L)}^{S} C_s$$

N_E = number of species expected at the observed site, S = number of species in the reference database, C_sL = optional low limit of C_s (0.5).

Finally, index B is computed as:

$$B = \frac{N_O}{N_E}$$

Table 3. Results of automatic forward selection of environmental variables (Monte-Carlo permutation test, 9999 permutations, CANOCO for Windows)

Marginal effects

Variable	Lambda1
Distance from source	0.23
Order of stream after Strahler	0.21
BOD	0.19
Mean width	0.18
Mean depth	0.18
Catchment area	0.16
Slope	0.15
COD_{Cr}	0.13
$N–NO_2$	0.12
P_{tot}	0.11
Mean annual air temperature	0.10
TOC	0.10
$N–NO_3$	0.09
Altitude	0.09
Conductivity	0.08
total hardness	0.07
SO_4^{2-}	0.07
Mean substratum roughness	0.07
Latitude	0.07
DO	0.06
Longitude	0.05
pH	0.04
$N–NH_4$	0.02

Conditional effects

Variable	LambdaA	*p*-value	*F*-ratio
Distance from source	0.23	0.000	19.69
Conductivity	0.07	0.000	6.64
Order of stream after Strahler	0.06	0.000	5.78
Latitude	0.06	0.000	4.85
BOD	0.04	0.000	4.42
Mean substratum roughness	0.04	0.000	3.73
$N–NO_3$	0.04	0.000	3.46

N_O = number of species with $C_s \geq C_sL$ found at the observed site.

The low limit of C_s is an essential number. The B index computed using the limit is a type of similarity index. When the C_sL is set to zero, only the simple number of taxa is compared and the result is not very useful.

Besides index B and the basic ecological indices, the ASPT, BMWP, saprobic index, EPT, and other indices were incorporated into the software. Their expected values can be predicted. It makes it possible to express these metrics in the form of ecological quality ratios (EQR).

The computation of expected values of some indices, for instance the saprobic index, needs a prediction of abundances. A mathematical method for an abundance prediction does not exist. Hobent, therefore, predicts abundances for each species in each group using pseudorandom numbers as follows:

It generates a pseudorandom value in a range from 0 to 1. If the value is smaller or equal to the probability of occurrence of the taxon in the group (the probabilities are a part of the reference database), taxon "occurs", if the value is higher or the probability is zero, taxon "does not occur". If taxon "occurs", Hobent generates a pseudorandom value in the range from the minimum to the maximum abundance in the group (minimum and maximum abundances of each taxon for each group are also a part of the reference database). The value is "abundance" of the species. By the way, Hobent predicts an "artificial sample" for each group. Consecutively, the saprobic index is computed for each group. Finally, the predicted saprobic index is computed as the sum of products of group indices and probabilities of the observed site that belongs to that group. The procedure is repeated; the number of repetitions is optional. The final predicted index is then computed as an average of all the predicted indices and the EQR_{Si} index as a quotient of the final predicted and observed indices.

The computation of the ecological profile also needs a prediction of abundances, which is done in the same way as in the case of the saprobic index. Computation of the ecological profile follows the method described in Schmedtje (1998), and individual species profiles published in Fauna Aquatica Austriaca (Moog, 1995) are used. The profiles of the two categories (trophic guilds and the biocenotic region) can be computed. Every category has ten subcategories: shredders, scrapers, active filtrators, passive filtrators, detritivores, miners, xylophagous taxa, predators, parasites, and other; and eucrenal, hypocrenal, epirhithral, metarhithral, hyporhithral, epipotamal, metapotamal, hypopotamal, littoral, and profundal. $EQR_{EkoProf}$

Table 4. Results of manual forward selection of environmental variables (Monte-Carlo permutation test, 9999 permutations, CANOCO for Windows

Variable	*p*-value	*F*-ratio	Variance explained by the selected variable
Distance from source	0.000	19.69	0.23
Order of stream	0.000	6.20	0.29
Altitude	0.000	5.01	0.35
Longitude	0.000	4.98	0.40
Latitude	0.000	4.72	0.46
Slope	0.000	2.84	0.49
Catchment area	0.000	2.20	0.51

Variance explained by all variables: 0.85.

for each category is computed as the sum of absolute values of differences between observed and predicted subcategories divided by 2. The index measures a difference between the observed and predicted states, but not the direction of the change.

The differences in the EQR indices in the classification groups were computed using the Kruskal–Wallis nonparametrical analysis of variance (Sokal & Rohlf, 1995).

One of the goals of HOBENT software is easier data treatment. It contains a list of synonyms, computes general indices, and makes data exchange between HOBENT and EXCEL possible.

Discussion

Differences in classification results between seasons

The number of end groups and their composition as results of the classification of biota by TWINSPAN differ slightly in spring, summer and autumn. It is caused, among other factors, by the different number of sites in the seasons and by the absence of Chironomidae in the summer and autumn data sets.

Selection of environmental variables used for the categorisation of sites

During the forward selection of environmental variables, conductivity and total hardness were omitted regardless of their significance. These variables are influenced both by geology and organic pollution. The importance of geological factors is unquestionable, but no relevant information on the geology of the area investigated is available. The geological classification is either very detailed or very rough at present. It is very difficult to distinguish slight organic enrichment from geological influence under these conditions. This is a task that remains to be solved in the near future.

Relations to abiotic stream typology

The HOBENT software and the whole PERLA system were not primarily oriented towards abiotic stream typology. Because Water Framework Directive (WFD) requires abiotic stream typology, this typology was also derived for the Czech Republic (http://heis.vuv.cz/_english/default.asp), and the PERLA dataset was subsequently used for the validation of typology, closely corresponding to typology A of WFD. The WFD A typology leads to many types which are often in a very small number of sites. According to our analysis, the stream types derived by typology A do not agree with the results of classification of benthic macroinvertebrate assemblages (see also Zahrádková et al., in press; Davy-Bowker et al., 2006). The large overlap of environmental variable values in classification groups exists, which is not in concordance with the strict division of environmental variable values in WFD A typology. The prediction models based on the RIVPACS approach are believed to provide a better solution than using abiotic typology, (especially A typology). Discriminant analysis seems to be a better tool for the ordering of a observed site into the groups of the reference database. In fact, the RIVPACS approach also contains a typology, but a more complicated one and not as evident as in the case of WFD A abiotic typology. The concordance of "typology" and groups of benthic assemblages can be easily computed as, for instance, the percentage of correctly ordered sites.

Relations to the multimetric assessment systems and stressor specific approach

The multimetric system is related to the type specific and stressor specific approach, which require the definition of class boundaries for each type and each metric. Dahl & Johnson (2004) stated the major differences between the use multimetric versus multivariate approaches was that the first one requires assumption regarding the expected response of indicator taxa, whilst multivariate approaches require no such a priori assumptions. In the case of a RIVPACS type models based on discriminant analysis, the construction of class boundaries for each group is not necessary and one common class boundary set suffices. It allows for a less complicated interpretation of results. One of the main arguments against the use of multivariate or predictive approaches in bioassessment is that they are consider to be complex to use (required expert knowledge in computer software) and the information is difficult to convey to managers. These shortcomings can be overcome by interactive computer software (Dahl & Johnson, 2004).

The PERLA system is assigned to the site specific and stressor non-specific approaches in principle. Nevertheless, the HOBENT software enables predictions of stressor specific indices like the saprobic index or ASPT; the EQR's of these metrics then also enable stressor specific assessment.

Interrelations with international research projects

Some parts of the PERLA system have interrelations with the STAR project (a research project supported by the European Commission under the Fifth Framework Programme, Contract No: EVK1-CT 2001-00089). The sampling method, sample processing, and assessment of predictive modelling results by HOBENT software were intercalibrated with AQEM-STAR methods within the project. The data of the PERLA system are partially shared by the STAR database.

Conclusion

The PERLA system is a complex of biological methods of ecological status assessment of running waters and interrelated activities in the Czech Republic. It involves (i) a network of reference sites, (ii) a database of reference sites involving both respective biotic and abiotic data, (iii) a prediction model using HOBENT software and iv) TRITON assessment software interrelated to the SALAMANDER information system. The most important tool of this system is the HOBENT software (Kokeš, 2002), which includes the prediction model comparing reference and observed status. The model generally follows the published mathematical principles of RIVPACS (Clarke et al., 1996). Due to this fact, the PERLA system is assigned to the site specific and stressor non-specific approaches in principle. The TRITON software (Jarkovský et al., 2003) represents a multivariate approach (multivariate comparisons based on Gower metric) which is alternative to the HOBENT software and interrelated to SALAMANDER – an information database system developed for the Agricultural Water Management Authority (AWMA). This organisation manages small-sized watercourses; both TRITON and SALAMANDER are restricted to these types of streams.

Methodical support is an inseparable part of the system – an instructional handbook was written (Kokeš & Vojtíšková, 1999); identification courses of benthic macroinvertebrates are regularly organised by Masaryk University and the Water Research Institute; a training course of the sampling method was organised for hydrobiologists participating in the monitoring programmes.

The practical application of the PERLA system started in 2001. Large streams were evaluated by WRI (about 20 sites per year in spring season) (Bernardová et al., 2003) and by AWMA (more than 300 sites a year in spring and autumn season).

The prediction model of the PERLA system enables a more sophisticated evaluation of an observed site than the assessment systems used in the Czech Republic in the past. It cannot be regarded as a universal means sufficient for ecological quality assessment, but as one of the tools which can help to fulfill the demands of the Water Framework Directive.

Acknowledgements

The Council of Government of the Czech Republic for Research and Development (VaV 510/2/96 and VaV 510/7/99), the Ministry of the Environment of the Czech Republic, and the Ministry of

Agriculture of the Czech Republic funded projects in which the PERLA system were developed. The study was supported also by the project of the Ministry of Education, Youth and Sports MSM 0021622416 and Grant Agency of the Academy of Sciences of the Czech Republic No. 1QS500070505. We thank Prof. František Kubíček, Prof. Rudolf Rozkošný, Prof. Jan Knoz, Jana Schenková, Michal Horsák, Jana Holasová, Bohumil Losos, Jan Špaček, Aleš Mergl, Petr Komzák for identifying of benthic macroinvertebrates. We thank Eva Strašáková and Marcela Růžičková for technical aid. We would like to thank our departed colleague Michal Fiala, initiator of the project.

References

Anonymous, 1997. SPSS© 7.5 Statistical Algorithms. SPSS inc.: 541 pp.

AQEM consortium, 2002. Manual for the application of the AQEM method. A comprehensive method to assess European streams using benthic macroinvertebrates, developed for the purpose of the Water Framework Directive. Version 1.0, February 2002.

Armitage, P. D., D. Moss, J. F. Wright & M. T. Furse, 1983. The performance of a new biological water quality score system based on macroinvertebrates over a wide range of unpolluted running-water sites. Water Research 17(3): 333–347.

Bernardová, I., K. Mrázek & M. Fiala, 1996. Water quality monitoring, modelling and protection projects in the Czech Republic. In Ganoulis, J. (ed.), Transboundary Water Resources Management: Technical and Institutional Issues. NATO ASI series. 2. Environment. vol. VII, Springer, Berlin, 150–158.

Bernardová, I., S. Zahrádková, J. Kokeš, J. Zahrádka & M. Rozkošný, 2003. Analýza hodnocení ekologického stavu řek Dyje a Bečvy. [An analysis of ecological state assessment of the Dyje and Bečva Rivers]. In Bitušík, P. & M. Novikmec (eds), Proc 13th Conference of Slovak Limnological Society and Czech Limnological Society, Banská Štiavnica, June 2003. Acta Facultatis Ecologiae, 10, Suppl. 1: 135–139.

Clarke, R. T., M. T. Furse, J. F. Wright & D. Moss, 1996. Derivation of a biological quality index for river sites: comparison of the observed with expected fauna. Journal of Applied Statistics 23(2&3): 311–332.

Culek, M., 1996. Biogeografické členění České republiky. [Biogeographical classification of the Czech Republic]. Enigma, Praha, 347 pp.

Dahl, J. & R. K. Johnson, 2004. Detection of ecological change in stream macroinvertebrate assemblages using single metric, multimetric and multivariate approaches. In Dahl, J., Detection of Human-Induced Stress in Streams Comparison of bioassessment approaches using macroinvertebrates. Doctoral dissertation. Acta Universitatis Agriculturae Sueciae, Silvestria 332, part III, 22 pp.

Davy-Bowker, J., R.T. Clarke, R.K. Johnson, J. Kokeš, J.F. Murphy, & S. Zahrádková, 2006. A comparison of the European Water Framework Directive physical typology and RIVPACS-type models as alternative methods of establishing reference conditions for benthic macroinvertebrates. Hydrobiologia 566: 91–105.

Deichsel, C. D. & H. J. Trampisch, 1985. Clusteranalyse und Diskrinimanzanalyse. Gustav Fischer Verlag, Stuttgart, New York, 135 pp.

Demek, J., 1987. Zeměpisný lexikon ČSR. Hory a nížiny. [Geographical lexicon of Czech Republic. Mountains and lowlands]. Academia, Praha, 584 pp.

European Commission, 2000. Directive 2000/60/EC. Establishing a framework for community action in the field of water policy. European Commission PE-CONS 3639/1/100 Rev 1, Luxembourg.

Furse, M.T., D. Moss, J. F.Wright, P. D. Armitage, & R. J. M Gunn, 1986. A practical manual to the classification and prediction of macroinvertebrate communities in running water in Great Britain. Freshwater Biological Association, River laboratory, 147 pp.

Hellawell, J. M., 1986. Biological Indicators of Freshwater Pollution and Environmental Management. Elsevier Applied Sciences Publishers, London and New York, 546 pp.

Hill, M. O., 1979. TWINSPAN – A FORTRAN program for arranging multivariate in an ordered two-way table by classification of the individuals and attributes. Cornell University Press, Ithaca, New York, 99 pp.

Illies, J. (ed.), 1978. Limnofauna Europaea. Gustav Fischer Verlag Stuttgart, 637 pp.

Jarkovský, J., L. Dušek, P. Pavliš, J. Hodovský, S. Zahrádková, P. Kukleta & R. Šmíd, 2003. Vícerozměrná pravděpodobnostní typologie říčních lokalit na základě biotických a abiotických dat – návrh uživatelsky přístupného řešení. [Multivariate classification of stream sites based on abiotic and biotic data – suggestion of a robust solution]. In Bitušík P. & M. Novikmec (eds), Proc. 13th Conference of Slovak Limnological Society and Czech Limnological Society, Banská Štiavnica, June 2003. Acta Facultatis Ecologiae, 10, Suppl. 1: 157–160.

Klecka,, W. R., 1980. Discriminant analysis. Series: Quantitative Applications in the Social Sciences. Sage University paper, 19: 70 pp.

Kokeš, J., 2002. Predikční modely říčních ekosystémů. [Predictions models of river ecosystems]. Final report of the grant No. 510/7/99 of the Council of the Government of the Czech Republic for Research and Development, T.G.M. Water Research Institute Prague, 85 pp.

Kokeš, J. & D. Vojtíšková, 1999. Nové metody hodnocení makrozoobentosu tekoucích vod. [New Methods of Running Waters Evaluation Using Benthic Macroinvertebrates.] Výzkum pro praxi, 39. T.G.M. Water Research Institute, Prague, 84 pp.

Landa, V. & T. Soldán, 1989. Rozšíření jepic v ČSSR a jeho změny v souvislosti se změnami kvality vody v povodí Labe. [The distribution of the order Ephemeroptera in Czechoslovakia with respect to water quality]. Studie ČSAV, 17, Academia, Praha: 172 pp.

Moog, O. (ed.), 1995. Fauna Aquatica Austriaca. Katalog zur autökologischen Einstufung aquatischer Organismen Österreichs. Wien: 206 pp.

Nijboer, R. C. & P. F. M. Verdonschot, 2000. Taxonomic adjustment affects data analysis: often forgotten error. Verh. Int. Ver. Limnol. 27: 1–4.

Schmedtje, U., 1998. Die Ökologische Bewertung von Fließgevässern in Bayern. In Integrierte Ökologische Gewässerbewertung: Inhälte unf Möglichkeiten. Oldenburg Verlag, 199–215.

Sokal, R. R. & J. F. Rohlf, 1995. Biometry. Freeman and Company, New York, 887 pp.

Soldán, T., S. Zahrádková, J. Helešic, L. Dušek & V. Landa, 1998. Distributional and quantitative patterns of Ephemeroptera and Plecoptera in the Czech Republic: a possibility of detection of long-term environmental changes of aquatic biotopes. Folia Fac. Sci. Nat. Masaryk. Brun., Biol. 98: 305.

Strahler, A. N., 1952. Hypsometric (area-altitude) analysis of erosional topography. Bulletin of the Geological Society of America 63: 1117–1142.

Braak, C. J. F. ter, 1986. Canonical correspondence analysis: a new eigenvector technique for multivariate direct gradient analysis. Ecology 67: 1167–1179.

WRI, 1993. Směrný vodohospodářský plán ČR [Master water management plan of the CR]. Vodohospodářský věstník 1992. MŽP ČR, Publ. SVP č. 40, Praha, 157 pp.

WRI, 2002. Směrný vodohospodářský plán ČR [Master water management plan of the CR]. Vodohospodářský věstník 2001. MŽP ČR, Publ. SVP č. 51, Praha, 171 pp.

Wright, J. F., P. D. Armitage & M. T. Furse, 1989. Prediction of invertebrate communities using stream measurements. Regulated Rivers, Research, Managements vol IV: 147–155.

Wright, J. F., M. T. Furse & P. D. Armitage, 1993. RIVPACS – a technique for evaluating the biological quality of rivers in the U.K. European Water Pollution Control 3(4): 15–25.

Wright, J. F., 1995. Development and use of a system for predicting the macroinvertebrate fauna in flowing waters. Australian Journal of Ecology 20: 181–197.

Zahrádková, S., K. Brabec, J. Kokeš, D. Němejcová, T. Soldán, J. Jarkovský, P. Pařil & O. Hájek. Abiotic stream types and species assemblages: is there any simple linkage? Czech streams and benthic macroinvertebrates as an example. International Association of Theoretical and Applied Limnology, Proc. 29th Congress, Lahti, 2004 (in press).

Intercalibration and Comparison

Hydrobiologia (2006) 566:357–364
M.T. Furse, D. Hering, K. Brabec, A. Buffagni, L. Sandin & P.F.M. Verdonschot (eds), The Ecological Status of European Rivers: Evaluation and Intercalibration of Assessment Methods
DOI 10.1007/s10750-006-0084-5

Intercalibration and comparison – major results and conclusions from the STAR project

Andrea Buffagni[1,*] & Mike Furse[2]
[1]*CNR – IRSA, Water Research Institute, Via della Mornera, 25, 20047 Brugherio, (MI), Italy*
[2]*Centre for Ecology & Hydrology, Winfrith Technology Centre, DT2 8ZD Dorchester, Dorset, UK*
*(*Author for correspondence: E-mail: buffagni@irsa.cnr.it)*

Key words: comparison, intercalibration, macroinvertebrates, macrophytes, sampling methods, Water Framework Directive

Abstract

The main results of the STAR project on the intercalibration of boundaries of European assessment systems and comparison between assessment methods are summarized here. The main findings are outlined in the context of the Water Framework Directive that requires reliable instructions to be set up on how to use and harmonise assessment systems and methods for European rivers. The main papers published on these subjects by STAR partners are reviewed, with focus on major questions addressed and approaches used for investigation. The need for broad coverage of geographic ranges and pressure gradients, together with the goal of providing outcomes appropriate to the effective application of the WFD are emphasized. Extensive datasets from a wide range of countries, stream types and sites and a large number of methods, metrics and approaches are compared and tested and various cross-cutting themes emerged. Among these, the value of the use of benchmarking systems for comparison and intercalibration is highlighted. Two ways of looking for comparability of assessment systems results were analyzed: a) by adopting identical sampling techniques across Europe and b) by harmonizing the classification results of the national assessment systems. In addition, the need, in the intercalibration process, for a proper definition of the criteria for reference conditions is underlined. This is because their imprecision now represents one of the major weaknesses of the whole intercalibration process. Direct and indirect approaches to intercalibration are considered and commented on for their potential use in distinct circumstances. Finally, the use common metrics for the intercalibration process, which make comparability across Europe valid, is tested and indeed encouraged.

Introduction

One of the major aims of the Water Framework Directive – WFD (European Commission, 2000) is to assess the ecological quality of European surface waters in a way that is comparable across Europe. This objective is co-ordinated through a Common Implementation Strategy (Heiskanen et al., 2004), a key feature of which is an intercalibration (IC) process (European Commission, 2003). This requires Member States of the European Union to perform a comparison of the classification results of their assessment systems. The intercalibration process is achieved through a series of Geographical Intercalibration Groups (GIGS), each of which is responsible for a set of river types in defined geographic areas.

An important aspect to be taken into account during this process is the comparison of sampling methods used by different countries. This would provide information on their inherent variability (Clarke & Hering, 2006), aid a more transparent interpretation of monitoring data and check the suitability of different methods in dissimilar geographical contexts. However, in order to achieve intercalibration, it is far easier and faster to harmonize the classification results of the national assessment systems than the assessment systems themselves. Indeed, while the WFD defines the general principles that have to be taken into account when assessing ecological status, it leaves Member States the flexibility to outline the specific details of their own assessment system. That is why the purpose of the European intercalibration exercise is not to harmonize assessment systems and methods, but only their results.

Research undertaken by STAR, and summarized here, has informed and supported the Common Implementation Strategy process and provided the GIGS with practical mechanisms to perform the actual intercalibration (Buffagni et al., 2005).

Results

Background and practical examples

Detailed accounts of the research on intercalibration undertaken by the STAR project consortium and practical examples of its application are provided by Buffagni et al. (2005, in press).

Buffagni et al. (2005) is a published version of one of the core project deliverables and incorporates most of the analysis performed to support the European Commission and Joint Research Centre in the intercalibration process. The publication specifically concentrates on macroinvertebrates but the principles elaborated have general applicability to the other Biological Quality Elements (BQEs). The background and history of the intercalibration process up to March 2005 provide the context in which practical methodologies for intercalibration are developed and tested by STAR using data from eleven European Member States.

The central tool developed and advocated by Buffagni and colleagues is the use of ICMi (Intercalibration Common Metric index) approach. The ICMi used for exemplary purposes by Buffagni et al. (2005) is a multimetric that comprises six metrics used in one or more European countries for bioassessment. In accordance with the requirements of the WFD, the six metrics are divided into three categories representing community tolerance, abundance and richness and diversity. The weighted contributions of each of the three categories are set as equal but the specific weightings of the six individual metrics vary according to their perceived importance in defining the ecological status of the site. The use of Intercalibration Common Metrics for the IC process is described, a pilot application of some of the possible IC options to a relatively large range of data from different GIGs and river types is presented and the delineation of many possible alternatives, including the so-called 'hybrid options' for the intercalibration process are compared and evaluated (Buffagni et al., 2005).

A more detailed example of one of the possible harmonization options, which includes the use of an ICMi in conjunction with data obtained from both the national bioassessment programme and an external benchmarking dataset, was presented for a sample stream type by Buffagni et al. (in press). In this example the national data were obtained from the Italian Environment Protection Agency using the Indice Biotico Esteso (IBE) sampling method (Ghetti, 1997) whilst the benchmark data were obtained from the AQEM (Hering et al., 2004) and STAR projects. Buffagni et al. (in press) compared and harmonized the existing boundaries for the standard Italian IBE assessment protocol, a method not yet adapted to the WFD requirements, with the classification of STAR and AQEM project data. By means of the calculation of the ICM metrics and index, the invertebrate communities obtained from nearly 400 samples collected in lowland rivers in Northern Italy were compared to those obtained in a wider geographical context (i.e., across Europe). In the intercalibration process, the IBE method class boundaries were re-allocated until no more statistically significant differences were observed between the Italian 'test' samples and the European 'benchmarking' ones. A major benefit of the paper is the practical demonstration it provides of the applicability of the ICMi approach for intercalibration purposes, right down to the final harmonization phase It offers a detailed, step by step exposition of the procedure and illustrates the changes in the attribution of samples to the five ecological quality classes after each boundary adjustment The procedure represents a crucial practical option for the sensible implementation of the WFD implementation process across Europe.

The aim of both papers (Buffagni et al., 2005, in press) was to provide examples of very pragmatic environmental research that involved key stakeholders and was compatible with the scientific research requirements of the WFD (e.g., Mostert, 2003).

Synopsis of papers in 2006

Issues addressed

One of the central objectives of the STAR project (Furse et al., 2006) was: 'How can data from different assessment methods and taxonomic groups be compared and intercalibrated and how can the results of the STAR programme be used to assist the WFD intercalibration exercise?'

The key questions addressed by the papers presented in 2006 of *Hydrobiologia* on comparison of methods (Friberg et al., 2006) and the intercalibration of results (Birk & Hering, 2006; Birk et al., 2006; Buffagni et al., 2006) therefore included:

(1) Are the invertebrate samples collected around Europe with different field methods comparable in terms of the invertebrate metrics used for the assessment of the ecological status of running water?
(2) Can any of the methods in use be improved in order to increase their suitability for meeting the aims of the WFD?
(3) Is the approach of using common metrics (e.g., ICMs) for the intercalibration process applicable to data collected in rivers from different geographical areas so that they can be compared easily and with respect of the WFD requirements?
(4) Can the comparison and harmonization of European class boundaries benefit from the use of external, international benchmarking?
(5) How large are the differences between the class boundaries we should expect for invertebrate and macrophytes methods?
(6) Is the direct comparison of existing methods scientifically reliable and feasible for the intercalibration process, at least in homogeneous European areas and when methods are similar in their structure and application?
(7) Can direct and indirect comparison be simultaneously utilized to enhance data interpretation before the final harmonization phase?

Comparison

The WFD is not prescriptive about the sampling and assessment methods to be used for evaluating the ecological status of surface water bodies. This will inevitably lead to many Member States retaining the own traditional methods of bioassessment. In order to compare results obtained by the different national protocols used to sample macroinvertebrate assemblages of running waters, Friberg et al. (2006) compared the results obtained by each method with the results obtained using a common method, STAR-AQEM, simultaneously in each country. The study covered a wide geographical range, including more than twenty stream types from eleven European ecoregions and compared eight different sampling protocols with the STAR-AQEM method. Comparisons included the sampling process and the results obtained from the sampling, as expressed in terms of the values of 12 widely applied metrics. The metrics were subdivided into four categories: structural (abundance) ($n=1$) structural (sensitivity/insensitivity) ($n=4$), structural (diversity) ($n=3$) and functional ($n=4$).

Comparison of sampling protocols showed them to have many features in common, particularly the use of hand-net sampling, a priori habitat assessment, multi-habitat sampling and dead sorting. However, there were exceptions in each case and only the STAR-AQEM procedure incorporated sub-sampling as an integral component of the method. There was greater variation between methods in the use of field or laboratory sorting and the area or duration of sampling. Comparison of metric values indicated that there were no consistent patterns of differences between values obtained using STAR-AQEM and national methods. The Normalized Sampling Effort (NSE) was quantified but offered poor support for the interpretation of between-method variability in metric values. However, differences between metric values could be related to some aspects of the field and laboratory methods being compared. Despite the observed differences, the values of the most of the twelve metrics analyzed correlated significantly and positively with each other when samples collected by the STAR-AQEM and the various national methods were compared. The study indicated that some of the assessments resulting from existing national methods render them relatively easy to intercalibrate. Nonetheless, most methods could be improved in several aspects and their revision should be an ongoing process.

Intercalibration

Buffagni et al. (2006), compared and harmonized class boundaries of three European assessment systems based on macroinvertebrates by means of the ICM index approach of Buffagni et al. (2005, in press). The topics of WFD-compliancy and Best Available Classification (Buffagni et al., 2005) were also briefly considered. In their analyses, ICMi values were calculated for test datasets from a single stream type in three European countries. Three different approaches to comparison were used, however only one was considered useful for the harmonization of boundaries (Buffagni et al., 2006). ICMi values were also calculated for samples included in a benchmark dataset assumed to be strictly WFD-compliant. The ICMi values for the ecological status classes Good and High for the test and benchmark datasets were then statistically compared and, when significant differences were observed in the harmonization phase, the boundaries of the national method were refined until no further differences were observed in the allocation of national sites to ecological status classes of both the national and benchmark assessment systems. In general, only minor or no refinements of the boundaries between High/Good and Good/Moderate classes were needed to remove the differences from the benchmark dataset. The suitability of ICMi for the harmonization phase is discussed and the use of external, benchmarking datasets is recommended in order to make the European intercalibration process more transparent and objective.

Birk & Hering (2006), in a different but complementary approach to Buffagni et al. (2006), focussed on a direct comparison of existing river quality assessment methods for two European stream types. Based on benthic macroinvertebrate data, national class boundaries of eight countries were compared. Comparisons of national methods were based on two common scales; (1) the national method showing the highest mean correlation of all indices and (2) the 'Integrative Multimetric Index for Intercalibration' (IMI-IC), an artificial index defined as the mean of all index values calculated with a sample. Comparisons of methods were achieved via a range of regression analyses. In general, simple nonlinear models provided higher coefficients of determination. Using this approach, the authors demonstrated that the Good quality status boundaries of the national methods deviated up to 25%, confirming the need for harmonization of class boundaries. Analyses showed that assessment methods of the same type (Saprobic Indices, BMWP/ASPT scores) showed best correlation results. Birk & Hering (2006) therefore recommended using the intercalibration approach described in their paper only for comparison of methods addressing similar components of the biocoenosis. The authors further recommend that knowledge of the relationships between assessment methods and abiotic pressure gradients should be integral to the process of boundary comparison. Birk & Hering (2006) concurred with the conclusion of Buffagni et al. (2006) that some form of 'benchmarking' system can help overcome differences in definitions of the reference state in different countries. Finally, the authors recognize that intercalibration is of political interest since the definition of quality boundaries sets the environmental standard to be achieved and implies agreement on a level of anthropogenic degradation acceptable for our freshwater systems.

Birk et al. (2006) applied some of the concepts proposed by Buffagni et al. (2005, 2006) and Birk & Hering (2006) to the comparison of class boundaries of European macrophyte methods. Four assessment methods were compared for one stream type in Central Europe and two alternative procedures were applied to compare the Good quality class boundaries. Both the direct comparison and the 'common metrics' approach were used. In direct comparisons, coefficients of determination between the methods tested were generally weaker than for the macroinvertebrates. The authors ascribed this in part to differences between countries/methods in the definitions of reference conditions and partly to the different aspects of ecological quality information that the methods have been designed to summarize.

In applying the common metrics approach of Buffagni et al. (2005), 70 metrics were tested by Birk et al. (2006) for their suitability to act as Intercalibration Common Metrics. These metrics cover the categories 'richness and diversity', 'composition and abundance', 'sensitivity and tolerance' and 'ecosystem function'. Of these, only one, Ellenberg et al. (1992) correlated significantly with all four assessment methods and thus represented a potential common metric for intercalibration. The difficulty in applying intercalibration approaches commonly used for invertebrates to macrophyte assessment methods was one of the major findings of the paper. Consequently, the authors advocated further research to produce suitable common macrophyte assessment metrics. They also note the paucity of 'reference conditions' data and surmised that further attention to the concept of reference condition – and related aspects – is crucial to the European intercalibration process.

Conclusions

General remarks

A general limitation to the validity of the results of any approach to test assessment methods for intercalibration purposes is the number of samples available. Where these are few, it is possible that even where differences exist they will not be revealed by statistical comparisons. Furthermore, an important requirement for a successful application of most intercalibration procedures is the availability of datasets covering the whole degradation gradient (Buffagni et al., 2005). Non-continuity of quality conditions within the data-set can cause problems in the interpretation of, for example, the regression model to test the outputs from different quality classification options. Thus, it was one of the more stringent guidelines of the STAR project to try to cover a geographic and pressure gradient as wide as possible and to provide research outcomes appropriate to the effective application for the WFD implementation strategy. The four papers in the 'Intercalibration and Comparison' section of this volume are therefore based on extensive data from a wide range of countries, stream types and sites and commensurately large numbers of methods, metrics and approaches are compared and tested (Table 1). Collectively, a number of cross-cutting themes and conclusions emerge from these papers and others emanating from STAR.

One such theme is the value to comparison and intercalibration of the use of benchmarking systems to standardize the interpretation of results. Standardization of the assessment of the ecological status of European running waters is intrinsic to the STAR project (Furse et al., 2006). Standardization can be achieved by both the adoption of identical sampling techniques across Europe and by the harmonization of the classification results of the national assessment systems. Thus, Friberg et al. (2006) compared invertebrate samples collected with national method to a standard, trans-national sampling method, the STAR-AQEM protocol first developed in the AQEM project (Hering et al., 2003) and refined for use during STAR. Similarly project members working on the intercalibration of class boundaries of assessment systems often looked for 'common metrics' to make comparison easier (Buffagni & Erba, 2004; Buffagni et al., 2005, 2006; Birk & Hering, 2006; Birk et al., 2006). Recommendations for the use of trans-national benchmarking datasets for intercalibration purposes is another recurrent theme of many of these papers (Buffagni et al., 2005, 2006, in press; Birk & Hering, 2006).

A third recurrent theme is the need, by the intercalibration process, for a proper definition of reference conditions for rivers, encompassing each of the BQEs used. Validation of the reference state through analysis of the biota is fundamental to ascribing the site to a natural or nearly natural condition. Birk & Hering (2006) and Buffagni et al. (2005) suggested that the establishment of biological reference conditions could be further refined by the combined analysis of pressures and biological data. Currently, the shortage of reference biological data is a significant restriction to achieving full European intercalibration. This is compounded by not all countries having established reference condition criteria for sites in their country. For example, the use of samples from the best available sites – as opposed to 'true' reference sites – can lead to inconsistent values for the High status class be-

Table 1. Overview of the BQEs, geographic gradient, number of sites/samples and methods/metrics compared in the four papers of this section

Main issue addressed	Paper	BQE	Number of ecoregions	Number of countries	Number of stream types	Number of sites	Number of samples	Number of approaches	Number of national methods	Number of metrics
Comparison of national sampling methods against the STAR-AQEM method	Friberg et al.	MI	11	11	22	300*	600*	1	9	12
Comparison and intercalibration of class boundaries of assessment systems via common metrics	Buffagni et al.	MI	3	3	1	1105	1587	3	3	12
Direct comparison of class boundaries of assessment systems	Birk & Hering	MI	4	8	2	196	466	1	5	16
Comparison of class boundaries of assessment systems direct and via common metrics	Birk et al.	MPHY	1	6	1	108	108	2	4	70

MI: macroinvertebrates; MPHY: macrophytes, (*circa).

tween countries. Where this applies, the correct intercalibration of datasets and methods may be strongly compromised. The extent of knowledge of true reference conditions also varies between the BQEs and these elements may themselves vary in their response to particular stresses and their ability to detect them (Johnson et al., 2006a, b). This further complicates the intercalibration process and, in the absence of adequate information on more than one element, a monitoring programme targeted at a single element's response to degradation may be preferable strategy (Buffagni et al., 2005).

The metrics included in the analysis of the papers presented in this section were usually selected in order to cover some of the WFD requirements and be applicable over a wide geographical range. Consequently, because the availability of comprehensive taxonomic keys and expertise varies from country to country, an issue identified as needing resolution prior to intercalibration is the level of identification required for each of the contributory BQEs. Friberg et al. (2006) and Buffagni et al. (2005, 2006), in analyses of macroinvertebrate data, resolved this by applying family level identification. Each concluded that this level of identification adequately met the objectives of their analyses. This approach is necessarily a pragmatic one and elsewhere Verdonschot (2006) clearly identified the consequences of information loss when macroinvertebrate data are analyzed at family rather than species level.

Approaches to compare class boundaries for the Intercalibration process

In the papers summarized here, comparison of European class boundaries and assessment systems led to different results for different stream types and options used, but showed how systems and boundaries can be actually compared in the short term.

The direct comparison approach i.e., not using 'common metrics' can easily be used to demonstrate apparent discrepancies between the class boundaries of the assessment systems of different Member States. Observed differences can be due to the fact that the existing methods have different sampling strategies and laboratory procedures (Friberg et al., 2006) and/or are based on different concepts (Birk et al., 2006). Intrinsic differences between, for example, the taxonomic richness of similar stream types in different ecoregions may even lead to disparity between their apparent ecological status when, in reality, their 'true' quality is the same. Despite this, the direct comparison approach has considerable potential for use in European IC harmonization exercise, especially when the systems being compared are quite similar (Birk & Hering, 2006), such as for the bilateral, fine tuning of class boundaries (see also Buffagni et al., 2005). A prerequisite for the successful application of the direct comparison approach to intercalibration is that large numbers of samples are available, each collected according to the requirements of the national systems being harmonized. Any options of averaging the values of class boundaries of Member States' assessment methods (e.g., IMI-IC in Birk & Hering, 2006) are only applicable when all of the incorporated biological methods can be demonstrated to be fully WFD-compliant.

Buffagni & Erba (2004) and Buffagni et al., (2005, 2006) therefore propose the use of common metrics, i.e., the ICMi approach incorporating the use of reference conditions and data normalization for each of the datasets under comparison, in order to achieve pan-European harmonization of assessment systems. As already seen for the direct comparison approach, the option of averaging (or using the median) of, for example, the ICMi values of the class boundaries of Member States' assessment methods, is applicable when all the biological methods are WFD-compliant (Buffagni et al., 2006). However, comparison and harmonization using a benchmark dataset (see also Birk & Hering, 2006, for discussion) handles the problem of not having fully WFD-compliant systems presently available for all Member States. By using an entirely external benchmarking system, the STAR-AQEM WFD-compliant dataset, Buffagni et al. (2005, 2006, in press) showed that the ICMi can be used to harmonize class boundaries within and between GIGs, achieving a full comparability and lack of ambiguity in results. Furthermore, Buffagni et al. (2005, in press) concluded that if the comparison of the tested datasets with the benchmark datasets does not show significant differences, then it indicates that the tested method can be considered to provisionally fulfil WFD requirements for the establishment of class boundaries for ecological status allocation.

The research summarized in this paper has therefore provided the Common Implementation Strategy and the Geographic Intercalibration Groups with the practical mechanisms that are needed to achieve intercalibration of European assessment systems. This approach allows for intrinsic differences in national sampling protocols (Friberg et al., 2006) and even for national systems that may be regarded as not fully WFD-compliant. The use of common metrics and the ICMi approach is now an integral part of the activities of European harmonization process (European Commission, 2005).

References

Birk, S. & D. Hering, 2006. Direct comparison of assessment methods using benthic macroinvertebrates: a contribution to the EU Water Framework Directive intercalibration exercise. Hydrobiologia 566: 401–415.

Birk, S., T. Korte & D. Hering, 2006. Intercalibration of assessment methods for macrophytes in lowland streams: direct comparison and analysis of common metrics. Hydrobiologia 566: 417–430 .

Buffagni, A. & S. Erba, 2004. A simple procedure to harmonize class boundaries of European assessment systems. Discussion paper for the intercalibration process – WFD CIS WG 2.A ECOSTAT, 6 February 2004, 21 pp.

Buffagni, A., S. Erba, S. Birk, M. Cazzola, C. Feld, T. Ofenböck, J. Murray-Bligh, M. T. Furse, R. Clarke, D. Hering, H. Soszka & W. van de Bund, 2005. Towards European Inter-calibration for the Water Framework Directive: Procedures and Examples for Different River Types from the E.C. Project STAR. Quaderni Istituto di Ricerca sulle Acque, Roma 123, Rome (Italy), IRSA, 468 pp.

Buffagni, A., S. Erba, M. Cazzola, J. Murray-Bligh, H. Soszka & P. Genoni, 2006. The STAR common metrics approach to the WFD intercalibration process: Full application for small, lowland rivers in three European countries. Hydrobiologia 566: 379–399.

Buffagni, A., S. Erba & M. T. Furse, (in press). A simple procedure to harmonize class boundaries of assessment systems at the pan-European scale. Environmental Science and Policy.

Clarke, R. T. & D. Hering, 2006. Errors and uncertainty in bioassessment methods – major results and conclusions from the STAR project and their application using STARBUGS. Hydrobiologia 566: 433–439.

Ellenberg, H., H. E. Weber, R. Düll, V. Wirth, W. Werner & D. Paulißen, 1992. Indicator Values of Plants in Central Europe. Erich Goltze, Göttingen.

European Commission, 2000. Directive 2000/60/EC of the European Parliament and of the Council of 23 October 2000 establishing a framework for Community action in the field of water policy. Official Journal of the European Communities L 327, 22.12.2000, 72 pp.

European Commission, 2003. Common Implementation Strategy for the Water Framework Directive (2000/60/EC). Guidance document no. 6. Towards a guidance on establishment of the intercalibration network and the process on the intercalibration exercise. Produced by Working Group 2.5 – Intercalibration, 54 pp.

European Commission, 2005. Common Implementation Strategy for the Water Framework Directive (2000/60/EC) – Guidance Document No. 14. Guidance on the Intercalibration Process 2004–2006, 26 pp.

Friberg, N., L. Sandin, M. T. Furse, S. E. Larsen, R. T. Clarke & P. Haase, 2006. Comparison of macroinvertebrate sampling methods in Europe. Hydrobiologia 566: 365–378.

Furse, M., D. Hering, O. Moog, P. Verdonschot, R. K. Johnson, K. Brabec, K. Gritzalis, A. Buffagni, P. Pinto, N. Friberg, J. Murray-Bligh, J. Kokes, R. Alber, P. Usseglio-Polatera, P. Haase, R. Sweeting, B. Bis, K. Szoszkiewicz, H. Soszka, G. Springe, F. Sporka & I. Krno, 2006. The STAR project: context, objectives and approaches. Hydrobiologia 566: 3–29.

Ghetti, P. E., 1997. Indice Biotico Esteso (IBE). I macroinvertebrati nel controllo della qualità degli ambienti di acque correnti. Provincia Autonoma di Trento, 222 pp.

Heiskanen, A.-S., W. van de Bund, A. C. Cardoso & P. Nõges, 2004. Towards good ecological status of surface waters in Europe – interpretation and harmonisation of the concept. Water Science & Technology 49: 169–177.

Hering, D., A. Buffagni, O. Moog, L. Sandin, M. Sommerhäuser, I. Stubauer, C. Feld, R. K. Johnson, P. Pinto, N. Skoulikidis, P. F. M. Verdonschot & S. Zahrádková, 2003. The development of a system to assess the ecological quality of streams based on macroinvertebrates – design of the sampling programme within the AQEM project. Internationale Revue der gesamten Hydrobiologie 88: 345–361.

Hering, D., O. Moog, L. Sandin & P. F. M. Verdonschot, 2004. Overview and application of the AQEM assessment system. In Hering, D., P. F. M. Verdonschot, O. Moog & L. Sandin (eds), Integrated Assessment of Running Waters in Europe. Kluwer Academic Publishers, Printed in the Netherlands, Hydrobiologia 516: 1–20.

Johnson, R. K., D. Hering, M. T. Furse & R. T. Clarke, 2006a. Detection of ecological change using multiple organism groups: metrics and uncertainty. Hydrobiologia 566: 115–137.

Johnson, R. K., D. Hering, M. T. Furse & P. F. M. Verdonschot, 2006b. Indicators of ecological change: comparison of the early response of four organism groups to stress gradients. Hydrobiologia 566: 139–152.

Mostert, E., 2003. The European Water Framework Directive and water management research. Physics and Chemistry of the Earth 28: 523–527.

Verdonschot, P. F. M., 2006. Data composition and taxonomic resolution in macroinvertebrate stream typology. Hydrobiologia 566: 59–74.

Hydrobiologia (2006) 566:365–378

M.T. Furse, D. Hering, K. Brabec, A. Buffagni, L. Sandin & P.F.M. Verdonschot (eds), The Ecological Status of European Rivers: Evaluation and Intercalibration of Assessment Methods
DOI 10.1007/s10750-006-0083-6

Comparison of macroinvertebrate sampling methods in Europe

Nikolai Friberg[1,*], Leonard Sandin[2], Mike T. Furse[3], Søren E. Larsen[1], Ralph T. Clarke[3] & Peter Haase[4]

[1]*Department of Freshwater Ecology, National Environmental Research Institute, Vejlsøvej 25, DK-8600 Silkeborg, Denmark*

[2]*Department of Environmental Assessment, Swedish University of Agricultural Sciences, P.O. Box 7050, SE-750 07 Uppsala, Sweden*

[3]*Centre for Ecology and Hydrology, CEH Dorset, Winfrith Technology Centre, Winfrith Newburgh, DT2 8ZD, Dorchester Dorset, UK*

[4]*Research Institute and Natural History Museum Senckenberg, Lochmuehle 2, 63599, Biebergemuend, Germany*
*(*Author for correspondence: E-mail: nfr@dmu.du)*

Key words: macroinvertebrates, metrics, bioassessment, Water Framework Directive, STAR

Abstract

The aim of this study was to describe in detail the national macroinvertebrate sampling methods used and to compare them with a common standard, the STAR-AQEM sampling method. Information on national methods and field data were collected from 11 countries (Austria, Czech Republic, Denmark, France, Germany, Greece, Italy, Latvia, Portugal, Sweden, and UK). The sampling included 22 stream types situated in 11 different Ecoregions. Within each country samples were taken in spring and one additional season (summer or autumn) using both the national method and the STAR-AQEM method. A single anthropogenic stressor was also defined for each stream type sampled within the project, with the three main stressor types being organic pollution (including eutrophication), toxic pollution and habitat degradation. In addition, not impacted reference sites were sampled in each country. A common set of metrics was calculated and compared between the methods. The majority of national methods employed had many features in common. Most of the 12 metrics analysed using the values derived from the STAR-AQEM method and the various national methods correlated significantly, and positively to each other. There was no clear pattern with respect to the differences between metric results obtained using STAR-AQEM and national methods. For some metrics, number of EPT-taxa and families, the value obtained was higher when using the majority of national methods when compared to the STAR-AQEM method. Variability in metric results between methods could not be explained from differences in sampling effort. Sorting in the field and sub-sampling appeared to affect e.g., number of taxa found negatively. The results of the present study supports that inter-calibration in Europe can be undertaken using samples collected with the existing national methods.

Introduction

Macroinvertebrates are the most frequently used organism group in biomonitoring of streams and rivers worldwide (e.g., Metcalfe-Smith, 1996). Currently more than 50 different approaches for biomonitoring using macroinvertebrates exist (De Pauw & Vanhooren, 1983; Metcalfe-Smith, 1996) and most countries in Europe have national and/or regional monitoring programmes that use macroinvertebrates (Birk & Hering, 2003). In most cases each country has developed individual sampling methodology and assessment systems. Among these are RIVPACS (Wright et al., 2000) in the UK, IBGN (AFNOR, 1982) in France and BBI in Belgium (De Pauw & Vanhooren, 1983). As the majority of methods have evolved from the same ancestors, the Saprobic index and the Trent index (Metcalfe-Smith, 1996), and use a hand net for sampling in accordance with CEN Standard EN 27 828 they should have many features in common. However, until now there has been no direct inter-comparison of the performance of various national sampling methods at a European scale. Johnson et al. (2001) found in an inter-country comparison that national sampling methods in the Nordic countries yielded very similar results.

The EU Water Framework Directive (Directive 2000/60/EC – Establishing a Framework for Community Action in the Field of Water Policy) defines a framework for assessing water bodies including streams and rivers. One of the indicator groups to be used in WFD monitoring of stream and rivers are macroinvertebrates. Intercalibration of the various methods used is essential in providing a consistent picture of the ecological quality within the EU. If existing methods cannot be intercalibrated within an acceptable range of precision, a common standard on sampling methodology should be developed. This is one of the key questions addressed by the STAR project (www.eu-star.at) and the main focus of the present study which was based on data collected from 11 countries (Austria, Czech Republic, Denmark, France, Germany, Greece, Italy, Latvia, Portugal, Sweden, and UK). The sampling included 22 stream types, where five were defined as being of the STAR project type 'Core stream type 1' (mid altitude, 200–500 m.a.s.l., and with a 'small' catchment area 10–100 km^2), seven were of the STAR project type 'Core stream type 2' (lowland, <200 m.a.s.l., and 'medium' catchment areas 100–1000 km^2), whereas ten other stream types were defined as STAR project type 'Additional stream type' (having a different characterisation). These stream types are situated in 11 Ecoregions according to Illies definition (Illies, 1978; as used in the Water Framework Directive), these were regions 3, 4, 6, 7, 8, 9, 10, 14, 15, 16 and 18.

The aim of this study was to describe in detail the national sampling methods used and to compare them against a common standard, the STAR-AQEM sampling method. We hypothesise that sampling effort and subsequent sampling treatment will have a clear impact on the final assessment result. Our additional aim was therefore to elucidate what components of the various methods affected their overall performance.

Material and methods

Sampling strategy

Within each country one or several STAR project stream types were sampled (see Introduction). Macroinvertebrate samples were taken in two different sampling seasons (all partners sampled in spring and one additional season [summer or autumn]). A single anthropogenic stressor was also defined for each stream type sampled within the project, with the two main stressor types being organic pollution (including eutrophication) and habitat degradation (Furse et al., 2006). For each stream type in each country a pre-defined number of sites were selected to cover all ecological classes from high to bad quality (poor quality when the stressor was habitat degradation).

For all sites investigated it were not always possible to apply both the national and the STAR-AQEM sampling method. However, in the comparison of sampling methods data were only included in the analysis where both sampling methods for macroinvertebrates were used at the same site in the same stream and in the same season. The number of samples used for these comparisons therefore differed between types, seasons, and methods used. The analysis undertaken in the present paper combines season and stream types.

Taxonomic adjustment

Each country has adjusted all of its own taxonomical data, so that there are no biases within each country's dataset caused by differences in taxonomic resolution used (e.g., between sampling seasons, where during some seasons it might be more difficult to identify certain taxa because they are in early instars). The taxonomic adjustments were made using common rules within the project.

Comparison of methods

When samples were obtained using a hand net, the area sampled cannot be completely fixed. However, as the sampling effort should be similar as long as the sampling protocol is followed, the number of individuals obtained should be directly comparable among samples. In addition, the area sampled can be roughly estimated from the area disturbed in front of net multiplied with net width. With respect to the RIVPACS and the PERLA method it is assumed that sampling distance is 1 m per 20 seconds and the area was calculated by multiplying total sampling length (=sampling time/20 s) with net width. Using this approximation the area sampled can be compared among methods. To further enable an inter-method comparison, a 'normalised sampling effort' (NSE) was calculated for each method using sampled area and mesh size. The STAR-AQEM was used as base line and was set to give a NSE value of 1. Consequently, NSE was calculated using the following formula:

$$NSE = (samplearea/1.25m^2)/(meshsize/0.5mm)$$

as the STAR-AQEM method samples an approximate area of 1.25 m^2 with a 0.5 mm mesh hand net. The NSE is dimensionless. If the method included a pick sample it was not used in the estimation of NSE.

To further allow an inter-comparison of methods used, a handling-processing score was calculated divided into a field and laboratory component. The score is subjective and based on giving the value 1 to each of the handling-processing steps which are considered by the authors to be positive for overall assessment quality (0 if negative), i.e. a high score indicates a high quality method (8 is maximum). In the field, the score 1 is given if field sorting is not undertaken, if no species are removed (1) and if no excess material is removed (1). In the laboratory, the score 1 is given if no live sorting is undertaken (1), if no subsampling is undertaken (1), if sorting is done using magnification (1), if all individuals are enumerated (1) and if identification is done to the species level (1). A sampling method was judged as enumerating all individuals either if it was an actual total count of all individuals in the sample or putting them into abundance classes. In the latter case the actual number will often be based on estimation. Identification to the species level means that all taxa are identified to the best attainable level and that the subsequent index calculation, to which the sampling method was developed, is at least partly based on species information. Project partners supplied all information on the national sampling methods for each country to ensure the most up-dated information. More details on the various methods can be found in Deliverable 8 of the STAR project, which is published on the STAR homepage (www.eu-star.at).

Metrics used

A group of metrics was selected which was generally applicable and covers various types of stress (e.g., Metcalfe-Smith, 1996; AQEM manual, 2002; Birk & Hering, 2003). The metrics vary in intrinsic properties as to which features of the macroinvertebrate community they respond to, i.e. structural (incl. sensitivity) or functional properties (Table 1). Metric values were calculated from species data using the various national methods and the STAR-AQEM method. This allows for a direct comparison of the national method with the STAR-AQEM method for each country individually.

Statistical analysis

The 12 metrics were calculated from samples obtained using the various national methodologies and the STAR-AQEM method. Only main samples were used (as opposed to replicate samples, where a second sample was taken in some streams to estimate sample variability) so that each site was represented by one sample per season. The national method and the STAR-AQEM method was tested using pair-wise comparisons for each country individually. This was accomplished by performing a Students t-test, or a non-parametric Sign test (Sokal & Rohlf, 1995) if the differences in metric values between the STAR-AQEM and national method for a given site and season were not normally distributed. Furthermore the correlation between the STAR-AQEM and national method was investigated by Spearman's rank correlation. For metrics with high correlations, the functional relationship between the STAR-AQEM and national method was investigated and estimated. For a number of selected metrics we plotted their dependence on NSE and the handling-processing score and tested for significant

Table 1. Common metrics used for the comparison of national methods and the STAR-AQEM method

Metric	Type
Saprobic Index (Zelinka & Marvan, 1961)	Structural (sensitivity)
Abundance	Structural – total number of individuals
ASPT (Armitage et al., 1983)	Structural (sensitivity)
Shannon–Wiener index (Shannon & Weaver, 1949)	Structural (diversity)
EPT-taxa	Structural (sensitivity) – total number of taxa belonging to Ephemeroptera, Plecoptera and Trichoptera
No. of taxa	Structural (diversity) – total number of taxa
No. of families	Structural (diversity) – total number of families
Oligochaeta [%]	Structural (insensitivity) – percentage of Oligochaeta in the sample
RETI (Schweder, 1992)	Functional
% Grazers	Functional – percentage of the individuals belonging to functional feeding group grazers in the sample
% Gatherers	Functional – percentage of the individuals belonging to functional feeding group gatherers in the sample
% Shredders	Functional – percentage of the individuals belonging to functional feeding group shredders in the sample

differences using one-way ANOVA followed by a t-test (pair-wise comparisons). Box and whisker plots were used for plotting NSE and the handling-processing score versus selected metrics.

Results

Comparison of sampling methods

The majority of sampling methods employed by the different countries have many features in common (Table 2). The majority of methods involve an *a priori* assessment of habitats at the sampling site, exceptions being the RIVPACS method and the DSFI method. In RIVPACS, habitats are sampled in proportion to their occurrence, which is subjectively assessed by the surveyor while sampling. DSFI uses a fixed sampling grid that should cover most habitats without introducing a sampling bias due to variability in how surveyors assess the number of habitats present. All methods, except the Swedish method, use a multi-habitat sampling approach. In contrast, it is the only method, which takes replicate samples to assess inter-sample variability. Most methods use standard hand nets with a width of 25 cm and mesh bag with a 500 μm mesh size in accordance with the CEN standard EN 27 828. The samples are therefore semi-quantitative. A Surber sampler can be used when employing the STAR-AQEM method, while it is obligatory when using the French IBGN protocol with the exception of sampling in lentic areas. Mesh sizes used vary between 475 and 1000 μm. Three of the methods (RIVPACS, DSFI and PERLA) include a pick sample of attached macroinvertebrates.

The smallest area sampled is 0.4 m^2 (IBGN) and the largest is 2.25 m^2 (RIVPACS and PERLA). NSEs ranged from 0.32 (IBGN) to 1.8 (PERLA). Three methods used field sorting of the whole sample (IBE, PERLA and the Latvian method), four collected some species for further identification in the field (STAR-AQEM, IBE, PERLA and the Latvian method) and excess material was removed using most methods. Only when using DSFI, IBGN and PMP is removal of excess material in the field not allowed.

Field sorting is only standard when applying the Italian IBE protocol and the Latvian method (Table 2). In addition, if samples are brought back to the laboratory only the IBE method has live sorting as standard. When using RIVPACS, Portuguese PMP and the Latvian sampling method live sorting is optional, but dead sorting is recommended. All other methods rely on the sorting of dead material. Only the STAR-AQEM method allows sub-sampling of the entire sample. With regard to sorting under magnification, enumeration of all individuals collected and identification to the best attainable taxonomic level, the methods investigated are highly variable. Enumeration of all individuals and identification to the best attainable level increase the biological information in the sample and hence potentially the quality of the assessment. The handling-processing score ranges between 1 (IBE) and 7 (the Swedish method) with most methods obtaining scores of either 4 or 5.

Correlation between STAR-AQEM and National methods

The majority of the 12 metrics analysed using values derived from the STAR-AQEM method and the various national methods correlated significantly and positively to each other (Table 3). Only a few correlations were negative. However, despite being significant a substantial number of correlations had coefficients below 0.7. Overall, number of EPT-taxa was the metric that was most highly correlated when compared among countries. Also the RETI index was highly correlated in most countries. The metric with the overall weakest correlation in an inter-country comparison was abundance. Especially four countries exhibited strong correlations between their national method and the STAR-AQEM method. These were the Czech Republic, Germany, Sweden and the UK. In contrast, especially Italy, but also Denmark and Portugal, had many weak correlations, although some lack of significance can be explained from the low number of sites in these countries. Strong correlations do not necessarily mean that methods will provide identical results. However, they show that results from the different method can be compared.

Table 2. Comparison of the methods used (Y = yes, N = no, V = variable). Calculation of NSE and handling-processing score are explained in the text

		STAR-AQEM	RIVPACS	Nordic methods		The French (IBGN) sampling method	The Italian (IBE) sampling method	The Czech (PERLA) sampling method	The Portugeuse (PMP) sampling method	The Latvian sampling method
				The Swedish method	The Danish (DSFI) method					
Strategy	*A priori* habitat assessment	Y	N	Y	N	Y	Y	Y	Y	Y
	Multi-habitat/number of habitats	Y/V	Y/V	N/1	Y/V	Y/8	Y/V	Y/V	Y/V	Y/V
	No. of samples/replicates	20/none	1/none	5/5	12/none	8/none	1/none	1/none	10/none	20/none
Effort	Sampling device	Hand net or Surber	Hand net	Hand net	Hand net	Surber	Hand net	Hand net	Hand net	Hand net
	Width of sampling device (m)	0.25	0.25	0.25	0.25	0.25	0.25	0.25	0.25	0.205
	Mesh size (μm)	500	1000	500	500	500	475	500	500	1000
	Kicking technique (area sample^{-1}, time used, distance sampled)	Fixed area (0.0625 m^2)	Time (3 min)	Time/distance (1 m 1 min^{-1})	Fixed area (0.1 m^2)	Fixed area (0.05 m^2)	Variable	Time (3 min)	Fixed area 1 m sampled (0.205 m^2)	Fixed area 1 m sampled (0.205 m^2)
	Pick sampling (effective sampling time)	N	Y (1 min)	N	Y (5 min)	N	N	N	N	Y
	Area covered (m^2)	1.25	2.25	1.25	1.20	0.4	$\approx 0.9^d$	2.25	2.05	4.1
	Sampling effort (NSE)	1	0.9	1	0.96	0.32	0.77	1.8	1.64	1.64

Field	Field sorting	N	N	N	N	N	Y	N	N	Y
	Some species collected from sample in the field	Y	N[a]	N	N	N	Y	Y	N	Y
	Excess material removed	Y	Y	Y	N	N	Y	Y	N	Y
Labora-tory	Live sorting	N	N[b]	N	N	N	Y	N	N[b]	N[b]
	Sub-sampling	Y	N	N	N	N	N	N	N[c]	N
	Use of magnification sorting	N	N	Y	N	Y	N	N	N	N
	Enumeration of all individuals	Y	Y	Y	N	N	N	Y	Y	N
	Identification to species level	Y	N	Y	N	N	N	Y	N	N
	Handling/processing score	4	5	7	5	6	1	5	6	2

[a]Except rare species which are released into the stream or river again, in the case where it is possible to identify them properly in the field and no biomass data is necessary.
[b]Dead sorting is recommended, but live sorting is optional.
[c]Only the fine fraction (>0.5 mm and <1 mm) can be subsampled.
[d]As the number of kick samples varies, the area is set using expert judgement (CNR-IRSA).

Table 3. Correlation matrix between the STAR-AQEM method and the respective national methods

	Abundance	Number of taxa	Saprobic	ASPT	Shannon – Wiener	Grazers (%)	Shredders (%)	Gatherers (%)	RETI	Oligochaeta (%)	EPT – taxa	Number of families
Austria	0.34	0.80***	0.62**	0.63**	0.84***	0.87***	0.80***	0.71**	0.85***	0.69**	0.74***	0.73***
	0.66**	0.71***	0.84***	0.51*	0.70**	0.71***	0.79***	0.72***	0.76***	0.29	0.77***	0.65**
Czech Rebublic	0.63**	0.86***	0.96***	0.89***	0.73***	0.71***	0.93***	0.85***	0.82***	0.80***	0.94***	0.87***
	0.39	0.90***	0.83***	0.90***	0.60**	0.65**	0.94***	0.72***	0.89***	0.87***	0.94***	0.87***
Denmark	0.67*	0.71*	0.82**	0.91***	0.14	0.55	0.28	0.72*	0.70*	0.49	0.72*	0.85**
	0.17	0.85**	0.68*	0.95***	0.76**	0.73*	0.61*	0.87***	0.87***	0.54	0.95***	0.90***
France	0.76*	0.70*	Not possible	0.83**	0.83**	0.87**	0.85**	0.86**	0.83**	0.59	0.85**	0.44
	0.13	0.51	Not possible	0.78**	0.82**	0.81**	0.80**	0.55	0.84**	0.73*	0.83**	0.57
Germany	0.66***	0.51*	0.93***	0.78***	0.68***	0.92***	0.83***	0.76***	0.70***	0.45*	0.86***	0.43*
	0.77***	0.66**	0.89***	0.74***	0.82***	0.87***	0.82***	0.60**	0.72***	0.19	0.65**	0.53*
Greece	0.05	0.93***	Not possible	0.61	0.78*	0.61	0.59	0.77*	0.66*	0.76*	0.79*	0.92***
	0.13	0.60***	Not possible	0.70***	0.79***	0.84***	0.34*	0.67***	0.47*	−0.18	0.64***	0.58***
Italy	0.10	0.49	0.52	0.39	0.56	0.18	0.50	0.35	0.38	No data	0.58	0.35
	−0.26	0.37	0.55	0.73*	0.22	0.63*	0.52	0.67*	0.37	No data	0.65*	0.33
Latvia	0.30	0.62**	0.64**	0.20	0.40	0.49*	0.65**	0.55*	0.73***	0.28	0.46*	0.63**
	0.49*	0.66**	0.79***	0.73***	0.38	0.37	0.85***	0.59**	0.63**	0.58**	0.25	0.69***
Portugal	0.37	−0.18	0.88**	0.75*	0.59	0.30	0.88**	0.88**	0.09	0.67*	0.45	-0.09
	0.54	0.82*	0.84**	0.63	0.72*	0.55	0.92***	0.66	0.76*	0.55	0.56	0.86**
Sweden	0.50*	0.80***	0.85***	0.84***	0.85***	0.82***	0.72***	0.70***	0.82***	0.59**	0.83***	0.74***
	0.67***	0.64***	0.93***	0.91***	0.51*	0.85***	0.68***	0.78***	0.81***	0.53**	0.83***	0.69***
UK	0.82***	0.77***	0.80***	0.88***	0.77***	0.89***	0.83***	0.77***	0.77***	0.62**	0.95***	0.88***
	0.60**	0.90***	0.93***	0.92***	0.66***	0.58**	0.85***	0.64**	0.64**	0.39	0.95***	0.89***

In the top panel for each country are correlations calculated on spring samples and in the lower panel correlations are calculated using summer/autumn samples. Significant correlations are denoted: $*p<0.05$; $**p<0.005$: $***p<0.0005$. Note that the number of samples varies among countries and between seasons. Therefore, similar correlation coefficients might not have the same p-value.

Table 4. Significant differences between the AQEM-STAR method (S) and the respective national methods (N)

	Abundance	Number of taxa	Saprobic index	ASPT	Shannon–Wiener	%Grazers	%Shredders	%Gatherers	RETI	%Oligochaeta	EPT – taxa	Number of families
Austria					S > N*	S > N*	S < N**				S < N*	S < N*
Czech	S < N*	S < N*	S > N*		S > N**			S > N*		S > N***	S < N***	S < N*
Denmark	S > N***	S < N***									S < N***	S < N***
France	S < N***		S < N***		S > N***						S < N**	S < N***
Germany	S > N*										S < N**	S < N*
Greece		S > N***	N/A									S > N***
Italy	S > N***	S > N***					S < N*			N/A	S > N**	S > N***
Latvia	S > N***	S > N*			S < N***			S > N***		S > N***		S > N***
Portugal		S < N**									S < N*	S < N*
Sweden		S < N***				S < N*	S < N*				S < N*	S < N***
UK	S > N***		S > N*		S > N**	S < N*	S < N***	S > N***	S < N***	S > N***		S < N*

Analyses are based on all sites and seasons within each country. $^{*}p < 0.05$; $^{**}p < 0.005$; $^{***}p < 0.0005$.

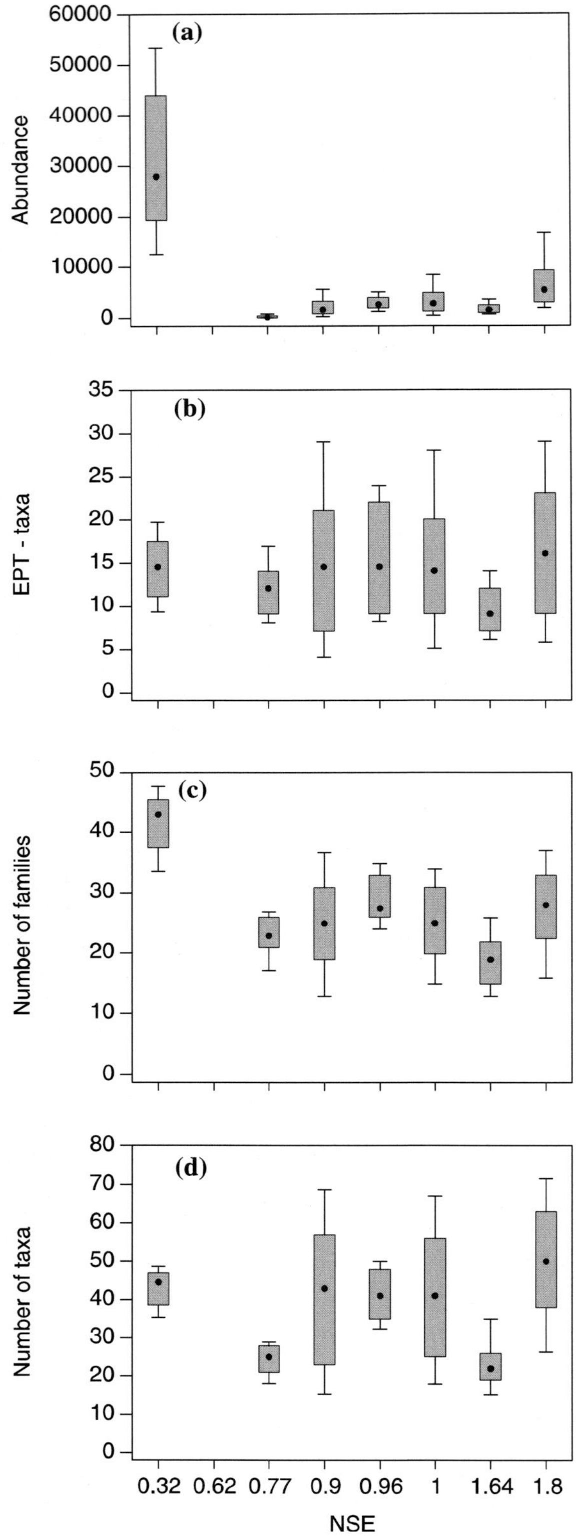

Figure 1. Relationship between normalised sampling effort (NSE) and abundance (a), EPT-taxa (b), number of families (c) and number of taxa (d). Median values (circle), 75th and 25th percentile (top and bottom edge of box, respectively) and 90th and 10th percentile (top and bottom of error bars, respectively) are shown.

Comparison of AQEM-STAR and national methods

No overall clear pattern emerged with respect to the differences between metric results obtained using STAR-AQEM and national methods (Table 4). Within countries, there was, in most cases, not a consistent pattern when comparing metrics: some metrics would score higher when calculated using data obtained by the national method while other would score lower than the STAR-AQEM method.

In most cases (64% of the countries) the various national methods yielded significantly higher EPT-taxa values than the STAR-AQEM method. A similar pattern was evident with respect to number of families. In 73% of the countries significantly more families were found using the national method. In contrast, the STAR-AQEM method yielded significantly more EPT-taxa and families in 9 and 27% of the countries, respectively.

The STAR-AQEM method yielded in general higher values (e.g. more taxa) than the national methods in Italy and Latvia when the methods were significantly different whereas the opposite was the case in Sweden and Portugal where the national method consistently yielded higher metric values than the STAR-AQEM method. In Denmark and Germany, significantly more individuals were found when employing the STAR-AQEM method whereas the opposite was true with respect to number of EPT-taxa and families.

Several countries used the RIVPACS method as their national method (Austria, Germany, Greece and UK; Table 2). In addition, the Czech PERLA system is very closely related to the RIVPACS method (Table 2). Overall, there were no clearly consistent results among these countries.

Inter-country comparison of metric performance

There was no relationship between abundance of macroinvertebrates in samples and NSE (Fig. 1a).

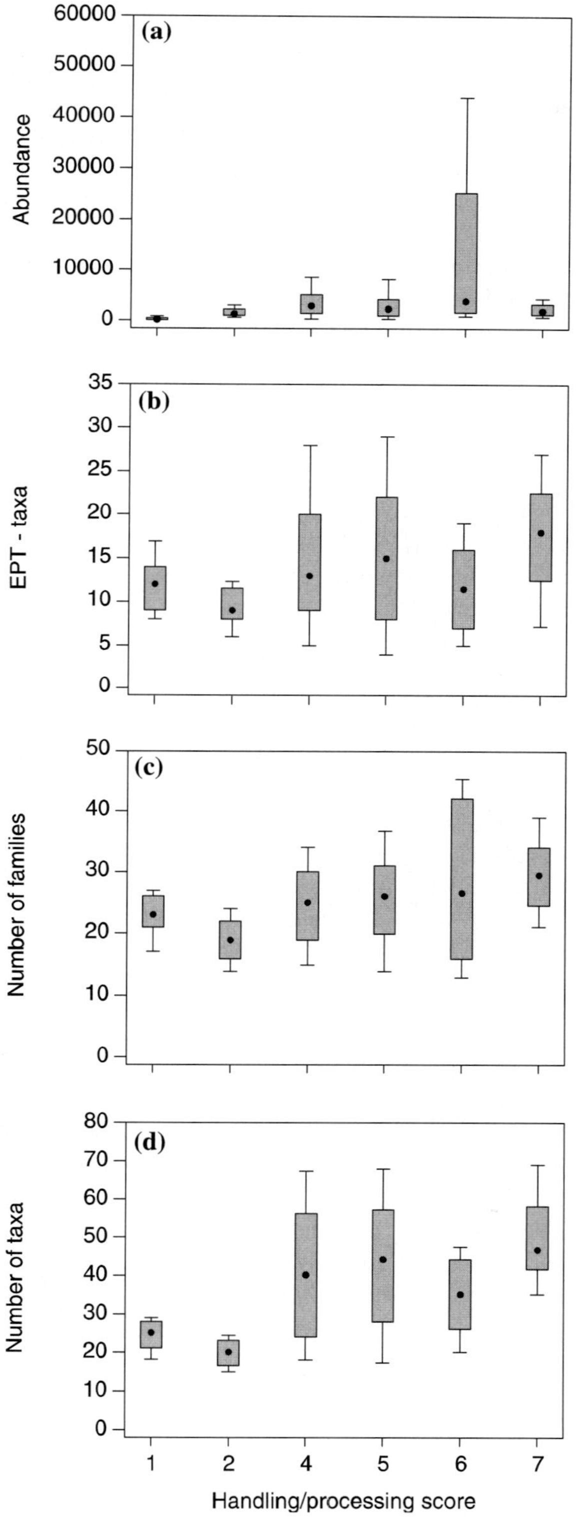

Figure 2. Relationship between handling/processing score and abundance (a), EPT-taxa (b), number of families (c) and number of taxa (d). Median values (circle), 75th and 25th percentile (top and bottom edge of box, respectively) and 90th and 10th percentile (top and bottom of error bars, respectively) are shown.

The French methods IBGN had a significant higher number of individuals than all methods and at the same time the lowest NSE ($p < 0.0001$). If the IBGN method is omitted from the data set there is a tendency for an increase in number of individuals caught with increasing NSE. There was no clear relationship between the number of EPT-taxa and NSE (Fig. 1b). The IBGN method caught a similar number of EPT-taxa as the other methods despite the low NSE. The number of families found was not related to NSE (Fig. 1c). As with abundance, the method with the smallest sampling area and NSE caught significantly the largest number of families (the IBGN method, $p < 0.0001$). The number of taxa was, as for the other metrics tested, not related to NSE (Fig. 1d). There was a high degree of variability, which appears to be method specific and cannot be explained from single variables as NSE.

Abundance was lower in samples with a handling/processing score of 1 (the IBE method) whereas abundance varied independently of the score in the range 4–7 (Fig. 2a), except for the IBGN method catching much more individuals than the other methods with a handling score of 6. There was a tendency that the number of EPT-taxa found increased with increasing handling/processing score, indicating that these taxa are lost during sample treatment (Fig. 2b). There was no effect of the handling/processing score on the number of families found (Fig. 2c) whereas there was significantly fewer taxa found when scores were 1 and 2 compared with scores 4–7 (Fig. 2d, $p < 0.0001$).

Discussion

The national methods compared in the present study had many features in common. All methods, except the French IBGN method, used a hand net and in most cases the mesh size was 500 μm in accordance with the CEN standard EN 27 828. It is therefore not surprising that the various methods yielded comparable results. That different sampling methods will provide almost identical results have previously been demonstrated in the Nordic countries (Johnson et al., 2001). They found that sampling methods from four countries (Denmark, Finland, Norway and Sweden) yielded very similar results when sampling was done in one perturbed and one unperturbed stream in south-central Sweden. Another study also shows similar results between RIVPACS and STAR-AQEM (Haase et al. 2004a). The STAR-AQEM method appeared to collect more individuals and taxa than the national methods in Italy and Latvia. This could reflect the very low handling-processing score obtained for both countries compared with the STAR-AQEM method as well as the other national methods. With respect to Latvia, a further explanation could be that a number of taxa are not considered in the national method, and consequently they will not appear in the taxa list. In Sweden and Portugal, the national method yielded consistently more taxa, EPT-taxa and families than the STAR-AQEM method. This could relate to the use of subsampling in the STAR-AQEM methodology, which might reduce the number of taxa found. In the case of Sweden, the higher number of taxa (all and EPT) and families might reflect that the sampling effort is concentrated in riffles which are the most species rich in stream ecosystems (e.g., Brown & Brussock, 2001). In Denmark and Germany, significantly more individuals were found when employing the STAR-AQEM method whereas the opposite was true with respect to number of EPT-taxa and families. Again, this might reflect that taxa are lost when subsampling the large STAR-AQEM sample.

Several countries used the RIVPACS method as their national method (Austria, Germany, Greece and UK; Table 2). In addition, the Czech PERLA system is very closely related to the RIVPACS method. Overall, there were no clearly consistent results among these countries. As the differences cannot be attributed to protocol itself, they might reflect the way samples were taken in the individual countries. It has previously been shown that sampling potentially is a major source of variation when employing the RIVPACS techniques (Clark, 2000; Dines & Murray-Bligh, 2000). In the STAR project, a workshop was undertaken prior to the start of the sampling programme in which the various methods, including the RIVPACS methodology, were demonstrated in order to reduce sampling variability among countries. This might not have been sufficient in reducing the variability as our results indicate that differences among countries in how sampling is undertaken are as important as the intrinsic differences in the methods employed.

Handling in the field and processing of samples in the laboratory will affect the quality of the assessment result. Field sorting, collection of some species from the sample in the field and removal of excess material can all potentially reduce sample quality by the loss of species (Haase et al., 2004b). Field handling is extremely dependent on the surveyors' abilities and is affected by weather conditions, time pressure etc. However, removal of fragile or endangered species can be necessary in certain cases and any negative impacts on sample quality should be reduced through training of the surveyors (e.g., Dines & Murray-Bligh, 2000). Obligatory live sorting is likely to affect quality negatively as it introduces a time constraint on the sorting procedure. Even though this is not obvious from the NSE value of the STAR-AQEM method, it collects large amounts of inorganic material, organic debris and plants, which makes sub-sampling necessary. Sub-sampling can potentially reduce the number of species found and hence affect sample quality negatively and increase sampling variance (Vinson & Hawkins, 1996). Sorting under magnification increases the likelihood of finding all species present in the sample, even the smaller specimens.

In conclusion, the STAR-AQEM method appears to collect fewer taxa (all and EPT) and families than the majority of the national methods. The most likely explanation of this finding is that species are lost during the sub-sampling procedure employed by the STAR-AQEM method. However, an additional explanation could be that the STAR-AQEM was developed to take samples habitat proportional, ignoring rare habitats which might contain additional species (AQEM, 2002). The advantage of this approach is that it limits sampling variability by reducing the subjective element introduced by the surveyor and that it is likely to be more sensitive towards hydromorphological degradation. Therefore, the lower number of taxa found in the STAR-AQEM samples than in the national methods might to some degree reflect a higher sensitivity to hydromorphological degradation. Two methods, the Italian IBE method and the Latvian method, appear to lose information about the macroinvertebrate community to a degree that might affect the assessment of ecological stream quality. Laboratory processing (IBE and Latvian method) and identification of more species (Latvian method) would probably improve their performance. Despite these differences it is difficult to estimate the effects of different methods on assessment results. Differences in single metrics might be covered by a multi-metric approach. In Germany, for example, the assessment results (multi-metric system with scores between 0 and 1) are highly correlated (Spearman $R = 0.92$) and their differences very small (mean difference −0.01), when comparing STAR-AQEM and RIVPACS (Haase et al., 2004a).

The results of the present study are promising as it clearly indicates that existing national sampling methods can be relatively easily intercalibrated as they are in general based on similar principles. It consequently supports that intercalibration among European countries at presently is undertaken by calculating a common set of metrics on species lists collected by the individual countries using their national methodology (Buffagni et al. 2005). However, it is important to keep in mind that many aspects of sampling methodology, such as sensitivity to different stressors and stability in time and space, was not covered by the present analysis. Many of the national methods were developed to detect organic pollution, focusing on the collection of indicator species that might occur in rare habitats. However, pressures on stream ecosystems change in time and consequently also the stressors acting on the biota. As examples, hydromorphology and introduction of exotic species are increasingly important and will affect assessment results using existing methods/systems in an inconsistent manner (e.g., Olsen & Friberg, 1999; Gabriels et al., 2005). Consequently, most methods should be improved in several aspects and the revision and improvement of methods should be an ongoing process.

Acknowledgement

STAR was partially funded by the European Commission, 5th Framework Program, Energy, Environment and Sustainable Development, Key Action Water, Contract no. EVK1-CT-2001-00089. The authors thank the whole STAR team for making this paper possible, especially the colleagues who collected the field samples and sorted and identified the taxa.

References

AFNOR, 1982. Essais des euax. Détermination de l'indice biologique global normalisé (IBGN) Association Francaise de Normalisation NF T 90–350 France.

AQEM, 2002. Manual for the application of the AQEM system, version 1.0. www.aqem.de.

Armitage, P. D., D. Moss, J. F. Wright & M. T. Furse, 1983. The performance of a new biological water quality score system based on macroinvertebrates over a wide range of unpolluted running-water sites. Water Research 17: 333–347.

Birk S. & D. Hering, 2003. Waterview Web-Database: a comprehensive review of European assessment methods for rivers. FBA News, 20 (winter 2002): 4.

Brown, A. V. & P. P. Brussock, 2001. Comparisons of benthic invertebrates between riffles and pools. Hydrobiologia 220: 99–108.

Buffagni, A., S. Erba, S. Birk, M. Cazzola, C. K. Feld T. Ofenböck, J. Murray-Bligh, M. T. Furse, R. Clarke, D. Hering, H. Soszka & W. van der Bund, 2005. Torwards European Inter-Calibration for the Water Framework Directive: Procedures and Examples for the Different River Types from the E.C. Project STAR. 11th STAR Deliverable. STAR contract no: EVK1-CT 2001–00089. Quaderni Instituto di Ricerca sulle Acque 123, IRSA, Rome, Italy 460 pp.

Clarke, R., 2000. Uncertainty in estimates of biological quality based on RIVPACS. In Wright, J. F., D. W. Sutcliffe, M. T. Furse (eds), Assessing the Biological Quality of Freshwaters: RIVPACS and Similar Techniques. Freshwater Biological Association, Ambleside, UK, 39–54.

De Pauw, N. & G. Vanhooren, 1983. Method for biological quality assessment of watercourses in Belgium. Hydrobiologia 100: 153–168.

Dines, R. A. & J. A. D. Murray-Bligh, 2000. Quality assurance and RIVPACS. In Wright, J. F., D. W. Sutcliffe & M. T. Furse (eds), Assessing the Biological Quality of Freshwaters: RIVPACS and Similar Techniques. Freshwater Biological Association, Ambleside, UK, 71–78.

Gabriels, W., P. L. M. Goethals & N. De Pauw, 2005. Implications of taxonomic modifications and alien species on biological water quality assessment as exemplified by the Belgian Biotic Index method. Hydrobiologia 542: 137–150.

Haase, P., S. Lohse, S. Pauls, K. Schindehütte, A. Sundermann, P. Rolauffs & D. Hering, 2004a. Assessing streams in Germany with benthic invertebrates: development of a practical standardised protocol for macroinvertebrate sampling and sorting. Limnologica 34: 349–365.

Haase, P., S. Pauls, A. Sundermann & A. Zenker, 2004b. Testing different sorting techniques in macroinvertebrate samples from running waters. Limnologica 34: 366–378.

Johnson, R. K., K. Aagaard, K. J. Aanes, N. Friberg, G. M. Gislason, H. Lax & L. Sandin, 2001. Macroinvertebrates. In Skriver, J. (ed.), Biological Monitoring in Nordic Rivers and Lakes. TemaNord 2001:513, Nordic Council of Ministers, Copenhagen, Denmark: 43–52.

Metcalfe-Smith, J. L., 1996. Biological water-quality assessment of rivers: use of macroinvertebrate communities. In Petts, G. & P. Calow (eds), River Restoration. Blackwell Science, Oxford, UK, 17–59.

Olsen, H. -M. & N. Friberg, 1999. Biological stream assessment in Denmark: The importance of physical factors. In Friberg, N. & J. D. Carl (eds), Proceedings from the Second Meeting of the Nordic Benthological Society, Silkeborg November 1997. National Environmental Research Institute, Denmark.

Schweder, H., 1992. Neue indices für die Bewertung des ökologischen Zustandes von Fliessgewässern, abgeleitet aus der Makroinvertebraten-Ernährungstypologie. Limnologie Aktuell 3: 353–377.

Shannon, C. E. & W. Weaver, 1949. The Mathematical Theory of Communication. The University of Illinois Press, Urbana, IL, U.S.

Sokal, R. R. & J. R. Rohlf, 1995. Biometry. Freeman and Company, New York, U.S.

Vinson, M. R. & C. P. Hawkins, 1996. Effects of sampling area and subsampling procedure on comparisons of taxa richness among streams. Journal of the North American Benthological Society 15: 392–399.

Wright, J. F., D. W. Sutcliffe & M. T. Furse (eds), 2000. Assessing the Biological Quality of Freshwaters: RIVPACS and Similar Techniques. Freshwater Biological Association, Ambleside, UK.

Zelinka, M. & P. Marvan, 1961. Zur Präzisierung der biologischen Klassifikation der Reinheit fließender Gewässer. Archiv Fur Hydrobiologie 57: 389–407.

Hydrobiologia (2006) 566:379–399
M.T. Furse, D. Hering, K. Brabec, A. Buffagni, L. Sandin & P.F.M. Verdonschot (eds), The Ecological Status of European Rivers: Evaluation and Intercalibration of Assessment Methods
DOI 10.1007/s10750-006-0082-7

The STAR common metrics approach to the WFD intercalibration process: Full application for small, lowland rivers in three European countries

Andrea Buffagni[1,*], Stefania Erba[1], Marcello Cazzola[1], John Murray-Bligh[2], Hanja Soszka[3] & Pietro Genoni[4]
[1]*CNR – IRSA, Water Research Institute, Via della Mornera, 25, 20047 Brugherio (MI), Italy*
[2]*Environment Agency, Manley House, Kestrel Way, Exeter EX2 7LQ, UK*
[3]*IOEP, Institute of Environmental Protection, Kolektorska, 401-692 Warszawa, Poland*
[4]*ARPA Lombardia – Regional Environment Protection Agency, Parabiago (MI), Italy*
(**Author for correspondence: Fax: +39-039-2004692; E-mail: buffagni@irsa.cnr.it*)

Key words: intercalibration, harmonization, boundaries, multimetric, invertebrates

Abstract

Class boundaries of three European assessment systems based on macroinvertebrates were compared and harmonized. Three different approaches to comparison, one based on regression analysis and the other two on statistical testing, were described and used, however only one was considered useful for the harmonization of boundaries. In all cases, the calculations were based on a set of six Intercalibration Common Metrics, combined into a simple multimetric index (ICMi). The ICMi was calculated for three test datasets from Italy, Poland and the UK, all belonging to the same stream type (small lowland siliceous sand rivers). For comparison, a regression model was employed to convert national assessment boundary values into ICMi values. The ICMi was also calculated on samples included in a strictly WFD-compliant benchmark dataset. The values of the ICMi obtained for the quality classes Good and High for the test and benchmark datasets were statistically compared. When significant differences were observed in the harmonization phase, the boundaries of the national method were refined until no further differences were observed. For the test datasets and assessment systems of Italy (IBE index) and Poland (Polish BMWP index) small refinements of the boundaries between High/Good and Good/Moderate classes were sufficient to remove the differences from the benchmark dataset. After harmonization, in the studied stream type, the percentage of samples requiring restoration to Good quality increased by 22 and 6% for Italy and Poland, respectively. For the UK dataset (EQI ASPT) the comparison to benchmark dataset showed no significant differences, thus no harmonization was proposed. A general discussion of the options used to compare boundaries based on the ICMi and their potential for harmonization is provided. Lastly, the option of harmonizing class boundaries through comparison to an external, benchmarking dataset and then re-setting them until no differences are found is supported.

Introduction

The Water Framework Directive (WFD) (European Commission, 2000) is intended to achieve a sustainable management of water resources, to attain good ecological quality and to prevent further deterioration of surface and groundwater by ensuring the sustainable functioning of aquatic ecosystems (and dependent wetlands and terrestrial systems). The environmental objectives of the WFD (i.e. the good ecological quality of natural water bodies and the good ecological potential of heavily modified and artificial water bodies) should be reached by 2015. The overall complexity of the Water Framework Directive and a very tight timetable for its implementation creates challenges for the fulfilment of its requirements. Therefore the European Commission and the Member States (MSs) instigated a Common Implementation Strategy (CIS) in 2001. This has resulted in a number of guidance documents, where various technical issues related to the WFD implementation requirements are interpreted according to the common understanding of Member States. They are not legally binding but present examples of best practises and a common understanding of the legal requirements. The general framework of the WFD, the underlying principles and the major steps needed for its implementation have been recently summarized (Heiskanen et al., 2004).

The WFD requires the harmonization of national ecological assessment systems and classifications through an intercalibration exercise (European Commission, 2003a), in order to ensure a uniform interpretation of the 'good ecological quality' of surface waters all over Europe. The aim is consistency and comparability in the classification results of the monitoring systems operated by each Member State (MS) for the biological quality elements. More specifically, the intercalibration exercise must establish values for the boundary between classes of High and Good status and the boundary between Good and Moderate status, consistent with the normative definitions of those class boundaries given in Annex V of the WFD (European Commission, 2004).

A general approach for carrying out such an intercalibration has been proposed by Buffagni & Erba (2004) and described in details in Buffagni et al. (2005). It is based on the concept of Intercalibration Common Metrics (ICMs) and ICM index (ICMi) (Buffagni et al., 2005). An ICM is a biological metric, widely applicable at European or regional scale, that can be used to derive comparable information between different countries/stream types and the ICMi is a simple multi-metric combination of ICMs (see Buffagni et al., 2005). The selection of these metrics for the IC process was performed to fit with WFD definitions. They have been demonstrated to show a high correlation with river quality classifications over a range of stream types (Buffagni et al., 2005). Also, they were chosen on the basis of recent metric selection experiences (e.g. AQEM Consortium, 2002; Buffagni et al., 2004; Hering et al., 2004a; Pinto et al., 2004). The potential application of the metrics over a wide geographical scale was taken into account in order that they can be applied to most datasets in Europe (i.e. to gain comparability).

It is not the aim of the present paper to describe in detail the procedure, widely explained elsewhere (Buffagni et al., 2005). In the present paper the applicability of the methodology based on ICMi for the intercalibration of assessment methods, in terms of the harmonization of their resulting classification (i.e. class boundaries) is demonstrated. By harmonization – used here as a surrogate for intercalibration – we mean: 'The process by which the class boundaries of MS National methods should be adjusted to be consistent with a common trans-National benchmarking' (Buffagni et al., 2005). The two steps of (a) comparison, i.e. to match data from different countries and look for discrepancies in terms of class boundaries and (b) the harmonization of boundaries, where needed on the basis of the studied data, are kept clearly distinct.

The example of harmonization of the national class boundaries presented here is intended to demonstrate the possibility of identifying and eliminating the possible differences arising out of the use of different approaches and methods. The main aims of the present paper are (a) in terms of their overall application, to compare the approach by regression analysis via ICMi to the indirect approach via the use of an external, benchmarking dataset; (b) to provide an example of the full application of the STAR ICMi harmonization procedure across Europe, namely in Italy, Poland and United Kingdom for invertebrates in a common stream type.

Methods

The concept of harmonization

In the present paper, the exploratory harmonization was carried out by 'shifting' boundaries – High/Good (HG) and Good/Moderate (GM) – in order to reduce/eliminate differences among samples grouped into the same quality classes for the methods which are being tested. The basis for the comparison and harmonization of boundaries was the calculation of the ICM index. The procedure adopted involved the comparison of three test datasets to a WFD-compliant, trans-national dataset (benchmark dataset) for which a Best Available Classification (BAC) is provided (i.e. based on STAR/AQEM data, see Buffagni et al., 2005).

Selection and calculation of the Intercalibration Common Metrics

For both comparison and harmonization, a range of metrics were calculated and combined into an Intercalibration Common Metric index (Buffagni & Erba, 2004; Buffagni et al., 2005). The Intercalibration Common Metrics used in the present paper were recently selected for various pilot exercises for European Intercalibration (Buffagni et al., 2005). The metrics used are reported in Table 1 (see also Buffagni et al., 2005). The identification level chosen for the calculation of the metrics is family (Buffagni et al., 2005).

After the normalization of metric values – obtained by dividing each value by the 75th percentile value of the High status class, classified according to the national method (Buffagni et al., 2005), – the metrics were combined into an Intercalibration Common Metric index (ICMi). The individual metrics were clustered into three groups, providing information on three major response areas: Tolerance, Abundance/Habitat and Richness/Diversity (Buffagni et al., 2004). A different weight was attributed to the metrics within each group, giving greater importance to the metrics based on the whole community (Buffagni et al., 2004, 2005). In order to obtain the final multimetric score, the same weight (0.333) was attributed to each of the three metric groups.

The values obtained by ICMi and the national classification methods were converted to EQRs by normalization, dividing them by the 75th percentile value for the high status samples, as previously done for single metrics (Buffagni et al., 2005). Also, the class boundaries were converted from values of the national classification to values of ICMi for comparison with boundaries of other countries' national systems. A regression between EQRs of ICMi and national method values (after normalization) was derived and class boundary values expressed in terms of ICMi were then obtained for all methods/countries.

The ASPT, Shannon, N_families, and EPT taxa were calculated by means of the ASTERICS assessment software (AQEM/STAR Ecological RIver Classification System. The remaining metric values were calculated by using electronic sheet macros.

Comparison, harmonization procedure and statistical testing

Simple linear regression models were used to describe the relationships between the original metric values of the national methods of Poland, Italy and the UK and the ICMi metric values for the same samples. Based on the ICMi approach, a simple option for the harmonization of the class boundaries of assessment systems was to search for common values for the thresholds between classes. This was achieved by calculating regression formulae describing the relationships between the metric values for national methods and the equivalent ICMi values. The starting point for the procedure was to set the HG boundary for each of the national methods/metrics equal to 1. These values were then converted into an ICMi value. Harmonization could thus be based on the selection of, for example, the average ICMi value calculated from a set of national boundary values. The harmonized national value for each of the relevant class boundaries could thus calculated from the ICMi value (e.g. the average between the three countries) by using the related regression formulae (Table 2). Following this approach, countries are recommended to increase or allow to decrease the original boundary values of their assessment systems. If the regression model is statistically robust, this approach does not necessarily imply further statistical testing. A check of original and 'harmonized' boundaries can be done by direct statistical testing the ICMi values obtained for each quality class as defined by the national protocols (test datasets) before and after any boundary adjustments. The ICMi values obtained for each quality class as defined by the national protocols (test datasets) were compared by means of the Mann–Whitney *U* test (Helsel & Hirsch, 1992).

A second option for comparing – and later harmonizing – boundaries, also based on the ICMi approach, is to contrast indirectly the national

Table 1. Metrics included in the ICM index

Metric name	Taxa considered in the metric/info on taxa	Literature reference
ASPT	Whole community	e.g. Armitage et al. (1983); Brabec et al. (2004); Buffagni et al. (2004); Dahl et al. (2004); Logan (2001); Morais et al. (2004); Pinto et al. (2004)
Log_{10} Sel_EPTD	Log_{10} (sum of Heptageniidae, Ephemeridae, Leptophlebiidae, Brachycentridae, Goeridae, Polycentropodidae, Limnephilidae, Odontoceridae, Dolichopodidae, Stratiomyidae, Dixidae, Empididae, Athericidae & Nemouridae + 1)	Based on Buffagni et al. (2004)
1-GOLD	1 − (relative abundance of Gastropoda, Oligochaeta and Diptera)	Pinto et al. (2004)
Total number of families	Sum of all Families present at the site	e.g. Thorne & Williams, (1997)
Number of EPT families	Sum of Ephemeroptera, Plecoptera and Trichoptera taxa	e.g. Barbour et al. (1999); Morais et al. (2004); Thorne & Williams (1997)
Shannon-Wiener diversity	$D_{S-W} = -\sum_{i=1}^{s}\left(\frac{n_i}{A}\right) \cdot \ln\left(\frac{n_i}{A}\right)$	e.g. Böhmer et al. (2004); Hering et al. (2004a); Lorenz et al. (2004); Morais et al. (2004); Shannon & Weaver (1949); Thorne & Williams (1997)

Table 2. Relationship between ICMi and National MS indices: original, normalized and harmonized boundaries (High/Good (HG) and Good/Moderate (GM) classes), conversion formulae, regression coefficient and level of significance

		Original HG boundary	Normalized HG boundary	HG boundary converted in ICMi	Harmonized HG boundary	Original GM boundary	Normalized GM boundary	GM boundary converted in ICMi	Harmonized G/M boundary	Conversion formula, R^2, *p*-level
ITA	IBE	9.6	1	0.883	9.6	7.6	0.792	0.662	8.0	IBE = 0.942*ICMi + 0.178 $R^2 = 0.72$; *p*-level < 0.001
POL	BMWP-POL	100	1	0.876	100	70	0.700	0.591	77	BMWP-POL = 1.051*ICMi + 0.074 $R^2 = 0.74$; *p*-level < 0.001
	Margalef-Ind	5.5	1	0.633	nc	4	0.727	0.381	nc	Marg-Ind = 1.079*ICMi + 0.292 $R^2 = 0.40$; *p*-level < 0.001
UK	EQI ASPT	1	1	0.844	1	0.89	0.890	0.738	0.8	ASPT_EQI = 1.046*ICMi + 0.112 $R^2 = 0.83$; *p*-level < 0.001
	EQI N-taxa	1	1	0.812	nc	0.78	0.78	0.628	nc	EQI N-taxa = 1.193*ICMi + 0.025 $R^2 = 0.72$; *p*-level < 0.001

nc indicates that the values were not calculated.

datasets (and therefore classification methods) with an external, WFD-compliant dataset. In this case the WFD compliancy relates mainly to the criteria for the identification of reference sites. Thus, a statistical comparison was undertaken between the ICMi values of samples in the benchmark dataset and those of the samples in the test dataset. Samples of Good status were considered first. If the ICM index values based on the two classification schemes significantly differed, the class boundary Good/Moderate for the national dataset was moved in order to eliminate the differences. This resulted in samples being removed from the pool of that quality class until the values of the remaining samples were no longer different between datasets. After the adjustment of the national Good/Moderate boundary, so that there were no significant differences between the two classification systems, the value of the national High/Good boundary, and the possible need for its revision was considered using the same statistical comparisons. The new, harmonized boundaries for the national classification system were thus set for High/Good and Good/Moderate classes.

In the two cases where significant differences existed and the test data had values significantly lower than benchmark data, the national Good/Moderate boundary was moved up and successive samples were removed from the Good class (i.e. those with results below the new national boundary), until no more statistically significant differences were observed. Thus, the harmonization option illustrated here involves the re-positioning of the boundary – and consequent sample exclusion from the dataset – until no more differences are found for ICMi values by statistically comparing the two datasets.

The Mann–Whitney *U* test was used for all the above mentioned statistical tests to compare ICMi values calculated for a given STAR/AQEM biological class (benchmark dataset) with the values observed for the corresponding national system classes (test dataset).

Datasets description

The assessment methods used for the samples under consideration were the Italian IBE, the Polish BMWP & Margalef index and the British BMWP system. The data presented in this paper were grouped into (a) three test datasets, containing data collected by Italian, Polish and British environment agencies as part of their national monitoring networks and (b) a benchmark dataset, which incorporates data mainly collected within the EU co-funded AQEM and STAR projects.

Test datasets

The three test datasets comprised samples from the River Intercalibration site type R-C1: small lowland siliceous sand rivers (European Commission, 2003b). Their catchment areas varied within the range 10–100 km2 and the altitude for all the sites was lower than 200 m. The data presented were collected for different purposes e.g. standard monitoring, methodology testing, local impact assessment etc. In all cases the adopted method was the national standard in use in the MS at the time of sampling (Environment Agency, 1997; Genoni, 2003; Genoni et al., 1998; Kownacki et al., 2002).

The streams studied in Southern Europe often had re-sectioned banks and channels and were located in intensively cultivated or urban areas. A total of 361 samples from 39 sites were analyzed for Italy. The main causes of degradation included morphological alteration, organic pollution, and pesticide contamination. The 49 Polish samples (1 per site) were mainly affected by organic pollution/eutrophication. In the UK, the 789 sites yielded 789 combined samples formed by merging the two separate season samples collected at each site. Here, general degradation was considered to be the main cause of degradation. In practice, these sites were affected by multiple stresses, of which organic pollution probably produced the greatest impact.

As far as national classification systems are concerned, in Italy the official method is IBE (Indice Biotico Esteso) (APAT-IRSA/CNR, 2004). The sampling method applied requires the collection of the sample in a representative reach of the river, along a transect in a riffle area. The final index score is obtained via a two-entry table, by comparison of two metrics: the total number of taxa collected and the Faunistic Group (ordered by an increasing scale of tolerance). Values of the index can theoretically vary from 0 to 14. In the studied dataset, from 361 applications, the minimum and maximum observed values are 2.4 and 13.

The Polish method of assessment was based on two components: a BMWP score system, (BMWP-PL), adapted to Polish conditions and a modified form of Margalef's diversity index (Kownacki et al., 2004). The sampling method applied was based on the collection of four quantitative replicates from dominant substrates plus 1 qualitative sample from all habitats. The standard BMWP system (Armitage et al., 1983) was modified (Kownacki et al., 2004), in order to represent the ecological gradient in Polish rivers better. These modifications included (a) verification of usefulness of taxa scored in the original British system in Polish conditions, supplementing the list of families with several taxa not occurring in Great Britain due to zoogeographical isolation but present in Poland and having a role as indicators of water quality; (b) modification of the scores assigned to several taxa (in comparison with the original BMWP). Margalef was modified according to the formula $S/\text{Log}\ N$ (where S is the number of families and N the total abundance). If the quality class of the site obtained from the two Polish assessment metrics differed, the final classification is based on the worst value. The minimum value for BMWP-PL is 0, the maximum is open-ended. In fact, for the data presented here, the Margalef index never determined the worst classification status.

The classification for UK sites is undertaken through the combination of two indices: EQI ASPT and EQI N-taxa. The sampling was done according to the RIVPACS procedure (Wright, 1995; Murray-Bligh, 1999), which required 3 min active sampling in all the habitats present and a minute's search using a hand net. The EQI ASPT (and the EQI N-taxa) corresponds to the observed ASPT (or Number of families) for combined spring and autumn samples, divided by the RIVPACS prediction for the same seasonal combination. Both indices give a classification. The poorest class indicated by either EQI ASPT or EQI N-taxa is the overall quality class for a site. Minimum and maximum values can vary according to the dataset being analysed (Table 2).

For the Italian and UK methods rough abundance classes were recorded. For the IBE calculation, for taxa present with less than 10 individuals a real count was usually undertaken. For more abundant taxa, the field surveyor, indicated the relative abundance of taxa by use of terms 'present', 'abundant' and 'dominant'. In this paper, these classes were converted into numbers, thus present was considered to be 20, abundant was set at 60 and dominant as 180 specimens. Logarithmic abundance classes recorded in the UK method were converted according to the following criteria: for category 1–9, a value of 4 was used; for category 10–99, a value of 40 was used; for category 100–999, a value of 400 was used and for category 1000+, a value of 4000 was used. For the Polish method the actual number of individuals found was recorded.

A summary of the investigated national methods and test datasets can be found in Table 3.

Benchmark dataset

A dataset assembled for the purposes of the WFD (benchmark dataset), including quality classification of sites, was identified (i.e. STAR/AQEM data), which is independent from the national monitoring datasets under examination (Table 4). The relationships between the environmental quality of rivers (e.g. water pollution, habitat degradation, acidification) and the biological response have already been examined for single subsets of data (e.g. Hering et al., 2004b) and for the dataset as a whole (Buffagni et al., 2005), in order to interpret the observed range of metric values and check the proposed ecological classification criteria. For each site, the 'Best Available Classification' (BAC) that fulfils WFD requirements, was provided (Buffagni et al., 2005).

Benchmark data were mainly collected within the AQEM and the STAR projects. A supplement to these data was provided by CEMAGREF (Lyon, France), which demonstrated the feasibility of merging information from non-STAR institutes and countries for the IC process and aims. The sites whose data was provided by France are included in the National monitoring network and regularly investigated for the assessment of river quality. These data were included in the benchmark set because an ecological classification fulfilling WFD requirements was provided, as well as a set of reference sites. The stressor investigated

Table 3. Main features of national methods investigated and tested datasets

Country	Method	Main degradation factor detected	Calculation	Min/Max observed values	#Years/seasons	Number of samples				Quality range
						Total	High status	Good status	Mod. status	
Italy	Indice Biotico Esteso (Extended Biotic Index) – IBE	General degradation (stressor not specified, but mainly organic pollution)	Two entries table (2 metrics: Number of taxa and Faunistic Group)	2.4/13	6 years, 4 seasons	361	84	176	69	High to Bad
Poland	BMWP-Polish version & Margalef index	Organic pollution/general degradation	Worst classification between BMWP-POL and Margalef Diversity Index	BMWP-POL: 5/158 Margalef: 0.7/11.7	1 year, 1 season	49	11	15	12	High to Bad
UK	General Quality Assessment (GQA) scheme EQI ASPT and EQI N-taxa	General degradation (stressor not specified, but mainly organic pollution)	Worst classification between EQI ASPT (observed ASPT/RIVPACS predicted ASPT) and NFAM_EQI	EQI ASPT 0.4/1.16 EQI N-taxa 0.11/1.54	1 year, 2 seasons combined data	789	202	345	113	High to Bad

Table 4. Summary description of the sub-sets of data that constitute the benchmark dataset

Country	IC river type	Data owner	Number of samples		Quality range	Best available classification criteria
			Tot	Ref. (BAC)		
Austria	R-C3/4	BOKU-Wien	24	5	Ref. to Bad	Multimetric classification. Range between 25th percentile of high status and 75th of bad divided by three. PCA and cluster-analysis performed on abiotic parameters to confirm the significant gradients in the dataset and thus the multimetric classification. AQEM analysis
Czech Republic	R-C3/4	Masaryk University	24	2	Ref. to Poor	Post-classification, community structure and thresholds for saprobic value. AQEM analysis
Czech Republic	R-C3	Masaryk University	22	7	Ref. to Poor	Post-classification, community structure and thresholds for saprobic value. AQEM analysis
France	R-M1	Direction Regionale de l'Environment	77	17	Ref. to Bad	IBGN classification, with WFD compliant reference definition and detailed pressure analysis. CEMAGREF analysis
Italy	R-C1	CNR-IRSA	33	9	Ref. to Bad	Ecological breakpoint between reference and good class along multivariate axis. Remaining classes equally spaced. AQEM analysis
Italy	R-M1	CNR-IRSA	33	8	Ref. to Bad	Ecological breakpoint between reference and good class along multivariate axis. Remaining classes equally spaced. AQEM analysis
Italy	R-M2	CNR-IRSA	33	7	Ref. to Mod.	Ecological breakpoint between reference and good class along multivariate axis. Remaining classes equally spaced. AQEM analysis
Italy	R-M1	CNR-IRSA	16	2	Ref. to Mod.	Ecological breakpoint between reference and good class along multivariate axis. Remaining classes equally spaced. STAR analysis
UK	R-C1/2	CEH Dorset	70	18	Ref. to Bad	RIVPACS classification (multivariate predictive system). STAR analysis
UK	R-C4	CEH Dorset	66	18	Ref. to Bad	RIVPACS classification (multivariate predictive system). STAR analysis
		Total	388	93		

was general degradation. These sites belonged to the hydro-ecoregion "Méditerrannée" (HER 6) of the French typology (Wasson et al., 2003). Hydrologic seasonality is high, but the streams are not regularly intermittent. Altitude ranged from 0 to 600 m and catchment area between 10 and 100 km^2.

At sites investigated for the AQEM project (Hering et al., 2003, 2004a), samples were collected with the aim of developing and testing macroinvertebrate-based assessment methods that satisfy WFD requirements (e.g. type specific, derived from comparison with reference conditions). Following the principles of the AQEM project, a STAR project objective (Furse et al., 2006) was to develop a framework method for calibrating different biological survey results against ecological quality classifications in accordance with the terms of the Water Framework Directive. For both projects, the collection of all invertebrate data used in this paper followed a multi-habitat sampling approach derived from Barbour et al. (1999) (see also Hering et al., 2004a, b), with the collection of 20 sample units proportionally distributed among the microhabitats present in the river. Data included in the benchmark dataset belonged to five countries and included 10 different stream types. The altitude of the stream types selected varies between sea level and 800 m a.s.l., with a predominance of mid-altitude sites. Catchment area was between 10 and 1000 km^2, with about half of the sites having a catchment smaller than 100 km^2. The main impacts were organic pollution and habitat degradation even if other perturbation factors were locally present (e.g. pesticides at northern Italian sites). For some stream types, only one stressor was active at the sites: for two stream types (Italy – Emilia Romagna and Austria) degradation in stream morphology only and in rivers from Czech Republic organic pollution only.

The variety of pressures acting across the investigated types and for some, their simultaneous occurrence, (i.e. multi-pressure systems) in certain types is advantageous for the aims of the present paper. In fact, the large scale IC process is aimed at inter-calibrating the final classification of sites without necessarily focusing on single pressures. The essential question addressed by the European IC process is whether or not the five classes of ecological status classification are equivalent across Europe and not the causes of possible discrepancies. Nevertheless, detailed information is needed to interpret the response pattern of biological elements in relation to acting pressures. In general terms, the availability of comprehensive and high-quality ecological and environmental data (i.e. data adequately covering the biological aspects and the different pressures acting at a site) differentiates the benchmark dataset used here from the three test datasets, which show a poorer compilation of supporting data. The total number of sites included in the benchmark dataset is 137. The sampling seasons investigated were mainly two, with the exception of Italy for which three seasons were investigated; the total number of samples is 388 (further details can be found in Buffagni et al., 2005). Each set of data collected for the AQEM and STAR projects included a set of reference sites, selected according to the demands of the WFD. Criteria for the selection of the reference sites are specified elsewhere (Buffagni et al., 2001; Hering et al., 2003; Nijboer et al., 2004). More details for each of the datasets included in the benchmarking systems and corresponding reference condition criteria are reported in Hering et al. (2004b).

The quality classification used here for benchmark datasets is referred to as a Best Available Classification. The BAC is an ecological classification obtained by applying WFD compliant procedures and all the available, relevant information on a site. For example, depending on the kind of the main acting pressures, the BAC may result from the integration of biological, physico-chemical and hydromorphological information. It must be based on detailed community analysis (e.g. by multivariate analysis) and not simply on the standard national method of classification. Agreed BACs at the European or GIG scale will be produced on the basis of the criteria outlined in the guidance document on the intercalibration process (European Commission, 2004; Buffagni et al., 2005).

The BAC concept was put into practice within the AQEM project by outlining an agreed classification based on a common framework for collecting data, developing assessment systems and setting final quality classes (e.g. Hering et al., 2004a). For each country, the BAC corresponds to the post (multivariate analysis, including pressures analysis) or final (multimetric including pressures) classification used in the AQEM project, depending on which of the two is better represented by the quality gradient of the sites described by macroinvertebrates' community and related pressures. For the STAR datasets in Table 4 the criteria for the BAC are specified. In Table 4, the sub-sets of data selected for inclusion in the benchmark dataset (Buffagni et al., 2005) are reported, with a general indication of the criteria used to define the BAC classification.

Results

Relationship between ICMi and national methods

The relationships between single metrics (ICMs), the ICM index and national assessment systems were analyzed. Table 5 shows the coefficient of determination, R^2, (i.e. how well the regression line represents the data), together with its statistical significance (p-level), between ICMs, ICMi and the five standard indices used in the three national assessment systems considered.

Apart from Margalef's index vs. Shannon and 1-GOLD, all the relationships analyzed were highly significant. Among single ICMs and the national indices, the highest correlation values were observed for the metric Number of families (R^2 higher than 0.60 with all National indices). EPT-taxa and ASPT also showed good correlations, with R^2 always higher than 0.50, with the exception of Margalef's Index for ASPT. For the metric based on the abundance of sensitive taxa (Log_EPTD), R^2 is always greater than 0.40, again with the exception of Margalef's index. The Shannon metric exhibits the lowest correlation values with respect to the two Polish indices. The highest R^2 value for Shannon was observed with respect to IBE (0.58). In general, 1-GOLD showed low correlation values with all the national methods (maximum R^2 value 0.21). For the ICMi, the R^2 values were greater than 0.70 for all the national indices included in comparisons, with the exception of Margalef's Index. This demonstrates the general increase of performance when single metrics are combined into a multi-metric index.

As far as UK data is concerned, finding a R^2 value of 1 between EQI ASPT and ASPT and between NFAM_EQI vs. N_families was not expected, for two different reasons. Firstly, EQI ASPT and NFAM_EQI were derived from RIVPACS and represent the deviation of observed invertebrate communities from those expected i.e. communities typical of the reference condition for the studied stream type. Secondly, the calculation of the metric ASPT was undertaken with the ASTERICS assessment software, which in its present version provides a calculation metric which is slightly different from the UK standard method.

Comparison of existing class boundaries

Comparison through linear regression

The three National assessment systems – IBE, BMWP-POL, EQI ASPT – and respective boundaries, were compared indirectly through linear regression with ICMi values (Table 2; Fig. 1).

Table 5. R^2 values and p-level between ICMs, ICMi and national indices

			ASPT	SHAN	1-GOLD	Log_EPTD	EPT_taxa	N_families	ICMi
ITA	IBE	R^2	0.59	0.58	0.21	0.51	0.55	0.80	0.72
		p-level	<0.001	<0.001	<0.001	<0.001	<0.001	<0.001	<0.001
POL	BMWP-POL	R^2	0.66	0.21	0.19	0.40	0.78	0.94	0.74
		p-level	<0.001	<0.001	0.001	<0.001	<0.001	<0.001	<0.001
	Margalef-Ind	R^2	0.44	0.01	0.12	0.14	0.59	0.79	0.40
		p-level	<0.001	0.42	0.02	0.008	<0.001	<0.001	<0.001
UK	EQI ASPT	R^2	0.88	0.32	0.20	0.62	0.77	0.62	0.83
		p-level	<0.001	<0.001	<0.001	<0.001	<0.001	<0.001	<0.001
	EQI N-taxa	R^2	0.57	0.31	0.39	0.53	0.72	0.87	0.72
		p-level	<0.001	<0.001	<0.001	<0.001	<0.001	<0.001	<0.001

In Figure 1, the ICM index value is reported on the *x*-axis and national method EQR on the *y*-axis for each sample included in the dataset. For this comparison, the values obtained by the national assessment methods were normalized by dividing every observed value by the value fixed nationally for the High/Good boundary, subtracting the minimum achievable value in order to get effectively 0 as minimum. In fact, until contrasting evidence is provided, it can be argued that the methods and corresponding class boundaries can be WFD compliant (i.e. the H/G boundary effectively indicates the transition from potential Reference Condition to Good status). In addition, information on WFD-compliant reference conditions is quite often unavailable across Europe (Buffagni et al., 2005).

The minimum values used were 5 for BMWP-POL and 2 for IBE. For EQI ASPT the minimum observed value (0.40) was used assuming that, due to the high number of samples in the British dataset (789), this would correspond to the actual minimum for the method in the stream type studied. For each set of data, the linear regression line between ICMi and national index is reported. The High/Good (dotted lines) and Good/Moderate (unbroken lines) boundary values are indicated. By using regression formulae, it is possible to derive the values for the ICMi (*x*-axis) boundaries, from the normalized EQR values of the boundaries of the national method and *vice versa*. Values of the boundaries, regression formulae used, R^2 coefficient and *p*-level are reported in Table 2 for all the National indices. The three regression lines between ICMi and National indices (Fig. 1) show similar slopes, all bearing angular coefficient close to 1 (see Table 2). The indices BMWP-POL and EQI ASPT show a noticeably linear trend and constantly increase with increasing values of ICMi. In contrast, the IBE index shows a less constant growth, i.e. its relationship, while roughly linear, is not ideal. With low ICMi values (ca from 0.15 to 0.40), IBE grows with a coefficient close to 1, while with ICMi values higher than 0.40 a small increase of IBE values corresponds to a wider variation of ICMi values.

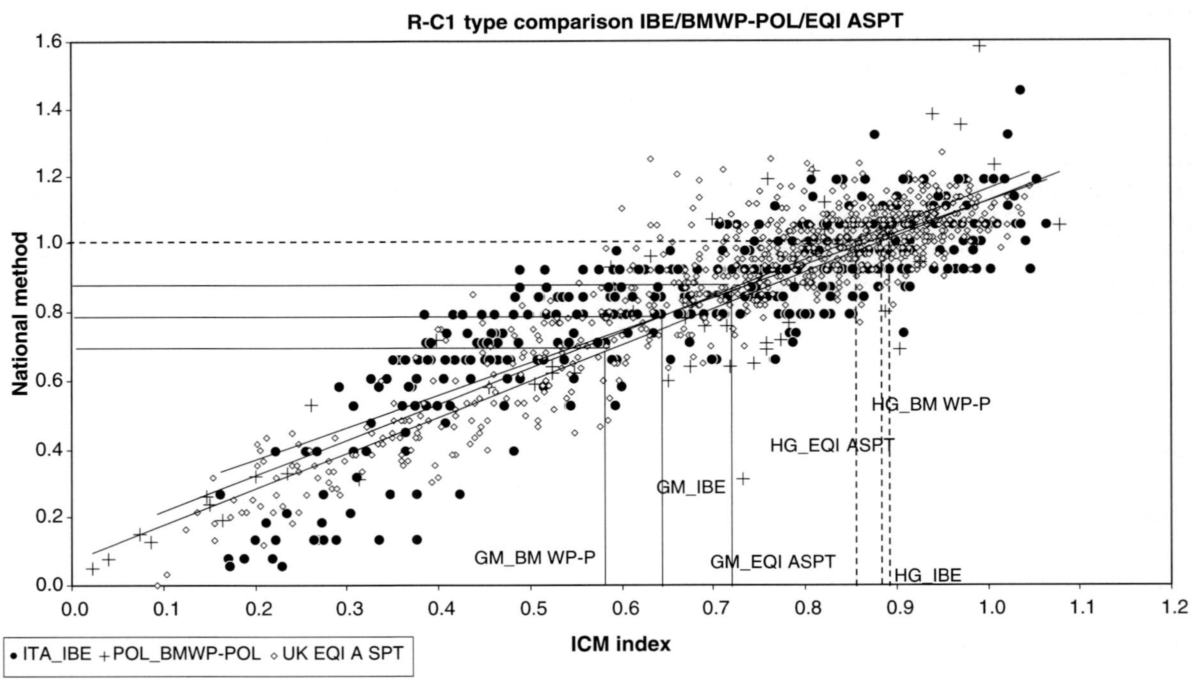

Figure 1. Relationship between ICM index and national indices (IBE, BMWP-POL and EQI ASPT) (linear regression). Lines in the diagram show the High/Good (dotted line) and Good/Moderate (unbroken line) boundaries, obtained after setting the High/Good boundary of national methods (EQR value) equal to one and normalizing all individual metric values as described in the text.

Direct comparison among test datasets

ICMi values were calculated for all test sites in each of the three countries. On the basis of these values, samples were allocated to High, Good or Moderate status using their national class boundaries. Direct statistical comparisons could then be made, using the Mann–Whitney *U*-test of the ICMi values of, Good status samples in one country with the values of ICMi of Good sites in another country. The summary results of the Mann–Whitney *U* test are reported in Table 6. In the direct comparison of ICMi values as classified by National systems, significant differences were found between UK Good status samples and both Polish and Italian samples. High status samples were not significantly different between countries. In Figure 2 the variation of the ICMi values obtained from the Good status samples of the test datasets is presented in the form of Box-and-Whisker, together with the values observed in the benchmark samples (left side of Fig. 2). The same kind of demonstration of ICMi value ranges is presented for High status samples (Fig. 3). For both Good and High status samples lower median values were observed for Italy and Poland with respect to both the benchmark and the UK data. In particular, for Good status the median ICMi value of UK samples is the highest observed.

Indirect comparison via benchmark dataset

In order to assess differences among datasets against a common comparison system, statistical tests were also performed between the ICMi values obtained for each MS's samples and the benchmark pool of samples belonging to the same quality class. The procedure followed was the same as in the previous section, with Good status samples compared first and High status samples afterwards. While UK data was found to be not statistically different from the benchmark samples, ICMi values from the Italian standard classification differ from benchmark values both for Good and High status quality classes (Table 6). Polish ICMi values differ significantly from benchmark ones only for Good status samples.

Harmonization of Good/Moderate and High/Good class boundaries among countries

For the datasets considered here, the average ICMi value corresponding to the national EQR boundaries on the basis of the regression formulae between national methods and ICMi is 0.87 for the H/G boundary and 0.66 for G/M. To give an example, the harmonized boundaries were calculated by regression from these average values (Table 2), even if the correct procedure would require the verification of WFD compliance of the methods. Only after demonstrating the WFD compliance of the methods used in all MSs that must be harmonized, can a meaningful average (e.g. median) value be derived. This ICMi value, converted into the national metric, corresponds to H/G boundaries very close to the original ones. As far as the G/M boundary for Italy is concerned this may have to be moved from 7.6 to 8 and for Poland from 70 to 77.

The results of the direct statistical comparison between the ICMi values obtained for the High and Good quality classes as defined by the national protocols in the three Countries are reported in Table 6 (original boundaries – first three

Table 6. Results of the statistical comparison – before and after harmonization – among the ICMi values obtained for the High and Good status classes as defined by the national protocols in the three Member States (national datasets) and between them and benchmark data (Mann–Whitney *U* test)

	High (original boundaries) *p*-level	High (after Harmonization) *p*-level	Good (original boundaries) *p*-level	Good (after Harmonization) *p*-level
IT vs. PL	0.53	0.43	0.60	0.64
IT vs. UK	0.11	0.27	$3*10^{-18}$	$7*10^{-4}$
PL vs. UK	0.29	nc	0.009	0.06
IT vs. Benchmark	0.04	0.09	$1.9*10^{-8}$	0.053
PL vs. Benchmark	0.22	–	0.035	0.12
UK vs. Benchmark	0.29	–	0.33	–

nc indicates that the value was not calculated;– means not calculated, because of no harmonization needed.

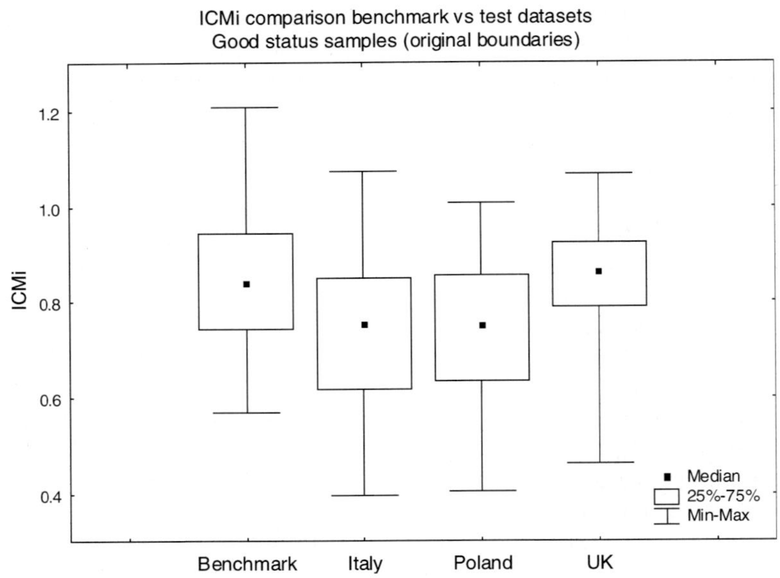

Figure 2. Box-and-Whisker representation of the variation of the ICMi values observed for the Good status samples of the benchmark dataset (left side) and of the test datasets (Italy, Poland and UK).

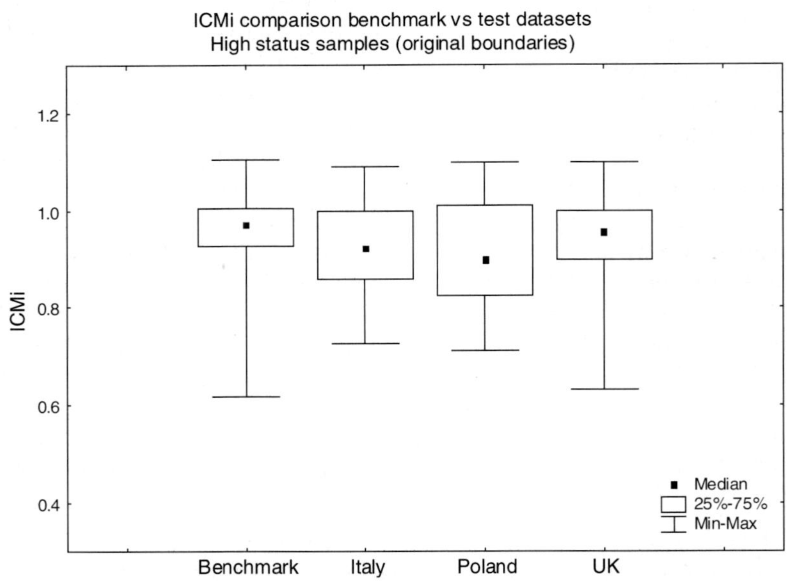

Figure 3. Box-and-Whisker representation of the variation of the ICMi values observed for the High status samples of the benchmark dataset (left side) and of the test datasets (Italy, Poland and UK).

comparisons). The outcome of testing them against the benchmark dataset, for the same classes, is also described (original boundaries, bottom three comparisons).

The level of significance for the harmonized classes is also shown in Table 6. As before, the comparison of national and benchmark data suggests that boundaries need to be adjusted for Good

Table 7. Boundaries for test datasets before and after harmonization via comparison with a benchmark dataset

Limit	Italy		Poland			UK	
	IBE score	IBE harmonized	BMWP score	BMWP harmonized	Marg-DI score	EQI ASPT	EQI N-taxa
High–Good	9.6	10	100	100	5.5	1.00	1.00
Good–Moderate	7.6	8.6	70	75	4	0.89	0.78
Moderate–Poor	5.6	nc	40	nc	2.5	0.76	0.57
Poor–Bad	3.6	nc	10	nc	1	0.65	0.36

status in Italy and Poland where respective *p* value were 1.9×10^{-8} and 0.035.

As a result of the procedure of harmonization against a benchmark, the IBE Good/Moderate status boundary was moved up, because the median value in the test dataset was significantly lower than in the benchmark dataset (Table 7). The threshold value was re-positioned step-by-step (i.e. from 7.6 to 8, from 8 to 8.4, etc.; see Spaggiari & Franceschini, 2000 for IBE original thresholds), until no more differences were found between the values of the ICMi according to the benchmark (STAR/AQEM) and IBE classifications. For the Italian dataset, statistical differences were found until the boundary was moved to 8.6, after which they were no longer significant ($p=0.053$). The new Good/Moderate boundary was thus fixed at 8.6. After having compared and tested ICMi values for the Good status class, the High status class was compared and tested. Because the observed ICMi values for the High status samples were different from benchmark values, the boundary High/Good was moved up step by step as was done for the Good/Moderate boundary. To remove differences, it was enough to move the High/Good boundary from an IBE value of 9.6 to 10, when a non-significant *p*-value of 0.09 was found. Figures 4 and 5 show the ICMi variation after harmonization respectively for Good and High status classes. The harmonized boundaries for the three countries are reported in Table 7. In Poland's case, because the Margalef index did not determine the site classification, only the original boundaries are reported, i.e. its values were not adapted. For the UK, because harmonization was not necessary, only the original boundaries are reported. After harmonization, the ICMi values were again compared between countries and against the benchmark dataset (Table 6). The results showed that the differences between countries were generally diminished after harmonization, the only exception being in the High status samples of Poland vs. Italy. In fact, the process of harmonization involved the H/G IBE boundary being moved upwards, resulting afterwards in higher median ICMi values, which increased its distance from the ICMi median value of the Polish dataset.

In the Polish dataset, significant differences were found for Good status samples compared to benchmark values ($p=0.035$). In Poland, site classification was derived from the combination of BMWP and Margalef's index classification, on the basis of the 'one-out all-out' principle. For the samples included in the dataset studied, BMWP always determined the final class. The re-positioning of the boundaries was thus done by moving up the threshold Good/Moderate for the BMWP score only. As a rule, the boundary was moved up by steps of five scores (e.g. from 70 to 75). The *p*-level for the comparison between the test data and the newly defined BMWP boundary for the Good class and the benchmark data was 0.12, indicating that significant differences no longer existed. Nevertheless, the median ICMi value still remains the lowest of the datasets considered. With regard to the comparison of High status samples, the test did not reveal any significant differences ($p=0.22$).

Discussion

The definition of class boundaries is a crucial step for implementing the Water Framework Directive. This process will need to involve political and ecological considerations. Ecological judgements will need to be based on a variety of messages

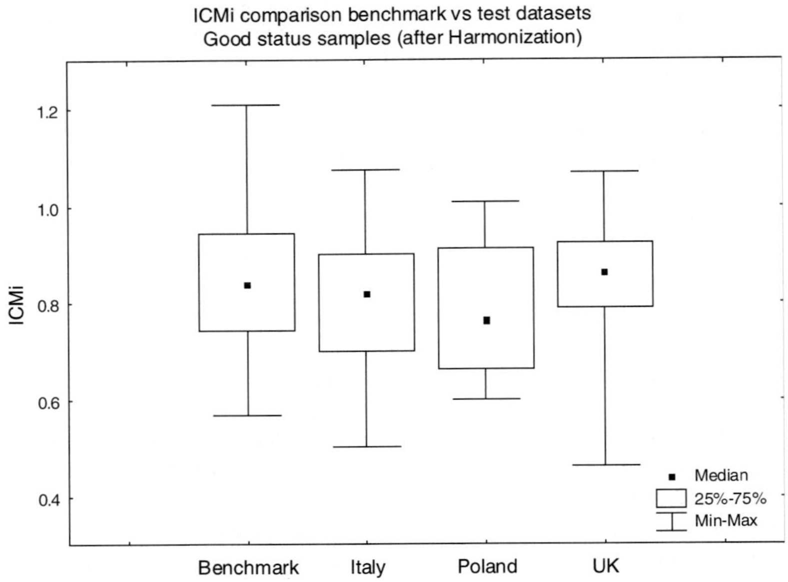

Figure 4. Box-and-Whisker representation of the variation of the ICMi values observed for the Good status samples of the benchmark dataset (left side) and of the test datasets (Italy, Poland and UK) after harmonization. The asterisk indicates countries for which harmonization was carried out.

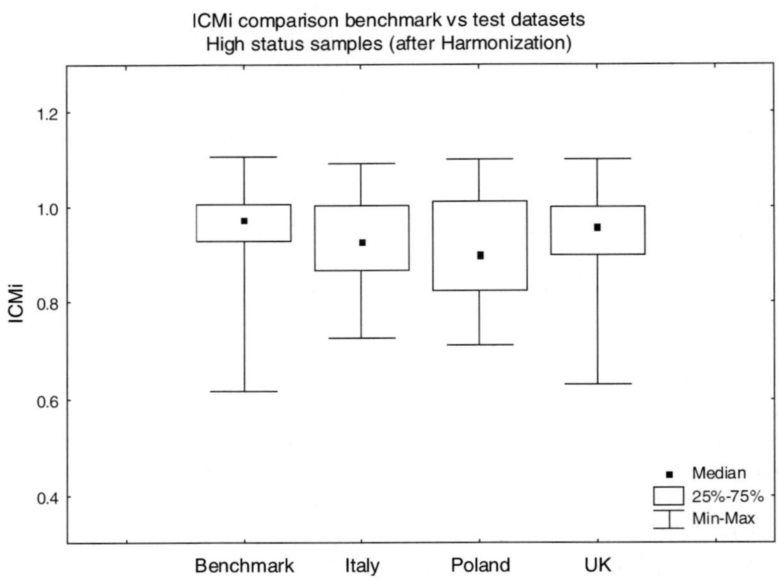

Figure 5. Box-and-Whisker representation of the variation of the ICMi values observed for the High status samples of the benchmark dataset (left side) and of the test datasets (Italy, Poland and UK) after harmonization. The asterisk indicates countries for which harmonization was carried out.

emanating from a variety of different taxonomic groups and hydromorphological conditions. At present there is no sound scientific basis for integrating these different sources of information.

It was the intention of STAR (Furse et al., 2006) to provide the background science needed to link classes defined by the use of different organism groups. It is necessary to advise the European

Commission and the countries involved in the WFD intercalibration process on how this information could be used, in conjunction with political considerations, in assisting the process of defining and delimiting the five grades of ecological status.

In the AQEM and STAR Projects, a multimetric approach was adopted for assessing river quality based on biological indicators. Ecological assessments, were integrated at a higher level with quality evaluations based on water chemistry and hydromorphological information. The same approach, although simplified has been used in this paper to make it compatible with the timing and scope of the IC process. Following the STAR ICMi approach (Buffagni et al., 2005), a simple multimetric index based on the selection of a set of common metrics has thus been developed for the purposes of European intercalibration (Buffagni et al., 2005). The main criteria for the selection of metrics to be included in this multimetric index (Buffagni et al., 2005) were:

- their consistency with WFD definitions, i.e. they have to deal with the three main aspects outlined for aquatic invertebrates in the WFD (tolerance, richness/diversity and abundance);
- their ability to describe degradation gradients and discriminate different quality classes based on existing literature and AQEM/STAR projects' experience;
- the feasibility of calculating them from a wide range of geographical contexts, i.e. where different effort is placed on monitoring and different expertise is available for taxonomic identifications.

The concept underlying the selection and combination of different metrics into a unique index is the assumption that each of the metrics can give different information on the status of the biological community (Barbour et al., 1999). Some of the metrics selected, such as ASPT and EPT taxa, are widely used for assessment and well known to be representative of rivers' ecological quality (e.g. Armitage et al., 1983; Kerans & Karr, 1994; Verdonschot, 2000). The remaining metrics were chosen from those selected for the multimetric assessment systems developed within the AQEM project (e.g. Buffagni et al., 2004; Ofenböch et al., 2004; Lorenz et al., 2004; Pinto et al., 2004), as they were defined or adapted to different levels of identification and wider geographical range (Buffagni et al., 2005). Some metrics like EPT taxa are in fact based on a family ID level for the purpose of the intercalibration, while they are normally used for monitoring at more detailed ID levels. The values obtained for each metric constituting the intercalibration index were compared to those of national assessment methods. All the metrics showed a generally good relationship with the national assessment methods of all the countries, with R^2 values higher than 0.4. The ability of single metrics to separate the different quality classes was also evaluated. In general, the metrics selected were proven to adequately reflect the gradient expressed by national assessment methods, being in most cases well able to discriminate between the different quality classes (especially Good and Moderate) (see Buffagni et al., 2005). The high correlation of the ICMi with national assessment metrics has confirmed the greater ability of a combined i.e. multimetric index to represent the integrity of the biotic community compared to single metrics (Cao et al., 1997). The good performance of the ICMi in representing the ecological quality gradient of the different test datasets is confirmed here by the relationships shown in Fig. 1, where the regression between ICMi and National methods is presented. Regression lines for the three methods are very close to each other with similar angular coefficients. This confirms its ability to translate MSs' results into a single format, which makes different methods truly comparable. R^2 values are high (about 0.70) between each National assessment system and ICMi.

The IBE method had a comparatively low potential to discriminate samples positioned in the Good/Moderate range. The trend in the IBE values is in fact not entirely linear (Fig. 1) and this is most probably because of the way that the presence of sensitive taxa is recorded. This, together with the total number of taxa, determines the IBE value and is not linear itself because it is based on a binary information (presence/absence of selected taxa) and not on a semi-continuous scale. The effect is that to a broad quality range well discriminated by the ICMi corresponds a low gradient for the IBE method, especially at sites characterized by fair/moderate ecological status (i.e. from Moderate to High). In turn, this forewarns of a potentially high uncertainty in the ecological status classification based on the IBE method in this quality range. A non-linear response of biological metrics has also been found by Cao et al. (1996). The model used here is based on a linear regression, but other non-linear models can offer a better performance (i.e. higher R^2). However, the low capacity of discriminating slightly impacted sites did not depend on the type of model used. The two other assessment systems compared (Polish and British), showed a better linearity with respect to ICMi.

In this paper, we dealt only indirectly with the problem of defining reference conditions, by referring to the BAC concept (Buffagni et al., 2005). The final cross-validation of the results of different MS's assessment methods will ultimately depend on the adequacy of the protocol used to derive reference conditions, i.e. criteria to accept or reject sites as reference sites (European Commission, 2004).

Comparatively few scientific papers have dealt with the problem of intercalibration of biological methods and results. Most of them compared different biological methods directly, e.g. by using regression and correlation analyses on the results (e.g. Ghetti & Bonazzi, 1977; Rico et al., 1992). However, care must be taken when considering the regression approach. For instance, the application of different metrics, derived by different assessment systems, to a single dataset (i.e. to the same samples) can lead to the comparison of misallocated class boundaries, even if the various metric values are transposed to EQR values. In fact, the sampling and sorting phases are often an integral part of the assessment system, and the allocation of class boundaries is closely linked to the approach and effort placed on sampling. Because most methods rely on taxon richness, if a class boundary is derived from samples collected from an area of 0.5 m^2, this boundary will differ from one based on samples collected from an area of 1 m^2, because of the different number of taxa likely to be accumulated. For further details on the regression approach, see Birk & Hering (2006). Such a methodology usually provides conversion formulae between different methods and metrics, which are nevertheless subject to some conceptual restraints. For instance, Buffagni et al. (2005) demonstrated that a certain degree of misclassification is inevitable when directly boundaries translating via regression formulae.

Apart from the few examples reported, very few authors approached the subject of intercalibrating the results of whole assessment systems, i.e. including the boundaries between quality classes. As far as this last kind of harmonization is concerned, a procedure based on the averaging of class boundaries after translating them into an ICMi value was briefly presented in this paper, mainly to provide a picture of the comparison of National methods *via* ICMi using a regression approach. It was not our intention to emphasize this approach and, for example, no mention of the estimation of confidence intervals was made. This is a principal concern when considering a regression approach. The pre-requisite for the harmonization of class boundary values by any kind of averaging is the inclusion of all assessment methods (i.e. MSs) for a given intercalibration stream type. Before averaging, by taking the median or other values, all methods must be demonstrated to be WFD-compliant, especially in terms of the definition of Reference conditions. The results of the harmonization can be highly influenced by the number and kind of methods and datasets considered (Buffagni et al., 2005). In addition, the procedure for averaging class boundaries (directly or by using an ICMi) gives MSs greater opportunity to look for a simple agreement on boundaries. In our opinion, no simple 'averaging' of existing class boundaries should be considered for European intercalibration, at least until all MSs' assessment systems are proved to be fully WFD-compliant. In contrast, the comparison against an external and agreed benchmarking dataset/classification supports a more objective (i.e. scientifically sound) definition of classes because it is less dependant on a single MS' view of boundary setting. In any case, when boundaries of different assessment systems are alike, the problem of averaging is overcome. Nonetheless, the central question is how big does the difference have to be for it not to be considered acceptable?

The harmonization option recommended here is the one that includes the comparison and adjustment with an external dataset. Based on this approach, the results presented here for a very common European stream type show that only small refinements would be required to the methods considered. In fact, the G/M boundary was only adjusted for Poland and Italy. No changes seem to be needed for the UK system. The percentage of samples moving from Good to Moderate status after harmonization is around 9% for Poland and 22% for Italy. The re-positioning of the H/G boundary involves very minor changes for Italy only, and no change for Poland and UK. Poland still presents quite a low median value but no significant differences were found. This is almost certainly because of the relatively high variability of ICMi values in the Polish dataset and to the number of samples tested, which is quite low, i.e. only 11 High status samples. It is probable that when more samples are included (increasing statistical power), the results of the statistical comparison might change, determining the need for future adjustments of the Polish High/Good boundary. To cope with this problem, we will recommend a minimum number of samples to be included in each dataset, especially for the High status class (Buffagni et al., 2005). The situation observed in Poland, where a low number of samples was available, provides a crucial warning of the impossibility of realistically performing the IC exercise for water body types for which there is a paucity of data: quite a common situation in some European areas. Those countries where available data are scarce should plan explicit collection activities to fill the existing gap so that a reliable pan-European comparison can be performed.

As far as large-scale studies on intercalibration are concerned, an exploratory study related some chemical indicators of organic pollution to the ASPT and Saprobic metrics in selected stream types in Europe (Sandin & Hering, 2004). Common class boundaries for High/Good and Good/Moderate Ecological Status were set, based on a multivariate gradient. More recently, an extensive study was undertaken to provide support to the E.C. intercalibration process by applying and discussing different approaches to compare class boundaries and test their similarity (Buffagni et al., 2005). The present paper is a further contribution to discussions on this issue, which will play a vital role in the WFD implementation process (Heiskanen et al., 2004) which is also likely to affect forthcoming European social and economic progress. Crucial next steps in the European intercalibration process are to describe agreed criteria to accept/reject reference sites having as starting point the work of REFCOND and to define criteria to check the validity of the possible options in setting a class boundary, including pressure analysis. In our view, it is essential that a benchmark classification is agreed and understood at a European level. This will support the validation of, for example, the STAR/AQEM benchmark dataset (or equivalent) and the development of a common procedure at the pan-European or GIG level for the comparison and harmonization of all national assessment systems, so that the WFD deadlines for the IC process can be met effectively.

Acknowledgements

We thank Jean-Gabriel Wasson (CEMAGREF, Lyon, France), Karel Brabec (Masaryk University, Brno, Czech Republic), Mike T. Furse (CEH, Dorset, UK) and Otto Moog (BOKU, Vienna, Austria) for providing the data included in the benchmark dataset. The STAR Project (Standardisation of River Classifications: Framework method for calibrating different biological survey results against ecological quality classifications to be developed for the Water Framework Directive) was co-funded by the European Commission, 5th Framework Program, Energy, Environment and Sustainable Development, Key Action Water, E.U. Contract number: EVK1-CT 2001-00089). We are also very grateful to M. T. Furse and to an anonymous referee for their review work.

References

APAT-IRSA/CNR, 2004. Indice Biotico Esteso (I.B.E). In APAT, Manuali e linee guida 29/2003. APAT-IRSA/CNR, Metodi analitici per il controllo della qualità delle acque 3: 1115–1136.

AQEM Consortium, 2002. Manual for the Application of the AQEM system. A Comprehensive Method to Assess European streams using Benthic Macroinvertebrates, Developed for the Purpose of the Water Framework Directive. Version 1.0, February 2002, 202 pp.

Armitage, P. D., D. Moss, J. F. Wright & M. T. Furse, 1983. The performance of a new biological water quality score system based on macroinvertebrates over a wide range of unpolluted running-water sites. Water Research 17: 333–347.

Barbour, M. T., J. Gerritsen, B. D. Snyder & J. B Stribling, 1999. Rapid Bioassessment Protocols for Use in Wadeble Streams and Rivers: Periphyton, Benthic Macroinvertebrates and Fish (2nd ed.). EPA 841-B-99–002. USEPA, Office of Water, Washington, D.C.

Birk, S. & D. Hering, 2006. Direct comparison of assessment methods using benthic macroinvertebrates: a contribution to the EU Water Framework Directive intercalibration exercise. Hydrobiologia 566: 401–415.

Böhmer, J., C. Rawer-Jost & A. Zenker, 2004. Multimetric assessment of data provided by water managers from Germany: assessment of several different types of stressors with macrozoobenthos communities. In Hering D., P. F. M. Verdonschot, O. Moog & L. Sandin (eds), Integrated Assessment of Running Waters in Europe. Kluwer Academic Publishers, The Netherlands. Hydrobiologia 516: 215–228.

Brabec, K., S. Zahrádková, D. Nějcová, P. Páril, J. Kokeš & J. Jarkovsk, 2004. Assessment of organic pollution effect considering differences between lotic and lentic stream habitats. In Hering D., P. F. M. Verdonschot, O. Moog & L. Sandin (eds), Integrated Assessment of Running Waters in Europe.Kluwer Academic Publishers, The Netherlands. Hydrobiologia 516: 331–346.

Buffagni, A., J. Kemp, S. Erba, C. Belfiore, D. Hering & O. Moog, 2001. A Europe-wide system for assessing the quality of rivers using macroinvertebrates: the AQEM Project and its importance for southern Europe (with special emphasis on Italy). Journal of Limnology 60(Suppl. 1): 39–48.

Buffagni, A. & S. Erba, 2004. A simple procedure to harmonize class boundaries of European assessment systems. Discussion paper for the intercalibration process – WFD CIS WG 2.A ECOSTAT, 6 February 2004, 21 pp.

Buffagni, A., S. Erba, M. Cazzola & J. L. Kemp, 2004. The AQEM multimetric system for the southern Italian Apennines: assessing the impact of water quality and habitat degradation on pool macroinvertebrates in Mediterranean rivers. In Hering D., P. F. M. Verdonschot, O. Moog & L. Sandin (eds), Integrated Assessment of Running Waters in Europe. Kluwer Academic Publishers, The Netherlands. Hydrobiologia 516: 313–329.

Buffagni A., S. Erba, S. Birk, M. Cazzola, C. Feld, T. Ofenböck, J. Murray-Bligh, M. T. Furse, R. Clarke, D. Hering, H. Soszka & W. van de Bund, 2005. Towards European Intercalibration for the Water Framework Directive: Procedures and examples for different river types from the E.C. project STAR. 11th STAR deliverable. STAR Contract No: EVK1-CT 2001-00089. Rome (Italy), Quaderni Istituto di Ricerca sulle Acque 123, Rome (Italy), IRSA, 468 pp.

Cao, Y., W. A. Bark & W. P. Williams, 1996. Measuring the response of macroinvertebrate communities to water pollution: a comparison of multivariate approaches, biotic and diversity indices. Hydrobiologia 341: 1–19.

Cao, Y., W. A. Bark & W. P. Williams, 1997. Analysing benthic macroinvertebrate community changes along a pollution gradient: a framework for the development of biotic indices. Water Research 31: 884–892.

Dahl J., R. K. Johnson & L. Sandin, 2004. Detection of organic pollution of streams in southern Sweden using benthic macroinvertebrates. In Hering, D., P. F. M. Verdonschot, O. Moog & L. Sandin (eds), Integrated Assessment of Running Waters in Europe. Kluwer Academic Publishers, The Netherlands. Hydrobiologia 516: 161–172.

Environment Agency, 1997. Assessing Water Quality – General Quality Assessment (GQA) Scheme for Biology. Fact Sheet. Environment Agency, Bristol.

European Commission, 2000. Directive 2000/60/EC of the European Parliament and of the Council of 23 October 2000 establishing a framework for Community action in the field of water policy. Official Journal of the European Communities L 327, 22.12.2000, 1–72.

European Commission, 2003a. Common Implementation Strategy for the Water Framework Directive (2000/60/EC). Guidance Document No. 6. Towards a Guidance on Establishment of the Intercalibration Network and the Process on the Intercalibration Exercise. Produced by Working Group 2.5 – Intercalibration, 54 pp.

European Commission, 2003b. Overview of common Intercalibration Types and Guidelines for the selection of intercalibration sites. Water Framework Directive Common Implementation Strategy Working Group 2A 'Ecological Status' Version 3. 9 October, 2003, 70 pp.

European Commission, 2004. Guidance on the Intercalibration Process. Water Framework Directive Common Implementation Strategy Working Group 2. A 'Ecological Status' Final Draft 4.1. October, 14th 2004.

Furse, M., D. Hering, O. Moog, P. Verdonschot, R. K. Johnson, K. Brabec, K. Gritzalis, A. Buffagni, P. Pinto, N. Friberg, J. Murray-Bligh, J. Kokes, R. Alber, P. Usseglio-Polatera, P. Haase, R. Sweeting, B. Bis, K. Szoszkiewicz, H. Soszka, G. Springe, F. Sporka & I. Krno, 2006. The STAR project: context, objectives and approaches. Hydrobiologia 566: 3–29.

Genoni, P., P. Beati, F. Buzzi, P. Casarini, M. Girami, E. Gozio, V. Mafessoni, P. Roella & A. Sarzilla, 1998. Intercalibrazione del metodo Indice Biotico Esteso I.B.E. (IRSA-CNR, 1995) per la valutazione della qualità dei corsi d'acqua. Regione Lombardia Direzione Generale Sanità Servizio Prevenzione Sanitaria. 34 pp.

Genoni, P., 2003. Influenza di alcuni fattori ambientali sulla composizione delle cenosi macrobentoniche dei corsi d'acqua planiziali minori. Biologia Ambientale 17: 9–16.

Ghetti, P. F. & G. Bonazzi, 1977. A comparison between various criteria for the interpretation of biological data in the analysis of running waters sites. Water Research 11: 819–831.

Heiskanen, A-S., W. van de Bund, A. C. Cardoso & P. Nõges, 2004. Towards good ecological status of surface waters in Europe – interpretation and harmonisation of the concept. Water Science & Technology 49: 169–177.

Helsel, D. R. & R. M. Hirsch, 1992. Statistical Methods in Water Resources. Elsevier (ed.) 522 pp.

Hering, D., A. Buffagni, O. Moog, L. Sandin, M. Sommerhäuser, I. Stubauer, C. Feld, R. K. Johnson, P. Pinto, N. Skoulikidis, P. F. M. Verdonschot & S. Zahrádková, 2003. The development of a system to assess the ecological quality of streams based on macroinvertebrates – design of the sampling programme within the AQEM project. Int. Rev. Hydrobiol. 88: 345–361.

Hering, D., O. Moog, L. Sandin & P. F. M. Verdonschot, 2004a. Overview and application of the AQEM assessment system. In Hering D., P. F. M. Verdonschot, O. Moog & L. Sandin (eds), Integrated Assessment of Running Waters in Europe. Kluwer Academic Publishers, Printed in the Netherlands. Hydrobiologia 516: 1–20.

Hering, D., P. F. M. Verdonschot, O. Moog & L. Sandin (eds), 2004b. Integrated Assessment of Running Waters in Europe. Developments in Hydrobiology, 175. Kluwer Academic Publishers. Reprinted from Hydrobiologia 516 (2004), 379 pp.

Kerans, B. L. & J. B. Karr, 1994. A benthic index of biotic integrity (B-IBI) for rivers of the Tennessee valley. Ecological Application 4: 768–785.

Kownacki, A., H. Soszka, T. Fleituch & D. Kudelska (eds), 2002. River biomonitoring and benthic invertebrate communities. Institute of Environmental Protection, Karol Starmach Institute of Freshwater Biology PAS, Warszawa-Krakow, 88 pp.

Kownacki, A., H. Soszka, D. Kudelska & T. Fleituch, 2004. Bioassessment of Polish rivers based on macroinvertebrates. In Geller W. et al. (eds), Proceedings of the International 11th Magdeburg Seminar on Waters in Central and Eastern Europe: Assessment, Protection, Management. 18–22 October 2004, UFZ Leipzig: 250–251.

Logan, P., 2001. Ecological quality assessment of rivers and integrated catchment management in England and Wales. Journal of Limnology 60(Suppl.1): 25–32.

Lorenz, A., D. Hering, C. K. Feld & P. Rolauffs, 2004. A new method for assessing the impact of hydromorphological degradation on the macroinvertebrate fauna of five German stream types. In Hering D., P. F. M. Verdonschot, O. Moog & L. Sandin (eds), Integrated Assessment of Running Waters in Europe. Kluwer Academic Publishers. The Netherlands. Hydrobiologia 516: 107–127.

Morais, M., P. Pinto, P. Guilherme, J. Rosado & I. Antunes, 2004. Assessment of temporary streams: the robustness of metric and multimetric indices under different hydrological conditions. In Hering D., P. F. M. Verdonschot, O. Moog & L. Sandin (eds), Integrated Assessment of Running Waters in Europe. Kluwer Academic Publishers, The Netherlands. Hydrobiologia 516: 229–249.

Murray-Bligh, J. A. D., 1999. Procedures for collecting and analysing macro-invertebrate samples. Quality Management Systems for Environmental Monitoring: Biological Techniques, BT001. (Version 2.0, 30 July 1999). Environment Agency, Bristol.

Nijboer, R. C., R. K. Johnson, P. F. M. Verdonschot, M. Sommerhäuser & A. Buffagni, 2004. Establishing reference conditions for European streams. In Hering D., P. F. M. Verdonschot, O. Moog & L. Sandin (eds), Integrated Assessment of Running Waters in Europe. Kluwer Academic Publishers, The Netherlands. Hydrobiologia 516: 91–105.

Ofenböck, T., O. Moog, J. Gerritsen & M. Barbour, 2004. A stressor specific multimetric approach for monitoring running waters in Austria using benthic macro-invertebrates. In Hering D., P. F. M. Verdonschot, O. Moog & L. Sandin (eds), Integrated Assessment of Running Waters in Europe. Kluwer Academic Publishers, The Netherlands. Hydrobiologia 516: 251–268.

Pinto, P., J. Rosado, M. Morais & I. Antunes, 2004. Assessment methodology for southern siliceous basins in Portugal. In Hering D., P. F. M. Verdonschot, O. Moog & L. Sandin (eds), Integrated Assessment of Running Waters in Europe. Kluwer Academic Publishers, The Netherlands. Hydrobiologia 516: 191–214.

Rico, E., A. Rallo, M. A. Sevillano & M. L. Arretxe, 1992. Comparison of several biological indices based on river macroinvertebrate benthic community for assessment of running water quality. Annales de Limnologie 28: 147–156.

Sandin, L. & D. Hering 2004. Comparing macroinvertebrate indices to detect organic pollution across Europe: a contribution to the EC Water Framework Directive intercalibration. In D. Hering, P. F. M. Verdonschot, O. Moog & L. Sandin (eds), Integrated Assessment of Running Waters in Europe. Kluwer Academic Publishers, The Netherlands. Hydrobiologia 516: 55–68.

Shannon, C. E. & W. Weaver, 1949. The Mathematical Theory of Communication. The University of Illinois Press, Urbana, IL.

Spaggiari, R. & S. Franceschini, 2000. Procedure di calcolo dello stato ecologico dei corsi d'acqua e di rappresentazione grafica delle informazioni. Biologia Ambientale 14(2): 1–6.

Thorne, R. S. & W. P. Williams, 1997. The response of benthic macroinvertebrates to pollution in developing countries: a multimetric system of bioassessment. Freshwater Biology 37: 671–686.

Verdonshot, P. F. M., 2000. Integrated ecological assessment methods as a basis for sustainable catchment management. Hydrobiologia 442/443: 389–412.

Wasson, J. G., A. Chandesris, H. Pella, & L. Blanc, 2003. Typologie des eaux courantes pour la Directive Cadre Européenne sur l'Eau : l'approche par Hydro-écorégion. "5ème séminaire du réseau d'animation REGLIS (Représentation et Gestion de l'Information Spatialisée)", Cemagref UMR Structures et Systèmes Spatiaux, Montpellier, 13–14 novembre, 7 p.

Wright, J. F., 1995. Development and use of a system for predicting the macroinvertebrate fauna in flowing waters. Australian Journal of Ecology 20: 181–198.

Hydrobiologia (2006) 566:401–415
M.T. Furse, D. Hering, K. Brabec, A. Buffagni, L. Sandin & P.F.M. Verdonschot (eds), The Ecological Status of European Rivers: Evaluation and Intercalibration of Assessment Methods
DOI 10.1007/s10750-006-0081-8

Direct comparison of assessment methods using benthic macroinvertebrates: a contribution to the EU Water Framework Directive intercalibration exercise

Sebastian Birk* & Daniel Hering
Department of Hydrobiology, University of Duisburg-Essen, Universitätsstr. 5, D-45117, Essen, Germany
*(*Author for correspondence: E-mail: sebastian.birk@uni-essen.de)*

Key words: ecological quality classification, biological assessment index, macroinvertebrates, common intercalibration type, STAR project, EU Water Framework Directive

Abstract

The aim of the intercalibration exercise presently performed by the EU is to identify and resolve significant inconsistencies between the ecological quality classifications of EU Member States and the normative definitions of the EU Water Framework Directive. Based on benthic macroinvertebrate data of two European stream types (small siliceous mountain streams and medium-sized lowland streams in Central and Western Europe) we correlated the indices of 10 river quality assessment methods (ASPT, BMWP, DSFI, German Multimetric Index, Saprobic Indices) applied in Austria, Czech Republic, Denmark, Germany, Poland, Slovak Republic, Sweden and United Kingdom. National class boundaries were compared via regression analysis. Assessment methods of the same type (Saprobic Indices, BMWP/ASPT scores) showed best correlation results ($R^2 > 0.7$). The good quality status boundaries of the national methods deviated up to 25%; thus indicating the necessity to harmonize the national classification schemes. Prerequisites of the presented intercalibration approach are (1) a sufficiently large and consistent dataset representative of the respective common intercalibration types and (2) agreement on common type specific reference conditions.

Introduction

In the individual European countries the practice of evaluating ecological river quality is very different (Metcalfe-Smith, 1994; Knoben et al., 1995; Birk & Hering, 2002). Although river monitoring programmes in most countries are based on the benthic macroinvertebrate community, design and performance of individual methods to assess rivers with this organism group vary significantly. On the one hand this is due to different traditions in stream assessment. While in many Central and Eastern European countries modifications of the Saprobic System have been applied for decades as standard methods (Birk & Schmedtje, 2005), other countries rely on the Biological Monitoring Working Party score (BMWP, 1978), which has been adjusted for the use in various countries (Armitage et al., 1983; Just et al., 1998; Alba-Tercedor & Pujante, 2000; Kownacki et al., 2004). On the other hand the EU Water Framework Directive had a great effect on European freshwater management, since it outlines an innovative concept of bioassessment: not the impact of single pressures on individual biotic groups but the deviation of the community from undisturbed conditions is decisive for ecological status classification. In many EU Member States efforts are being made to adapt the national programmes to these new requirements; however, different approaches are being used, since in some countries a single stressor (e.g. organic pollution) is overwhelming, while in other regions different stressors are of equal importance and simultaneously affect river inhabiting communities.

To overcome the difficulties in comparing the various national assessment methods the Directive outlines an intercalibration procedure of the methods' outputs. Member States are enabled to establish or to maintain their own methods; a definition of high, good or moderate biological quality is provided centrally through the intercalibration exercise. The aim of the intercalibration exercise is to identify and to resolve significant inconsistencies between the quality class boundaries established by Member States and indicated by the normative definitions of the Directive (CIS WG 2.A Ecological Status, 2004).

The first efforts to compare different national assessment methods in Europe go back to 1975. Three intercalibration campaigns organized by the Commission of the European Communities included comparisons of field sampling, sample treatment and quality assessment applied in Germany, Italy and United Kingdom (Tittizer, 1976; Woodiwiss, 1978; Ghetti & Bonazzi, 1980). These early studies established strong correlations between the individual assessment methods and compared the methods directly. This approach towards intercalibration was then followed by various authors both to demonstrate the relationship of methods and to point out discrepancies between national quality classifications (Ghetti & Bonazzi, 1977; Rico et al., 1992; Friedrich et al., 1995; Biggs et al., 1996; Morpurgo, 1996; Stubauer & Moog, 2000). In their preparatory study for the Water Framework Directive Nixon et al. (1996) explicitly recommended direct comparison to be used for the intercalibration of assessment methods.

However, the official intercalibration exercise for the Water Framework Directive has adopted an alternative approach due to the lack of a sufficiently large and consistent international database covering all of Europe: indirect comparison via intercalibration common metrics, thus, generating a 'common' multimetric assessment procedure, which is more or less applicable in most of Europe and comparing national assessment methods against this common method (Buffagni et al., 2006).

In this paper we

(1) evaluated the principal suitability of directly comparing assessment methods for intercalibration procedures;
(2) tested a variety of different regression techniques to refine the practical application of direct comparison for intercalibration purposes;
(3) directly compared assessment methods frequently applied for two broadly defined European river types and suggest steps for harmonising class boundaries.

Methods

Overview

This study was based on a two-step analysis: first, different assessment methods, which are presently being used in national water management, were calculated with the same taxa lists. The results of the individual assessment methods were then directly compared by regression analysis.

All data used in this study resulted from the AQEM project (Hering et al., 2004) and the STAR project (Furse et al., 2006). Only data on invertebrate samples restricted to two broadly defined stream types were used. With the data from each stream type up to 10 national assessment systems were calculated, which were first normalized by calculating ecological quality ratios (EQR) (i.e., transferring the results into a common scale ranging from 0 to 1). These normalized assessment results were fed into a regression analysis, to translate the index results of country A into the index results of country B. Comparison of more than two methods was enabled by including the index of country C and translating these results into the index results of country B ('common scale'). In addition, the assessment results were correlated to environmental gradients. In a second step, the class boundaries between the individual quality classes, as applied by the national assessment systems, were compared.

To test the impact of different regression techniques on the results, linear and nonlinear techniques were compared.

Samples and sites

This study was based on benthic invertebrate data sampled in the EU projects AQEM and STAR with standardized field and laboratory protocols (Furse et al., 2006). The data were limited to two broadly defined stream type groups: small, siliceous mountain streams and medium-sized lowland streams in Central and Western Europe. In the official intercalibration exercise for the Water Framework Directive, these stream types were named 'small-sized, mid-altitude brooks of siliceous geology' (R-C3) and 'medium-sized, lowland streams of mixed geology' (R-C4) in Central Europe (Table 1).

Two hundred ninety four samples taken at 125 sites located in four different countries in spring and summer were analysed for the small mountain streams. The lowland stream type embraced a total of 217 samples taken at 71 sites in four different countries in spring, summer and autumn.

The ecological quality of each sampling site was pre-classified based on expert judgement of the field researchers having sampled the streams and, if available, additional knowledge derived from previous studies. Each site was assigned to one of five quality classes ('high', 'good', 'moderate', 'poor', 'bad') referring to the estimated main stressor's degree of impairment. For the AQEM sites, the pre-classification of most sites was replaced by the post-classification after sampling due to additional environmental parameters gained during the field work (physical–chemical and hydromorphological variables).

National assessment methods and quality classifications

Altogether ten biological assessment indices were compared in this analysis (Table 2), all of which are either in current usage in certain European countries or are about being implemented into water management as standard techniques. Most represented biotic index or score methods (Saprobic Index (SI), Biological Monitoring Working Party (BMWP) Score, Average Score Per Taxon (ASPT), Danish Stream Fauna Index (DSFI)). All indices were part of the respective national method

Table 1. Overview of samples included in the analysis

Stream type	Country	Stream type	Ecoregion no.	Number of samples
Small siliceous mountain streams	Austria	Small-sized shallow mountain streams	9	36
	Czech Republic	Small-sized shallow mountain streams	9, 10	40
		Small-sized streams in the Central Sub-alpine mountains	9	32
		Small-sized streams in the Carpathians	10	28
	Germany	Small streams in lower mountainous areas of Central Europe	9	86
		Small-sized Buntsandstein-streams	9	24
	Slovak Republic	Small-sizes siliceous mountains streams in the West Carpathians	10	48
Medium-sized lowland streams	Denmark	Medium-sized deeper lowland streams	14	46
	Germany	Mid-sized sand bottom streams in the German lowlands	14	86
	Sweden	Medium-sized deeper lowland streams	14	14
		Medium-sized streams on calcareous soils	14	35
	United Kingdom	Medium-sized deeper lowland streams	18	36

Table 2. Overview of national assessment methods

Stream type	Country	Assessment index	Category	Abundance	Reference
Small siliceous mountain streams	Austria	SI (AT) – Austrian Saprobic Index	BI	Y	Moog et al. (1999)
	Czech Republic	SI (CZ) – Czech Saprobic Index	BI	Y	CSN 757716 (1998)
	Germany	SI (DE) – German Saprobic Index	BI	Y	Friedrich & Herbst (2004)
	Poland	BMWP (PL) – Polish Biological Monitoring Working Party score	BI	N	Kownacki et al. (2004)
	Slovak Republic	SI (SK) – Slovak Saprobic Index	BI	Y	STN 83 0532-1 to 8, (1978/79)
	United Kingdom	ASPT (UK) – Average Score Per Taxon	BI	N	Armitage et al. (1983)
Medium-sized lowland streams	Denmark	DSFI (DK) – Danish Stream Fauna Index	BI	N	Skriver et al. (2000)
	Germany	GD (DE) – Module 'General Degradation' of the German Assessment System Macrozoobenthos	MI*	Y	Böhmer et al. (2004)
	Sweden	ASPT (SE) – Average Score Per Taxon applied in Sweden	BI	N	Swedish Environmental Protection Agency (2000)
		DSFI (SE) – Danish Stream Fauna Index applied in Sweden	BI	N	
	United Kingdom	ASPT (UK) – Average Score Per Taxon	BI	N	Armitage et al. (1983)

BI, biotic index; MI, multimetric index. *Includes the following single metrics: relative abundance of ETP taxa, German Fauna Index Type 15, number of Trichoptera taxa, Shannon–Wiener diversity, share of rheobiontic taxa, share of shredders (%).

planned for biological monitoring in the context of the Water Framework Directive. With the exception of DSFI and ASPT, applied in Sweden, calculation of index values was based on a nationally adjusted indicator species list.

For the indices applied in Austria, the Czech Republic, Germany and Denmark, stream type specific reference values existed; these described the value of an index to be expected under 'undisturbed conditions'. The system used in the United Kingdom predicted site specific reference values, Sweden defined reference conditions for broad-scale natural geographical regions but in Poland and the Slovak Republic reference values have not yet been established. All indices distinguished between five classes of biological quality. The British and Swedish methods and the German multimetric index defined class boundary values as EQR. The Polish BMWP and the Saprobic Systems used quality classes given as absolute index values. The Austrian, Czech and German quality bands were stream type specific. An overview of nationally defined reference conditions and class boundaries is given in Table 3.

Data preparation

National assessment methods were calculated to the taxa lists of each sample. Absolute index values were converted into EQR by dividing the calculated (observed) value by the index specific reference value. Since, for the Saprobic Indices, biological quality decreased with increasing index values these were converted by the following equation:

Table 3. Original reference and class boundary values of the national assessment methods

Index	SI (AT)	SI (CZ)	SI (DE)	BMWP (PL)	SI (SK)	ASPT (UK)
Small siliceous mountain streams						
Reference (abs)	≤1.50	≤1.20	≤1.25	n.a.	n.a.	≥6.62*
High-good	1.50	1.20	1.40	100	1.79	1.00
Good-moderate	2.10	1.50	1.95	70	2.30	0.89
Moderate-poor	2.60	2.00	2.65	40	2.70	0.77
Poor-bad	3.10	2.70	3.35	10	3.20	0.66
Lit. source	–	Brabec et al. (2004)	Rolauffs et al. (2003)	Kownacki et al. (2004)	STN 83 0532-1 to 8 (1978/79)	National Rivers Authority (1994)

Index	DSFI (DK)	GD (DE)	BMWP (PL)	ASPT (SE)	DSFI (SE)	ASPT (UK)
Medium-sized lowland streams						
Reference (abs)	7	1	n.a.	≥4.7	≥5	≥6.38*
High-good	7	0.80	100	0.90	0.90	1.00
Good-moderate	5	0.60	70	0.80	0.80	0.89
Moderate-poor	4	0.40	40	0.60	0.60	0.77
Poor-bad	3	0.20	10	0.30	0.30	0.66
Lit. source	–	Böhmer et al. (2004)	Kownacki et al. (2004)	Swedish Environmental Protection Agency (2000)	Swedish Environmental Protection Agency (2000)	National Rivers Authority (1994)

Abs, absolute value. *Values were derived by RIVPACS predictions for the corresponding stream type group based on averaged environmental parameter values and combined season information for the analysed samples.

$$EQR_{SI} = 1 - \frac{\text{observed SI value} - \text{reference SI value}}{\text{maximum SI value} - \text{reference SI value}}$$

To validate the national reference values, an index specific reference value was calculated as the 75th percentile of all samples taken at sites pre- or post-classified as high quality status (excluding outliers). For the small mountain streams, sampling sites located in Austria (6 samples), Czech Republic (14 samples), Germany (13 samples) and Slovak Republic (1 sample) were used. For the lowland type sites from Denmark (13 samples), Germany (26 samples), Sweden (2 samples) and United Kingdom (9 samples) were the basis of this calculation.

Conversion into the EQR scale resulted in values ranging from 0 to >1 since several samples revealed biological index values representing higher quality than the respective reference value. These values were not transformed into the value '1' in order to improve the correlation and regression analysis by enlarging the quality gradient.

Correlation and regression analysis

The magnitude of the relation between two assessment methods was specified by the 'coefficient of determination'. Beside linear regression, we applied nonlinear modelling via automatic curve-fitting using the software TableCurve 2D (SYSTAT Software Inc., 2002).

Comparison of quality class boundaries

In order to compare the national quality classes the boundary values of the different assessment methods were transformed into a 'common scale'. In this study two common scales were used: (1) The national method showing the highest mean correlation of all indices. (2) The 'integrative multimetric index for intercalibration' (IMI-IC), an artificial index designed here for the purpose of intercalibration. This index was defined as the mean of all index values calculated for a sample.

The transformation was done based on the results of linear regression analyses, in which the predictor variables were represented by the national indices and the response variables by the 'common scale'. Each boundary value transformed by regression was given including its 95% confidence interval. Class boundaries showing overlapping ranges (translated class boundary ± confidence interval) were considered as being equal.

Based on environmental variables, abiotic gradients were generated for each stream type and the pressure gradients best correlating to the methods analysed in this intercalibration approach were identified. Indirect gradient analysis was aimed at the identification and quantification of physical–chemical and hydromorphological gradients that can be assigned to human impairment. Therefore, Principle Component Analysis (PCA) was run separately on correlation matrices of physical–chemical, catchment land use, hydromorphological and microhabitat variables of the mountain and lowland dataset (see Feld et al., in prep.). A dimensionless value of abiotic pressure, including the 95% confidence interval, was assigned to each national class boundary via regression analysis. These pressure data were used to support class boundary comparisons.

Results

Definition of reference values

The 75th percentiles of reference values were specified in Table 4. Each reference was based on a slightly different number of samples due to the elimination of outliers. Except for the German indices and the assessment methods for which no reference was nationally defined (Polish BMWP and Slovak SI), the 75th percentile, as calculated in this study, generally represented higher biological quality than the minimum values of the national reference.

Descriptive statistics of national indices calculated from the AQEM–STAR datasets

The overall mean of normalized index values (0–1) for the small mountain streams amounted to 0.87, while the same statistic for medium-sized lowland streams was 0.77 (Table 5). The maximum values of all indices except DSFI exceeded 1.0. This was due to the selection of the 75th percentile of AQEM–STAR high status sites as the reference value. The values of the Polish BMWP and the German GD covered ranges of more than 1.0, while the Austrian and German SI, and the British and Swedish ASPT showed value ranges of less than 0.65.

Correlation and regression of national assessment methods

The correlation analysis revealed differences between assessment methods (Table 6). The linear equations of the regression analysis of national methods against methods representing a common scale (best correlating national index, IMI-IC) were displayed in Table 7.

For small mountain streams coefficients of determination ranged from 0.20 (Slovak SI and

Table 4. Reference values of national assessment methods derived by using the 75th percentile of index values calculated from samples taken at high status sites

Index	SI (AT)	SI (CZ)	SI (DE)	BMWP (PL)	SI (SK)	ASPT (UK)
Small siliceous mountain streams						
75th percentile	1.46 (32)	0.91 (34)	1.44 (33)	187 (33)	1.21 (30)	7.26 (33)
Index	DSFI (DK)	GD (DE)	BMWP (PL)	ASPT (SE)	DSFI (SE)	ASPT (UK)
Medium-sized lowland streams						
75th percentile	7	0.67	150	6.57	7	6.57

For small mountain streams the number of high status sites' samples is individually specified in brackets. Values of lowland streams are based on 50 samples.

Table 5. Descriptive statistics of national indices calculated from the AQEM–STAR datasets (normalized index values)

	Mean	Minimum	Maximum	25th percentile	75th percentile	Range	Quartile range
Small siliceous mountain streams ($n=294$)							
SI (AT)	0.902	0.526	1.112	0.833	0.972	0.585	0.138
SI (CZ)	0.853	0.374	1.112	0.761	0.963	0.739	0.202
SI (DE)	0.920	0.444	1.055	0.895	0.984	0.611	0.088
BMWP (PL)	0.768	0.102	1.273	0.636	0.936	1.171	0.299
SI (SK)	0.890	0.444	1.281	0.798	0.984	0.837	0.186
ASPT (UK)	0.908	0.448	1.077	0.869	0.988	0.629	0.119
Medium-sized lowland streams ($n=217$)							
DSFI (DK) and DSFI (SE)	0.767	0.286	1.000	0.571	1.000	0.714	0.429
GD (DE)	0.709	0.090	1.149	0.552	0.896	1.060	0.343
BMWP (PL)	0.741	0.173	1.480	0.580	0.900	1.307	0.320
ASPT (SE) and ASPT (UK)	0.869	0.457	1.091	0.797	0.956	0.634	0.159

Table 6. Coefficients of determination based on linear and nonlinear regression ($p<0.05$)

Index	SI (AT)		SI (CZ)		SI (DE)		BMWP (PL)		SI (SK)		ASPT (UK)	
	Linear	Nonl.	Linear	Nonl.	Linear	Nonl.	Linear	nonl.	Linear	Nonl.	Linear	Nonl.
Small siliceous mountain streams ($n=294$)												
SI (AT)	1.00	–	0.62	–	0.70	0.74	0.36	0.39	0.73	0.77	0.45	0.46
SI (CZ)	0.62	–	1.00	–	0.62	0.64	0.31	0.35	0.55	–	0.38	–
SI (DE)	0.70	0.73	0.62	0.70	1.00	–	0.53	0.63	0.48	0.56	0.69	0.73
BMWP (PL)	0.36	0.37	0.31	0.34	0.53	–	1.00	–	0.20	0.23	0.69	0.70
SI (SK)	0.73	–	0.55	–	0.48	0.51	0.20	0.21	1.00	–	0.24	0.26
ASPT (UK)	0.45	0.50	0.37	0.45	0.69	0.70	0.69	0.75	0.24	0.36	1.00	–
IMI-IC$_{R\text{-}C3}$	0.79	0.80	0.72	0.74	0.86	0.87	0.72	0.75	0.62	0.66	0.75	–
PE1	0.31	0.33	0.23	0.27	0.46	–	0.37	0.38	0.19	0.23	0.53	–

Index	DSFI (DK) and DSFI (SE)		GD (DE)		BMWP (PL)		ASPT (SE) and ASPT (UK)	
	Linear	Nonl.	Linear	Nonl.	Linear	Nonl.	Linear	Nonl.
Medium-sized lowland streams ($n=217$)								
DSFI (DK) and DSFI (SE)	1.00	–	0.61	–	0.53	0.54	0.65	–
GD (DE)	0.61	–	1.00	–	0.41	0.46	0.49	–
BMWP (PL)	0.53	0.54	0.41	–	1.00	–	0.51	–
ASPT (SE) and ASPT (UK)	0.65	0.67	0.49	0.50	0.51	0.57	1.00	–
IMI-IC$_{R\text{-}C4}$	0.90	–	0.76	–	0.73	0.75	0.80	–
HY1	0.23	–	0.35	–	0.12	0.13	0.24	0.26

IMI-IC, integrative multimetric index for intercalibration (see text for explanation); PE1, pollution/eutrophication gradient; HY1, hydromorphological gradient.

Polish BMWP) to 0.77 (Austrian SI and Slovak SI). Nonlinear regression gained higher R^2 values in 23 out of 36 relations. The mean difference in R^2 values between linear and nonlinear regressions was 0.04. The maximum difference in R^2 values of 0.12 was between linear and nonlinear equations for the relationship between SI (SK) and ASPT (UK). German SI had the highest average

Table 7. Coefficients of linear regression equations (*a* – slope, *b* – intercept) for the common scales and the abiotic gradients

Index	SI (AT)		SI (CZ)		SI (DE)		BMWP (PL)		SI (SK)		ASPT (UK)	
Parameter	*a*	*b*	*a*	*b*	*a*	*b*	*a*	*b*	*a*	*b*	*a*	*b*
Small siliceous mountain streams												
SI (DE)	0.784	0.212	0.562	0.440	1.000	0	0.319	0.675	0.511	0.465	0.687	0.296
IMI-IC$_{R\text{-}C3}$	0.992	−0.021	0.717	0.261	1.100	−0.138	0.441	0.535	0.688	0.261	0.850	0.102
PE1	−0.845	1.000	−0.567	0.720	−1.089	1.236	−0.450	0.577	−0.542	0.721	−0.976	1.120

Index	DSFI (DK) and DSFI (SE)		GD (DE)		BMWP (PL)		ASPT (SE) and ASPT (UK)	
Parameter	*a*	*b*	*a*	*b*	*a*	*b*	*a*	b
Medium-sized lowland streams								
DSFI	1.000	0.000	0.579	0.356	0.344	0.570	1.349	−0.405
IMI-IC$_{R\text{-}C4}$	0.825	0.154	0.566	0.386	0.357	0.580	1.301	−0.343
HY1	−0.627	0.934	−0.583	0.857	−0.360	0.720	−1.078	1.396

IMI-IC, integrative multimetric index for intercalibration (see text for explanation); PE1, pollution/eutrophication gradient; HY1, hydromorphological gradient.

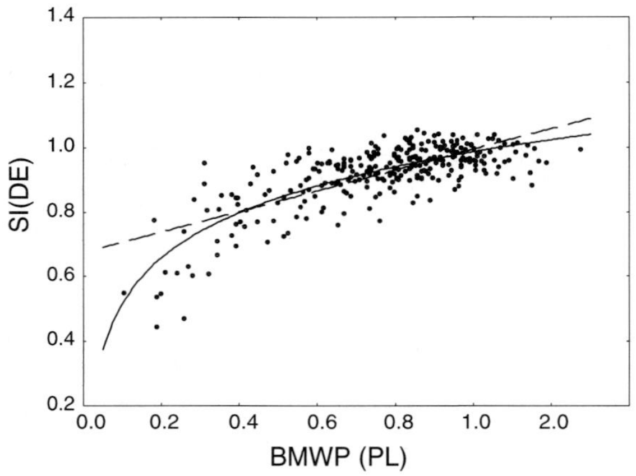

Figure 1. Regression of BMWP (PL) against SI (DE). Both linear ($R^2=0.53$, dashed) and nonlinear ($R^2=0.63$) regression lines are plotted.

correlation to the other assessment methods ($R^2=0.67$). The IMI-IC for this stream type was characterized by coefficients of determination ranging from 0.62 (Slovak SI) to 0.87 (German SI). In Figure 1 regression lines of BMWP (PL) against SI (DE) were exemplarily plotted for linear and nonlinear regression.

R^2 values for regressions of methods for the lowland streams varied between 0.41 (German GD and Polish BMWP) and 0.67 (British and Swedish ASPT, and Danish and Swedish DSFI). In 6 out of 16 correlations, nonlinear regression provided a higher proportion of the variance explained. Mean difference of the linear and nonlinear coefficients of determination was $R^2=0.02$ and the maximum difference was $R^2=0.06$ (Polish BMWP and British ASPT). DSFI showed the highest mean correlation for the lowland samples ($R^2=0.60$). The IMI-IC had coefficients of correlation ranging from 0.73 (Polish BMWP) to 0.90 (Danish and Swedish DSFI). All correlations were significant at $p<0.05$. Since none of the differences between the linear and nonlinear coefficients of determination were significant, we assumed linear relationships between indices in the following analyses.

Correlation to environmental gradients (PCA)

Index values of the small mountain streams showed the strongest relationship with the PCA gradient reflecting nutrient enrichment and organic pollution. Determination coefficients of this gradient and the assessment methods varied from 0.19 (Slovak SI) to 0.53 (British ASPT). Index values of the lowland streams showed highest correlations with the main hydromorphological gradient that comprised physical features of the river channel, its banks and immediate vicinity,

Table 8. EQR values of the high-good (H/G) and good-moderate (G/M) quality class boundaries transferred into 'common scale'

Class boundary	Common scale	SI (AT)		SI (CZ)		SI (DE)		BMWP (PL)		SI (SK)		ASPT (UK)	
		Boundary value	95% confid.	Boundary value	95% confid.	Boundary value	95% confid.	Boundary value	95% confid.	Boundary value	95% confid.	Boundary value	95% confid.
Small siliceous mountain streams													
H/G	SI (DE)	0.984	0.008	0.949	0.008	1.016	–	0.846	0.011	0.870	0.010	0.983	0.008
	IMI-$IC_{R\text{-}C3}$	0.955	0.008	0.911	0.008	0.979	0.008	0.771	0.010	0.806	0.011	0.952	0.009
	PE1	0.169	0.023	0.206	0.019	0.130	0.023	0.336	0.022	0.291	0.023	0.144	0.019
G/M	SI (DE)	0.799	0.012	0.895	0.007	0.801	–	0.794	0.016	0.776	0.020	0.907	0.006
	IMI-$IC_{R\text{-}C3}$	0.721	0.012	0.842	0.008	0.743	0.009	0.700	0.015	0.680	0.021	0.858	0.007
	PE1	0.368	0.032	0.262	0.019	0.364	0.025	0.409	0.032	0.391	0.045	0.251	0.014
Class boundary	Common scale	DSFI (DK)		GD (DE)		BMWP (PL)		ASPT (SE)		DSFI (SE)		ASPT (UK)	
		Boundary value	95% confid.	Boundary value	95% confid.	Boundary value	95% confid.	Boundary value	95% confid.	Boundary value	95% confid.	Boundary value	95% confid.
Medium-sized lowland streams													
H/G	DSFI	1.000	–	1.048	0.012	0.724	0.018	0.809	0.016	0.900	–	0.944	0.025
	IMI-$IC_{R\text{-}C4}$	0.979	0.012	1.061	0.008	0.744	0.012	0.827	0.011	0.897	0.009	0.958	0.017
	HY1	0.307	0.054	0.162	0.021	0.480	0.036	0.426	0.035	0.370	0.042	0.318	0.054
G/M	DSFI	0.714	–	0.875	0.016	0.610	0.016	0.674	0.019	0.800	–	0.795	0.016
	IMI-$IC_{R\text{-}C4}$	0.744	0.008	0.892	0.011	0.628	0.011	0.697	0.012	0.814	0.007	0.814	0.010
	HY1	0.486	0.035	0.335	0.030	0.552	0.034	0.534	0.041	0.432	0.035	0.437	0.034

In addition, the values of the abiotic gradients (PE1, HY1) corresponding to the national class boundaries are displayed. For each value derived by regression the 95% confidence interval is specified. IMI-IC, integrative multimetric index for intercalibration (see text for explanation); PE1, pollution/eutrophication gradient; HY1, hydromorphological gradient.

including information on the degree of impairment. The coefficients of determination ranged between 0.12 (Polish BMWP) and 0.35 (German GD).

Comparison of national quality classes

The comparison of biological quality classes was based on the transformation of boundary values of the assessment methods into a common scale. This allowed for a direct juxtaposition of class boundaries in Table 8.

Small-sized siliceous mountain streams

The common scales used in the comparison procedure for the mountain streams were SI (DE) and IMI-$IC_{R\text{-}C3}$ (multimetric index composed of all national assessment methods). In SI (DE) scale, the high-good boundaries of SI (AT) and ASPT (UK) were similar considering the 95% confidence interval. ASPT (UK) and SI (CZ) showed overlapping good-moderate boundary intervals and thus shared equal class boundaries. The same applied for the group of indices SI (AT), SI (DE), BMWP (PL) and SI (SK). Based on IMI-$IC_{R\text{-}C3}$ the high-good boundaries of SI (AT) and ASPT (UK) shared common intervals. For the good-moderate boundary the comparison showed similar values for SI (AT), BMWP (PL) and SI (SK).

The pollution/eutrophication gradient showed similar pressure between high-good boundaries of SI (AT), SI (CZ), SI (DE), ASPT (UK), and BMWP (PL) and SI (SK). For the good-moderate boundary corresponding levels of chemical impairment were between SI (AT) and SI (DE), SI (SK) and BMWP (PL), and SI (CZ) and ASPT (UK). The average confidence interval amounted to 0.025 units.

Medium-sized, lowland, mixed geology

The DSFI and IMI-$IC_{R\text{-}C4}$ (multimetric index composed of all national assessment methods) were used as common scales for the boundary comparisons of the lowland stream type. Using DSFI as the common scale, none of the national indices showed similar high-good class boundaries but the good-moderate boundaries of DSFI (SE) and ASPT (UK) were corresponding. The average confidence interval amounted to 0.017 DSFI units.

In the IMI-$IC_{R\text{-}C4}$ scale, the high-good boundaries of DSFI (DK) and ASPT (UK) had similar values and the good-moderate boundaries of DSFI (SE) and ASPT (UK) corresponded closely. Confidence intervals showed an average value of 0.011 units.

Boundary comparisons using the hydromorphological gradient were difficult because the large confidence intervals (0.038 units in average) resulted in overlapping boundary ranges. Both good quality boundaries of GD (DE) showed the lowest level of pressure. For the good-moderate boundary, levels of pressure were similar between DSFI (DK), DSFI (SE) and ASPT (UK), and between BMWP (PL) and ASPT (SE).

Discussion

Role of reference conditions in the intercalibration exercise

Within the intercalibration exercise, class boundaries of national assessment methods need to be defined as EQR. The position of each boundary on this relative scale is dependent on (1) the definition of reference conditions and (2) the procedure of setting class boundaries. If the former is not properly dealt with in the intercalibration process, the different nationally defined reference values may strongly impact upon comparability.

In this study we have defined a common reference, which is based on sites in several countries. As a result of this common reference, it was possible to include several methods in the comparison, even if countries have not yet defined reference values for a specific method. A further advantage of common references is that differences in national approaches to define references are avoided. On the other hand, common references are in danger of not adequately accounting for the differences between more specific streams types.

More importantly, countries have applied different procedures to define reference values and quality classification schemes. While this study is restricted to the analysis of national class boundary settings, it must be an objective of the official intercalibration exercise to overcome differences in the references too.

Relations between assessment methods

In this study, the calculation of national assessment metric values is based on taxa lists derived by application of the standardized STAR–AQEM field and laboratory protocol. Thus, the correlation analyses of index values mainly reveal the numerical relation between these indices and is less biased by differences in field and laboratory procedures. The character of these relations depends on the architecture of the individual indices, e.g. number and indicative value of taxa included in the evaluation, type of abundance information used and the assessment formula. The effect of different national sampling methods on the comparability of taxa lists and metric results as a major constraint of intercalibration is investigated by Friberg et al. (2006). Buffagni et al. (2006) present a practical approach enabling the use, in intercalibration, of datasets derived by the national monitoring programmes.

An additional factor, impacting on the relationships, is the dataset itself, in particular the number of samples, the biogeographical gradient, the types of pressures influencing sampling sites and the range of degradation covered. The different ranges of index values (cf. Table 5) indicate a larger degradation gradient being covered by the lowland dataset. This is, in particular, obvious from the Polish BMWP and British ASPT values, which have been calculated for both datasets.

For the mountain stream data, relationships are strongest between the values of the different Saprobic Indices of Austria, Czech Republic, Germany and Slovak Republic and between the score methods applied in Poland and the United Kingdom. In general, the strength of correlations between the different Saprobic Indices results from similarities in indicator taxa and their indication values (Table 9). For instance, the Austrian and Slovak Saprobic Indices ($R^2 > 0.73$) share the largest number of indicator taxa and are most closely related concerning indicator taxa value and weight. Schmidt-Kloiber et al. (2006) provide a comprehensive analysis of saprobic indicator taxa applied in Europe.

For the lowland stream dataset, BMWP (PL) and ASPT (UK) correlate less strongly ($R^2 < 0.60$), which can be explained by the different taxonomic composition of the lowland dataset compared to that of the mountain streams. The two indices have 66 indicator taxa in common, amounting to a share of 73% (Polish BMWP) and 80% (British ASPT), respectively. BMWP indicator values of the common taxa in the Polish and UK systems are correlated with $R^2 = 0.73$.

Method comparisons of earlier studies show similar results. Based on 232 samples from various lowland and mountain stream types in Germany, Friedrich et al. (1995) found correlations of $R^2 = 0.71$ between ASPT (UK) and a previous version of the German Saprobic Index. The weak relation of ASPT and the Austrian Saprobic Index has already been demonstrated by Stubauer & Moog (2000), who used a large dataset covering all Austrian stream types ($n = 588$; $R^2 = 0.52$). Analyses of Birk & Rolauffs (2003) revealed strong correlations between the Austrian and German Saprobic Indices ($n = 262$; $R^2 = 0.75$).

Several indices revealed higher coefficients of determination when applying a nonlinear fit, in

Table 9. Comparison of the saprobic indicator taxa lists of Austria, Czech Republic, Germany and Slovak Republic: Share of common taxa and coefficients of determination derived from correlation analysis of indicator values and indicator weights

	SI (AT)			SI (CZ)			SI (DE)			SI (SK)		
	Share of common taxa (%)	Indicator value	Indicator weight	Share of common taxa (%)	Indicator value	Indicator weight	Share of common taxa (%)	Indicator value	Indicator weight	Share of common taxa (%)	Indicator value	Indicator weight
SI (AT)	–	1.00	1.00	56	0.64	0.14	72	0.74	0.04	77	0.88	0.53
SI (CZ)	36	0.64	0.14	–	1.00	1.00	54	0.74	0.14	53	0.73	0.31
SI (DE)	35	0.74	0.04	41	0.74	0.14	–	1.00	1.00	41	0.73	0.04
SI (SK)	45	0.88	0.53	48	0.73	0.31	49	0.73	0.04	–	1.00	1.00

particular if BMWP (PL) was involved. This index combines the parameters taxon richness and sensitivity into a single value which may cause the observed relationship. Also, due to the large range of values covered by the method, the nonlinearity of the relationships became evident (cf. Fig. 1). Nevertheless, these difference of the coefficients of determination are not significant. Therefore, the simple model of linear relationship between indices is most appropriate in this example of direct comparison.

Comparison of class boundary values

While earlier intercalibration studies focussed on the comparison of quality class bands (Ghetti & Bonazzi, 1977; Friedrich et al., 1995; Morpurgo, 1996), the Water Framework Directive specifically requires the comparability of the high-good and good-moderate quality class boundaries. Thus, the intercalibration exercise is focussed on the range medium to high biological quality. The original procedure outlined in the Directive is restricted to the use of just a few intercalibration sites, selected because they represent the boundary status between quality classes. However, this approach seems not to be feasible, since sites known to be on class boundaries cannot be selected prior to the intercalibration is completed and those boundaries are defined. Furthermore, the uncertainty of intercalibration results is high if the analysis is based on insufficient data.

Therefore, the primary step, in comparing national class boundary values and best identifying the type and magnitude of the relationship between national assessment methods, should be based on a large number of samples covering the entire quality gradient. In a further step, regression analysis should be used to transform boundary values into other assessment scales. By applying an acceptable level of uncertainty (e.g., confidence interval of 95% derived from regression analysis), ranges of index values can be compared.

The comparison of assessment methods has revealed discrepancies between national classification schemes of more than 25% in particular cases (e.g. high-good boundary of German SI and Polish BMWP translated in German SI scale). The extent of differences between class boundaries is largely dependent on the common scale used for comparison. While class boundaries clearly differ if compared through the German Saprobic Index scale, no differences occur between the same boundaries if compared through a multimetric index. Each method used as a common scale is somewhat related to other assessment methods as expressed by the correlation coefficient and the regression equation.

Based on these findings we recommend using the intercalibration approach described in this paper only for comparison of methods addressing similar components of the biocoenosis, e.g. for methods that are closely related such as ASPT, BMWP and the Saprobic Indices, or methods that are fully compliant with the requirements of the Water Framework Directive (i.e., methods evaluating taxonomic composition and abundance, ratio of disturbance sensitive to insensitive taxa and diversity of the macroinvertebrate community). This principle makes sure that 'like with like' comparisons are applied in intercalibration and minimizes errors in the comparison analysis due to the selection of inappropriate common scales. Furthermore, the relation between assessment methods needs to be carefully evaluated. Nonlinear correlations yielding significantly better fit and smaller confidence intervals are to be favoured over weaker linear relations.

When shall boundaries be considered as different?

Intercalibration encompasses two steps: Firstly, national quality boundaries are compared. If this analysis discovers major differences in classification schemes, they need to be harmonized in a second step. For the first step, we have described a possible procedure to translate boundary values into a common scale, which determines whether or not boundary values are corresponding. According to our results only a few class boundaries are similar, which thus requires the remaining boundaries to be harmonized.

The use of abiotic pressure data in intercalibration allows for additional interpretation of results. Sandin & Hering (2004) applied organic pollution gradients to set intercalibration class boundaries defining a standard level of pollution. We particularly propose to use pressure information for the process of boundary comparison. Figure 2 displays the relative position of the national good-moderate

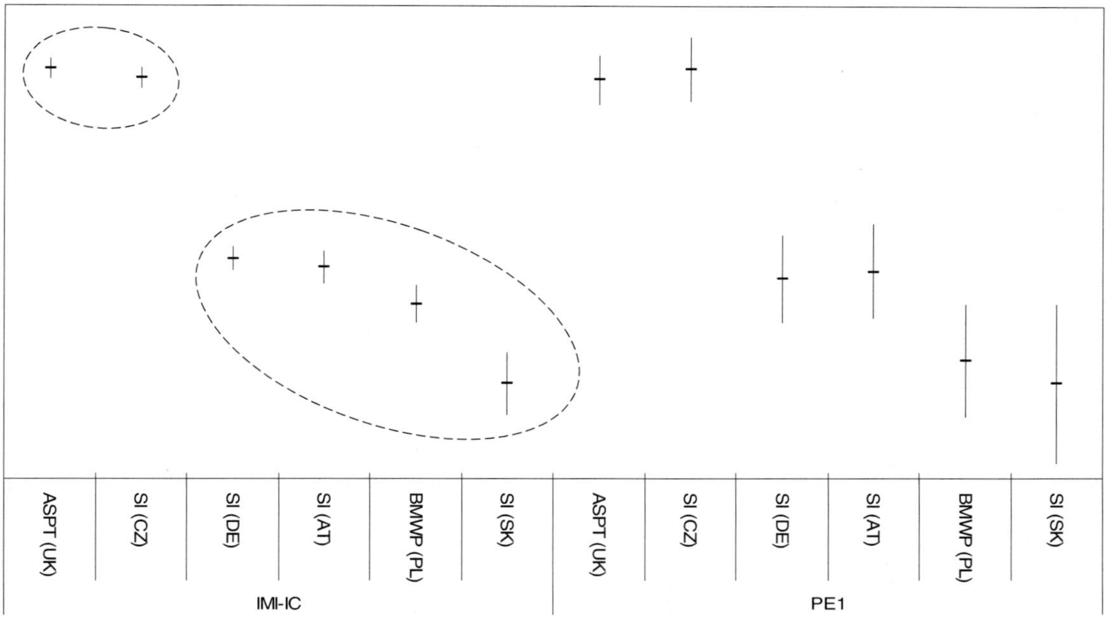

Figure 2. Relative comparison of good-moderate class boundary values (incl. 95% confidence intervals) using IMI-IC$_{R\text{-}C3}$ and corresponding chemical pressure values of the small siliceous mountain streams. Based on the results of the pressure data analysis two groups of similar boundaries are highlighted by dashed circles.

boundaries, including confidence intervals translated into a common biotic scale and an abiotic pressure scale (pollution/eutrophication gradient). Comparisons based on the interpretation of biotic data indicate that four out of six class boundaries are deviating (cf. Table 8), while the consideration of pressure data (Fig. 2) reveals only two groups of boundaries with overlapping pressure intervals. Thus, harmonization is only needed between the two groups of boundaries.

Conclusions

Intercalibration represents a crucial step towards the implementation of a pan-European water quality standard. Besides scientific issues, which we partly addressed in this paper, it holds a major social challenge. Although assessment methods are in general scientifically sound instruments, the element of quality classification is a concession to the practical requirements of decision making in water policy. According to the Water Framework Directive the quality assigned to a site can decide on restoration efforts to be spent or saved. Therefore, intercalibration is of political interest since the definition of quality boundaries sets the environmental standard to be achieved. Furthermore, intercalibration holds an ethical component: By selecting certain quality criteria we agree on a level of anthropogenic degradation acceptable for our freshwater systems. Although beyond its scope science needs to consider all these aspects in the preparation of reasonable and tenable results.

Acknowledgements

STAR was funded by the European Commission, 5th Framework Programme, Energy, Environment and Sustainable Development, Key Action Water, Contract no. EVK1-CT-2001-00089. The authors gratefully acknowledge the fruitful discussions held with Andrea Buffagni, Stefania Erba and Marcello Cazzola on intercalibration topics. Thanks are due to Peter Rolauffs and Christian Feld for their support in data preparation and analysis, and to Jean-Nicolas Beisel, Wouter van de Bund and Mike Furse for their valuable comments on the manuscript.

References

Alba-Tercedor, J. & A. M. Pujante, 2000. Running-water biomonitoring in Spain: opportunities for a predictive approach. In Wright, J. F., D. W. Sutcliffe & M. T. Furse (eds), Assessing the Biological Quality of Fresh Waters - RIVPACS and Other Techniques. FBA, Ambleside, 207–216.

Armitage, P. D., D. Moss, J. F. Wright & M. T. Furse, 1983. The performance of a new biological water quality score system based on macroinvertebrates over a wide range of unpolluted running-water sites. Water Research 17: 333–347.

Biggs, J., A. Corfield, D. Walker, M. Whitfield & P. Williams, 1996. A preliminary comparison of European methods of biological river water quality assessment. NRA Thames Region Operational Investigation. Environment Agency Technical Report No. 0I/T/001. National Rivers Authority Thames Region, Reading.

Birk, S. & D. Hering, 2002. Waterview web-database: a comprehensive review of European assessment methods for rivers. FBA News 20: 4.

Birk, S. & P. Rolauffs, 2003. A preliminary study comparing the results between the Austrian, Czech and German saprobic systems for the intercalibration of cross-border river basin districts. In Deutsche Gesellschaft für Limnologie (DGL) – Tagungsbericht (Köln). DGL, Werder, 74–79.

Birk, S. & U. Schmedtje, 2005. Towards harmonization of water quality classification in the Danube River Basin: overview of biological assessment methods for running waters. Archiv für Hydrobiologie, Supplement Large Rivers 16: 171–196.

BMWP (Biological Monitoring Working Party)., 1978. Final Report of the Biological Monitoring Working Party: Assessment and presentation of the biological quality of rivers in Great Britain. Department of the Environmental Water Data Unit, London.

Böhmer, J., C. Rawer-Jost, A. Zenker, C. Meier, C. K. Feld, R. Biss & D. Hering, 2004. Assessing streams in Germany with benthic invertebrates: development of a multimetric invertebrate based assessment system. Limnologica 34: 416–432.

Brabec, K., S. Zahradkova, D. Nemejcova, P. Paril, J. Kokes & J. Jarkovsky, 2004. Assessment of organic pollution effect considering differences between lotic and lentic stream habitats. Hydrobiologia 516: 331–346.

Buffagni, A., S. Erba, M. Cazzola, J. Murray-Bligh, H. Soszka & P. Genoni, 2006. The STAR common metrics approach to the WFD intercalibration process: Full application for small, lowland rivers in three European countries. Hydrobiologia 566: 379–399.

CIS WG 2.A Ecological Status (ECOSTAT), 2004. Guidance on the intercalibration process. Agreed version of WG 2.A Ecological Status meeting held 7–8 October 2004 in Ispra. Version 4.1. 14. October 2004. ECOSTAT, Ispra.

CSN 757716., 1998. Water quality, biological analysis, determination of saprobic index. Czech Technical State Standard, Czech Standards Institute, Prague.

Feld, C. K., T. Ofenböck, O. Moog & D. Hering, in prep. Assessing hydromorphological degradation and organic pollution in European rivers – selection of suited metrics derived from benthic macroinvertebrates. Manuscript.

Friberg, N., L. Sandin, M. T. Furse, S. E. Larsen, R. T. Clark & P. Haase, 2006. Comparison of macroinvertebrate sampling methods in Europe. Hydrobiologia 566: 365–378.

Friedrich, G. & V. Herbst, 2004. Eine erneute Revision des Saprobiensystems – weshalb und wozu?. Acta Hydrochimica et Hydrobiologica 32: 61–74.

Friedrich, G., E. Coring & B. Küchenhoff, 1995. Vergleich verschiedener europäischer Untersuchungs- und Bewertungsmethoden für Fließgewässer. Landesumweltamt Nordrhein-Westfalen, Essen.

Furse, M., D. Hering, O. Moog, P. Verdonschot, R. K. Johnson, K. Brabec, K. Gritzalis, A. Buffagni, P. Pinto, N. Friberg, J. Murray-Bligh, J. Kokes, R. Alber, P. Usseglio-Polatera, P. Haase, R. Sweeting, B. Bis, K. Szoszkiewicz, H. Soszka, G. Springe, F. Sporka & I. Krno, 2006. The STAR project: context, objectives and approaches. Hydrobiologia 566: 3–29.

Ghetti, P. F. & G. Bonazzi, 1977. A comparison between various criteria for the interpretation of biological data in the analysis of the quality of running waters. Water Research 11: 819–831.

Ghetti, P. F. & G. Bonazzi, 1980. Biological water assessment methods: Torrente Parma, Torrente Stirone, Fiume Po. 3rd Technical Seminar. Final Report. Commission of the European Communities, Brussels.

Hering, D., O. Moog, L. Sandin & P. F. M. Verdonschot, 2004. Overview and application of the AQEM assessment system. Hydrobiologia 516: 1–20.

Just, I., F. Schöll & T. Tittitzer, 1998. Versuch einer Harmonisierung nationaler Methoden zur Bewertung der Gewässergüte im Donauraum am Beispiel der Abwässer der Stadt Budapest. Umweltbundesamt, Berlin.

Knoben, R. A. E., C. Roos & M. C. M. van Oirschot, 1995. Biological Assessment Methods for Watercourses. UN/ECE Task Force on Monitoring and Assessment, Lelystad.

Kownacki, A., H. Soszka, D. Kudelska & T. Fleituch, 2004. Bioassessment of Polish rivers based on macroinvertebrates. In Geller, W. et al. (eds), Proceedings of the International 11th Magdeburg Seminar on Waters in Central and Eastern Europe: Assessment, Protection, Management. 18–22 October 2004, UFZ Leipzig, 250–251.

Metcalfe-Smith, J. L., 1994. Biological water-quality assessment of rivers: Use of macroinvertebrate communities. In Calow, P. & G. E. Petts (eds), The Rivers Handbook – Hydrological and Ecological Principles. Blackwell Scientific Publications, Oxford, 144–170.

Moog, O., A. Chovanec, J. Hinteregger & A. Römer, 1999. Richtlinie zur Bestimmung der saprobiologischen Gewässergüte von Fliessgewässern. Bundesministerium für Land- und Forstwirtschaft, Wien.

Morpurgo, M., 1996. Confronto fra Indice Saprobico (Friedrich e DIN, 1990) e Indice Biotico Esteso (Ghetti e IRSA, 1995). Biologia Ambientale 14: 30–36.

National Rivers Authority, 1994. The Quality of Rivers and Canals in England and Wales (1990 to 1992) as Assessed by a New General Quality Assessment Scheme. HMSO, London.

Nixon, S. C., C. P. Mainstone, T. Moth Iversen, P. Kristensen, E. Jeppesen, N. Friberg, E. Papathanassiou, A. Jensen & F. Pedersen, 1996. The harmonized monitoring and classification of ecological quality of surface waters in the European Union. Final Report. European Commission Directorate General XI, Brussels.

Rico, E., A. Rallo, M. A. Sevillano & M. L. Arretxe, 1992. Comparison of several biological indices based on river macroinvertebrate benthic community for assessment of running water quality. Annales de Limnologie 28: 147–156.

Rolauffs, P., D. Hering, M. Sommerhäuser, S. Rödiger & S. Jähnig, 2003. Entwicklung eines leitbildorientierten Saprobienindexes für die biologische Fließgewässerbewertung. Umweltbundesamt, Berlin.

Sandin, L. & D. Hering, 2004. Comparing macroinvertebrate indices to detect organic pollution across Europe: a contribution to the EC water framework directive intercalibration. Hydrobiologia 516: 55–68.

Schmidt-Kloiber, A., W. Graf, A. Lorenz & O. Moog, 2006. The AQEM/STAR taxalist – a pan-European macro-invertebrate ecological database and taxa inventory. Hydrobiologia 566: 325–342.

Skriver, J., N. Friberg & J. Kirkegaard, 2000. Biological assessment of running waters in Denmark: introduction of the Danish stream fauna index (DSFI). Verhandlungen der Internationalen Vereinigung für theoretische und angewandte Limnologie 27: 1822–1830.

STN (Slovenská Technická Norma) 83 0532-1 to 8, 1978/79. Biologický rozbor povrchovej vody. (Biological analysis of surface water quality.) Slovak Standardisation Institute, Bratislava.

Stubauer, I. & O. Moog, 2000. Taxonomic sufficiency versus need for information – comments based on Austrian experience in biological water quality monitoring. Internationale Vereinigung für theoretische und angewandte Limnologie: Verhandlungen 27: 1–5.

Swedish Environmental Protection Agency, 2000. Environmental quality criteria: lakes and watercourses. Swedish Environmental Protection Agency, Stockholm.

SYSTAT Software Inc., 2002. TableCurve 2D – Version 5.01. SSI, Richmond CA.

Tittizer, T., 1976. Comparative study of biological–ecological water assessment methods. Practical demonstration on the river Main. 2–6 June, 1975 (summary report). In Amavis, R.-J. (ed.) Principles and Methods for Determining Ecological Criteria on Hydrobiocoenosis: Proceedings of the European Scientific Colloquium Luxembourg, Nov. 1975. Pergamon Press, Oxford, 403–463.

Woodiwiss, F. S., 1978. Comparative study of biological–ecological water quality assessment methods. Second practical demonstration. Summary Report. Commission of the European Union, Brussels.

Hydrobiologia (2006) 566:417–430
M.T. Furse, D. Hering, K. Brabec, A. Buffagni, L. Sandin & P.F.M. Verdonschot (eds), The Ecological Status of European Rivers: Evaluation and Intercalibration of Assessment Methods
DOI 10.1007/s10750-006-0080-9

Intercalibration of assessment methods for macrophytes in lowland streams: direct comparison and analysis of common metrics

Sebastian Birk*, Thomas Korte & Daniel Hering
Department of Hydrobiology, University of Duisburg-Essen, Universitätsstr. 5, D-45117, Essen, Germany
*(*Author for correspondence: E-mail: sebastian.birk@uni-essen.de)*

Key words: ecological quality classification, STAR project, EU Water Framework Directive, macrophytes, Mean Trophic Rank, Indice Biologique Macrophytique en Rivière

Abstract

The results of four macrophyte assessment methods (French Indice Biologique Macrophytique en Rivière, German Reference Index, British Mean Trophic Rank and Dutch Macrophyte Score) were compared, based on plant survey data of medium-sized lowland streams in Central Europe. To intercalibrate the good quality class boundaries two alternative methods were applied: direct comparison and the use of "common metrics". While the French and British methods were highly related ($R^2 > 0.75$), the German RI showed less ($0.20 < R^2 < 0.55$) and the Dutch DMS least correlation ($R^2 < 0.10$) with other methods. Of 70 macrophyte metrics tested only Ellenberg_N was considerably related to three of the national assessment methods, thus representing a potential common metric for intercalibration. Comparison of quality class boundaries via regression analysis using both intercalibration approaches revealed major differences between classifications of the French, German and British methods, which are, in addition, related in a nonlinear way.

Introduction

According to the EU Water Framework Directive (WFD; European Commission, 2000) European surface waters must achieve good ecological quality by the year 2015. Responsiblity for the quality assessment lies with the individual Member States, which have developed or modified assessment methods at the national level. To ensure the comparability of the national methods, an intercalibration exercise is stipulated by the Directive, in which quality class boundaries are checked for comparability and consistency with normative requirements.

Although benthic macroinvertebrates are presently most commonly applied for the quality assessment of rivers (Birk & Hering, 2002), macrophytes are also surveyed in some countries to monitor the effects of anthropogenic pressures, especially eutrophication (Kelly & Whitton, 1998; Birk & Schmedtje, 2005). Macrophytes were first used in water quality assessment in relation to various modifications of the saprobic system. Several indicator catalogues (e.g., Sládeček, 1973) included single macrophyte species to evaluate the degree of organic pollution. More generally, the monitoring of macrophyte communities was confined to the description of the vegetation without inferring water quality (e.g., Holmes & Whitton, 1975; Janauer et al., 2003). With the increasing awareness of the effects of nutrient enrichment the community assessment of phototrophs gained in importance. The Mean Trophic Rank (MTR, Holmes et al., 1999), for instance, focuses on the impact of nutrient enrichment only, since it was elaborated and tested specifically for the application of the EU Urban Waste Water Treatment Directive (Council of the European Communities, 1991). The Water Framework Directive recently led to the development of national methods aimed at assessment of ecological quality of the aquatic flora (e.g., van der Molen et al., 2004; Leyssen et al., 2005; Meilinger et al., 2005). These methods differ in design and performance from macroinvertebrate-based systems and, thus, require a separate intercalibration process.

The intercalibration procedure as outlined in the Directive comprises the comparison of intercalibration sites whose individual biological quality, in the opinion of the Member States, represents the boundary between quality classes. Recent studies on intercalibration of macroinvertebrate methods are based on data representing a broad quality gradient, and class boundaries are compared via correlation and regression analysis. While Birk & Hering (2006) directly compare national methods by the help of a "common scale" (method best correlating with all other methods), Buffagni et al. (2006) use "common metrics" as a general scale. Common metrics are defined as biological metrics widely applicable within a geographical region, which can be used to derive comparable information among different countries and stream types (Buffagni et al., 2005).

In this paper we apply the two above outlined approaches of boundary comparison to macrophyte data from lowland rivers covering a broad spectrum of anthropogenic disturbance from reference to heavily impacted sites. Furthermore, we test both techniques for their applicability in the intercalibration of four assessment methods for macrophytes.

Methods

Samples and sites

This study is based on river macrophyte survey data collected at medium-sized lowland streams in six countries in the framework of the EU project STAR (Furse et al., 2006). In the official intercalibration exercise for the Water Framework Directive, this stream type is named "medium-sized, lowland streams of mixed geology" (R-C4) in Central Europe (CIS WG2.A ECOSTAT, 2004). Data used here were limited to 108 sites at which macrophytes covered at least 1% of the total channel area investigated (Table 1).

Macrophytes were sampled using a single survey in late summer or early autumn. A 100 m stream length was surveyed in each stream by wading in a zigzag manner across the channel. Macrophytes of non-wadable sites were observed by boat or by walking along the banks. All macrophytes species were recorded as well as the percent cover of the overall macrophyte growth. Species were normally identified in the field, but if identification was uncertain a representative sample was collected for later identification. In addition, physico-chemical data were sampled. Table 2 lists statistical descriptors for the sampling site's trophic status.

National assessment methods and quality classifications

Four methods to assess the quality of streams, which are being used in France, Germany, the Netherlands and United Kingdom were compared (Table 3). All methods are based on species-level data and integrate specific indicator values and abundance information. Except for the German Reference Index, abundance is specified in classes of relative plant coverage. Abundance data used by the German index is an estimation of the three-dimensional structure of the instream vegetation (Kohler, 1978). Table 4 compares the different macrophyte abundance schemes. While the French and British methods were used alone,

Table 1. Overview of the sites surveyed at medium-sized lowland streams

Country	Number of surveys
Denmark	11
Germany	11
Latvia	36
Poland	24
Sweden	20
United Kingdom	6

Table 2. Range of trophic status covered by the dataset ($n = 108$): descriptive statistics of the chemical parameters nitrate and total phosphorus

	Min	25th	Median	75th	Max
Nitrate (mg l^{-1})	0.03	0.25	1.50	2.00	12.10
Total phosphorus (mg l^{-1})	0.01	0.09	0.21	0.28	15.40

Table 3. Overview of macrophyte assessment methods

Country	Assessment method	Reference
France	IBMR (FR) – Indice Biologique Macrophytique en Rivière	NF T90-395 (2003)
Germany	RI (DE) – Reference Index	Schaumburg et al. (2004)
The Netherlands	DMS (NL) – Dutch Macrophyte Score ("Soortensamenstelling macrofyten")	Molen et al. (2004)
United Kingdom	MTR (UK) – Mean Trophic Rank	Holmes et al. (1999)

the German and Dutch indices are part of generic methods to assess the "aquatic flora", which is defined as including macrophytes and phytobenthos.

The Dutch and German methods aim at assessing the degree of deviation from the reference state and are, thus, based on stream type specific reference conditions. It is therefore necessary to classify the streams sampled here into specific stream types: for the German method sampling sites were assigned to the stream type "medium sized lowland rivers of northern Germany" (Schaumburg et al., 2004). Since the stream typology of the Netherlands is more complex, sites have been allocated to eight different national types for the Dutch index (Elbersen et al., 2003).

The French, German and Dutch methods distinguish between five classes of ecological quality (Table 5). Since the British MTR was developed to illustrate responses to urban discharges by surveying two physically similar sites upstream and downstream the method is not designed for classifying the ecological quality of rivers. For interpretation purposes only, Holmes et al. (1999) suggest MTR boundary values to determine if the investigated site is (1) 'unlikely to be eutrophic', (2) 'likely to be either eutrophic or at risk of becoming eutrophic' or (3) 'badly damaged by either eutrophication, organic pollution, toxidity or physically damaged'. Here, the MTR value discriminating between (1) and (2) was exemplarily used as the good–moderate ecological status boundary.

Description of biotic metrics analysed to provide "common macrophyte metrics"

Seventy macrophyte metrics were analysed to detect "common metrics" enabling intercalibration of national assessment methods (Table 6). These metrics cover the categories "richness and diversity", "composition and abundance", "sensitivity and tolerance", and "ecosystem function". The basic criterion for the selection of common metrics was a correlation ($R^2 > 0.5$; $p < 0.05$) of the metric with all assessment methods evaluated in this study. As an additional criterion, redundant metrics were excluded from further analysis. Of metric pairs with a coefficient of determination of >0.65,

Table 4. Comparison of macrophyte abundance schemes

IBMR (FR)		RI (DE)		DMS (NL)		MTR (UK)	
Abundance class	Cover [%]	Abundance class (Kohler, 1978)	Plant quantity (Schneider, 2000)	Abundance class	Cover [%]	Abundance class	Cover [%]
1	<0.1	1	1	1	<5	1	<0.1
2	0.1–1	2	8			2	0.1–1
3	1–10	3	27			3	1–2.5
						4	2.5–5
				2	5–50	5	5–10
4	10–50	4	64			6	10–25
						7	25–50
5	>50			3	>50	8	50–75
		5	125			9	>75

Table 5. Class boundaries of the national assessment methods and derived reference values using the 95th percentile value of all survey sites (n.a. – not applicable).

Index	IBMR (FR)	RI (DE)*	DMS (NL)	MTR (UK)
High–good	15	0.5	0.8	n.a.
Good–moderate	12	0.25	0.6	66**
Moderate–poor	9	0.15	0.4	n.a.
Poor–bad	7	0	0.2	n.a.
Literature source	NF T90-395 (2003)	Schaumburg et al. (2005)	Van den Berg et al. (2004)	Holmes et al. (1999)
Reference (95th percentile)	13.2	0.86	0.42	60.4

*Classification scheme relates to sites where only the Reference Index provides validated results within the assessment method for aquatic flora.

**Boundary based on recommendations for the interpretation of MTR scores to evaluate the trophic state (Holmes et al., 1999; see text for details).

the metric showing the lesser correlation with the assessment methods was omitted.

Data preparation

The national assessment methods were manually calculated for each macrophyte sample, with the exception of DMS (NL), which was calculated by the software QBWat (Pot, 2005). Due to the minimum criteria for confidence specified by the German and Dutch indices, they could not be determined for 15 and 9 sites, respectively. The index values were converted into Ecological Quality Ratios, i.e. dividing the observed score of each site by a reference value to normalise the output. The 95th percentile value of all samples was chosen as index reference assuming that approximately 5% of surveyed sites hold macrophyte communities in reference state.

Correlation and regression analysis: macrophyte assessment methods, potential common metrics and pressure gradients

The relationships between the four assessment indices were analysed and the strength of correlation was specified by the "coefficient of determination" (R^2). This measure was also used to determine common macrophyte metrics suitable for intercalibration. Both linear and nonlinear regression was tested using the software TableCurve 2D (SYSTAT Software Inc., 2002).

Physical–chemical, hydromorphological and land use/type data were used to construct complex stressor gradients by means of principle components analysis (PCA). General degradation gradients were derived from physical–chemical, hydromorphological and land use data. In addition, separate degradation gradients were constructed via PCA, using water chemistry, hydromorphological and microhabitat data. (see Hering et al., in prep.). The results were used to test the response of the macrophyte methods to individual pressure groups. Gradients best correlating to the macrophyte assessment methods were determined.

Comparison of quality class boundaries

Two intercalibration approaches were applied in this study: (1) National quality classes of the macrophyte methods were compared directly following the procedure described by Birk & Hering (2006). The assessment method showing the highest correlation to all other indices was used as a "common scale". (2) The approach of indirect boundary comparison (Buffagni et al., 2006) employed "common metrics" as response variables in the regression analysis.

Results

Comparison of classification schemes

The classification results of the four methods applied differed noticeably. According to the German method more than 50% of sites were in high and good status. The French, British and Dutch

Table 6. Metrics tested with the macrophyte dataset

Name of metric	Metric type	Reference
Proportion of community with preference for certain amount of water supply		
Typical macrophytes (# taxa and %)	f/rd/ca	Holmes et al. (1999)
Species submerged (# taxa and %)	f/rd/ca	Szoszkiewicz et al. (2006)
Species amphibious (# taxa and %)	f/rd/ca	
Mosses and liverworts (# taxa and %)	f/rd/ca	
Species terrestrial (# taxa and %)	f/rd/ca	
Diversity indices		
Shannon diversity	rd	Shannon & Weaver (1949)
Simpson diversity	rd	Simpson (1949)
Evenness	rd	Pielou (1966)
Shannon diversity (growth forms)	rd	Following Wiegleb (1991) and
Evenness (growth forms)	rd	Van de Weyer (2003)
Morphological groups according to growth forms		
Species anchored but with floating leaves or heterophyllus (# taxa and %)	f/rd/ca	Szoszkiewicz et al. (2006)
Species floating free (# taxa and %)	f/rd/ca	
Growth forms (# taxa and %)	f/rd/ca	Wiegleb (1991), Van de Weyer (2003)
Growth form Myriophyllids (# taxa and %)	f/rd/ca	
Growth form Parvopotamids (# taxa and %)	f/rd/ca	
Growth form Peplids (# taxa and %)	f/rd/ca	
Growth form Vallisnerids (# taxa and %)	f/rd/ca	
Reference and disturbance indicating taxa and growth forms of lowland streams		
Disturbance indicating taxa (# taxa and %)	st	Van de Weyer (2003)
Reference taxa (# taxa and %)	st	Following Van de Weyer (2003)
Reference growth forms (# taxa and %)	st	
Disturbance indicating growth forms (# taxa and %)	st	
Ratio: reference taxa to disturbance indicating taxa (# taxa and %)	st	
Disturbance indicating growth form: Elodeids (# taxa and %)	st	
Disturbance indicating growth form: Lemnids (# taxa and %)	st	
Disturbance indicating growth form: Myriophyllids (# taxa and %)	st	
Disturbance indicating growth form: Parvopotamids (# taxa and %)	st	
Disturbance indicating growth form: Peplids (# taxa and %)	st	
Reference growth form: Batrachids (# taxa and %)	st	
Reference growth form: Ceratophyllids (# taxa and %)	st	
Reference growth form: Magnonymphaeids (# taxa and %)	st	
Reference growth form: Magnopotamids (# taxa and %)	st	
Reference growth form: Myriophyllids (# taxa and %)	st	
Reference growth form: Parvopotamids (# taxa and %)	st	
Reference growth form: Peplids (# taxa and %)	st	
Selected reference taxa (*Potamogeton natans*, *P. polygonifolius*, *Nuphar lutea*, *Sagittaria sagittifolia*, *Sparganium emersum*, *Berula erecta*) (# taxa and %)	st	
Ratio: reference growth forms to disturbance indicating growth forms (# taxa and %)	st	

Continued on p. 422

Table 6. (Continued)

Name of metric	Metric type	Reference
Nitrogen indicating metric		
Ellenberg_N	st	Ellenberg et al. (1992)

For taxa assignment to growth forms refer to Table 9 (# taxa – number of taxa, % – relative abundance, ca – composition/abundance, f – functional, rd – richness/diversity, st – sensitivity/tolerance).

methods assessed nearly all sites as of moderate or worse quality (Fig. 1).

Due to the different range of quality covered by the individual methods, the 95th percentile value chosen as the reference value was allocated to different quality classes for each of the four national classification schemes (Table 5): The reference value was allocated as high quality in the German RI system, good quality in the French IBMR system, and moderate quality in the Dutch DMS and British MTR systems. Nevertheless, the reference obtained in the analysis for the British MTR corresponded to the mean of top 10% MTR values for similar British lowland river types given by Holmes et al. (1999).

Correlation and regression analysis

Macrophyte assessment methods

The coefficients of determination given in Table 7 revealed the differences between the four assessment methods. The French and British methods were most closely related ($R^2 > 0.75$). The German RI showed lower correlations with these methods, especially with the French IBMR, while DMS (NL) was negatively correlated to all other methods.

Nonlinear regression generally resulted in higher coefficients of determination. Between RI (DE) and MTR (UK) the difference between the two regression models was $R^2 = 0.12$.

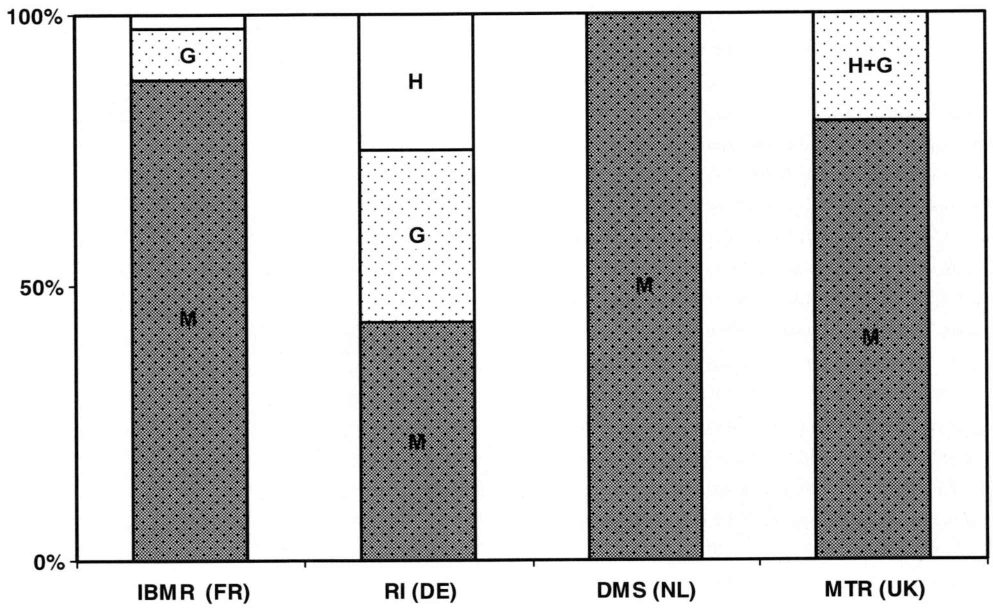

Figure 1. Distribution of quality classes in the dataset resulting from four macrophyte assessment methods (H – high; G – good; M – moderate and worse). Quality classes of RI (DE) are based on the analysis of the Reference Index and additional criteria (Schaumburg et al., 2004). The class boundary between high/good (H + G) and moderate quality of MTR (UK) is based on recommendations for the interpretation of MTR scores to evaluate the trophic state (Holmes et al., 1999; see text for details).

Potential common macrophyte metrics

Of the 70 biotic macrophyte metrics tested, only Ellenberg_N correlated significantly to all 4 assessment methods. For all four assessment methods nonlinear regression yielded higher coefficients of determination to Ellenberg_N than linear regression. While IBMR (FR), RI (DE) and MTR (UK) were negatively correlated to this metric, the Dutch index values were positively related to Ellenberg_N.

None of the other biotic metrics showed strong correlations with all four macrophyte assessment methods. For example, the richness measure "number of species" was strongly related to the German and Dutch methods (Table 7). However, due to the type of relation to the German RI (Fig. 2) it could not be considered as a common macrophyte metric, since the regression function was nonmonotonic. Thus, for each normalised value for "number of species", two values of the German RI were possible.

The DMS (NL) showed coefficients of determination of $R^2 > 0.5$ with several functional metrics (e.g. "relative abundance of disturbance indicating growth forms", "relative abundance of disturbance indicating growth form: Lemnids" and "number of selected reference taxa").

Environmental gradients (PCA)

The French, German and British methods related most strongly to the PCA gradient reflecting water chemistry ("pollution/eutrophication", PCA axis 1, Eigenvalue: 0.527; Table 7). The Dutch method was correlated with "general degradation" including chemical, hydromorphological and land use parameters (PCA axis 1, Eigenvalue: 0.287). Coefficients of determination of the regression analysis are listed in Table 8 (see Hering et al., in prep. for details of the gradients).

Direct comparison of quality class boundaries

British MTR correlated best with all other methods and was therefore used as "common scale" according to Birk & Hering (2006). Due to its weak relationship with any of the other macrophyte methods, the Dutch DMS was not included in direct class boundary comparison. Considering the 95% confidence intervals, direct comparison revealed large differences in national definitions of the high-good quality boundary (>0.6 MTR units, Table 8). The differences between the good-moderate boundaries were smaller (<0.3 MTR units on average). The mean value of confidence intervals amounted to 0.079 MTR units.

The nonlinear regression graph (Fig. 3) shows decreasing slope values with increasing deviation of IBMR (FR) and RI (DE) from the reference state. Especially in the lower range of the RI (DE), the British MTR was not responding to changes of the German method. Therefore, the high–good and good–moderate class boundary intervals of RI (DE) transferred into MTR scale were overlapping (cf. Table 8).

Indirect comparison of quality class boundaries using Ellenberg_N as common macrophyte metric

The high–good boundary comparison of the French and German method using Ellenberg_N resulted in a difference of >0.4 units. For the German and British method, confidence intervals of the good–moderate class boundaries shared similar ranges when compared via Ellenberg_N. The average confidence interval amounted to 0.141 units.

As observed in the "direct comparison approach" the quality class boundaries of RI (DE) showed overlapping confidence ranges using Ellenberg_N (Table 8). Regression analysis disclosed a similar type of relation between the German method and each of MTR and Ellenberg_N (Fig. 4).

Discussion

The obviously different quality classes of the sites assessed with the four methods (Fig. 1) reveal that intercalibration efforts for macrophyte methods are indispensable. Starting from this conclusion we applied analytical methods currently used in intercalibration of benthic invertebrate systems (Birk & Hering, 2006; Buffagni et al., 2006) to compare quality class boundaries of macrophyte assessment methods.

Testing of intercalibration approaches

This study discloses difficulties in adopting commonly used intercalibration approaches to macrophyte based assessment methods. Direct

Table 7. Correlation and regression analysis of macrophyte assessment methods, selected macrophyte metrics and environmental gradients

	IBMR (FR)			RI (DE)			DMS (NL)			MTR (UK)		
	Type	Linear	Nonlinear	Type	Linear	Nonlinear	Type	Linear	Nnonlinear	Type	Linear	Nonlinear
Macrophyte assessment methods												
IBMR (FR)	pos.	1.00	1.00	pos.	0.22	0.31	neg.	0.06	–	pos.	0.76	0.79
RI (DE)	pos.	0.22	0.26	pos.	1.00	1.00	–	n.s.	n.s.	pos.	0.41	–
DMS (NL)	neg.	0.06	0.10	–	n.s.	0.15	pos.	1.00	–	neg.	0.05	0.07
MTR (UK)	pos.	0.76	0.77	pos.	0.41	0.53	neg.	0.05	–	pos.	1.00	–
Selected macrophyte metrics												
Ellenberg_N	neg.	0.46	0.56	neg.	0.46	0.58	pos.	0.05	0.11	neg.	0.69	0.70
Number of species	neg.	0.04	0.06	–	n.s.	0.28	pos.	0.56	0.59	–	n.s.	0.06
Disturbance-indicating growth forms (%)	neg.	0.07	–	–	n.s.	n.s.	pos.	0.56	–	neg.	0.05	n.s.
Environmental gradients												
Pollution/eutrophication	neg.	0.46	–	neg.	0.14	0.22	pos.	0.09	0.10	neg.	0.51	0.52
General degradation	–	n.s.	n.s.	–	n.s.	n.s.	neg.	0.41	0.42	–	n.s.	n.s.

Type of correlation (pos. – positive, neg. – negative) and coefficients of determination (R^2) based on linear and nonlinear regression. Nonlinear R^2 is only given if providing higher coefficients of determination ($p < 0.05$; n.s. – not significant).

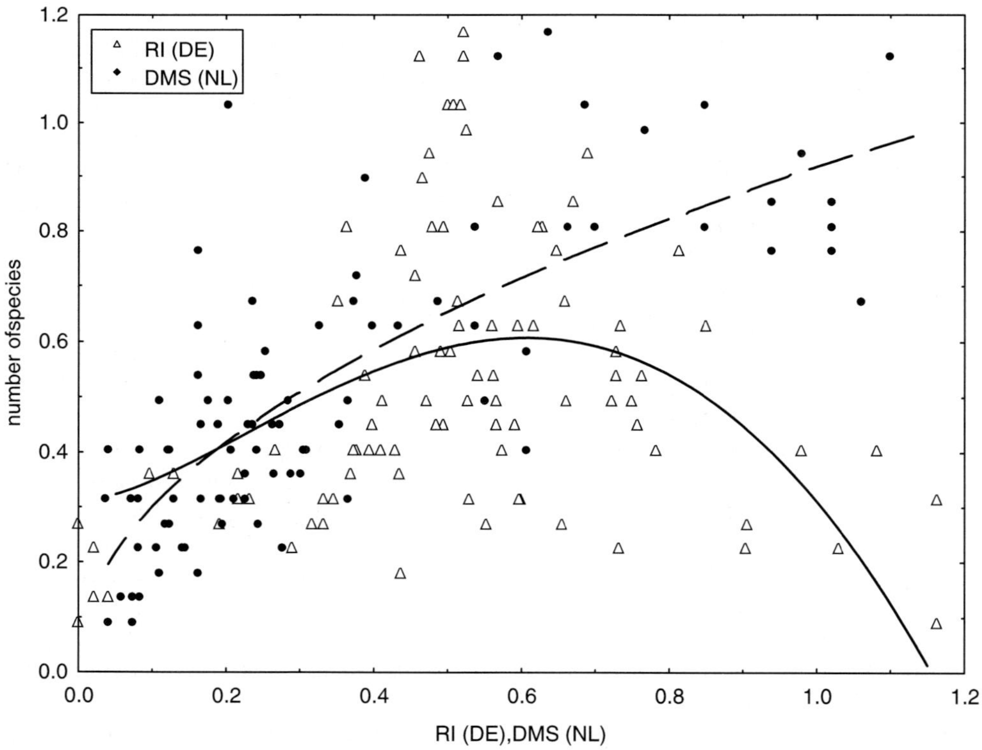

Figure 2. Nonlinear regression of German RI (solid line; $R^2 = 0.28$) and Dutch DMS (dashed line; $R^2 = 0.59$) against the number of species.

Table 8. EQR values of the high–good (H|G) and good–moderate (G|M) quality class boundaries transfered into MTR and Ellenberg_N scales via nonlinear regression analysis. For each value derived by regression the 95 % confidence interval is specified (n.a. – not applicable)

Class boundary	Common scale	IBMR (FR)			RI (DE)			MTR (UK)		
		Equation	Boundary value	*95% confid.*	Equation	Boundary value	*95% confid.*	Equation	Boundary value	*95% confid.*
H|G	MTR	(1)	1.497	0.150	(2)	0.638	0.056	n.a.	n.a.	n.a.
	Ellenberg_N	(2)	1.185	0.287	(2)	0.394	0.079	n.a.	n.a.	n.a.
G|M	MTR	(1)	0.820	0.044	(2)	0.565	0.067	–	1.094	–
	Ellenberg_N	(2)	0.638	0.103	(2)	0.294	0.094	(1)	0.911	0.143

(1) $f(x) = a + b{\cdot}x^{1.5}$.
(2) $f(x) = a + b{\cdot}x^{3}$.

comparison of class boundaries (Birk & Hering, 2006) only yielded sound results between the closely related methods IBMR (FR) and MTR (UK). These two indices share many common indicator species (Szoszkiewicz et al., 2006) whose indicator values correlate strongly ($R^2 = 0.61$). The sound correlation of the German RI with MTR (UK) seems to allow for direct boundary comparison. However, the specific nonlinear character of this relationship impedes significant resolution between the good quality boundaries, thus making direct comparison of these methods impossible. The low correlations of DMS (NL) with all other national assessment methods exclude this index from further intercalibration analysis.

Against this background we tested whether intercalibration could be accomplished using common metrics (Buffagni et al., 2006). Our

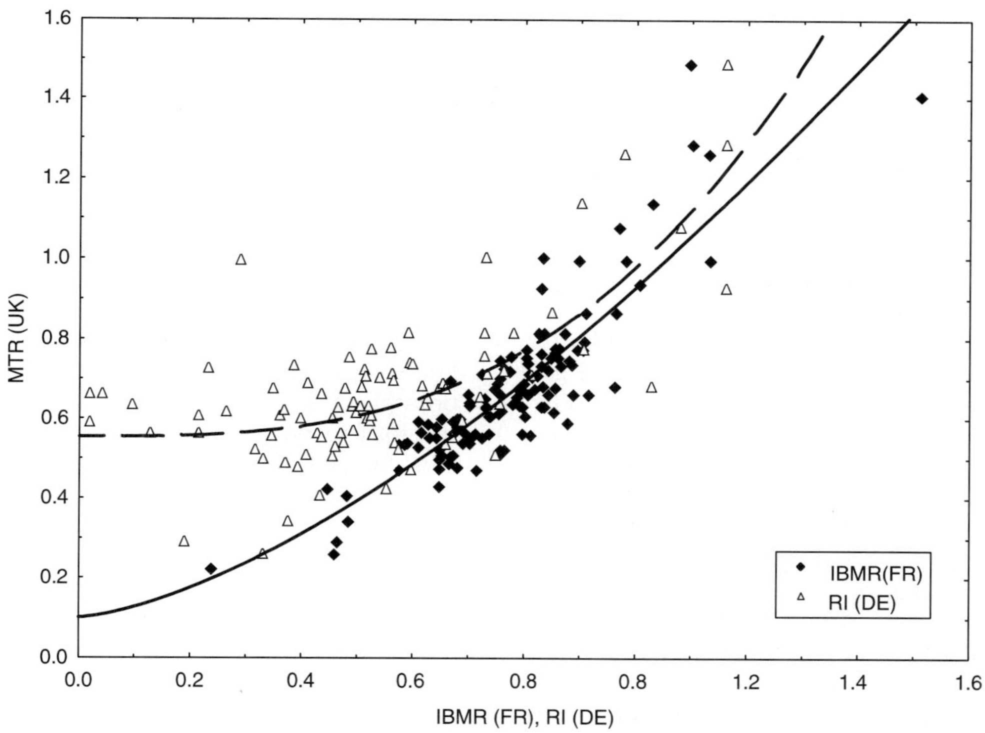

Figure 3. Nonlinear regression of French IBMR (solid line; $R^2 = 0.77$) and German RI (dashed line; $R^2 = 0.53$) against British MTR.

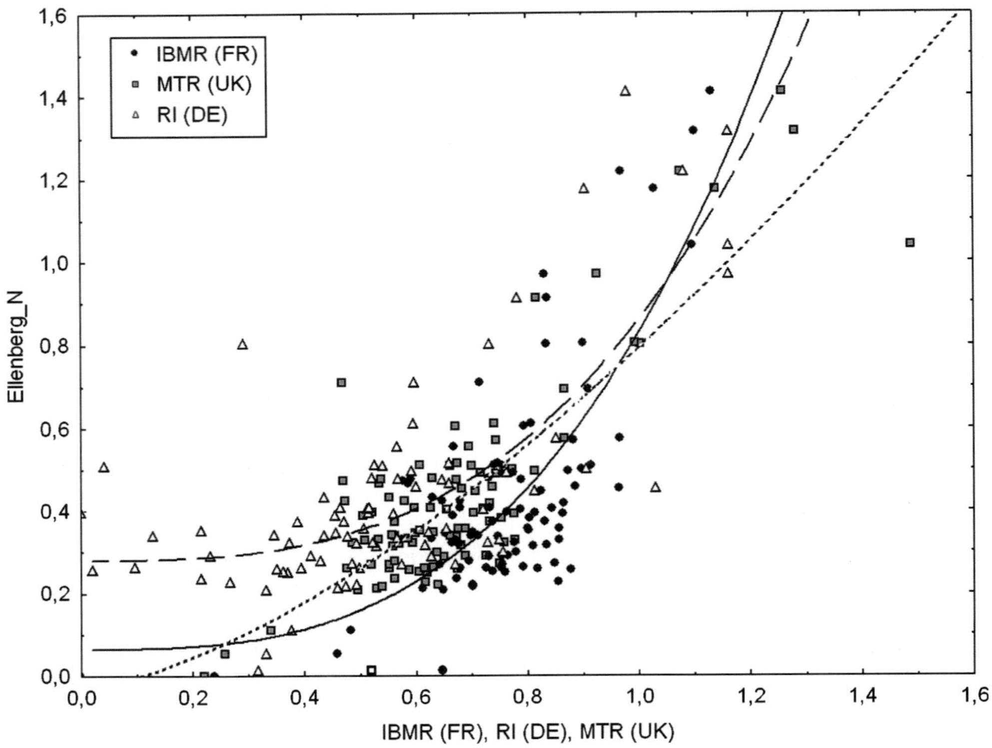

Figure 4. Nonlinear regression of French IBMR (solid line; $R^2 = 0.56$), German RI (dashed line; $R^2 = 0.58$) and British MTR (dotted line; $R^2 = 0.70$) against Ellenberg_N.

analysis showed that none of the tested macrophyte metrics met our common metric criteria for all assessment methods. Only Ellenberg_N showed strong relationships with at least three methods. The metric is based on the response of higher plants to nitrogen compounds (nitrate and/or ammonium) and, thus, corresponds to trophic categories and to general nutritional conditions in the rivers that are represented with a broad gradient in the dataset (Table 2). The French, German and British methods relating to this potential common metric also respond significantly to the abiotic PCA gradient reflecting organic pollution and eutrophication. These findings underline the general ability of macrophyte methods to assess the trophic status of rivers. While Holmes et al. (1999) designed the British MTR for this specific purpose, the German method in particular is aimed at detecting "general degradation", i.e., the level of deviation from a reference community (Schaumburg et al., 2004).

Like in direct comparison, DMS (NL) cannot be included in the intercalibration analysis using common metrics. Although it shares objectives with the German method (unspecific pressure assessment based on type specific macrophyte communities), we found no biotic metric suitable for intercalibration. Either different types of relation (cf. Fig. 2) or no common relationship at all, limits the applicability of the common metric approach. DMS (NL) is characterised by strong relations to richness and diversity measures and, most remarkably, by positive correlations with metrics indicating disturbance in lowland streams of North-Rhine Westphalia (Western Germany; Table 9; Van de Weyer, 2003). In this respect, the broad spectrum of environmental factors influencing the occurrence of macrophytes in streams on various spatial scales (Wiegleb, 1988) may confine the validity of indicator species to narrow geographic regions. Furthermore, Korte & Van de Weyer (2005) observed that, in two separate

Table 9. Reference taxa and disturbance indicating taxa of lowland streams and their growth forms (following Van de Weyer 2003)

Reference taxa	Growth form
Chara fragilis Desvaux	Charids
Chara sp. L. ex Vaillant	
Nitella flexilis C. A. Ag.	
Nitella sp. C. A. Ag.	
Ceratophyllum submersum L.	Ceratophyllids
Berula erecta (Huds.) Coville	Herbids
Juncus bulbosus L.	
Nuphar lutea (L.) Sibth. & Sm.	Magnonymphaeids
Nymphaea alba L.	
Persicaria amphibia (L.) Gray	
Ranunculus flammula L.	
Potamogeton alpinus Balbis	Magnopotamids
Potamogeton gramineus L.	
Potamogeton natans L.	
Potamogeton polygonifolius Pourret	
Potamogeton lucens L.	
Potamogeton obtusifolius Mert. & Koch	
Potamogeton perfoliatus L.	
Potamogeton praelongus Wulfen	
Myriophyllum alterniflorum DC.	Myriophyllids
Myriophyllum verticillatum L.	
Utricularia intermedia Hayne	
Utricularia vulgaris L.	
Potamogeton berchtoldii Fieber	Parvopotamids
Potamogeton compressus L.	
Potamogeton filiformis Pers.	
Callitriche cophocarpa Sendtn.	Peplids
Callitriche hamulata Kutz ex W.D.J. Koch	
Callitriche platycarpa Kütz.	
Alisma plantago-aquatica L.	Vallisnerids
Sagittaria sagittifolia L.	
Sparganium emersum Rehmann	
Sparganium erectum L.	
Sparganium sp. L.	
Disturbance indicating taxa	
Ceratophyllum demersum L.	Ceratophyllids

Table 9. (Continued)

Reference taxa	Growth form
Ceratophyllum demersum var. *apiculatum* Cham.	
Elodea canadensis Michx.	Elodeids
Lemna gibba L.	Lemnids
Lemna minor L.	
Spirodela polyrhiza (L.) Schleid	
Myriophyllum spicatum L.*	Myriophyllids
Ranunculus fluitans Lamk.*	
Potamogeton crispus L.	Parvopotamids
Potamogeton pectinatus L.	
Potamogeton pusillus L.	
Potamogeton trichoides Cham. & Schltdl.	
Zannichellia palustris L.	
Callitriche obtusangula Le Gall	Peplids

*According to Van de Weyer (2003) these species indicate increased current velocity (e.g., due to channel straightening).

methods for the assessment of German lowland streams, indicative characteristics of macrophyte species are evaluated differently. Nevertheless, the weak but significant positive correlation of the Dutch method with Ellenberg_N points at basic differences in the conception of the reference state. This is also indicated by the negative correlation of DMS (NL) to the "general degradation" gradient. However, since the dataset analysed covers only sites of moderate or worse status according to the Dutch classification system the validity of our findings is limited to a restricted range of quality.

Further incomparability results from the different calculation methods. The French, German and British indices are calculated by weighted average equations, yielding values less influenced by the species richness of the site. Abundance scores are accounted by multiplication by the indicator values. Results of DMS (NL) are obtained by summation of taxa scores, whose values depend on the relative abundance of the species. For certain species in specific river types this score value decreases with increasing abundance and *vice versa*.

National assessment methods for all biological quality elements will need to assess ecological quality in a general way; therefore, the intercalibration exercise of macrophyte-based methods has to simultaneously target the effects of different types of degradation. The selection of common intercalibration metrics should thus respond to general degradation (see also Buffagni et al., 2005 using common metrics for intercalibration of invertebrate-based methods). We tested a broad range of general and specific macrophyte metrics covering biotic parameters like taxonomic composition and abundance, richness and diversity, and functional groups (Table 6). Since none of the metrics analysed qualified for intercalibration purposes, further research to produce suitable common macrophyte assessment metrics is indispensable.

Implications for the macrophyte intercalibration exercise

This study presents preliminary results which may become relevant in the further discussion of macrophyte intercalibration. Nevertheless, several procedural requirements of the official intercalibration exercise are not met: (1) The international STAR dataset covers different biogeographical regions. Therefore, the applicability of national methods may be affected because the assessment is adjusted to the regional flora and the indicative characteristics of its macrophyte species. (2) Since macrophytes were surveyed according to a standard procedure (Furse et al., 2006), national survey techniques and their effect on the taxa list are neglected. (3) We based our analyses on macrophytes only, whereas two of the methods examined, RI (DE) and DMS (NL) are designed to assess the broader "aquatic flora", including phytobenthos. Considering these items, the following implications for the macrophyte intercalibration exercise can, however, be stated.

Comparison of quality classes for European river assessment methods using benthic invertebrates (Buffagni et al., 2005) showed promising outcomes for the success of the intercalibration exercise. This can substantially be attributed to the strong relationships of the methods (Birk & Hering, 2006), their focus on similar pressures and their common tradition. In the view of the present study, the practicability of the analytical approaches applied to the intercalibration of macrophyte methods (direct comparison, use of common metrics) is questionable. Two main factors that complicate comparisons between methods are (1) differently defined reference conditions and (2) gaps in knowledge about pressure–impact relationships. The delineation of reference communities, particularly for the medium-sized lowland rivers of Central Europe, is difficult due to the lack of existing reference sites. Therefore, expert opinion is used to estimate natural conditions in the lowlands. Furthermore, the Dutch and German methods both define reference states via index scores, but include diverse macrophyte species and apply different formulae. The lack of knowledge about pressure–impact relationships may generally impede the intercalibration of macrophyte methods. While higher plants are well known for their response to nutrient pollution, the effect of other impairments on the community has been little studied (Kelly & Whitton, 1998; Janauer, 2001). This also delimits the availability of appropriate common assessment metrics. This study demonstrates on the one hand that intercalibration of methods specifically addressing eutrophication is possible but, on the other hand, it also highlights deficiencies for the coming macrophyte intercalibration exercise.

Since the EU Water Framework Directive stipulates intercalibration of national methods by end of 2006, scientific activities at the European level are currently being carried out to fulfil these legal requirements. Thus, the intercalibration task has initiated a process of Europe-wide discussion on ecological quality and the harmonisation of its assessment. Tailor-made approaches for each biological element are required relying on national expertise and international coordination. As a first step towards intercalibration of macrophyte methods in Central Europe, we propose to compile an international database including national data on macrophytes and abiotic pressures taken from sites at common intercalibration types. Since field procedures of the countries involved are very similar (visual survey of 100 m stream sections), this will enable more extensive analyses of the relation between the assessment indices and the definition of the reference state. The outcome may necessitate detailed bilateral discussion on the assessment results at individual sites. This time consuming approach has already yielded results in a preliminary intercalibration study between macrophyte methods of Austria and Germany (Pall et al., 2005).

With regard to the multitude of issues to be addressed in the near future, intercalibration represents a major chance for the implementation of harmonised quality standards at the European level beyond the short timeframe given by the Directive. For macrophyte-based ecological quality assessment in particular, which is still in its early stages in Europe, communality can be gained by maintaining and extending international collaboration to enhance scientific exchange and trigger common outputs.

Acknowledgements

STAR was funded by the European Commission, 5th Framework Program, Energy, Environment and Sustainable Development, Key Action Water, Contract No. EVK1-CT-2001-00089. The authors like to thank Nigel Holmes, Mike Furse, Rebi Nijboer and an anonymous reviewer for their valuable comments on earlier versions of the manuscript.

References

NF T90-395, 2003. Water quality – Determination of the Macrophytes biological index for rivers (IBMR). Association Française de Normalisation (AFNOR), Saint Denis La Plaine.

Birk, S. & D. Hering, 2006. Direct comparison of assessment methods using benthic macroinvertebrates: a contribution to the EU Water Framework Directive intercalibration exercise. Hydrobiologia 566: 401–415.

Birk, S. & D. Hering, 2002. Waterview Web-Database: a comprehensive review of European assessment methods for rivers. FBA News 20: 4.

Birk, S. & U. Schmedtje, 2005. Towards harmonisation of water quality classification in the Danube River Basin: overview of biological assessment methods for running waters. Archiv für Hydrobiologie, Supplement "Large Rivers" 16: 171–196.

Buffagni, A., S. Erba, S. Birk, M. Cazzola, C. Feld, T. Ofenböck, J. Murray-Bligh, M. T. Furse, R. T. Clark, D. Hering, H. Soszka & W. v. d. Bund, 2005. Towards European Inter-calibration for the Water Framework Directive: Procedures and examples for different river types from the E.C. project STAR. 11th STAR deliverable. STAR Contract No: EVK1-CT 2001-00089. Quaderni Istituto di Ricerca sulle Acque 123: 1–468.

Buffagni, A., S. Erba, M. Cazzola, J. Murray-Bligh, H. Soszka & P. Genoni, 2006. The STAR common metrics approach to the WFD intercalibratin process: Full application for small, lowland rivers in three European countries. Hydrobiologia 566: 379–399.

Council of the European Communities, 1991. Urban Waste Water Treatment Directive 91/271/EEC. Official Journal of the European Communities, L135/40–52, 30 May 1991, Brussels.

CIS WG 2.A Ecological Status (ECOSTAT), 2004. Guidance on the intercalibration process. Agreed version of WG 2.A Ecological Status meeting held 7–8 October 2004 in Ispra. Version 4.1. 14. October 2004. ECOSTAT, Ispra.

Elbersen, J. W. H., P. F. M. Verdonschot, B. Roels & J. G. Hartholt, 2003. Definitiestudie KaderRichtlijn Water (KRW). I. Typologie Nederlandse Oppervlaktewateren. Alterra-rapport 669. ALTERRA, Wageningen.

Ellenberg, H., H. E. Weber, R. Düll, V. Wirth, W. Werner & D. Paulißen, 1992. Indicator Values of Plants in Central Europe. Erich Goltze, Göttingen.

European Commission, 2000. Directive 2000/60/EC. Establishing a framework for community action in the field of water policy. European Commission PE-CONS 3639/1/100 Rev 1, Luxembourg.

Furse, M., D. Hering, O. Moog, P. Verdonschot, R. K. Johnson, K. Brabec, K. Gritzalis, A. Buffagni, P. Pinto, N. Friberg, J. Murray-Bligh, J. Kokes, R. Alber, P. Usseglio-Polatera, P. Haase, R. Sweeting, B. Bis, K. Szoszkiewicz, H. Soszka, G. Springe, F. Sporka & I. Krno, 2006. The STAR project: context, objectives and approaches. Hydrobiologia 566: 3–29.

Hering, D., R. K. Johnson, S. Kramm, S. Schmutz, K. Szoszkiewicz & P. F. M. Verdonschot, in prep. Assessment of European rivers with diatoms, macrophytes, invertebrates and fish: A comparative metric-based analysis.

Holmes, N. T. H., J. R. Newman, S. Chadd, K. J. Rouen, L. Saint & F. H. Dawson, 1999. Mean Trophic Rank: A User's Manual. R & D Technical Report E38. Environment Agency, Bristol.

Holmes, N. T. H. & B. A. Whitton, 1975. Macrophytes of the river Tweed. Transactions of the Botanical Society of Edinburgh 42: 369–381.

Janauer, G. A., 2001. Is what has been measured of any direct relevance to the success of the macrophyte in its particular environment? Journal of Limnology 60(Suppl.): 33–38.

Janauer, G. A., P. Hale & R. Sweeting (eds), 2003. Macrophyte inventory of the river Danube: A pilot study. Archiv für Hydrobiologie, Supplement "Large Rivers" 147 (1–2): 1–229.

Kelly, M. G. & B. A. Whitton, 1998. Biological monitoring of eutrophication in rivers. Hydrobiologia 384: 55–67.

Kohler, A., 1978. Methoden der Kartierung von Flora und Vegetation von Süßwasserbiotopen. Landschaft & Stadt 10: 73–85.

Korte, T. & K. Van de Weyer, 2005. Die Bewertung von Fließgewässern mit Makrophyten gemäß EU-WRRL - Ergebnisse des Vergleichs von zwei Bewertungsverfahren. Wasser und Abfall 9/2005: 46–49.

Leyssen, A., P. Adriaens, L. Denys, J. Packet, A. Schneiders, K. van Looy & L. Vanhecke, 2005. Toepassing van verschillende biologische beoordelingssystemen op Vlaamse potentiële interkalibratielocaties overeenkomstig de Europese Kaderrichtlijn Water – partim "Macrofyten". Instituut voor Natuurbehoud in opdracht van VMM, Brussels.

Meilinger, P., S. Schneider & A. Melzer, 2005. The reference index method for the macrophyte-based assessment of rivers – a contribution for the implementation of the European

Water Framework Directive in Germany. International Review of Hydrobiology 90: 322–342.

Pall, K., V. Moser, J. Schaumburg, C. Schranz, & P. Meilinger, 2005. Ergebnisse zur Interkalibrierung der Fließgewässerbewertung mit Makrophyten (Option 3: Vergleich Deutschland-Österreich). Oral presentation held at the conference of the "Deutsche Gesellschaft für Limnologie" in Karlsruhe, 28 September 2005.

Pielou, E. C., 1966. The measurement of diversity in different types of biological collections. Journal of Theoretical Biology 13: 131–144.

Pot, R., 2005. QBWat – ecologische beoordeling van waterkwaliteit conform de Europese Kaderrichtlijn Water. Version 1.01.

SYSTAT Software Inc., 2002. TableCurve 2D – Version 5.01. SSI, Richmond CA.

Schaumburg, J., C. Schranz, J. Foerster, A. Gutowski, G. Hofmann, P. Meilinger, S. Schneider & U Schmedtje, 2004. Ecological classification of macrophytes and phytobenthos for rivers in Germany according to the Water Framework Directive. Limnologica 34: 283–301.

Schaumburg, J., U. Schmedtje, B. Köpf, C. Schranz, S. Schneider, P. Meilinger, D. Stelzer, G. Hofmann, A. Gutowski & J. Foerster, 2005. Makrophyten und Phytobenthos in Flüssen und Seen. Leitbildbezogenes Bewertungsverfahren zur Umsetzung der EG-Wasserrahmenrichtlinie. Informationsbericht Heft 1/05. Bayerisches Landesamt für Wasserwirtschaft, München.

Schneider, S., 2000. Entwicklung eines Makrophytenindex zur Trophieindikation in Fließgewässern. Shaker Verlag, Aachen.

Shannon, C. E. & W. Weaver, 1949. Mathematical Theory of Communication. University of Illinois Press, Urbana.

Simpson, E. H., 1949. Measurement of diversity. Nature 163: 688.

Sládeček, V., 1973. System of water quality from the biological point of view. Archiv für Hydrobiologie Beiheft Ergebnisse der Limnologie 7: 1–218.

Szoszkiewicz, K., T. Ferreira, T. Korte, A. Baattrup-Pedersen, J. Davy-Bowker & M. O'Hare, 2006. European river plant communities: the importance of organic pollution and the usefulness of existing macrophyte metrics. Hydrobiologia 566: 211–234.

Van de Weyer, K., 2003. Kartieranleitung zur Erfassung und Bewertung der aquatischen Makrophyten der Fließgewässer in Nordrhein-Westfalen gemäß den Vorgaben der EU-Wasser-Rahmenrichtlinie. LUA-Merkblätter Nr. 39. Landesumweltamt (LUA) NRW, Düsseldorf.

Van den Berg, M. S., H. C. Coops, R. Pot, W. Altenburg, R. Nijboer, T. v. d. Broek, M. Fagel, G. Arts, R. Bijkerk, H. v. Dam, T. Ietswaart, J. v. d. Molen, K. Wolfstein, D. d. Jong & H. Hartholt, 2004. Achtergronddocument referenties en maatlatten waterflora. RIZA, Lelystad.

van der Molen, D. T., M. Beers, M. S. v. d. Berg, T. v. d. Broek, R. Buskens, H. C. Coops, H. v. Dam, G. Duursema, M. Fagel, T. Ietswaart, M. Klinge, R. A. E. Knoben, J. Kranenbarg, J. d. Leeuw, R. Noordhuis, R. C. Nijboer, R. Pot, P. F. M. Verdonschot & T. Vriese, 2004. Referenties en maatlatten voor rivieren ten behoeve van de Kaderrichtlijn Water – version July 2004. Alterra, Wageningen.

Wiegleb, G., 1988. Analysis of flora and vegetation in rivers: concepts and applications. In Symoens, J. J. (ed.) Vegetation of Inland Waters. Kluwer Academic Publishers, Dordrecht: 311–340.

Wiegleb, G., 1991. Die Lebens- und Wuchsformen der makrophytischen Wasserpflanzen und deren Beziehung zur Ökologie, Verbreitung und Vergesellschaftung der Arten Tuexenia 11: 135–147.

Errors and Uncertainty in Bioassessment Methods

Hydrobiologia (2006) 566:433–439
M.T. Furse, D. Hering, K. Brabec, A. Buffagni, L. Sandin & P.F.M. Verdonschot (eds), The Ecological Status of European Rivers: Evaluation and Intercalibration of Assessment Methods
DOI 10.1007/s10750-006-0079-2

Errors and uncertainty in bioassessment methods – major results and conclusions from the STAR project and their application using STARBUGS

Ralph T. Clarke[1,]* & Daniel Hering[2]
[1]*Centre for Ecology and Hydrology, Winfrith Technology Centre, Winfrith Newburgh, DT2 8ZD Dorchester, Dorset, UK*
[2]*Department of Hydrobiology, University of Duisburg-Essen, D-45117 Essen, Germany*
*(*Author for correspondence: E-mail: rtc@ceh.ac.uk)*

Key words: replicate, sampling variation, sub-sampling, macroinvertebrates, multi-metric, uncertainty, STARBUGS, simulations, software, audit, quality assurance, sample coherence

Abstract

The STAR project's extensive replicated sampling programmes have provided the first ever quantitative comparative studies of the susceptibility of a wide range of national macroinvertebrate sampling methods and taxonomic metrics to uncertainty resulting from the effects of field sampling variability and subsequent sub-sampling and laboratory (or bank-side) procedures and protocols. We summarise six STAR project papers examining various aspects of the potential sources of uncertainty in the observed fauna and observed metric values. The use of new simulation software STARBUGS (STAR Bioassessment Uncertainty Software System) to incorporate the effects of these potential errors into quantitative assessments of the uncertainty in assigning water bodies to WFD ecological status classes is discussed.

Introduction

Any indices of freshwater biological quality are of little value without some knowledge and quantitative estimates of their precision and of the confidence in assigning individual water bodies (river sites or lakes) to ecological status classes. This is a requirement of the Water Framework Directive (WFD), which states that 'Estimates of the confidence and precision attained by the monitoring system used shall be stated in the river basin monitoring plan' (European Commission, 2000). In particular, given the importance being placed in the WFD on determining whether a water body is in 'good' or better status class, we would like to be able to estimate the probability that a water body could actually be of 'moderate' or worse status.

The WFD requires each Member State to express bioassessment results as Ecological Quality Ratios (EQRs), where the ratios represent the relationship between the values of the biological parameters observed for a water body and the values for these parameters in the reference conditions applicable to that water body. The directive also requires each country to use these EQRs to classify water bodies into five ecological status classes and to monitor any changes in the status of water bodies (European Commission, 2000). When the ecological condition of a river site is assessed in two different years, the observed estimates of site quality will usually differ and the ecological status class may also have changed. We need to be able to place some confidence on the likelihood that a real change in quality or change in status class has occurred or whether the observed changes are just due to the inherent errors and sampling variation in the whole site assessment process.

As a general guide to the likely levels of uncertainty in assignment of sites to WFD ecological status classes, Figure 1 and Table 1 show the probability of misclassifying a site of any particular true quality (i.e., EQR value) according to the size of the errors or uncertainty in the

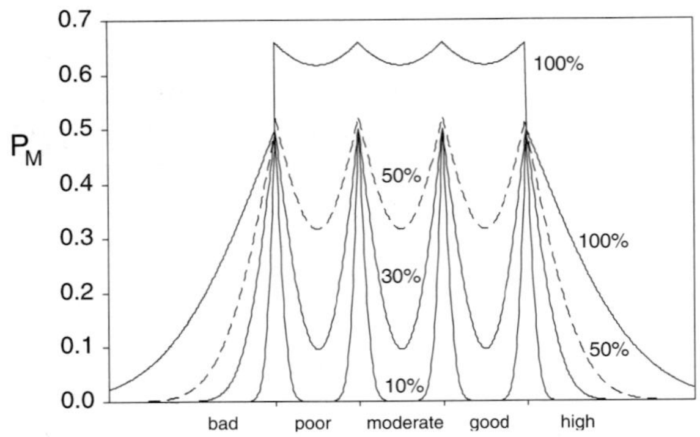

Figure 1. Plot of the probability (P_M) of classifying a site into a different status class versus its true Environmental Quality Ratio (EQR) value for a range of error/uncertainty standard deviations (σ) in the observed sample EQR value. The EQR range has been divided into the five WFD classes (high, good, moderate, poor and bad) with the middle three classes each of width W. Plots are shown for $\sigma = 10$, 30, 50 and 100% of W, where the broken line indicates the 50% plot.

Table 1. Mean and range of misclassification rates (P_M) for sites with true qualities in a middle (i.e., not top or bottom) ecological status class for each of a range of error/uncertainty standard deviations (σ) in their observed sample EQR values, where σ is expressed as a percentage of the EQR range of each middle class

σ (%)	Mean % misclassification (%)	Range (%)
10	8	0–50
30	24	10–50
50	39	32–52
100	63	62–66

metric's EQR values expressed as a percentage of the width of the status classes for the EQR (see Clarke et al., 1996 for the mathematical derivation). When the uncertainty standard deviation of the EQR values is only 10% of the width of status class, then sites whose true quality lies in the centre of a status class would never be misclassified. Sites whose true quality lies on the border of two classes will always have at least a 50% chance of being placed in the wrong class. With uncertainty standard deviations of 10% of class width, the overall misclassification rate for sites in a middle class (i.e., 'good', 'moderate' or 'poor') (assuming an even spread of true qualities across the class) is only 8% (Table 1). If however, the error standard deviation is 50% of the class width, then even sites in the centre of a middle class have a roughly one in three chance of being placed in the wrong class and roughly 40% of all sites in the class will be misplaced into either a higher or lower class. If the error standard deviation is equal to the class width (i.e., 100%), as if possible for metrics with high sampling variability, then all sites whose true quality lies within a middle class will more likely than not be placed in the wrong class (Fig. 1 and Table 1). Sites with EQR values either well above the high/good boundary or well below the poor/bad boundary will obviously have the lowest probabilities of being misclassified.

Sources of uncertainty in the observed biota

The sources of variation in the fauna observed at a site are due to:

(i) *Sampling variation and sampling method.* Within each site there will still be spatial heterogeneity in the microhabitats and distributions of macroinvertebrates and other organisms. Thus taxonomic richness and composition will vary between samples taken during the same period. The precision of a sampling method will therefore be influenced by the number of sampling units, the range of habitats and/or the total area sampled at a site.

(ii) *Sample processing and taxonomic identification errors.* Sub-sampling in the laboratory (or in the field) will also lead to increased uncertainty. In sorting the material in a test site's sample and identifying the taxa, some taxa may be missed or misidentified by less experienced staff. This may lead to biases and under-estimation of any index involving some from of taxonomic richness.

(iii) *Natural temporal variation.* There will also be what might be called 'natural' temporal variation whereby the taxa present at a site, not just in the sample, will vary 'naturally' over time for reasons other than stress or pollution.

(iv) *The effects of pollution or environmental stress on the biota.* This is what we are trying primarily to detect and quantify.

It may be difficult to distinguish between (iii) and (iv). For example, biological effects of a reduction in river discharge due to the weather may be considered natural, but reductions in river flow when abstraction is present may be considered a man-induced stress.

The potential sources of error in estimates of the expected fauna and Reference Condition (RC) for a site include having an inadequate set of reference sites, the choice of statistical prediction method or modelling technique, not involving all relevant environmental predictor variables and errors in measuring these variables for new sites (for further details see Clarke et al., 1996). For example, the WFD permits the determination of RC for a site from the average biota of the reference sites in the same stream type, where WFD System A types are based on only 3–4 classes of altitude, catchment area and geology. System B types and site-specific predictive models such as RIVPACS (Clarke et al., 2003) which use more site variables, might be expected to give more precise target RC, as recently shown for RIVPACS-type models in the UK, Sweden and the Czech Republic (Davy-Bowker et al., 2006).

The STAR project's extensive replicated sampling programme and the subsequent analysis of results has provided the first ever quantitative comparative study of the susceptibility of each of a wide range of established and 'national' macroinvertebrate sampling methods and a wide range of metrics to uncertainty resulting from the effects of field sampling method variability and subsequent subsampling and laboratory (or bank-side) procedures and protocols (Furse et al., 2006). We provide an integrated summary of six STAR project papers examining various aspects of the potential sources of uncertainty in the observed fauna and observed metric values.

Sampling method and sample size

Most commonly used macroinvertebrate sampling methods for rivers involving sampling each of the major habitats at a site and combining these basic sampling units into one overall composite sample for the site. Usually only one composite sample is obtained and thus there is no replication. Vlek et al. (2006) examined the effect of varying the number of sampling units involved in the composite sample on the precision of six commonly used macroinvertebrate metrics. They took repeated random subsets of 20 sampling units (each 25 cm sampling length using a 25 cm wide pond-net) from the dominant habitat type at each of four sites in the Netherlands and from each of two different habitats in each two streams in Slovakia. Although, as expected, the precision of all metrics increased with sample size (i.e., number units), the typical number of sampling of units required to achieve a 10% coefficient of variation (CV) for the composite sample varied from 1–2 (e.g., Saprobic index), to 3–8 (ASPT and 'Number of taxa'), while '% EPT-taxa' and 'total number of individuals' often required 10–17 sampling units. Accuracy was measured by treating the metric values based on all 20 sampling units combined as the 'truth', this was not ideal as it forces the any systematic bias to decrease as sample size approaches the maximum 20 units. The two most precise metrics also showed no systematic bias or trends with increasing sample size. However, ASPT values tended to under-estimate the 'true' ASPT when based on very few sampling units (<4–10) (Supplementary material in Vlek et al., 2006). This example reminds us that the sampling methods and protocols used to estimate the observed values of metrics should be exactly the same as those used at the reference sites involved in setting the target RC values; otherwise there may be systematic biases in the EQR values.

Vlek et al. (2006) found that, for a fixed sample size, precision was fairly similar across most habitat types for most metrics. However, there were exceptions, especially for '% EPT-taxa', which suggests caution in extrapolating estimates of sampling precision from one habitat or stream type to another.

Sample processing time was also found to increase linearly with sample size. Although the number of sampling units needed to achieve a target precision for a particular metric was similar for many stream types and metrics, the costs in terms of sample processing time for a given sample size varied significantly between habitats (Vlek et al., 2006). Samples from habitats, which had the most individuals per sample (and often the most taxa) (e.g., Fine particulate organic matter

(FPOM) riverine habitats in the Netherlands) tended to take longer to process, as might be expected with a method, which identifies all of the individuals in a sample.

Sampling variation

The STAR project involved the first ever-extensive replicated sampling programme to estimate and compare the overall effects of sampling variation on a wide range of 27 commonly used metrics for nine macroinvertebrate sampling methods across Europe. Replicate samples were taken in each of two seasons at a subset of 2–6 sites of varying pre-classified ecological status within each of 18 stream types spread over 12 countries, using both the STAR-AQEM method and a national sampling method or, where unavailable, the RIVPACS sampling protocol.

Clarke et al. (2006a) analysed these data to provide the first comparative estimates of the susceptibility to sampling variability of a range of macroinvertebrate sampling methods and metrics, including the six metrics involved in the proposed Inter-calibration Common Metric multi-metric index (ICMi, Buffagni et al., 2006). Clarke and colleagues determined the transformation scale for each metric, which made the replicate sampling standard deviation (SD) the most homogeneous, enabling a single best estimate of sampling SD of a metric to be determined for any particular method and stream type. These estimates can be then used to simulate the likely uncertainty in metric values associated with any other single sample taken from the same stream type using the same method (Table 2); as incorporated in the STAR project's STARBUGS software (see below).

Clarke et al. (2006a) estimated the precision of the combination of method and metric by expressing the replicate sampling variance as a percentage P_{samp} of the total variance in metric values with a stream type. High percentages indicate low sampling precision and low repeatability and hence that such a combination of sampling method and metric is unlikely to have much power to detect differences in ecological status class. The national methods used in the Czech Republic, Denmark, France, Poland and the RIVPACS method used in the UK and Austria all had percentage sampling variances <10% for most metrics. Because two methods were used on the same set of sites within a stream type, the P_{samp} values provided a valid comparison of their relative sampling precision. Most national methods, including RIVPACS, had sampling precisions at least as good as those for the STAR-AQEM method. In contrast, none of the metrics had percentage sampling variances <10% when based on either the Italian (IBE) method, which used bank-side sorting, or the Latvian national method, which identified only a limited set of taxa. When averaged over all stream types and methods, the three Saprobic metrics had the lowest average percentage sampling variances (3–6%). Obviously, metrics with high sampling precision and repeatability may still not be good ecological metrics or accurate indicators of ecological status class.

Lorenz & Clarke (2006) assessed the taxonomic community similarity of all pairs of samples taken within a stream type. They introduced the new concept of sample 'coherence' as a measure of the relative strength of within-site, within-season and within-method similarity. Site-coherence (i.e., the percentage of samples which are most similar to another sample from the same site) amongst sites

Table 2. Mathematical procedure used to simulate random sampling values of metrics with sampling SD (σ), which are constant on a particular transformation scale. X denotes the user-supplied untransformed observed value for a site. Z denotes a random standard normal deviate with a mean of zero and SD of σ

Transformation	Mathematical notation	Simulated value of metric in untransformed units
None	x	$X+Z$
Square root	$\sqrt{x}$	$(\sqrt{X}+Z)^2$
Double square root	$\sqrt{\sqrt{x}}$	$(\sqrt{\sqrt{X}}+Z)^4$
Arcsine square root for proportions	$\text{arcsine}(\sqrt{x})$	$\text{sine}(\text{arcsine}(\sqrt{X})+Z))^2$
Arcsine square root for percentages	$\text{arcsine}(\sqrt{(x/100)})$	$\text{sine}(\text{arcsine}(\sqrt{(X/100)})+Z))^2$

with replicate samples varied between 83% and 100%. Season-coherence of samples was nearly 100% even if different sampling methods were compared; indicating that time of year has a major influence on in-stream fauna. The STAR-AQEM method is most comparable in relative community similarity to the Nordic, Portuguese and Czech (PERLA) national methods and less comparable to the Italian (IBE) and Latvian methods. Samples collected by these latter methods had higher similarities to other sites sampled with the same methods than to samples from the same site obtained using the STAR-AQEM method, thus there was low site-coherence. Lorenz & Clarke (2006) found that replicate samples are less coherent within site, within season or within sampling method if the taxonomic resolution is family rather than species.

Sample processing and taxonomic identification errors

Having obtained a sample in the field, the procedures used to process the sample can all influence the overall reliability of the recorded taxonomic information. For example, the STAR-AQEM method requires the sub-sampling and taxonomic identification of at least one-sixth of the sample and at least 700 individuals. To assess the effect of this on the precision of results, replicate STAR-AQEM sub-samples were taken at most of the STAR sites where replicate samples were taken. Clarke et al. (2006b) found that STAR-AQEM sub-sampling effects caused more than 50% of the overall variance between replicate samples values for 12 of the 27 macroinvertebrate metrics analysed and was generally greatest for metrics that depend on the number of taxa present.

Sorting and identifying a larger fraction of the sample would reduce this source of variation (in the extreme, sorting the whole sample would eliminate it); but at increased costs. Vlek (2004) found that, on average across the sampled sites, STAR-AQEM samples took 18 h to process (including sorting and identification, whilst RIVPACS samples took only 9 h – half the amount of time. As the RIVPACS method led to no more than marginally higher average percentage sampling variances within the four countries where both methods were used, the RIVPACS method may be more cost-effective than the STAR-AQEM method.

Since the early 1990s, the UK government's environment agencies have used internal quality assurance and external auditing schemes to monitor the quality of their processing and taxonomic identification of RIVPACS macroinvertebrate samples (Dines & Murray-Bligh, 2000). Using this experience, a sample-auditing scheme involving 10 countries was implemented within the STAR project to assess the joint impact of sorting and identification errors of macroinvertebrate samples collected and analysed using different methods (notably STAR-AQEM and RIVPACS) (Haase et al., 2006). Haase and colleagues analysed differences in terms of 'gains' and 'losses' of taxa between the original and audited recorded lists of taxa for a sample. They found a surprising degree of sorting and identification errors, the total impact of which was reflected in many functional metrics and in metrics indicative of taxonomic richness. The results stress the importance of implementing quality control mechanisms in macroinvertebrate assessment schemes to monitor, improve and maintain sample-processing performance.

Natural temporal variation

Within STAR, Šporka et al. (2006) made an assessment of the effect of natural temporal seasonal variability on macroinvertebrate community composition and metric values. They took replicate multi-habitat samples at two-monthly intervals for a year from two stretches of a calcareous stream in the Carpathian Mountains and found major seasonal distinct differences in community composition. Moreover, seasonal differences were detected for many metrics, often related to the amount of organic material present. This study re-enforces the problem of deciding when to sample a stream for biomonitoring. Ignoring natural seasonal variability can confound the detection of anthropogenic environmental change. In the context of the WFD, it is important that the RC value of one or more metrics for a water body are not only appropriate for that type of site, but are determined from samples taken at roughly the same type of year as the sampling season(s) used in the monitoring programme. Sampling in more than one season (e.g., spring and autumn) and perhaps combining the samples (to determine both observed and RC metric values) can lead to more reliable estimates of ecological status, as shown by Clarke et al. (2002).

Implications for uncertainty in ecological status assessments and STARBUGS

As part of the STAR project, a new simulation software package called STARBUGS (STAR Bioassessment Uncertainty Guidance Software, Clarke, 2004) has been produced to help assess the effect of the various sources of variation and errors in the observed and RC values of one or more metrics on the overall uncertainty in assignment of water bodies to ecological status classes. See www.eu-star.at for further details about downloading the software and user manual.

Within STARBUGS, the ecological status class assessment for individual metrics can be based on just the observed (O) values of metrics or on Ecological Quality Ratios (EQRs) involving the ratio of the observed metric values to the RC (or RIVPACS-type Expected) values (E_1) of the metric. More generally, EQRs are determined by:

$$\text{EQR} = \frac{O - E_0}{E_1 - E_0} \qquad (1)$$

where O = observed value, E_1 = Reference Condition value (= value of metric for which EQR = 1) and E_0 = value of metric for which EQR = 0.

Statistical distributions of the uncertainty in the estimated EQR values are obtained in STARBUGS using stochastic simulations, as follows

$$\text{Simulated EQR} = \frac{O + S + B - E_0}{E_1 + R - E_0} \qquad (2)$$

where S = random sampling (+ sub-sampling) variation term, B = random sorting and identification bias and variation term, R = random error in estimating RC value E_1.

The estimates of overall replicate sampling (including perhaps sub-sampling) SD for the term S can be obtained from Clarke et al. (2006a), the Deliverable 8 report on the STAR web-site www.eu-star.at, or elsewhere as appropriate for the metric(s), stream type and sampling methods. The sample sorting and identification term B is more complex as such errors can lead to both additional variances and systematic biases. For example, inexperienced staff tends to miss some taxa present and under-estimate metrics involving taxonomic richness. An estimate of error SD for the RC values could, for example, be obtained from the standard error of the simple or weighted mean of the metric values for the reference sites' samples on which the RC value was based. STARBUGS uses these various estimates of the components of uncertainty to generate many random simulations of the potential metric values for a site, from which the pre-defined metric-based classification rules and class boundaries are used repeatedly on each simulation to build up estimates of the probabilities that a particular water body belongs to each of the WFD ecological status classes.

It should always be remembered that there is no absolute truth. The uncertainty in any approach can only be assessed using the limited information available.

References

Buffagni, A., S. Erba, M. Cazzola, J. Murray-Bligh, H. Soszka & P. Genoni, 2006. The STAR common metrics approach to the WFD intercalibration process: Full application for small, lowland rivers in three European countries. Hydrobiologia 566: 379–399.

Clarke, R. T., 2004. 9th STAR Deliverable. Error/Uncertainty Module Software STARBUGS (STAR Bio Assessment Uncertainty Guidance Software) User Manual. www.eu-star.at.

Clarke, R. T., M. T. Furse, J. F. Wright & D. Moss, 1996. Derivation of a biological quality index for river sites: comparison of the observed with the expected fauna. Journal of Applied Statistics 23: 311–332.

Clarke, R. T., M. T. Furse, R. J. M. Gunn, J. M. Winder & J. F. Wright, 2002. Sampling variation in macroinvertebrate data and implications for river quality indices. Freshwater Biology 47: 1735–1751.

Clarke, R. T., J. F. Wright & M. T. Furse, 2003. RIVPACS models for predicting the expected macroinvertebrate fauna and assessing the ecological quality of rivers. Ecological Modelling 160: 219–233.

Clarke, R. T., J. Davy-Bowker, L. Sandin, N. Friberg, R. K. Johnson & B. Bis, 2006a. Estimates and comparisons of the effects of sampling variation using 'national' macroinvertebrate sampling protocols on the precision of metrics used to assess ecological status. Hydrobiologia 566: 477–503.

Clarke, R. T., A. Lorenz, L. Sandin, A. Schmidt-Kloiber, J. Strackbein, N. T. Kneebone & P. Haase, 2006b. Effects of sampling and sub-sampling variation using the STAR-AQEM sampling protocol on the precision of macroinvertebrate metrics. Hydrobiologia 566: 441–459.

Dines, R. A. & J. A. D. Murray-Bligh, 2000. In J. F. Wright, D. W. Sutcliffe & M. T. Furse (eds), Assessing the Biological Quality of Freshwaters: RIVPACS and Similar Techniques, Freshwater Biological Association, Ambleside, pp. 71–78.

European Union, 2000. Directive 2000/60/EC. Establishing a Framework for Community Action in the Field of Water Policy. European Commission PE-CONS 3639/1/100 Rev 1, Luxemburg.

Furse, M., D. Hering, O. Moog, P. Verdonschot, R. K. Johnson, K. Brabec, K. Gritzalis, A. Buffagni, P. Pinto, N. Friberg, J. Murray-Bligh, J. Kokes, R. Alber, P. Usseglio-Polatera, P. Haase, R. Sweeting, B. Bis, K. Szoszkiewicz, H. Soszka, G. Springe, F. Sporka & I. Krno, 2006. The STAR project: context, objectives and approaches. Hydrobiologia 566: 3–29.

Haase, P., J. Murray-Bligh, S. Lohse, S. Pauls, A. Sundermann, R. Gunn & R. Clarke, 2006. Assessing the impact of errors in sorting and identifying macroinvertebrate samples. Hydrobiologia 566: 505–521.

Lorenz, A. & R. T. Clarke, 2006. Sample coherence – a field study approach to assess similarity of macroinvertebrate samples. Hydrobiologia 566: 461–476.

Šporka, F., H. E. Vlek, E. Bulánková & I. Krno, 2006. Influence of seasonal variation on bioassessment of streams using macroinvertebrates. Hydrobiologia 566: 543–555.

Vlek, H. E., 2004. Comparison of cost effectiveness between various macroinvertebrate field and laboratory protocols. European Commission, STAR (Standardisation of river classifications), Deliverable, N1, 78 pp, www.eu-star.at.

Vlek, H. E., F. Šporka & I. Krno, 2006. Influence of macroinvertebrate sample size on bioassessment of streams. Hydrobiologia 566: 523–542.

Hydrobiologia (2006) 566:441–459
M.T. Furse, D. Hering, K. Brabec, A. Buffagni, L. Sandin & P.F.M. Verdonschot (eds), The Ecological Status of European Rivers: Evaluation and Intercalibration of Assessment Methods
DOI 10.1007/s10750-006-0078-3

Effects of sampling and sub-sampling variation using the STAR-AQEM sampling protocol on the precision of macroinvertebrate metrics

Ralph T. Clarke[1,*], Armin Lorenz[2], Leonard Sandin[3], Astrid Schmidt-Kloiber[4], Joerg Strackbein[2], Nick T. Kneebone[1] & Peter Haase[5]
[1]*Centre for Ecology & Hydrology, Winfrith Technology Centre, Dorchester, Dorset DT2 8ZD, United Kingdom*
[2]*University of Duisburg-Essen, 45117 Essen, Germany*
[3]*Department of Environmental Assessment, Swedish University of Agricultural Sciences, P.O. Box 7050, SE-750 07 Uppsala, Sweden*
[4]*Department of Hydrobiology, Fisheries Management and Aquaculture, Max-Emanuel-Strasse 17, A-1180 Vienna, Austria*
[5]*Department of Limnology and Conservation Research, Senckenberg – Research Institute and Natural History Museum, Lochmuehle 2, 63599 Biebergemuend, Germany*
*(*Author for correspondence: E-mail: rtc@ceh.ac.uk)*

Key words: replicate, sampling variation, sub-sampling, uncertainty, macroinvertebrate metrics

Abstract

As part of the extensive field sampling programme within the European Union STAR project, replicate macroinvertebrate samples were taken using the STAR-AQEM sampling method at each of 2–13 sites of varying ecological quality within each of 15 stream types spread over 12 countries throughout Europe. The STAR-AQEM method requires the sub-sampling and taxonomic identification of at least one-sixth of the sample and at least 700 individuals. Replicate sub-samples were also taken at most of these sites. Sub-sampling effects caused more than 50% of the overall variance between replicate samples values for 12 of the 27 macroinvertebrate metrics analysed and was generally greatest for metrics that depend on the number of taxa present. The sampling precision of each metric was estimated by the overall replicate sampling variance as a percentage P_{samp} of the total variance in metric values within a stream type. Average over all stream types, the three Saprobic indices had the lowest percentage sampling variances with median values of only 3–6%. Most of the metrics had typical replicate sampling variances of 8–18% of the total variability within a stream type; this gives rise to estimated rates of mis-classifying sites to ecological status class of between 22 and 55% with an average of about 40%. This suggests that the precision of such metrics based on the STAR-AQEM method is only sufficient to indicate gross changes in the ecological status of sites, but there will be considerable uncertainty in the assignment of sites to adjacent status classes. These estimates can be used to provide information on the effects of STAR-AQEM sampling variation on the expected uncertainty in multi-metric assessments of the ecological status of sites in the same or similar stream types, where only one sample has been taken at a point in time and thus there is no replication.

Introduction

The original AQEM macroinvertebrate sampling method and protocol was developed during the AQEM project within the 5th Framework Programme of the European Union (Hering et al., 2004). The protocol was designed as a possible approach to providing a standardised sampling and assessment methodology across Europe (see special issue volume 516 of Hydrobiologia). The

AQEM method was subsequently modified slightly for use within the STAR project of the 5th Framework Programme, and is now referred to as the STAR-AQEM method and described in detail in Furse et al. (2006).

Assessments of the ecological status of rivers and streams using macroinvertebrate sampling are usually based on the values of one or more indices or metrics based on all or specific aspects of the macroinvertebrate community present at the site (e.g., Karr & Chu, 1999; Wright, 2000; Herring et al., 2004). The metrics are often designed to measure the ecological response to some specific form of stress, such as organic or toxic pollution, acid stress, or degradation in stream morphology and the diversity of habitats. The multimetric assessment system developed within the AQEM project were based on a wide range of different metrics (Dahl et al., 2004; Lorenz et al., 2004; Sandin et al., 2004; Ofenbock et al., 2004; Vlek et al., 2004). In total, the values of over 200 metrics can be calculated for any sample by the AQEM/STAR Ecological River Classification System ASTERICS (for a full list of the metrics, see Table 3 in Hering et al., 2004).

All assessments of the ecological status of a river site using macroinvertebrate sampling are subject to uncertainty and errors due to a range of factors (Ostermiller & Hawkins, 2004). Any measure of ecological quality or status is of little value without some knowledge of its levels of uncertainty (Clarke, 2000; REFCOND, 2003). In particular, it is important to have quantitative estimates of the effects of sampling variation on the value of any biotic index or metric used to assess the ecological status of a river site. Field sampling and sample processing methods and the derived biotic metrics which are very prone to high levels of variation between replicate samples will tend to provide less reliable estimates of ecological quality ratios and ecological status for a site and have less power and confidence to detect changes in ecological quality (Clarke, 2000).

A STAR-AQEM field sample is based on 20 sampling units taken in proportion to the estimated percentage cover of each major habitat type at the site. However, the taxonomic composition of replicate field samples will still vary because of small-scale spatial heterogeneity in habitat and patchiness in macroinvertebrate distribution and density within a site. Different samples will involve taking sampling units from different locations within the site. These differences are likely to be even greater when the replicate samples are taken completely independently by different personnel who may estimate the percentage cover of the micro-habitats differently and thus take different numbers of sampling units from several habitats. Thus there will be real differences in taxonomic composition and derived metric values between replicate field samples.

The STAR-AQEM method protocol involves a standardised method of laboratory sub-sampling of the macroinvertebrate field sample. The sample material is spread out as evenly as possible on a tray marked out with a 6 by 5 grid of cells. The STAR-AQEM protocol requires the biologist to randomly select five of the 30 grid cells and identify and count all of the macroinvertebrate specimens in these five cells. If necessary additional cells are randomly selected until at least 700 individuals have been identified. This sub-sampling procedure will introduce an additional source of variation in the recorded taxonomic composition for the site and hence in the values of metrics for the site at that time. Thus, variation in taxonomic composition and metric values between replicate field samples taken from the same site at the same time will be due to both sampling spatial variation in the field and laboratory sub-sampling effects.

Within the AQEM project, there was insufficient time to include any replicated sampling or sub-sampling study to assess the impact of either sampling or sub-sampling procedures on the precision of metric values and the estimates of ecological status. However, Lorenz et al. (2004) used computer-simulated sub-sampling of AQEM samples from Germany to assess the effect of using sub-sample sizes of 100, 200, 300, 400, 500 and the 700 individuals on the values of 45 metrics. Not surprisingly, they found that metrics which increase with the total number of taxa present (e.g., number of families, total BMWP score) increase with sample count size, whereas those that are based on relative abundance (e.g., % gatherers/collectors) are less dependent on, and sensitive to, the number of individuals counted. This highlights the fact that the critical or reference condition values for metrics in an assessment system for any particular type of river site should be based on exactly the same sampling and sample processing procedures, including the sub-sample count size.

This paper summarises the results of an extensive replicated sampling programme within the STAR field sampling programme (Furse et al., 2006), designed to assess the individual and combined effects of field sampling variation and subsequent laboratory sub-sampling procedures for the STAR-AQEM method on the variability and hence relative precision of the various metrics.

Methods

Replicate sampling and sub-sampling

As part of the STAR field sampling programme, STAR-AQEM samples were taken at all sites by each participating partner. Samples were collected in two seasons – spring and either summer or autumn (Furse et al., 2006); the precise months involved varied because of climatic differences across Europe. (At each site in nearly all of the main stream types, each partner also collected samples using a "national" method – see Furse et al., 2006 for further details).

Most STAR project partners took a second replicate STAR-AQEM field sample in each sampling season at a subset of their sites (Table 1). These sites were carefully selected within each sampled stream type to cover a range of the perceived (i.e., pre-classified) qualities of sites from 'high' and 'good' to 'moderate' or 'poor'/'bad'. This was important because the sampling variability of one or more metrics may depend on the quality of a site; poorer quality sites with fewer taxa present might be less variable in some taxonomic richness/diversity metrics, but more variable in metrics based on some form of average stress-tolerance score of the taxa present (such as a saprobic index or Average Score Per Taxon (ASPT; Armitage et al., 1983)).

To quantify the size of the sub-sampling source of variation, especially in relation to field sampling variability, most partners took a second replicate sub-sample from one of the replicate STAR-AQEM samples for all or most of the sites at which two replicate samples were taken (Table 1).

Taxonomic resolution and calculation of metric values

After storage in the AQEMdip database system (www.eu-star.at), macroinvertebrate data were taxonomically adjusted to a consistent national level. The values of all metrics for all sub-samples of all samples were then calculated using AQEM-STAR assessment software ASTERICS (www.eu-star.at). The analyses reported here are for 27 ecological quality metrics intended to represent a wide range of aspects and responses of the macroinvertebrate fauna (Table 2).

Most taxa in all macroinvertebrate samples taken from stream types in Italy, Greece and Latvia were identified to family level. In France and Portugal, samples were identified to mainly genus level. In other countries, the macroinvertebrates were identified mainly to species level. The three Saprobic indices, which require data identified to species or genus level, were therefore not calculated for sites from stream types with predominantly family level data. The metrics measuring percentage or proportional abundance of specific guilds (% Rheophilic, % Littoral, % Grazers/scrapers, % Shredders, % Gatherer/collectors and Rhithron Feeding Type Index (RETI)) were calculated using the best available data for each country. The metrics measuring the total 'Number of taxa' and the Shannon–Wiener diversity index will be dependent on the taxonomic resolution of the data; higher taxonomic resolution will obviously lead to more individual taxa being recorded and probably more variability in results. The sampling standard deviation (SD) of these metrics for the stream types based on family data may not be comparable with those based on species and genus level data.

Two new metrics, 'Log(Sel_EPTD + 1)' based on the logarithm of the total abundance of selected Ephemeroptera, Plecoptera, Tricoptera and Diptera, and '1-GOLD' based on the proportion of all individuals which are not Gasteropoda, Oligochaeta or Diptera (Pinto et al., 2004), were included because they are two of the six proposed Inter-calibration Common Metrics (ICMs) for use throughout Europe (Buffagni et al., 2006). Their values were calculated separately, but replicate sub-samples values were only available for sites in the Czech Republic. The replicate variability of the Italian national IBE metric (Indice Biotico Esteso) and of four species trait metrics (m1, m2, m7 and m12; Bis & Usseglio-Polatera, 2004) was also assessed.

Statistical methods

The statistical analyses aimed to estimate and compare the sampling variances in the observed values of each metric within stream types. Analysis of variance (ANOVA) and hierarchical nested ANOVA techniques (calculated using Minitab Release 14 statistics package (http://www. minitab.com)) were used to estimate the various sources of variation and variance components contributing towards the total variance in values of a metric within each stream type. Specifically, if Y_{ijrs} is the value of the metric for sub-sample *s* of replicate sample *r* from site *j* in season *i*, then Y_{ijrs} can be expressed in terms of the sum of the components contributing towards the overall variation in its values, namely:

$$Y_{ijrs} = \mu + a_i + b_{ij} + c_{ijr} + d_{ijrs}$$

where

$\mu=$ overall mean value of Y within the stream type;

Table 1. Number of sites in each stream type and country with replicate STAR-AQEM field samples and/or replicate sub-samples taken in at least one season; (small-sized = 10–100 km^2, medium sized = 100–1000 km^2, lowland = < 200 m asl, 'mountain' and 'alpine' streams cover a very wide range of altitudes mostly >200 m asl)

Country	Stream Type	Description	Replicate sampling		Replicate sub-sampling	
			Sites	Sites × seasons	Sites	Sites × seasons
Austria	A05	small-sized, shallow mountain streams	5	8	5	8
	A06	small-sized crystalline streams of the ridges of the Central Alps	6	8	4	6
Czech Republic	C04	small-sized, shallow mountain streams	3	6	3	6
	C05	small-sized streams in the Central sub-alpine mountains	3	6	3	6
Germany	D03	medium-sized lowland streams	2	4	2	4
	D04	small-sized, shallow mountain streams	2	4	2	4
	D06	small-sized Buntsandstein-streams	2	4	2	4
France	F08	small-sized, shallow headwater streams in Eastern France	6	12	6	11
Greece	H04	small-sized calcareous mountain streams in Western, Central and Southern Greece	6	12	0	0
Italy	I05	small-sized streams in the southern calcareous Alps	3	6	3	6
	I06	small-sized calcareous streams in the Central Apennines	6	11	0	0
Denmark	K02	medium-sized lowland streams	6	12	6	12
Latvia	L02	medium-sized lowland streams	13	19	0	0
Poland	O02	medium-sized lowland streams	7	12	5	8
Portugal	P04	medium-sized streams in lower mountainous areas of S. Portugal	6	11	3	6
Sweden	S05	medium-sized lowland streams	3	5	5	6
	S06	medium-sized streams on calcareous soils	3	6	3	6
UK	U15	small-sized, shallow lowland streams	3	6	3	6
	U23	medium-sized lowland streams	3	6	3	6

Table 2. STAR-AQEM method: Estimate of the standard deviation (SD_U) in transformed ($f(x)$) metric values due to sub-sampling, separately for each STAR stream type and the median (Med) values across stream types

Metric	$f(x)$	Stream Type A05	A06	C04	C05	D03	D04	D06	F08	I05	K02	O02	P04	S05	S06	U15	U23	Med
Abundance [ind/m^2]	$\sqrt{\sqrt{x}}$	0.320	0.243	0.256	0.229	0.281	0.425	0.288	0.804	0.134	0.390	0.544	0.129	0.928	0.347	0.486	0.282	0.304
Number of taxa	$\sqrt{x}$	0.212	0.255	0.298	0.149	0.216	0.356	0.205	0.338	0.195	0.422	0.257	0.250	0.460	0.404	0.150	0.201	0.253
Number of Families	$\sqrt{x}$	0.203	0.202	0.273	0.251	0.217	0.188	0.155	0.272	0.204	0.286	0.176	0.214	0.321	0.191	0.244	0.226	0.216
Number of EPT taxa	$\sqrt{x}$	0.261	0.212	0.335	0.283	0.245	0.320	0.230	0.288	0.138	0.348	0.280	0.356	0.260	0.213	0.232	0.134	0.261
Saprobic Index	x	0.022	0.035	0.021	0.023	0.011	0.055	0.054	0.155		0.110	0.059		0.023	0.023	0.037	0.020	0.023
German Saprobic new	x	0.035	0.024	0.076	0.087	0.041	0.051	0.013	0.025		0.046	0.073		0.056	0.040	0.079	0.034	0.041
Czech Saprobic	x	0.047	0.049	0.045	0.034	0.022	0.044	0.049	0.041		0.110	0.088		0.065	0.034	0.084	0.058	0.046
ASPT	x	0.154	0.237	0.315	0.251	0.178	0.255	0.221	0.282	0.145	0.247	0.420	0.461	0.172	0.214	0.311	0.206	0.242
IBE	x	0.304	0.700	0.770	0.548	0.458	0.854	0.357	0.742	0.356	0.735	0.760	0.796	1.098	0.600	0.735	0.404	0.718
Shannon Diversity	x	0.112	0.066	0.108	0.058	0.162	0.145	0.071	0.107	0.036	0.168	0.101	0.107	0.225	0.102	0.129	0.086	0.107
% Rheophilic	asin	0.033	0.029	0.032	0.013	0.026	0.036	0.058	0.034	0.036	0.036	0.031	0.090	0.100	0.030	0.042	0.070	0.035
% Rheophilic (ab-class)	asin	0.022	0.026	0.037	0.028	0.038	0.030	0.022	0.046		0.049	0.035	0.077	0.031	0.031	0.031	0.040	0.031
% Littoral	asin	0.018	0.015	0.017	0.015	0.014	0.017	0.016	0.037		0.019	0.015	0.034	0.069	0.015	0.010	0.031	0.017
% Grazers/Scrapers	asin	0.023	0.016	0.022	0.004	0.019	0.040	0.017	0.028	0.013	0.025	0.022	0.067	0.084	0.016	0.009	0.023	0.022
% Shredders	asin	0.022	0.014	0.019	0.011	0.023	0.013	0.018	0.019		0.025	0.011	0.007	0.044	0.017	0.022	0.015	0.018
% Gatherers/Collectors	asin	0.014	0.012	0.013	0.010	0.005	0.017	0.010	0.044		0.036	0.043	0.073	0.059	0.011	0.026	0.013	0.014
% Oligochaeta	asin	0.025	0.029	0.019	0.016	0.025	0.027	0.026	0.042	0.011	0.030	0.069	0.069	0.022	0.025	0.039	0.037	0.027
% EPT individuals	asin	0.035	0.022	0.046	0.014	0.048	0.042	0.022	0.032	0.042	0.026	0.014	0.044	0.178	0.036	0.017	0.033	0.034
% EPT (ab-class)	asin	0.028	0.010	0.029	0.059	0.052	0.017	0.014	0.039		0.026	0.033	0.088	0.035	0.027	0.032	0.038	0.032
% EPT Taxa	asin	0.037	0.022	0.049	0.081	0.057	0.015	0.025	0.070	0.034	0.037	0.043	0.117	0.022	0.028	0.042	0.044	0.040
RETI	asin	0.020	0.017	0.016	0.015	0.022	0.020	0.018	0.035	0.014	0.029	0.025	0.081	0.094	0.020	0.023	0.013	0.020
Log(Sel_EPTD + 1)	x			0.081	0.086													0.084
1-GOLD	asin			0.038	0.011													0.025
Trait m1: max size ≤ 1 cm	asin	0.018	0.019	0.023	0.022	0.028	0.012		0.017	0.011	0.023		0.031	0.030	0.010	0.008	0.018	0.019
Trait m2: >1 cycle	asin	0.017	0.014	0.025	0.024	0.021	0.006		0.013	0.018	0.027		0.057	0.024	0.021	0.021	0.022	0.021
Trait m7: crawler loco.	asin	0.011	0.016	0.029	0.019	0.026	0.010		0.010	0.008	0.020		0.038	0.012	0.013	0.014	0.015	0.015
Trait m12: current <25 cm s^{-1}	asin	0.013	0.012	0.015	0.016	0.017	0.019		0.017	0.009	0.021		0.019	0.028	0.008	0.006	0.012	0.016

a_i = deviation of mean value for season i from the overall mean value μ

b_{ij} = deviation of mean value for site j in season i from the mean for season i

c_{ijr} = deviation of replicate r for site j in season i from the mean for site j in season i

d_{ijrs} = deviation of sub-sample s of replicate r for site j in season i from the mean for replicate r for site j in season i

The total variance (σ_T^2) in metric values over all sampled sites within any one stream type is given by:

$$\sigma_T^2 = \sigma_I^2 + \sigma_J^2 + \sigma_R^2 + \sigma_U^2$$

where

σ_I^2 = variance of the a_i = variance due to inter-season differences in mean value

σ_J^2 = variance of the b_{ij} = variance due to inter-site differences within a season

σ_R^2 = variance of the c_{ijr} = variance due to differences between replicate samples within a site and season

σ_U^2 = variance of the d_{ijrs} = variance due to differences between replicate sub-samples within a sample.

This approach correctly identifies that part of the overall variance between replicate samples which is merely the consequence of sub-sampling (namely σ_U^2) from that due to real differences between the two samples in the fauna obtained (namely σ_R^2). The overall true variance (denoted σ_E^2) between replicate samples taken using the STAR-AQEM method is the sum of the two components, namely:

$$\sigma_E^2 = \sigma_U^2 + \sigma_R^2.$$

The percentage of the overall variance (σ_E^2) in metric values between replicate field samples which is due specifically to sub-sampling variation is therefore estimated by:

$$P_{\text{sub}} = 100\sigma_U^2/\sigma_E^2.$$

If a particular metric and sampling method are to be effective in discriminating the ecological status classes of river sites within a stream type, then the overall replicate sampling variance (σ_E^2) should be small relative to the total variance (σ_T^2) in metric values across the range of ecological qualities within the stream type. This is measured by the statistic:

$$P_{\text{samp}} = 100\sigma_E^2/\sigma_T^2$$

referred to as the 'percentage sampling variance'. This is a better practical measure of the precision of each metric than using the usual coefficient of variation (CV) determined as the ratio of the replicate SD to the replicate mean. This is because the typical actual range of values many metrics take with real samples rarely includes values near zero, so a low CV may not indicate high precision in practice. As an example, a metric may have a sampling SD of say 0.5 on replicate means ranging from around 5.0 to 6.0; giving a CV of 10% or less. However, because of the limited range of values of the metric (roughly 4–7), the percentage P_{samp} of total variance due to sampling is much higher at around 40%.

The variance components are usually quoted in the tables in their standard deviation (SD) form (e.g., $SD_U = \sqrt{\sigma_U^2}$ denotes the SD due to STAR-AQEM sub-sampling and $SD_E = \sqrt{\sigma_E^2}$ denotes the overall SD due to variability between replicate sample values. When a SD is based on only two values (x_1 and x_2), the SD is equal to the absolute value of their difference divided by the square root of two (i.e., $|x_1 - x_2|/\sqrt{2} = 0.71|x_1 - x_2|$).

Frequently in ecology, the replicate sampling variability in a biotic index of taxonomic abundance, richness or composition varies with the value of the index. For example, Clarke et al. (2002) found that the variance in the number of macroinvertebrate taxa found in replicate RIVPACS samples increased roughly in proportion to the average number of taxa found in samples from the same site, but that by transforming the data, the replicate variability in the square root of the number of taxa was roughly constant and did not depend on the physical type or ecological quality of the sites.

Using a similar approach for the STAR dataset, Taylor's Power Law regressions, Spearman rank correlations and plots of replicate variance against replicate mean value were used to estimate the best data transformation to reduce the systematic variability in the replicate standard deviation of the values of each metric (Taylor, 1961; Elliott, 1977; Clarke et al., 2002). Many of the selected metrics are percentages (range 0–100) or proportions (0–1) which are based on the fraction of all individuals or of all taxa which are in a particular group or have particular characteristics. Such metrics were often most variable at intermediate values (20–80%) and the arcsine transformation (i.e., arcsine ($\sqrt{x}$) for proportions and arcsine ($\sqrt{(x/100)}$) for percentages (Sokal & Rohlf, 1995)) was used to make their sampling variance more equitable. For reasons of consistency and robustness, only one transformation was used for any single metric regardless of stream type. All estimates of variance components are based on the appropriately transformed metric values (Table 2).

These transformations enable us to derive a single estimate of the sampling and sub-sampling SD for all sites within a stream type regardless of site quality. Estimates for a stream type only depend on the sample data for that stream type and so are not influenced by potentially inappropriate sample data from other stream types. However, because it was only possible to take replicate samples at a few (2–13) sites in each stream type of each STAR partner, estimates of the above variance components for individual stream types may be imprecise. Therefore, more robust overall estimates of SD_E and P_{samp} were also derived by taking the median of the stream-specific estimates. The median is not influenced or biased by very high values for stream types for which the metric values may be invalid or inappropriate.

Results

Estimates of sub-sampling variability

The effect of only identifying the individuals within a sub-sample fraction of a STAR-AQEM sample was measured by the average variability in metric values between replicate sub-samples taken from the same sample. The difference between sub-samples in the number of families recorded was zero or one in 33% of cases and no more than three in 75% of cases, but there was a difference of nine families for one Swedish sample (stream type S05) and of 10 families for one Danish sample (stream type K02) (Fig. 1a). The difference between replicate sub-samples in the 'Number of Taxa' recorded was two or less for 40% of samples and five or less in 77% of samples, but 10% of sub-samples differed by more than seven taxa, with 25 and 44 taxa found in the two sub-samples from the same sample from Denmark.

Table 2 gives the estimates of the standard (SD_U) deviation in (appropriately transformed) metric values due to STAR-AQEM sub-sampling, with separate estimates for each STAR stream type and the median value across all stream types. Samples from some stream types might be expected to have more 'nuisance' material of small-scale debris than others, which might influence the ability to distribute the macroinvertebrates evenly between the grid cells. Kruskal–Wallis tests of ranked SD highlighted higher sub-sampling variability for stream types P04 and S05 in some of the metrics based on percentage composition of selected taxa ('% Rheophilic' to '% Oligochaeta' in Table 2). However, after metric values had been suitably transformed, there were few other systematic consistent differences between stream types in the pattern and extent of sub-sampling variation.

For the Saprobic index calculated according to Zelinka & Marvan (1961), the standard deviation (SD_U) in values due to sub-sampling was less than 0.023 for half of the stream types (Table 2), but considerably larger for others, as seen from the differences between replicate sub-sample values in Figure 1(b). The overall median sub-sample SD in ASPT across all stream types was 0.242, and although half of the differences in ASPT between two sub-samples were less than 0.2 and 75% less than 0.4, a few large differences of 1.0–1.25 were recorded.

Relative importance of sub-sampling variation to field sampling variation

Estimates of the relative importance (P_{sub}) of STAR-AQEM sub-sampling on the overall variance in metric values obtained for replicate samples were calculated independently for each stream type (Table 3). STAR-AQEM sub-sampling variation caused a relatively large part of the overall variance between replicate sample values for many metrics, and was estimated on average (see median values in Table 3) to contribute more than 50% of the overall variance between replicate samples for 12 of the 27 metrics analysed. In general, sub-sampling variance has a large effect on those metrics which are based on the number of taxa

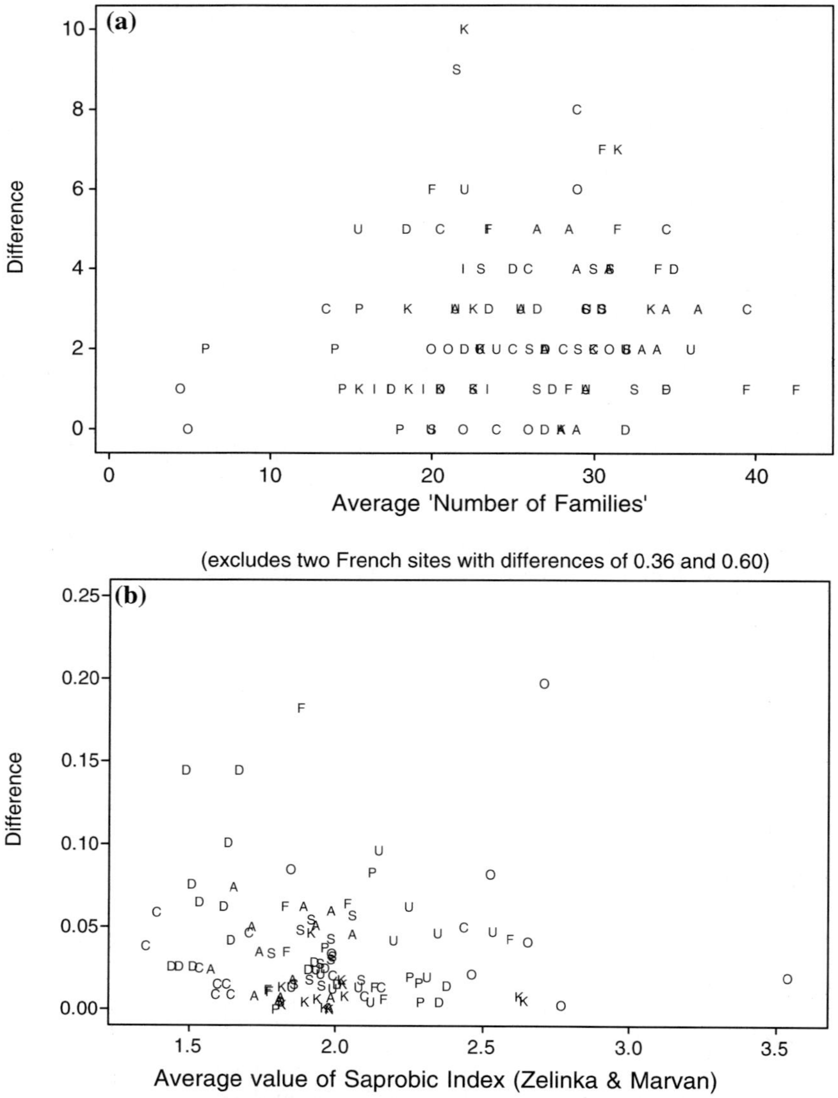

Figure 1. Difference between two replicate STAR-AQEM sub-samples plotted against the average of the two values for the metrics (a) 'Number of Families' and (b) 'Saprobic Index (Zelinka & Marvan)'; letters indicate country (Austria, Czech Republic, Germany(D), France, Italy, Denmark, Poland, Portugal, Sweden, UK).

present, such as number of families and number of EPT taxa.

Sub-sampling variation was estimated to be responsible for more than half of the overall replicate sampling variance in ASPT for the vast majority of stream types (Table 3). ASPT only depends on the presence, rather than abundance, of BMWP families (Biological Monitoring Working Party; Armitage et al., 1983). Several taxa which occur in the sample at very low abundances may be found in one sub-sample but not another. The percentage (%EPT Taxa) of all taxa which belong to the EPT group (Ephemeroptera, Plecoptera and Trichoptera), which was also dependent only on the presence of each taxon, was also relatively variable between sub-samples. The metrics based on relative abundance (i.e., percentage composition) of one of more taxonomic groups seem to be less prone to the effects of sub-sampling with less than one third of overall sampling

Table 3. STAR-AQEM method: Estimates of the percentage (P_{sub}) of the overall variance between replicate samples which is due to sub-sampling for transformed ($f(x)$) metric values, separately for each STAR stream type and the median (Med) values across stream types

Metric	Stream Type																	
	$f(x)$	A05	A06	C04	C05	D03	D04	D06	F08	I05	K02	O02	P04	S05	S06	U15	U23	Med
Abundance [ind/m^2]	$\sqrt{\sqrt{x}}$	15	18	14	9	100	100	51	49	6	40	38	3	100	5	77	20	29
Number of Taxa	$\sqrt{x}$	12	50	72	13	30	100	100	87	95	60	54	46	80	100	5	29	57
Number of Families	$\sqrt{x}$	44	40	100	65	100	74	83	61	100	52	94	41	71	72	38	93	72
Number of EPT Taxa	$\sqrt{x}$	29	32	100	100	30	93	100	75	53	50	100	100	58	100	42	100	84
Saprobic Index	x	50	46	14	15	15	100	100	50		94	98		30	37	13	39	42
German Saprobic new	x	100	46	66	100	100	100	37	6		100	65		100	74	75	28	74
Czech Saprobic	x	66	25	8	5	10	90	41	18		100	16		100	13	32	5	21
ASPT	x	20	86	100	100	26	100	75	100	25	75	100	76	28	100	45	52	76
IBE	x	41	82	100	50	49	100	85	54	57	63	97	100	76	25	63	25	63
Shannon Diversity	x	50	15	16	12	43	52	20	44	4	16	20	9	78	21	25	12	20
% Rheophilic	asin($\sqrt{(x/100)}$)	39	17	12	2	6	51	100	9	35	6	1	27	100	12	13	9	12
% Rheophilic (ab-class)	asin($\sqrt{(x/100)}$)	60	37	60	43	63	80	40	61		66	2	100	23	75	79	53	60
% Littoral	asin($\sqrt{(x/100)}$)	12	4	15	10	7	16	16	20		4	4	15	100	24	3	8	12
% Grazers/Scrapers	asin($\sqrt{(x/100)}$)	36	16	20	1	13	57	24	60	10	10	12	95	100	9	4	4	14
% Shredders	asin($\sqrt{(x/100)}$)	52	14	26	2	4	91	40	20		18	7	4	83	18	9	5	18
% Gatherers/Collectors	asin($\sqrt{(x/100)}$)	8	16	3	5	0	47	34	37		10	20	48	100	5	9	1	10
% Oligochaeta	asin($\sqrt{(x/100)}$)	15	77	2	16	66	99	65	14	38	3	9	32	50	26	35	13	29
% EPT individuals	asin($\sqrt{(x/100)}$)	29	11	28	3	59	45	12	9	25	4	9	24	100	13	26	4	18
% EPT (ab-class)	asin($\sqrt{(x/100)}$)	52	6	57	100	59	42	35	68		21	100	100	79	56	46	53	56
% EPT Taxa	asin($\sqrt{(x/100)}$)	56	19	82	100	91	17	50	100	70	28	100	100	65	45	65	100	68
RETI	asin($\sqrt{x}$)	12	14	6	1	3	15	100	61	10	7	10	100	100	24	6	1	11
Log(Sel_EPTD + 1)	x			34	12													23
1-GOLD	asin($\sqrt{x}$)			45	1													23
Trait m1: max size $\leq$ 1 cm	asin($\sqrt{x}$)	90	56	39	48	34	23		14	27	36		73	53	45	27	69	42
Trait m2: > 1 cycle	asin($\sqrt{x}$)	35	47	76	100	92	16		50	53	45		100	83	100	29	100	64
Trait m7: crawler loco.	asin($\sqrt{x}$)	23	80	100	85	100	37		24	31	39		100	64	82	27	100	72
Trait m12: current < 25 cm s^{-1}	asin	63	75	78	100	100	91		40	90	87		39	100	27	6	64	76

variability in metric values usually due to sub-sampling (Table 3).

Estimates of overall replicate sampling variability

Figures 2–3 show the pattern of differences between replicate STAR-AQEM field samples in values of selected metrics for individual sites (and seasons) in relation to the average of the replicate values at that site and season. The differences are based on just two sample values (i.e., excluding the second replicate sub-sample for the samples with two sub-samples). Estimates of the average overall replicate sampling SD (SD_E) for each metric,

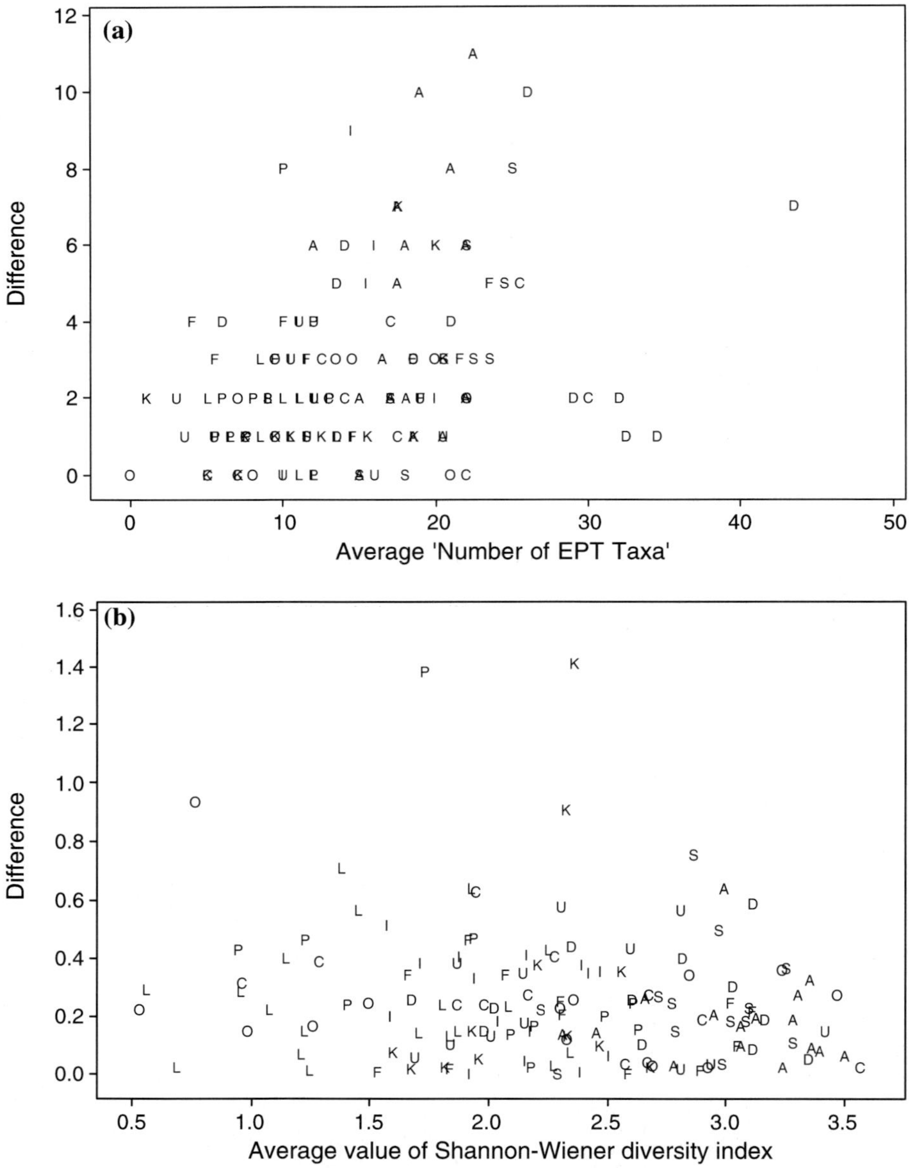

Figure 2. Difference between two replicate STAR-AQEM samples plotted against the average of the two values of (a) 'Number of EPT Taxa' and (b) 'Shannon–Weiner Diversity Index'; letters indicate country (Austria, Czech Republic, Germany(D), France, Italy, Denmark, Poland, Portugal, Sweden, UK).

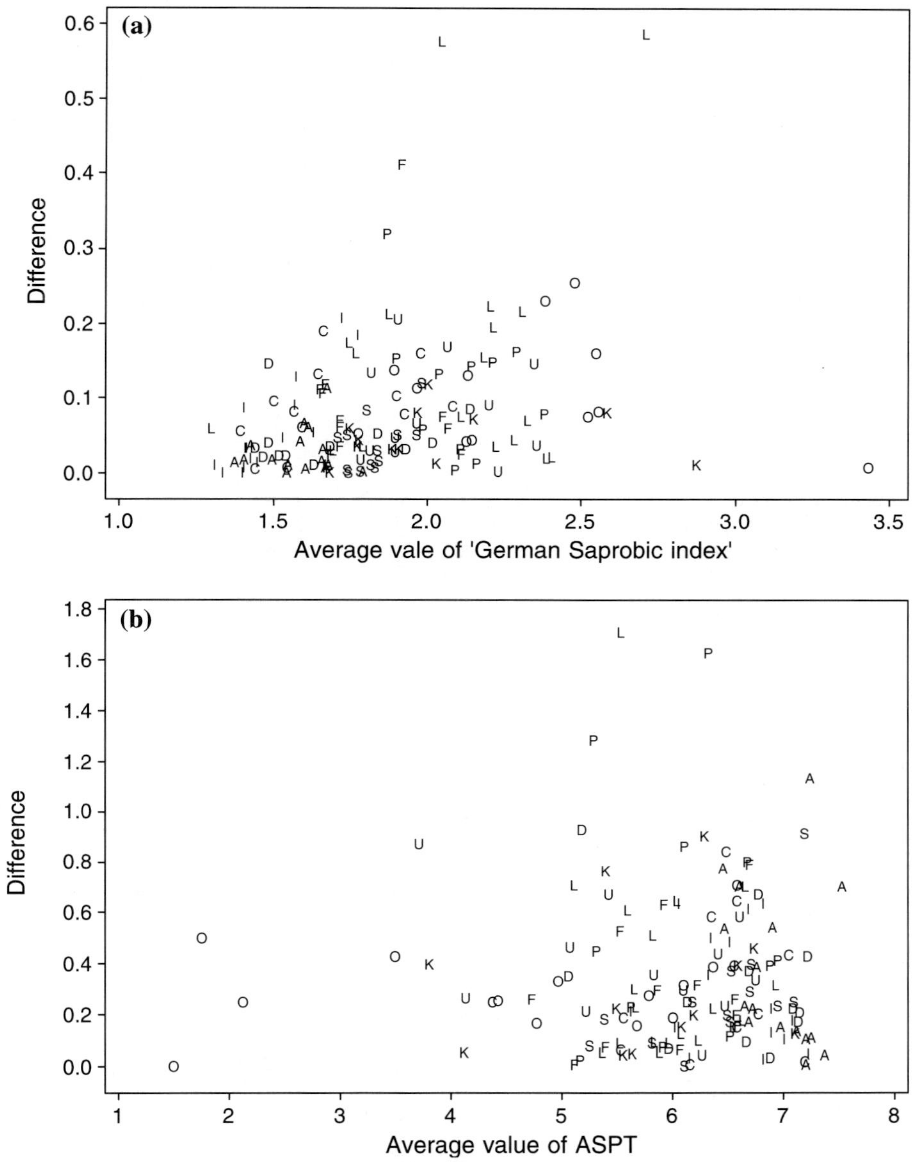

Figure 3. Difference between two replicate STAR-AQEM samples plotted against the average of the two values for the metrics (a) 'German Saprobic index (new version)' and (b) 'Average Score per Taxon (ASPT); letters indicate country (Austria, Czech Republic, Germany(D), France, Italy, Denmark, Poland, Portugal, Sweden, UK).

transformed where appropriate, are given separately for each stream type, together with the median value across all stream types (Table 4).

At sites where, on average, more than 40 taxa are recorded per STAR-AQEM sample (usually involving genus and species level identification), the difference between replicates in the number of taxa recorded was less than five is 53% of cases but was greater than 10 in 14% of cases and two replicate samples from one Austrian site had 84 and 53 taxa, a difference of 31. Half of all differences between two replicate samples in the number of families were two or less and 90% were six or less. When restricted to number of EPT taxa, differences between replicate samples were two or less for 63% of cases, but more than five in 11% of cases (Fig. 2).

Table 4. STAR-AQEM method: Estimate of the overall standard deviation (SD_E) in transformed ($f(x)$) metric values due to sampling, separately for sites in each STAR stream type and the median (Med) values across stream types

		Stream Type																		
Metric	$f(x)$	A05	A06	C04	C05	D03	D04	D06	F08	H04	I06	K02	L02	O02	P04	S05	S06	U15	U23	Med
Abundance [ind/m^2]	$\sqrt{\sqrt{x}}$	0.746	0.511	0.659	0.656	0.258	0.222	0.529	1.185	0.257	0.995	0.673	0.676	0.984	0.711	0.311	1.498	0.430	0.622	0.659
Number of taxa	$\sqrt{x}$	0.620	0.378	0.324	0.398	0.432	0.334	0.186	0.321	0.282	0.496	0.357	0.185	0.317	0.358	0.449	0.247	0.664	0.383	0.358
Number of Families	$\sqrt{x}$	0.352	0.261	0.166	0.250	0.144	0.258	0.191	0.314	0.299	0.478	0.300	0.175	0.172	0.331	0.317	0.177	0.406	0.274	0.261
Number of EPT taxa	$\sqrt{x}$	0.489	0.417	0.212	0.199	0.477	0.377	0.253	0.358	0.343	0.430	0.377	0.191	0.185	0.346	0.342	0.196	0.333	0.134	0.342
Saprobic Index	x	0.025	0.053	0.062	0.058	0.032	0.017	0.032	0.243			0.030	0.215	0.102		0.045	0.044	0.085	0.035	0.049
German Saprobic new	x	0.019	0.036	0.081	0.063	0.040	0.055	0.021	0.097			0.041	0.159	0.094		0.028	0.044	0.079	0.068	0.063
Czech Saprobic	x	0.046	0.110	0.163	0.159	0.077	0.032	0.054	0.105			0.102	0.185	0.206		0.039	0.095	0.121	0.265	0.105
ASPT	x	0.381	0.303	0.309	0.262	0.363	0.217	0.258	0.265	0.437	0.211	0.293	0.375	0.247	0.534	0.333	0.108	0.339	0.299	0.299
IBE	x	0.485	0.634	0.424	0.858	0.660	0.543	0.255	1.009	0.545	0.897	0.727	0.525	0.869	0.618	1.057	1.314	0.580	0.785	0.660
Shannon Diversity	x	0.141	0.186	0.254	0.169	0.205	0.253	0.163	0.168	0.272	0.236	0.362	0.230	0.240	0.354	0.262	0.192	0.223	0.224	0.224
% Rheophilic	asin	0.053	0.079	0.096	0.100	0.105	0.056	0.038	0.109	0.146	0.130	0.143	0.092	0.313	0.176	0.062	0.080	0.112	0.244	0.100
% Rheophilic (ab-class)	asin	0.029	0.044	0.036	0.038	0.054	0.034	0.025	0.065	0.096		0.057	0.042	0.311	0.068	0.051	0.032	0.030	0.069	0.043
% Littoral	asin	0.052	0.081	0.048	0.048	0.054	0.034	0.044	0.090	0.022		0.085	0.079	0.084	0.089	0.033	0.033	0.061	0.116	0.058
% Grazers/Scrapers	asin	0.047	0.040	0.050	0.033	0.051	0.044	0.038	0.041	0.052	0.063	0.077	0.042	0.070	0.067	0.029	0.051	0.041	0.112	0.047
% Shredders	asin	0.021	0.038	0.035	0.072	0.127	0.016	0.024	0.043	0.053		0.059	0.038	0.042	0.035	0.040	0.035	0.069	0.073	0.039
% Gatherers/Collectors	asin	0.049	0.030	0.075	0.046	0.090	0.025	0.012	0.079	0.075		0.109	0.036	0.102	0.104	0.034	0.050	0.089	0.112	0.063
% Oligochaeta	asin	0.067	0.028	0.120	0.036	0.037	0.020	0.029	0.112	0.028	0.057	0.158	0.063	0.248	0.121	0.030	0.046	0.059	0.108	0.059
% EPT individuals	asin	0.066	0.066	0.080	0.079	0.047	0.068	0.062	0.098	0.094	0.095	0.126	0.086	0.049	0.094	0.130	0.097	0.039	0.174	0.080
% EPT (ab-class)	asin	0.036	0.040	0.033	0.054	0.061	0.031	0.022	0.047	0.083		0.055	0.036	0.031	0.060	0.043	0.033	0.038	0.054	0.039
% EPT Taxa	asin	0.048	0.054	0.058	0.065	0.054	0.041	0.043	0.064	0.100	0.049	0.067	0.034	0.032	0.068	0.030	0.040	0.045	0.033	0.048
RETI	asin	0.057	0.049	0.066	0.053	0.145	0.054	0.015	0.053	0.046	0.039	0.104	0.056	0.081	0.075	0.049	0.039	0.084	0.163	0.056
Log(Sel_EPTD + 1)	x	0.182		0.114	0.268			0.127	0.277	0.354	0.333	0.215			0.439	0.186			0.087	0.182
1-GOLD	asin	0.084	0.061	0.091	0.100	0.040	0.088	0.077	0.117	0.094	0.076	0.135	0.099	0.097	0.093	0.133	0.103	0.165	0.241	0.097
Trait m1: max size $\leq$ 1 cm	asin	0.020	0.023	0.040	0.024	0.039	0.026		0.048	0.042	0.028	0.037	0.031	0.086	0.037	0.036	0.014	0.012	0.028	0.030
Trait m2: > 1 cycle	asin	0.028	0.020	0.023	0.017	0.018	0.016		0.019	0.037	0.017	0.038	0.028	0.102	0.040	0.018	0.019	0.035	0.024	0.022
Trait m7: crawler loco.	asin	0.026	0.013	0.023	0.019	0.018	0.013		0.022	0.035	0.020	0.035	0.016	0.083	0.029	0.013	0.010	0.021	0.016	0.020
Trait m12: current < 25 cm s^{-1}	asin	0.019	0.015	0.013	0.021	0.013	0.020		0.031	0.039	0.021	0.024	0.019	0.020	0.030	0.008	0.013	0.022	0.019	0.020

Sampling variability in the percentage of the total abundance in a sample comprised by key taxonomic or feeding groups tended to be less when the relative abundance of that group at a site was low (percentages greater than 50% were rare). High Spearman rank correlations (r_s) between replicate variance and replicate mean for the untransformed metrics '% Oligochaeta' (r_s = 0.82) and '% Shredders' (r_s = 0.74), together with 'Abundance' (r_s = 0.77) and 'Number of taxa' (r_s = 0.45), were all reduced by half or more on the appropriate transformed scales.

The only diversity index analysed was the Shannon–Wiener index (Shannon & Weaver, 1949) for which the best single estimate of sampling SD was 0.224 (Table 4) and for the vast majority of sites, the difference in Shannon–Wiener diversity between the two replicate samples was less than 0.4 (Fig. 2b).

Although inter-replicate variability for the original Saprobic index and the Czech Saprobic index shows no pattern, the German Saprobic index (which weights taxa by their abundances category rather than raw abundance) does show some tendency to vary less between replicates for sites with replicate mean values less than about 1.6 (Fig. 3a). The variation is generally higher for sites from France (stream type F08) and Latvia (L02), but this may be because the Saprobic indices are not valid for the taxonomic level of identification used by these STAR partners, and highlights the more general problem of only using metrics in situations for which they are appropriate.

Although the Spearman rank correlations between sampling SD and replicate mean of ASPT is very weak, there is some suggestion that the very low quality sites with very low ASPT values (<3) may be less variable between samples, but such STAR sites are all in Poland and too few to make wider inferences (Fig. 3b). The median value of the sampling SD for ASPT across all sites and stream types was 0.299 (Table 4).

Relative precision of different metrics derived from STAR-AQEM samples

The relative precision of each metric was estimated its percentage sampling variance (P_{samp}), separately within each stream type variance components (Table 5). To compare the overall relative susceptibility of metrics to sampling variation, the median value of P_{samp} across stream types was calculated for each metric (Table 5).

The STAR field sampling programme included sites from 'high' or 'good' quality to 'poor'/'bad' quality. For a fixed size of replicate sampling variance, the percentage variance P_{samp} will be less in stream types for which a wider range of qualities of sites were sampled. This should be remembered when comparing values of P_{samp} across stream types for any particular metric. However, comparisons of the values of P_{samp} between metrics within stream types are completely valid because they are all based on the same set of sites. The values of P_{samp} for the 27 metrics were therefore ranked within each stream type and then the ranks averaged across stream types to give the column 'mean rank' in Table 5. (For stream types where estimates were not calculated for some metrics, the ranks were re-scaled to give a comparable range of 1–27).

The estimates of percentage sampling variance for any particular metric varies considerably between stream types. However, overall patterns in P_{samp} are detectable. The original Saprobic index, the German new Saprobic index and the Czech Saprobic index appear to have the lowest percentage sampling variance with median values of only 3, 5 and 6% respectively. This suggests that these Saprobic indices have amongst the lowest susceptibilities to sampling variation and can be estimated with the greatest relative precision within a stream type. Sampling variance tends to be less than 10% of the total variance in Saprobic metric values within any one stream type. In contrast, ASPT, another indicator of organic pollution stress (but based on only the presence-absence of families), has higher levels of percentage sampling variance within most, but not all, stream types and a median value of 17% (Table 5.) Interestingly, P_{samp} is very high for ASPT, but not for the saprobic indices, for Austrian stream types A05 and A06 whose sites were chosen to assess stress from degradation in stream morphology rather than eutrophication/organic pollution and so did not have a great range of values of ASPT (Fig. 3b), making sampling variability a greater proportion of total variability (Table 5).

The highest median percentage sampling variances were for the metric 'species trait m1'

Table 5. STAR-AQEM method: ANOVA estimates of the percentage of overall variance (P_{samp}) within each stream type due to the overall replicate sampling variance (SD^2_E). Estimates are based on, and applicable to, transformed ($f(x)$) values of metrics as indicated and only given for stream types with at least three site/season combinations with replicate samples

Metric	$f(x)$	A05	A06	C04	C05	D03	D04	D06	F08	H04	I05	I06	K02	L02	O02	P04	S05	S06	U15	U23	Med	Mean rank
Abundance [ind/m^2]	$\sqrt{\sqrt{x}}$	21	11	16	38	4	6	34	55	11	6	48	60	30	23	22	3	70	7	15	21.5	18
Number of taxa	$\sqrt{x}$	15	16	6	7	29	18	3	9	11	54	66	20	13	3	31	32	13	32	9	15.5	16
Number of Families	$\sqrt{x}$	11	17	10	9	8	24	7	18	13	46	63	22	12	2	35	28	14	26	9	15.5	17
Number of EPT taxa	$\sqrt{x}$	22	28	4	2	56	16	7	15	27	60	77	13	16	1	28	20	7	6	1	15.5	15
Saprobic Index	x	2	10	3	4	3	1	2	60				1	46	5		6	3	10	1	3	8
German Saprobic new	x	2	7	6	3	6	14	2	29				2	16	5		3	4	10	5	5	8
Czech Saprobic	x	2	6	4	8	6	1	2	29				6	29	18		4	13	14	24	6	9
ASPT	x	42	40	15	4	31	16	19	15	36	57	18	9	31	3	43	29	3	11	7	17	17
IBE	x	6	9	4	18	18	12	3	23	16	92	47	15	27	7	18	40	74	8	13	16.5	16
Shannon Diversity	x	3	14	9	5	15	19	7	9	18	20	36	41	13	7	35	37	14	14	17	14	15
% Rheophilic	asin	2	8	8	13	7	11	6	13	38	6	32	13	11	74	21	7	15	17	60	12	12
% Rheophilic (ab-class)	asin	3	14	5	9	5	39	13	9	29			7	14	94	15	15	11	5	22	12	12
% Littoral	asin	9	35	6	6	9	7	26	40	3			25	14	44	36	4	5	17	51	15.5	15
% Grazers/Scrapers	asin	5	13	16	8	16	10	6	3	10	8	24	23	34	41	21	5	17	18	80	16	14
% Shredders	asin	6	5	2	13	64	1	2	10	9			18	16	7	7	46	10	18	19	10	11
% Gatherers/Collectors	asin	8	4	12	7	37	2	1	8	30			27	21	16	19	7	19	43	61	14	14
% Oligochaeta	asin	25	1	12	3	16	9	21	40	50	25	36	48	16	56	38	8	13	5	10	16	16
% EPT individuals	asin	7	14	17	18	4	11	13	25	11	16	31	19	13	4	9	36	29	6	44	15	14
% EPT (ab-class)	asin	7	24	5	9	33	7	8	25	36			13	9	2	15	17	9	6	10	9	11
% EPT Taxa	asin	15	44	21	16	45	13	28	44	52	64	20	21	10	2	26	11	13	9	5	18	18
RETI	asin	8	17	8	8	99	12	1	3	12	7	10	24	39	20	20	12	6	29	97	12	14
Log(Sel_EPTD + 1)	x	6		3	16			3	14	27		35	6			29	8			1	7	11
1-GOLD	asin	7		7	20			6	16	17		20	45			9	41			17	16.5	17
Trait m1: max size ≤ 1 cm	asin	2	17	60	10	25	45	55	64	21	26	45	41	28	99	27	32	6	4	11	27.5	20
Trait m2: > 1 cycle	asin	6	5	5	3	9	5	9	10	15	20	10	12	33	73	18	12	10	18	5	10	11
Trait m7: crawler loco.	asin	27	4	7	6	18	6	18	31	24	11	28	21	18	72	17	10	8	18	4	17.5	15
Trait m12: current < 25 cm s^{-1}	asin	12	10	6	11	2	12	15	19	16	22	14	11	15	14	20	2	6	17	10	12	12

Med = median P_{samp} ; mean rank = average across stream types of the ranks of the P_{samp} values within each stream type (percentage of all individuals from taxa with maximum body size ≤ 1 cm) (P_{samp} = 27.5%) and for 'total abundances' (P_{samp} = 21.5%), highlighting the enormous variability between replicate STAR-AQEM samples in the total number of individuals found, even with the STAR-AQEM protocol sampling of a fixed area and restriction that the sub-sample of five of more sampling units must contain at least 700 individuals.

Although values were not available for many stream types, the new metric 'Log(Sel_EPTD + 1), proposed as an Inter-calibration Common Metric (ICM) (Buffagni et al., 2006), appears to have relatively low replicate sampling variance with a median value of P_{samp} of only 7%. The other proposed ICM metrics of 'Number of EPT taxa' (median P_{samp} = 18%), ASPT (P_{samp} = 17%), Shannon–Wiener diversity (P_{samp} = 14%) and '1-GOLD' (P_{samp} = 16.5%) all have highly variable estimates of P_{samp} but with similar intermediate size median percentage sampling variances all less than 20% (Table 5). The metric '%EPT individuals' had a slightly lower median percentage sampling variance when based on abundance classes (P_{samp} = 9%) than when based on raw abundances (P_{samp} = 15%).

Amongst the four species trait metrics analysed, those with the lowest percentage sampling variance and greatest precision were 'Trait m2' (median P_{samp} = 10%) and 'Trait m12' (median P_{samp} = 12%) which are based on the proportions of all individuals in the sample from taxa with, respectively, more than one generation per year or preferring slow-flowing water (<25 cm s^{-1}). For example, individuals sample values of trait metric 'm12' varied between 0.3 and 0.8 but most differences in values between replicates were less than 0.04.

Discussion

The study has provided the first quantitative estimates and comparisons of the sampling and sub-sampling variances of a wide range of commonly used macroinvertebrate metrics based on the STAR-AQEM sampling and sample processing method. Replicate sampling variances for many metrics showed a tendency to increase or otherwise vary with the replicate mean value for the metric for a site. To overcome this and derive a more representative single estimate of sampling variance for all sites within a stream type for a particular metric and method, the metric values were often transformed and variance components and sampling SD were estimated and compared on these transformed scales.

Effects of sub-sampling

STAR-AQEM sub-sampling variation causes a major part of the overall variance between replicate sample values for many commonly used metrics, and is estimated, on average, to contribute more than 50% of the overall variance between replicate samples for 12 of the 27 metrics analysed. In general, sub-sampling variance is largest for those metrics which are based on the numbers of taxa present, such as number of families and number of EPT taxa.

Sorting and identifying a larger fraction of the sample would reduce this source of variation; in the extreme, sorting the whole sample would eliminate it. However, all extra identification increases costs and the aim is use the most cost-effective sub-sample size and effort (Barbour & Gerritsen, 1996; King & Richardson, 2002). Although costs were not assessed in this study, Vlek (2004) and Vlek et al. (2006) made some comparisons of the time and costs associated with both STAR-AQEM and RIVPACS sample processing. It is only possible to determine the cost effectiveness of extra sub-sampling effort by sorting all 30 tray cells of a STAR-AQEM sample and doing repeated computerised random combination of increasing numbers of cells macroinvertebrates to assess the rate of reduction in sub-sorting variance. These results also highlight the importance of always trying to spread and distribute the sample material as evenly as possible amongst the 30 grid cells on a sorting tray for any STAR-AQEM macroinvertebrate sample.

In a well-designed study of several sources of variation in site assessments in western Oregon and Washington in the USA, Ostermiller and Hawkins (2004) assessed the effect of using sub-samples fixed-size counts of 50, 100, ..., 450 individuals on variability of observed to expected ratios (O/E) of number of taxa for reference sites and concluded that at least 350 individuals should be counted. The STAR-AQEM protocol requires that a minimum of 700 individuals must be counted.

Although sub-sampling contributes a major part of the overall inter-replicate variance in numerous metrics, overall inter-replicate variance may still be small compared to the range in metric values amongst sites of varying quality and thus such metrics may still have high precision to detect differences between sites.

Metric precision and implications for uncertainty in assessments of ecological status

The practical size and importance of overall replicate variance was estimated by expressing the variance as a percentage (P_{samp}) of the total variance in metric values amongst all sites within a stream type. A low percentage indicates that the combination of sampling method and metric has high statistical precision compared to variability amongst sites of differing quality. High percentages indicate low sampling precision and low repeatability and hence that such a combination of sampling method and metric is unlikely to have much power to detect differences in ecological status class.

The following approach was used in an attempt to convert a value of P_{samp} for a metric into the approximate level of uncertainty it may cause in assigning sites to ecological status classes. The European Union Water Framework Directive (WFD) (European Union, 2000) requires the assessment of ecological quality of water bodies to be based on Ecological Quality Ratios (EQRs), representing a comparison of the observed (O) value of a metric for a site with the Reference Condition (E) value for that type of site. Assuming the sampled sites within a stream type are all given the same E value for a metric (based on the metric values for the reference or high quality sites) and the EQR is the O/E ratio of values, then all observed values for the metric for sites within the same stream type will be divided by a constant and therefore the percentage sampling variance of the EQR values will be the same as the percentage sampling variance of the observed (O) values of the metric, namely P_{samp}. Furthermore, assume the EQR values within the sampled sites in a stream type are roughly evenly spread over the range 0–1, which implies a total variance of EQR values of 1/12; and assume that the ecological status class limits are evenly spread at 0.2, 0.4, 0.6 and 0.8. Then the sampling SD of EQR values is a percentage $P_U = 500\sqrt{P_{samp}/1200}$ of the width (0.2) of the status classes. Equations (13) and (14) of Clarke et al. (1996) can then be used to calculate the overall mis-classification rates for all sites in the middle status classes (overall rates for the 'high' or 'bad' classes are half those of the middle classes). Table 6 shows the average and range of estimated mis-classification rates for metrics grouped by the median percentage sampling variance across stream types (as given in Table 5).

The three Saprobic indices and the new ICM metric 'Log(Sel_EPTD + 1)' have the lowest median percentage sampling variances (P_{samp}) of only 3–7%, which leads to estimated mis-classifications rates of 5–50% according to position within a status class and average mis-classification rates (P_B) over all sites within a middle class of 25%. The proposed ICM metrics of Number of EPT taxa, ASPT, Shannon–Wiener diversity and (1-GOLD) have highly variable estimates of P_{samp} across stream types, but with similar intermediate size median values of 13–18%, causing estimated average mis-classification rates of 43%. This suggests that the precision of such metrics based on the STAR-AQEM method is sufficient to indicate gross changes in the ecological status of sites, but there will be considerable uncertainty in the assignment of sites to particular status classes.

A more realistic approach could be to use the estimates of overall replicate sampling SD (SD_E) reported in Table 4 in the STARBUGS simulation software package (STAR Bioassessment Uncertainty Guidance Software; Clarke, 2004) to assess the effect of sampling variability of metric values based on the STAR-AQEM sampling method on the uncertainty of single-metric or multi-metric assessments of the ecological status of sites. The STARBUGS program copes with sampling SD for observed values of metrics based on either untransformed values or any of the transformations used here (for further details of STARBUGS, see Clarke & Hering, 2006).

Perhaps most importantly, the estimates of sampling SD derived here can also be used to provide information on the expected uncertainty

Table 6. Grouping of metrics in terms of their median percentage sampling variance (P_{samp}) across all stream types, together with an rough estimate of sampling variance expressed as a percentage (P_U) of the EQR status class widths and the estimated average and range of rates of mis-classifying sites whose true EQR value lies within the 'good', 'moderate' or 'poor' class (Mis-classification rates are approximately half for sites in the extreme 'high' or 'bad' classes. (Calculations based on assumptions of uniform distribution of true EQR values for the sampled sites over the full EQR range 0–1 and using status class limits of EQR values at 0.2, 0.4, 0.6 and 0.8.)

Sampling precision group (range of P_{samp})	Metrics	P_U	Mis-classification rates	
			Average (P_B)	Range
3–7%	Saprobic Index German Saprobic new Czech Saprobic Log(Sel_EPTD + 1)	25–38%	25%	5–50%
8–12%	% EPT (ab-class) % Shredders, RETI Trait m2: > 1 cycle % Rheophilic % Rheophilic (ab-class) Trait m12: current < 25cm s^{-1}	41–50%	36%	22–52%
13–18 %	Shannon Diversity % Gatherers/Collectors Number of taxa, ASPT, Number of Families Number of EPT taxa % EPT individuals % Grazers/Scrapers, IBE % Oligochaeta, % Littoral % EPT Taxa, 1-GOLD Trait m7: crawler loco.	52–61%	43%	33–55%
19–28%	Abundance [ind/m^2] Trait m1: max size ≤ 1 cm	63–76%	51%	43–59%

in metrics values for other sites in the same stream types, sampled using exactly the same STAR-AQEM field sampling and laboratory sorting and identification procedures, but where only one sample has been taken at a point in time and thus there is no replication. For any new site, it may be most reliable to use the estimates of overall sampling SD for the same stream type, if available. This may be most appropriate when the metric values and variability are highly dependent on either the stream type or the precise taxonomic resolution used by each STAR partner (e.g., as for the metric 'Number of taxa'). However, if there are no obvious major differences in the SD between stream types, it may be more robust to use estimates based on the information from a combination of stream types, or even the median SD value across all stream types, as given in the right-hand column of Table 4.

Adding metrics with very low precision to a multi-metric index (MMI) based on the average of the component metrics' EQR values can reduce the precision of the MMI. Specifically adding any metric with an EQR sampling variance more than three times the average EQR sampling variance of the metrics already involved in a MMI will always increase the sampling variance of the MMI and reduce its precision (Clarke et al., 2006).

Precision not accuracy

It must be remembered that, metrics which do not vary much between replicate samples and have high statistical precision may not necessarily be good informative indicators of the ecological quality and status of a water body. More subtly, although adding an extra metric may increase the precision of a MMI, it may not lead to a more meaningful assessment of site condition if the newly added metric is not a good ecological measure of some aspect of ecological quality of sites. In other words, increased precision does not necessarily imply increased "accuracy" of site assessment. Determining "accuracy" requires some understanding of the true quality and ecological status of sites. In other studies, "truth" is usually set by a pre-classification of sites based on independent non-biological information about the condition of each site (Hering et al., 2004; Ofenböck et al., 2004). However, the aim of this study was merely to provide estimates of the susceptibility of metrics to the combined effects of sampling and sub-sampling variation.

Acknowledgements

STAR was partially funded by the European Commission, 5th Framework Program, Energy, Environment and Sustainable Development, Key Action Water, Contract no. EVK1-CT-2001-00027. The authors acknowledge the crucial overall project coordination of Mike Furse and the support of all their project colleagues who collected the field samples and sorted and identified the taxa. We also thank the referees for constructive criticisms which have improved the paper.

References

Armitage, P. D., D. Moss, J. F. Wright & M. T. Furse, 1983. The performance of a new biological water quality score system based on macroinvertebrates over a wide range of unpolluted running-water sites. Water Research 17: 333–347.

Barbour, M. T. & J. Gerritsen, 1996. Sub-sampling of benthic samples: a defense of the fixed-count method. Journal of the North American Benthological Society 15: 386–391.

Bis, B. & P. Usseglio-Polatera, 2004. Species Traits Analysis. European Commission, STAR (Standardisation of river classifications), Deliverable N2, 134 pp.

Buffagni, A., S. Erba, M. Cazzola, J. Murray-Bligh, H. Soszka & P. Genoni, 2006. The STAR common metrics approach to the WFD intercalibration process: Full application for small, lowland rivers in three European countries. Hydrobiologia 566: 379–399.

Clarke, R. T., 2000. Uncertainty in estimates of river quality based on RIVPACS. In Wright, J. F., D. W. Sutcliffe & M. T. Furse (eds), Assessing the biological quality of freshwaters: RIVPACS and similar techniques. Freshwater Biological Association, Ambleside, 39–54.

Clarke, R. T., 2004. 9th STAR deliverable. Error/uncertainty module software STARBUGS (STAR Bio Assessment Uncertainty Guidance Software) User Manual.

Clarke, R. T., M. T. Furse, R. J. M. Gunn, J. M. Winder & J. F. Wright, 2002. Sampling variation in macroinvertebrate data and implications for river quality indices. Freshwater Biology 47: 1735–1751.

Clarke, R. T. & D. Hering, 2006. Errors and uncertainity in bioassessment methods – major results and conclusions from the STAR project and their application using STARBUGS. Hydrobiologia 566: 433–439.

Clarke, R. T., J. Davy-Bowker, L. Sandin, N. Friberg, R. K. Johnson & B. Bis, 2006. Estimates and comparisons of the effects of sampling variation using 'national' macroinvertebrate sampling protocols on the precision of metrics used to assess ecological status. Hydrobiologia 566: 477–503.

Dahl, J., R. K. Johnson & L. Sandin, 2004. Detection of organic pollution of streams in southern Sweden using benthic macroinvertebrates. Hydrobiologia 516: 161–172.

Elliott, J. M., 1977. Some methods for the statistical analysis of samples of benthic invertebrates. Scientific Publication No. 25 (2nd edn.). Freshwater Biological Association, Ambleside 160 pp.

European Union, 2000. Directive 2000/60/EC. Establishing a framework for community action in the field of water policy. European Commission PE-CONS 3639/1/100 Rev 1, Luxemburg.

Furse, M., D. Hering, O. Moog, P. Verdonschot, R. K. Johnson, K. Brabec, K. Gritzalis, A. Buffagni, P. Pinto, N. Friberg, J. Murray-Bligh, J. Kokes, R. Alber, P. Usseglio-Polatera, P. Haase, R. Sweeting, B. Bis, K. Szoszkiewicz, H. Soszka, G. Springe, F. Sporka & I. Krno, 2006. The STAR project: context, objectives and approaches. Hydrobiologia 566: 3–29.

Hering, D., O. Moog, L. Sandin & P. Verdonschot, 2004. Overview and application of the AQEM assessment system. Hydrobiologia 516: 1–20.

Karr, J. R. & E. W. Chu, 1999. Restoring Life in Running Waters: Better Biological Monitoring. Island Press, Washington, DC.

King, R. S. & C. J. Richardson, 2002. Evaluating sub-sampling approaches and macroinvertebrate taxonomic resolution for wetland bioassessments. Journal of the North American Benthological Society 21: 150–171.

Lorenz, A., L. Kirchner & D. Hering, 2004. 'Electronic subsampling' of macrobenthic samples: how many individuals are needed for a valid assessment result? Hydrobiologia 516: 299–312.

Ofenböck, T., O. Moog, J. Gerritsen & M. Barbour, 2004. A stressor specific multimetric approach for monitoring running waters in Austria using benthic macro-invertebrates. Hydrobiologia 516: 251–268.

Ostermiller, J. D. & C. P. Hawkins, 2004. Effects of sampling error on bioassessments of stream ecosystems: application to RIVPACS-type models. Journal of the North American Benthological Society 23: 363–382.

Pinto, P., J. Rosado, M. Morais & I. Antunes, 2004. Assessing methodology for southern siliceous basins in Portugal. Hydrobiologia 516: 191–214.

REFCOND, 2003. Guidance on establishing reference conditions and ecological status class boundaries for inland surface waters. Final version, 30 April 2003, produced by WG 2.3.

Sandin, L., J. Dahl & R. K. Johnson, 2004. Assessing acid stress in Swedish boreal and alpine streams using benthic macroinvertebrates. Hydrobiologia 516: 129–148.

Shannon, C. E. & W. Weaver, 1949. The Mathematical Theory of Communication. The University of Illinois Press, Urbana, IL.

Sokal, R. R. & J. R. Rohlf, 1995. Biometry (3rd edn.). Freeman and Company, New York.

Taylor, L. R., 1961. Aggregation, variance and the mean. Nature 189: 732–735.

Vlek, H. E., 2004. Comparison of (cost) effectiveness between various macroinvertebrate field and laboratory protocols. European Commission, STAR (Standardisation of river classifications), Deliverable N1, 78 pp.

Vlek, H. E., P. F. M. Verdonschot & R. C. Nijboer, 2004. Towards a multimetric index for the assessment of Dutch streams using benthic invertebrates. Hydrobiologia 516: 173–189.

Vlek, H. E., F. Šporka & I. Krno, 2006. Influence of macroinvertebrate sample size on bioassessment of streams. Hydrobiologia 566: 523–542.

Wright, J. F., 2000. An introduction to RIVPACS. In Wright, J. F., D. W. Sutcliffe & M. T. Furse (eds), Assessing the biological quality of fresh waters: RIVPACS and other techniques. Freshwater Biological Association, Ambleside, UK, 1–24.

Zelinka, M. & P. Marvan, 1961. Zur Präzisierung der biologischen Klassifikation der Reinheit fließender Gewässer. Archiv für Hydrobiologie 57: 389–407.

Hydrobiologia (2006) 566:461–476
M.T. Furse, D. Hering, K. Brabec, A. Buffagni, L. Sandin & P.F.M. Verdonschot (eds), The Ecological Status of European Rivers: Evaluation and Intercalibration of Assessment Methods
DOI 10.1007/s10750-006-0077-4

Sample coherence – a field study approach to assess similarity of macroinvertebrate samples

Armin Lorenz[1,*] & Ralph T. Clarke[2]
[1]*Department of Hydrobiology, University of Duisburg-Essen, D-45117 Essen, Germany*
[2]*Centre for Ecology & Hydrology, Winfrith Technology Centre, DT2 8ZD Dorchester, Dorset, UK*
*(*Author for correspondence: Tel.: +49-201-183-2442; Fax: +49-201-183-4442; E-mail: armin.lorenz@uni-essen.de)*

Key words: macroinvertebrates, replicate sampling, similarity, sampling methods, coherence

Abstract

The EU-funded STAR-project provided an opportunity to analyse 1418 macroinvertebrate samples from 310 sampling sites throughout Europe. At most of the sites, samples were taken in two seasons using both national protocols and the project's STAR-AQEM protocol. At a subset of sites (86), two replicate samples were taken by each method in each of the two seasons. The resulting taxalists were analysed in terms of community similarity using the Bray–Curtis Index, Jaccard, and Renkonen Indices. A new concept of sample 'coherence' is used to measure the relative strength of within-site, within-season and within-method similarity and to determine their importance on variability in community composition. Site-coherence (i.e., highest similarity to another sample from the same site) was much higher where replicate samples were available. Season-coherence of samples was nearly 100% even if different methods were compared. Season appeared to be one of the major determinants of in-stream fauna. The STAR-AQEM method is most comparable to the Nordic, Portuguese and Czech (PERLA) national methods and less comparable to the Italian (IBE) and Latvian methods. Samples collected by these latter methods had higher similarities to other sites sampled with the same methods than to samples of the same site using the STAR-AQEM method, thus there was low site-coherence. In three stream types from Italy, Latvia and Greece 28–38% of the samples were most similar to a sample from a different site than to a replicate sample from the same site. This fact could have serious consequences for follow up bioassessments or impact assessments by cluster analysis based on similarity measures. Replicate samples are less coherent within site, season or method if the taxonomic resolution is family rather than species.

Introduction

Bioassessment based on macroinvertebrate taxa has been used for more than a century (Kolkwitz & Marsson, 1902). Many methods and protocols have been proposed for standardisation of macroinvertebrate sampling (e.g., Zelinka & Marvan, 1961; DEV, 1992; Ghetti, 1997; Barbour et al., 1999; Wright et al., 2000; Hering et al., 2003; Furse et al., 2006).

For assessment systems to be effective, the sampling protocol must provide a sample which adequately represents the status of the biota at a site and which gives repeatable results in the sense that there is relatively little variation between replicate samples taken from the same site at the same time. This representativeness and 'coherence' of a sample taken at a site is often assumed implicitly. The coherence of different methods and samples can be assessed by comparing different samples from a site in terms of their (i) similarity of taxonomic composition (presence-absence and/or abundance) Friberg et al., (2006) (ii) values of derived biological indices or metrics (Feld, 2004;

Birk & Hering, 2006), or (iii) ecological status as determined by bioassessment systems (Hering et al., 2003). The first method, a coefficient of similarity of taxonomic composition between different samples taken at a site provides a direct measure of the extent to which samples are replicable. This is independent of any specific derived metrics, indices or bioassessment system.

Small scale replicate programs and electronic simulation of replicate similarity have frequently been performed (e.g., Wolda, 1981; Cao et al., 2002; Clarke et al., 2002). But designs with a large number of replicates in different regions have been restricted to simulation experiments (Wolda, 1981; Boyle et al., 1990). The STAR-project (Furse et al., 2006) is the first European-wide field study of the replicate sampling variability and inter-method comparability of a wide range of national sampling methods.

Many different indices of community similarity and their analysis have been published (see Washington, 1984 for an overview). Each of the indices express different aspects of similarity in taxonomic composition. This paper concentrates on three common ones, the Bray–Curtis Index (Bray & Curtis, 1957), the Jaccard Index (Jaccard, 1912) and the Renkonen Percent similarity Index (Renkonen, 1938).

In this paper, we try to answer the following questions:

- How similar are replicate macroinvertebrate samples from the same site?
- How similar are samples taken with the same protocol in different seasons?
- How similar are samples taken with two different protocols?
- Does the protocol influence the similarity of replicate samples at a site?

Methods

Sampling sites and protocol

A total of 310 stream sites from 23 stream types spread over 13 countries across Europe were sampled using two different protocols in two seasons (Fig. 1, Table 1). Although an overall sampling protocol refers strictly to all aspects of obtaining macroinvertebrate sample data, namely the combined effect of the sampling method, sorting and optional sub-sampling methods and the identification level used, the protocol is often referred to as the sampling method. In the STAR-project each stream type in each country was given a unique code; in this paper they are hereafter only referred to as stream types. On each sampling occasion at each site, both the STAR-AQEM sampling method and protocol (Furse et al., 2006) and a national sampling protocol were applied. For streams in the Czech Republic and the Slovak Republic the PERLA method was used (Birk & Hering, 2002), in Denmark and Sweden the Nordic sampling procedure (Skriver et al., 2000), in France the IBGN (AFNOR, 1982), in Great Britain the RIVPACS method (Murray-Bligh et al., 1997), in Italy the IBE (Ghetti, 1997), in Latvia the Latvian national method (Latvian Standard Ltd., 1999), in Poland the Polish sampling method (Birk & Hering, 2002) and in Portugal the PMP (Birk & Hering, 2002) (for a summary on the sampling methods see www.eu-star.at). Austria, Germany and Greece used the RIVPACS method as the second method. As exceptions, the Italian stream type I05 was only sampled using the STAR-AQEM method and replicate samples were not taken in the Slovak stream type V01.

The samples were sorted and identified to the best possible taxonomic resolution, which was country dependent (Table 2). As an exception, the French partner identified the samples taken with the STAR-AQEM method to a lower resolution (mainly genus) than their national method (IBGN, mainly family), thus their taxalists were not directly comparable. Prior to statistical analysis, all the taxalists were taxonomically adjusted (Nijboer & Verdonschot, 2000) on the national level to ensure compatibility and standardisation within a country.

Similarity measures

Three similarity indices were calculated from the macroinvertebrate samples: the Bray–Curtis (*BC*) Index, the Renkonen (*R*) Index and the Jaccard (*J*) Index (see supplementary material for equations).[1]

[1] Electronic supplementary material is available for this article at <http://dx.doi.org/ 10.1007/s10750-006-0077-4> and accessible for authorised users.

Figure 1. Number of sampling sites and samples taken in each country in the STAR-project.

They express different features of the macroinvertebrate taxalists. The Jaccard Index compares the presence/absence of taxa within a pair of samples. The Bray–Curtis Index is a widely used abundance-based similarity index (Bloom, 1981; Field et al., 1982; Hruby, 1987) with a well-established reputation for its use with multivariate ordination and clustering methods to recover and summarise patterns in ecological space (Faith et al., 1987; Clarke & Gorley, 2001). The Renkonen Index compares the proportional dominance of taxa in samples. This index was recommended by Wolda (1981) following his assessment of 22 diversity and similarity indices in simulation experiments and by Hruby (1987), who found the Bray–Curtis and Renkonen similarity indices based on log transformed abundances to be the best for benthic impact studies, out of 11 coefficients tested. In our analyses, logarithmic transformed abundances were used for the Bray–Curtis Index because the different protocols being compared involved different sampling and sample processing procedures and efforts and thus yielded different abundances.

Data analysis

All the data analyses below were carried out separately for each of the three similarity indices. The average similarity between replicate samples taken at the same site in the same season using the same method was calculated separately for each method and season. Three other average similarities between samples from the same site were calculated, using samples with (i) same method but different season, (ii) same season but different method and (iii) different method and season.

Table 1. Methods applied and number of samples in the stream types (spr = spring, sum = summer, aut = autumn, win = winter)

Country/stream type	National method	Seasons	Replicates of the same method				No replicates of the same method			
			STAR-AQEM method		National method		STAR-AQEM method		National method	
			1st season	2nd season	1st season	2nd season	1st season	2nd season	1st season	2nd season
Austria stream type A05	RIVPACS	Spr/sum	12	12	4	8	9	10	11	10
Austria stream type A06	RIVPACS	Spr/sum	9	13	6	8	4	3	4	3
Czech Republic stream type C04	PERLA	Spr/sum	9	9	6	6	11	11	11	11
Czech Republic stream type C05	PERLA	Spr/sum	9	9	6	6	7	7	7	7
Denmark stream type K02	Nordic	Spr/sum	18	18	12	12	5	5	5	5
France stream type F08	IBGN	Spr/aut	17	18	12	12	0	0	0	0
Germany stream type D03	RIVPACS	Spr/sum	6	6	4	4	10	10	10	10
Germany stream type D04	RIVPACS	Spr/sum	6	6	4	4	10	10	10	10
Germany stream type D06	RIVPACS	Spr/sum	8	8	4	4	4	4	4	4
Greece stream type H04	RIVPACS	Spr/sum/win	12	12	12	12	4	4/win 10	4	4
Greece stream type H05	RIVPACS	Sum	0	0	0	0	0	10	0	10
Greece stream type H06	RIVPACS	Sum	0	0	0	0	0	10	0	10
Greece stream type H07	RIVPACS	Spr/sum	0	0	0	0	0	5	0	5
Italy stream type I05	n.a.	Spr/sum	12	12	0	0	0	0	0	0
Italy stream type I06	IBE	Spr/sum	12	12	12	12	5	5	5	5
Latvia stream type L02	L	Spr/aut	35	12	12	12	11	18	18	18
Poland stream type O02	P	Spr/aut	14	18	12	14	19	19	19	19
Portugal stream type P04	PMP	Spr/aut	16	17	12	12	2	2	4	4
Slovak Republic stream type V01	PERLA	Spr/aut	0	0	0	0	6	6	6	6
Sweden stream type S05	Nordic	Spr/aut	9	11	6	6	13	11	13	13
Sweden stream type S06	Nordic	Spr/aut	9	9	6	6	8	9	8	9
UK stream type U15	RIVPACS	Spr/aut	9	9	6	6	10	10	10	10
UK stream type U23	RIVPACS	Spr/aut	9	9	6	6	9	9	9	9
Sum			231	220	142	150	147	188	158	182

Table 2. Taxonomic resolution of the macroinvertebrate samples in the different countries

Country	Taxonomic resolution
Austria	Mainly species
Czech Republic	Mainly species
Denmark	Mainly species
France IBGN	Family
France STAR-AQEM	Mainly genus
Germany	Mainly species
Greece	Mainly family
Italy	Mainly family
Latvia	Mainly family
Poland	Mainly species
Portugal	Mainly genus
Slovak Republic	Mainly species
Sweden	Mainly species
UK	Mainly species

The similarity of each sample to all other samples (regardless of site, season or method) within the same stream type was calculated. The sample with the highest similarity to the original sample was selected and details about the site, the season and the sampling method of the two samples were compared. This was repeated for each sample in turn and the results were summarised for each stream type.

As a measure of the extent to which samples with attributes (site, season, and/or method) in common are most similar to each other, we introduce the new concept of sample 'coherence'. We followed the hypothesis that each sample should be most similar to another sample from the same site. Thus, we calculated the percentage of samples from the same stream type for which the highest community similarity was with another sample from the same site; this is referred to as 'site-coherence'. Hundred percent site-coherence in a stream type indicates that all samples are most similar to another sample from the same site. A lower site-coherence indicates that at least some samples had higher similarities with samples from different sites than to the one they derive from.

The same analyses were used to derive the 'season-coherence' which was estimated by the percentage of samples most similar to a sample in the same stream type taken in the same season. Similarly, the 'method-coherence' in a stream type was measured by the percentage of samples most similar to another sample obtained using the same sampling method and protocol within the same stream type. Low method-coherence indicates that the two methods used within a stream type result in relatively similar communities.

The site-coherence, season-coherence and method-coherence for each stream type were calculated separately for two subsets of samples, namely (i) 'replicated' samples, for which there was another replicate sample of the same method from the same site in the same season, and (ii) 'un-replicated' samples for which there was no replicate sample of the same method from the same site in the same season.

The range of errors in sampling variation is discussed in detail in Clarke et al. (2006a, b).

Results

Within-site similarity

The average similarity between samples taken from the same site in the same season by any particular method was always higher than the average similarity between samples taken from the same site in different seasons (Table 3). For example, using the STAR-AQEM method, the average Bray–Curtis similarity between replicate samples taken in the same season is 71.5% in spring and 71.7% in the second season, compared to an average similarity of only 48.9% when taken from the same site in different seasons. The overall average within-site similarity is roughly the same for the STAR-AQEM method as that averaged over all national methods, for individual seasons and also for samples from different seasons (but same site) (Table 3). However, the average similarity between a STAR-AQEM sample and a national method sample from the same site was 7–16% less than for two equivalent samples taken by the same method (Table 3). For example, the average within-site Renkonen similarity between a STAR-AQEM sample and national method sample taken in spring is 56.7% compared to equivalent values of 70.3 and 73.2% within the two methods. The average between-seasons within-sites similarity is between 17 and 25% less than the equivalent within-seasons-similarity for the

Table 3. Average of the *n* similarity index values between pairs of samples within the same site, separately for each season and method, and for when the two samples are from the same site but different seasons and/or different methods

Season	Method	*n*	Bray–Curtis similarity (%)	Jaccard similarity (%)	Renkonen similarity (%)
Spring	National	71	72.6	57.5	70.3
	STAR-AQEM	85	71.5	57.5	73.2
	Different	223	62.5	50.3	56.7
Second season	National	75	70.6	57.0	66.4
(summer or autumn)	STAR-AQEM	83	71.7	55.5	73.3
	Different	251	60.1	47.8	57.1
Different season	National	235	47.1	37.7	36.4
	STAR-AQEM	276	48.9	37.6	36.9
	Different	228	43.1	34.0	31.7
	Overall mean		53.7	42.8	45.0

three similarity indices (Table 3). Using different methods in different seasons reduces the average similarity by another 3–6% for all three indices.

Within any season, the average within-site similarity values for the Bray–Curtis and Renkonen indices are about the same and the Jaccard Index is on average about 13% less. In the different season analysis, the Renkonen and Jaccard Index are about the same and Bray–Curtis Index has about 10% higher similarities in all three method variables.

Coherence amongst replicated samples

Where two samples were taken from a site in the same season by the same method, between 90 and 100% of the samples from most stream types were most similar to another sample from the same site (see site-coherence in Tables 4–6). Only the Greek stream type H04, the Italian stream type I06 and the Latvian stream type had substantially lower site-coherence, especially when based on the Renkonen Index, with values of 64.6, 62.5 and 64.8% respectively (Table 6).

The season-coherence is 100% in many stream types and over 90% in the vast majority for all three indices (Tables 4–6). The lowest season-coherences occur using the Jaccard Index based on taxa presence–absence for the French stream type F08 IBGN and the Italian stream type I05 (Table 5).

The method-coherence ranged between 40 and 98% over all three indices with average values of 82.2, 71.3 and 77.6% for the Bray–Curtis, Jaccard and Renkonen Indices (Tables 4–6). Using the Bray–Curtis Index (Table 4) lowest method-coherences occurred for the Greek stream type H04, the Portuguese stream type P04 and the Swedish stream types S05 (all three below 70%) and S06 (73.3%). When based on the Jaccard Index, method-coherence below 70% was recorded for Greek stream type H04, the two Swedish stream types S05 and S06, the German stream type D04, the UK stream type U23, the Latvian stream type L02 and the Austrian stream type A06 (Table 5). Equivalent low values using the Renkonen Index were obtained for again Greek stream type H04, Swedish stream types S05 and the Portuguese stream type P04 (all of them below 60%) (Table 6). There was no stream type for which all samples were most similar to another sample obtained using the same method (i.e., with method-coherence equal to 100%).

Coherence amongst un-replicated samples

If only one sample was taken by each method in each season from each site, the site-, season- and method-coherence were significantly lower (Tables 7–9) than for cases with replicate samples. The site-coherence is strongly correlated to the method-coherence and is up to 60% less than for the situations with replicate samples (e.g., based on the Bray–Curtis similarity for the Polish stream type O02, site-coherence is reduced from 93.1 to 31.6% if no second replicate samples are available).

Table 4. Site-, season- and method-coherence based on the Bray–Curtis similarity Index for the STAR sampling program; only replicated samples are analysed; each of *n* such samples is compared to all other samples within the stream type (Methods: F = IBGN, I = IBE, L = Latvian, N = Nordic, O = Polish, P = PERLA, PMP = Portuguese, R = RIVPACS, S = STAR-AQEM)

Country/stream type	Methods	*n*	Site-coherence (%)	Season-coherence (%)	Method-coherence (%)
Austria stream type A05	R/S	36	93.1	100.0	86.1
Austria stream type A06	R/S	36	94.4	100.0	86.1
Austria all samples	R/S	72	91.7	100.0	86.1
Czech Republic stream type C04	P/S	30	100.0	100.0	80.0
Czech Republic stream type C05	P/S	30	100.0	100.0	96.7
Czech Republic all samples	P/S	60	100.0	100.0	88.3
Denmark stream type K02	N/S	60	100.0	100.0	90.0
France stream type F08 IBGN	F	24	100.0	100.0	n.a.
France stream type F08 STAR-AQEM	S	35	97.1	100.0	n.a.
Germany stream type D03	R/S	20	100.0	100.0	90.0
Germany stream type D04	R/S	20	100.0	100.0	85.0
Germany stream type D06	R/S	24	100.0	100.0	95.8
Germany all samples	R/S	64	100.0	100.0	90.6
Greece stream type H04	R/S	48	85.4	95.8	54.2
Italy stream type I05	S	24	100.0	100.0	n.a.
Italy stream type I06	I/S	48	83.3	91.7	93.8
Latvia stream type L02	L/S	71	95.8	98.6	85.9
Poland stream type O02	O/S	58	93.1	100.0	98.3
Portugal stream type P04	PMP/S	57	100.0	100.0	64.9
Sweden stream type S05	N/S	32	100.0	100.0	53.1
Sweden stream type S06	N/S	30	96.7	100.0	73.3
Sweden all samples	N/S	62	98.4	100.0	62.9
UK stream type U15	R/S	30	96.7	100.0	83.3
UK stream type U23	R/S	30	100.0	100.0	80.0
UK all samples	R/S	60	98.3	100.0	81.7
Overall mean			96.8	99.3	82.2

In only three stream types (Portuguese stream type P04, Swedish stream type S06 and UK stream type U15) did all samples have the highest Bray–Curtis similarities with another sample from the same site (i.e., 100% site-coherence in Table 7). This was also true for the stream type U15 when based on the Jaccard Index (Table 8). Using the Renkonen Index, highest site-coherence occurred for the Swedish stream type S06 with 92% (Table 9). The lowest site-coherences were obtained for the Italian stream type I06 ($BC = 25.0\%$, $J = 35.0\%$ and $R = 5.0\%$), the Polish stream type O02 ($BC = 31.6\%$ and $R = 26.3\%$) and the Greek stream type H05 ($BC = 20.0\%$ and $R = 30.0\%$). Averaged over all stream types, the means of the site-coherence were 72.2% (*BC*), 66.6% (*J*), and 60.4% (*R*), which were 24.5% (*BC*), 23.2% (*J*) and 29.4% (*R*) lower than for situations where two replicate samples were available.

High season-coherence (>80%) was found for most of the stream types and all three indices. Outliers were only the Greek stream type H04, the Italian stream type I06 and the Austrian stream type A06. The Czech and the Slovakian un-replicated samples were all most similar to another sample from the same season for all three similarity indices (100% season-coherence).

The method-coherence varied considerably between stream types and methods used. The highest percentages were calculated for the Italian stream type I06 ($BC = 80.0\%$, $J = 70.0\%$, $R = 80.0\%$) and the Polish stream type O02 ($BC =$

Table 5. Site-, season- and method-coherence based on the Jaccard similarity Index for the STAR sampling program; only replicated samples are analysed; each of *n* such samples is compared to all other samples within the stream type (Methods: F = IBGN, I = IBE, L = Latvian, N = Nordic, O = Polish, P = PERLA, PMP = Portuguese, R = RIVPACS, S = STAR-AQEM)

Country/stream type	Methods	*n*	Site-coherence (%)	Season-coherence (%)	Method-coherence (%)
Austria stream type A05	R/S	36	83.3	100.0	91.7
Austria stream type A06	R/S	36	80.6	100.0	69.4
Austria all samples	R/S	72	81.9	100.0	80.6
Czech Republic stream type C04	P/S	30	96.7	100.0	76.7
Czech Republic stream type C05	P/S	30	93.3	100.0	83.3
Czech Republic all samples	P/S	60	95.0	100.0	80.0
Denmark stream type K02	N/S	60	100.0	100.0	71.7
France stream type F08 IBGN	F	24	87.5	75.0	n.a.
France stream type F08 STAR-AQEM	S	35	97.1	97.1	n.a.
Germany stream type D03	R/S	20	100.0	100.0	80.0
Germany stream type D04	R/S	20	100.0	100.0	55.0
Germany stream type D06	R/S	24	100.0	100.0	83.3
Germany all samples	R/S	64	100.0	100.0	73.4
Greece stream type H04	R/S	48	72.9	89.6	41.7
Italy stream type I05	S	24	79.2	79.2	n.a.
Italy stream type I06	I/S	48	72.9	95.8	75.0
Latvia stream type L02	L/S	71	60.6	85.9	69.0
Poland stream type O02	O/S	58	93.1	98.3	98.3
Portugal stream type P04	PMP/S	57	86.0	98.3	79.0
Sweden stream type S05	N/S	32	93.8	100.0	40.6
Sweden stream type S06	N/S	30	100.0	100.0	60.0
Sweden all samples	N/S	62	96.8	100.0	50.0
UK stream type U15	R/S	30	100.0	96.7	70.0
UK stream type U23	R/S	30	100.0	100.0	66.7
UK all samples	R/S	60	100.0	98.3	68.3
Overall mean			89.9	95.8	71.3

68.4%, $J = 50.0\%$, $R = 82.9\%$). Methods-coherence of less than 10% were found in the Czech stream types C04 and C05 as well as in the Swedish stream types S05 and S06 and the UK stream type U15 in the Bray–Curtis Index but also for some of these stream types in the other two similarity indices.

Discussion

The results integrate different aspects of the analysis of community similarity between samples, namely the level of similarity and the extent to which samples are most similar to other samples from the same site, season or method.

Site-coherence and within-site similarity

The estimate of site-coherence is generally much higher if replicate samples are available because samples are often most similar to the other replicate for the same site, season and method. Although there is high site-coherence (>90%) amongst the replicate subset of samples for the majority of stream types, the stream types I06 (Italy), L02 (Latvia) and H04 (Greece) all have low site-coherence of 62–64% when based on the Renkonen Index and 60–72% using the Jaccard Index. This indicates that 28–38% of samples were most similar to a sample from a different site than to a replicate sample from the same site. This fact could have serious consequences for follow up site

Table 6. Site-, season- and method-coherence based on the Renkonen similarity Index for the STAR sampling program; only replicated samples are analysed; each of *n* such samples is compared to all other samples within the stream type (Methods: F = IBGN, I = IBE, L = Latvian, N = Nordic, O = Polish, P = PERLA, PMP = Portuguese, R = RIVPACS, S = STAR-AQEM)

Country/stream type	Methods	*n*	Site-coherence (%)	Season-coherence (%)	Method-coherence (%)
Austria stream type A05	R/S	36	97.2	100.0	88.9
Austria stream type A06	R/S	36	94.4	97.2	91.7
Austria all samples	R/S	72	95.8	98.6	90.3
Czech Republic stream type C04	P/S	30	93.3	100.0	86.7
Czech Republic stream type C05	P/S	30	93.3	100.0	86.7
Czech Republic all samples	P/S	60	93.3	100.0	86.7
Denmark stream type K02	N/S	60	96.7	96.7	80.0
France stream type F08 IBGN	F	24	100.0	100.0	n.a.
France stream type F08 STAR-AQEM	S	35	82.9	97.1	n.a.
Germany stream type D03	R/S	20	95.0	100.0	70.0
Germany stream type D04	R/S	20	100.0	100.0	90.0
Germany stream type D06	R/S	24	100.0	100.0	87.5
Germany all samples	R/S	64	98.4	100.0	82.8
Greece stream type H04	R/S	48	64.6	87.5	50.0
Italy stream type I05	S	24	95.8	100.0	n.a.
Italy stream type I06	I/S	48	62.5	95.8	77.1
Latvia stream type L02	L/S	71	64.8	97.2	74.7
Poland stream type O02	O/S	58	89.7	100.0	96.6
Portugal stream type P04	PMP/S	57	91.2	93.0	59.7
Sweden stream type S05	N/S	32	93.8	96.9	50.0
Sweden stream type S06	N/S	30	93.3	100.0	70.0
Sweden all samples	N/S	62	93.6	98.4	59.7
UK stream type U15	R/S	30	90.0	100.0	90.0
UK stream type U23	R/S	30	96.7	100.0	70.0
UK all samples	R/S	60	93.3	100.0	81.7
Overall mean			89.8	98.1	77.6

assessments or impact assessments by cluster analysis based on similarity measures (Hruby, 1987).

One of the main causes of low similarity is the taxonomic resolution of the samples. Samples from stream types with low taxonomic resolution (i.e., family for the stream types of Greece, Latvia or Italy) have lower coherence than samples from stream types with high taxonomic resolution (e.g., species in stream types of Czech Republic, Germany or UK). Another reason could be the sampling and sorting methods used in these national protocols. The Italian and the Latvian protocol require life sorting in the field, furthermore according to the Latvian protocol the species are mainly identified in the field. This could cause a higher error and variability in the resulting taxalists than standardised lab sorting and identification methods. The Greek stream types are generally very poor in abundance terms. In simulation and field studies, Cao et al. (2002) found that equal-sized samples and methods in which only a relatively small number of individuals are counted may not characterise the structure of communities adequately or enable reliable comparison of samples (see also Vlek et al., 2006). Cao et al. (1997) pointed out that the similarity of samples from the same site increases with the sample size (i.e., number of individuals counted) whereas the similarity of samples of two different sites varies unpredictably with samples size depending on individual community structure. Thus, if the

Table 7. Site-, season- and method-coherence based on the Bray–Curtis similarity Index for the STAR sampling program; only un-replicated samples are analysed; each of *n* such samples is compared to all other samples within the stream type (Methods: F = IBGN, I = IBE, L = Latvian, N = Nordic, O = Polish, P = PERLA, PMP = Portuguese, R = RIVPACS, S = STAR-AQEM)

Country/stream type	Methods	*n*	Site-coherence (%)	Season-coherence (%)	Method-coherence (%)
Austria stream type A05	R/S	40	62.5	95.0	35.0
Austria stream type A06	R/S	14	85.7	85.7	14.3
Austria all samples	R/S	54	68.5	92.6	29.6
Czech Republic stream type C04	P/S	44	97.7	100.0	2.3
Czech Republic stream type C05	P/S	28	92.9	100.0	7.1
Czech Republic all samples	P/S	72	95.8	100.0	4.2
Denmark stream type K02	N/S	20	90.0	100.0	10.0
Germany stream type D03	R/S	40	90.0	100.0	10.0
Germany stream type D04	R/S	40	57.5	100.0	42.5
Germany stream type D06	R/S	16	93.8	100.0	6.3
Germany all samples	R/S	96	77.1	100.0	22.9
Greece stream type H04	R/S	26	53.9	69.2	46.2
Greece stream type H05	R/S	20	20.0	95.0	75.0
Greece stream type H06	R/S	20	75.0	90.0	15.0
Greece stream type H07	R/S	10	60.0	100.0	0.0
Greece all samples	R/S	76	51.3	85.5	39.5
Italy stream type I06	I/S	20	25.0	70.0	80.0
Latvia stream type L02	L/S	65	56.9	83.1	50.8
Poland stream type O02	O/S	76	31.6	96.1	68.4
Portugal stream type P04	PMP/S	12	100.0	100.0	0.0
Slovak Republic stream type V01	P/S	24	37.5	100.0	54.2
Sweden stream type S05	N/S	50	98.0	96.0	2.0
Sweden stream type S06	N/S	34	100.0	97.1	2.9
Sweden all samples	N/S	84	98.8	96.4	2.4
UK stream type U15	R/S	40	100.0	97.5	0.0
UK stream type U23	R/S	36	88.9	91.7	19.4
UK all samples	R/S	76	94.7	94.7	9.2
Overall mean			72.2	93.6	25.8

methods rely overly on small sample sizes, a low degree of site-coherence is achieved. Here, small size relates to the size of area sampled as well as the number of individuals picked and identified from the sample.

The overview of within-sites similarity (Table 3) gave no indication of consistent differences in the precision/coherence of the sampling methods applied at all sites. On average, the similarities were equally high. But the precision declines if different methods were compared for each site and even more if different methods were compared in different seasons.

This was also supported by the analysis of the un-replicated samples (Tables 7–9) for which site-coherence depends on the comparability of the sampling methods. Site-coherence is reduced in cases of strongly differing methods (e.g., Italian IBE method compared to STAR-AQEM method in stream type I06 or Polish national method compared to STAR-AQEM method in stream type O02) but remains very high if the two methods are comparable (e.g., Czech PERLA method and STAR-AQEM method in the Czech stream types).

The two German stream types D03 and D04 display the problems of unique sampling protocols applied to totally different stream types. D03 are mid-sized lowland streams with sand as the main and often only bottom substrate, a low current

Table 8. Site-, season- and method-coherence based on the Jaccard similarity Index for the STAR sampling program; only un-replicated samples are analysed; each of *n* such samples is compared to all other samples within the stream type (Methods: F = IBGN, I = IBE, L = Latvian, N = Nordic, O = Polish, P = PERLA, PMP = Portuguese, R = RIVPACS, S = STAR-AQEM)

Country/stream type	Methods	*n*	Site-coherence (%)	Season-coherence (%)	Method-coherence (%)
Austria stream type A05	R/S	40	47.5	95.0	40.0
Austria stream type A06	R/S	14	57.1	92.9	21.4
Austria all samples	R/S	54	50.0	94.4	35.2
Czech Republic stream type C04	P/S	44	81.8	100.0	15.9
Czech Republic stream type C05	P/S	28	85.7	100.0	14.3
Czech Republic all samples	P/S	72	83.3	100.0	15.3
Denmark stream type K02	N/S	20	85.0	100.0	5.0
Germany stream type D03	R/S	40	85.0	100.0	12.5
Germany stream type D04	R/S	40	60.0	100.0	30.0
Germany stream type D06	R/S	16	56.3	100.0	25.0
Germany all samples	R/S	96	69.8	100.0	21.9
Greece stream type H04	R/S	26	50.0	80.8	61.5
Greece stream type H05	R/S	20	55.0	90.0	15.0
Greece stream type H06	R/S	20	55.0	85.0	35.0
Greece stream type H07	R/S	10	60.0	100.0	10.0
Greece all samples	R/S	76	54.0	86.8	35.5
Italy stream type I06	I/S	20	35.0	75.0	70.0
Latvia stream type L02	L/S	65	41.5	80.0	47.7
Poland stream type O02	O/S	76	54.0	94.7	50.0
Portugal stream type P04	PMP/S	12	75.0	83.3	33.3
Slovak Republic stream type V01	P/S	24	62.5	100.0	33.3
Sweden stream type S05	N/S	50	78.0	92.0	14.0
Sweden stream type S06	N/S	34	94.1	85.3	11.8
Sweden all samples	N/S	84	84.5	89.3	13.1
UK stream type U15	R/S	40	100.0	87.5	5.0
UK stream type U23	R/S	36	80.6	86.1	30.6
UK all samples	R/S	76	90.8	86.8	17.1
Overall mean			66.6	91.8	27.7

diversity and a poor taxa richness (Feld, 2004). D04 are small-sized mountain streams with a diverse substrate ranging from large boulders to sand, a high current diversity and a high taxon richness (Lorenz et al., 2004). The RIVPACS protocol (Murray-Bligh et al., 1997) requires proportional time-related sampling of the riffle areas of a site, whereas the STAR-AQEM protocol (Hering et al., 2004; Furse et al., 2006) requires a habitat specific proportional area sampling in riffles and in pools. In the stream type D04 these differences result in a low site-coherence for the un-replicated samples because many pool/slow current inhabiting species are the cause of the difference between the two methods' sample communities at a site. In stream type D03 with the moderate diversity in habitats and current velocities the same areas are sampled by both methods and thus the site-coherence is high.

Season-coherence

A high season-coherence, over 85% and often 100%, was found for all stream types and countries. Even if different protocols/methods are applied, the most similar samples were usually from the same season. This indicates strong differences and changes between seasons in the macroinvertebrate communities dependent on the emergence of the insect taxa (see also Sporka et al.,

Table 9. Site-, season- and method-coherence based on the Renkonen similarity Index for the STAR sampling program; only un-replicated samples are analysed; each of *n* such samples is compared to all other samples within the stream type (Methods: F = IBGN, I = IBE, L = Latvian, N = Nordic, O = Polish, P = PERLA, PMP = Portuguese, R = RIVPACS, S = STAR-AQEM)

Country/stream type	Methods	*n*	Site-coherence (%)	Season-coherence (%)	Method-coherence (%)
Austria stream type A05	R/S	40	47.5	100.0	45.0
Austria stream type A06	R/S	14	78.6	64.3	50.0
Austria all samples	R/S	54	55.6	90.7	46.3
Czech Republic stream type C04	P/S	44	63.6	100.0	27.3
Czech Republic stream type C05	P/S	28	71.4	100.0	21.4
Czech Republic all samples	P/S	72	66.7	100.0	25.0
Denmark stream type K02	N/S	20	75.0	80.0	20.0
Germany stream type D03	R/S	40	67.5	87.5	17.5
Germany stream type D04	R/S	40	45.0	92.5	52.5
Germany stream type D06	R/S	16	87.5	93.8	12.5
Germany all samples	R/S	96	61.5	90.6	31.3
Greece stream type H04	R/S	26	46.2	34.6	23.1
Greece stream type H05	R/S	20	30.0	85.0	45.0
Greece stream type H06	R/S	20	65.0	90.0	20.0
Greece stream type H07	R/S	10	50.0	80.0	10.0
Greece all samples	R/S	76	47.4	68.4	26.3
Italy stream type I06	I/S	20	5.0	95.0	80.0
Latvia stream type L02	L/S	65	27.7	78.5	72.3
Poland stream type O02	O/S	76	26.3	76.3	82.9
Portugal stream type P04	PMP/S	12	75.0	75.0	8.3
Slovak Republic stream type V01	P/S	24	79.2	100.0	16.7
Sweden stream type S05	N/S	50	92.0	100.0	6.0
Sweden stream type S06	N/S	34	70.6	91.2	14.7
Sweden all samples	N/S	84	83.3	96.4	9.5
UK stream type U15	R/S	40	87.5	90.0	7.5
UK stream type U23	R/S	36	77.8	94.4	16.7
UK all samples	R/S	76	82.9	92.1	11.8
Overall mean			60.4	86.1	30.9

2006). However, it is also apparent that stream types identified to family (e.g., Italian stream types I05 and I06, Latvian stream type L02, Greek stream type H04) are less replicable within seasons. Most unusually, low season-coherences can be found using the Renkonen Index on the un-replicated samples (e.g., Greek stream type H04: 34.6% or the Austrian stream type A06: 64.3%). These two stream types were sampled with the RIVPACS and the STAR-AQEM protocol. Differences in their sampling and sorting methods could have resulted in shifts in the dominance structure of the taxalists and therefore lower community similarity between methods and within seasons, resulting in a low season-coherence and a higher method-coherence.

The average similarity between samples taken from the same site in the same season is equally high regardless of season, but the average similarity between samples from the same site in different seasons is 20–30% less. Thus, season is one of the major determinants of stream macroinvertebrate fauna (see also Furse et al., 1984; Sporka et al., 2006). The Renkonen similarity for samples from different methods and seasons gave the lowest within-site similarity (31.7%) and highlights the overwhelming effect of seasonality. This index relies directly on the abundance and dominance of taxa and indicates that the relative abundance of taxa changes considerably between seasons in many stream types.

Method-coherence

This analysis highlights the diversity of the various sampling methods but also the commonalities and similarities between them (see also Friberg et al., 2006). The average within-site similarity of samples is about equally high for the STAR-AQEM method as the average for the national methods, whether based on samples from any one season or from different seasons (Table 3). Thus, the proposed methods have a good integrity within the sites and the stream types. But where samples from two different methods were compared at one site, the community similarity between them was about 10% lower than the within-method similarity for each of the three similarity indices. This reduction in similarity depended on the national method involved.

Where replicate samples of the same method were analysed, a low method-coherence in connection with a high season- and site-coherence was conspicuous in certain stream types. For the stream types P04 (Portugal), S05 and S06 (Sweden), this means that the two methods applied (PMP or Nordic respectively, and STAR-AQEM) seem to be comparable in terms of taxonomic similarity. For species presence/absence according to the Jaccard Index, this was also true for the Austrian stream type A06, the German stream type D04 and the UK stream types U15 and U23; for all of which the STAR-AQEM and RIVPACS methods were applied. However, the correspondence between the STAR-AQEM and RIVPACS methods was not apparent in all stream types; for example, the method-coherence in the Austrian stream type A05 and the other two German stream types D03 and D06 was high (>80%). If the method-coherence and the site-coherence are both low (e.g., as in Greek stream type H04 and Latvian stream type L02) then the recorded taxa are not sufficiently different between sites, and the methods are not necessarily comparable.

Comparing the Jaccard Index of replicated (Table 5) and un-replicated (Table 8) methods leads to the conclusion that the sampling and sample processing of some national methods produces sample taxalists that are different to those obtained using the STAR-AQEM protocol; several taxa were found only by one of the methods. Thus, similarities are often higher between samples from different sites with the same method than between different methods at the same site. Neither the Latvian or Italian (IBE) national methods give samples which are comparable to those from the STAR-AQEM method, even though taxonomic resolution is mainly only to family in both countries.

When the abundance and dominance of taxa are involved by using the Renkonen Index, the method-coherence is high for the Italian (80.0%), Latvian (72.3%) and Polish (82.9%) stream types (Table 9), indicating that their national methods are not comparable to the STAR-AQEM method. The site-coherence for the un-replicated samples from these stream types is also very low (5.0, 27.7 and 26.3%, respectively; Table 9).

In contrast, the Portuguese, Nordic and Czech PERLA methods are comparable with the STAR-AQEM method. For every similarity index, the site-coherence was very high (BC>90%; J>75%; R > 63%) and the method-coherence was low (BC: <10%; J: <35%; R: <30%). The comparability is caused by many overlaps in the sampling protocols. The Czech PERLA method and the Portuguese method allocate sampling units in proportion to the percentage cover of all habitats in a stream site, as is the case in the STAR-AQEM method. The main difference in the PERLA method to the STAR-AQEM method is that the samples are time related and not area related. The difference in the Portuguese method is that it uses a slightly larger sampling area (1.5 m^2 instead of 1.25 m^2). Thus, by processing the samples in the laboratory, there are sorting or sub-sampling steps that even-up the resulting taxalists.

Another feature detected with the Renkonen Index is the difference between the STAR-AQEM method and the RIVPACS method. Small-scale comparisons for Austria, Germany and Greece showed that taxon- and organism-rich stream types like A05 and D04 have even lower site-coherences than less diverse stream types like D03. This can also be seen in the United Kingdom (taxon-rich: U23; taxon-poor: U15; difference in site-coherence: 9.7%).

There could be an additional reason for the two-sided character of the results of the comparison of RIVPACS method and STAR-AQEM method. The British project partner is far more familiar with the RIVPACS sampling approach, the other three partners used it for the first time in the STAR-project. In the UK stream types the site-coherence was always more than 90% in the replicate sample analysis and always more than 78% in the un-replicated sample analyses. Thus, the two methods seem to be comparable in the UK. This suggests that the experience of the people actually doing the sampling influences the coherence of samples. A second possibility is that the differences between the communities are larger in the UK streams.

Despite that, the coherence of samples could depend on the ecological status of the sampling sites, which influences taxon richness and thus inter-sample similarity. However, this is beyond the scope of this paper.

Similarity indices

Average values of the Jaccard similarity Index are generally lower than for the Bray–Curtis or Renkonen indices; the former only involves the presence–absence of taxa and thus may be more prone to sampling variation. Cao et al. (2002) used the same Jaccard coefficient, assuming it to be a more strict measure for sample representativeness and more closely related to the taxon richness used in their bioassessments.

There are obvious differences between the three similarity indices in the estimated coherence of samples between methods. The Jaccard Index emphasises the presence of species, indicating that e.g., the different methods used in the UK stream type U15 or the Swedish stream type S06 had sampled the same habitats (high site-coherences). In contrast, the Italian and Latvian sampling and sorting methods resulted in strongly differing taxalists compared to the STAR-AQEM protocol and thus in low Jaccard similarity and low site-coherences. The UK stream type U15 ($J=100\%$; $R=87.5\%$) and Swedish stream type S06 ($J=94.1\%$; $R=70.6\%$) had much lower site-coherence when based on the abundance and dominance of taxa according to the Renkonen Index. Thus, the sampling protocols seem to result in similar lists of taxa, but with differing relative densities. This was underlined by the intermediate Bray–Curtis Index which led to 100% site-coherence for both of these stream types. The results showed that the three indices can be used to concentrate on different aspects of the sample similarity of macroinvertebrate communities.

Consequences for bioassessment

An overall measure of coherence of all samples based on taxonomic similarity cannot be given. However, some methods applied in this project lead to a very good coherence within sites and seasons (e.g., Czech PERLA method), whereas others have a low coherence (e.g., Italian method on stream type I06 and Latvian method on stream type L02). A very high site-coherence is desirable for any bioassessment or impact assessment involving cluster analysis of samples (and thus sites) based on community similarity indices from single samples per site. Few stream types have a site-coherence of 100%. Low site-coherence does not necessarily mean that (assessment) results become arbitrary using those methods. Recent assessment systems for streams in Europe are often composed of several metrics (e.g., Buffagni et al., 2004; Lorenz et al., 2004) which depend on the ecological, biological or compositional parameters of macroinvertebrate communities. Samples can still give similar values for such metrics even though they include different taxa. A serious problem would occur if the impact assessment is done by similarity or dissimilarity analysis or with multivariate methods using similarity indices as the separating measures.

The Italian IBE, the Latvian national and the Polish national method are least comparable to the STAR-AQEM method, probably because the habitat sampling design is too different. Some others like the Nordic method (Sweden and Denmark) and the Portuguese PMP method are more like the STAR-AQEM method. The similarity and comparability of samples obtained with RIVPACS method with those obtained with the STAR-AQEM method seemed to depend on the stream type and the people applying the methods.

Finally, if the taxonomic resolution is low (e.g., family), the discrimination of samples and sites and thus the site-coherence is worse than if the taxonomic resolution is high (species).

Acknowledgement

This paper is based on samples taken for the EU-funded STAR-project (Standardisation of River Classifications: Framework method for calibrating different biological survey results against ecological quality classifications to be developed for the Water Framework Directive; Contract No: EVK1-CT 2001-00089). We are grateful to Jörg Strackbein for illustrations and provision of data and to Sascha Weyers for programming the Jaccard and Renkonen Index.

References

AFNOR, 1982. Essais des euax. Détermination de l'indice biologique global normalisé (IBGN). Association Francaise de Normalisation NF T 90-350.

Barbour, M. T., J. Gerritsen, B. D. Snyder & J. B. Stribling, 1999. Rapid Bioassessment Protocols for Use in Wadeable Streams and Rivers. Periphyton, Benthic Macroinvertebrates, and Fish (2nd edn). U.S. Environmental Protection Agency, Office of Water, Washington, D.C. EPA 841-B-99-002.

Birk, S. & D. Hering, 2002. Waterview Web-Database: a comprehensive review of European assessment methods for rivers. FBA News 20: 4.

Birk, S. & D. Hering, 2006. Direct comparison of assessment methods using benthic macroinvertebrates: a contribution to the EU Water Framework Directive intercalibration exercise. Hydrobiologia 566: 401–415.

Bloom, S. A., 1981. Similarity Indices in community studies: Potential pitfalls. Marine Ecology Progress Series 5: 125–138.

Boyle, T. P., G. M. Smillie, J. C. Anderson & D. R. Beeson, 1990. A sensitivity analysis of nine diversity and seven similarity indices. Research Journal of the Water Pollution Control Federation 62: 749–762.

Bray, J. R. & J. T. Curtis, 1957. An ordination of the upland forest communities of South Wisconsin. Ecological Monographs 27: 325–347.

Buffagni, A., S. Erba, M. Cazzola & J. L. Kemp, 2004. The AQEM multimetric system for the southern Italian Appenines: assessing the impact of water quality and habitat degradation on pool macroinvertebrates in Mediterranean rivers. Hydrobiologia 516: 315–331.

Cao, Y., D. D. Williams & D. P. Larsen, 2002. Comparisons of ecological communities: the problem of sample representativeness. Ecological Monographs 72: 41–56.

Cao, Y., W. P. Williams & A.W. Bark, 1997. Effects of sample size (number of replicates) on similarity measures in river Aufwuchs community analysis. Water Environment Research 69: 107–114.

Clarke, K. R. & R. N. Gorley, 2001. PRIMER v5: User Manual/Tutorial. PRIMER-E Ltd, Plymouth, UK.

Clarke, R. T., M. T. Furse, R. J. M. Gunn, J. M. Winder & J. F. Wright, 2002. Sampling variation in macroinvertebrate data and implications for river quality indices. Freshwater Biology 47: 1735–1751.

Clarke, R. T., A. Lorenz, L. Sandin, A. Schmidt-Kloiber, J. Strackbein, N. T. Kneebone & P. Haase, 2006a. Effects of sampling and sub-sampling variation using the STAR-AQEM sampling protocol on the precision of macroinvertebrate metrics. Hydrobiologia 566: 441–459.

Clarke, R. T., J. Davy-Bowker, L. Sandin, N. Friberg, R. K. Johnson & B. Bis, 2006b. Estimates and comparisons of the effects of sampling variation using 'national' macroinvertebrate sampling protocols on the precision of metrics used to assess ecological status. Hydrobiologia 566: 477–503.

DEV (Deutsches Institut für Normung e.V.), 1992. Biologisch-ökologische Gewässergüteuntersuchung: Bestimmung des Saprobienindex (M2). In Deutsche Einheitsverfahren zur Wasser-, Abwasser- und Schlammuntersuchung. VCH Verlagsgesellschaft mbH, Weinheim, 1–13.

Faith, D. P., P. R. Minchin & L. Belbin, 1987. Compositional dissimilarity as a robust measure of ecological distance. Vegetatio 69: 57–68.

Feld, C., 2004. Identification and measure of hydromorphological degradation in Central European lowland streams. Hydrobiologia 516: 69–90.

Field, J. G., K. R. Clarke & R. M. Warwick, 1982. A Practical Strategy for analysing multispecies distribution patterns. Marine Ecology Progress Series 8: 37–52.

Friberg, N., L. Sandin, M. T. Furse, S. E. Larsen, R. T. Clarke & P. Haase, 2006. Comparison of macroinvertebrate sampling methods in Europe. Hydrobiologia 566: 365–378.

Furse, M. T., D. Moss, J. F. Wright & P. D. Armitage, 1984. The influence of seasonal and taxonomic factors on the ordination and classification of running-water sites in Great Britain and on the prediction of their macro-invertebrate communities. Freshwater Biology 14: 257–280.

Furse, M., D. Hering, O. Moog, P. Verdonschot, R. K. Johnson, K. Brabec, K. Gritzalis, A. Buffagni, P. Pinto, N. Friberg, J. Murray-Bligh, J. Kokes, R. Alber, P. Usseglio-Polatera, P. Haase, R. Sweeting, B. Bis, K. Szoszkiewicz, H. Soszka, G. Springe, F. Sporka & I. Krno, 2006. The STAR project: context, objectives and approaches. Hydrobiologia 566: 3–29.

Ghetti, P. F., 1997. Manuale di applicazione Indice Biotico Esteso (I.B.E.). I macroinvertebrati nel controllo della qualità degli ambienti di acque correnti. Provincia Autonoma di Trento, Agenzia provinciale per la protezione dell'ambiente.

Hering, D., A. Buffagni, O. Moog, L. Sandin, M. Sommerhäuser, I. Stubauer, C. K. Feld, R. Johnson, P. Pinto, N. Skoulikidis, P. F. M. Verdonschot & S. Zahrádková, 2003. The development of a system to assess the ecological quality of streams based on macroinvertebrates – design of the sampling programme within the AQEM project. International Review of Hydrobiology 88: 345–361.

Hering, D., O. Moog, L. Sandin & P. F. M. Verdonschot, 2004. Overview and application of the AQEM assessment system. Hydrobiologia 516: 1–20.

Hruby, T., 1987. Using similarity measures in benthic impact assessments. Environmental Monitoring and Assessment 8: 163–180.

Jaccard, P., 1912. The distribution of flora in the alpine zone. New Phytologist 11: 37–50.

Kolkwitz, R. & M. Marsson, 1902. Grundsätze für die biologische Beurteilung des Wassers nach seiner Flora und Fauna. Mitt. Prüfungsanstalt Wasserversorgung Abwasserreinigung 1: 33–72.

Latvian Standard Ltd., 1999. LVS 240:1999. Water quality – Operative evaluation biological quality of small stream by saprobity index of macrozoobenthos community. In Catalogue of Latvian standards, Riga, Latvian Standard Ltd, 1999: Group 13.060, 1–11.

Lorenz, A., D. Hering, C. K. Feld & P. Rolauffs, 2004. A new method for assessing the impact of hydromorphological degradation on the macroinvertebrate fauna of five German stream types. Hydrobiologia 516: 107–127.

Murray-Bligh, J. A. D., M. T. Furse, F. H. Jones, R. J. M. Gunn, R. A. Dines & J. F. Wright, 1997. Procedure for collecting and analysing macroinvertebrate samples for RIVPACS. Joint publication by the Institute of Freshwater Ecology and the Environment Agency, 162 pp.

Nijboer, R. C. & P. F. M. Verdonschot, 2000. Taxonomic adjustment affects data analysis: an often forgotten error. Verhandlungen Internationale Vereinigung der Limnologie 27: 2546–2549.

Renkonen, O., 1938. Statistisch-Oekologische Untersuchungen ueber die terrestrische Kaeferwelt der Finnischen Bruchmoore. Annales zoologici Societatis Zoologicae Botanicae Fennicae Vanamo 6: 1–231.

Skriver, J., N. Friberg & J. Kirkegaard, 2000. Biological assessment of running waters in Denmark: Introduction of the Danish Stream Fauna Index (DSFI). Verhandlungen Internationale Vereinigung der Limnologie 27: 1822–1830.

Šporka, F., H. E. Vlek, E. Bulánková & I. Krno, 2006. Influence of seasonal variation on bioassessment of streams using macroinvertebrates. Hydrobiologia 566: 543–555.

Vlek, H. E., F. Šporka & I. Krno, 2006. Influence of macroinvertebrate sample size on bioassessment of streams. Hydrobiologia 566: 523–542.

Washington, H. G., 1984. Diversity, biotic, and similarity indices. Water Research 18: 653–694.

Wolda, H., 1981. Similarity indices, samples size and diversity. Oecologia 50: 296–302.

Wright, J. F., D. W. Sutcliffe & M. T. Furse (eds), 2000. Assessing the Biological Quality of Fresh Waters – RIVPACS and Other Techniques. Freshwater Biological Association, Ambleside, Cumbria, U.K, 373 pp.

Zelinka, M. & P. Marvan, 1961. Zur Präzisierung der biologischen Klassifikation der Reinheit fließender Gewässer. Archiv für Hydrobiologie 57: 389–407.

Hydrobiologia (2006) 566:477–503

M.T. Furse, D. Hering, K. Brabec, A. Buffagni, L. Sandin & P.F.M. Verdonschot (eds), The Ecological Status of European Rivers: Evaluation and Intercalibration of Assessment Methods
DOI 10.1007/s10750-006-0076-5

Estimates and comparisons of the effects of sampling variation using 'national' macroinvertebrate sampling protocols on the precision of metrics used to assess ecological status

Ralph T. Clarke[1,*], John Davy-Bowker[1], Leonard Sandin[2], Nikolai Friberg[3], Richard K. Johnson[2] & Barbara Bis[4]

[1]*Centre for Ecology & Hydrology, Winfrith Technology Centre, DT2 8ZD, Dorchester, Dorset, United Kingdom*

[2]*Department of Environmental Assessment, Swedish University of Agricultural Sciences, P.O. Box 7050, SE-750 07, Uppsala, Sweden*

[3]*Department of Freshwater Ecology, National Environmental Research Institute, Vejlsøvej 25, DK-8600 Silkeborg, Denmark*

[4]*Institute of Ecology and Nature Protection, Department of Invertebrate Zoology and Hydrobiology, University of Łodź, Banacha 12/16, 90-237 Łodź, Poland*

*(*Author for correspondence: E-mail: rtc@ceh.ac.uk)*

Key words: replicate sampling variation, uncertainty, macroinvertebrate metrics, Water Framework Directive, RIVPACS, STAR-AQEM, PERLA

Abstract

The Water Framework Directive (WFD) of the European Union requires all member countries to provide information on the level of confidence and precision of results in their river monitoring programmes to assess the ecological status class of river sites. As part of the European Union project STAR, the overall effects of sampling variation for a wide range of commonly used metrics and sampling methods were assessed. Replicate samples were taken in each of two seasons at 2–6 sites of varying ecological status class within each of 18 stream types spread over 12 countries, using both the STAR-AQEM method and a national sampling method or, where unavailable, the RIVPACS sampling protocol. The sampling precision of a combination of sampling method and metric was estimated by expressing the replicate sampling variance as a percentage P_{samp} of the total variance in metric values with a stream type; low values of P_{samp} indicate high precision. Most metrics had percentage sampling variances less than 20% for all or most stream types and methods. Most national methods including RIVPACS had sampling precisions at least as good as those for the STAR-AQEM method as used in their country at the same sites; the main exceptions were the national methods used in Latvia and Sweden. The national methods used in the Czech Republic, Denmark, France, Poland and the RIVPACS method used in the UK and Austria all had percentage sampling variances of less than 10% for the majority of metrics assessed. In contrast, none of the metrics had percentage sampling variances less than 10% when based on either the Italian (IBE) method, which used bank-side sorting, or the Latvian national method which identifies only a limited set of taxa. P_{samp} was lowest on average for the two stream types sampled in the Czech Republic using either the PERLA national method or the STAR-AQEM method. Averaged over all stream types and methods, the three Saprobic-based metrics had the lowest average percentage sampling variances (3–6%) amongst the 26 metrics assessed. These estimates of sampling standard deviation can be used to help assess the uncertainty in single or multi-metric systems for estimating site ecological status using the general STAR Bioassessment Guidance Software (STARBUGS) developed within the STAR project.

Introduction

Most quantitative assessments of the biological status of water bodies are based on the values of one or more biological indices or metrics derived from the taxonomic composition of the sample (Smith et al., 1999; Wright et al., 2000). The metrics are often designed to measure the ecological response to some specific form of stress, such as organic or toxic pollution, acid stress, or degradation in stream morphology and the diversity of habitats (Herring et al., 2004). Any measure of ecological quality or status is of little value without some knowledge of its levels of uncertainty (Clarke, 2000; REFCOND, 2003). All assessments of the ecological status of river sites using macroinvertebrate sampling are subject to uncertainty and errors due to a range of factors (Ostermiller & Hawkins, 2004). Replicate sample values will vary because of inherent natural small-scale spatial heterogeneity in the fauna at a site. Subsequent sample processing, perhaps involving sub-sampling or identifying only a fixed number of individuals, and the taxonomic resolution (family, genus or species) will all affect the precision and uncertainty in site assessments. Assessment methods which are prone to high levels of variation between replicate sample values will tend to provide less reliable estimates of the ecological status for a site and provide less power and confidence to detect changes in ecological quality (Clarke, 2000). Therefore, it is important to have quantitative estimates of the effects of sampling variation on the values of any biotic index or metric used to assess the ecological status of a river site (Clarke et al., 1996).

There have, surprisingly, been few extensive quantitative field studies of the susceptibility of freshwater biotic metrics to sampling variation. Clarke et al. (2002) quantified the sampling variation in BMWP score, number of BMWP families (taxa) and Average Score Per Taxon (ASPT) for the RIVPACS sampling method (Murray-Bligh et al., 1997) by taking three replicate samples in each of three seasons at each of 16 UK sites selected to cover a balanced range of ecological qualities and physical types. They showed that with trained staff, inter-personnel differences contributed less than 12% of the total variance between replicate RIVPACS samples in each of the three metrics. Johnson (1998) took five or six replicate macroinvertebrate samples from littoral, sub-littoral and profundal habitats in each of 16 Swedish lakes and assessed sampling variation in six metrics. In contrast, Hose et al. (2004) assessed the within-season short-term temporal variation in observed number of taxa at two reference quality sites in New South Wales, Australia by sampling the same riffle and edge habitats on three occasions over a 6 week autumn period and found that the O/E ratio of the observed to AUSRIVAS-predicted expected number of taxa did not vary significantly within the season.

Within Europe, the EU Water Framework Directive (WFD) (Council of the European Union, 2000) requires all partner countries to provide information in their river basin management plans on the level of confidence and precision of results in their river monitoring programmes to assess the ecological status class of river sites (WFD: Annexe V, Section 1.3). The STAR project of the European Union 5th Framework Programme included an extensive field sampling programme across 13 countries using existing nationally used macroinvertebrate sampling methods and protocols, together with a standard STAR-AQEM method (Furse et al., 2006). The STAR-AQEM method and protocol was developed originally during the AQEM project within an earlier EU 5th Framework Programme (Herring et al., 2004). The protocol was designed as a possible approach to providing a standardised sampling and assessment methodology across Europe (see special issue volume 516 of Hydrobiologia). The AQEM method has subsequently been modified for use within the STAR project, and is now referred to as the STAR-AQEM method and described in detail in Furse et al. (2006).

This paper summarises the results of an extensive replicated sampling programme within the main STAR field sampling programme. It quantifies the overall effects of sampling variation and subsequent laboratory sample processing procedures on the sampling variability of a wide range of macroinvertebrate metric values for 'national' sampling methods (or, where unavailable, the RIVPACS method) and compares these with the estimates of variability obtained for the STAR-AQEM samples taken at the same sites at the same times.

Methods

Replicate field sampling programme

The replicated sampling study covered 12 countries encompassing 22 stream types spread over 11 Ecoregions (*sensu* Illies, 1978; as used in the WFD). Within the STAR field sampling programme, STAR-AQEM samples were taken at all sites by each participating partner (Clarke et al., 2006; Furse et al., 2006). At each site in nearly all of the main stream types, each partner also collected samples using a notional 'national' method. This was normally a widely used protocol within the individual partner's Member State, but in Germany, Austria and Greece where there were no existing common 'national' sampling protocols, the UK RIVPACS protocol was used (Murray-Bligh et al., 1997) (Table 1). Both STAR-AQEM and 'national' or RIVPACS samples were collected in two seasons – spring and either summer or autumn. Further details of all of the sampling methods and protocols are given in Furse et al. (2006).

Most STAR project partners took a second replicate field sample at a subset of their sites, usually by both methods and usually in both sampling seasons (Table 1). These sites were carefully selected within each sampled stream type to cover a range of perceived (i.e., pre-classified) ecological status classes (*sensu* WFD) of sites from 'high' and 'good' to 'moderate' or 'poor'/'bad'. This was important because the sampling variability of many metrics may depend on the quality of a site; poorer quality sites with fewer taxa present might be less variable in some taxonomic richness/diversity metrics, but more variable in metrics based on some form of average stress-tolerance score of the taxa present (e.g., ASPT or a Saprobic Index).

Taking replicate samples at the same set of sites using both methods provided valid direct comparisons of the sampling standard deviation (SD) for individual metrics between the 'national' or RIVPACS method and the STAR-AQEM method, because both sampling methods were then based on the same range of site qualities and within-site

Table 1. Number of sites in each stream type and country with replicate samples obtained using either the RIVPACS or 'national' sampling method in at least one season; (small-sized = 10–100 km^2, medium sized = 100–1000 km^2, lowland = < 200 m asl)

Country	Method	Stream type	Description	Sites	Sites× seasons
Austria	RIVPACS	A05	Small-sized, shallow mountain streams	4	6
		A06	Small-sized crystalline streams of the ridges of the Central Alps	4	7
Czech Republic	PERLA	C04	Small-sized, shallow mountain streams	3	6
		C05	Small-sized streams in the Central sub-alpine Mountains	3	6
Germany	RIVPACS	D03	Medium-sized lowland streams	2	4
		D04	Small-sized, shallow mountain streams	2	4
		D06	Small-sized Buntsandstein-streams	2	4
France	IBGN	F08	Small-sized, shallow headwater streams in Eastern France	6	12
Greece	RIVPACS	H04	Small-sized calcareous mountain streams in Western, Central and Southern Greece	6	12
Italy	IBE	I06	Small-sized calcareous streams in the Central Apennines	6	11
Denmark	DSFI	K02	Medium-sized lowland streams	6	12
Latvia	LVS 240:1999	L02	Medium-sized lowland streams	6	12
Poland	National	O02	Medium-sized lowland streams	7	13
Portugal	PMP	P04	Medium-sized streams in lower mountainous areas of S. Portugal	6	12
Sweden	National	S05	Medium-sized lowland streams	3	6
		S06	Medium-sized streams on calcareous soils	3	6
UK	RIVPACS	U15	Small-sized, shallow lowland streams	3	6
		U23	Medium-sized lowland streams	3	6

habitat heterogeneities – both of which could influence sampling variability in macroinvertebrate composition and derived metric values at a site.

Taxonomic resolution and calculation of metric values

For the majority of countries and STAR partners, most taxa were identified to species level for all samples obtained using both methods. However, in Italy, Greece and Latvia samples were identified to mainly family level, while in Portugal, samples were identified to mainly genus level. In France, the STAR-AQEM samples were identified to mainly genus level, but the 'national' IBGN method samples were only identified to mainly family level. Prior to calculating metric values, the macroinvertebrate data were taxonomically adjusted to a consistent level, separately for each country.

Metrics values for all samples were then calculated using the AQEM/STAR Ecological River Classification System ASTERICS, which has the ability to calculate over 200 macroinvertebrate-based metrics (www.eu-star.at). The analyses reported here are for up to 26 metrics chosen for their generally applicability and coverage of a variety of aspects of the macroinvertebrate fauna (structural, functional or life cycle types) and its response to particular environmental stresses (Table 2 onwards). A new metric '1-GOLD' based on the proportion of all individuals which are not Gastropoda, Oligochaeta or Diptera (Pinto et al., 2004) was included because it is one the six proposed Inter-calibration Common Metrics (ICMs) (Buffagni et al., 2006); its values were calculated separately from ASTERICS by Andrea Buffagni, but only for both methods for samples from the Czech Republic, Greece and Italy. The Italian 'national' IBE metric is also included, although it may be inappropriate for many non-Mediterranean stream types. Values for the new species trait metrics (Bis & Usseglio-Polatera, 2004), including trait metrics m1, m2, m7 and m12 analysed here, were available for both STAR-AQEM and 'national' method samples from most countries, but only for RIVPACS samples from the UK.

The three Saprobic indices, which require identifications to species or genus level, were not calculated for stream types and methods with data at mainly family level. Metric values will often depend on the taxonomic resolution of the samples. This is especially true for the metrics measuring some form of taxonomic richness or diversity, such as the total 'Number of taxa' and the Shannon–Wiener diversity index; higher taxonomic resolution will obviously lead to more individual taxa being recorded and probably more variability in results. The sampling SD of these metrics for the stream types based on family data may not be comparable those based on species and genus level data.

Statistical methods

The statistical analyses aimed to estimate and compare the sampling variances in the observed values of each metric within and between stream types, countries and sampling methods. Analysis of variance (ANOVA) and hierarchical nested ANOVA techniques (calculated using Minitab Release 14 statistics package (http://www.minitab.com)) were used to estimate the various sources of variation and variance components contributing towards the total variance in values of a metric within each stream type. Specifically, if Y_{ijr} is the value of the metric for replicate sample r of site j in season i, then Y_{ijr} can be expressed in terms of the sum of the components contributing towards the overall variation in its values, namely:

$$Y_{ijr} = \mu + a_i + b_{ij} + c_{ijr}$$

where

μ = overall mean value of Y within the stream type

a_i = deviation of mean value for season i from the overall mean value μ

b_{ij} = deviation of mean value for site j in season i from the mean for season i

c_{ijr} = deviation of sample r for site j in season i from the mean for site j in season i

The total variance in metric values over all sampled sites within any one stream type is given by

$$\sigma_T^2 = \sigma_I^2 + \sigma_J^2 + \sigma_E^2$$

where

σ_I^2 = variance of the a_i = variance due to inter-season differences in mean value

Table 2. Comparisons of RIVPACS (R) and STAR-AQEM (S-A) methods used in Austria for overall standard deviations (SD_E) and percentage variance (P_{samp}) due to replicate sampling for transformed ($f(x)$) values of metrics; asin = arcsine($\sqrt{(x/100)}$), $asin^1$ = arcsine($\sqrt{x}$)

Metric	$f(x)$	SD_E for RIVPACS within stream type		Averaged across stream types			
				SD_E		P_{samp}	
		A05	A06	R	S-A	R	S-A
Abundance (ind/m^2)	$\sqrt{\sqrt{x}}$	0.891	0.518	0.715	**0.640**	24	**17**
Number of taxa	$\sqrt{x}$	0.353	0.404	**0.038**	0.513	**8**	14
Number of families	$\sqrt{x}$	0.164	0.251	**0.215**	0.310	**6**	12
Number of EPT taxa	$\sqrt{x}$	0.272	0.159	**0.219**	0.454	**5**	23
Saprobic Index	x	0.072	0.087	0.080	**0.042**	13	**6**
German Saprobic new	x	0.052	0.044	0.048	**0.029**	**4**	5
Czech Saprobic	x	0.123	0.099	0.111	**0.084**	6	**5**
ASPT	x	0.266	0.124	**0.202**	0.344	**9**	40
IBE	x	0.936	0.626	0.784	**0.565**	13	**7**
Diversity SW	x	0.224	0.263	0.246	**0.165**	19	**5**
% Rheophilic	asin	0.060	0.104	0.087	**0.068**	8	**5**
% Rheophilic (ab-class)	asin	0.031	0.043	0.038	**0.037**	7	**6**
% Littoral	asin	0.032	0.022	**0.027**	0.068	**3**	19
% Grazers/scrapers	asin	0.049	0.044	0.046	**0.043**	10	**6**
% Shredders	asin	0.029	0.056	0.046	**0.031**	10	**6**
% Gatherers/collectors	asin	0.038	0.026	**0.032**	0.041	**4**	6
% Oligochaeta	asin	0.067	0.045	0.056	**0.051**	7	**6**
% EPT individuals	asin	0.099	0.064	0.082	**0.066**	13	**9**
% EPT (ab-class)	asin	0.038	0.022	**0.031**	0.038	**8**	11
% EPT taxa	asin	0.037	0.029	**0.033**	0.051	**11**	22
RETI	$asin^1$	0.056	0.042	**0.049**	0.053	**7**	10
Average (range)						9 (3–24)	11 (5–40)
Metrics (%) with $P_{samp}<10\%$						13 (62%)	12 (57%)
Metrics with smaller (bold) value of SD_E or P_{samp}				9	12	10	11

σ_J^2 = variance of the b_{ij} = variance due to inter-site differences within a season

σ_E^2 = variance of the c_{ijr} = variance between replicate samples within a site and season

The inter-season variance σ_I^2 represents the extent of systematic seasonal changes in metric values.

If a particular metric and sampling method are to be effective in discriminating the ecological status classes of river sites within a stream type, then the overall replicate sampling variance (σ_E^2) should be small relative to the total variance (σ_T^2) in metric values within the stream type. This is measured by the statistic:

$$P_{samp} = 100\sigma_E^2/\sigma_T^2$$

and referred to as the percentage sampling variance. A low value of P_{samp} indicates that the combination of sampling method and metric has high statistical precision compared to variability amongst sites of differing quality. High values of P_{samp} indicate low sampling precision and low repeatability and hence that such combinations of sampling method and metrics are unlikely to have much power to detect differences in ecological status class. Obviously, metrics with high sampling precision and repeatability may still not be meaningful ecological metrics or reliable indicators of ecological status class. P_{samp} is a better practical measure of the precision of each metric than using the usual coefficient of variation (CV) determined as the ratio of the replicate SD to the replicate mean (see Clarke et al., (2006) for an explanation).

This method of estimating P_{samp} is the best approach if the observed metric values are subsequently to be compared against a single reference condition value for a stream type or site, regardless of season, because in such cases the relevant average total variance in metric values within any one stream type should include the variance σ_I^2 due to systematic differences between seasons in average metric value. This is the approach used throughout this study to estimate and compare the relative sizes (P_{samp}) of the sampling variances for each metric and sampling method. However, if the observed metric values are subsequently to be compared against a season-specific reference condition value for a stream type or site (as done in the UK RIVPACS bioassessment system and software), then the relevant average total variance σ_T^2 in metric values within any one stream type should exclude the variance σ_I^2. Our analyses indicated that σ_I^2 was usually much less than inter-site variance σ_J^2 such that the estimates of P_{samp} provided in this study will also give a reasonable guide to the relative precision of metrics in the case where inter-season variance is excluded.

Because it was only possible to take replicate samples at a few (2–7) sites in each stream type of each STAR partner (albeit usually in two seasons), estimates of the above variance components for individual stream types may be imprecise (Table 1). Therefore, to obtain more robust estimates for a particular sampling method, the variance components and values of P_{samp} for a particular metric were also derived by combining data across all stream types for which the method was used in a particular country. For the RIVPACS method, estimates were also derived by averaging across all four countries where the method was used. The sampling variance (σ_E^2) is usually quoted in the tables in its SD form (i.e., $SD_E = \sqrt{\sigma_E^2}$).

The differences in the values of some metrics between replicate samples varied systematically with the average of the two replicate values. The replicate sampling variability of metrics involving crude abundances or some form of taxonomic richness (i.e., number of taxa, number of families and number of EPT taxa) increased with the average of the two replicate values, as is common in ecology (Taylor, 1961). Several of the selected metrics were percentages (range 0–100) or proportions (0–1) based on the fraction of all individuals or of all taxa which are in a particular group or have particular characteristics. The replicate sample values of such metrics tended to be less variable when their values for a site are very low (near zero) or very high (near 100%) and most variable at intermediate values (30–70%). Visual plots and Spearman rank correlations between replicate variance and replicate mean value across all sites were used to determine the most appropriate transformation of each metric's values to make sampling variances less variable between sites within a stream type and across stream types and increase the validity of a single estimate of sampling variance for all sites within a stream type and even across stream types . For reasons of consistency, compatibility and robustness, only one transformation was used for any single metric regardless of sampling method or stream type. The decision of which single transformation, if any, to use for each metric, was based on assessments of patterns in the overall replicate sampling variability for the STAR-AQEM and RIVPACS methods, as the other methods were only used on relatively few sites and in a single region. Subsequent analyses to estimate sampling and other variances and values of P_{samp} were conducted on the transformed values of each metric, using the transformations indicated in the tables. Where appropriate, Kruskal–Wallis analyses of the ranks of the replicate sampling SD (SD_E) were used to test for statistically significant difference in SD_E between stream types (Sokal & Rohlf, 1995).

Sampling the same (or almost the same) set of sites in any one stream type by two sampling methods, enabled and justified a direct comparison of replicate sampling variability in metric values for the two methods, both in numerical terms (SD_E) and as a percentage (P_{samp}) of the total variance in metric values with a stream type.

Results

Sampling variability using the RIVPACS method

The difference in (untransformed) metric values between two replicate RIVPACS samples taken from the same site were plotted against the average of the two values for a range of metrics (Figs. 1–4). Each point denotes an individual site in one season. The differences rather than the SD of the two replicate samples are shown in these initial plots to aid visual understanding.

The difference between replicate RIVPACS samples in both the total number of taxa and the number of families recorded tends be less when there are fewer taxa and families present at a site (i.e., as measured by the average of the two replicate values) (Fig. 1). This is a similar pattern to that found for STAR-AQEM samples (Clarke et al., 2006), supporting the overall decision to

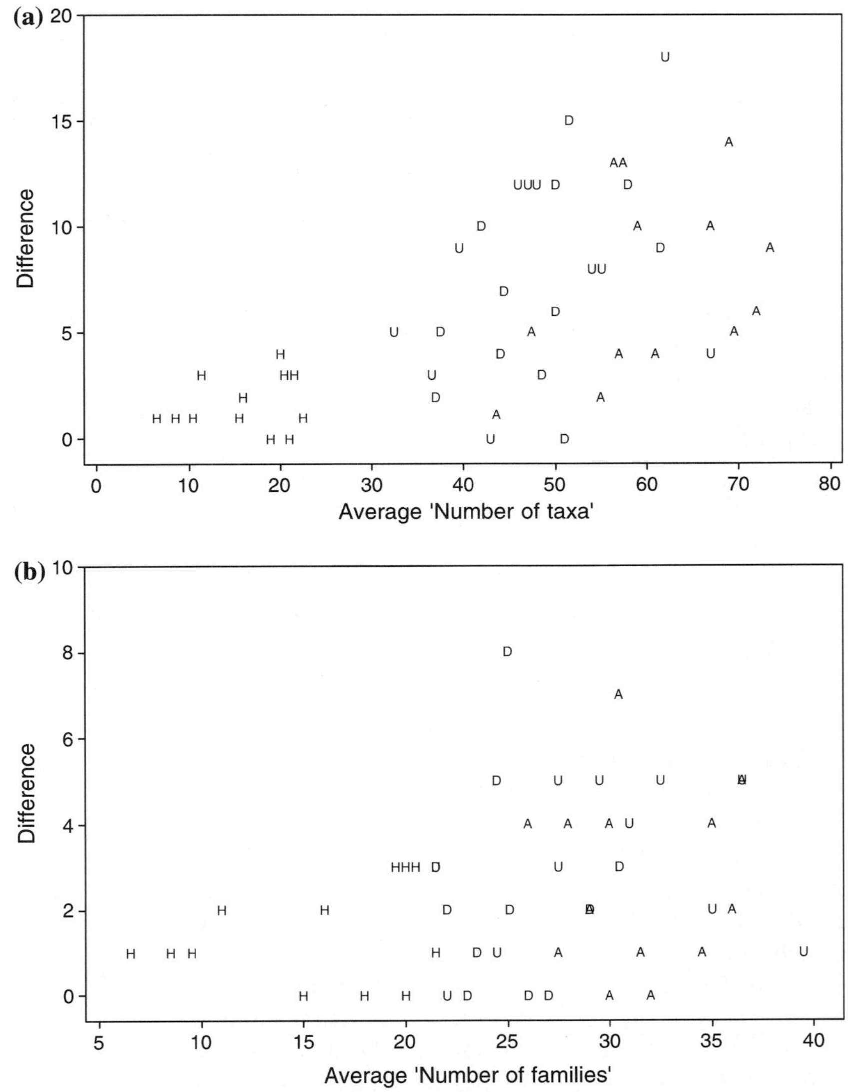

Figure 1. Difference between two replicate RIVPACS samples plotted against the average of the two values for untransformed values of the metrics (a) 'Number of taxa' and (b) 'Number of families'; letters indicate country (Austria, Germany(D), Greece(H) and UK); plot (but not analyses) excludes one outlier UK site with 28 taxa and 18 families for one replicate sample and 53 taxa and 35 families for the other.

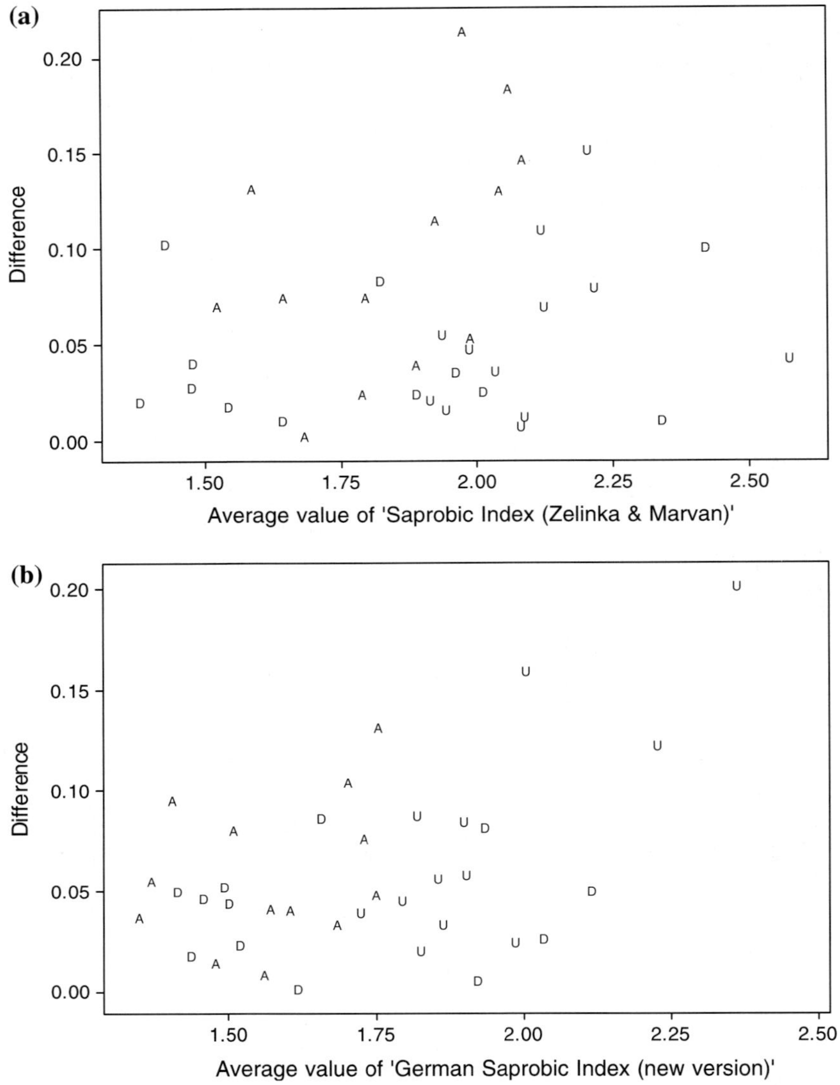

Figure 2. Difference between two replicate RIVPACS samples plotted against the average of the two values for the metrics (a) 'Saprobic Index (Zelinka & Marvan, 1961)' and (b) German Saprobic Index (new version); letters indicate country (Austria, Germany(D) and UK).

estimate variance components based on the square root transformed values for 'Number of taxa' and other taxonomic richness metrics. Before transformation, a Kruskal–Wallis analysis of the ranks of the individual site replicate sampling SD in number of taxa showed statistically significant differences between stream types (test $p=0.005$). After transformation, the systematic differences in sampling SD were greatly reduced (test $p=0.044$) but were still marginally lower for the Greek sites from stream type H04 which had much lower recorded taxonomic richness (Fig. 1b).

Differences (and SD) between replicate RIVPACS sample values of the original Saprobic Index of Zelinka and Marvan (1961) (Fig. 2a) and either the new version of the German Saprobic Index (Fig. 2b) or the Czech Saprobic index showed no major trends or statistically significant differences between stream types. (Kruskal–Wallis test $p=0.157$, 0.699 and 0.468 respectively). However,

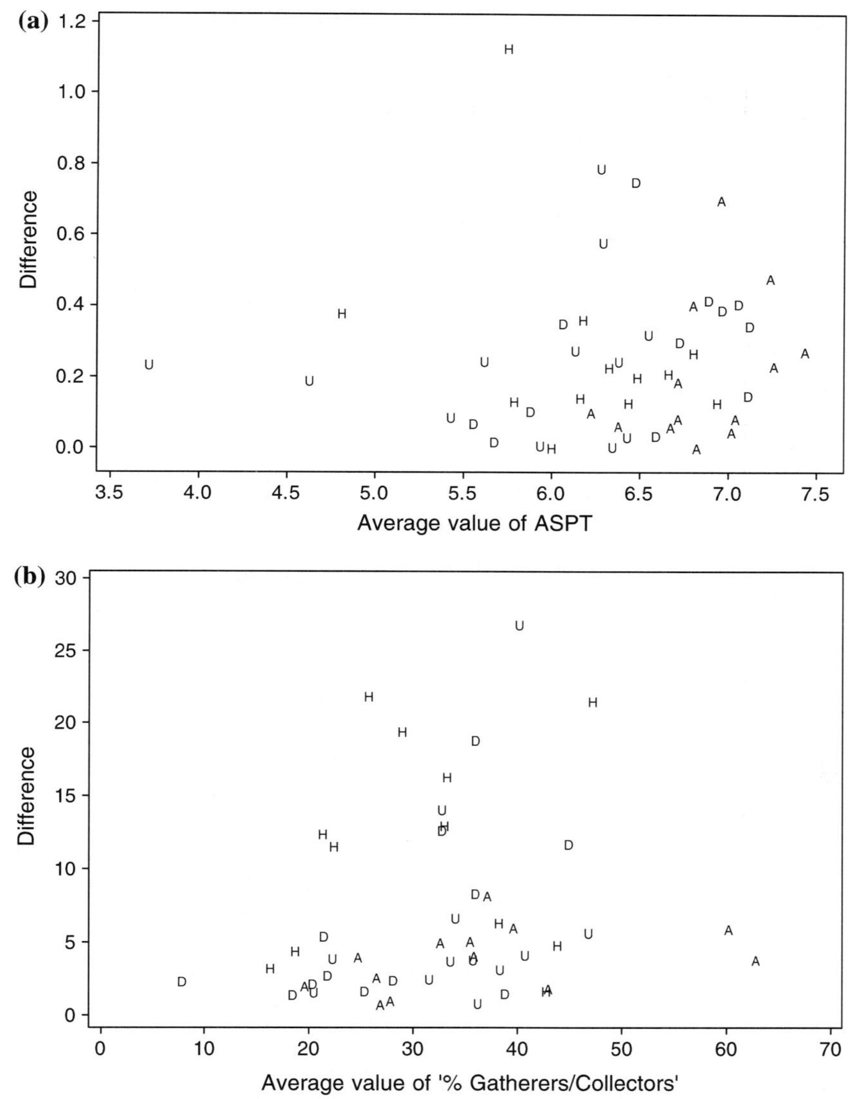

Figure 3. Difference between two replicate RIVPACS samples plotted against the average of the two values for the metrics (a) ASPT and (b) '% Gatherers/collectors'; letters indicate country (Austria, Germany (D), Greece (H) and UK).

this may be partly because replicate RIVPACS samples were only taken from a few (2–3) sites in each stream type. Although, ASPT varied considerably between sites and stream types, differences in ASPT between replicate samples show no systematic patterns with either the stream type or general level of quality (Fig. 3a). After transformation, the only other metric for which the RIVPACS sampling SD differed significantly between stream types was for '% Gatherers/Collectors' (Kruskal–Wallis test $p = 0.010$), which was most variable for replicate samples from sites in Greece (Fig. 3b).

Differences in the Shannon–Wiener diversity index (Shannon & Weaver, 1949) between replicate samples are often large relative to the actual values of the index, especially for some UK and German sites (Fig. 4), suggesting that this diversity index is highly susceptible to sampling variation and hence may be of low precision. The general precision of metrics derived from samples based on the RIVPACS method is discussed below.

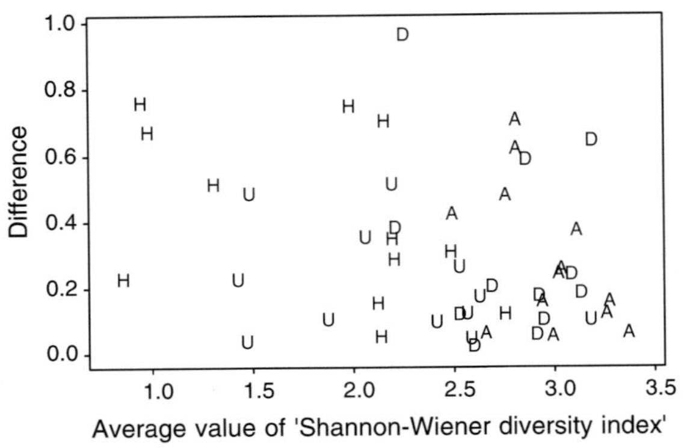

Figure 4. Difference between two replicate RIVPACS samples plotted against the average of the two values for the Shannon–Wiener diversity index; letters indicate country (Austria, Germany(D), Greece (H) and UK).

Estimates of the average sampling SD (SD_E) for the transformed (where appropriate) values of each metric were calculated separately for each stream type and country for which replicates RIVPACS samples were taken (Tables 2–5). The sampling variance expressed as a percentage (P_{samp}) of the total variance within a stream type was also estimated for each stream type separately. The corresponding estimates of SD_E and P_{samp} for replicate samples taken using the STAR-AQEM method from (mostly) the same sites as used for the replicate RIVPACS sampling were also calculated for each country (Tables 2–5). These provide direct comparisons of the relative precision of the two methods in terms of susceptibility to sampling variability in each metric.

RIVPACS method in Austria and Germany

Using the RIVPACS method to sample the streams in Austria gave high precision to estimates of most metrics with nearly two-thirds (62%) of metrics having replicate sampling variances of less than 10% of the total variance in metric values within any one stream type (Table 2). The average value of P_{samp} was only 9% and no metric had a value of P_{samp} greater than 24% – this suggests high sampling repeatability of all aspects of the macroinvertebrate community structure. For these Austrian stream types, the RIVPACS and STAR-AQEM methods were (on average) equally precise (Table 2).

The RIVPACS method gave variable levels of sampling precision of metric values for sites in the three STAR stream types (D03, D04 and D06) in Germany. Six metrics (29%), including the three Saprobic metrics, had replicate sampling variances of less than 10% of the total variance in metric values within any one stream type – indicating very high sampling repeatability and precision for these metrics (Table 3). The RIVPACS method gave even higher precision (in terms of P_{samp}) for the German Saprobic Index (new version) than the STAR-AQEM method (Table 3). In Germany, the average sampling variability in metric values for the RIVPACS and STAR-AQEM methods was similar (average values of P_{samp} of 17 and 13% respectively), although the STAR-AQEM method gave the lower value of SD_E and P_{samp} for slightly more metrics. The percentage sampling variance of ASPT for the German stream types was much higher (23–27%) than that of any of the Saprobic metrics for both the RIVPACS and STAR-AQEM methods. However, the main stress operating within the stream types sampled in Germany was degradation of stream morphology rather than organic pollution. Therefore both ASPT and the Saprobic indices, designed primarily to indicate biological impacts of organic pollution, did not vary greatly within these stream types. Within that constraint, the Saprobic species-based metrics appear to be relatively less susceptible to sampling variation.

RIVPACS method in Greece

In Greece, macroinvertebrates were identified to family level, rather than genus or species level – this is why the Saprobic metrics are excluded from analyses (Table 4). The RIVPACS method led to higher percentage sampling variance (P_{samp}) for several metrics than was found for the other countries where the RIVPACS method was used. The average value of P_{samp} for Greek sites was 28% and only two metrics ‘Number of taxa’ and ‘Number of families’ had percentage sampling variances of less than 10% (Table 4). In comparison with the STAR-AQEM method applied to the same sites in Greece, only 5 of the 19 metrics assessed had smaller sampling standard deviations (SD_E) for the RIVPACS method. However, the STAR-AQEM method had similar average levels

Table 3. Comparisons of RIVPACS (R) and STAR-AQEM (S-A) methods used in Germany for overall standard deviations (SD_E) and percentage variance (P_{samp}) due to replicate sampling for transformed ($f(x)$) values of metrics

Metric	$f(x)$	SD_E for RIVPACS within stream type			Averaged across stream types			
					SD_E		P_{samp}	
		D03	D04	D06	R	S-A	R	S-A
Abundance (ind/m^2)	$\sqrt{\sqrt{x}}$	0.486	0.502	0.334	0.447	**0.363**	17	**12**
Number of taxa	$\sqrt{x}$	0.519	0.382	0.324	0.416	**0.333**	24	**14**
Number of families	$\sqrt{x}$	0.323	0.139	0.182	0.228	**0.203**	18	**12**
Number of EPT taxa	$\sqrt{x}$	0.341	0.276	0.248	**0.291**	0.380	**12**	20
Saprobic Index	x	0.039	0.032	0.040	0.037	**0.028**	2	2
German Saprobic new	x	0.035	0.039	0.025	**0.034**	0.041	**2**	7
Czech Saprobic	x	0.090	0.087	0.031	0.075	**0.057**	4	**2**
ASPT	x	0.131	0.314	0.246	**0.242**	0.286	27	**23**
IBE	x	0.686	0.255	0.381	**0.476**	0.515	**10**	11
Diversity SW	x	0.418	0.098	0.246	0.286	**0.210**	25	13
% Rheophilic	asin	0.094	0.049	0.093	0.081	**0.072**	12	**7**
% Rheophilic (ab-class)	asin	0.027	0.054	0.020	**0.037**	0.040	7	7
% Littoral	asin	0.047	0.040	0.011	**0.036**	0.045	**4**	11
% Grazers/scrapers	asin	0.055	0.032	0.039	**0.043**	0.045	**6**	10
% Shredders	asin	0.162	0.041	0.049	0.100	**0.075**	38	**21**
% Gatherers/collectors	asin	0.093	0.045	0.018	0.061	**0.054**	14	**11**
% Oligochaeta	asin	0.045	0.079	0.040	0.057	**0.030**	55	**15**
% EPT individuals	asin	0.074	0.058	0.074	0.069	**0.060**	11	**8**
% EPT (ab-class)	asin	0.054	0.019	0.041	**0.040**	0.042	**14**	16
% EPT taxa	asin	0.046	0.045	0.049	0.047	**0.046**	**21**	24
RETI	asin[1]	0.122	0.050	0.009	**0.077**	0.090	**27**	32
Average (range)							17 (2–55)	13 (2–32)
Metrics (%) with P_{samp} <10%							6 (29%)	6 (29%)
Metrics with smaller (bold) value of SD_E or P_{samp}					9	12	8	11

of percentage sampling variance (23%) and only two (different) metrics with sampling variances less than 10% of total variance within the stream type. Thus the RIVPACS and STAR-AQEM give roughly equal moderately high sampling percentage sampling variances as applied within the STAR project to this stream type in Greece.

Values for the new proposed ICM 'Log (Sel_EPTD + 1)', based on the logarithm of the total abundance of selected Ephemeroptera, Plecoptera, Tricoptera and Diptera, were only available for sites in Greece for samples obtained using the RIVPACS method. It is encouraging that its estimated percentage sampling variance was only 8%, indicating high precision and repeatability, albeit only tested on these six sites in one stream type. The corresponding value of P_{samp} for this metric based on the STAR-AQEM method for the same sites is 27%.

RIVPACS method in the UK

In the UK, the RIVPACS method has been used by the government environment agencies to assess the biological quality of UK rivers for 15 years. However, most national surveys and monitoring programmes have concentrated on the use of the ratios of the observed to RIVPACS expected values of the metrics 'Number of BMWP families (taxa)' and the BMWP system ASPT. The STAR project is the first time such a wide range of metrics have been calculated for samples obtained using

Table 4. Comparisons of RIVPACS (R) and STAR-AQEM (S-A) methods used in Greece (stream type H04) for overall standard deviations (SD_E) and percentage variance (P_{samp}) due to replicate sampling for transformed ($f(x)$) values of metrics

Metric	$f(x)$	SD_E		P_{samp}	
		R	S-A	R	S-A
Abundance (ind/m^2)	$\sqrt{\sqrt{x}}$	0.513	**0.257**	21	**11**
Number of taxa	$\sqrt{x}$	**0.182**	0.282	**5**	11
Number of families	$\sqrt{x}$	**0.157**	0.299	**4**	13
Number of EPT taxa	$\sqrt{x}$	**0.291**	0.343	**20**	27
ASPT	x	**0.274**	0.437	**19**	36
IBE	x	0.742	**0.545**	32	**16**
Diversity SW	x	0.332	**0.272**	26	**18**
% Rheophilic	asin	0.160	**0.146**	**27**	38
% Rheophilic (ab-class)	asin	0.118	**0.096**	**25**	29
% Littoral	asin	0.065	**0.022**	15	**3**
% Grazers/scrapers	asin	0.108	**0.052**	34	**10**
% Shredders	ain	0.056	**0.053**	42	**9**
% Gatherers/collectors	asin	0.105	**0.075**	44	**30**
% Oligochaeta	asin	0.047	**0.028**	75	**50**
% EPT individuals	asin	0.116	**0.094**	17	**11**
% EPT (ab-class)	asin	0.083	0.083	40	**36**
% EPT taxa	asin	**0.089**	0.100	**51**	52
RETI	asin[1]	0.061	**0.046**	18	**12**
1-GOLD	asin[1]	0.107	**0.094**	**16**	17
Average (range)				28 (4–75)	23 (3–52)
Metrics (%) with $P_{samp} < 10\%$				2 (10%)	2 (10%)
Metrics with smaller (bold) value of SD_E or P_{samp}		5	14	8	11

the RIVPACS method. It is therefore encouraging that, of the 26 metrics assessed, 15 (58%) had sampling variances which formed less than 10% of the total variance in each metric's values within any one stream type in the UK (Table 5). ASPT, the three Saprobic indices and most of the metrics based on the percentage abundance of selected taxa all had very low sampling variability and hence high precision. However, the two taxonomic richness metrics 'Number of taxa' and 'Number of families' both had much higher sampling variances with estimated values of P_{samp} of 32 and 36% respectively. This low precision and repeatability of these two richness metrics was not found for the RIVPACS method for either the Austrian or Greek streams and merits further investigation. Values of the UK LIFE index (Extence et al., 1999), designed to measure flow-related biological stress, were calculated for the UK STAR samples and provided the first estimates of the susceptibility of LIFE (family level) to sampling variation (Table 5).

In the UK, when compared with the STAR-AQEM method used at the same sites, the RIVPACS method tended to give both lower sampling SD (SD_E) and lower percentage sampling variance for the majority of metrics. This suggests that the STAR-AQEM method would not offer any improvement in precision for stream assessment in the UK over the current RIVPACS method.

Over the four countries where the RIVPACS method was used, it gave slightly lower sampling precision (i.e., higher P_{samp}) than the STAR-AQEM method for more than half of the metrics. However, the differences between the two methods in estimated values of both SD_E and P_{samp} is small for many metrics and probably within the

Table 5. Comparisons of RIVPACS (R) and STAR-AQEM (S-A) methods used in the UK for overall standard deviations (SD_E) and percentage variance (P_{samp}) due to replicate sampling for transformed ($f(x)$) values of metrics

Metric	$f(x)$	SD_E for RIVPACS within stream type		Averaged across stream types			
				SD_E		P_{samp}	
		U15	U23	R	S-A	R	S-A
Abundance (ind/m^2)	$\sqrt{\sqrt{x}}$	0.353	0.812	0.626	**0.535**	15	**11**
Number of taxa	$\sqrt{x}$	0.702	0.564	0.637	**0.542**	32	**20**
Number of families	$\sqrt{x}$	0.509	0.267	0.406	**0.346**	36	**16**
Number of EPT taxa	$\sqrt{x}$	0.288	0.326	0.308	**0.254**	6	**4**
Saprobic Index	x	0.061	0.029	**0.048**	0.065	**3**	5
German Saprobic new	x	0.083	0.044	**0.067**	0.074	**6**	7
Czech Saprobic	x	0.106	0.074	**0.091**	0.206	**3**	21
ASPT	x	0.114	0.314	**0.236**	0.320	**5**	9
IBE	x	1.115	0.854	0.993	**0.699**	19	**11**
Diversity SW	x	0.231	0.108	**0.180**	0.223	**10**	15
% Rheophilic	asin	0.046	0.064	**0.056**	0.190	**2**	41
% Rheophilic (ab-class)	asin	0.049	0.040	**0.045**	0.053	**7**	14
% Littoral	asin	0.016	0.039	**0.030**	0.093	**3**	38
% Grazers/scrapers	asin	0.026	0.040	**0.034**	0.085	**9**	57
% Shredders	asin	0.035	0.027	**0.031**	0.071	**3**	18
% Gatherers/collectors	asin	0.094	0.030	**0.070**	0.101	**23**	53
% Oligochaeta	asin	0.134	0.084	0.112	**0.087**	14	**8**
% EPT individuals	asin	0.030	0.078	**0.059**	0.126	**5**	33
% EPT (ab-class)	asin	0.036	0.049	**0.043**	0.046	**7**	8
% EPT taxa	asin	0.041	0.051	0.046	**0.039**	9	**6**
RETI	asin[1]	0.050	0.031	**0.041**	0.130	**6**	65
Trait m1: max size ≤ 1 cm	asin[1]	0.021	0.025	0.023	**0.022**	14	**9**
Trait m2: > 1 cycle	asin[1]	0.040	0.020	0.031	**0.030**	12	**10**
Trait m7: crawler locomotion	asin[1]	0.032	0.026	0.029	**0.019**	26	**8**
Trait m12: current < 25 cm/s	asin[1]	0.015	0.013	**0.014**	0.020	**4**	12
LIFE	x	0.260	0.241	0.251	**0.226**	11	**8**
Average (range)						11 (2–36)	20 (4–65)
Metrics (%) with $P_{samp} < 10\%$						15 (58%)	10 (38%)
Metrics with smaller (bold) value of SD_E or P_{samp}				15	11	15	11

estimation error of this dataset. The three Saprobic metrics have values of P_{samp} less than 10% for each of the eight stream types sampled using the RIVPACS method. This suggests that these metrics are highly robust to sampling effects using the RIVPACS method. The high sampling precision for these saprobic metrics based on RIVPACS samples is generally about the same as that obtained when the STAR-AQEM method is used at the same sites and stream types (Tables 2–5). The general level of percentage sampling variance for metric values obtained using the RIVPACS method tends to be highest for the Greek sites in stream type H04, for which macroinvertebrates were only identified to family rather than species or genus level.

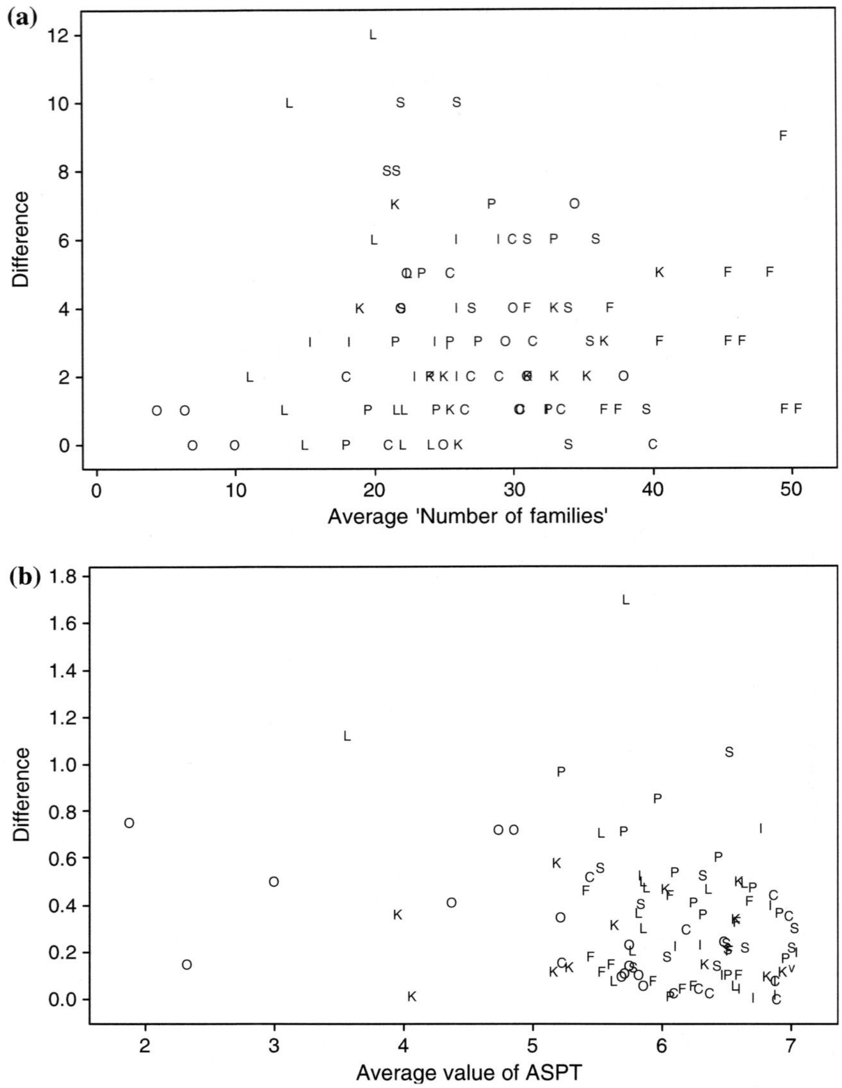

Figure 5. Difference between two replicate 'national' samples plotted against the average of the two values for the metrics (a) 'Number of families' and (b) 'ASPT; letters indicate country (Czech Republic, France, Italy, Denmark, Latvia, Poland, Portugal, Sweden).

Sampling variability in metric values for other 'national' methods

For most of the stream types where a 'national' sampling method was used, replicate STAR-AQEM samples were taken at the same time at all, or nearly all, of the same selected subset of sites at which replicate 'national' method samples were taken (Table 1).

The difference in (untransformed) metric values between two replicate samples taken by the various 'national' methods was plotted against the average of the two values for a range of metrics (Figs. 5–7). All of the 'national' methods are shown together on the same plot for compactness, but differences in replicate sampling variability and sampling SD (SD_E) should be interpreted with caution as the different methods may collect and identify varying amounts of macroinvertebrates. For example, methods recording larger numbers of taxa (or families) at a site are likely to record greater inter-sample variation in number of taxa (or families).

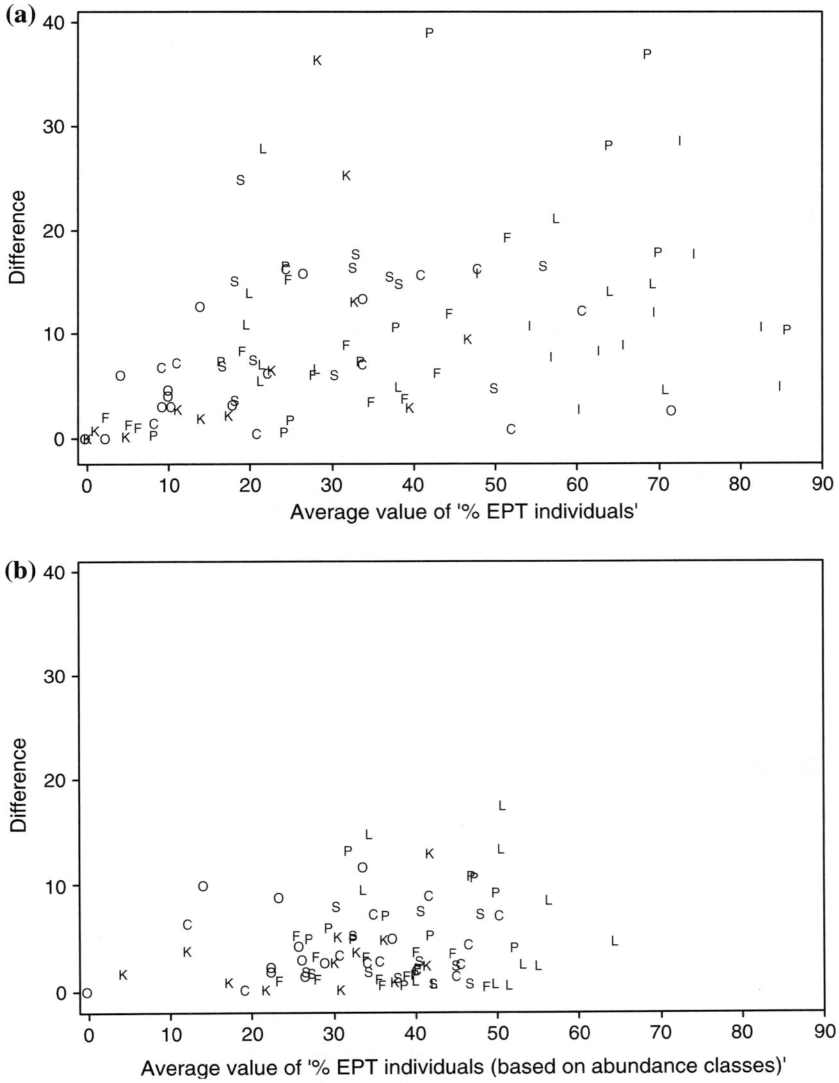

Figure 6. Difference between two replicate 'national' method samples plotted against the average of the two values for untransformed values of metrics (a) '% EPT individuals' and (b) '% EPT individuals (based on abundance classes)' – not available for Italian sites; letters indicate country (Czech Republic, France, Italy, Denmark, Latvia, Poland, Portugal, Sweden); (a) excludes one outlier Latvian site with replicate values of 8 and 64.

The sampling variance expressed as a percentage (P_{samp}) of the total variance in the metric's values within a stream type is more appropriate for comparing the sampling precision of methods.

Estimates of the replicate sampling SD (SD_E) and percentage sampling variance (P_{samp}) for (transformed) metric values within a stream type are given for each 'national' method in Tables 6–9. For each country, there is a direct comparison of these estimates for the 'national' method with those obtained for samples taken and processed using the STAR-AQEM method from all (or almost all) of the same sites at the same time.

The difference between replicates in the 'Number of families' recorded was less than four in the majority of sites for all 'national' methods, except for the Swedish method (Fig. 5a). For the Swedish 'national' method, the difference in recorded 'Number of families' was between 4 and 10 for many sites, leading to the highest replicate

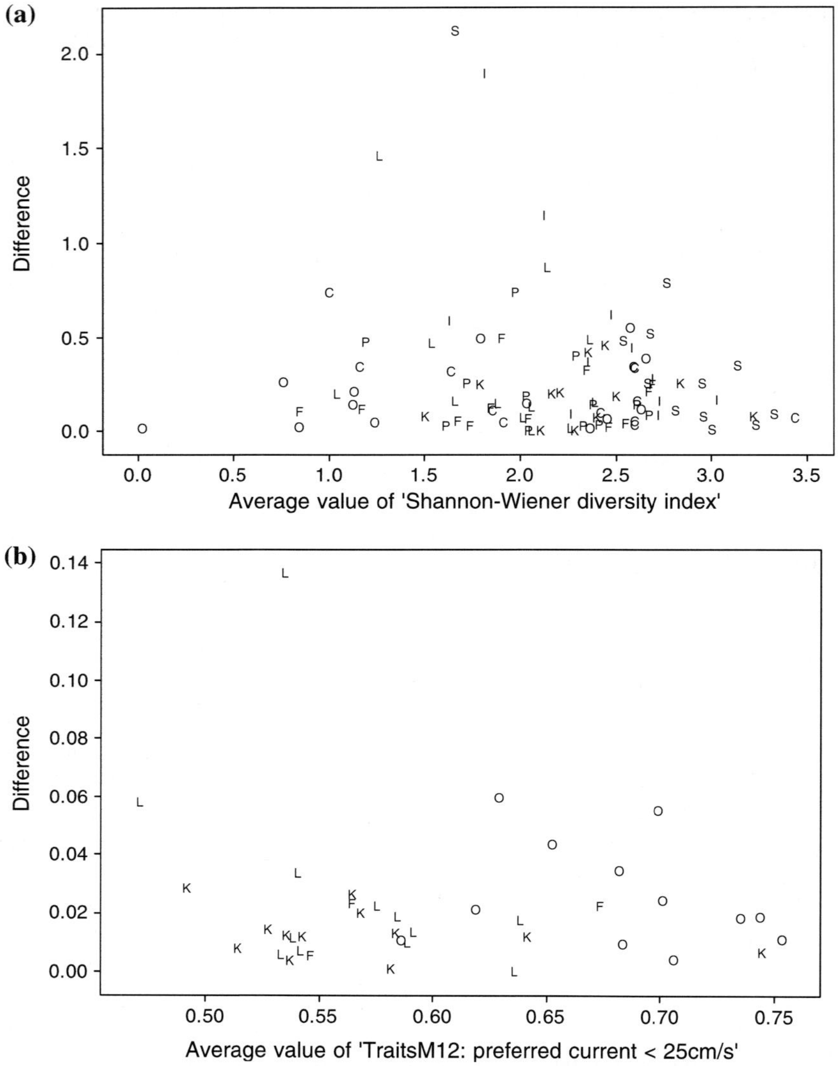

Figure 7. Difference between two replicate 'national' method samples plotted against the average of the two values for untransformed values of metrics (a) 'Shannon–Wiener diversity index' and (b) 'Trait m12: preferred current <25 cm/s'; letters indicate country (Czech Republic, France, Italy, Denmark, Latvia, Poland, Portugal, Sweden).

sampling variance with SD_E for untransformed 'Number of families' of 11.94 and 6.47 for stream types S05 and S06 respectively. Even after transformation to the square root scale, the sampling SD in 'Number of families' is still higher than that for any other stream type and 'national' method (Tables 6–9) except for stream type U15 in the UK when sampled by the RIVPACS method (Table 5). Sampling variability in 'Number of families' was also very high for two sites sampled using the Latvian 'national' method, with replicate values of 9 and 19 for one site and 14 and 26 for the other (Fig. 5a).

Figure 6 shows the distribution of differences between replicate samples taken using the 'national' methods in each of two metrics measuring the percentage of all individuals at the site which are EPT taxa (Emphemeroptera + Plecoptera + Trichoptera). It is clear that sampling variability is usually less when the metric is based on abundance classes than when based on raw abundances (Fig. 6a and b respectively). A similar result was found for the two

Table 6. Estimates of the average replicate sampling SD (SD_E) and percentage sampling variance (P_{samp}) in transformed ($f(x)$) metric values within stream types for (a) Czech 'national' method (PERLA) and (b) Danish 'national' method (DFSI) and comparisons with the STAR-AQEM (S-A) method

Metric	$f(x)$	(a) Czech Republic				(b) Denmark			
		SD_E		P_{samp}		SD_E		P_{samp}	
		Nat	S-A	Nat	S-A	Nat	S-A	Nat	S-A
Abundance (ind/m^2)	$\sqrt{\sqrt{x}}$	**0.594**	0.600	**8**	18	**0.411**	0.673	**18**	60
Number of taxa	$\sqrt{x}$	**0.284**	0.342	**5**	6	**0.252**	0.357	**12**	20
Number of families	$\sqrt{x}$	**0.187**	0.274	**5**	16	**0.233**	0.300	**17**	22
Number of EPT taxa	$\sqrt{x}$	**0.198**	0.281	**3**	6	**0.214**	0.377	**5**	13
Saprobic Index	x	0.056	**0.048**	3	**2**	0.043	**0.030**	3	**1**
German Saprobic new	x	**0.066**	0.079	**3**	5	0.059	**0.041**	4	**2**
Czech Saprobic	x	**0.125**	0.132	3	3	0.116	**0.102**	9	**6**
ASPT	x	**0.181**	0.271	**3**	8	**0.226**	0.293	**5**	9
Diversity SW	x	0.210	**0.191**	8	**6**	**0.164**	0.362	**13**	41
% Rheophilic	asin	0.086	**0.080**	**6**	7	**0.080**	0.143	**4**	13
% Rheophilic (ab-class)	asin	0.053	**0.041**	8	8	**0.045**	0.057	**6**	7
% Littoral	asin	0.042	**0.039**	5	**4**	**0.049**	0.085	**10**	25
% Grazers/scrapers	asin	**0.022**	0.035	**2**	8	**0.054**	0.077	**16**	23
% Shredders	asin	0.052	**0.048**	4	4	**0.053**	0.059	18	18
% Gatherers/collectors	asin	**0.034**	0.054	**3**	7	**0.096**	0.109	30	**27**
% Oligochaeta	asin	**0.032**	0.074	**1**	5	**0.132**	0.158	**33**	48
% EPT individuals	asin	0.078	**0.069**	**10**	12	**0.112**	0.126	22	**19**
% EPT (ab-class)	asin	**0.038**	0.043	**5**	7	**0.036**	0.055	**7**	13
% EPT taxa	asin	**0.040**	0.059	**7**	17	**0.025**	0.067	**4**	21
RETI	asin	**0.046**	0.049	**4**	5	**0.072**	0.104	**18**	24
1-GOLD	asin	0.092	**0.078**	7	7				
Trait m1: max size ≤ 1 cm						**0.018**	0.037	**9**	41
Trait m2: > 1 cycle						**0.031**	0.038	**9**	12
Trait m7: crawler loco.						**0.024**	0.035	**10**	21
Trait m12: current < 25 cm/s						**0.011**	0.024	**6**	11
Average (range)				5 (1–10)	8 (2–18)			13 (3–39)	22 (1–60)
Metrics (%) with $P_{samp} < 10\%$				20 (95%)	17 (81%)			12 (50%)	5 (21%)
Metrics with smaller (bold) value of SD_E or P_{samp}		13	8	14	3	21	3	18	5

metrics '% Rheophilic' and '% Rheophilic (ab-class)' measuring the percentage of all individuals which prefer fast-flowing water. Figure 6 also highlights the general tendency for replicate sampling variability to be less for percentage abundance metrics when the values are very low (near zero) or very high (>80%), supporting the decision to analyse this type of metric on the arcsine transformed scale to reduce the systematic heterogeneity in the sampling variance within a stream type.

Most sites have replicate differences in 'ASPT' less than 0.6, but the observed range of ASPT values is limited for most stream types (Fig. 5b). There was some statistical evidence of differences between 'national' methods (and/or stream types) in the size of the replicate sampling variability of ASPT (Kruskal–Wallis test $p = 0.038$), with greatest sampling SD (SD_E) occurring for sites sampled using the Portuguese 'national' method (Table 7b).

Czech 'national' (PERLA) method

The Czech PERLA method has a very high sampling precision within the two sampled stream types (C05 and C06) with values of P_{samp} of 10% or less for all of the metrics analysed (Table 6a). The precision is usually roughly the same as for the STAR-AQEM method; where they differ the PERLA method appears to be more precise. For example, the value of P_{samp} for 'Number of families' is estimated to be 16% for the STAR-AQEM method, but only 5% for the PERLA method. Overall, the Czech 'national' method gives very low variability in metric values between replicate samples and hence high sampling precision, mostly as good or better than that for the STAR-AQEM method.

Danish 'national' (DFSI) method

The sampling SD (SD_E) for the Danish 'national' method was less than that based on STAR-AQEM method at the same sites and time for 21 of the 24 metrics analysed (Table 6b).

More impressively, 75% (18) of the metrics had lower percentage sampling variances within the studied stream type (K02) for the Danish 'national' method than for the STAR-AQEM method. Half of the 24 metrics assessed had replicate sampling variances which formed less than 10% of the total variance in the metric values amongst all sites within the stream type. Only three metrics measuring percentage abundance of selected taxa (% Gatherers/Collectors, % Oligochaeta and % EPT individuals) had values of P_{samp} greater than 20%. Thus the Danish (DFSI) method seems to lead, in most cases, to metric values with low sampling variances and thus high sampling precision and repeatability.

The Danish Stream Fauna Index (DSFI), although used nationally in stream bioassessments within Denmark, was not included amongst the analysed metrics partly because it only takes seven possible integer values (1–7), so its sampling variation needs to be summarised in a different way. In addition, the current DSFI metric might be revised to conform directly to five class system of the WFD (Friberg *pers comm.*).

French 'national' (IBGN) method

The French 'national' (IBGN) method gives high sampling precision (i.e., $P_{samp} < 10\%$) for just over half (57%) of the metrics assessed and especially for those metrics based on percentage abundance of taxa with selected characteristics (Table 7a). Sampling SDs were less using the 'national' method than using the STAR-AQEM method for 20 of the 21 metrics assessed; and this was also true to a lesser extent when expressed in terms of percentage (P_{samp}) of total variance within stream type (F08). However, it should be noted that within STAR, samples taken using the apparently more precise IBGN method were identified to family level for most groups, whereas samples taken from the same French sites using the STAR-AQEM method were identified to a more detailed level.

Portuguese 'national' (PMP) method

Using the Portuguese 'national' method gave replicate sampling variance in metric values which on average contributed 22% (range 6–50%) of the total variance in metric values amongst all sampled sites within the studied stream type (P04) (Table 7b). When compared with the STAR-AQEM method used at the same time at the same sites, the 'national' (PMP) method gave smaller estimates of sampling SD (SD_E) for 75% of the 20 metrics assessed. However, once expressed as percentage sampling variance (P_{samp}), the Portuguese 'national' method appears to give about the same average level of sampling variance and precision as the STAR-AQEM method (Table 7b). The new metric '1-GOLD' had an estimated percentage sampling variance of 19% when based on samples obtained using the Portuguese method. When based on the STAR-AQEM method at the same sites, the value of P_{samp} for '1-GOLD' was only 9% (the second lowest value amongst all metrics).

Italian 'national' (IBE) method

The Italian *Indice Biotico Esteso* (IBE) 'national' method of sampling and subsequent sorting and processing of samples appears to lead to relatively

Table 7. Estimates of the average replicate sampling SD (SD_E) and percentage sampling variance (P_{samp}) in transformed ($f(x)$) metric values within stream types for (a) French 'national' method (IBGN) and (b) Portuguese 'national' method and comparisons with the STAR-AQEM (S-A) method

Metric	$f(x)$	(a) France				(b) Portugal			
		SD_E		P_{samp}		SD_E		P_{samp}	
		Nat	S-A	Nat	S-A	Nat	S-A	Nat	S-A
Abundance (ind/m^2)	$\sqrt{\sqrt{x}}$	**0.983**	1.185	**34**	55	**0.566**	0.711	31	**22**
Number of taxa	$\sqrt{x}$	**0.201**	0.321	18	**9**	**0.295**	0.358	**15**	31
Number of families	$\sqrt{x}$	**0.213**	0.314	19	**18**	**0.237**	0.331	**15**	35
Number of EPT taxa	$\sqrt{x}$	**0.164**	0.358	**7**	15	**0.305**	0.346	**11**	28
Saprobic Index	x					**0.078**	0.091	18	**17**
German Saprobic new	x								
Czech Saprobic	x								
ASPT	x	**0.186**	0.265	**14**	15	**0.384**	0.534	**29**	43
IBE						1.185	**0.618**	49	**18**
Diversity SW	x	**0.146**	0.168	**6**	9	**0.213**	0.354	**13**	35
% Rheophilic	asin	**0.060**	0.109	**5**	13	0.192	**0.176**	30	**21**
% Rheophilic (ab-class)	asin	**0.029**	0.065	**8**	9	**0.067**	0.068	37	**15**
% Littoral	asin	**0.030**	0.090	**5**	40	**0.086**	0.089	50	**36**
% Grazers/scrapers	asin	**0.029**	0.041	5	**3**	**0.051**	0.067	**11**	21
% Shredders	asin	**0.032**	0.043	**6**	10	**0.031**	0.035	**6**	7
% Gatherers/collectors	asin	**0.058**	0.079	10	**8**	**0.086**	0.104	**15**	19
% Oligochaeta	asin	**0.078**	0.112	**17**	40	**0.085**	0.121	**22**	38
% EPT individuals	asin	**0.011**	0.098	**10**	25	0.152	**0.094**	22	**9**
% EPT (ab-class)	asin	**0.020**	0.047	**5**	25	**0.055**	0.060	**11**	15
% EPT taxa	asin	**0.025**	0.064	**8**	44	**0.053**	0.068	**12**	26
RETI	asin	**0.049**	0.053	5	**3**	0.078	**0.075**	23	**20**
1-GOLD	asin					0.144	**0.093**	19	**9**
Trait m1: max size ≤ 1 cm		**0.011**	0.048	**11**	64				
Trait m2: > 1 cycle		0.020	**0.019**	**8**	10				
Trait m7: crawler loco.		**0.010**	0.022	**25**	31				
Trait m12: current < 25 cm/s		**0.014**	0.031	**5**	19				
Average (range)				11 (5–34)	22 (3–64)			22 (6–50)	23 (7–43)
Metrics (%) with P_{samp} < 10%				12 (57%)	6 (29%)			1 (5%)	3 (15%)
Metrics with smaller (bold) value of SD_E or P_{samp}		20	1	16	5	15	5	11	9

high variability in metric values between replicate samples, at least relative to the total variance in metric values amongst the sampled sites within the stream type I06 for which data were available (Table 8a). The replicate sampling variance was not less than 10% of the total variance within the stream type for any of the 13 metrics analysed. Fewer metrics were assessed because all Italian macroinvertebrate samples (both IBE and STAR-AQEM) were only identified to family level, and some metrics required lower level identification.

However, it is very interesting to note that on average, the Italian IBE method appeared to give as repeatable results and the same replicate sampling precision as the STAR-AQEM sampling method in terms of both the sampling SD (SD_E) and the percentage sampling variance (P_{samp}) of metrics (Table 8a). Both methods, at least as

Table 8. Estimates of the average replicate sampling SD (SD_E) and percentage sampling variance (P_{samp}) in transformed ($f(x)$) metric values within stream types for (a) Italian 'national' method (IBE) and (b) Lavian 'national' method and comparisons with the STAR-AQEM (S-A) method

Metric	$f(x)$	(a) Italy				(b) Latvia			
		SD_E		P_{samp}		SD_E		P_{samp}	
		Nat	S-A	Nat	S-A	Nat	S-A	Nat	S-A
Abundance (ind/m^2)	$\sqrt{\sqrt{x}}$	1.311	**0.995**	73	**48**	0.657	**0.676**	31	**30**
Number of taxa	$\sqrt{x}$	**0.258**	0.496	**21**	66	0.475	**0.185**	57	**13**
Number of families	$\sqrt{x}$	**0.245**	0.478	**25**	63	0.436	**0.175**	55	**12**
Number of EPT taxa	$\sqrt{x}$	**0.211**	0.430	**12**	77	0.401	**0.191**	70	**16**
Saprobic Index						**0.134**	0.215	**14**	46
ASPT	x	0.263	**0.211**	38	**18**	0.494	**0.375**	43	**31**
IBE	x	**0.831**	0.897	**37**	47				
Diversity SW	x	0.520	**0.236**	56	**36**	0.380	**0.230**	56	**13**
% Rheophilic						0.141	**0.092**	32	**11**
% Rheophilic (abund classes)						0.070	**0.042**	43	**14**
% Littoral						0.098	**0.079**	20	**14**
% Grazers/scrapers	asin	0.100	**0.063**	49	**24**	0.050	**0.042**	**19**	34
% Shredders						0.066	**0.038**	46	**16**
% Gatherers/collectors						0.064	**0.036**	39	**21**
% Oligochaeta	asin	**0.052**	0.057	**14**	36	0.096	**0.063**	30	**16**
% EPT individuals	asin	0.110	**0.095**	**25**	31	0.173	**0.086**	47	**13**
% EPT (abund. classes)						0.063	**0.036**	37	**9**
% EPT taxa	asin	**0.038**	0.049	**18**	20	0.062	**0.034**	54	**10**
RETI	asin	0.115	**0.039**	67	**10**	0.057	**0.056**	**19**	39
1-GOLD	asin	0.098	**0.076**	22	**20**				
Trait m1: max size ≤ 1 cm						0.050	**0.031**	61	**28**
Trait m2: > 1 cycle						0.049	**0.028**	48	**33**
Trait m7: crawler loco						0.046	**0.016**	98	**18**
Trait m12: current < 25 cm/s						0.032	**0.019**	47	**15**
Average (range)				35 (12–73)	38 (10–77)			43 (14–98)	21 (9–96)
Metrics (%) with P_{samp} < 10%				0	0			0 (0%)	1 (5%)
Metrics with smaller (bold) value of SD_E or P_{samp}		6	7	7	6	1	21	3	19

carried out in Italy for the Italian sites within the STAR project, give amongst the highest overall percentage sampling variances and lowest precisions of all countries, stream types and methods.

Latvian 'national' (LVS 240:1999) method

When using the Latvian 'national' method (LVS 240:1999), all of the 22 metrics assessed had replicate sampling variances which contributed more than 10% of the total variance in metrics values amongst all sites within the sampled stream type L02 (Table 8b). The average percentage sampling variance (P_{samp}) was 43% – this was the highest average amongst all countries, stream types and methods. Replicate samples were taken using the STAR-AQEM standard method at all of the Latvian sites at which replicate samples were taken by the 'national' method. Although the number of families recorded in 'national' method samples was on average only slightly less than that recorded in STAR-AQEM samples, the 'national' method gave higher values of sampling SD (SD_E) for all except one metric and higher values of P_{samp} for all

Table 9. Estimates of the average replicate sampling SD (SD_E) and percentage sampling variance (P_{samp}) in transformed ($f(x)$) metric values within stream types for (a) Polish 'national' method and (b) Swedish 'national' method and comparisons with the STAR-AQEM (S-A) method

Metric	$f(x)$	(a) Poland				(b) Sweden			
		SD_E		P_{samp}		SD_E		P_{samp}	
		Nat	S-A	Nat	S-A	Nat	S-A	Nat	S-A
Abundance (ind/m²)	$\sqrt{\sqrt{x}}$	1.202	**0.984**	**20**	23	**0.692**	1.028	**21**	30
Number of taxa	$\sqrt{x}$	**0.312**	0.317	4	**3**	0.716	**0.415**	60	**32**
Number of families	$\sqrt{x}$	0.214	**0.172**	3	**2**	0.427	**0.296**	39	**30**
Number of EPT taxa	$\sqrt{x}$	0.332	**0.185**	4	**1**	0.445	**0.253**	26	**10**
Saprobic Index	x	0.103	**0.102**	**3**	5	0.071	**0.035**	10	**3**
ASPT	x	0.296	**0.247**	6	**3**	0.304	**0.232**	19	**8**
Diversity SW	x	**0.184**	0.240	**6**	7	0.495	**0.218**	93	**22**
% Rheophilic	asin	**0.069**	0.313	**11**	74	0.171	**0.086**	49	**14**
% Rheophilic (ab-class)	asin	**0.073**	0.311	**8**	94	0.067	**0.046**	24	**14**
% Littoral	asin	**0.038**	0.084	**7**	44	0.069	**0.047**	21	**9**
% Grazers/scrapers	asin	**0.026**	0.070	**4**	41	0.072	**0.054**	28	**18**
% Shredders	asin	0.047	**0.042**	9	**7**	0.045	**0.041**	26	**16**
% Gatherers/collectors	asin	**0.060**	0.102	**4**	16	0.112	**0.044**	64	**12**
% Oligochaeta	asin	**0.119**	0.248	**7**	56	0.067	**0.036**	21	**10**
% EPT individuals	asin	0.070	**0.049**	10	**4**	**0.116**	0.124	**30**	32
% EPT (ab-class)	asin	0.046	**0.031**	5	**2**	**0.032**	0.036	**6**	7
% EPT taxa	asin	0.056	**0.032**	6	**2**	**0.026**	0.032	**5**	6
RETI	asin	**0.068**	0.081	**12**	20	0.095	**0.055**	**7**	13
Trait m1: max size ≤ 1 cm	asin	**0.075**	0.086	**63**	99				
Trait m2: > 1 cycle	asin	**0.054**	0.102	**21**	73				
Trait m7: crawler loco	asin	0.083	0.083	**38**	72				
Trait m12: current < 25 cm/s	asin	0.024	**0.020**	21	**14**				
Average (range)				12 (3–63)	30 (1–99)			30 (5–93)	16 (3–32)
Metrics (%) with $P_{samp} < 10\%$				15 (68%)	10 (45%)			3 (17%)	5 (28%)
Metrics with smaller (bold) value of SD_E or P_{samp}		12	10	11	9	4	14	5	13

except three of the 22 metrics analysed (Table 8b). Overall, the current Latvian 'national' method gives metrics values which are highly susceptible to sampling variation and hence of low precision – so it may benefit from refinement.

Polish 'national' method

The Polish 'national' method for macroinvertebrate sampling appears to give replicate sampling SD for metric values of about the same absolute size as for many other 'national' methods (Tables 6–9). However, once expressed as a percentage of the total variances in metrics values across all sites within the sampled stream type (O02), the Polish 'national' method appears to have high relative precision with values of P_{samp} of less than 10% for 15 of the 20 metrics assessed (Table 9a). The four trait metrics had the highest percentage sampling variances with values of P_{samp} ranging from 21 to 63%.The Polish 'national' method gave dramatically higher estimates precision than the STAR-AQEM method for some of the metrics based on selective percentage abundances (e.g., % Rheophilic, % Littoral, % Grazers/Scrapers).

Swedish 'national' method

The replicate sampling SD, averaged over the two sampled stream types (S05 and S06) was higher using the Swedish 'national' method than using the STAR-AQEM method for all except four of the 18 metrics assessed (Table 9b). However, especially for some richness-related metrics, this could be partly because the Swedish 'national' method tends to record more taxa (average = 48 and 53 taxa per sample in stream types S05 and S06) than the STAR-AQEM method (average = 43 and 44 respectively). After sampling variance is standardised as a percentage of total variance for that method within each stream type, the Swedish 'national' method still appears to be less precise than the STAR-AQEM method for the majority of metrics. Only five of the 18 metrics had percentage sampling variances less than 20% using the 'national' method and the average value was 30%. This compares with an average value of P_{samp} of 16% when metric values are based on samples obtained using the STAR-AQEM method at the same time at the same sites.

Discussion

This study has provided the first ever quantitative comparison of the susceptibility of each of a wide range of metrics and established 'national' macroinvertebrate sampling methods to uncertainty resulting from the combined effects of field sampling variability and subsequent sub-sampling and laboratory (or bank-side) procedures and protocols.

Comparison of precision of methods

It is obviously difficult to compare the relative precision of different methods when they are sampled in different stream types and countries. However, within any one STAR stream type, samples have been obtained at the same sites using both the 'national' or RIVPACS method and the STAR-AQEM method. The precision of a given method relative to the STAR-AQEM method can therefore be compared in terms of their sampling variance expressed as a percentage (P_{samp}) of the total variance in a metric's values within a stream type (Table 10). Most 'national' methods had sampling precisions at least as good as those for the STAR-AQEM method (as used in their country at the same sites); the main exceptions being Latvia and Sweden (Fig. 8). Perhaps more importantly, the national methods used in the Czech Republic, Denmark, France, Poland and the RIVPACS method used in the UK and Austria all had percentage sampling variances of less than 10% for the majority of metrics assessed. In contrast, none of the metrics had percentage sampling variances less than 10% when based on the Italian (IBE) or Latvian 'national' method and less than one-fifth of metrics when based on the Greek, Portuguese or Swedish 'national' methods (Table 10).

As used in Italy, both the 'national' (IBE) method and the STAR-AQEM method gave amongst the highest percentage sampling variances (averages 37 and 38%) of any stream type and method, although the analyses were based on only those metrics considered to be valid for community data based on identification to family level. Part of this explanation may be that the sites sampled in Italy may have covered a narrower

Table 10. Overall precision ($P_{samp(N)}$) of the 'national' sampling method (or RIVPACS) for each country; $P_{samp(N)}$ and $P_{samp(STAR\text{-}AQEM)}$ denote the percentage sampling variance P_{samp} for the 'national'/RIVPACS method and STAR-AQEM method respectively

Country	% of metrics with $P_{samp(N)}$ < 10% (%)	% of metrics with $P_{samp(N)} \leq P_{samp(STAR\text{-}AQEM)}$ (%)
Austria – RIVPACS	65	48
Germany – RIVPACS	29	48
Greece – RIVPACS	14	45
UK – RIVPACS	58	58
Czech Republic (PERLA)	95	86
Denmark (DSFI)	50	75
France (IBGN)	57	76
Italy (IBE)	0	54
Latvia (LVS 240:1999)	0	14
Poland	68	60
Portugal (PMP)	5	55
Sweden	17	28

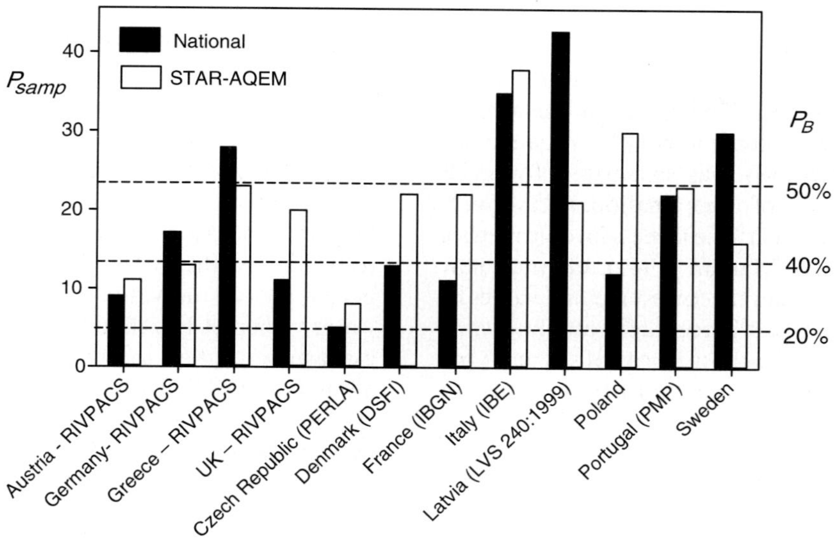

Figure 8. Average percentage sampling variance (P_{samp}) over all metrics for the 'national' and STAR-AQEM methods for each country. Dashed lines indicate the levels of P_{samp} which may correspond to average mis-classification rates (P_B) for sites of good, moderate or poor status class of approximately 20, 40 and 50%. (Based on the assumption that the STAR sites within a stream type uniformly cover the full range (0–1) of EQR values for a metric and assuming the EQR class limits are set at 0.2,0.4,0.6 and 0.8 – see Clarke et al. (2006) for further details).

range of ecological qualities so that total variability amongst the sampled sites within the stream type is relatively low, and thus replicate sampling variance is a relatively large percentage of the total variance. Moreover, in Italy, all samples are sorted on the bank-side (rather than in the laboratory) and this may cause extra variability between replicate samples in their recorded faunal composition.

The Latvian 'national' method was the most prone to sampling variability amongst all of the national methods and led to higher values of P_{samp} than the STAR-AQEM method for almost all metrics based on the same set of sampled sites. This may be because most samples were sorted as live material on the stream bank and only a limited list of taxa are identified using the 'national' method – precision might be improved by sorting in the laboratory and by identifying a larger range of taxa within each sample. The greatest general sampling precision of all methods and stream types was obtained for the stream types (C04 and C05) sampled in the Czech Republic using both the 'national' PERLA method and the STAR-AQEM method, for which average values of P_{samp} were only 5 and 8% respectively.

Overall, the RIVPACS method and STAR-AQEM methods gave similar sized percentage sampling variances when used in both Austria and Germany; differences between the two methods in estimated values of both SD_E and P_{samp} was small for many metrics and probably within the estimation error of these datasets. The percentage sampling variance for metric values obtained using the RIVPACS method tended to be highest for the Greek sites and, on average, slightly higher than for the STAR-AQEM method. In Greece, macroinvertebrates were only identified to family rather than species or genus level. In the UK, the RIVPACS method gave, on average, higher sampling precision (i.e., lower P_{samp}) than the STAR-AQEM method (Fig. 8). The RIVPACS method tended to lead to higher sampling variances and percentage sampling variances than the STAR-AQEM method for both total abundances and percentage abundance of Oligochaeta. It may be that the RIVPACS method has a less precisely standardised sampling area and proportional sampling of fine sediments. In the UK, abundances of common taxa in RIVPACS samples are often determined from a fraction of the sample and only recorded and used as log abundance categories.

Comparison of precision of metrics

As an attempt to assess which metrics were overall most susceptible to sampling variation, the median value of P_{samp} across all of the sampled stream types (i.e., from Tables 2–9) was calculated for each metric, regardless of the 'national/RIVPACS method used (Table 11). The three Saprobic metrics (original Zelinka & Marvan (1961), German new version and Czech) had the lowest median values of P_{samp} (3–4%) and had percentage sampling variances less than 10% for the majority of stream types and methods. These metrics are highly robust to sampling effects using STAR-AQEM, RIVPACS and most 'national' methods. The taxonomic richness metrics are of average general sampling precision using 'national' methods, as are most of the percentage composition metrics, ranging from P_{samp} of 7% for '% Littoral' and '% EPT individuals based on abundance classes' to 19% for '% Oligochaeta'. Amongst the newly developed species trait metrics, the metric 'Trait m12' based on the percentage of individuals from species preferring water velocities less than 25 cm s^{-1} had the lowest average percentage sampling variance (6%) across all 'national'/RIVPACS samples.

Table 11. Median values of the percentage sampling variance (P_{samp}) across all stream types and countries for the 'national'/RIVPACS method and for the STAR-AQEM method

Metric	National/ RIVPACS	STAR-AQEM
Saprobic Index	3	3
German Saprobic new	4	5
Czech Saprobic	4	6
Trait m12: preferred current < 25 cm/s	6	12
% Littoral	7	15.5
% EPT (abundance-classes)	7	9
% Rheophilic (abundance-classes)	8	12
Number of EPT taxa	9	15.5
% Shredders	10	10
% EPT taxa	10	18
% Grazers/scrapers	10.5	16
% Rheophilic	11	12
Trait m2: > 1 cycle per annum	12	10
Trait m1: max size ≤ 1 cm	14	27.5
% Gatherers/collectors	15	14
% EPT individuals	15	15
RETI	15	12
Diversity SW	16	14
Number of taxa	16.5	15.5
ASPT	16.5	17
Number of families	17.5	15.5
1-GOLD	17.5	16.5
% Oligochaeta	19	16
Abundance (ind/m^2)	21	21.5
IBE	25.5	16.5
Trait m7: crawler locomotion	26	17.5

The proposed ICM '1-GOLD' had a median percentage sampling variance of about 17% for both the national/RIVPACS and STAR-AQEM methods. Another ICM metric, 'Log (Sel_EPTD + 1)' based on the logarithm of the total abundance of selected Ephemeroptera, Plecoptera, Tricoptera and Diptera (Buffagni et al., 2006), for which replicate sample values were only available for six Greek sites in each of two seasons, had moderate P_{samp} values of 8 and 17% when based RIVPACS and STAR-AQEM samples respectively.

The metrics 'number of EPT-Taxa' and '% EPT-Taxa' tended to have higher sampling variances for the STAR-AQEM method than for the RIVPACS and several other 'national' methods. This may be because many EPT taxa occur at relatively low numbers in whole samples and there is a large degree of chance whether they are in the one sixth (minimum) sub-sample of a STAR-AQEM sample which is actually identified. Other methods have more tendency to scan all of the sample for all of the EPT taxa present.

By working with the transformed values of many metrics, the replicate sampling SD was less variable between sites of differing quality and stream types. For example, the sampling SD of the square root of 'Number of Families' did not vary systematically between stream types – a Kruskal–Wallis one-way ANOVA of the ranks of the individual site sampling SD showed no statistically significant differences between stream types (test $p = 0.274$). This suggests using the overall ANOVA estimate of sampling SD of 0.268 based on all of the sites sampling using the RIVPACS method. This estimate is similar but slightly higher than the equivalent estimate of 0.228 derived by Clarke et al. (2002) based on single season replicate sampling of a set of 16 UK sites covering a wide range of stream types and qualities. The average sampling SD for ASPT based on the RIVPACS method was 0.236 for the 6 UK sites, and 0.239 when averaged over all four countries where RIVPACS samples were taken. These values are very close to the value of 0.249 obtained by Clarke et al. (2002). The consistency of these estimates demonstrates the repeatability and robustness of our uncertainty analyses.

Obviously, these comparisons of sampling methods excluded information on the relative costs of taking and processing a sample, which is highly relevant to the cost-effectiveness of a method, but beyond the scope of this study. However the component costs of obtaining and processing RIVPACS and STAR-AQEM method samples were assessed and compared for some sites in a separate study within the STAR project. Vlek (2004) found that, on average across the sampled sites, STAR-AQEM samples took 18 h to process (including sorting and identification), whilst RIVPACS samples took only 9 h – half the amount of time. As the RIVPACS method led to no more than marginally higher average percentage sampling variances within the four countries where both methods were used, the RIVPACS method may be more cost-effective than the STAR-AQEM method, at least when the aim is to base site assessments on one or more of the metrics assessed here.

No attempt has been made here to identify which components of the whole sampling method and sample processing protocol cause the most variation in recorded community composition and metric values. Clarke et al. (2006) found that sub-sampling of STAR-AQEM samples caused more than 50% of the overall replicate sampling variance for nearly half of all metrics analysed. In a companion study, Haase et al. (2006) found significant effects of laboratory sampling sorting and taxonomic identification errors on the values of several macroinvertebrate metrics for the STAR-AQEM and RIVPACS methods.

Precision of multi-metric indices

Assessments of the ecological condition or status of a water body are often based on some form of average of the normalised values of several biological metrics measuring complementary aspects of the biological community composition and diversity within a sample and site (Barbour et al., 1999; Ofenböck et al., 2004). The sampling precision of such MMIs depends on the sampling variances of the individual component metrics (and potentially their sampling covariances).

If a set of metrics all have the roughly the same size sampling variance (σ^2) in their EQR values, then a MMI based on the simple average of the EQR values of a subset of m of these metrics will have sampling variance equal to σ^2/m (assuming uncorrelated sampling variability of the metrics). Therefore, involving more such metrics into the MMI will reduce the sampling variance of the MMI.

Adding metrics with very low precision to a MMI based on the average of the component metrics' EQR values can reduce the precision of the MMI. As an example, suppose a MMI is based on the average of the EQR values of m metrics with an average sampling variance of σ^2, and the additional metric has a variance of $C\sigma^2$. Adding this extra metric to the MMI will increase its variance from σ^2/m to $\sigma^2(m+c)/(m+1)^2$ (assuming independent sampling variability of metrics) if $C > 2 + (1/m)$. This means that adding any metric with an EQR sampling variance more than three times the average EQR sampling variance of the metrics already involved in a MMI will always reduce the precision of the MMI.

Implications for uncertainty in assessments of ecological status

These estimates of sampling variability will provide valuable provisional estimates for use in any assessment of uncertainty in any single metric or multi-metric assessments of river quality based on any of the national macroinvertebrate sampling methods and metrics tested within the STAR project. Any estimate of sampling SD derived here can be used to provide information on the expected uncertainty in a metric's values due to sampling variation for other sites in the same stream type sampled using the same method, but where only one sample has been taken at a point in time and thus there is no replication. Of course, the validity of their use depends on the assumption that exactly the same field sampling and laboratory sorting and identification procedures have been used by staff with a similar level of training.

Figure 8 gives estimates of what a given value of P_{samp} might mean in terms of average rates of mis-classifying sites in one of the middle WFD status classes, namely good, moderate or poor (Mis-classification rates are, on average, generally lower for sites of extreme high or bad status). These conversions from percentage sampling variance to average mis-classification rates are based on the assumptions used by Clarke et al. (2006); the resulting estimates are very approximate and only intended as rough guidelines to aid interpretation.

A more precise approach would be to use the estimates of overall replicate sampling standard deviations (SD_E) in the STARBUGS simulation software package (STAR Bioassessment Uncertainty Guidance Software, www.eu-star.au, Clarke, 2004) to assess the effect of sampling variability in individual metric values using a particular sampling method on the uncertainty of single- or multi-metric assessments of the ecological status of sites based on user-specified class limits and multi-metric rules. The STARBUGS software has options to deal with user-supplied estimates of sampling SD for metrics on all of the transformation scales used here (see Clarke & Herring, 2006 for further details).

In summary, this study has provided comparative estimates of the susceptibility of a wide range of macroinvertebrate metrics to the combined effects of field sampling variation and sample processing procedures for a large number of existing nationally used macroinvertebrate sampling methods across Europe. The results can be used to help river monitoring agencies assess the uncertainty in their assignment of river sites to ecological status classes based on any of these metrics and sampling protocols. Finally, this study has only assessed the relative precision of metrics, not their accuracy; the latter depends on the true, but unknown, ecological status of the site. A single metric or MMI could give consistent and repeatable results for a site, but may not be a informative measure of its ecological condition. However, methods, metrics and MMI which together have been found to give high sampling variability cannot provide reliable and precise measures of site ecological status.

Acknowledgements

STAR was partially funded by the European Commission, 5th Framework Program, Energy, Environment and Sustainable Development, Key Action Water, Contract no. EVK1-CT-2001-00027. The authors thank Mike Furse for his leadership of the STAR project and its field sampling programme and acknowledge the support of all their project colleagues who collected the field samples and sorted and identified the taxa.

References

Barbour, M. T., J. Gerritsen, B. D. Snyder & J. B. Stribling, 1999. Rapid Bioassessment Protocols for Use in Streams and Wadeable Rivers: Periphyton, Benthic Macroinvertebrates and Fish. 2nd edn. EPA/841-B-98-010. U.S. EPA. Office of Water. Washington, DC.

Bis, B. & P. Usseglio-Polatera, 2004. Species Traits Analysis. European Commission, STAR (Standardisation of river classifications), Deliverable N2, 134 pp.

Buffagni, A., S. Erba, M. Cazzola, J. Murray-Bligh, H. Soszka & P. Genoni, 2006. The STAR common metrics approach to the WFD intercalibration process: Full application for small, lowland rivers in three European countries. Hydrobiologia 566: 379–399.

Clarke, R. T., 2000. Uncertainty in estimates of river quality based on RIVPACS. In: Assessing the biological quality of freshwaters: RIVPACS and similar techniques. Wright, J. F., D. W. Sutcliffe & M. T. Furse (eds), Freshwater Biological Association, Ambleside: 39–54.

Clarke, R. T., 2004. 9th STAR deliverable. Error/uncertainty module software STARBUGS (STAR Bio Assessment Uncertainty Guidance Software) User Manual.

Clarke, R. T., M. T. Furse, J. F. Wright & D. Moss, 1996. Derivation of a biological quality index for river sites: comparison of the observed with the expected fauna. Journal of Applied Statistics 23: 311–332.

Clarke, R. T., M. T. Furse, R. J. M. Gunn, J. M. Winder & J. F. Wright, 2002. Sampling variation in macroinvertebrate data and implications for river quality indices. Freshwater Biology 47: 1735–1751.

Clarke, R. T. & D. Hering, 2006. Errors and uncertainty in bioassessment methods – major results and conclusions from the STAR project and their application using STARBUGS. Hydrobiologia 566: 433–439.

Clarke, R. T., A. Lorenz, L. Sandin, A. Schmidt-Kloiber, J. Strackbein, N. T. Kneebone & P. Haase, 2006. Effects of sampling and sub-sampling variation using the STAR-AQEM sampling protocol on the precision of macroinvertebrate metrics. Hydrobiologia 566: 441–459.

Council of the European Communities, 2000. Directive 2000/60/EC of the European Parliament and of the Council of 23

October 2000 establishing a framework for Community action in the field of water policy. Official Journal of the European Communities L327(43): 1–72.

Extence, C. A., B. M. Balbi & R. P. Chadd, 1999. River flow indexing using British benthic macroinvertebrates: a framework for setting hydroecological objectives. Regulated Rivers: Research and Management 15: 543–574.

Furse, M., D. Hering, O. Moog, P. Verdonschot, R. K. Johnson, K. Brabec, K. Gritzalis, A. Buffagni, P. Pinto, N. Friberg, J. Murray-Bligh, J. Kokes, R. Alber, P. Usseglio-Polatera, P. Haase, R. Sweeting, B. Bis, K. Szoszkiewicz, H. Soszka, G. Springe, F. Sporka & I. Krno, 2006. The STAR project: context, objectives and approaches. Hydrobiologia 566: 3–29.

Haase, P., J. Murray-Bligh, S. Lohse, S. Pauls, A. Sundermann, R. Gunn & R. Clarke, 2006. Assessing the impact of errors in sorting and identifying macroinvertebrate samples. Hydrobiologia 566: 505–521.

Hering, D., O. Moog, L. Sandin & P. Verdonschot, 2004. Overview and application of the AQEM assessment system. Hydrobiologia 516: 1–20.

Hose, G., E. Turak & N. Wadell, 2004. Reproducibility of AUSRIVAS rapid bioassessments using macroinvertebrates. Journal of the North American Benthological Society 23: 126–139.

Illies, J. (ed.), 1978. Limnofauna Europaea (2nd edn). Gustav Fischer Verlag Stuttgart, New York; Swets and Zeitlinger B.V., Amsterdam, 532 pp.

Johnson, R.K., 1998. Spatiotemporal variability of temperate lake macroinvertebrate communities: detection of impact. Ecological Applications 81: 61–70.

Murray-Bligh, J. A. D., M. T. Furse, F. H. Jones, R. J. M. Gunn, R. A. Dines & J. F. Wright, 1997. Procedure for collecting and analysing macroinvertebrate samples for RIVPACS. Joint publication by the Institute of Freshwater Ecology and the Environment Agency, 162 pp.

Ofenböck, T., O. Moog, J. Gerritsen & M. Barbour, 2004. A stressor specific multimetric approach for monitoring running waters in Austria using benthic macro-invertebrates. Hydrobiologia 516: 251–268.

Ostermiller, J. D. & C. P. Hawkins, 2004. Effects of sampling error on bioassessments of stream ecosystems: application to RIVPACS-type models. Journal of the North American Benthological Society 23: 363–382.

Pinto, P., J. Rosado, M. Morais & I. Antunes, 2004. Assessing methodology for southern siliceous basins in Portugal. Hydrobiologia 516: 191–214.

REFCOND, 2003. Guidance on establishing reference conditions and ecological status class boundaries for inland surface waters. Final version, 30 April 2003, produced by WG 2.3.

Shannon, C. E. & W. Weaver, 1949. The Mathematical Theory of Communication. The University of Illinois Press, Urbana, IL.

Smith, M. J., W. R. Kay, D. H. D. Edward, P. J. Papas, St. J. Richardson, J. C. Simpson, A. M. Pinder, D. J. Cale, P. H. J. Horwitz, J. A. Davis, F. H. Yung, R. H. Norris & S. A. Halse, 1999. AusRivAS: using macroinvertebrates to assess ecological condition of rivers in Western Australia. Freshwater Biology 41: 269–282.

Sokal, R. R. & J. R. Rohlf, 1995. Biometry (3rd edn). Freeman and Company, New York.

Taylor, L. R., 1961. Aggregation, variance and the mean. Nature 189: 732–735.

Vlek, H. E., 2004. Comparison of (cost) effectiveness between various macroinvertebrate field and laboratory protocols. European Commssion, STAR (Standardisation of river classifications), Deliverable N1, 78 pp.

Wright, J. F., D. W. Sutcliffe & M. T. Furse (eds), 2000. Assessing the Biological Quality of Freshwaters: RIVPACS and Similar Techniques. Freshwater Biological Association, Ambleside, 373 pp.

Zelinka, M. & P. Marvan, 1961. Zur Präzisierung der biologischen Klassifikation der Reinheit fließender Gewässer. Arch. Hydrobiol. 57: 389–407.

Hydrobiologia (2006) 566:505–521
M.T. Furse, D. Hering, K. Brabec, A. Buffagni, L. Sandin & P.F.M. Verdonschot (eds), The Ecological Status of European Rivers: Evaluation and Intercalibration of Assessment Methods
DOI 10.1007/s10750-006-0075-6

Assessing the impact of errors in sorting and identifying macroinvertebrate samples

Peter Haase[1,*,†], John Murray-Bligh[2], Susanne Lohse[1], Steffen Pauls[1], Andrea Sundermann[1,†], Rick Gunn[3] & Ralph Clarke[3]

[1]*Department of Limnology and Conservation Research, Senckenberg – Research Institute and Natural History Museum, Clamecystraße 12, 63571 Gelnhausen, Germany*
[2]*Environment Agency, Manley House, Kestrel Way, EX6 8EX Exeter, UK*
[3]*CEH Dorset, Winfrith Technology Centre, Winfrith Newburgh, Dorchester, DT2 8ZD Dorset, UK*
*(*Author for correspondence: E-mail: peter.haase@senckenberg.de)*

Key words: stream assessment, error estimation, sample sorting, macroinvertebrate identification

Abstract

This study assesses the impact of errors in sorting and identifying macroinvertebrate samples collected and analysed using different protocols (e.g. STAR-AQEM, RIVPACS). The study is based on the auditing scheme implemented in the EU-funded project STAR and presents the first attempt at analysing the audit data. Data from 10 participating countries are analysed with regard to the impact of sorting and identification errors. These differences are measured in the form of gains and losses at each level of audit for 120 samples. Based on gains and losses to the primary results, qualitative binary taxa lists were deducted for each level of audit for a subset of 72 data sets. Between these taxa lists the taxonomic similarity and the impact of differences on selected metrics common to stream assessment were analysed. The results of our study indicate that in all methods used, a considerable amount of sorting and identification error could be detected. This total impact is reflected in most functional metrics. In some metrics indicative of taxonomic richness, the total impact of differences is not directly reflected in differences in metric scores. The results stress the importance of implementing quality control mechanisms in macroinvertebrate assessment schemes.

Introduction

All assessments of the ecological status of a river site based on biological samples are subject to uncertainty and errors. Biological surveys can only detect a change in river quality when the difference in the results before and after change is greater than uncertainty caused by natural variability and human error. In this paper, we explore the size of these errors in survey and analytical methods for assessing river quality that are used throughout Europe. Error is rarely measured in monitoring surveys or considered negligible as it is often assumed to be small and constant. If this assumption is incorrect, there is a high risk that conclusions drawn from such surveys could be wrong.

Most quantitative assessments of the biological status of water bodies are based on the values of biological indices or metrics derived from the taxonomic composition of the sample, where the metric is intended to measure some specific aspect or general feature of the biota (Cao et al., 2003; Böhmer et al., 2004; Hering et al., 2004a, b). These measures are of little value without knowing their degree of uncertainty (Clarke, 2000; Clarke et al., 2002). This is because differences in river quality can only be confirmed when they exceed the

[†]These authors contributed equally to this work.

uncertainty inherent in the data. Uncertainty is caused both by the natural variability of the biota used to evaluate river quality and by human error introduced by the analyst. It arises from every stage of data collection, from sampling (e.g. Carter & Resh, 2001; Clarke et al., 2002; Ostermiller & Hawkins, 2004) to sample analysis and data handling (e.g. Doberstein et al., 2000; Haase et al., 2004a, b). The sources of these must be identified so that they can be reduced and can be accounted for when results are evaluated. This study focuses on the two major sources of analytical error: sorting error and identification error.

The EU Water Framework Directive (EU-WFD) (European Union, 2000) requires the level of confidence and precision of results provided by monitoring programmes to be given in the River Basin Management Plans (EU-WFD: Annexe V, Section 1.3). As with all ecological analyses, it is more important to have moderate errors that have been quantified than to have small errors but no estimate of their magnitude. The former allows the significance of any differences to be determined whereas the latter does not. In most member states, monitoring for the EU-WFD will be undertaken by environmental protection agencies or commercial environmental laboratories, and in same cases by research laboratories. Irrespective of who does the analysis, it is impossible to eliminate all errors from data based on field survey and laboratory analysis. Therefore it is essential to understand how error can be quantified and minimised and to provide tools to assess these errors.

One aim of the EU-funded STAR[1] project was to identify and quantify different sources of error that affect metrics and thus assessment results (Furse et al., 2006). Within the project the uncertainty entailed to site selection, natural variability within a site or between seasons, different sub-sampling strategies, and human error caused in sample processing were studied (Furse et al., 2006). To evaluate the error of sample processing, a sorting and identification audit was implemented for two major biological quality assessment components, macroinvertebrates and diatoms. For invertebrates, the emphasis was placed on qualitative sorting and identification errors inherent in the laboratory treatment of invertebrate samples.

In the present study, pre audit and post audit macroinvertebrate taxa lists and resulting metric values are compared, based on samples collected following standardised sampling and processing protocols. The focus of this study thus lies on determining differences between individual sources of error and how sorting and identification errors effect metrics commonly used in river quality assessment. The results from other sources of error, such as replicate sampling (Clarke et al., 2006a), sub-sampling (Clarke et al., 2006a; Vlek et al., 2006) and natural variability like seasonal change (Šporka et al., 2006) are presented in other papers of this issue.

[1] Standardisation of river classifications: Framework method for calibrating different biological survey results against ecological quality classifications to be developed for the Water Framework Directive, STAR. Contract No: EVK1-CT 2001-00089

Materials and methods

Audit design

The auditing approach for macroinvertebrate samples applied in the STAR project involved two separate components: (1) a sorting audit at family level, undertaken by a single auditing laboratory, to assess sorting errors across the whole project in a consistent and unified way, and (2) an identification audit undertaken by partners familiar with analysing invertebrates from similar environments. Generally, these were laboratories from neighbouring countries from the same Ecoregion. This approach was chosen for the identification audit, because no laboratory involved in the project had sufficient experience in analysing all the species found in the geographic area covered by the project to undertake the identification audit for the whole project. Because of this, each partner's identification audit was done by one or more partners from neighbouring countries. Although this caused the quality of the identification audit to vary between partners, it ensured that audit results were more accurate. Throughout this paper, the terms primary sample, primary analyst and primary data relate to the main analysis of a sample, the terms audit sample and audit data to the re-analysis of a sample in the audit.

Selecting audit samples

Macroinvertebrate samples of the STAR project sampling programme were taken at all sites by each participating partner using two different methods: (a) the STAR-AQEM method, a multi-habitat-sampling protocol developed within the STAR project (Furse et al., 2006) and (b) a 'national' method, which was normally a widely used protocol within the individual partner's member state (Furse et al., 2006). In Germany, Austria and Greece there were no existing common 'national' sampling protocols. Alternativly, the UK RIVPACS protocol was used (Murray-Bligh et al., 1997). Twelve samples from 10 countries ($N_{sort} = 120$) were analysed in the sorting audit: 6 STAR-AQEM samples ($N_{SA} = 60$) and 6 collected and analysed by the corresponding national method ($N_{Nat} = 60$) (Table 1). A subset of these samples was used for the identification audit. This subset comprised the 12 samples from the 6 countries for which RIVPACS or the RIVPACS comparable PERLA was their national method (from hereon referred to as RIVPACS/PERLA), to complement the data available from the STAR-AQEM data sets. This allowed for a comparison of different methods based on a reasonable sample size for both STAR-AQEM and RIVPACS/PERLA samples ($N = 72$ for both methods) (Table 1).

Primary analysts were aware that *all* 1090 invertebrate samples collected for the STAR project were potentially subject to audit. A partner not involved in any primary analyses selected the samples for the audit randomly. Audit samples were selected roughly evenly between seasons and included samples representing high, good and moderate ecological quality (pre-classification based on expert judgement). For each combination of site and season, one sample collected by the STAR-AQEM protocol and one sample collected by the national survey protocol was chosen for audit. Partners were not told which samples were selected for audit until all the primary data had been entered into the STAR database, AQEMdip (AQEM Consortium, 2004) so that the primary data could not be altered after the audit samples had been selected. This ensured that primary analysts could not give any special attention to audit samples and the audit results would therefore reflect the quality of all the primary analyses.

Audit procedure

Sorting audit

When sorting STAR-AQEM samples, the primary analyst had to remove all specimens from the sub-sample. After analysis, the sorted specimens were placed in a labelled vial or jar containing preservative and stored for the identification audit (see Furse et al., 2006 for a detailed sampling and sorting protocol). All organic and inorganic material from the sorted STAR-AQEM sub-sample, together with any animals remaining in it, was returned to a jar with preservative for the sorting audit. In the sorting audit, the auditors re-sorted the whole sub-sample removing any animals they found and placing them in a new, labelled vial. The only identification undertaken by the sorting auditors was to identify any additional families that were not recorded by the primary analyst, which were then recorded as gains.

For national methods protocols that did not demand that all specimens were removed during sorting, the primary analysts had to remove up to three representatives (but not every specimen) of every taxon for the identification audit (see Furse et al., 2006 for a detailed protocols of sampling and sorting procedures used in the STAR project). The taxa were based on the taxonomic level of the primary analysis: if the identification was to family level, the taxa removed were families; if the sample was analysed to species level, the taxa removed were species (Table 1). The specimens removed had to be good quality examples and not simply the first ones that the analysts found in the sample. The sorting auditors re-sorted the sample and removed from it all specimens of families missed by the primary analyst. They also removed up to three good quality specimens of every potentially different species that they found in the sample. The auditors put these specimens in a vial with preservative.

Table 1. Samples used in the present study. "National Method" refers to the sampling and sorting protocol applied in the respective country for the national method samples, "National method category" refers to the method category into which protocol was placed for the selection of the identification audit subset of samples

Country	STAR-AQEM sample*	National method sample*	National method	National method category	Season	River	Site	STAR Site No.	Sample date	Sorting audit	ID audit analyses	
Austria	A0500261	A0500431	RIV	RIV/PER	Spring	Sarmingbach	Wolfsschlucht	600	16.04.2003	F	S	X
Austria	A0500291	A0500461	RIV	RIV/PER	Spring	Grosse Ysper	near Altenmarkt	603	16.04.2003	F	S	X
Austria	A0500332	A0500502	RIV	RIV/PER	Summer	Sarmingbach	Waldhausen	607	09.07.2002	F	S	X
Austria	A0600141	A0600341	RIV	RIV/PER	Spring	Wildbach	near Kramermirtl	701	28.05.2003	F	S	X
Austria	A0600192	A0600392	RIV	RIV/PER	Summer	Stullneggbach	near Aichegg	706	30.07.2002	F	S	X
Austria	A0600232	A0600432	RIV	RIV/PER	Summer	Stullneggbach	near Mainsdorf	708	30.07.2002	F	S	X
Czech Republic	C0401621	C0403561	PERLA	RIV/PER	Spring	Velka Hana	Rychtarov	614	04.04.2003	F	S	X
Czech Republic	C0401701	C0403631	PERLA	RIV/PER	Spring	Nectava	Brezinky	620	27.03.2003	F	S	X
Czech Republic	C0401172	C0403152	PERLA	RIV/PER	Summer	Umori	Zbraslavec	625	19.07.2002	F	S	X
Czech Republic	C0501212	C0503182	PERLA	RIV/PER	Summer	Huntava	Valsovsky dul	713	26.07.2002	F	S	X
Czech Republic	C0501272	C0503232	PERLA	RIV/PER	Summer	Luha	Sloup	717	22.07.2002	F	S	X
Czech Republic	C0501941	C0503831	PERLA	RIV/PER	Spring	Trebuvka	Borsov	722	09.04.2003	F	S	X
Denmark	K0201011	K0201021	DSFI	Nat	Spring	Karstoft	Noerre Grene	662	01.04.2003	F	S	
Denmark	K0202012	K0202022	DSFI	Nat	Summer	Mattrup	Stids Moelle	663	06.08.2002	F	S	
Denmark	K0206011	K0206021	DSFI	Nat	Spring	Kastbjerg	Edderup	667	01.04.2003	F	S	
Denmark	K0207012	K0207022	DSFI	Nat	Summer	Fjederholt	Okkels	668	08.08.2002	F	S	
Denmark	K0209011	K0209021	DSFI	Nat	Spring	Skibsted	Skibstedbro	670	07.04.2003	F	S	
Denmark	K0210012	K0210022	DSFI	Nat	Summer	Skals	Faarup	671	12.08.2002	F	S	
France	F0800013	F0800193	IBGN	Nat	Autumn	Aube	Aubepierre-sur-Aube	724	25.09.2002	F	F	
France	F0800021	F0800201	IBGN	Nat	Spring	Seine	Ermitage du Val de Seine	725	15.04.2003	F	F	
France	F0800041	F0800221	IBGN	Nat	Spring	Aujon	upstream of Giey-sur-Aujon	726	25.05.2003	F	F	
France	F0800063	F0800243	IBGN	Nat	Autumn	Ornain	downstream of Abainville	728	30.09.2002	F	F	
France	F0800073	F0800253	IBGN	Nat	Autumn	Meuse (Bassoncourt)	between Daillecourt & Bassoncourt	729	10.10.2002	F	F	
France	F0800111	F0800291	IBGN	Nat	Spring	Mouzon	Sartes	733	09.04.2003	F	F	
Germany	D0400392	D0400512	RIV	RIV/PER	Summer	Wehebach	Wehebachtalsperre	627	29.06.2002	F	S	X
Germany	D0400461	D0400581	RIV	RIV/PER	Spring	Salwey	Niedersalwey	634	25.03.2003	F	S	X
Germany	D0300202	D0300352	RIV	RIV/PER	Summer	Stepenitz	near Putlitz	649	15.07.2002	F	S	X
Germany	D0300201	D0300351	RIV	RIV/PER	Spring	Stepenitz	near Putlitz	649	10.04.2003	F	S	X

Germany	D0600022	D0600122	RIV	RIV/PER	Summer	Ilme	above Relliehausen	816	21.06.2002	F	S	X
Germany	D0600071	D0600171	RIV	RIV/PER	Spring	Klingbach	above Hausen	821	11.03.2003	F	S	X
Greece	H0400222	H0400282	RIV	RIV/PER	Summer	Peristeria	Artiki	735	29.07.2002	F	F	X
Greece	H0400242	H0400302	RIV	RIV/PER	Summer	Tsouraki	Tsouraki	737	01.08.2002	F	F	X
Greece	H0400051	H0400151	RIV	RIV/PER	Spring	Tsouraki	SL 98	738	21.05.2003	F	F	X
Greece	H0400031	H0400131	RIV	RIV/PER	Spring	Krathis	Tsivlos	739	22.05.2003	F	F	X
Greece	H0400262	H0400322	RIV	RIV/PER	Summer	Gadouras	Gadouras	753	24.08.2002	F	F	X
Greece	H0400011	H0400111	RIV	RIV/PER	Spring	Gorgopotamos	Gorgopotamos Bridge	756	18.05.2003	F	F	X
Portugal	P0411313	P0431313	PMP	Nat	Autumn	Xévora	Xévora	863	03.02.2003	F	G	
Portugal	P0411133	P0431133	PMP	Nat	Autumn	Tripeiro	Tripeiro	864	06.12.2002	F	G	
Portugal	P0411121	P0431121	PMP	Nat	Spring	Taveiró	Taveiró	865	13.05.2003	F	G	
Portugal	P0411213	P0431213	PMP	Nat	Autumn	Alpreade	Alpreade	866	05.12.2002	F	G	
Portugal	P0411321	P0431321	PMP	Nat	Spring	Ponsul	Ponsul	867	14.05.2003	F	G	
Portugal	P0411221	P0431221	PMP	Nat	Spring	Baságueda	Baságueda	868	14.05.2003	F	G	
Slovakia	V0100473	V0100483	PERLA	RIV/PER	Autumn	Bystrica	pod Vel'kou skalou	984	17.09.2003	F	S	X
Slovakia	V0100493	V0100503	PERLA	RIV/PER	Autumn	Bystrica	Horná domovina	986	17.09.2003	F	S	X
Slovakia	V0100513	V0100523	PERLA	RIV/PER	Autumn	Bystrica	Bystrièany	987	17.09.2003	F	S	X
Slovakia	V0100423	V0100433	PERLA	RIV/PER	Autumn	Hostiansky potok	pri Pod Javorom	988	16.09.2003	F	S	X
Slovakia	V0100443	V0100453	PERLA	RIV/PER	Autumn	Hostiansky potok	pod Obecným vrchom	989	16.09.2003	F	S	X
Slovakia	V0100463	V0100563	PERLA	RIV/PER	Autumn	Hostiansky potok	nad Topolèiankami	990	18.09.2003	F	S	X
Sweden	S0501351	S0502311	Swedish	Nat	Spring	Nittälven	downstream of Nordtjärnsälven	685	04.06.2003	F	S	
Sweden	S0501063	S0502023	Swedish	Nat	Autumn	Sävälven	upstream of Sävefors	689	23.10.2002	F	S	
Sweden	S0501431	S0502391	Swedish	Nat	Spring	Hörksälven	Brattforsen	691	04.06.2003	F	S	
Sweden	S0601193	S0602153	Swedish	Nat	Autumn	Forsmarksan	Johannisfors	875	30.10.2002	F	S	
Sweden	S0601293	S0602253	Swedish	Nat	Autumn	Hågaån	Lurbo	876	19.11.2002	F	S	
Sweden	S0601561	S0602521	Swedish	Nat	Spring	Strömaran	Hillebola	878	22.05.2003	F	S	
United Kingdom	U1510011	U1510321	RIV	RIV/PER	Spring	Ecchinswell Brook	Headley	639	07.04.2003	F	S	X
United Kingdom	U1510663	U1510973	RIV	RIV/PER	Autumn	Westbury Brook	Westbury	642	08.10.2002	F	S	X
United Kingdom	U1510101	U1510411	RIV	RIV/PER	Spring	Cliff Brook	Crowton	648	13.04.2003	F	S	X
United Kingdom	U2310763	U2311073	RIV	RIV/PER	Autumn	Clun	Marlow	674	28.09.2002	F	S	X
United Kingdom	U2310181	U2310491	RIV	RIV/PER	Spring	Ogmore	Bridgend	678	09.04.2003	F	S	X
United Kingdom	U2310833	U2311143	RIV	RIV/PER	Autumn	Sirhowy	Ynysddu	681	27.09.2002	F	S	X
No. of samples	N_{SA} = 60	N_{Nat} = 60		$N_{RIV/PER}$ = 32							N_{sort} = 120	N_{ID} = 72

"STAR Site No." refers to the unique site code used throughout the STAR project. (SA = STAR-AQEM, RIV = RIVPACS, IBGN = French national method, DSFI = Danish Stream Fauna Index protocol, PERLA = Czech national method, Swedish = Swedish national method, PMP = Portuguese national method, NAT = other national methods; further information on the different methods see Furse et al., 2005). Taxonomic level indicated for sorting audit and identification audit (F = family level, G = mainly genus level, S = mainly species level). "X" indicates samples used for subsequent analyses in the identification audit. *Refers to STAR samples code.

Identification audit

The identification audit was undertaken at the taxonomic level used for the calculation of the metrics by the primary analyst partner. For some partners, this was species, for others it was mixed taxonomic level (Table 1). Vials of specimens and material mounted on permanent microscope slides by the primary analysts were sent to the identification auditors. Temporary mounts could not be sent. The identification auditors used the same method of identification that they used for their primary analysis. Partners that used experts for their primary analyses used the same experts for auditing identifications. The identification auditors recorded a new list of taxa based on their identification of the vial(s) and slide mounts from the primary analyst. They recorded gains and losses, compared to the primary analyst's taxa list. Because it was impractical to undertake the identification audit quantitatively, the metrics used to compare the two samples were based on presence/absence data. A binary taxa list was created to allow a qualitative comparison of the results. The binary taxa lists comprised the primary analysts' results (from hereon referred to as "primary" or "P"), the primary analysts' results plus further taxa observed at the sorting audit level (from hereon referred to as "sorting audit" or "A_S") and a taxa list based on the identification auditor's results (from hereon referred to a "identification audit" or "A_{ID}"). A_{ID} is only based on taxa found in the primary analysis and does not consider taxa gained through the sorting audit. A last taxa list represents the combined results of the sorting and identification audits (from hereon referred to as "total audit " or "A_{TOT}"). By this design, it was possible to establish the effects of errors at each audit level independently and the cumulative error of both the sorting and identification audit.

Audit analyses

Chironomidae, Nematoda and Oligochaeta taxa were not included in the audit. All sorting and identification audit results were based on qualitative errors only. Two parameters were used to measure analytical quality in these audits: the number of gains (taxa that were not recorded as being present in the sample but which the auditors found in the sample) and the number of losses (taxa that are recorded as being present but which were not found in the sample by the auditor). Gains (*G*) and losses (*L*) were identified by comparing the auditor's taxa list to that of the primary analyst. Only gains were recorded in the sorting audit. Losses and gains were recorded in the identification audit.

Neither the primary analyst's nor the identification auditor's species lists were considered to be definitive – they were considered simply as two views of the same data. Audit results were not used to correct the primary data.

The primary analysts calculated a range of metrics separately for the primary and audit sample analyses using the AQEM-STAR assessment software ASTERICS (www.eu-star.at). The differences in metric results for primary and audit samples were calculated and used to determine the effect of analytical errors on a selection of metrics, that are commonly used in the member states to classify river quality (Hering et al., 2004). The audit results were qualitative and the metrics were calculated from the binary taxa lists, i.e. presence/absence data. For some of the selected metrics, which are normally based on quantitative data, e.g. Shannon–Wiener diversity index, this approach could only reflect the qualitative component of the error. Depending on the abundance structure of a sample, this approach may overestimate or underestimate the impact of differences in some metrics. However, this approach made it possible not only to test the effect of uncertainty on "counting" metrics, such as Number of Taxa, but also to get an idea of the uncertainty related to sorting and identification errors based on commonly used richness measures and functional metrics. The similarity between the primary and audit results was investigated by Jaccard similarity (Jaccard, 1901), which was calculated using PC-ORD Version 4.25 (McCune & Mefford, 1999). The statistical analysis of metric results included mean deviation and the spread of differences between primary and audit results. It was also possible to compare the performance of the STAR-AQEM and RIVPACS methods in regard to the deviation of metric results between primary and audit samples. Mann–Whitney *U*-tests (Mann & Whitney, 1947) were used to see whether deviation was larger using one particular method. Wilcoxon

Test (Wilcoxon, 1945) was used to see if differences between primary and audit samples were significant. All statistical analyses were performed in Statistica 6.1 (StatSoft, 2002).

Results

Absolute differences in resulting taxa lists

The results of the sorting audit are summarised as gains and the identification audit results as a combination of gains and losses. Figure 1 shows the number of differences between taxa lists based on gains and losses after each audit by different methods. In STAR-AQEM samples, the number of differences identified in the sorting audit is significantly smaller than those observed during identification audit (Wilcoxon Test, $p < 0.001$). In both the national methods and the RIVPACS/PERLA methods complex there is no significant difference between the number of differences identified in the sorting or the identification audit (Wilcoxon Test, $p < 0.5$). The number of gains observed at the sorting level is significantly higher for RIVPACS/PERLA than for national methods or for STAR-AQEM (Mann–Whitney *U*-Test, $p < 0.003$). The number of gains observed at the identification audit and the total audit level is significantly lower in the national methods than in RIVPACS/PERLA or STAR-AQEM (Mann–Whitney *U*-Test, $p < 0.001$). This could be because many of the national methods determine taxa at the family or genus level, where there is a very low error. In both RIVPACS/PERLA and STAR-AQEM identification is generally to species level, presumably leading to higher levels of identification difference. The results further indicate that in national method samples and RIVPACS/PERLA complex samples, both sorting and identification differences contribute about equally to the total number of differences, while in STAR-AQEM samples the differences are mainly caused by varying identification results.

The qualitative similarity between taxa lists based on primary results, sorting audit results, identification audit results and total audit results were also tested by calculating Jaccard similarity between the different fractions. This similarity value was calculated for samples from those countries where STAR-AQEM and RIVPACS/PERLA methods were applied (c.f. Table 1). Figure 2 shows the Jaccard similarity values by method between the primary taxa lists (P) and the sorting audit taxa lists (A_S) (P/A_S), between the primary and identification audit taxa lists (A_{ID}) (P/A_{ID}) and between the primary and the total audit taxa lists (P/A_{TOT}). In both methods, there is a significant difference in Jaccard similarity between sorting audit and identification audit: in STAR-AQEM Jaccard similarity is significantly higher after sorting audit, while in RIVPACS/PERLA it is significantly lower after the sorting audit (Wilcoxon Test for both, $p < 0.01$). There is no significant difference in Jaccard similarity after identification audit or total between the two methods (Mann–Whitney *U*-Test, $p > 0.36$). However, in RIVPACS/PERLA samples the Jaccard similarity between primary and sorting audit samples is significantly lower than in STAR-AQEM samples (Mann–Whitney *U*-Test, $p < 0.01$). This indicates that in RIVPACS/PERLA samples the sorting error contributes more to the total error than differences in identification, while in STAR-AQEM samples the effect of the sorting audit is much less than that of the identification audit. In both methods, the Jaccard similarity is significantly lower after the total audit (P/A_{TOT}: RIVPACS/PERLA: median = 0.58; STAR-AQEM: median = 0.63) than after the sorting audit or identification audit (Wilcoxon Test, $p < 0.01$). This shows that in both methods there is a cumulative effect of both errors with respect to Jaccard similarity.

Metric results

For the same subset of samples, qualitative taxa lists were also used to calculate 12 metrics commonly used in river quality assessments, to examine the impact of sorting and identification error on the metrics and thus the assessment results. Six of the metrics examined were richness measures: number of taxa (No. Taxa), number of families (No. Families), number of genera (No. Genera), number of Ephemeroptera, Plecoptera, Trichoptera, Coleoptera, Odonata and Bivalvia taxa (EPTCOB Taxa), number of Ephemeroptera, Plecoptera, Trichoptera taxa (EPT Taxa), Shannon–Wiener Diversity index (Diversity Shannon–Wiener)

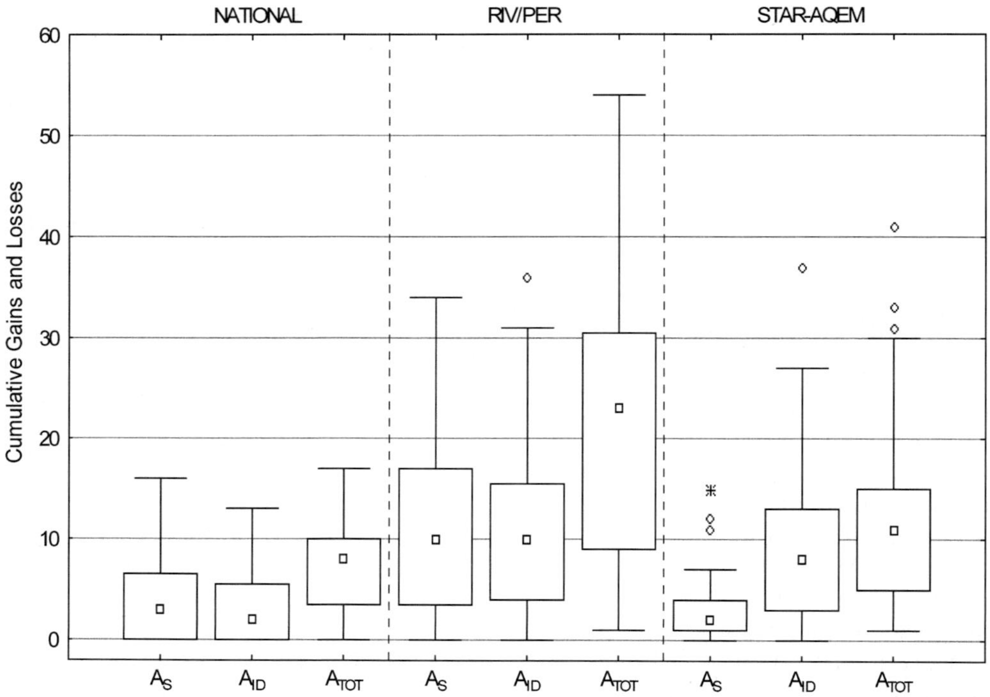

Figure. 1 Box Plots showing the number of differences observed during the sorting (A_S: gains only) and identification audit (A_{ID}: gains plus losses) and cumulative number of differences (A_{TOT}: gains plus losses) for 24 samples collected and analysed following national methods protocols, 36 RIVPACS/PERLA (RIV/PER) samples and 60 STAR-AQEM samples. Box Plots: □ indicates median; box indicates 25–75th percentile range; ○ indicate outliers; ◇ indicate extreme values. $N = 120$.

(Shannon & Weaver, 1949). Two were relative measures of composition: number of Ephemeroptera, Plecoptera, Trichoptera taxa to the number of Diptera taxa (EPT/Diptera Taxa) and the number of taxa scored as r-strategists compared to the number of taxa scored as K-strategists (r/K relationship). Four were functional metrics: Biological Monitoring Working Party score (BMWP), the average score per taxon (ASPT) (both Armitage et al., 1983), the Rhithron Typie Index (RTI) (Biss et al., 2002) and the Rhithron-Feeding types index (RETI), which analyses the proportion of shredders and grazers (Schweder, 1992). An explanation of all these metrics can be found in the AQEM-STAR assessment software ASTERICS (www.eu-star.at).

Table 2 gives the absolute differences in metric values compared to the primary result. A Wilcoxon Test was used to see if there are significant differences between metric values scored for the primary taxa list and those scored after the sorting audit, the identification audit and the total audit. In STAR-AQEM samples there are significant differences between the primary results and results after sorting audit and identification audit for eight metrics (Wilcoxon Test $p < 0.05$). In RIVPACS/PERLA samples eight metrics showed significant differences between the primary metric results and the results after sorting and seven metrics showed significant differences after the identification and total audit (Wilcoxon Test $p < 0.05$). In both methods, six of these significant differences were observed in metrics that measure taxonomic richness. BMWP was significantly different at all levels of the audit in both methods (Wilcoxon Test, $p < 0.05$). In STAR-AQEM samples only two metrics were significantly different at the total audit level (BMWP and ASPT).

The absolute differences between primary metric results and those metric results scored after sorting (P-A_S), identification (P-A_{ID}) and total audit (P-A_{TOT}) were also calculated (Fig. 3).

In STAR-AQEM samples differences in metric results observed after identification audit are

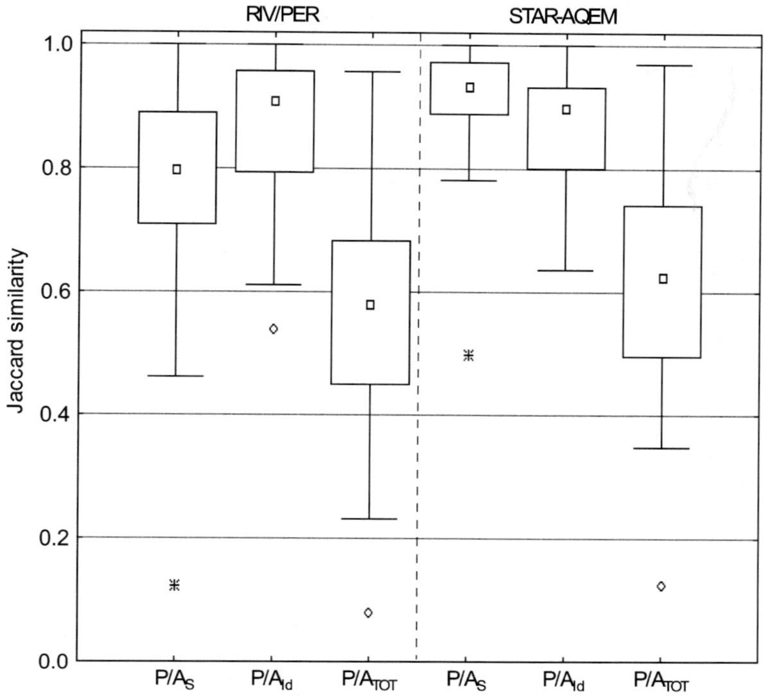

Figure. 2 Box Plots showing Jaccard similarity values between the primary taxa lists (P) and the taxa lists based on the sorting audit (A_S) (P/A_S), between the primary taxa lists and the identification audit (A_{ID}) taxa lists (P/A_{ID}) and the difference between the primary taxa lists and the total audit (A_{TOT}) taxa lists (P/A_{TOT}) for 36 RIVPACS/PERLA (RIV/PER) and 36 STAR-AQEM samples. Box Plots: □ indicates median; box indicates 25–75th percentile range; ○ indicate outliers; ◇ indicate extreme values. $N = 72$.

higher than after sorting audit for all metrics but ASPT and RETI. For RTI, No. Taxa, EPTCOB Taxa, EPT Taxa and Diversity Shannon–Wiener the difference is significant (Wilcoxon Test, $p < 0.02$) (Table 2). In RIVPACS/PERLA samples differences in metric results are higher after sorting audit than after identification audit, the only exception is r/k relationship (Table 2). The differences after sorting audit are significantly higher in eight metrics. These eight metrics are all richness measures, BMWP score and ASPT score (Wilcoxon Test, $p < 0.02$). These results suggest that in RIVPACS/PERLA samples, sorting error causes more analytical error than identification, while in STAR-AQEM differences in identification appear to be more important than sorting error.

The differences after both audits are generally lower in STAR-AQEM samples than they are in RIVPACS/PERLA samples (Fig. 3). For all functional metrics the differences observed are largest after the total audit, suggesting a cumulative effect of the two audit levels. For richness measures the situation is different. In RIVPACS/PERLA samples the largest differences are observed at the sorting audit level (Table 2, Fig. 3). In STAR-AQEM samples the largest differences are observed at the identification audit level for richness measures (Table 2, Fig. 3).

Discussion

In this paper, we make a first attempt to evaluate the analytical error observed in stream assessments based on sorting and identification components of laboratory sample treatment. The errors entailed in these processes are important for providing confidence in assessment results. There are many other sources of variation in macroinvertebrate sampling and sample analysis. These include natural variability (e.g. McElravy et al., 1989; Weatherby & Ormerod, 1990; Boulton & Lake, 1992), operator dependent sampling variability (Clarke et al., 2002), variability in sample

Table 2. Mean values (Ø) and standard deviation (SD) of absolute differences in metric values

	Primary (P) Ø ± SD	Sorting audit ($P\text{-}A_S$) Ø ± SD	Identification audit ($P\text{-}A_{ID}$) Ø ± SD	Total audit ($P\text{-}A_{TOT}$) Ø ± SD	P/A_S p*	P/A_{ID} p*	P/A_{TOT} p*
STAR-AQEM							
BMWP	118.36 ± 44.89	4.92 ± 7.90	7.92 ± 11.73	9.53 ± 8.65	**< 0.001**	**< 0.001**	**0.031**
ASPT	6.70 ± 0.98	0.10 ± 0.32	0.10 ± 0.23	0.30 ± 0.50	**0.041**	**0.011**	**< 0.001**
RTI	11.10 ± 4.15	0.26 ± 0.48	0.55 ± 0.55	0.94 ± 1.00	0.600	0.789	0.307
RETI	0.58 ± 0.11	0.03 ± 0.05	0.02 ± 0.02	0.05 ± 0.08	0.175	0.926	0.598
EPT/Diptera Taxa	3.12 ± 2.17	0.34 ± 0.64	0.55 ± 0.86	0.99 ± 1.41	0.218	0.939	0.304
r/K relationship	0.06 ± 0.05	0.01 ± 0.01	0.01 ± 0.01	0.02 ± 0.03	0.260	0.125	0.092
No. Taxa	34.53 ± 15.12	3.00 ± 3.22	5.42 ± 4.06	3.14 ± 2.98	**< 0.001**	**< 0.001**	0.352
No. Families	21.58 ± 6.51	1.08 ± 1.36	1.31 ± 1.85	1.42 ± 1.25	**< 0.001**	**< 0.001**	0.737
No. Genera	27.56 ± 10.46	1.78 ± 2.27	2.58 ± 2.89	2.56 ± 2.47	**< 0.001**	**< 0.001**	0.596
EPTCOB Taxa	24.92 ± 13.04	2.17 ± 2.62	3.89 ± 3.47	2.81 ± 2.75	**< 0.001**	**< 0.001**	0.244
EPT Taxa	17.83 ± 10.27	1.33 ± 1.96	3.14 ± 2.87	2.36 ± 2.32	**< 0.001**	**< 0.001**	0.110
Diversity Shannon–Wiener	3.41 ± 0.60	0.09 ± 0.12	0.15 ± 0.11	0.10 ± 0.09	**< 0.001**	**< 0.001**	0.214
RIVPACS							
BMWP	124.72 ± 51.88	18.83 ± 18.49	6.83 ± 12.54	19.39 ± 18.18	**< 0.001**	**0.034**	**< 0.001**
ASPT	6.90 ± 0.96	0.22 ± 0.28	0.09 ± 0.16	0.28 ± 0.29	1.000	0.438	0.797
RTI	11.54 ± 3.43	0.59 ± 0.88	0.48 ± 0.66	1.01 ± 0.90	0.265	0.514	0.838
RETI	0.58 ± 0.15	0.04 ± 0.04	0.02 ± 0.03	0.05 ± 0.05	0.299	0.355	0.528
EPT/Diptera Taxa	3.64 ± 2.38	0.52 ± 0.74	0.46 ± 0.48	0.69 ± 0.88	**0.027**	0.688	0.188
r/K relationship	0.04 ± 0.05	0.01 ± 0.02	0.01 ± 0.02	0.02 ± 0.03	0.207	0.614	0.432
No. Taxa	36.19 ± 17.64	10.89 ± 8.86	5.28 ± 6.39	8.17 ± 7.21	**< 0.001**	**< 0.001**	**< 0.001**
No. Families	21.86 ± 8.41	4.08 ± 3.86	1.39 ± 2.23	4.06 ± 3.88	**< 0.001**	**0.003**	**< 0.001**
No. Genera	28.53 ± 12.96	6.56 ± 6.12	2.83 ± 4.31	5.83 ± 5.46	**< 0.001**	**< 0.001**	**< 0.001**
EPTCOB Taxa	26.56 ± 14.12	7.78 ± 6.49	3.36 ± 4.03	5.78 ± 5.54	**< 0.001**	**< 0.001**	**< 0.001**
EPT Taxa	20.33 ± 11.77	4.53 ± 4.65	2.44 ± 3.08	3.58 ± 3.78	**< 0.001**	**< 0.001**	**0.001**
Diversity Shannon–Wiener	3.39 ± 0.76	0.29 ± 0.35	0.15 ± 0.16	0.26 ± 0.36	**< 0.001**	**< 0.001**	**< 0.001**

Metrics values based on the sorting audit ($P\text{-}A_S$), identification audit ($P\text{-}A_{ID}$) and total audit ($P\text{-}A_{TOT}$) compared to the values observed using primary taxa list (P). Significance of differences between $P\text{-}A_S$ and $P\text{-}A_{ID}$ and significance between average primary metric result and the average metric results at the different audit levels were tested using Wilcoxon Test (p indicates level of significance). *Wilcoxon Test: boldface values. significant at $p < 0.05$.

treatment (Haase et al., 2004b) and sub-sampling variability (Lorenz et al., 2004; Clarke et al. 2006b). The present study provides a first approximation of the quality and degree of error that may be observed from sorting errors and operator-dependent differences in macroinvertebrate identifications. Aspects of variability and sources of error related to replicate sampling (Clarke et al., 2006a), sub-sampling (Clarke et al., 2006a; Vlek et al., 2006) and natural variability (Šporka et al., 2006) are discussed in other essays in this issue.

Absolute differences in resulting taxa lists

Our study provides some interesting insights into the two sources of error examined in this study: sorting and identification error. The two components of error play a different role in STAR-AQEM and RIVPACS/PERLA samples. While sorting error seems to be more important in RIVPACS/PERLA samples, identification error seems to be more important in STAR-AQEM samples (Figs. 1–3). The effect of identification error should, by its nature, be similar or equal in both methods because it is caused by two analysts looking at the same set of specimens. The difference in taxonomic expertise or interpretations of distinguishing morphological characters – e.g. relative bristle length or coloration – is the same for the operators, independent of the method used to obtain the set of specimens. Therefore, it is not surprising that no significant differences in identification error between the two methods were detected. STAR-AQEM samples seem to be less affected by sorting error than RIVPACS/PERLA samples. This could be the result of differences in the sorting procedures for the methods. While both methods apply a sub-sampling procedure, the sub-sampling approach is very different. In STAR-AQEM, a defined fraction of the sample is completely sorted and all animals are removed from the sub-sample for identification. In RIVPACS, the whole sample is sorted sequentially by transferring small aliquots of sample material into a dish and sorting a defined fraction of this dish (e.g. ¼ or ⅛), depending on the number of specimens in the total sample (see Furse et al., 2006 for a detailed sampling and sorting protocol for the methods used in the STAR project). This defined fraction will from hereon be referred to as the "sorted fraction". The rest of the material in the dish is scanned and only taxa, which have not been observed in the sorted fraction or any of the sorted fractions in previous dishes are picked and recorded. Also, instead of removing all individuals of abundant taxa from the sorted fraction, they are left in the tray and counted instead. This inevitably leads to a more variable sorting protocol in RIVPACS, which also requires a higher level of taxonomic expertise from the person sorting the sample than the sorting protocol in STAR-AQEM (Haase et al. 2004a, b and references therein). This source of error may be overestimated in the present study, because for many partners, this was the first time they applied the RIVPACS protocol. Although the same is true for most partners with respect to the STAR-AQEM protocol, the complexity of the RIVPACS protocol may make it more prone to mistakes by novices.

Our results indicate that, for RIVPACS samples the sorting error is equally as important as the identification error, while in STAR-AQEM samples, the sorting error is less severe. This is supported by both the Jaccard similarity analyses as well as the number of gains and losses observed at each level of the audit procedure (Figs. 1 and 2).

There appears to be a cumulative effect of analytical error in the two levels of audit. The single effect of sorting error and differences in identification still maintain a high Jaccard similarity (>0.8) between samples regardless of method. The cumulative effect is much more severe in both methods. In both methods the cumulative error decreases Jaccard similarity to about 0.6; 60% shared taxa between pre and post audit taxa lists is a very poor value. In both methods there are numerous samples in which differences based only on analytical error exceed this value. Similar values are e.g. observed when comparing caddis fly assemblages from different regions (Wiberg-Larsen et al., 2000) or differently impacted sites along a river stretch (Ganasan & Hughes, 1998). This suggests that the differences between taxa lists caused by the analytical errors assessed in this study are severe. These results stress the need for a high degree of standardisation of methods and raise the issue of increasing confidence in assessment results through independent sample auditing. Our results suggest that errors caused during sorting and identification procedures cannot be ignored in river quality assessment. Several questions concerning the effect of error on quantitative data sets and the practical implementation of auditing schemes should be subject to further study. For example, are the errors observed in the present study more or less pronounced in quantitative data sets? If they are as pronounced, are the errors constant or stable and can the degree of error be estimated in a one-off survey or research project? If not, must they be measured continually as an integral part of the survey?

Metric results

This study provides a preliminary view of how metrics are affected by sorting and identification errors, but can only quantify these errors on single qualitative metrics. How quantitative metrics or multimetric assessment results are affected can only be estimated. The reason for this lies in the qualitative nature of the audit approach. The indices of the BMWP-score system (National Water Council, 1981) were ideally suited to the sorting audit because they are based on family-level presence-absence data and included both an index of organic pollution (ASPT) and of general stress (No. Taxa). However, other metrics which make use of the absolute or relative abundances of taxa or involve species are less suited to this qualitative approach. Many of the metrics used for stream assessment rely not only on presence/absence data, but also on abundance data for each taxon. Four of the twelve metrics investigated in this study normally use quantitative data. The interpretation of the results obtained for these metrics based on qualitative data must therefore be interpreted with caution. Also, all of the multimetric assessment schemes implemented to date for the countries whose samples were analysed in this study are based on at least one metric that requires quantitative data (Böhmer et al., 2004; Hering et al. 2004a; Ofenböck et al., 2004). Therefore, we cannot estimate the effect of sorting and identification error on a multimetric assessment result. It is important that this subject is addressed by a quantitative audit scheme in future studies. Then error ranges can be assigned to assessment results and confidence in assessment results increased.

Despite the difficulties related to the qualitative nature of our audit design, some interesting observations were made regarding the effect of analytical error on certain metrics. Intuitively, one would expect the total number of differences or the total error (observed in both the sorting and identification audit) to be higher than that observed in the sorting or the identification audit only, independent of the method. For the number of gains and losses and Jaccard similarity values, this was the case. This was also the case for functional metrics. In qualitative metrics that are indicative of richness, the results presented in this study are different and somewhat counter-intuitive. In RIVPACS/PERLA samples especially, the differences in richness metrics are more pronounced in the sorting audit than in the total audit result where both sorting and identification differences are considered (Fig. 3, Table 2). One explanation for this could be that in RIVPACS samples, most gains were identified in the sorting audit. These gains (G_S) plus the number of gains observed in the identification audit (G_{ID}) are not eliminated by the losses (L) observed in the identification audit ($G_S + G_{ID} - L = 8.03$). Therefore, in metrics that measure taxonomic richness – i.e. are qualitative, counting taxa – the number of differences increases more strongly between original and audited samples in RIVPACS samples than it does in STAR-AQEM samples. This is because in STAR-AQEM samples, sorting and identification gains are more or less eliminated by identification losses ($G_S + G_{ID} - L = 0.44$). The higher number of differences observed in the sorting audit compared to the total audit can be explained by the number of losses. On average, $G_{ID} - L$ is −2.83 for RIVPACS/PERLA and −2.55 for STAR-AQEM, so the effect is about the same for both methods. However, because G_S is much greater in RIVP-

→

Figure. 3 Box Plots showing the absolute differences in metric values between the primary taxa lists (P) and the taxa lists based on the sorting audit (A_S) (P-A_S), between the primary taxa list and the identification audit (A_{ID}) taxa lists (P-A_{ID}) and the difference between the primary taxa lists and the total audit (A_{TOT}) taxa lists (P-A_{TOT}) for 36 RIVPACS/PERLA (RIV/PER) and 36 STAR-AQEM samples. Box Plots: (indicates median; box indicates 25–75th percentile range; ○ indicate outliers; ◇ indicate extreme values. N = 72.

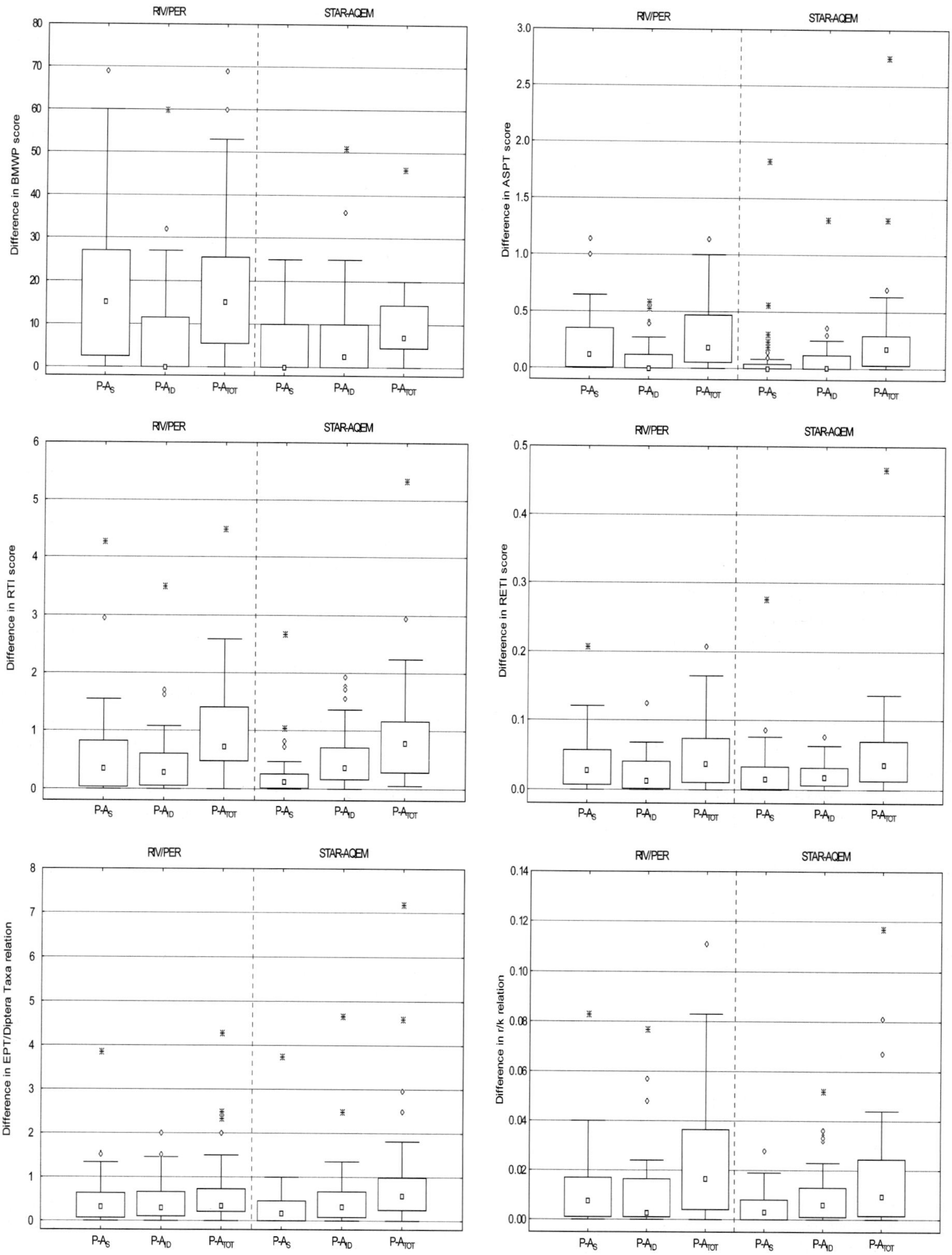
RIV/PER
STAR-AQEM
Difference in BMWP score
Difference in ASPT score
Difference in RTI score
Difference in RETI score
Difference in EPT/Diptera Taxa relation
Difference in r/k relation
P-A$_S$
P-A$_{ID}$
P-A$_{TOT}$

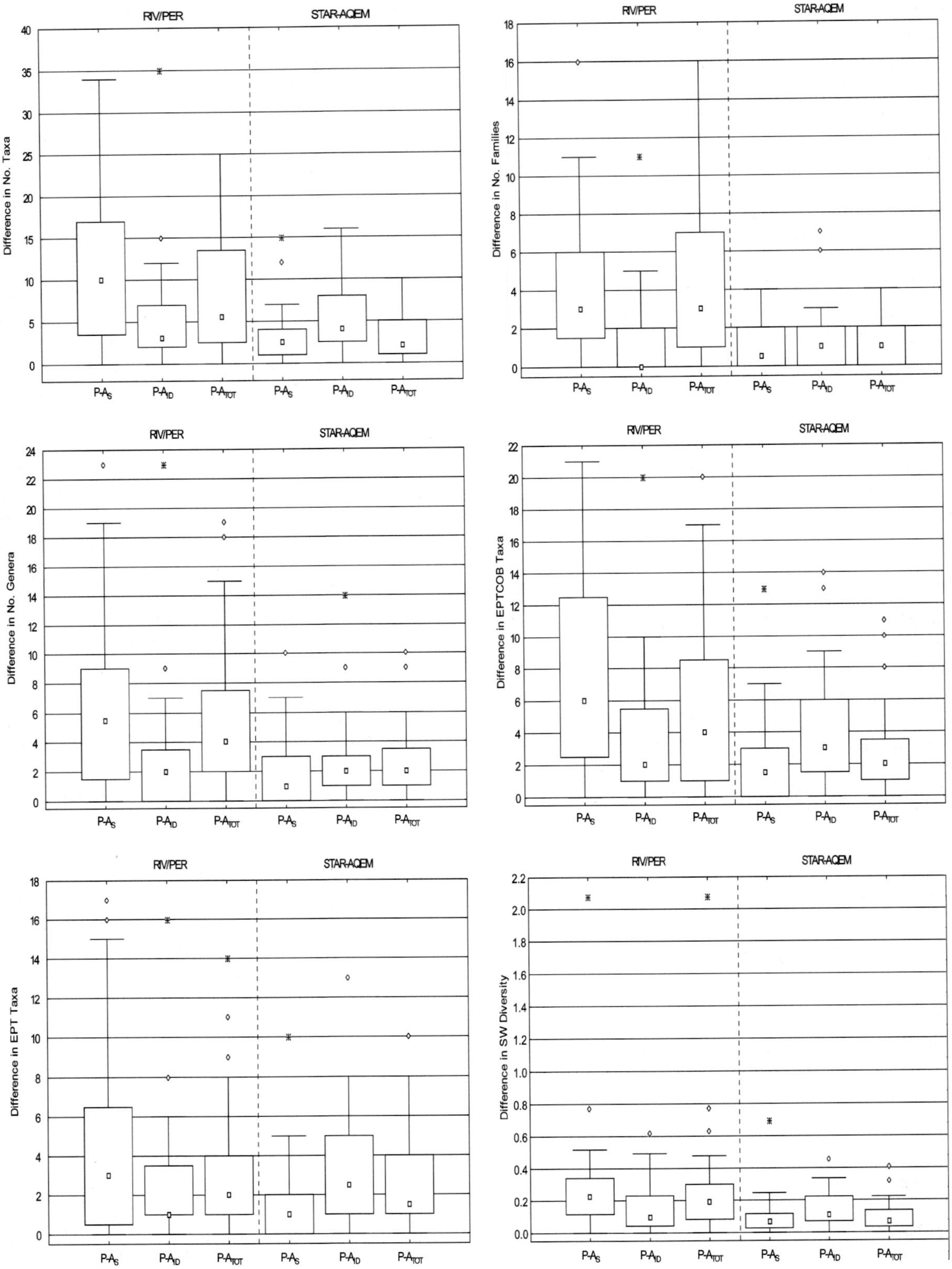

Figure 3. (Continued)

ACS than in STAR-AQEM samples, the differences in number of taxa between the primary results and the sorting audit results are greater than those between the primary result and the sorting plus identification audit result. This effect is maximised in metrics with the highest level of taxonomic differentiation, i.e. highest probability of observing differences, e.g. No Taxa. The effect is reduced as the number of taxa observed is reduced. This is, for example, the case when only certain taxonomic groups are considered (effect in EPT-COB Taxa > EPT Taxa) or when the level of identification is lowered (effect in No. Taxa > No. Genera > No. Families).

It thus appears, that some metrics will hardly be affected by the cumulative error in sorting and identification of samples. These are the metrics that count taxa as measures of species richness. While affected by both sorting and identification errors, the overall number of taxa and therefore the number of taxa belonging to a taxonomic group is hardly influenced in the overall assessment as the errors do not act cumulatively but cancel each other out. For example, if the primary analyst identifies *Drusus annulatus* and another analyst identifies the same individuals as *Drusus destitutus*, one would have two differences in the taxa lists, but no differences in the number of taxa, genera, family or number of EPT taxa. Functional metrics may however respond to such differences, e.g. feeding types, r/K relationship, ASPT or BMWP scores, saprobic valences.

Sorting and identification audits and quality control

Regardless the methods used, a considerable amount of sorting and identification error could be shown. It also became evident that these errors affect metric results and therefore should be taken into account in stream assessment.

The performance of partners varied considerably, especially in the sorting audit. This could be the result of the limited experience of some partners with one or both of the protocols. Another reason might be the existence of an audit system. In our study the best performance (of all samples) in the sorting audit has been achieved by the UK partner, the only country which has established an audit system many years ago. It appears that experience and common auditing of samples leads to better quality of performance. Errors in the long-term auditing scheme in the UK were greatest in the first year of the audit and have decreased over time for various laboratories in the UK, including other government agencies and commercial contractors. Poor results are especially common in the first audit but improve very rapidly thereafter (Murray-Bligh et al., 2006). This is an effect of training and experience, but may also be an effect of operators knowing that samples can and will be audited. Those partners whose audit results were much poorer than expected are unlikely to have similarly poor results if they are audited again in the future. Conclusions made on the basis of the results in this study may therefore differ from those of potential future studies with a similar auditing scheme and the results may not be generally applicable to laboratories currently involved in operational monitoring schemes with auditing. However, the present study and past experience with auditing schemes shows that there is a considerable effect resulting from experience and the training that operators receive. Biologists often receive no formal training, particularly in sorting, and unless someone points-out mistakes, they will remain unaware of shortcomings. Sorting is conceptually very simple and the task is sometimes left to the most junior and inexperienced biologists. The audit results demonstrate that sorting is in fact a task that requires more skill than has been recognised in the past. The audit results also point out the need for formal training and audit strategies for operators working to implement bioassessment schemes using macroinvertebrates. Extensive training is of utmost importance in the identification of macroinvertebrates. This is demonstrated by the large contribution identification error had on the total audit error (Figs. 1 and 2). In STAR-AQEM it was the main component of error, and was equally important in RIVPACS. Correct taxonomic identification is also very important when assessment strategies are based on metrics, because many functional metrics are based on species-specific autecological data. As less and less alpha-taxonomical skills are being taught within tertiary education programs around the globe, the need for specialist-based training and extra-curricular schooling for analysts dealing with the identification of stream biota will become increasingly important. Rigid training programs and auditing schemes will minimise analytical error related to sample sorting and species identification observed in this study. This will increase the precision of assessment results and strengthen water managers' confidence in assessment results. Such programs should therefore become an integral part of biological stream assessment in the future and seem vital for the successful implementation of the EU-WFD.

Acknowledgements

We would like to thank all project partners who contributed data to this study. Rebecca Bloch, Britta Gehenio and Jenny Schmidt are thanked for assistance in data formatting. This study was supported by the EU and presents results from the STAR project (Contract No: EVK1-CT 2001-00089).

References

AQEM consortium, 2004. AQEMdip: AQEM data input program. Downloadable from http://www.eu-star.at.

Armitage, P. D., D. Moss, J. F. Wright & M. T. Furse, 1983. The performance of a new biological water quality score system based on macroinvertebrates over a wide range of unpolluted running-water sites. Water Research 17: 333–347.

Biss, R., P. Kübler, I. Pinter & U. Braukmann, 2002. Leitbildbezogenes biozönotisches Bewertungsverfahren für Fließgewässer (aquatischer Bereich) in der Bundesrepublik Deutschland. Ein erster Beitrag zur integrierten ökologischen Fließgewässerbewertung – Final report on CD-ROM. UBA Texts 62/02, Berlin.

Böhmer, J., C. Rawer-Jost, A. Zenker, C. Meier, C. Feld, R. Biss & D. Hering, 2004. Development of a multimetric invertebrate based assessment system for German rivers. Limnologica 34: 416–432.

Boulton, A. J. & P. S. Lake, 1992. The ecology of two streams in Victoria, Australia. III. Temporal changes in species composition. Freshwater Biology 27: 123–138.

Cao, Y., C. P. Hawkins & M. R. Vinson, 2003. Measuring and controlling data quality in biological assemblage surveys with special reference to stream benthic macroinvertebrates. Freshwater Biology 48: 1898–1911.

Carter, J. L. & V. H. Resh, 2001. After site selection and before data analysis: sampling, sorting, and laboratory procedures used in stream benthic macroinvertebrate monitoring programs by USA state agencies. Journal of the North American Benthological Society 20: 658–682.

Clarke, R. T., 2000. Uncertainty in estimates of river quality based on RIVPACS. In Wright, J. F., D. W. Sutcliffe & M. T. Furse (eds), Assessing the Biological Quality of Freshwaters: RIVPACS and Similar Techniques. Freshwater Biological Association, Ambleside 39–54.

Clarke, R. T., M. T. Furse, R. J. M. Gunn, J. M. Winder & J. F. Wright, 2002. Sampling variation in macroinvertebrate data and implications for river quality indices. Freshwater Biology 47: 1735–1751.

Clarke, R. T., J. Davy-Bowker, L. Sandin, N. Friberg, R. K. Johnson & B. Bis, 2006a. Estimates and comparisons of the effects of sampling variation using 'national' macroinvertebrate sampling protocols on the precision of metrics used to assess ecological status. Hydrobiologia 566: 477–503.

Clarke, R. T., A. Lorenz, L. Sandin, A. Schmidt-Kloiber, J. Strackbein, N. T. Kneebone & P. Haase, 2006b. Effects of sampling and sub-sampling variation using the STAR-AQEM sampling protocol on the precision of macroinvertebrate metrics. Hydrobiologia 566: 441–459.

Doberstein, C., J. Karr & L. Conquest, 2000. The effect of fixed-count subsampling on macroinvertebrate biomonitoring in small streams. Freshwater Biology 44: 355–371.

European Union, 2000. Directive 2000/60/EC. Establishing a framework for community action in the field of water policy. European Commission PE-CONS 3639/1/100 Rev 1, Luxemburg.

Furse, M., D. Hering, O. Moog, P. Verdonschot, R. K. Johnson, K. Brabec, K. Gritzalis, A. Buffagni, P. Pinto, N. Friberg, J. Murray-Bligh, J. Kokes, R. Alber, P. Usseglio-Polatera, P. Haase, R. Sweeting, B. Bis, K. Szoszkiewicz, H. Soszka, G. Springe, F. Sporka & I. Krno, 2006. The STAR project: context, objectives and approaches. Hydrobiologia 566: 3–29.

Ganasan, V. & R. M. Hughes, 1998. Application of an index of biological integrity (IBI) to fish assemblages of the rivers Khan and Kshipra (Madhya Pradesh), India. Freshwater Biology 40: 367–383.

Haase, P., S. Lohse, S. Pauls, K. Schindehütte, A. Sundermann, P. Rolauffs & D. Hering, 2004a. Assessing streams in Germany with benthic invertebrates: development of a practical standardised protocol for macroinvertebrate sampling and sorting. Limnologica 34: 349–365.

Haase, P., S. Pauls, A. Sundermann & A. Zenker, 2004b. Testing different sorting techniques in macroinvertebrate samples from running waters. Limnologica 34: 366–378.

Hering, D., C. Meier, C. Rawer-Jost, R. Biss, C. Feld, A. Zenker, A. Sundermann, S. Lohse & J. Böhmer, 2004a. Assessing streams in Germany with benthic invertebrates: selection of candidate metrics. Limnologica 34: 398–415.

Hering, D., O. Moog, L. Sandin & P. F. M. Verdonschot, 2004b. Overview and application of the AQEM assessment system. Hydrobiologia 516: 1–20.

Jaccard, P., 1901. Étude comparative de la distribution florale dans une portion des Alpes et des Jura. Bulletin de la Société Vaudoise des Sciences Naturelles 37: 547–579.

Lorenz, A., L. Kirchner & D. Hering, 2004. 'Electronic subsampling' of macrobenthic samples: how many individuals are needed for a valid assessment result? Hydrobiologia 516: 299–312.

Mann, H. B. & D. R. Whitney, 1947. On a test of whether one of two random variables is stochastically larger than the other. Annals of Mathematical Statistics 18: 50–60.

McCune, B. & M. J. Mefford, 1999. PC-ORD. Multivariate Analysis of Ecological Data. Version 4.25. MjM Software, Gleneden Beach, Oregon, USA.

McElravy, E. P., G. A. Lamberti & V. H. Resh, 1989. Year-to-year variation in the aquatic macroinvertebrate fauna of a northern Californian Stream. Journal of the North American Benthological Society 8: 51–63.

Murray-Bligh, J. A. D., M. T. Furse, F. H. Jones, R. J. M. Gunn, R. A. Dines & J. F. Wright, 1997. Procedure for collecting and analysing macroinvertebrate samples for RIVPACS. Joint publication by the Institute of Freshwater Ecology and the Environment Agency, 162 pp.

Murray-Bligh, J., J. van der Molen & P. Verdonschot, 2006. STAR deliverable No. 7: Audit of Performance incorporating Results of the La Bresse sampling and analysis workshop. Unpublished report. www.eu-star.at.

National Water Council, 1981. River Quality: The 1980 Survey and Future Outlook. National Water Council, UK.

Ofenböck, T., O. Moog, J. Gerritsen & M. Barbour, 2004. A stressor specific multimetric approach for monitoring running waters in Austria using benthic macro-invertebrates. Hydrobiologia 516: 251–268.

Ostermiller, J. D. & C. P. Hawkins, 2004. Effects of sampling error on bioassessments of stream ecosystems: application to RIVPACS-type models. Journal of the North American Benthological Society 23: 363–382.

Shannon, C. E. & W. Weaver, 1949. Mathematical Theory of Communication. The University of Illinois Press, Urbana, IL.

Schweder, H., 1992. Neue Indices für die Bewertung des ökologischen Zustandes von Fließgewässern, abgeleitet aus der Makroinvertebraten-Ernährungstypologie. Limnologie Aktuell 3: 353–377.

Šporka F., H. E. Vlek, E. Bulánková & I. Krno, 2006. Influence of seasonal variation on bioassessment of streams using macroinvertebrates. Hydrobiologia 566: 543–555.

StatSoft, Inc., 2002. STATISTICA for Windows (Software-System for Data Analysis) Version 6.1. www.statsoft.com.

Weatherby, N. S. & S. J. Ormerod, 1990. The constancy of univoltine assemblages in soft water streams: implications for the publication and detection of environmental change. Journal of Applied Ecology 27: 952–964.

Wiberg-Larsen, P., K. P. Brodersen, S. Birkholm, P. N. Grøn & J. Skriver, 2000. Species richness and assemblage structure of Trichoptera in Danish streams. Freshwater Biology 43: 633–647.

Wilcoxon, F., 1945. Individual Comparisons by Ranking Methods. Biometrics 1: 80–83.

Vlek, H. E., F. Šporka & I. Krno, 2006. Influence of macroinvertebrate sample size on bioassessment of streams. Hydrobiologia 566: 523–542.

Hydrobiologia (2006) 566:523–542
M.T. Furse, D. Hering, K. Brabec, A. Buffagni, L. Sandin & P.F.M. Verdonschot (eds), The Ecological Status of European Rivers: Evaluation and Intercalibration of Assessment Methods
DOI 10.1007/s10750-006-0074-7

Influence of macroinvertebrate sample size on bioassessment of streams

Hanneke E. Vlek[1,*], Ferdinand Šporka[2] & Il'ja Krno[3]
[1]*Alterra, Green World Research, P.O.Box 47, 6700 AA Wageningen, The Netherlands*
[2]*Department of Hydrobiology, Institute of Zoology, Slovak Academy of Sciences, Dúbravská cesta 9, SK-84506 Bratislava, Slovakia*
[3]*Department of Ecology, Faculty of Natural Sciences of Comenius University, Mlynská dolina B-2, SK-84215 Bratislava, Slovakia*
*(*Author for correspondence: E-mail: hanneke.vlek@wurl.nl)*

Key words: sample size, costs, macroinvertebrates, metrics, bioassessment, streams, the Netherlands, Slovakia

Abstract

In order to standardise biological assessment of surface waters in Europe, a standardised method for sampling, sorting and identification of benthic macroinvertebrates in running waters was developed during the AQEM project. The AQEM method has proved to be relatively time-consuming. Hence, this study explored the consequences of a reduction in sample size on costs and bioassessment results. Macroinvertebrate samples were collected from six different streams: four streams located in the Netherlands and two in Slovakia. In each stream 20 sampling units were collected with a pond net (25×25 cm), over a length of approximately 25 cm per sampling unit, from one or two habitats dominantly present. With the collected data, the effect of increasing sample size on variability and accuracy was examined for six metrics and a multimetric index developed for the assessment of Dutch slow running streams. By collecting samples from separate habitats it was possible to examine whether the coefficient of variation (CV; measure of variability) and the mean relative deviation from the "reference" sample (MRD; measure of accuracy) for different metrics depended only on sample size, or also on the type of habitat sampled. Time spent on sample processing (sorting and identification) was recorded for samples from the Dutch streams to assess the implications of changes in sample size on the costs of sample processing. Accuracy of metric results increased and variability decreased with increasing sample size. Accuracy and variability varied depending on the habitat and the metric, hence sample size should be based on the specific habitats present in a stream and the metric(s) used for bioassessment. The AQEM sampling method prescribes a multihabitat sample of 5 m. Our results suggest that a sample size of less than 5 m is adequate to attain a CV and MRD of ≤10% for the metrics ASPT (Average Score per Taxon), Saprobic Index and type Aka + Lit + Psa (%) (the percentage of individuals with a preference for the akal, littoral and psammal). The metrics number of taxa, number of individuals and EPT-taxa (%) required a multihabitat sample size of more than 5 m to attain a CV and MRD of ≤10%. For the metrics number of individuals and number of taxa a multihabitat sample size of 5 m is not even adequate to attain a CV and MRD of ≤20%. Accuracy of the multimetric index for Dutch slow running streams can be increased from ≤20 to ≤10% with an increase in labour time of 2 h. Considering this low increase in costs and the possible implications of incorrect assessment results it is recommended to strive for this ≤10% accuracy. To achieve an accuracy of ≤10% a multihabitat sample of the four habitats studied in the Netherlands would require a sample size of 2.5 m and a labour time of 26 h (excluding identification of Oligochaeta and Diptera) or 38 h (including identification of Oligochaeta and Diptera).

Introduction

One of the objectives of the European Water Framework Directive (WFD; European Commission, 2000) is to standardise the biological assessment of surface waters in Europe. In the AQEM project assessment systems based on macroinvertebrates, which meet the requirements of the WFD (Hering et al., 2004), were developed. For example, an assessment system for slow running streams was developed in the Netherlands (Dutch AQEM assessment system; Vlek et al., 2004). For the development of the assessment systems data were collected in eight European countries using a standardised method for sampling, sorting and identification (Hering et al., 2004). This standardised AQEM method requires a pond net (width 25 cm) or kick sample collected over a length of 5 m, divided into 20 sampling units of 25 cm. The 20 sampling units are proportionally distributed over the habitats present in a stream consistent with their relative coverage. The AQEM method has proved to be relatively time-consuming, i.e., sample processing of Dutch samples can take 155 h per sample (Vlek, 2004). Before water managers are willing to apply the AQEM method for the purpose of biological monitoring the costs associated with the method will have to be drastically reduced.

Costs of monitoring can, among others, be reduced by reducing the sample size. The interpretation of the concept of sample size is variable. Cao et al. (1997) and Bartsch et al. (1998) interpreted sample size as the number of samples (replicates), while Metzling & Miller (2001) interpreted sample size as the physical size of a sample. In most cases a decrease in the costs of biological monitoring programs has been achieved by limiting the number of samples or restricting the number of organisms picked (Metzling & Miller, 2001). The implications of these measures to reduce costs have been the subject of many studies (e.g., Needham & Usinger, 1956; Chutter, 1972; Elliot, 1977; Barbour et al., 1996; Somers et al., 1998; Lorenz et al., 2004). The implications of reducing the physical sample size, however, have hardly been studied. Also, investigations concerning the number of replicate samples are not relevant in the context of biological monitoring by water managers, since water managers usually take only one multihabitat sample for the purpose of biological monitoring. This multihabitat sample consists of several sampling units from different habitats and all sampling units together form one composite multihabitat sample. In this study we, therefore, addressed the influence of physical sample size instead of the number of replicate samples.

Two important aspects of biological monitoring results should be considered in making decisions on the applied sample size: variability and accuracy. Biological monitoring usually has two purposes: (1) to estimate variables of interest at one site and (2) to make comparisons among sites or times. Variables of interest in biological monitoring are primarily metric values (e.g., the number of taxa, ASPT values, BMWP values) and ecological quality indications resulting from assessment systems. Accuracy is a very important aspect of estimating metric values, since accuracy refers to the closeness of a measurement to its true value (Norris et al., 1992). For the purpose of this study the definition of accuracy by Norris et al. (1992) has been adopted. The aspect of variability is very important in making comparisons, because the validity of conclusions depends on data variability (Norris et al., 1992). Higher variability and lower accuracy increase the risk of incorrect assessment results. In case the ecological quality at a site is incorrectly assessed as less than good, water managers will unnecessarily take costly restoration measures to reach a good ecological quality by 2015 (European Commission, 2000). From this point of view, the consequences of poor decision-making due to low accuracy and/or high variability potentially outweigh the savings associated with a smaller sample size (Doberstein, 2000).

Given the importance of accuracy, variability and costs in the process of decision-making, the aim of this study was to assess the implications of changes in sample size for different habitats on (1) the variability and accuracy in metric values, (2) the variability and accuracy of assessment results calculated with the Dutch AQEM assessment system and (3) the costs of sample processing.

Methods

Study site and data collection

The Netherlands

Streams dominated by a single habitat (coverage > 50%) were selected to enable sampling of that habitat over a total length of 5 m. In total, four sites at four different streams (the Oude beek, the Heelsumse beek, the Tongerensche beek and the Molenbeek) were sampled. Each stream is dominated by a different habitat. The streams represent slow flowing (current velocity < 50 cm/s) middle and downstream reaches of poor to moderate ecological quality in the Netherlands, except for the Oude Beek. The Oude Beek is an upstream reach of good ecological quality. The catchment area of all streams is smaller than 100 km^2 and is located between 0 and 200 m a.s.l. Fine to medium-sized gravel (0.2–2 cm; akal) was sampled in the Oude Beek (N 52° 9′ 47.9″ E 5° 57′ 30.1″), submerged macrophytes (*Callitriche* sp.) in the Heelsumse beek (N 51° 58′ 40.7″ E 5° 45′ 30.6″), sand in the Tongerensche beek (N 52° 20′ 22.9″ E 5° 55′ 47.3″) and FPOM (fine particulate organic matter) in the Molenbeek (N 51° 59′ 26.2″ E 5° 43′ 53.5″). The Heelsumse beek, the Tongerensche beek, and the Molenbeek were selected because they represent a stream type and ecological quality which frequently occurs in the Netherlands. The Oude Beek was selected because gravel is frequently found in streams of good ecological quality.

Sampling took place between June and September 2002. From each stream 20 sampling units of the dominant habitat were collected. A sampling unit was collected by pushing a rectangular pond net (25×25 cm, mesh size 500 μm) through the upper part of the substratum (2–5 cm) over a length of approximately 25 cm. A ruler was used to visually point out the length of approximately 25 cm. The 20 sampling units were collected in buckets, and kept separately during sample processing. In the laboratory the sampling units were stored overnight in a refrigerator, where they were oxygenated until sorting. The sampling units were washed through a 1000 and a 250 μm sieve prior to sorting. Live organisms were sorted from the sampling units by eye and preserved in 70% ethanol, except for Oligochaeta and Hydracarina. Oligochaeta were preserved in 4% formaldehyde and Hydracarina in Koenike fluid. Organisms were identified to the lowest taxonomic level possible, i.e., species level for almost all specimens. Literature used for identification purposes is listed in AQEM consortium (2002: p. 156, Appendix 8). Time spent on sorting and identification of all specimens in each sampling unit was recorded.

Slovakia

In Slovakia, four different habitats were sampled in two streams: Pokútsky potok (N 48° 34′ 14.8″ E 18° 40′ 16.5″) and Hostiansky potok (N 48° 29′ 36.3″ E 18° 28′ 40.1″). Both streams are siliceous mountain streams in the West Carpathian. Their catchment is smaller than 100 km^2 and is located between 200 and 500 m a.s.l. Pokútsky potok represents streams of high ecological quality and Hostiansky potok represents streams of good to moderate ecological quality. Two dominating habitats were sampled in both streams: macrolithal (20–40 cm) and mesolithal (6–20 cm) in Pokútsky potok, akal and microlithal (2–6 cm) in Hostiansky potok. The streams were selected because they represent a range in ecological quality that is frequently found in small siliceous mountain streams in the West Carpathian.

Sampling took place in June 2003. From each habitat 20 sampling units were collected as described for the Dutch streams. The 20 sampling units were collected in buckets, preserved in 4% formaldehyde, and kept separately during sample processing. The buckets were transported to the laboratory. The sampling units were washed through a 1000 μm and a 500 μm sieve in the laboratory prior to sorting. Preserved organisms were sorted from the sampling units by stereomicroscope and preserved in 70% ethanol. Organisms were identified to the lowest taxonomic level possible, i.e., species level for almost all specimens. Literature used for identification purposes is listed in AQEM consortium (2002: p. 143, Appendix 8).

Data analysis

In total 158 sampling units were collected from eight different habitats. The assumption was made that the 20 pooled sampling units from one habitat would accurately represent the macroinvertebrate community composition of the respective habitat. The 20 pooled sampling units (with a total sample size of 5 m) are therefore referred to as the "reference" sample. The sample size is expressed as the length over which the pond net was pushed through the substratum. This length can be easily converted into the sampled area by multiplying it by 0.25 m (width of the pond net). Different numbers and combinations of sampling units were pooled per habitat to "construct" composite samples of different sizes. To gain insight into the effect of sample size on variability and accuracy the sampling units from each habitat were randomly reordered 50 times. In case of one sampling unit or 19 sampling units it was only possible to reorder 20 times. For each sample size the randomly selected sampling units were pooled to form a composite sample. Sampling units were selected randomly without replacement because in the field the same area is normally not sampled twice. The described procedure resulted in 50 or 20 replicate (composite) samples per sample size with sample size ranging from 0.25 to 4.75 m. For example, 50 randomly selected combinations of eight sampling units were used to study a sample size of 2 m.

For evaluation, six metrics were selected from an extensive list of metrics that can be calculated with the program ASTERICS version 1.0 (AQEM/STAR Ecological RIver Classification System; http://www.aqem.de): the Saprobic Index (Zelinka & Marvan, 1961), the Average Score per Taxon (ASPT; Armitage et al., 1983), the number of individuals, the number of taxa, the percentage of Epehemeroptera, Plecoptera and Trichoptera taxa (EPT-taxa (%); Lenat, 1988), and the percentage of individuals with a preference for the akal, littoral and psammal (type Aka + Lit + Psa (%); Schmedtje & Colling, 1996). The first reason to select these metrics was that they represent a variety of metric types (taxon richness, community composition, tolerance-intolerance, habitat preference, population attributes). Second, some of these metrics are frequently used in Europe. Third, EPT-taxa (%), type Aka + Lit + Psa (%) and ASPT have proven to be well correlated to anthropogenic stress in Dutch slow running streams and are incorporated in a revised version of the multimetric index for the assessment of Dutch slow running streams described by Vlek et al. (2004). Fourth, EPT-taxa (%) and ASPT have proven to be well correlated to anthropogenic stress in streams with habitats similar to the habitats present in Slovakian mountain streams (Hering et al., 2004).

Metric values were calculated for all composite samples and plotted against the sample size (number of pooled sampling units) (Heyer & Berven, 1973; Bartsch et al., 1998). Species abundances in a sample of a certain size were always standardised to a sample size of 5 m (abundance × 5/sample size (m)), e.g., species abundances in a composite sample consisting of 10 sampling units (2.5 m) were multiplied by 2 to make them comparable to the species abundances in a composite sample consisting of 20 pooled sampling units (5 m). To compare accuracy between metrics, habitats and sample size, the relative deviation of the metric value for each composite sample from the "reference" sample (true value) was calculated. The information concerning accuracy was summarised by calculating the mean relative deviation (MRD) over all composite samples of a certain size.

The coefficient of variation (CV = SD/mean), a measure of variability, was calculated for the metric values of each sample size per habitat. The minimal sample size required to attain a CV and MRD of both $\leq 10\%$ and $\leq 20\%$ was graphically depicted to facilitate the comparison of the effect of sample size on accuracy and variability for different metrics and habitats. The minimal sample size, henceforth referred to as the sample size, required to achieve a certain level of variability or accuracy is used as a measure for variability and accuracy. This is possible because sample size is correlated with variability/accuracy; a larger sample size implies lower variability or higher accuracy. The sample sizes required to reach a CV or MRD of both $\leq 10\%$ and $\leq 20\%$ for the individual habitats (FPOM, sand akal and submerged macrophytes in the Netherlands; akal, macrolithal, mesolithal and microlithal in Slovakia) were summed per country to gain insight into the sample size required for a multihabitat sample.

For all composite samples from Dutch habitats, ecological quality classes were calculated with a revised version of the multimetric index described by Vlek et al. (2004), in order to determine the effects of sample size and habitat on the variability and accuracy in assessment results. The ecological quality class for the samples from

Slovakia was not calculated because no suitable multimetric index was available for the assessment of samples from Slovakian streams.

Sample processing time (time spent on sorting and identification) was recorded for each Dutch sampling unit. The mean sample processing time, including and excluding the time needed for the identification of Oligochaeta and Diptera, was plotted against sample size per habitat to study the consequences of an increase in sample size in terms of costs. A *t*-test ($\alpha = 0.05$) was performed per sample size to look for significant differences in sample processing time between habitats. Residuals were plotted against predicted values to check for normality in sample processing time. No deviations from normality in sample processing time were found.

Results

Variability and sample size

The mean and standard deviation for sample sizes ranging from 0.25 to 4.75 m are given for each metric and habitat in the supplementary material[1]. Depending on the metric, the effect of increasing sample size on metric values showed different types of responses (supplementary material). A decrease in variation with increasing sample size and a relative stable mean (e.g., Fig. 1) was observed for the following metrics: number of individuals, Saprobic Index, type Aka + Lit + Psa (%) and EPT-taxa (%) (supplementary material). A decrease in variation and an increase in the mean value with increasing sample size (e.g., Fig. 2) was observed for the number of individuals and the number of taxa (supplementary material). The type of metric response to increasing sample size was identical for all habitats and streams in both the Netherlands and Slovakia (supplementary material). The ASPT values showed either one of the two described responses or an intermediate response (Fig. 3), depending on the habitat (supplementary material).

The Saprobic Index and the metric type Aka + Lit + Psa (%) showed relatively low variability (Fig. 4). A sample size of 0.5 m or less was in all cases sufficient to reach a CV of ≤ 10%, with two exceptions: (1) in case of the habitat akal (NL) and the Saprobic Index a sample size of 2.5 m was required to reach a CV of ≤ 10% and (2) in case of the habitat submerged macrophytes (NL) and the metric type Aka + Lit + Psa (%) a sample size of 1.5 m was required to reach a CV of ≤ 10% (Fig. 4).

The ASPT and the number of taxa showed intermediate variability (Fig. 4). The sample size required to achieve a CV of ≤ 20% for the ASPT was 0.25 m. However, to achieve a CV of ≤ 10% for the ASPT the sample size had to be much larger for the habitats akal (1.25 m) and sand (1.75 m) in the Netherlands. For the other habitats the sample size required to achieve a CV of ≤ 10% varied between 0.25 and 0.75 m. For the number of taxa the sample size required to achieve a CV of ≤ 20 % was low (0.25–0.75 m). As for the ASPT, however, the sample size had to be much larger to achieve a CV of ≤ 10% (0.75–2 m) and differences between habitats became obvious. Variability in the number of taxa did not increase as a function of the number of taxa or the number of individuals collected from a habitat. For example, the metric number of taxa showed higher variability for sand samples than FPOM samples (Fig. 4), while the number of individuals and the number of taxa collected from the FPOM samples were higher than the number of individuals and taxa collected from the sand samples (Table 1).

The EPT-taxa (%) and the number of individuals showed high variability in most cases (Fig. 4). The sample size required to achieve a CV of ≤ 10% for the EPT-taxa (%) varied highly from 0.5 to 4.25 m in both countries, depending on the habitat. Results for the EPT-taxa (%) from the habitat FPOM are not depicted in Figure 4, because EPT-taxa were only found in three of the 20 sampling units and in very low percentages (3.4% on average). The sample size required to achieve a CV of ≤ 10% for the EPT-taxa (%) was 2.5 m on average, whereas it was 1 m on average to achieve a CV of ≤ 20%. To achieve a CV of ≤ 10%, all habitats required a sample size of at least 1.75 m, except for the habitats akal (NL) and macrolithal (S). The differences between habitats were

[1] Electronic supplementary material is available for this article at <http://dx.doi.org/ 10.1007/s10750-006-0074-7> and accessible for authorised users.

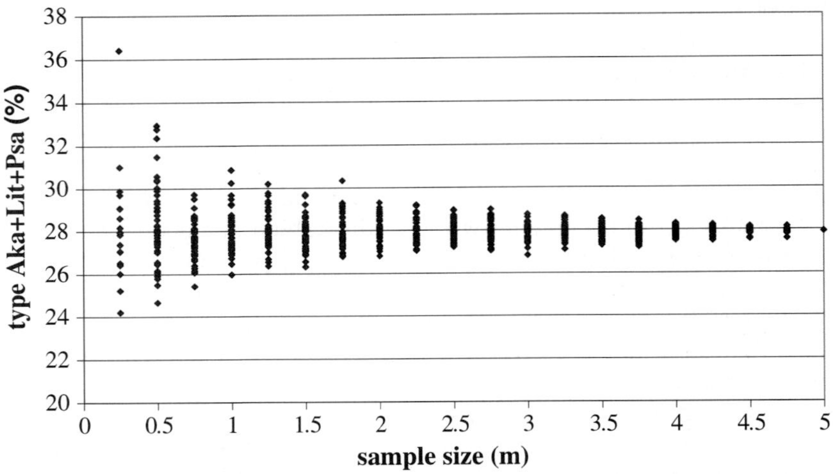

Figure 1. Response of type Aka + Lit + Psa (%) values to increasing sample size for composite FPOM samples from the Molenbeek.

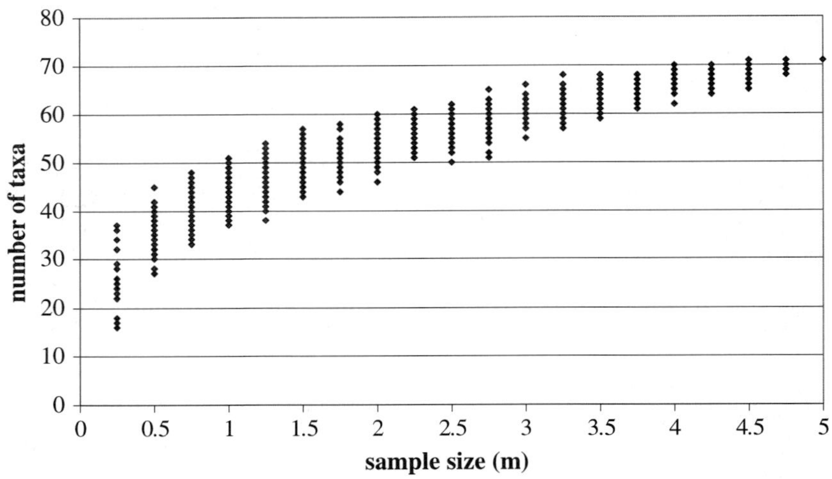

Figure 2. Response of the number of taxa to increasing sample size for composite FPOM samples from the Molenbeek.

somewhat smaller for the number of individuals than for the EPT-taxa (%) with the sample size required to achieve a CV of ≤10% ranging from 2.5 to 4 m. On average sampling of 3 m (CV of ≤20%) and 1.5 m (CV of ≤10%) was required for the number of individuals.

Akal was the only habitat sampled both in the Netherlands and in Slovakia. The difference in the sample size required to achieve a CV of ≤10% for this habitat between the Netherlands and Slovakia was less than 0.75 m for the number of individuals, the ASPT and the metric type Aka + Lit + Psa (%) (Fig. 4). The differences in the sample size required to achieve a CV of ≤10% were much higher for the number of taxa (1 m), the EPT-taxa (%) (2 m) and the Saprobic Index (2.25 m).

The sample size required to reach a CV of ≤10% and ≤20% for a multihabitat sample from streams in the Netherlands and Slovakia is shown in Table 2. The sample size required to attain a CV ≤10% for the Saprobic index, the metric type Aka + Lit + Psa (%) and the ASPT was considerable smaller than 5 m (between 1.5 and 3.75 m). The minimal sample size required to attain a CV of ≤10% for the metrics number of taxa, number of individuals and EPT-taxa (%)

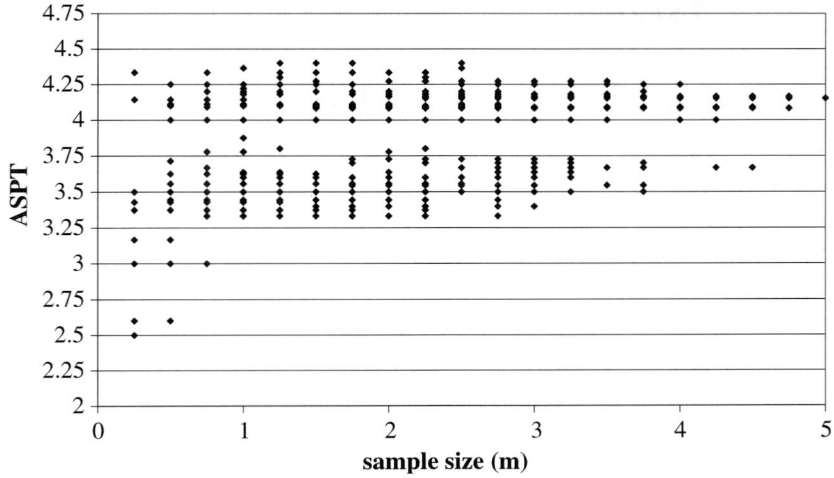

Figure 3. Response of the ASPT values to increasing sample size for composite FPOM samples from the Molenbeek.

varied between 4.5 and 13.75 m. To reach a CV of ≤20% the metrics ASPT, number of taxa, Saprobic Index and type Aka + Lit + Psa (%) required a considerable smaller minimal sample size compared to the EPT-taxa (%) and the number of individuals (between 2.75 and 6.5 m smaller). All metrics, except the number of individuals, required a minimal sample size of less than 5 m to attain a CV of ≤20%.

Accuracy and sample size

The same patterns were observed in the relative accuracy of metrics as in the relative variability of metrics: high accuracy corresponds to low variability. Like the differences in variability (Fig. 4), the differences in accuracy between metrics were high (Fig. 5). The Saprobic Index and the metric type Aka + Lit + Psa (%) showed relative high accuracy (Fig. 5). For both metrics a sample size of 0.25–0.5 m was sufficient to reach a MRD of ≤10%, with two exceptions: (1) in case of the habitat akal and the Saprobic index a sample size of 2.25 m was required and (2) in case of the habitat submerged macrophytes and the metric type Aka + Lit + Psa (%) a sample size of 1.5 m was required.

The ASPT showed intermediate accuracy (Fig. 5). The sample size required to achieve a MRD ≤20% for the ASPT was low (0.25–0.5 m). However, the sample size required to attain a MRD of ≤10% varied from 0.25 to 1.5 m depending on the habitat.

The EPT-taxa (%), number of individuals and number of taxa showed relatively low accuracy (Fig. 5). The sample size required to attain a MRD of ≤10% was 3 m on average for all three metrics. To attain a MRD of ≤20% this was 1.5 m on average. The pattern in relative accuracy for the number of taxa differed (Fig. 5) from the pattern in relative variability (Fig. 4). The metric showed intermediate variability compared to low accuracy.

The differences in accuracy and variability between habitats for the different metrics showed similar patterns (Figs. 4, 5). Differences in accuracy between habitats were larger when the deviation from the "reference" sample was higher, except for the number of taxa (Fig. 5). Differences in variability and accuracy between habitats were highest for the EPT-taxa (%) (Figs. 4, 5). Differences between habitats were minimal for the Saprobic Index values and the metric type Aka + Lit + Psa (%) for both variability and accuracy, with two exceptions: (1) the habitat akal showed low accuracy and high variability for the Saprobic Index and (2) the habitat submerged macrophytes showed low accuracy and high variability for the metric type Aka + Lit + Psa (%) compared to all other habitats (Figs. 4, 5). The difference in accuracy between habitats for the number of taxa was low compared to the differences in variability.

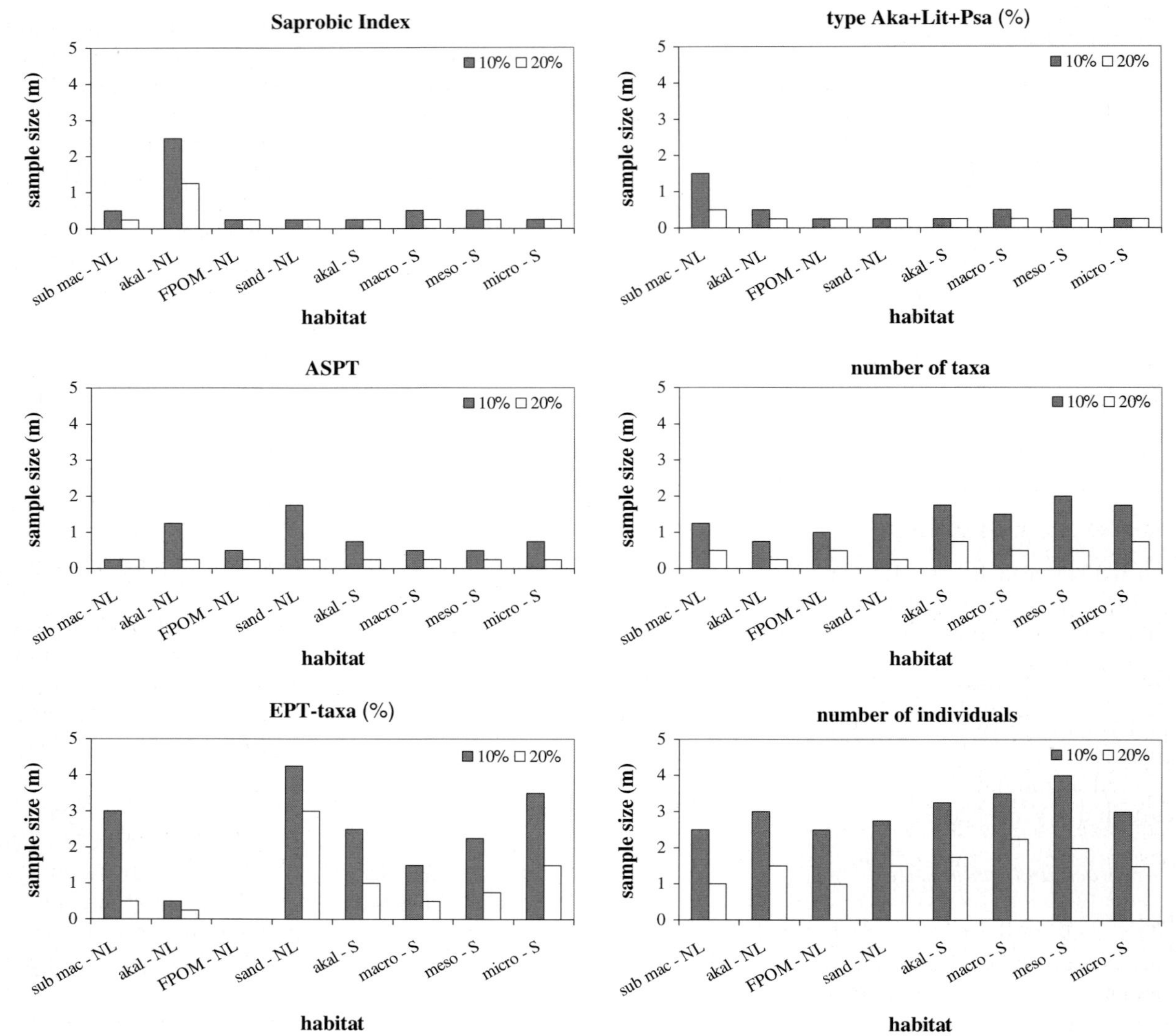

Figure 4. Overview of the minimal sample size required to attain a CV of ≤10% and ≤20 % (maximum) for each combination of habitat and metric (sub mac = submerged macrophytes; macro = macrolithal; micro = microlithal, meso = mesolithal; NL = Netherlands; S = Slovakia).

The differences in the sample size required to attain a MRD of ≤10% for the habitat akal between the Netherlands and Slovakia was less than 0.75 m for all metrics, except for the Saprobic Index (2 m; Fig. 5).

The sample size required to reach a MRD of ≤10% and ≤20% for a multihabitat sample from streams in the Netherlands and Slovakia is shown in Table 2. The sample size required to attain a MRD of ≤10% for the Saprobic index, the metric type Aka + Lit + Psa (%) and ASPT was smaller than 5 m (between 1.25 and 4 m). The sample size required to attain a MRD of ≤10% for the metrics number of taxa, number of individuals and EPT-taxa (%) varied between 6.75 and 15.5 m. To reach a MRD of ≤20% the metrics ASPT, Saprobic Index and type Aka + Lit + Psa (%) required a considerable smaller sample size compared to the EPT-taxa (%), the number of taxa and the number of individuals (between 1 and 10.25 m smaller). All metrics, except the EPT-taxa (%) from Dutch streams and the number of taxa,

Table 1. Overview of the number of individuals and number of taxa collected from the 20 sampling units per habitat and country

Habitat	Number of individuals	Number of taxa
The Netherlands		
Akal	2759	59
Submerged Macrophytes	3032	44
FPOM	7693	71
Sand	5404	63
Slovakia		
Akal	3246	54
Microlithal	2152	59
Mesolithal	1056	58
Macrolithal	1198	66

required a sample size of less than 5 m to attain a CV of ≤20%.

Assessment and sample size

The relation between sample size and the deviation from the ecological quality class associated with the "reference" sample differed between habitats. Assessment results for the habitat FPOM did not depend on sample size; a sample size of only 0.25 m resulted in all cases in an ecological quality class identical to that of the "reference" sample (Table 3). Assessment results for the habitat sand deviated from the "reference" samples for sample sizes varying between 1 and 1.75 m, but only in 4% of the cases (Table 3). In many cases small samples (0.25–0.75 m) from the habitats submerged macrophytes and akal showed a deviation in ecological quality class from the "reference" sample. To reduce the percentage of samples indicating an ecological quality class deviating from the "reference" sample to less than 10%, a sample size of at least 1 m is required when collecting samples from submerged macrophytes or akal (Table 3).

Sample processing costs

Mean sample processing time (or costs) increased with sample size for all habitats (Fig. 6). A twofold increase in sample size resulted in approximately a doubling of the costs. The relative increase in costs with an increase in sample size of 0.25 m (for sample sizes larger than 0.5 m) was relatively low (≤factor 1.3). The absolute increase in costs, however, was considerable, e.g., between 139 and 519 min for an increase in sample size from 0.75 to 1 m.

Costs varied considerably between habitats (Fig. 6). Irrespective of sample size, costs significantly differed between habitats ($p < 0.001$), except

Table 2. Overview of the minimal multihabitat sample size required to attain a CV of ≤10%, a CV of ≤20%, a mean relative deviation of ≤10% and mean relative deviation of ≤20% for each combination of metric and country (NL = The Netherlands; S = Slovakia)

Metric	Country	Sample size (m)			
		CV ≤10%	Mean relative deviation ≤10%	CV ≤20%	Mean relative deviation ≤20%
Type Aka + Lit + Psa (%)	NL	2.5	2.5	1.25	1.25
Type Aka + Lit + Psa (%)	S	1.5	1.25	1	1
EPT-taxa (%)	NL	7.75	9.75	3.75	5.25
EPT-taxa (%)	S	9.75	9.25	3.7	3.5
Number of individuals	NL	10.75	6.75	5	2.75
Number of individuals	S	13.75	12.25	7.5	6
ASPT	NL	3.75	3	1	1.5
ASPT	S	2.5	4.5	1	1
Number of taxa	NL	4.5	13.75	1.5	9.75
Number of taxa	S	7	15.5	2.5	11.25
Saprobic Index	NL	3.5	3.25	2	1.75
Saprobic Index	S	1.5	1.25	1	1

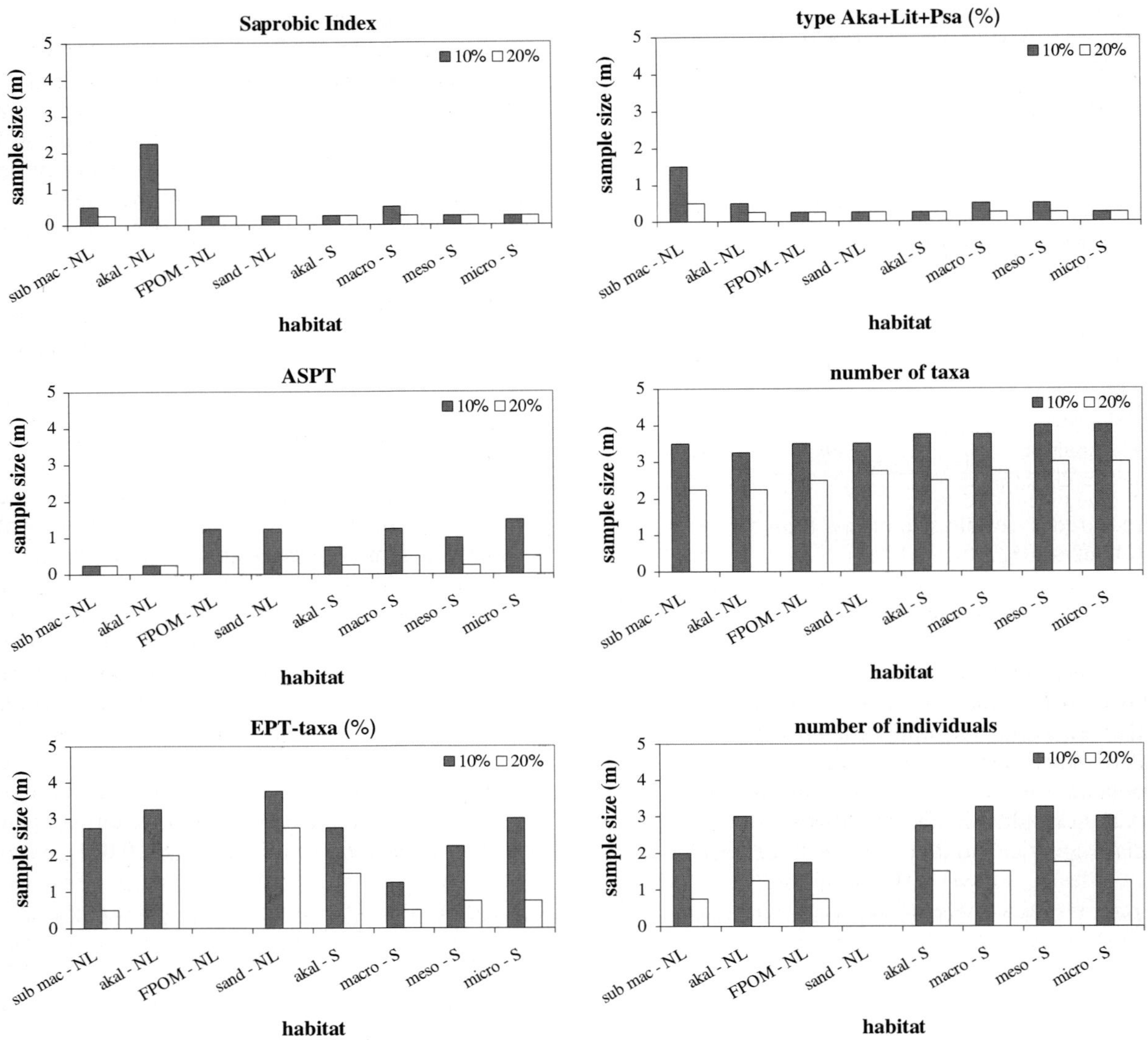

Figure 5. Overview of the minimal sample size required to attain a mean relative deviation of ≤10% and ≤20 % for each combination of habitat and metric (sub mac = submerged macrophytes; macro = macrolithal; micro = microlithal, meso = mesolithal; NL = The Netherlands; S = Slovakia).

for costs between sand and akal samples that did not differ significantly for a sample size of 0.25 m ($p = 0.053$). Processing of FPOM samples proved to be the most costly, followed by samples from the habitat sand, akal and submerged macrophytes, respectively (Fig. 6). The differences in costs between sand, akal and submerged macrophytes samples were relatively small compared to the differences in costs between FPOM samples and samples from all other habitats (Fig. 6).

Costs were related to the number of individuals collected from a sample. Costs for FPOM samples were relatively high, and so was the number of individuals collected from the FPOM samples (Fig. 6 and Table 1). The costs of FPOM samples were high compared to sand samples (factor 2.2 higher) and so was the number of individuals collected from FPOM samples (factor 1.4 higher). However, the differences in costs between FPOM and sand samples could not be completely

Table 3. Overview of the percentage of samples indicating an ecological quality class different from the "reference sample" per habitat (sampled in the Netherlands) and sample size

Sample size (m)	Habitat			
	Submerged macrophytes	Akal	FPOM	Sand
0.25	25	26	0	0
0.5	55	30	0	0
0.75	16	24	0	0
1	6	6	0	2
1.25	8	2	0	4
1.5	0	0	0	4
1.75	2	0	0	4

Percentages for sample sizes larger than 1.75 m are not listed, because these were zero.

explained by the differences in the number of individuals; the costs of FPOM samples were much higher than expected based on the number of individuals.

Costs were greatly reduced by not identifying Oligochaeta and Diptera (Figs. 6, 7). The costs of sand samples were reduced with a factor 2.7, of FPOM samples with a factor 1.9, of akal samples with a factor 1.3, and of submerged macrophytes with a factor 1.2. These reductions in costs were related to the number of Oligochaeta and Diptera individuals present in the samples. The FPOM and sand samples consisted for approximately 70% of Oligochaeta and Diptera individuals, while this percentage was only 40% for akal samples and 18% for submerged macrophytes samples. Even when Oligochaeta and Diptera were not identified the costs of FPOM samples were still the highest, followed by samples from the habitat akal, submerged macrophytes and sand (Fig. 7). Despite the decrease in costs associated with not identifying Oligochaeta and Diptera, a twofold increase in sample size still resulted in approximately a doubling of the costs.

The cost that had to be made to reach a CV of ≤10% and ≤20% for the individual habitats and the multihabitat samples are given in Table 4. The costs in Table 4 are directly related to the sample size. Only the costs related to variability are shown in Table 4 because results for accuracy and variability were similar (Figs. 4, 5). The costs of FPOM samples for the EPT-taxa (%) were not included in Table 4 because EPT-taxa were only found in 3 of the 20 sampling units, which means that the total costs for the EPT-taxa (%) were underestimated. The total costs (costs for a multihabitat sample) to achieve a CV of ≤20% were high for the number of individuals and the EPT-taxa (%), 96 and 62 h, respectively (Table 4). The total cost to achieve a CV of ≤20% for the other metrics varied between 20 and 34 h. To reduce CV from ≤20% to ≤10% an increase in total costs by a factor of 1.6 (19 h) for the Saprobic Index and by a factor of 1.5 (12 h) for the metric type Aka + Lit + Psa (%) was required (Table 4). The other metrics required an increase in total costs by a factor of 1.8 to a factor of 3.4, or an absolute increase in hours between 50 and 199.

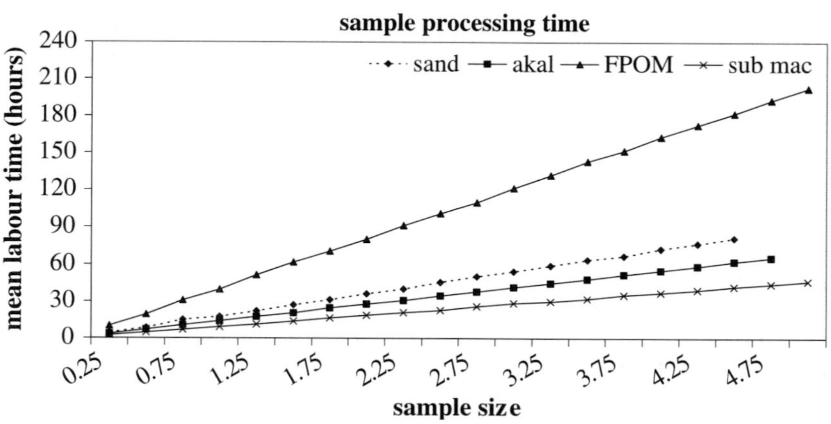

Figure 6. Mean sample processing time as a function of sample size for the habitats sand, akal, FPOM and submerged macrophytes from Dutch streams.

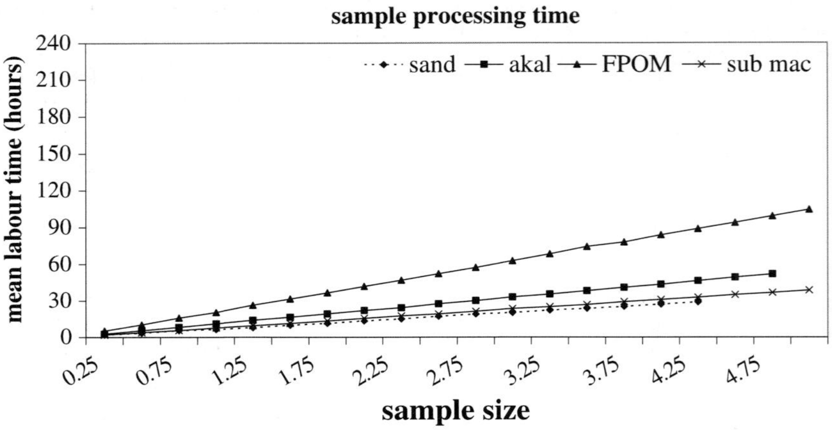

Figure 7. Mean sample processing time (excluding the identification of Oligochaeta and Diptera) as a function of sample size for the habitats sand, akal, FPOM and submerged macrophytes from Dutch streams.

The differences in total costs between metric to reach a CV of ≤10% were much larger than the differences in total costs between metrics to reach a CV of ≤20%. The total costs to reach a CV of ≤10% were low for the Saprobic Index (54 h) and the metric type Aka + Lit + Psa (%) (35 h) compared to the others metrics (between 70 and 215 h) (Table 4).

The absolute differences in total costs between metrics were lower when the costs for the identification of Oligochaeta and Diptera were not included, while the relative differences in total costs between metrics remained similar. When Oligochaeta and Diptera were not identified an increase in total costs by a factor of 1.7 for the Saprobic Index (16 h) and for the metric type Aka+-Lit+Psa (%) (10 h) was required to reduce CV from ≤20% to ≤10% (Table 4). The other metrics required an increase in costs by a factor of 2.1 to a factor of 3.1, or an absolute increase in hours between 26 and 69, when Oligochaeta and Diptera were not identified (Table 4).

To gain accuracy in assessment results, by reducing deviations from the ecological quality class with the "reference" sample, from ≤20% to ≤10% sample size (and costs) did not have to be increased for the habitats FPOM, sand and akal (Table 3). The habitat-submerged macrophytes required an increase in sample size from 0.75 to 1 m to achieve this gain in accuracy (Table 3), which is equal to an increase in labour time of 2 h (Fig. 6).

Discussion

Methodological approach

The optimal sample size is the largest possible (Green, 1979). One of the restrictions of this study was that variation and accuracy were studied based on the assumption that a sample size of 5 m would cover all variation of one habitat at a site. The data showed decreasing variation in metric values and increasing accuracy with increasing sample size. The decrease in variation with sample size might have been more gradual in reality. Samples of different sizes were created by randomly combining samples from the complete pool of 20 sampling units. The question is whether variation might have been higher if the samples of different sizes had been collected in the field. It is difficult to judge whether the 5 m sampled in this study covers all variation at a site. Compared to the sample sizes applied in biological surveillance monitoring an area sampled of 1.25 m^2 (=sampling over a length of 5 m) from *one habitat* is quite large, e.g., the mean area sampled in macroinvertebrate monitoring programs by USA state agencies is 1.7 m^2 for a *mulitihabitat* sample (Carter & Resh, 2001).

The sample size of the individual sampling units was approximately 25 cm. It was not possible to sample exactly 25 cm without disturbing the substrate prior to sampling. The small variation in sample size between the sampling units is not

Table 4. Overview of the sample processing time required to attain a CV of ≤ 10% and ≤ 20% including and excluding (labour time excl.) the identification of Oligochaeta and Diptera per habitat and metric

Metric	Habitat	Labour time (hours)		Labour time excl. (hours)	
		CV ≤ 10%	CV ≤ 20%	CV ≤ 10%	CV ≤ 20%
ASPT	Akal	17	3	14	3
	FPOM	19	10	10	5
	Sand	31	5	11	2
	Sub mac	2	2	2	2
	Total	**70**	**20**	**38**	**12**
EPT-taxa (%)	Akal	7	3	5	3
	FPOM	0	0	0	0
	Sand	77	54	29	20
	Sub mac	28	5	24	4
	Total	**112**	**62**	**58**	**27**
Number of individuals	Akal	41	20	33	16
	FPOM	101	40	52	20
	Sand	50	27	19	10
	Sub mac	23	9	19	8
	Total	**215**	**96**	**123**	**55**
Number of taxa	Akal	11	3	8	3
	FPOM	40	19	20	10
	Sand	27	5	10	2
	Sub mac	11	5	9	4
	Total	**88**	**32**	**48**	**19**
Saprobic Index	Akal	35	17	28	14
	FPOM	10	10	5	5
	Sand	5	5	2	2
	Sub mac	5	2	4	2
	Total	**54**	**34**	**39**	**23**
Type Aka + Lit + Psa (%)	Akal	7	3	5	3
	FPOM	10	10	5	5
	Sand	5	5	2	2
	Sub mac	14	5	11	4
	Total	**35**	**23**	**24**	**14**

Sample processing time was only recorded for habitat samples collected from streams in the Netherlands.

expected to have consequences regarding the applicability of the results of this study, since it will always be a problem to determine the exact sample size when sampling with a pond net in slow running streams.

Samples in this study have been collected between June and September. The fact that the habitats were not sampled simultaneously might have influenced the results. Studies performed in the Netherlands and in Slovakia, however, indicated that there are no significant differences in the number of individuals, the number of taxa, the EPT-taxa (%), ASPT values or Saprobic Index values between months (Šporka et al., 2006; Vlek, 2004). These findings make it unlikely that differences in variability between habitats were the result of differences between months.

In many European countries samples are preserved prior to sorting, while the samples (from the Netherlands) collected during this study were not preserved. Findings by Vlek (2004) suggest that the choice to preserve a sample or not will not influence variability and accuracy in metric values, i.e., Vlek (2004) detected no significant differences in the number of individuals, the number of taxa, the EPT-taxa (%), ASPT values or Saprobic Index values between preserved and unpreserved macroinvertebrate samples collected in the Netherlands.

The samples collected in this study came from different streams which makes it difficult to determine the effect of sample size on variability in metric values of a multihabitat sample. In this study the assumption was made that by reaching a CV (or MRD) of ≤10% for the individual habitats, a CV (or MRD) of ≤10% for the multihabitat samples would be guaranteed. Unfortunately, it was not possible to test this assumption since the habitats in this study came from different streams. Generally, macroinvertebrate community composition differs more among streams than within sites (e.g., Doberstein et al., 2000; Sandin & Johnson, 2000). Consequently, variability would be much higher in combining habitat samples from different streams than combining habitat samples from one stream. According to Beisel (1998) the variability in taxon richness and total abundance does not depend on the number of habitats sampled. This would suggest that metric values based on multihabitat samples would not be more variable than metric values based on single habitat samples, as was assumed in this study. Another difficulty was that the relation between variability/accuracy and multihabitat sample size was based on the four specific habitats sampled in the Netherlands and in Slovakia. This relation will have to be adjusted depending on the number and type of habitats present in the stream that is subjected to monitoring. Carter & Resh (2001) suggested that multihabitat samples would be more variable than single habitat samples, since sampling from multiple habitats in proportion to their cover is most likely to be operator-dependent and therefore more difficult to standardize than collecting from a single habitat samples. The variability in habitat coverage estimates is an extra source of variation that should be studied in the future.

Variability and sample size

High variability in metric values creates problems with assessment. As a result of high variability metric values will overlap between ecological quality classes. This overlap makes it impossible to distinguish between many ecological quality classes, complicating assessment (Doberstein et al., 2000).

When considering costs the metrics type Aka + Lit + Psa (%), Saprobic Index and ASPT should be preferred over the number of individuals, the number of taxa and EPT-taxa (%), for these showed relative low variability and high accuracy, which means that the required sample size to attain a certain degree of variability is smaller. For biological assessment it is important to know whether these metrics are also (highly) correlated to anthropogenic stress. Both the ASPT and the Saprobic Index are frequently applied in Europe and have proven to be highly correlated to organic pollution. The ASPT has been incorporated in multimetric indices in the Czech Rebuplic (Brabec et al., 2004), Greece (Skoulikidis et al., 2004), Italy (Buffagni et al., 2004), Sweden (Dahl et al., 2004) and the United Kingdom (Clarke et al., 2002). The Saprobic Index (or derivations from this index) has been incorporated in multimetric indices in Austria (Ofenböck et al., 2004), the Czech Republic (Brabec et al., 2004), Germany (Rolauffs et al., 2004), the Netherlands (Vlek et al., 2004) and Sweden (Dahl et al., 2004). A possible correlation between anthropogenic stress and type Aka + Lit + Psa (%) values are yet to be established.

The number of taxa and the number of individuals are notoriously poor metrics (Karr & Chu, 1999). The number of individuals showed high variation compared to the other metrics evaluated in this study. Apparently, significant variation in faunal densities occurs over small spatial scale, possibly caused by invertebrate aggregations (Downes et al., 1993).

Differences in variability between habitats depended on the metric studied, indicating that differences in variability between habitats could not be explained based on general assumptions about habitat heterogeneity. In general, metrics characterised by higher variability showed larger differences between habitats.

The large differences in variability for the number of taxa, the EPT-taxa (%) and the Saprobic Index between akal samples from the Netherlands and Slovakia might have been the result of regional differences or different sample processing protocols. The Slovakian samples were washed through a 500 μm mesh size sieve, while the Dutch samples were washed through a 250 μm mesh size sieve. It is not clear why the differences in variability are so high for the EPT-taxa (%) and the Saprobic Index compared to the other metrics.

Accuracy and sample size

As long as metric values are highly correlated to anthropogenic stress, high accuracy is not per definition required for assessment purposes, since class boundaries applied in an assessment system should always be calibrated based on data. In cases where scientists are interested in the 'true' community composition instead of biological assessment, accuracy (apart from variability) becomes very important. It is difficult to obtain accurate measurements of richness due to the collector's curve phenomenon (Colwell & Coddington, 1994; Fig. 2). This phenomenon resulted in high costs to establish accurate values for the number of taxa and the percentage of EPT-taxa. Colwell & Coddington (1994) stated that the number of taxa encountered in a sample increases asymptotically as a function of both the area sampled and the number of individuals in a sample. Lorenz et al. (2004) suggested that the curve is also a function of taxa diversity and that in streams with lower species diversity richness measures are likely to approach an asymptote at a smaller sample size. In this study no evidence was found to suggest that the number of taxa collected increased as a function of the number of individuals or the number of taxa in a sample. Cao et al. (2002) and Clarke et al. (2002) found that sampling variability in the number of taxa increased with the mean number of taxa recorded at a site. Doberstein et al. (2000) found low variances in metric values in streams with relatively few taxa. This study did not confirm the findings of Doberstein et al. (2000), Cao et al. (2002) and Clarke et al. (2002) because no evidence was found to suggest that the number of taxa collected increases as a function of the number of taxa in a sample and only minor differences were detected between habitats (determines the number of taxa in a sample) in variability and accuracy in the number of taxa compared to Cao et al. (2002). Where Cao et al. (2002) compared differences between habitats in the same river or site we compared habitats from different streams in different countries. Cao et al. (2002) detected differences in total taxon richness of more than 30% (based on one sampling unit). We detected differences in total taxon richness between Dutch habitats of 8% and between Slovakian habitats of 18%. An explanation for the differences between our study and that of Doberstein et al. (2000), Cao et al. (2002) and Clarke et al. (2002) might be the range in the number of taxa collected from the habitats in our study (between 44 and 71 taxa). This assumption is supported by Cao et al. (2002), who showed that relative differences in total taxon richness (%) are much larger when comparing a community of 20 taxa with a community of 60 taxa, than when comparing a community of 60 with a community of 100 taxa. So, caution should be taken in basing decisions concerning sample size on the results of this study when sampling habitats with less than 44 taxa.

Differences in accuracy between habitats depended on the metric studied, indicating that differences in accuracy between habitats could not be explained based on general assumptions about habitat characteristics. In general, metrics characterised by lower accuracy showed lager differences between habitats.

The large differences in accuracy for the Saprobic Index between akal samples from the Netherlands and Slovakia might have been the result of regional differences or different sample processing protocols.

Sample processing costs

Costs were based on identifications to species level and identification of all specimens. Some metrics, however, do not necessitate identification to species level or identification of all groups. For example, the calculation of the Saprobic Index, the metric type Aka + Lit + Psa (%), the ASPT or the EPT-taxa (%) does not require the identification of Oligochaeta and Diptera. In the Netherlands Oligochaeta and Diptera can make up a large part of the total number of individuals in a sample. Instead of determining the costs for the different metrics separately, which would be lengthy, the costs excluding the identification of Oligochaeta and Diptera were determined. This means that the costs for the ASPT and the EPT-taxa (%) are in reality lower than indicated in this study because these metrics do not necessitate the identification of other groups besides Oligochaeta and Diptera. The assumption made in this study was that often a combination of metrics (multimetric) will be used for assessment, thereby requiring the identification of the majority of the groups. For this reason, differences in costs between metrics were not taken into account. In case these differences in costs are taken into account the metrics ASPT and EPT-taxa (%) might still be calculated against reasonable costs, despite their high variability.

Apart from the groups that are identified, taxonomic resolution plays an important role in the costs associated with sample processing. All cost-related comparisons made in this study have been based on identifications to species level. The ASPT is a metric that requires identification to family level only. When the ASPT is the only metric used for bioassessment purposes and identifications can be performed at family level, the cost associated with the ASPT would probably be comparable to the costs associated with the Saprobic Index or the metric type Aka + Lit + Psa (%).

Differences in sample processing costs between habitats could not completely be related to the number of individuals collected. Other factors, e.g., the characteristics of the collected material sampled (large amounts of small dark particulate matter makes it more difficult to detect organisms) or previous experience of the analysts with the taxa collected also might have played a role.

The samples in this study were collected by pushing the net through the upper layer of the substratum, collecting the complete upper layer. The amount of material and the number of individuals collected through kick sampling or jabbing the substratum would have been much lower (Vlek, 2004). Since costs are directly related to the amount of material and the number of individuals collected (Barbour & Gerritsen, 1996), sample processing costs can expected to be much lower in case of kick sampling or jabbing the substratum instead of sampling the complete upper layer of the substratum.

Assessment and sample size

Reason for this study was the large amount of time that is needed for the processing of samples collected with the AQEM method. In the AQEM project multimetric indices were developed based on multihabitat samples collected according to the AQEM method (Hering et al., 2004). The assessment of anthropogenic stress with multimetric indices based on multihabitat samples has been frequently applied in the United States (Ohio EPA, 1987; Plafkin et al., 1989; Barbour et al., 1992; Kerans et al., 1992; Barbour et al., 1996; Major et al., 1998 and Maxted et al., 2000) and Europe (Hering et al., 2004). Arguments in favour of this approach are (1) by collecting macroinvertebrates from all the habitats present in proportion to their coverage a sample is a better representative of the habitats (and organisms) present in the sampled reach than when collecting from a single habitat (Carter & Resh, 2001); limiting sampling to a single habitat means that certain kinds of anthropogenic stress, which only influence specific habitats, may go undetected (Kerans et al., 1992); (2) multimetric indices provide detection capability over a broader range and nature of stressors and give a more complete picture about ecosystem health (Karr et al., 1986; Barbour et al., 1996).

The calculation of ecological quality classes in this study was based on samples from one habitat. However, the multimetric index used to calculate the classes was calibrated based on multihabitat samples (Vlek et al., 2004). Calculations of the ecological quality classes based on multihabitat samples would most likely have resulted in different classes compared to the calculations based on samples from one habitat. Still, the acquired information is very valuable in the sense that it gives an idea about the sensitivity of assessment results to reductions in sample size.

The differences in the percentage of misclassifications (a deviation in ecological quality class from the "reference" sample) between habitats could not be explained based on general assumptions about habitat heterogeneity; otherwise the variability in metric values would have been higher for samples from submerged macrophytes and akal than for samples from sand and FPOM for all metrics studied. Of the metrics evaluated in this study the metrics EPT-taxa (%), ASPT and type Aka + Lit + Psa (%) are incorporated in the multimetric index. The differences in misclassification between habitats could neither be explained by the variation in EPT-taxa (%) values. Variability in EPT-taxa (%), ASPT and type Aka + Lit + Psa (%) values together were not higher for submerged macrophytes and akal samples than for sand and FPOM samples. The differences in misclassification between habitats seemed to be related to other metrics incorporated in the multimetric index. The low number of misclassifications for the sand samples did not reflect the relatively high variation in EPT-taxa (%) values, two possible explanations can be (1) EPT-taxa (%) values did not happen to fall near a breakpoint in the scoring criteria (Fore et al., 2001) and/or (2) the combination of several metrics makes the multimetric index robust.

It is difficult to predict the influence of variability/accuracy for different individual metrics on the variability and accuracy of the final assessment result (Vlek, 2004). This is, among others, due to the fact that it is very important whether metric values for a single sample happen to fall near a breakpoint in the scoring criteria (Fore et al., 2001). Water managers will be interested in the probability that assessment results indicate less than good ecological quality while in reality ecological quality is good (false positives, type I error), because false positives will lead to unnecessary restoration measures (CIS working group 2.3, 2003). Organisations dealing with nature conservation will of course be interested in the probability that assessment results indicate good quality while in reality the ecological quality is less than good (false negatives, type II error). It is unlikely that water managers will take more than one multihabitat sample for the purpose of routine biological monitoring, due to costs considerations. So, instead of calculating the number of samples necessary to achieve a low error, they would be interested in knowing the error associated with taking only one sample. With information on the variability in individual metric values, the program STARBUGS (Clarke, 2004) can be used to calculate the effect of differences in estimates of habitat coverage and the effect of variability in individual metric values on the final assessment result of individual samples. The information on variability in the supplementary material can be used to perform the mentioned calculations for different multimetric indices. However, assumptions will have to be made about the variability of multihabitat samples based on single habitat variability. Because it is not clear whether the differences in variability and accuracy between samples from the Netherlands and Slovakia were caused by regional differences or different sample processing protocols, the application of the information in the supplementary material should be limited to the studied stream types in Slovakia and in the Netherlands.

The information in this paper gives scientists and water managers the opportunity of weighing a decrease in variability and an increase in accuracy on the one hand against the increase in costs on the other hand. Hopefully, the outlined approach shows water managers that the consequences of poor decision making potentially outweigh the savings associated with smaller sample area (Doberstein et al., 2000).

Conclusions and recommendations

Accuracy and variability varied depending on the habitat and the metric examined. This leads to the conclusion that sample size applied for biological monitoring should be based on the specific habitats present in a stream and the metric(s) used for bioassessment.

Assessment based on the number of taxa, the ASPT, the EPT-taxa (%) or the number of individuals is relative expensive compared to assessment based on the Saprobic Index or the metric type Aka + Lit + Psa (%), when specimens are identified to species level and a CV of 10% is aspired. These relative expensive metrics also require a high absolute increase in costs to realise a decrease in CV from ≤20% to ≤10%, while this decrease in costs requires (for most habitats) a relative low (or even no) increase in costs for the Saprobic Index and the metric type Aka + Lit + Psa (%). The increase in costs necessary to reduce variability for the Saprobic Index and the metric type Aka + Lit + Psa (%) is certainly justifiable given the possible implications of incorrect assessment results. When assessment of Dutch streams is based on the Saprobic Index or the metric type Aka + Lit + Psa (%) it is, therefore, recommended to strive for a CV of ≤10%. A CV of ≤10% can be achieved by sampling 3.5 m (54 h, including identification of Oligochaeta and Diptera) in case of the Saprobic Index or 2.5 m (35 h, Oligochaeta and Diptera) in case of the metric type Aka + Lit + Psa (%). The indicated sample sizes for multihabitat samples are based on streams in the Netherlands where the habitats FPOM, akal, submerged macrophytes and sand are present. For streams in Slovakia (small siliceous mountain streams in the West Carpathian) a CV of ≤10% can be achieved by sampling 1.5 m in case of both metrics. The indicated sample size is based on multihabitat samples from streams in Slovakia where the habitats akal, macrolithal, mesolithal and microlithal are present.

The recommended multihabitat sample sizes are based on a fixed sample size per habitat and do not depend on the coverage of the individual habitats in a stream. Results of this study suggested that a multihabitat sample size of less than 5 m is also adequate to attain a CV and MRD of ≤10% for the metric ASPT. The metrics number of taxa, number of individuals and EPT-taxa (%) require a multihabitat sample size of more than 5 m to attain a CV and MRD of ≤10%. For the metrics number of individuals and number of taxa a multihabitat sample size of 5 m is not even adequate to attain a CV and MRD of ≤20%.

Accuracy of the multimetric index for Dutch slow running streams depends on the sampled habitat(s). No extra costs are associated with an increase in accuracy from ≤20% to ≤10% for akal, FPOM and sand samples. However, the sample size of submerged macrophytes samples has to be increased from 0.75 to 1 m to achieve this increase in accuracy. This increase in sample sizes equals an increase in labour time of 2 h, which is no much considering the possible implications of incorrect assessment results. Hence, it is recommended to strive for a an accuracy of ≤10%, which requires a multihabitat sample size of 2.5 m (0.25 m FPOM, 0.25 m sand, 1 m akal and 1 m submerged macrophytes) and a labour time of 26 h (excluding Oligochaeta and Diptera) or 38 h (including Oligochaeta and Diptera).

Acknowledgements

This study was carried out within the STAR project funded by the European Commission, 5th Framework Program, Energy Environment and Sustainable Development, Key Action Water, Contract No. EVK1-CT-2001-00089. We are very grateful for the helpful comments made by Rebi Nijboer and two anonymous reviewers on an earlier version of the manuscript. We would like to thank Tjeerd-Harm van den Hoek, Martin van den Hoorn, Rink Wiggers, Tomáš Derka, Eva Bulánková, Daniela Illéšová, Zuzana Pastuchová and Zuzana Zaťovičová for their efforts in collecting the data on which this study was based.

References

AQEM consortium, 2002. Manual for the Application of the AQEM System. A Comprehensive Method to Assess European Streams using Benthic Macroinvertebrates, Developed for the purpose of the Water Framework Directive. Version 1.0, February, 2002.

Armitage, P. D., D. Moss, J. F. Wright & M. T. Furse, 1983. The performance of a new biological water quality score system based on macroinvertebrates over a wide range of unpolluted running-water sites. Water Research 17: 333–347.

Barbour, M. T. & J. Gerritsen, 1996. Subsampling of benthic samples: a defense of the fixed-count method. Journal of the North American Benthological Society 15: 386–391.

Barbour, M. T., J. Gerritsen, G. E. Griffith, R. Frydenborg, E. McCarron, J. S. White & M. L. Bastian, 1996. A framework for biological criteria for Florida streams using benthic macroinvertebrates. Journal of the North American Benthological Society 15: 185–211.

Barbour, M. T., J. L. Plafkin, B. P. Bradley, C. G. Graves & R. W. Wisseman, 1992. Evaluation of EPA's rapid bioassessment benthic metrics: metric redundancy and variability among reference stream sites. Environmental Toxicology and Chemistry 11: 437–449.

Bartsch, L. A., W. B. Richarson & T. J. Naimo, 1998. Sampling benthic macroinvertebrates in a large floodplain river: considerations of study design, sample size and cost. Environmental Monitoring and Assessment 52: 425–439.

Beisel, J. N., P. Usseglio-Polatera, S. Thomas & J. C. Moreteau, 1998. Effects of mesohabitat sampling strategy on the assessment of stream quality with benthic invertebrate assemblages. Archiv für Hydrobiologie 142: 493–510.

Brabec, K., S. Zahrádlová, D. Němejcová, P. Pařil, K. Kokeš & J. Jarkovský, 2004. Assessment of organic pollution effect considering differences between lotic and lentic stream habitats. Hydrobiologia 516: 331–346.

Buffagni, A., S. Erba, M. Cazzola & J. L. Kemp, 2004. The AQEM multimetric system for the southern Italian Apennines: assessing the impact of water quality and habitat degradation on pool macroinvertebrates in Mediterranean rivers. Hydrobiologia 516: 313–329.

Cao, Y., W. P. Williams & A. W. Bark, 1997. Effects of sample size (replicate number) on similarity measures in river benthic Aufwuchs community analysis. Water Environment Research 69: 107–114.

Cao, Y., D. Williams & D. P. Larsen, 2002. Comparison of ecological communities: the problem of sample representativeness. Ecological Monographs 72: 41–56.

Carter, J. L. & V. H. Resh, 2001. After site selection and before data analysis: sampling, sorting, and laboratory procedures used in stream benthic macroinvertebrate monitoring programs by USA state agencies. Journal of the North American Benthological Society 20: 658–682.

Chutter, F. M., 1972. A reappraisal of Needham and Usinger's data on the vriability of a stream fauna when sampled with a Surber sampler. Limnology and Oceanography 17: 139–141.

Common Implementation Strategy (CIS) Working Group 2.3 – REFCOND, 2003. Guidance on Establishing Reference Conditions and Ecological Status Class Boundaries for Inland Surface Waters. European Commission, Version 7.0, 93 pp.

Clarke, R. T., 2004. Error/Uncertainty Module Software STARBUGS (STAR Bioassessment Uncertainty Guidance Software) User Manual. STAR (Standardisation of River Classifications) Deliverable 9. Produced under European Union 5th Framework Programme Contract EVK1-CT 2001-00089.

Clarke, R. T., M. T. Furse, R. J. M. Gunn, J. M. Winder & J. F. Wright, 2002. Sampling variation in macroinvertebrate data and implications for river quality indices. Freshwater Biology 47: 1735–1751.

Colwell, R. K. & J. A. Coddington, 1994. Estimating terrestrial biology through extrapolation. Philosophical Transactions of the Royal Society (Series B) 345: 101–118.

Dahl, J., R. K. Johnson & L. Sandin, 2004. Detection of organic pollution of streams in southern Sweden using benthic macroinvertebrates. Hydrobiologia 516: 161–172.

Doberstein, C. P., J. R. Karr & L. L. Conquest, 2000. The effect of fixed-count subsampling on macroinvertebrate biomonitoring in small streams. Freshwater Biology 44: 355–371.

Downes, B. J., P. S. Lake & E. S. G. Schreiber, 1993. Spatial variation in the distribution of stream macroinvertebrates. Implications of patchiness for models of community organization. Freswater Biology 30: 119–132.

Elliot, J. M., 1977. Some Methods for the Statistical Analysis of Benthic Invertebrates, 2nd edn. Sci. Publ. No. 25, Freshwater Biological Association, Ferry House, U.K., 156 pp.

European Commission, 2000. Directive 2000/60/EC of the European Parliament and of the Council – Establishing a Framework for Community Action in the Field of Water Policy. Brussels, Belgium, 23 October 2000.

Fore, L. S., K. Paulsen & K. O'Laughlin, 2001. Assessing the performance of volunteers in monitoring streams. Freshwater Biology 46(1): 109–123.

Green, R. H., 1979. Sampling Design and Statistical Methods for Environmental Biologists. John Wiley and Sons, New York, 257 pp.

Heyer, R. W. & K. A. Berven, 1973. Sepcies diversity of herpetofauna samples from similar microhabitats at two tropical stations. Ecology 54: 642–645.

Hering, D., O. Moog, L. Sandin & P. F. M. Verdonschot, 2004. Overview and application of the AQEM assessment system. Hydrobiologia 516: 1–20.

Karr, J. R., K. D. Fausch, P. L. Angermeier, P. R. Yant & I. J. Schlosser, 1986. Assessing biological integrity in running waters: a method and its rationale. Illinois National History Survey, Champaign, Illinois, Special Publication 5.

Karr, J. R. & E. W. Chu, 1999. Restoring Life in Running Waters: Better Biological Monitoring. Island Press, Washington, DC.

Kerans, B. L., J. R. Karr & S. A. Ahlstedt, 1992. Aquatic invertebrate assemblages: spatial and temporal differences among sampling protocols. Journal of the North American Benthological Society 11: 377–390.

Lenat, D. R., 1988. Water quality assessment of streams using a qualitative collection method for benthic macroinvertebrates. Journal of the North American Benthological Society 7: 222–233.

Lorenz, A., L. Kirchner & D. Hering, 2004. 'Electronic subsampling' of macrobenthic samples: how many individuals are needed for a valid assessment result? Hydrobiologia 516: 299–312.

Major, E. B., M. T. Barbour, J. S. White & L. S. Houston, 1998. Development of a Biological Assessment Approach for Alaska Streams: A Pilot Study on the Kenai Peninsula. Environment and Natural Resources Institute, University of Alaska Anchorage, Anchorage, AK. Report for Alaska Department of Environmental Conservation, Anchorage, AK, 31 pp.

Maxted, J. R., M. T. Barbour, J. Gerritsen, V. Poretti, N. Primrose, A. Silvia, D. Penrose & R. Renfrow, 2000. Assessment framework for mid-Atlantic coastal plain streams using benthic macroinvertebrates. Journal of the North American Benthological Society 19: 128–144.

Metzeling, L. & J. Miller, 2001. Evaluation of sample size used for the rapid bioassessment of rivers using macroinvertebrates. Hydrobiologia 444: 159–170.

Needham, P. R. & R. L. Usinger, 1956. Variability in the macrofauna of a single riffle in Prosser Creek, California, as indicated by the Surber sampler. Hilgardia 24: 383–409.

Norris, R. H., E. P. McElravy & V. H. Resh, 1992. The sampling problem. In Calow, P. & G. E. Petts (eds), Rivers Handbook. Blackwell Scientific Publications, Oxford: 282–306.

Ofenböck, T., O. Moog, J. Gerritsen & M. Barbour, 2004. A stressor specific multimetric approach for monitoring running waters in Austria using benthic macro-invertebrates. Hydrobiologia 516: 251–268.

Ohio EPA (Enviromental Protection Agency), 1987. Biological Criteria for the Protection of Aquatic LifeI–IIIOhio EPA, Division of Water Quality Monitoring and Assessment, Surface Water Section, Columbus, Ohio.

Plafkin, J. L., M. T. Barbour, K. D. Porter, S. K. Gross & R. M. Hughes, 1989. Rapid Bioassessment Protocols for Use in Streams and Rivers: Benthic Macroinvertebrates and Fish. EPA/440/4–89/001. U. S. EPA Office of Water, Washington, DC.

Rolauffs, P., I. Stubauer, S. Zahrádlová, K. Brabec & O. Moog, 2004. Integration of the saprobic system into the European Union Water Framework Directive. Hydrobiologia 516: 285–298.

Sandin, L. & R. K. Johnson, 2000. The statistical power of selected indicator metrics using macroinvertebrates for assessing acidification and eutrophication of running waters. Hydrobiologia 422/423: 233–243.

Schmedtje, U. & M. Colling, 1996. Ökologische Typisierung der aquatischen Makrofauna. Informationsberichte des Bayerischen Landesamtes für Wasserwirtschaft 4/96.

Skoulikidis Th., N., K. C. Gritzalis, T. Kouvarda & A. Buffagni, 2004. The development of an ecological quality assessment and classification system for Greek running waters based on benthic macroinvertebrates. Hydrobiologia 516: 149–160.

Somers, K. M., R. A. Reid & S. M. David, 1998. Rapid biological assessments: how many animals are enough? Journal of the North American Benthological Society 17: 348–358.

Šporka, F., H. E. Vlek, E. Bulánková & I. Krno, 2006. Influence of seasonal variation on bioassessment of streams using macroinvertebrates. Hydrobiologia 566: 543–555.

Vlek, H. E. (ed.), 2004. Comparison of (Cost) Effectiveness between Various Macroinvertebrate Field and Laboratory Protocols. European Commssion, STAR (Standardisation of River Classifications), Deliverable N1, 78 pp.

Vlek, H. E., P. F. M. Verdonschot & R. C. Nijboer, 2004. Towards a multimetric index for the assessment of Dutch streams using benthic macroinvertebrates. Hydrobiologia 516: 173–189.

Zelinka, M. & P. Marvan, 1961. Zur Präzisierung der biologischen Klassifikation der Reinheit fließender Gewässer. Archiv für Hydrobiology 57: 389–407.

Hydrobiologia (2006) 566:543–555

M.T. Furse, D. Hering, K. Brabec, A. Buffagni, L. Sandin & P.F.M. Verdonschot (eds), The Ecological Status of European Rivers: Evaluation and Intercalibration of Assessment Methods
DOI 10.1007/s10750-006-0073-8

Influence of seasonal variation on bioassessment of streams using macroinvertebrates

Ferdinand Šporka[1,*], Hanneke E. Vlek[2], Eva Bulánková[3] & Il'ja Krno[3]
[1] *Department of Hydrobiology, Institute of Zoology, Slovak Academy of Sciences, Dúbravská cesta 9, SK-84506 Bratislava, Slovakia*
[2] *Alterra, Green World Research, P.O. Box 47, 6700 AA Wageningen, The Netherlands*
[3] *Department of Ecology, Faculty of Natural Sciences, Comenius University, Mlynská dolina B-2, SK-84215, Bratislava, Slovakia*
(**Author for correspondence: E-mail: ferdinand.sporka@savba.sk*)

Key words: bioassessment, macroinvertebrates, seasonal variation, Slovakia, stream

Abstract

The EU Water Framework Directive requires assessment of the ecological quality of running waters using macroinvertebrates. One of the problems of obtaining representative samples of organisms from streams is the choice of sampling date, as the scores obtained from macroinvertebrate indices vary naturally between seasons, confounding the detection of anthropogenic environmental change. We investigated this problem in a 4th order calcareous stream in the western Carpathian Mountains of central Europe, the Stupavský potok brook. We divided our 100 m study site into two stretches and took two replicate samples every other month alternately from each stretch for a period of 1 year, sampling in the months of February, April, June, August, October and December. Multivariate analysis of the macroinvertebrate communities (PCA) clearly separated the samples into three groups: (1) April samples (2) June and August samples (3) October, December and February samples. Metric scores were classified into two groups those that were stable with respect to sampling month, and those that varied. Of the metrics whose values increase with amount of allochthonous organic material (ALPHA_MESO, hyporhithral, littoral, PASF, GSI new, DSI, CSI), the highest scores occurred in February, April, October and December, while for metrics whose values decrease with content of organic material (DSII, DIS, GFI D05, PORI, RETI, hypocrenal, metarhithral, RP, AKA, LITHAL, SHRED, HAI) the highest values occurred in February, April, June and December. We conclude that sampling twice a year, in early spring and late autumn, is appropriate for this type of metarhithral mountain stream. Sampling in summer is less reliable due to strong seasonal influences on many of the metrics examined while sampling in winter is inappropriate for logistical reasons.

Introduction

With the implementation of the Water Framework Directive (WFD) every EU member state is obligated to assess the effects of human activities on the ecological quality of all water bodies (European Commission, 2000). Assessment of the ecological state of surface waters based on selected groups of living organisms as required by the Water Framework Directive (WFD) poses the problem of obtaining samples representative of the stream community. In collecting macroinvertebrate samples temporal and spatial changes in the community composition are two of the most important aspects that should be taken into account when collecting representative samples.

Temporal distributions of freshwater communities, both on the bottom and in the water column, are known to be influenced by the life histories of the various species (Hynes, 1972; Williams, 1981a). Ormerod (1987) showed that the most precise categorisation of assemblage type required a sampling strategy that combines both habitat and seasonal data. While many physical factors that have been shown to affect faunal assemblages are known to change seasonally (e.g., hydrological regime, water chemistry, light levels and temperature), lotic assemblages of invertebrates vary both seasonally and with spatial position within the stream (Matthews & Bao, 1991; Cowell et al., 2004). Setting a suitable time period for sampling a given habitat type is therefore a complex problem.

The establishment of reliable biomonitoring programmes is central to the effective implementation of the WFD for surface waters. Water managers prefer cost efficient methods, e.g. sampling in most cases only once a year for the purpose of surveillance monitoring. In contrast, studies aiming to assess conservation value normally require more than one sampling occasion within a given year to obtain adequate site evaluations (Furse et al., 1984). The choices made related to sampling strategies are always a trade off between biological reliability and economic considerations. When cost do not allow to take more than one sample a year at a site for the purpose of surveillance monitoring a higher level of standardisation and between site comparability could be reached if samples from the same area were collected in the same time period, thereby minimising variability in the observed communities due to natural seasonal differences. In many European countries there is an agreement about the period most suited for sampling macroinvertebrates, however in most cases scientific background to these agreements is lacking.

The aim of this study was therefore (1) to examine the variation in macroinvertebrate community composition between months (2) to assess the effects of natural seasonal community variation on metric values, and (3) to determine whether a preferred sampling period(s) could be identified for mountainous streams in Slovakia. A similar study in lowland streams (Heelsumse beek) was performed in the Netherlands (Vlek, in press). In combination these two studies combined make it possible to evaluate the influence of seasonal changes in macroinvertebrate community composition on metrics used for bioassessment purposes across two widely differing European stream types.

Materials and methods

Study site and data collection

Samples were collected from the Stupavský potok brook (N 48° 15′ 09.1″ E 17° 06′ 44.4″), a small, calcareous, 4th order stream in the Carpathian Mountains of central Europe (Fig. 1). The long-term discharge of Stupavský potok brook is characteristic of highland snowmelt streams (Šimo & Zaťko, 1980), with the highest discharges occurring at the beginning of spring (March and April; Fig. 2). It should be noted that the discharge during the study period was to some extent atypical, being generally lower than the long-term average and lacking a peak in the usual snow-melt period (gradual spring snow melt; Fig. 2).

The study site was a relatively uniform 100 m section of the stream (average width 5.1 m: average depth 0.16 m). This 100 m section was divided into two 50 m stretches. Two (replicate) samples were taken every other month in the last week of the month (April, June, August, October, December* and February, actually sampled 8th January), alternately from the two stretches (stretch 1 in April, stretch 2 in June etc.). Prior to sampling, habitat coverage was estimated for the complete 100 m section (AQEM consortium, 2002). For each habitat an area of 25×25 cm was sampled by kick-sampling using a 500 μm hand-net. Each habitat with a coverage of more than 5% was sampled separately. The area sampled per habitat was the same on all sampling occasions and the same operator collected all of the sub-samples. The samples were preserved in 4% formaldehyde prior to transportation to the laboratory for processing. In the laboratory the samples collected from the different habitats were sieved using 1000 and 500 μm sieves, and fully sorted under a stereomicroscope. Sorting was performed by a group of three people. The same specialist preformed all identifications of each major organism group. Macroinvertebrates were identified to the lowest taxonomic level possible (species level for almost all groups).

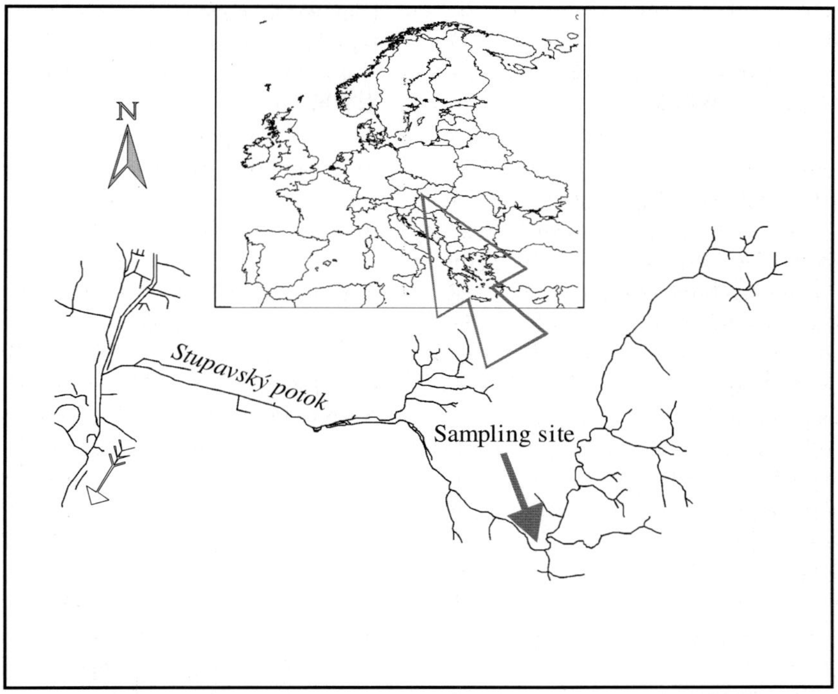

Figure 1. The catchment area of the Stupavský potok brook with sampling site.

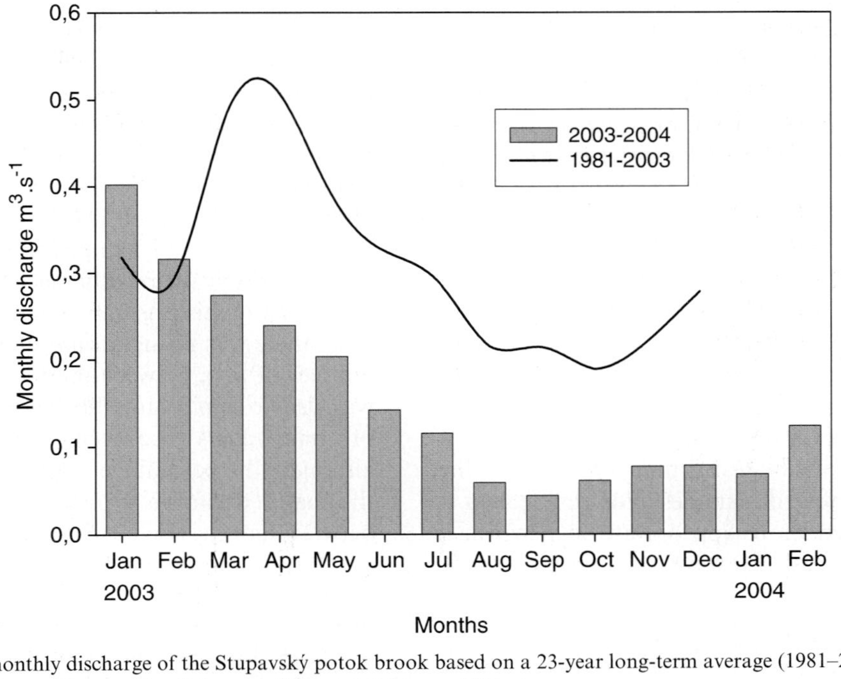

Figure 2. Average monthly discharge of the Stupavský potok brook based on a 23-year long-term average (1981–2003) and individual monthly averages between the months of January 2003 and February 2004.

Data analysis

Prior to analysis, samples from the different habitats were pooled together to form two composite samples. The number of individuals per taxon were standardised to a total sample area of 1.25 m^2 for each composite sample based on habitat coverage and sampled area (abundance* 1.25/area sampled). A Principal Components Analysis (PCA) using CANOCO 4.5 (ter Braak & Smilauer, 2002) was performed to examine variation in macroinvertebrate community composition between months. Species data were $\log_2 (x+1)$ transformed before analysis.

The effects of natural seasonal variation in community composition on metric values were assessed using a list of metrics commonly used in Europe (supplementary material).[1] The metrics were selected from an extensive list given by Hering et al. (2004). In addition to these metrics the number of taxa and the number of individuals for each major macroinvertebrate group (e.g. Diptera, Ephemeroptera, Plecoptera) was also evaluated. Some groups were only present at low abundances and in just a few samples. These groups were therefore excluded from our analyses because of the difficulties of finding appropriate transformations to normalise the data and the problems of having many zero values (Metzling et al., 2003). Metric values were calculated with the software ASTERICS version 1.0 (AQEM/STAR Ecological RIver Classification System; http://www.aqem.de) for all composite samples, except for the Slovak Saprobic index which is not included in the software. Slovak Saprobic index values were obtained from Šporka (2003). The coefficient of variation (CV = SD/mean), a measure of variability, was calculated for the different metrics. One-way analysis of variance (ANOVA) was used to identify significant differences between months ($\alpha = 0.05$) by SigmaStat 3.1 for Windows software.

Assumptions for normality and homogeneity of variance could not be tested in a reliable way due to the low number of samples. For this reason it might have been more appropriate to perform a non-parametric test. However, a non-parametric test would never be able to detect significant differences between protocols based on two replicates. Therefore it was decided to use the ANOVA and to transform metric values based on experiences in other studies. Abundance metrics were $\ln(x+1)$ transformed (Supplementary Material type 1). Taxa counts were not transformed and proportions were transformed $\ln(x+1)$–$\ln(y+1)$ (Supplementary Material type 2), where x = the number of individual taxa and y = the number of total taxa (Kerans et al., 1992). Biotic index data (e.g. Saprobic Index, BMWP, ASPT) were not transformed (Norris & Georges, 1993). Metrics like XENO (%), SHRED (%) and littoral (%) are not simple proportional metrics. The values for these metrics also depend on the strength with which a species prefers a certain category (AQEM consortium, 2002). The decision was made not to transform values of these metrics, since no information could be found to describe a suitable transformation. Acronym, metric description and type of transformation are given in Supplementary Material.

[1] Electronic supplementary material is available for this article at <http://dx.doi.org/ 10.1007/s10750-006-0073-8> and accessible for authorised users.

Results

Taxa analysis

In total 218 taxa were collected during this study. Each replicate contained on average 42% of the total number of taxa, and the total number of taxa occurring in both replicates from any 1 month varied between 56% and 70%. In macroinvertebrate community of the Stupavský potok brook the highest of number of taxa reached Diptera and Trichoptera (Fig. 3). Samples from different months did not exhibit major differences in the number of taxa per organism groups (Fig. 3, Table 1). Similarly, there was no significant difference in the total number of taxa between months ($p = 0.185$). There was also no significant difference in the total number of individuals between months ($p = 0.062$), although, the percentage of individuals for some of the major organism groups did vary significantly between months (Fig. 4, Supplementary Material).

During most months (except February and April) the Crustacea formed the largest proportion of the community (varying between 25 and 57%), followed by the Diptera (varying between 15 and 38%). In February however, the Diptera

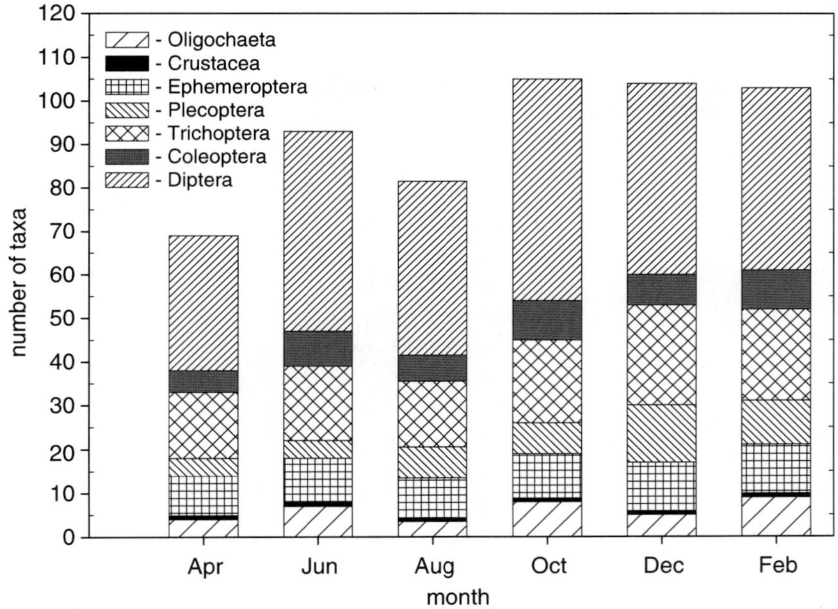

Figure 3. Between month variation in the number of taxa in the Stupavský potok brook based on the sum of both replicates. Only those groups that formed more than 5% of the total abundance are shown.

represented the largest part of the community, while Crustacea numbers were far lower and conversely represented the smallest proportion of the community (Fig. 4).

Multivariate analysis clearly divided the samples into three groups: (1) April samples (2) June and August samples (3) October, December and February samples (Fig. 5).

Dominant taxa that were found in high abundance more than 5% in at least 1 month are compiled in Table 2. *Gammarus fossarum* and species of the family Simuliidae predominated in the summer months. *Rhithrogena semicolorata* dominated in early spring, as did the caddisflies *Agapetus* sp., *Hydropsyche instabilis* and midges of the genus *Micropsectra*. Midges also formed a large proportion of the macroinvertebrates assemblage in October and December and *Hydraena gracilis* dominated in February.

Metric analysis

About 31 out of 76 metrics showed significant ($p < 0.05$) differences between months (Table 1). Between which months significant differences occurred depended on the metric. Metrics showing significant differences between individual months were classified into three groups – (a) those with values increasing with anthropogenic stress (e.g. organic pollution, general degradation, acidification) (b) those with values decreasing with anthropogenic stress and (c) those showing no direct relation to degradation (Hering et al., 2004) or being based on insufficient knowledge:

group a

Metrics that increase values with degradation – ALPHA_MESO, hyporhithral, littoral, PASF, GSI new, CSI. Five out of six metrics reached their lowest values in April and one in February.

group b

Metrics that decrease values with degradation – DSII, DIS, GFI D05, PORI, RETI, hypocrenal, RP, AKA, LITHAL, SHRED, HAI, EPHE, PLEC%, PLEC taxa, PLEC, TRIC. Five out of sixteen metrics reached their highest values in April, 3 out of 16 in August and October, 2 out of 16 in February, June and December.

group c

Metrics with unidentified or insignificant relationships with degradation: GRA + SCRA, metarhithral, DSI, COL taxa, RHYTI, CRUS,

Table 1. Months between which metrics values differed significantly ($p < 0.05$) in the Stupavský potok brook, based on the Least Significant Difference (LSD, $\alpha = 0.05$) and months when metrics reached minimal and maximal value

Acronym	p	Significant differences between	Min value	Max value
ALPHA-MESO (%)	0.003	Apr–other	Apr	Aug
GFI D03	0.045	None		
GFI D05	< 0.001	Apr–other	Jun	Apr
		Dec–other (except Feb)		
		Feb–Jun		
GSI new	0.018	Apr–Feb/Jun/Oct	Apr	Feb
DSI	< 0.001	Jun–other (except Aug)	Oct	Aug
		Aug–Feb/Oct/Dec		
		Apr–Feb/Oct		
		Dec–Feb		
CSI	0.013	Feb–Apr/Aug	Apr	Feb
		Apr–Oct		
MTS	0.049	None		
HAI	0.001	Feb–Jun/Oct/Dec	Jun, Oct	Feb, Aug
		Aug–Jun/Oct/Dec		
DSII	< 0.001	Feb–Jun/Aug	Aug	Feb
		Apr–Jun/Aug		
		Dec–Jun/Aug		
		Oct–Jun/Aug		
DIS	< 0.001	Dec–Jun/Aug	Aug	Feb
		Feb–Jun/Aug		
		Oct–Jun/Aug		
		Apr–Jun/Aug		
EVENNESS	< 0.001	Dec–Jun/Aug	Aug	Dec
		Apr–Jun/Aug		
		Feb–Jun/Aug		
		Oct–Jun/Aug		
RP (%)	0.004	Aug–Feb/Oct	Feb	Jun
		Jun–Feb		
		Dec–Feb		
AKA (%)	0.034	Jun–Apr	Apr	Jun
LITHAL (%)	0.026	Apr–Feb/Oct	Feb	Apr
Hypocrenal (%)	0.011	Jun–Feb/Dec	Feb	Jun
Littoral (%)	0.014	Apr–Jun/Aug/October	Apr	Jun
Metarhithral (%)	0.01	Apr–other	Feb	Apr
Hyporhithral (%)	0.018	Aug–Apr/Feb	Apr	Dec
SHRED (%)	0.008	Aug–Febr/April	Feb	Aug
		Jun–Feb		
PASF (%)	0.006	Aug–other (except Dec)	Feb	Aug
GRA + SCRA (%)	0.001	Apr–other	Aug	Apr
RETI	0.044	Apr–Feb	Feb	Apr
EPT taxa	0.05	None		
PLEC (%)	0.021	Dec–Apr/Jun	Feb	Apr
CRUS	0.006	Apr–other (except Feb)	Apr	Oct
EPHE	0.022	Oct–Jun/Aug	Aug	Oct

Continued on p. 549

Table 1. (Continued)

Acronym	p	Significant differences between	Min value	Max value
PLEC	0.018	Oct–Aug	Aug	Oct
PLEC taxa	0.03	Dec–Jun/Aug	Jun	Dec
TRIC	0.009	Oct–others (except April)	Jun	Oct
COL	0.005	Oct–Apr/Aug/Dec	Apr	Feb
		Feb–Apr/Aug		
COL taxa	0.032	Feb–Aug	Apr	Feb
DIP	0.02	Apr–Feb/Oct	Apr	Feb
PORI	0.012	Apr–Jun/Aug	Aug	Apr
RHYTI	0.032	Apr–Oct	Oct	Apr

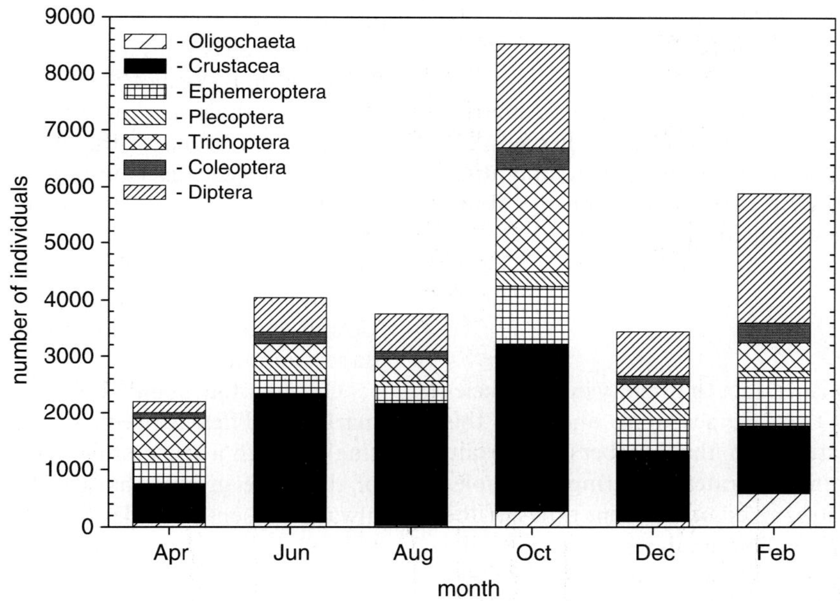

Figure 4. Between month variation in the number of individuals in the Stupavský potok brook, based on the average of both replicates. Only those groups that formed more than 5% of the total abundance are shown.

COL, DIP, Evenness. Among them, three metrics showed highest values in February and April and one in August, October and December, respectively. Four metrics reached the lowest values in April, two metrics in August and October and one in February.

Metrics that reached their maximum values in summer (group a) and differed significantly in value between summer and the other months were associated with poor water quality caused by low discharges (high CSI, PASF %, littoral %). Values of metrics indicating impairment of water quality in summer samples (June, August) are also influenced by summer emergence and the consequent absence of larval stages. The effects of summer emergence were also evident in the low values of the diversity (DIS, DSII) and evenness and low abundance values for certain taxonomic groups e.g., Plecoptera (Table 1). Percentage of dominant feeding types shows differences in individual months during the year (Fig. 6).

The coefficient of variation (CV) of significant metrics varied from 4.2 to 90.6% during the year

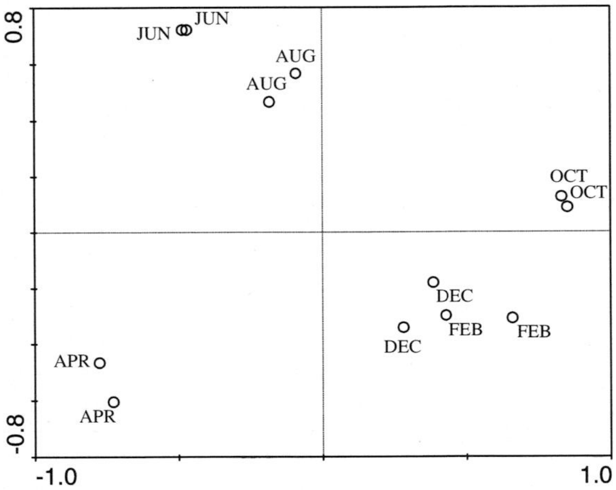

Figure 5. The first two axes of a PCA ordination of Stupavský potok brook macroinvertebrate samples from different seasons.

(Table 3). CV of the most of qualitative metrics does not exceed 20%. However, the highest CV values (above 40%) were found for the quantitative metrics that were based mainly on the abundance of a particular taxonomic group.

Discussion

It is a well-established fact that many insect species have life cycles that are seasonal, and that this results in fluctuations in the numbers of certain groups of macroinvertebrates occurring in samples taken from the streambed at different times of the year (Hynes, 1972). Our analyses show how the community as a whole is affected by macroinvertebrate seasonality and how individual bioassessment metrics can differ significantly between months as a consequence. We found that the majority of metrics exhibiting significant differences between months were quantitative metrics. So, when using quantitative metrics in assessment it is important to recognise that the season in which samples are taken can and often will have a strong influence on the results obtained. In terms of individual metrics, differences between months strongly depend on the metric under evaluation. This makes it difficult to give a general recommendation for a preferred sampling month or season. One option (although not a very practical one) might be to select a preferred season for each individual metric. For metrics directly related to the number of taxa or the number of individuals, the preferred sampling period might be the month in which their values are typically at their highest. In the Stupavský potok brook, the highest numbers of individuals of most major taxonomic groups were found at the end of October. Hynes (1972) showed that autumn is a period of egg hatching, and for many species it is a period of increasing or often of maximum, numbers, including many small individuals. Similarly, in lowland headwater streams of the Alafia River, Cowell et al. (2004) also found the highest abundances in autumn.

On the other hand, EPT metric values did not markedly differ between seasons because in any single month a reasonably representative selection of the three groups that make up this index was always present. Sprules (1947) similarly showed that while the number and diversity of Plecoptera decreases with increasing average summer temperature, the number and diversity of Ephemeroptera and Trichoptera increase, thereby

Table 2. Taxa with abundances more than 5% in one month. Percentage of individuals based on the average of both replicates

Months	Number of individuals (%)						
	Gammarus fossarum	*Rhithrogena semicolorata*	*Agapetus* sp.	*Hydropsyche instabilis*	Simuliidae Gen. sp.	*Micropsectra* sp.	*Hydraena gracilis*
Feb	20	7	1	3	1	23	8
Apr	31	3	16	7	1	0	0
Jun	56	1	1	3	3	3	0
Aug	57	0	7	1	12	2	0
Oct	35	0	12	5	1	9	0
Dec	36	6	1	5	4	7	0

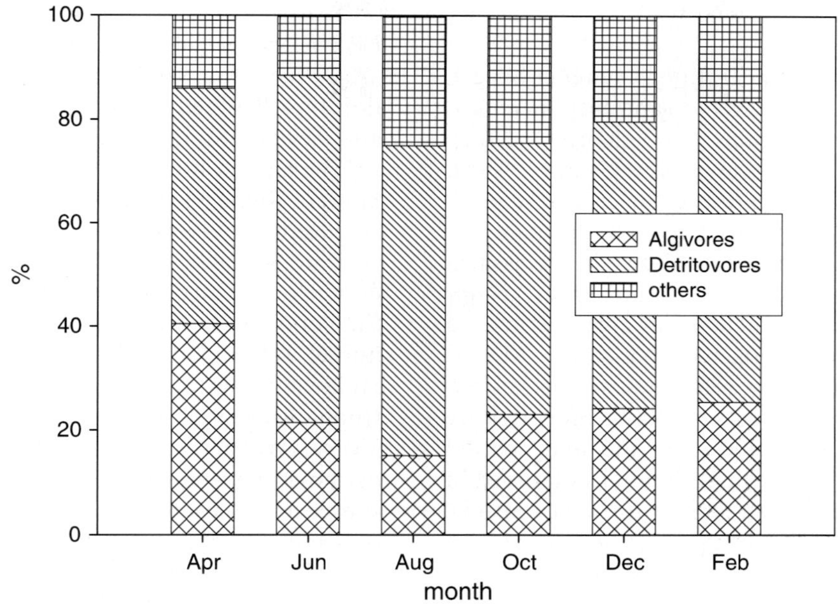

Figure 6. Between month variations in invertebrate food guilds of in the Stupavský potok brook. Percentage of functional feeding groups based on the average of both replicates. Only dominant food guilds are shown.

Table 3. The coefficient of variation (CV) of significant metrics for samples from the Stupavský potok

Metric	CV	Metric	CV	Metric	CV
GSI new	4.2	EPT-taxa	20.4	GRA+SCRA (%)	34.6
RHYTI	7.7	LITHAL (%)	22.8	PLEC	40.1
HAI	9.1	GFI D03	23.3	PLEC (%)	40.7
DSII	12.0	ALPHA-MESO 9%)	23.5	PLEC taxa	45.7
EVENNESS	13.2	RP (%)	23.7	CRUS	48.4
RETI	13.5	Littoral (%)	24.7	COL	52.8
DIS	14.2	Hypocrenal (%)	26.3	EPHE	54.2
MTS	15.4	Metarhithral (%)	26.8	PASF (%)	63.3
GFI D05	16.2	COL taxa	27.9	TRI	81.4
DSI	16.9	PORI	29.5	DIP	90.6
Hyporhithral (%)	17.8	SHRED (%)	29.8	–	–
AKA (%)	20.2	CSI	31.7	–	–

avoiding strong seasonal differences of EPT index scores. This effect has also been observed in the lowland stream Heelsumse beek in the Netherlands (Vlek, in press).

By examining the whole community using multivariate analyses we identified three distinct seasonal assemblages from spring (April), summer (June and August), and autumn and winter (October, December, and February). Individual metric results also indicated that macroinvertebrate community composition in the Stupavský potok brook in April differed from all other months. ALPHA-MESO (%) values were significantly lower in April than in all other months. The low values of ALPHA-MESO (%) in April indicate low amounts of allochthonous organic material. The significantly low CSI values can also be related to organic pollution. The low CSI values and the high values of RETI, GFI, PLEC (%), PORI in April suggest that the water quality of the

Stupavský potok is better in April than in all other months.

With increasing temperature in summer oxygen levels decrease and therefore saprobity increases. Under extreme conditions these changes become readily apparent, as shown by Coimbra et al. (1996) in their investigation of macroinvertebrate community in a temporary stream in Portugal. On the basis of multivariate analysis they classified macroinvertebrate communities into three groups according to environmental variables related to seasons and anthropogenic influences. Morais et al. (2004) studied the robustness of metrics under different hydrological conditions in temporary streams. Seasonal changes over the study period followed the general temporal pattern observed in other Mediterranean streams, with taxa sensitive to organic pollution being present under high discharge and more tolerant taxa under low discharge. The same pattern could be observed in the Stupavský potok brook. In summer due to low discharge the fauna consisted mostly of eurytopic species e.g., *Simulium* sp.

Several other studies have also shown that eurytopic species of the family Simuliidae are dominant in streams of the Small Carpathians Mts. in summer (Halgoš & Jedlička, 1974; Illéšová & Halgoš, 2003). Dahl et al. (2004) stated "However, though a summer sampling window may result in a better detection of oxygen stress, the summer emergence by aquatic insects often precludes the use of this season in bioassessment programmes in Sweden." Nijboer & Schmidt-Kloiber (2004) found that taxa indicating oligosaprobic conditions were taxa with small distribution ranges living in close proximity to stones and gravel (i.e., lithal). In the Stupavský potok brook, colonisation of the lithal substrate was at its greatest in April.

Many studies have shown that seasonal abundance of food may strongly influence the life cycles of the stream community (Ross, 1963; Neel, 1968; Williams & Hynes, 1973; Cummins, 1977; Moore 1977; Townsend & Hildrew, 1979; Williams 1981b). Based on the evaluation of energy flow, Krno (1996) distinguished two significantly different time periods within a year in terms of abiotic factors and food availability:

Cold season – High discharge, periphyton biomass and production of scrapers

Warm season – High temperature, biomass FPOM and production of filterers and collectors.

In the Stupavský potok brook similar relationships between abiotic factors, food resources, and the composition of trophic groups were found.

The highest values of the metrics GRA + SCRA % were found in April when discharge was highest. Representation of feeding types during the year in Stupavský potok brook shows a strong dominance of algophagous forms in spring and, on the contrary, dominance by detritophagous taxa during other parts of year. Similarly, Krno & Hullová (1988) found the largest proportion of this trophic group in the metarithral stretch of the Vydrica stream in the Carpathians in spring, when periphyton (representing an important food resource in this system) develops under the influence of increasing illumination. Krno (1996) also recorded the highest percentage of PASF % in summer when water temperatures were highest. These studies support the view that temperature is a key abiotic factor influencing macrozoobenthos structure (Sprules, 1947, Williams & Hynes, 1974). High temperatures result in high microbial activity and subsequently low oxygen concentrations (Dahl et al., 2004). The metrics reaching significantly higher values in August and June in relation to other months are typically regarded as indicators of poor water quality caused by reduced discharges and high temperatures (CSI, PASF %, hypocrenal % and littoral %).

In this study we have shown that seasonal changes in macroinvertebrate community composition have marked effects on many biotic indices. The life cycles of stream invertebrates, and the seasonal changes in community composition reflect on metric values, are caused primarily by the seasonal dynamics of variables such as temperature, light regime and the supply of nutrients and allochthonous organic material (Clifford, 1978; Bunn, 1986; Krno & Hullová, 1988; Doledec, 1989; Krno 1996). Spring is characterised by an increase in temperature, discharge, light and nutrient supply which results in an increase in primary production and abundance of algophagous invertebrates. This situation is accompanied by a stronger representation of lithophiles and rheophils, and the rapid development of spring forms of macrozoobenthos and emergence of water insects. In spring the metabolism of Small

Carpathian streams has been shown to be predominantly autotrophic (Krno & Hullová, 1988; Rodrigez & Derka, 2003). In the Stupavský potok brook this was confirmed by the highest values of the metric LITHAL % in April and the dominance of algophagous invertebrates (GRA + SCRA%). The progression to summer is characterised by relatively stable and high temperatures, reduced discharge and reduced illumination due to shadowing, and the concurrent development of summer forms of the macrozoobenthos. Signatures of these changes are readily apparent in the metrics littoral, hypocrenal and hyporhithral, which all peak in summer. In autumn and winter, a marked decrease in temperature, lower illumination, and (in contrast to earlier months of the year) a strong supply of allochthonous organic material result in the development of detritophagous invertebrates. Development of detritophagous invertebrates can however be slower than the onset of the preceding seasonal changes in the macroinvertebrate community and in winter it can be strongly inhibited or even stopped. During the winter, the metabolism of Small Carpathian streams has been shown to predominantly heterotrophic (Krno & Hullová, 1988; Rodriguez & Derka, 2003). The strong development detritophagous Crustacea, Plecoptera, Ephemeroptera, and Coleoptera in our study confirms these findings.

The question of determining an appropriate number of sampling occasions during the year is important. From an economic perspective there is a desire to minimise the frequency of sampling while biological studies tend to indicate the reverse. Several studies (e.g., Ormerod, 1987) have demonstrated the benefit of combining datasets from at least two seasons so that taxa rarely recorded in one season are gained from the additional season. Similarly, Furse et al. (1984) showed that combined season data enabled better categorisation and prediction of macroinvertebrate communities than single season data. They advocated sampling in three seasons wherever feasible to allow the characteristic annual pattern of change in the fauna of a site to be incorporated into the analyses. The advantage of taking more than one sample a year was also evident from this study. The complementary value of a late autumn or winter sample to a spring sample was obvious. The autumn and winter community consisted of many species that were uncommon in spring yet were found in high abundances in the later part of the year. It should be noted, however, that mid winter sampling is not suitable for purely logistic reasons (e.g., problems reaching and entering streams and sampling in ice and snow). Futhermore sampling three times a year can be very time-consuming, particularly if identifications are to be taken to species level.

Since seasonal changes are a natural phenomenon it is not possible to give advice on the time period most suited for sampling. For metrics that show high seasonal variation the best solution would be to always sample during the same month or to take into account seasonal variation in setting class boundaries for assessment purposes.

Many of the metrics evaluated in this study depend on indicator values. In many cases indicator values for these taxa were unknown and the influence of taxa with indicator values (and high abundance) and the sensitivity of the metrics to seasonal variation will be overestimated. Increasing the knowledge of autecology will help to reduce this problem. For metrics where the optimal sampling period is not directly related to the highest metric value, the best solution would be to sample in a comparable month or months or to take into account seasonal variation in setting class boundaries.

In this study only the effects of seasonal variation in macroinvertebrate community composition on metric values were evaluated. When selecting metrics for the development of a biological assessment system apart from variability and differences in values between months it is most important to know whether metrics are (highly) correlated to anthropogenic stress.

Acknowledgements

This study was carried out within the STAR project, a research project under the 5th Framework programme of the European Union (EVK1-CT-2001-00089) and grant 1/292/04 from the Slovak Grant Agency for Sciences. We would like to thank John Davy-Bowker for linguistic revision of the manuscript. We would like to thank Tomáš Derka, Daniela Illéšová and Zuzana Pastuchová for their efforts in collecting the data on which this study was based.

References

AQEM consortium, 2002. Manual for the application of the AQEM system. A comprehensive method to assess European streams using benthic macroinvertebrates, developed for the purpose of the Water Framework Directive. Version 1.0. February 2002.

Bunn, S. E., 1986. Spatial and temporal variation in the macroinvertebrate fauna of streams of the northern jarrah forest, Western Australia: functional organization. Freshwater Biology 16: 621–632.

Clifford, H. F., 1978. Descriptive phenology and seasonality of a Canadian Brown-water stream. Hydrobiologia 58: 213–231.

Coimbra, C. N., M. A. S. Graqa & R. M. Cortes, 1996. The effects of a basic effluent on macroinvertebrate community structure in a temporary Mediterranean river. Environmental Pollution 94: 301–307.

Cowell, B. C., A. H. Remley & D. M. Lynch, 2004. Seasonal changes in the distribution and abundance of benthic invertebrates in six headwater streams in central Florida. Hydrobiologia 522: 99–115.

Cummins, K. W., 1977. From headwater streams to river. American Biology Teacher (May), 305-312.

Dahl, J., R. K. Johnson & L. Sandin, 2004. Detection of organic pollution of streams in southern Sweden using benthic macroinvertebrates. Hydrobiologia 516: 161–172.

Dolédec, S., 1989. Seasonal dynamics of benthic macroinvertebrate communities in the Lower Ardeche River (France). Hydrobiologia 183: 73–89.

European Commission, 2000. Directive 2000/60/EC of the European Parliament and of the Council – Establishing a framework for Community action in the field of water policy. Brussels, Belgium, 23 October 2000.

Furse, M. T., D. Moss, J. F. Wright & P. D. Armitage, 1984. The influence of seasonal and taxonomic factors on the ordination and classification of running-water sites in Great Britain and on the prediction of their macro-invertebrate communities. Freshwater Biology 14: 257–280.

Halgoš, J. & L. Jedlička, 1974. The distribution of black flies (Diptera, Simuliidae) in the Little Carpathians. Acta Rerum Naturalium Musei Nationalis Slovaci, Bratislava 19: 173–193.

Hering, D., O. Moog, L. Sandin & P. F. M. Verdonschot, 2004. Overview and application of the AQEM assessment system. Hydrobiologia 516: 1–20.

Hynes, H. B. N., 1972. The Ecology of Running Waters. University of Toronto Press, Toronto, 555 pp.

Illéšová, D. & J. Halgoš, 2003. Phenology of Blackflies (Diptera, Simuliidae) in the Gidra River Basin. Acta Zoologica Universitatis Comenianae 45: 69–75.

Kerans, B. L., J. R. Karr & S. A. Ahlstedt, 1992. Aquatic invertebrate assemblages: spatial and temporal differences among sampling protocols. Journal of the North American Benthological Society 11: 377–390.

Krno, I. & D. Hullová, 1988. Influence of the water pollution on the structure and dynamics of benthos in the stream Vydrica (Small Carpathians). Biologia (Bratislava) 43: 513–526.

Krno, I. (ed) 1996. Limnology of the Turiec river basin (West Carpathians, Slovakia). Biologia (Bratislava) 51(Suppl. 2): 1–122.

Matthews, R. C. jr. & Y. Bao, 1991. Alternative instream flow assessment methodologies for warm water river systems. In Cooper, J. L. & R. H. Hamre (eds) Proceedings of Warm-water Fisheries Symposium 1. U.S. Forest Service (General Technical Report RM–207), Fort Collins, CO, 189–196.

Metzeling, L., B. Chessman, R. Hardwick & V. Wong, 2003. Rapid assessment of rivers using macroinvertebrates: the role of experience, and comparisons with quantitative methods. Hydrobiologia 510: 39–52.

Moore, J. W., 1977. Seasonal succession of algae in rivers II. Examples from Highland water, a small woodland stream. Archiv für Hydrobiologie 80: 160–171.

Morais, M., P. Pinto, P. Guilherme, J. Rosado & I. Antunes, 2004. Assessment of temporary streams: the robustness of metric and multimetric indices under different hydrological conditions. Hydrobiologia 516: 231–251.

Neel, J. K., 1968. Seasonal succession of benthic algae and their macroinvertebrate residents in head-water limestone stream. Journal Water Pollution Control Federation 40: 10–30.

Nijboer, R. C. & A. Schmidt-Kloiber, 2004. The effect of excluding taxa with low abundances or taxa with small distribution ranges on ecological assessment. Hydrobiologia 516: 349–366.

Norris, R. H. & A. Georges, 1993. Analysis and interpretation of benthic macroinvertebrate surveys. In Rosenberg, D. M. & V. H. Resh (eds), Freshwater Biomonitoring and Benthic Macroinvertebrates. Chapman & Hall, New York and London: 234–286.

Ormerod, S. J., 1987. The influences of habital and seasonal sampling regimes on the ordination and classification of macroinvertebrate assemblages in the catchment of the River Wye, Wales. Hydrobiologia 150: 143–151.

Rodriguez, A. & T. Derka, 2003. Physiographical and hydrobiological characteristics of the Gidra river basin. Acta Zoologica Universitatis Comenianae 45: 11–18.

Ross, H. H., 1963. Stream communities and terestrial biomes. Archiv für Hydrobiologie 59: 235–242.

Šimo, E. & M. Zaťko, 1980. Typy režimu odtoku, s. 65. In Mazúr, M. (ed.), Atlas Slovenskej socialistickej republiky, SAV, 296 pp.

Šporka, F. (ed.), 2003. Vodné bezstavovce (makroevertebráta) Slovenska. Súpis druhov a autekologické chrakteristiky. Slovak aquatic macroinvertebrates. Checklist and catalogue of autecological notes. Slovenský hydrometeorologický ústav, Bratislava, 590 pp.

Sprules, V. M., 1947. An ecological investigation of stream insects in Algonquin Park, Ontario. University Toronto Studies, Biology Series 56: 1–81.

ter Braak, C. J. F. & P. Smilauer, 2002. Canoco reference manual and CanoDraw for Windows Users guide: Software for Canonical Community Ordination (version 4.5). Microcomputer Power (Ithaca, NY, USA), 500 pp.

Townsend, C. R. & A. G. Hildrew, 1979. Foraging strategies and coexistence in a seasonal environment. Oecologia 38: 231–234.

Vlek, H. E., Influence of seasonal variation on bioassessment of streams using macroinvertebrates. Verhandlungen der Internationalen Vereinigung für Limnologie (in press).

Williams, D. D., 1981a. Emergence pathways of adult insects in the upper reaches of a stream. Internationale Revue der gesamten Hydrobiologie 67: 223–234.

Williams, D. D., 1981b. Migrations and distributions of stream benthos. In Lock, M. A. & D. D. Williams (eds), Perspectives in Running Water Ecology. Plenum Press, New York and London, 155–207.

Williams, D. D. & H. B. N. Hynes, 1974. The occurrence of benthos deep in the substratum of a stream. Freshwater Biology 4: 233–256.

Williams, N. E. & H. B. N. Hynes, 1973. Microdistribution and feeding of the net/spinning caddisflies (Trichoptera) of a Canadian stream. Oikos 24: 73–84.